2154;_

Arcadia

"Art,
The Imitation Of Life,
Life,
The Imitation Of Art."

V. V.

Lieu, Michael W.
廖, 米高 伍 o

...Unless, it is Not Safe For Work (N.S.F.W.) porn, parody, fan fiction, fan art...
that is okay, because Murphy's Law, Chaos Theory, String Theory, and Human Beings' Behavior Of Internet Rule 34 (thirty four).

Purpose of copyright is to encourage imaginative artist author designer writer to produce creative works that enrich human beings, Artificial Intelligence (A.I.), animal kingdom, and planet earth culture. Any actual names or likenesses of human(s), A.I., and animal kingdom, and planet earth culture are used in a science fictitious and parody manner

Stories, characters, location settings, plot points in this "Visual Story Text Book { Travel Drawings Audio Retcon }", *Science-Fiction Character Story Universe*" is a mix of make belief, fiction, and actual reality existence, non-fiction, plus composite character designs. Copyrighted © I.P. story, characters, plot points in "**Model References Work Inspiration; Bibliography**" page is plausible under fair use U.S. copyright conduct code law.

This product is from the imagination of the artist author designer writer; actual persons alive living, undead augmented, or deceased necrosis, business establishments, events, or locales is entirely coincidental.

Their real identities, actual names, characteristics have been changed in accordance to Privacy Protection aug made possible by Kay's Private Security Services Limited Liability Company (L.L.C.)

Print book, Electronic-Book (E-Book), made possible by Planet Earth's Evolved Earthlings

2154;_Arcadia (L.C.C.N., U.S.C.O., I.S.B.N.)
Library of Congress Control Number: 2022914885
United States Copyright Office: { X }
International Serial Book Number: "978-1-959211-00-6 = Softcover Paperback"
"978-1-959211-04-4 = Hardcover Hardback"
"978-1-959211-02-0 = Electronic Book E-Book"
"0 (zero) none = Audio Book"

Subject Topics, Genre Type;_
Science-Fiction, Hard Science-Fiction, Soft Science-Fiction = "The Literature Of Ideas"
Alternative History = non-fiction mixed fiction
Apocalyptic, Neo-Western { Macro Bigger Story Setting } = non-fiction mixed fiction
Utopia, Parody { Micro Smaller Story Setting }= non-fiction mixed fiction
Semi-Autobiography = non-fiction
Traditional Art Character Designer = non-fiction mixed fiction
Coloring-Book Line Art Style = non-fiction mixed fiction
Photo-Book = non-fiction
Travel Drawings Audio Retcon = non-fiction mixed fiction

No D.R.M. (Digital Rights Management) dog poo

2154visualstorytextbook@gmail.com

x

Story 1 "2154;_Arriving Coloring Book"+"2154;_Arriving" Concept Art Script Experiment
Story 1(one) On Amazon Lex Luthor Corp (L.L.C.) Insane Inspiration;_
Simon And Schuster + Tom Bissell, Greg Sestero's The **Disaster Artist Book**
Lionsgate + A24's The **Disaster Artist** <= **Movie** => The Room's Tommy Wiseau.
E.A. + DICE's **Mirror's Edge Catalyst remake sandbox game**
Warner Brothers + D.C., Christopher Nolan's *The Dark Knight* { Night } story 2(two)
With Greg Wiseau's Tommy Planet Parody "*The Mark Knight*"
Ubisoft's *Watch Dogs 1(one) original* sandbox game
Bethesda + Obsidian's **Fallout New Vegas** sandbox simulator game
2K + 2K Boston, 2K Australia, Blind Squirrel's **BioShock** sandbox simulator game
Valve's **Half-Life 2**, Lost Coast, Episode: One, Two, Left 4 Dead 2, Portal: 1, 2 game(s)

Original Art By;_
Lieu, Michael W.
廖, 米高 伍o

Original Book Design;_
Lieu, Michael W.
廖, 米高 伍o

Original Character Design;_
Lieu, Michael W.
廖, 米高 伍o

Original Concept By;_
Lieu, Michael W.
廖, 米高 伍o

Original Cover Design;_
Lieu, Michael W.
廖, 米高 伍o

Originally Edited By;_
Lieu, Michael W.
廖, 米高 伍o

Original Story By;_
Lieu, Michael W.
廖, 米高 伍o

Originally Written By;_
Lieu, Michael W.
廖, 米高 伍o

Improving Infrastructure;_
Lieu, Michael W.
廖, 米高 伍o
+(plus) Benavidez, Richard

"Elle K. Mcgee"
"BladeMaster 刀大師"
character conceptualized art by;_
Lieu, Michael W.
廖, 米高 伍o
and(&) naming by
Benavidez, Richard

"Big Ben"
"BigBoss 大老闆"
"Mexican-Man \ Mexican-Men"
characters conceptualized art by;_
Lieu, Michael W.
廖, 米高 伍o
and(&) naming by
Benavidez, Richard

"Xtra-Men"
"Jerry Duty"
"Dog-Name-Cat \ Rover"
"Social Worker Mars Hall"
characters conceptualized art by;_
Lieu, Michael W.
廖, 米高 伍o
and(&) naming by
Benavidez, Richard

Story 2(two) "2154;_Arcadia" = Travel Drawings Audio Retcon
"Art, The Imitation Of Life. Life, The Imitation Of Art."
Insane Inspiration;_

Anime eXpo (A.X.) + Comic Con(s) = "Anime Comic Con Expo(s)" Convention Center(s)
2006's KIPI { kosplayer } or Kipi cosplayer, Elizabeth Rage, Alex Drastal, Julia Valentine,
Lisa A. Pinelli, Tara Cosplay \ Recreation, Economics, Infrastructure (REI) \ Free Market
Imagining 2019's Convention Centers as the new world 22nd(twenty second) Century Cities
Economic Freedom Merchant Artists Author Alleyways \ Animal Human Cosplay Cafe
United States Dollars + Crypto Currency \ Anime Los Angeles \ Long Beach Comic Expo
Anifest \ Comic-Con's Wonder Con \ Anime Pasadena \ Ronin-Expo \ Anime eXpo (A.X.)
Animanga \ Anime California \ Stan Lee's Kamikaze, L.A. Comic Con \ "Animal eXpo"

"Martin Wong Photography", Linda Le A.K.A. "VampyBitMe", "YaYa Han", "D-Piddy",
"Alodia Gosiengfiao" A.K.A. BlackMage9 Eastern Society Of Asia's Cosplay Queen

Nisioisin, Vofan, Aniplex's MoNoGaTaRi story universe character(s) of
KissShot-AcerolaOrion-HeartUnderBlade { Best Vampire Lady And Or Girl } cosplayer(s);
@MakiEraclea, @SayaTheFox, @HaneAme_Cos, @Owldepot, @HaidyKnox, @Standesu

Party At The Sausage Castle, A.K.A. The Wildest House In America(s), @TheSausageCastle
"Arrive as strangers, depart as family."
Bill And Ted universe, "Be excellent to each other." / The Summer Of Love <3 Or Ice Cream

Warner Brothers(W.B.), Adult Swim, Robot Chicken T.V. show series comedy, parody, fan fic

Square Soft Enix, Dontnod, Deck 9(nine), Titan Books' Life Is Strange "travel mark book"
Rockstar's Read Dead Redemption 2, story 0(zero) origins, Arthur Morgan's travel journal

W.B., D.C., Rocksteady's **Batman Arkham**: Asylum, **City**, Knight. Kill The Justice League

Bethesda + Obsidian's **Fallout New Vegas' Dead Money Add-On**
Down Loadable Content (D.L.C.) sandbox simulator game

2K + Irrational Games' BioShock: Infinite story 0(zero) origins
BioShock Infinite: Burial At Sea Episode 1(one), Burial At Sea Episode 2(two)
BioShock 1(one) Original, BioShock 2(two) science-fiction violent video games

Valve's Half-Life 1(one) Original, Half-Life Blue Shift, Half-Life Opposing Force,
Half-Life 2, Half-Life 2 Lost Coast, Half-Life 2 Episode: 1(One),
Half-Life 2 Episode: 2(Two), Left 4(four) Dead 1(one) original, Left 4(four) Dead 2(two),
Portal 1(one) original, Portal 2(two) science-fiction violent video games

4(four) part arc(ark) acts

"Electronic-Book (E-Book) pages subject to change if not in P.D.F. type."

Quotes To Lieu By

Hero(Good) \ Vigilante(Neutral) \ Villain(Bad)

英雄(好) \ 義務警員(中性) \ 惡棍(壞)

"...There is nothing either good or bad, but thinking makes it so..."
"...All the world's a stage..."
-Shakespeare, William { Will.I.am }

Wiseau-Wisdom, "Respect equals success."

"...Without Checks And Balances..."
"...Power Corrupts, Absolute Power, Corrupts Absolutely..."

"A science-fiction story mixed with non-fiction reality"
2006-2016;_Final-Fan-Fiction-Fantasy-6-Versus-13-Equals=15

"Because { they } thought it was good sport. Because some { people } aren't looking for anything logical, like money. They can't be bought, bullied, reasoned or negotiated with. Some { people } just want to watch the world burn..."
-Michael Caine { Alter Ego, My Cocaine } as Sir Alfred Pennyworth Butler,
Warner Brothers + D.C., Chris Nolan's Batman story 2(two) The Dark Knight { Night }

"(cool { kool }) Whoa... I know kung fu... We need guns, lots of guns."
-{ Agent } Keanu Neo Women Infant Children (WIC)

"(laughing) Ha ah! ...What a story Mark!!" -Wiseau's The Room, Johnny { Jonny }

"You can't connect the dots looking forward; you can only connect them looking backwards. So you have to trust that the dots will somehow connect in your future. You have to trust in something – your gut, destiny, life, karma, whatever." -Jobs, Steve

"(actual silent protagonist) ..."
-Red, Pokémon Trainer from Pallet Town KaNTo region

"(Kawaii kute anime { Annie-May } voice over) Oh-Knee-Chan!"

"When the system fails you, you create your own system." Williams, Michael K.

Lieu's Hall Of Dedication

This *Science-Fiction Character Story Universe* is *dedicated to all* ***military veteran(s)*** of past, present, future... I know future ***military veteran(s)***, that sounds weird to think we all will know { string theory } who can be ***military veteran(s)***...

Also to include in this *narrative*, those who were *medically separated* as well as *discharged* under: '**Honorable**', '**General**', '**Other Than Honorable**', '**Bad Conduct**', '**Dishonorable Discharge**', plus ***military officer(s)*** *discharged* to include '**Dismissal Notice**'.

Not to forget: ***family***, ***friends***, and ***supporting staff(s)*** to ***military veteran(s)*** who *encourage*, *guide them* in times of needed assistance when they reached out.

For those all on ***military service***: ***active duty***, ***reserve members***, ***national guard***, ***with others*** and finally ***first responders vigilant heroes***... *Godspeed* to you all.

This *Science-Fiction Character Story Universe* is ***personally dedicated*** to;_ Issac Asimov, Phillip K. Dick, Sir Arthur C. Clarke, Robert A. Heinlein, Mary Shelley, H.G. Wells, Jules Verne, George Lucas, 8, MuGeN, infinity, no limit, Amy Lieu, Saxton Dee I.T. Man, my grandpa, Jonny Huynh Sushi Chef, Stan Lee The Man, Winner Twins, Todd Mccaffrey, Self-Publishing School With Chandler Bolt, Self-Publishing With Dale L. Roberts, 2003-2019, Andy Weir, Any Weird, E.L. James, J, my family, friends, my mom.

Hope you all enjoy this *Science-Fiction Character Story Universe*, that based on *non-linear narratives* and *subjective topics* that I personally enjoyed in the reality of 20th and 21st century time frame, ***blended together into this melting pot*** *concept, idea, characters, setting of a 22nd century science-fiction alternative history parody utopia...*

Along with ***cosplayers*** *modeling* themselves after their *favorite character(s)* in the *stories* they all personally *experienced* and or *enjoyed*.

...Finally, this story spoiler title directly to the **"2009 release date " 2154; Avatar Movie Director James Cameron" + "2013 release date" 2154; Elysium Movie Director Neil Blomkamp"** for the title of this story references the year their stories are set in as my "**Model References Work Inspiration; Bibliography" + "2001: A Space Odyssey, 2010: Year We Make Contact" universe.**

"2154" ;_ { Plus Steins;Gate }

P.S.;_Dedicated to all those who did not make it to live past 2019's C-Virus C.B.R. outbreak.

Science-Fiction Quotes To Lieu By

"The basic tool for the manipulation of reality is the manipulation of words. If you can control the meaning of words, you can control the people who must use the words."

"Reality is that which, when you stop believing in it, does not go away."

"Sometimes the appropriate response to reality is to go insane."

-Philip K. Dick { Soft Science-Fiction Alternative History Universe genre(s) }

The 3 (three) Laws Of { Artificial Intelligence (A.I.) Synthetic Human(s) } Robotics

Law 1 (one), An { A.I. } may not injure a { naturally unmodified birth } human being(s) or through inaction, allow a { naturally unmodified birth } human being to come to harm.

Law 2 (two), An { A.I. } must obey any orders given to it by a { naturally unmodified birth } human being(s), except where such orders would conflict with { Law 1 (one) }.

Law 3 (three), An { A.I. } must protect its own existence as long as such protection does not conflict with { Law 1 (one) or Law 2 (two) }. **-Issac Asimov**

Clarke's 1st (first) Law;
"When a distinguished but elderly scientist states that something is possible, { they are } almost certainly right. When { they } state that a something is impossible, { they are } very probably wrong."

Clarke's 2nd (second) Law;
"The only way of discovering the limits of the possible is to venture a little way past them into the impossible."

Clarke's 3rd (third) Law;
"Any sufficiently advanced { biotech } is indistinguishable from magic."

-Sir Arthur C. Clarke { Mythology Religious Spiritual Science-Fiction genre(s) }

"There are no dangerous weapons { or tools }; there are only dangerous { humans }."

"It takes 2 to create a heaven { utopia }, but hell { dystopia } can be accomplished by 1."

"Do not ever become a pessimist… a pessimist is correct oftener than an optimist, but an optimist has more fun, and neither can stop the march of events."

"A generation which ignores history has no past and no future."

"{ A human } is not a rational animal, { a human } is a rationalizing animal."

"Take big bites. Anything worth doing is worth overdoing."

-**Robert A. Heinlein** { Hard Military Science-Fiction genre(s) }

V. V.'s Travel Itinerary 2154;_ Happy 34th (thirty fourth) Built Day, Summer, Fall, Winter

2154;_Arcadia \ Valentina Vita (V. V.) 'Healthy-Strong Life' \ Left to Right(L->R)

V. V.'s Travel Itinerary 2154;_ Happy 34th (thirty fourth) Built Day, Summer, Fall, Winter

V. V.'s Travel Itinerary 2154;_ Happy 34th (thirty fourth) Built Day, Summer, Fall, Winter

2154;_11(Nov)-12(twelve) Revisit, Chicago Self-Governing Sovereignty (S), "**Animal** Sanctuary", East Side, All For One Shopping Center pages #402 - 431
2154;_11(Nov)-13(thirteen) Revisit, Chicago Self-Governing Sovereignty (S), "**Animal** Sanctuary", West Side, Chicago City Christian Church Belief Systems, And Other Religions Welcome <3 pages #432 - 446
2154;_11(Nov)-28(twenty eight) Arriving, Chicago Self-Governing Sovereignty (S), "**Animal** Sanctuary", Downtown District, **Elysium Corporation** Biotech Dome Center Point **Captain Rage's Public Po Po Place** **Chicago City Police Department (C.C.P.D.) Special Weapons And Tactics (SWAT)** pages #447 - 457
2154;_11(Nov)-29(twenty nine) Arriving, Chicago Self-Governing Sovereignty (S), @ 00;00_a.m. { **act 4(four) conclusion** } Beauty Implant+ "**Animal** Sanctuary", Downtown District, **Elysium Corporation** Biotech Dome Center Point **Captain Rage's Public Po Po Place** **Chicago City Police Department (C.C.P.D.) Special Weapons And Tactics (SWAT)** pages #458 - 459
2155;_01(Jan)-01(one) @ 00;00_a.m. "Maybe... New Years' Travel Resolution?" Pages #X

2154;_08(Aug)-01(one) @ 00;00_a.m. { act 1(one) back story } Beauty Implant+ Engineering Environment Biotech Dome, Interior, +024° C (twenty four degrees Celsius) "Naples, A **Human** Sanctuary In **Italy**. Utopian Sovereignty of Eur-Asia."
'Residence to a 22nd (twenty second) century human sanctuary biotech utopia.'
Naples ***City***, Valentina Vita (V. V.)'s Home Sweet Home, ***Elysium Corporation***
Naples (Nap) 80121 U.S. of E-A, Planet Earth, Solar System, Milky Way Galaxy

V. V. wakes up to the selected music "**Nier: Automata's Peaceful Sleep**" playing in the **Utopia A.R. Holo HUD** background on one third volume of 33% that was set on loop from the **Date-Time "2154;_02(Feb)-22 @ 00;00_a.m.**" on her biotech cryogenic stasis sleep.

Valentina Vita (V. V.); Traveling Cosplayers And Kosplayers;
"(@_@) ...Time to load up my Ai-Aug biotech travel recording red Holo HUD projection... **Twenty one fifty four**, I **am thirty four this year**. **Happy Built Day To** Me, V. V. Time to **beauty enhance**! Need to apply an ex-foiling mask to my head, face and lips!" (^_^) **Dear travel recording day one, twenty one fifty four, August fifteen**... Preparing for my travel trip adventure outside **Utopian** Sovereignty **of Eur-Asia** to the **Americas counter part**... on Earth. Lots of preparation... Time to get up from my biotech **beauty** stasis sleep... and figure out which **beauty implants to put on**. What **face mask to put on**... I **am not cosplaying yet**! After all, **being a cosplayer is like being mythological Super-Hero double life**. (O_O) ...Any cosplayer must have experienced situations like the **costume destroying itself through wear and tear for physical forms**... **Ugly threads popping up from middle of nowhere** when preparation or photo session... I must be prepared. (^_^) I always, always **prepare a handy cosplay emergency kit to solve those ugly threads** or in case a cosplay friend needs help. I guess this is cosplay's negative self-talk points, first the worst... What I dislike about cosplay... (@_@) **Wig headache, that moment when you take off that wig and a super strong powerful headache comes to visit you**... (>_<) ...**When** I **have to refuse a holo photo... hate to say no, but sometimes** I **have say no because** I **am going home, wearing half the cosplay or have a meeting**. (O_<) ...**When people photograph without asking permission. Cosplay... cosplay is not consent.** (>_<) Another one, **stress caused by deadlines,** I **do not want to pressure myself, but when** I **have a to do list... sewing for hours, days, a nightmare come true, but** (^_^) **the con meet up days is amazing**! (=_=) Another negative point here... **Holo photographers insisting on us cosplayers having sexy angle poses, if** I **do not want to do it,** I **will not do it, this aspect of cosplay** I **do not like**. Some holo photographers are really self-confident on this... part of quote sexy angle poses, and... (>_O) I **have stand up and push myself to be as self-confident as those holo photographers**... (^_^) Okay, done with the self-talk care for now... Time to check on my **travel face mask time, EGG passport**!"

V. V. selects the music "**Alphaville's Forever Young**" to play in the **Utopia A.R. Holo HUD** background on one third volume of 33% and is set on loop.

Reads = Left to Right (L -> R)

2154;_ Naples, Italy Intranet, Wireless Broadcasting Frequencies Online

<table>
<tr><td colspan="3"><u>Measuring System</u>;_ Metric degrees Celsius (°C) [<u>Switch to</u> <u>Imperial °F</u>, <u>Kelvin</u>]</td></tr>
<tr><td colspan="3"><u>Geolocation</u>;_ Planet Earth, Eur-Asia, Utopian Sovereignty of Eur-Asia (U.S. of E-A)</td></tr>
<tr><td colspan="3"><u>Language</u>;_ English, Western Accent, Augmented Reality (A.R.) Application (App)</td></tr>
<tr><td colspan="3"><u>Century</u>;_ 22nd (twenty second) [<u>Compare statistics to</u> <u>21st (twenty first)</u> <u>Aug+</u>]</td></tr>
<tr><td>Actual Legal Given Name</td><td>Height</td><td>Weight</td></tr>
<tr><td>Deoxyribo Nucleic Acid (D.N.A.)</td><td>Holographic Retina Biometric Body Projection</td><td>OCEAN 5 (five) Common Personality Traits</td></tr>
<tr><td>Alias Network Nickname</td><td>7 (seven) Deadly Sins</td><td>7 (seven) Heavenly Virtues</td></tr>
<tr><td>Birth Day Or Built Day</td><td>Blood Type Or Biotech Fluid</td><td>Physical Plus Psych Age</td></tr>
<tr><td>Hair Texture Color Tone</td><td>Eye Retina Color Tone</td><td>Skin Covering Color Tone</td></tr>
<tr><td>Ethnicity</td><td>Nationality</td><td>Alien Or Citizen</td></tr>
<tr><td>Natural Matter
Organic Photosynthesis</td><td>Augmented Compound
Enhanced Mutation</td><td>Percentage Rate
Synthetic Synthesis (S.S.)</td></tr>
<tr><td>Human Being</td><td>Animal Kingdom</td><td>Artificial Intelligence (A.I.)</td></tr>
<tr><td>Biotech Or Bio Or Tech</td><td>Exosuit Or Exoskeleton</td><td>Implant Or Explant</td></tr>
<tr><td>Alive = Heart-Rate</td><td>Undead = Flatline</td><td>Deceased = Necrosis</td></tr>
<tr><td>Chemical (C.) Toxicity</td><td>Biological (B.) Toxicity</td><td>Radiological (R.) Toxicity</td></tr>
<tr><td>Personal
Protective
Equipment (P.P.E.)</td><td>Ballistic Motion
Energy Transfer
Shielding Resistance</td><td>Electro
Magnetic
Pulse (E.M.P.)</td></tr>
<tr><td>Sexual Orientation</td><td>Sexual Preference</td><td>Virginity Test</td></tr>
<tr><td>Sexual Transmitted Diseases (S.T.D.)</td><td>Human Immunodeficiency Virus (H.I.V.)</td><td>Acquired Immuno Deficiency Syndrome (AIDS)</td></tr>
<tr><td>Known Address(es)</td><td>Community Network(s)</td><td>Language Level(s)</td></tr>
<tr><td>Family Member(s)</td><td>In-Law Relative(s)</td><td>Relationship(s)</td></tr>
<tr><td>Biological Genealogy</td><td>Psychological Traits</td><td>Social Status</td></tr>
<tr><td>Employment</td><td>Education</td><td>Training</td></tr>
<tr><td>Frequency
Hertz (Hz)</td><td>Frequency
Modulation (F.M.)</td><td>Amplitude
Modulation (A.M.)</td></tr>
<tr><td>Electromagnetic Spectrum</td><td>Intelligence Quotient (I.Q.)</td><td>Kinesis Module</td></tr>
<tr><td>Radio Wave Music</td><td>Broadcasting Film Movie</td><td>Electronic Book Genre</td></tr>
<tr><td>Hate(s) Or Dislike(s)</td><td>Neutral(s) Or No Comment</td><td>Love(s) Or Like(s) [<u>Aug+</u>]</td></tr>
</table>

2154;_ Naples, Italy Intranet, Wireless Broadcasting Frequencies Online

Reads = Left to Right (L -> R)

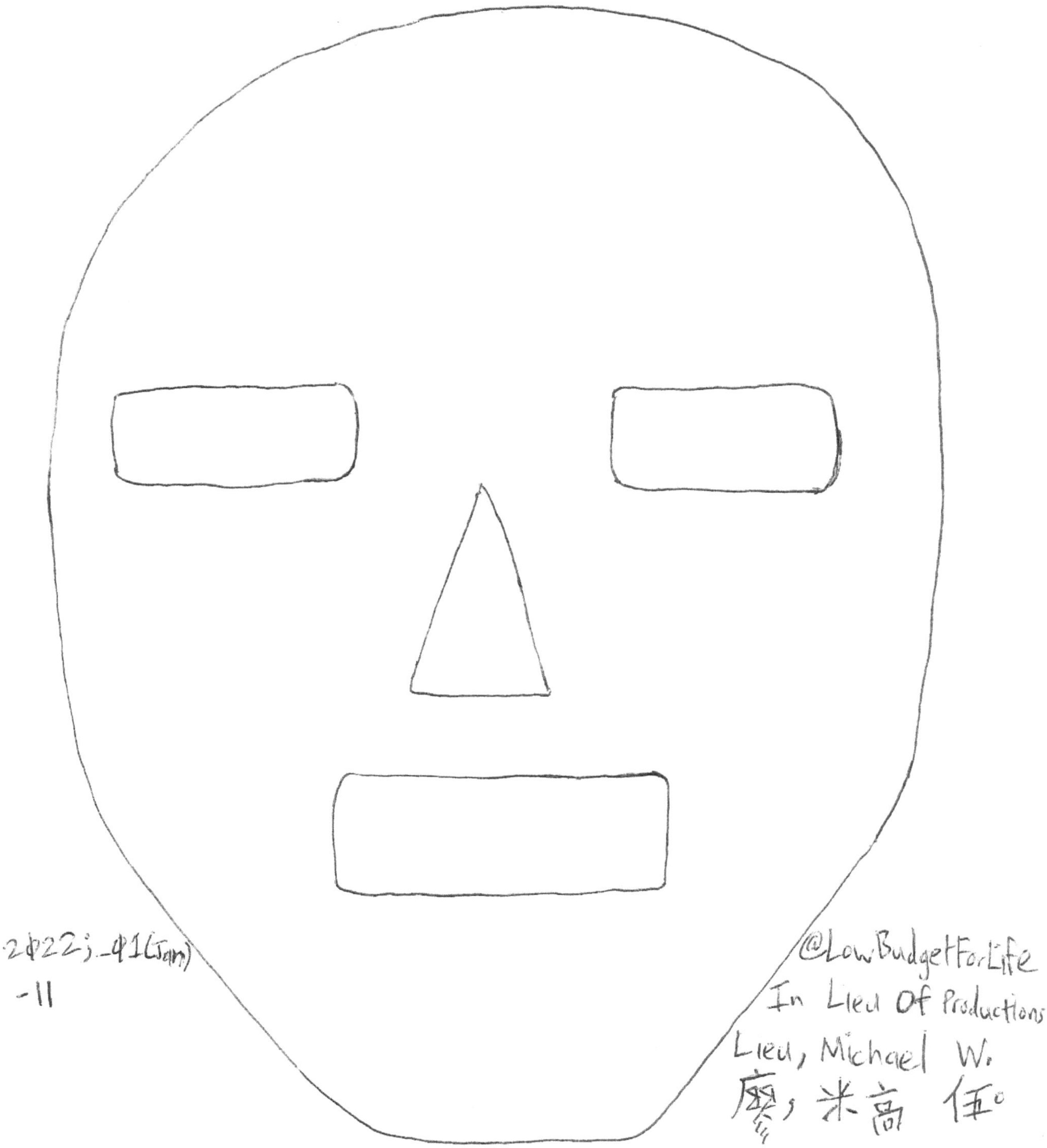
2Ø22;_Ø1(Jan)
-11
@LowBudgetForLife
In Lieu Of Productions
Lieu, Michael W.
廖, 米高 伍o

Earth's Global Government (EGG),
Planet Earth, Earthling Augmented And Natural Human(s), Eur-Asia Passport

Eur→ English	Asia→ Kanji
Actual Legal Surname(s) Vita	**實際法定姓氏** 維他
Actual Legal Given Name(s) Valentina	**實際法定名字** 瓦倫蒂娜
Actual Legal Middle Name(s) No Middle Name (N.M.N.)	**實際法定中間名** 沒有中間名
Nationality Eur-Asia	**國籍** 歐亞大陸
Birth Day Or Built Day 2120;_08(Aug)-15(fifteen)	**建造日或出生日** 2120;_08（八月）-15（十五）
Built Place Or Birth Place Italy Of Eur-Asia	**建地或出生地** 歐亞意大利
Issue Date 2154;_08(Aug)-15(fifteen)	**發行日期** 2154;_08（八月）-15（十五）
Expiration Date 2164;_08(Aug)-15(fifteen)	**截止日期** 2164;_08（八月）-15（十五）
Endorsement(s) Civilian, Citizen Of Eur-Asia	**背書** 平民, 歐亞公民
Biological Sex Natural Birth Human Female (F)	**生理性別** 自然出生人類女性 (F)
Biotech Augmentation(s) Beauty Augment+ Limb Enhancement(s) Breast Surgery Face Surgery Liposuction Tummy Tuck Anti-Aging Skin Bleaching +	**生物技術增強** 美容增強+ 肢體增強 乳房手術 面部手術 吸脂 腹部除皺 抗衰老 皮膚漂白 +

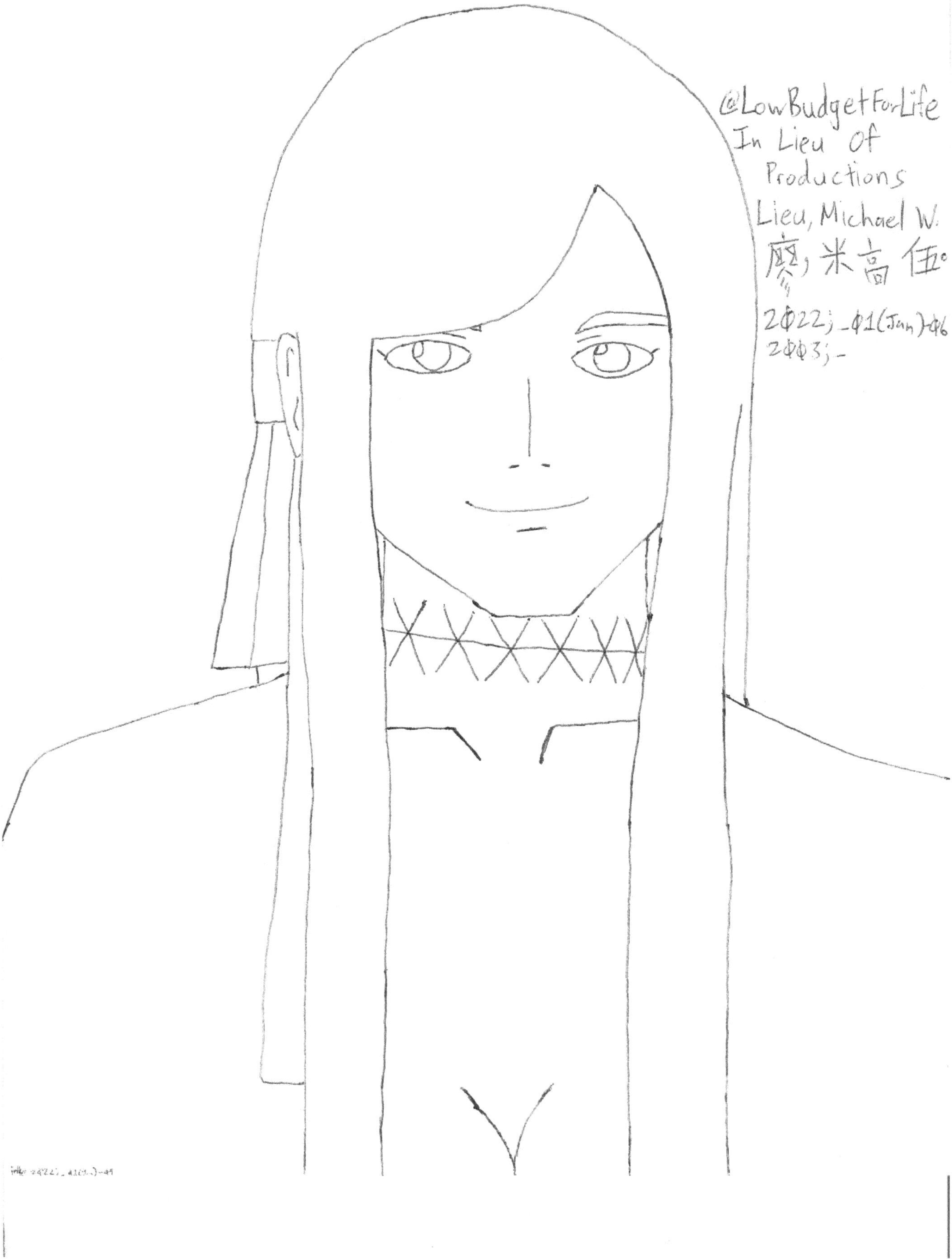
@LowBudgetForLife
In Lieu Of
Productions
Lieu, Michael W.
廖, 米高 伍o
2Φ22;_Φ1(Jan)-Φ6
2ΦΦ3;_

Beauty Implant Augmentations (Aug+);

Travel Size Biotech Cosmetic Product Enhancements
Beautiful Biotech Bodies And Or "The Beautiful Biotech Body"
Made Possible By

Elysium Corporation

Established 2077
The Magical Corporation Of Desire

Purifying Face Cleanser	Anti-Wrinkle Eye Treatment
High Potency Evening Oil	Anti-Wrinkle Overnight Cream
Clay Cleanser	Advanced Tightening Neck Cream Plus
Do It Yourself (D.I.Y.) Surgical Stitching Line	Essential Vitamins Serum Capsules
Acne Essentials	Bi-Facil Double-Action Eye Makeup Remover
Peeling Gel	No More Wrinkle Smoothing Cream
Deodorant	Recovery Moisturizer
Beauty Sleep	Pillow Cream
Foaming Face Wash	Eye Gels
Detox Charcoal Mask	Herbal Cleanser
Lip Sleeping Mask	Cleaning Balm
Face Finishing And Firming Tinted Moisturizer	Hydrating Serum
Ultra Repair Cream Minis	Travel Size Maracuja Oil
Acne Foam Cleanser	Peel-Off Mask
Oil Control Moisturizer	Skin Purifying Mask
Skin Refining Serum	Enzyme Cleansing
Original Skin Re-Texturing Mask	Facial Cleanser
Facial Radiance Pads	Facial Spray
Facial Oil	Firming Neck Gels
Moisturizer With Essential Vitamins	Anti-Aging Cleansing Gel
Anti-Aging Facial Moisturizer	Shampoo And Conditioner Hair Wash
Japanese Hair Straightening	Brazilian Blowout Hair Treatment
Skin Bleaching	Limb Enhancement Strengthening

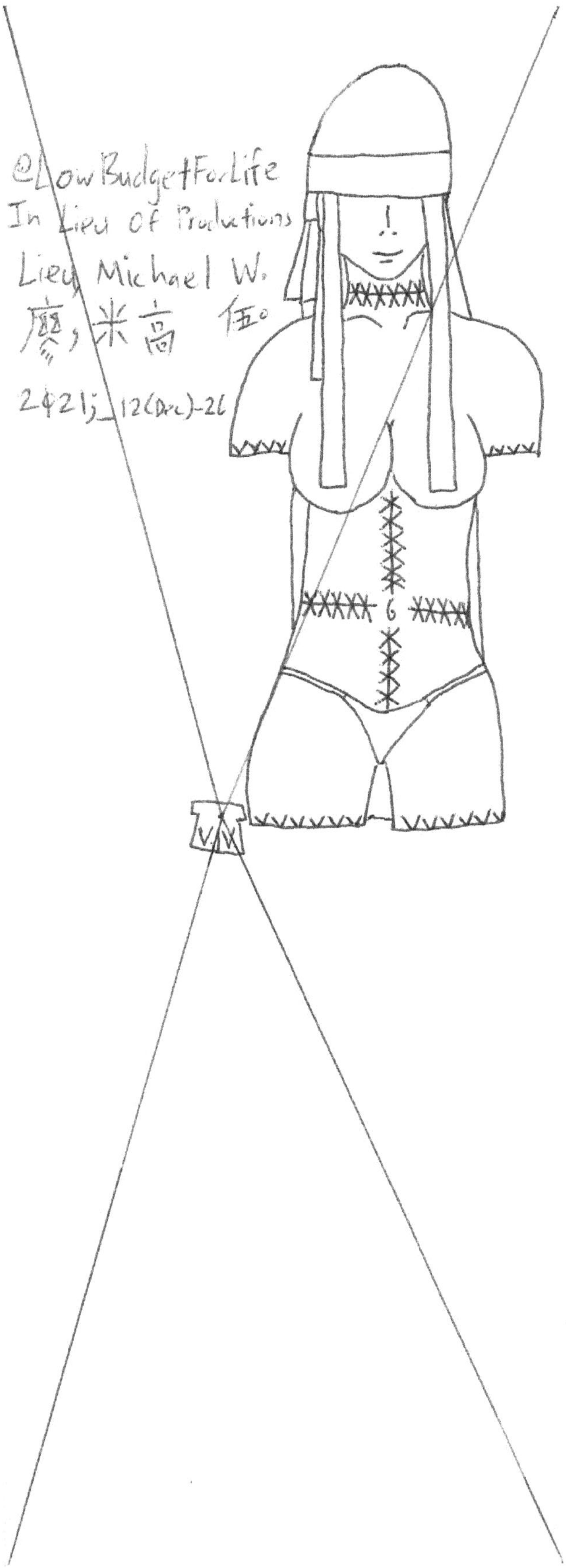
@LowBudgetForLife
In Lieu Of Productions
Lieu Michael W.
廖, 米高　伍o
2021j_12(Dec)-26

Beauty Implant Augmentations (Aug+);

Elysium Corporation

Established 2077

The Magical Corporation Of Desire

Biology-Biological + Technology-Technological = Bio Tech (B.T.);
BioTech "Biotech" Implants And Or Explants

Physical Plus Psychological

Beautiful Biotech Body Enhancement(s)

Physical Body Alterations, Modifications (Mods), And Splicing Junction

Biology = Organism	Technology = Cybernetics	BioTech; CybOrg
Biology = Reptitlian Rapid Regeneration Scaling + Chameleon Color Toner Shifting		
Technology = Eternal Electrons Electrical Tubes + Electro Magnetic Pulse Shielding		

Intelligence Amplification

Invisibility, Stealth Suit(s), Biotech Suit(s)

Life Extension, Existence Extension,
Biological Immortality, Technological Immortality, Biotech Immortality

Artificial Synthetic Organ(s)	Kloning, Klone(s)	Cryonics And Stasis
Augmented And Natural Organ Bank	Organ Harvesting	Organ Transplant(s)

Molecular Manufacturing And Or Molecular Machine(s), Nanities

Nano Biotech, Gray Goo Or Grey Goo, Self-Reproducing Tool(s)

Nano-Military	Nano-Materials	Nano-Ecology	Nano-Medicine
Nano-Computing	Nano-Space	Nano-Energy	Nano-Tattoo

Naples Controller City Mayor selects the music "**Crosses' Telepathy**" to play in the **Utopia A.R. Holo HUD** background on one third volume of 33% and is set on loop.

Elysium Corporation

亞

Established 2077

The Magical Corporation Of Desire

Psychological Mind Interfacing Research And Development (R&D)

Psychological Memory Removal / Editing	Psychological Mind Control, Persuasion Power
Psychological Mind Swap, Consciousness Transfer	Psychological Mind Uploading, Downloading, And Archival / Neural Network
Two Or More Psychological Minds Coexisting In One Body	Human(s), Animal Kingdom(s), And Or Space Alien(s) Coexisting In One Physical Platform
Neurological Implants And Or Explants To Directly Interface With Biotech(s)	
Magnetoencephalography: Measuring Magnetic Fields Produced By Electrical Brain Activity	
Neural Oscillations, Brain Waves, Neuro Science, Psi	
The Origins Of Consciousness, Awareness	
Additional Psychological R&D	

Broadcasting Frequencies Hertz Electromagnetic Spectrum Transmitter And Antenna, Signal Manipulation UltraViolet (U.V.) Rays City Controllers' City Mayor

Creating Consciousness, Activating Awareness, The Voice(s), Frequencies' Jamming	
Frequencies And Vibration (Vibe)	Biotech Kinesis, Psychokinesis
Clairvoyance, Extra Sensory Perception (E.S.P.), Parapsychology, Psychic, Paranormal	
Day Dreaming, Dreams, Visions, Creativity Curse, Latent Precognition, Mediumism	

String Theory, Multiple uniVerse (MultiVerse) Theory

Alternative: Universe(s), Timeline(s), History	Parallel: Universe(s), World(s), Earth(s)

Elysium Corporation

亞

Established 2077

The Magical Corporation Of Desire

Paradise Engineering, Creating Utopia, Environmental Habitats

Sovereign (S), Self-Governing (S), Animal Sanctuary (S), Human Sanctuary (S)

The 2121 Utopian Sovereignty (U.S.) Movement

Animal Zoo, Human Zoo	Architecture + Ecology = Arcologies	Domed City, Domed Sanctuary
Endemic, Epidemic	Extraterrestrial Space Alien Habitats	Floating City, Floating Fortress
Sea Steading, Aircraft Carriers	Ocean Colonization, Underwater Colonization	Pirate Utopia, Criminal Utopia, Terrorist Utopia
Reality Television Entertainment	Space Colonization, Transforming Planets To Earth "Terraforming"	Underground City, Underground Rail Road
Walking City	Walled Gated Community Enclave	X, Variable In Mathematics

Resizing

Size-Changing	Magnification (+)	Miniaturization (-)	Shrinking
Enlargement		Ooo.	

Shapeshifting

Visual Verbal Mimicking Copycat(s)	Artists, Models, Costume Play (Cosplay)
Method: Mime, Acting, Writing, Performer	Impersonators, Imitation, Theater Simulation

Teleportation

Augmented Biotech Human Body Host Hopping
Undead Utopian De-Assembling And Re-Assembling Consciousness Platform Transfer

V. V. selects the music "**Solar Fields' Mirror's Edge Catalyst, Anchor District Theme, Extended Ambient**" to play in the **Utopia A.R. Holo HUD** background on one third volume of 33% and is set on loop.

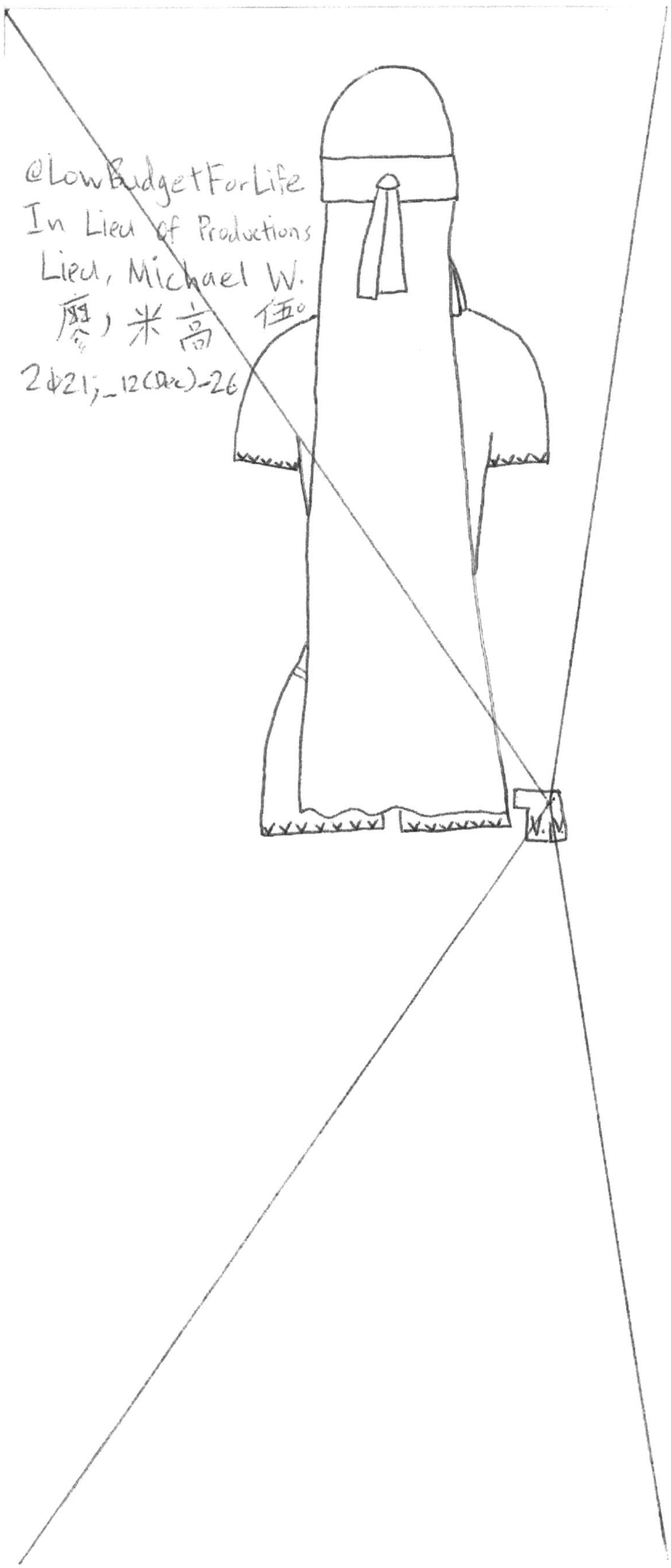
@LowBudgetForLife
In Lieu of Productions
Lieu, Michael W.
廖, 米高 伍o

2154;_08(Aug)-10(ten)

Valentina Vita (V. V.); Traveling Cosplayers And Kosplayers;
"(@_@) Okay, something to do while I wait for my **beauty implant augment and limb enhancements to take into effect... Time to study up and practice the good nutrition** healthy-strong life... as my name Valentina Vita implies..."

A Good Nutrition Guide Earth's Global Government (EGG) For Earthling Human Civilians

Eating Right Keeps You Healthy! "Good nutrition means eating the right foods in the right amounts to keep your body and mind healthy. It can help you fight infections, stay at a healthy weight and lower the risk of problems like heart disease and diabetes."
Key Nutrients; Water & Protein; "Water transports other nutrients, and helps us digest food and regulate temperature. Sources include tap water, purified water, milk and soup. Protein helps our bodies grow and repair themselves. Sources include meat, poultry, seafood and beans."
Key Nutrients; Carbohydrates & Fats; "Carbohydrates are our main source of energy. Sources include fruits, grains and beans. Fats also provide energy and help our bodies use certain vitamins. Healthy sources include olive and canola oils. Limit solid fats, like butter."
Key Nutrients; Vitamins & Minerals; "Vitamins help our bodies function properly. Sources include vegetables, fruits and whole grains. Minerals provide the building blocks for bones and other tissue. Sources include leafy greens, dairy, meat and poultry."
Healthy Eating Basics; Balance Calories; "One key to a healthy eating pattern, and a healthy weight is to take in only as many calories as you burn. Always check the calories in foods and drinks. To stay on track, adjust your intake to match your daily calorie needs."
Healthy Eating Basics; Get Needed Nutrients; "Choose foods and drinks high in fiber, vitamins and minerals, and low in solid fats, added sugar and sodium. Fill half your plate with vegetables and fruits. Round it out with whole grains, lean proteins and low-fat or fat-free dairy products."
Healthy Eating Basics; Know What To Limit; "Limit items high in calories, sodium, cholesterol, saturated or trans fats, or added sugars. For example, eat less red and processed meat, like sausage, full-fat cheese, fried food, soda and alcohol."

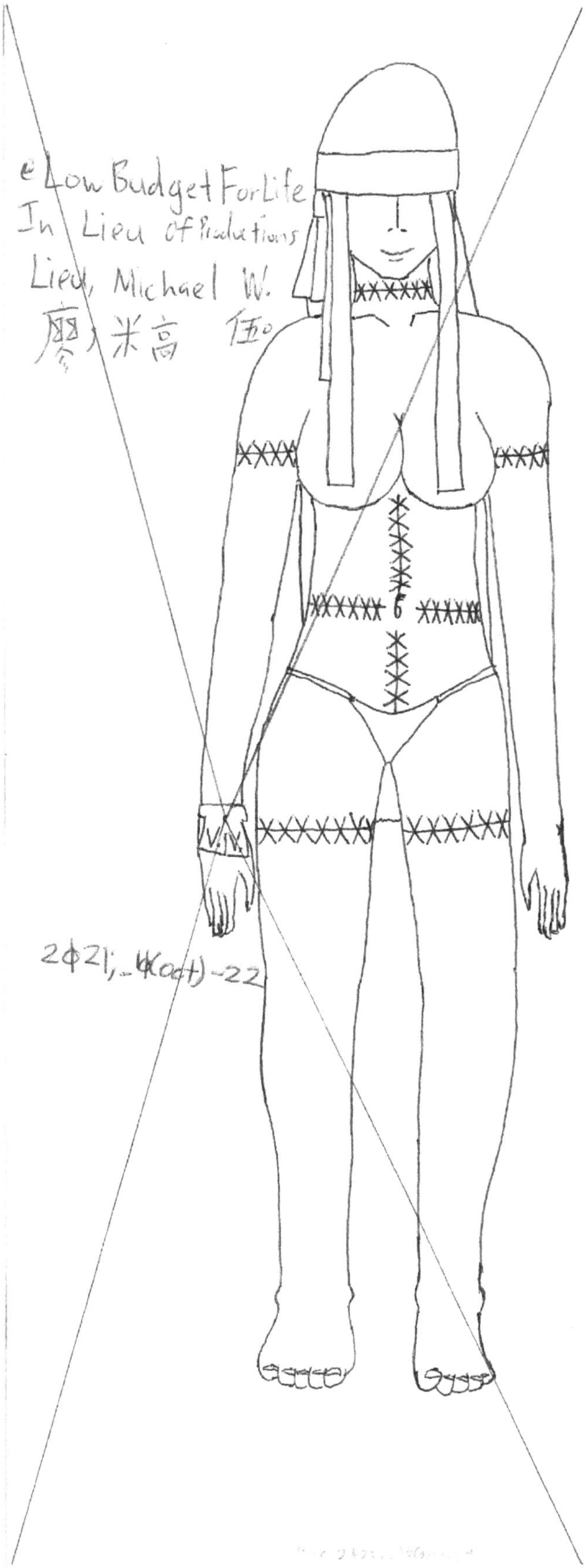

©Low Budget For Life
In Lieu Of Productions
Lieu, Michael W.
廖, 米高 伍o
2021-10(Oct)-22

A Good Nutrition Guide Earth's Global Government (EGG) For Earthlings Human Continued

Eat A Variety Of Foods;	
Vegetables	"Have plenty of beans, peas and red, orange and dark-green vegetables. Count 473 milli Liters (mL) of leafy greens or 236 mL of other raw, cooked or juiced vegetables as 236 mL. For a typical 2,000-calorie diet, you need 591 mL a day."
Fruits	"For a 2,000 calorie diet, you need 473 mL of fruits a day. Count 236 mL of raw, cooked or juiced fruit or 118 mL of dried fruit as 236 mL. Go for mostly whole or cut-up fruit. Limit juice."
Dairy Products	"Go for low-fat or fat-free dairy products, including milk, cheese and yogurt. For a 2,000-calorie diet, you need 710 mL a day. Court 44 mL of hard cheese or 236 mL of milk, yogurt or fortified soy "milk" as 0.23 Liters.
Grains	"For a 2,000 calorie diet, you need 177 mL of grains a day. Choose whole grains for at least half of your daily intake. Count 118 mL of cooked cereal, rice or pasta, 1-30 mL slice of bread or 236 mL of cold cereal as 29 mL."
Proteins	"Choose a variety of different proteins. Eat seafood at least twice a week. For a 2,000-calorie diet, you need 163 mL of protein a day. Count 30 mL of lean meat or seafood, 1 egg or 59 mL of cooked beans as 29 mL."
Get Plenty Of Exercise Too!	"Regular exercise is also key to good health- and a healthy weight. Most adults should get at least 2 ½ hours of moderate – or 1 ¼ hours of vigorous – exercise each week. Check with your health-care provider before starting."

Question, How many calories a day do you need?

Answer, It depends on your age, biological sex, any augmentation biological and or technological "biotech" enhancements you currently have equipped and or implanted, and activity level. The ranges below are a public general guidelines for Human Earthlings.

Males (M)	Females (F)	Hermaphrodite (H)
All children Males (M), Females (F), Hermaphrodites (H), Ages 2-3 = 1,000-1,400		
Ages 4-8 = 1,200-2,000	Ages 4-8 = 1,200-1,800	Ages 4-8 = 1,200-2,200
Ages 9-13 = 1,600-2,600	Ages 9-13 = 1,400-2,200	Ages 9-13 = 1,800-3000
Ages 14-18 = 2,000-3,200	Ages 14-30 = 1,800-2,400	Ages 14-18 = 2,200-3,600
Ages 19-30 = 2,400-3,000	Ages 31-50 = 1,800-2,200	Ages 19-30 = 2,400-3,800
Ages 31-50 = 2,200-3,000		Ages 31-50 = 2,600-4000

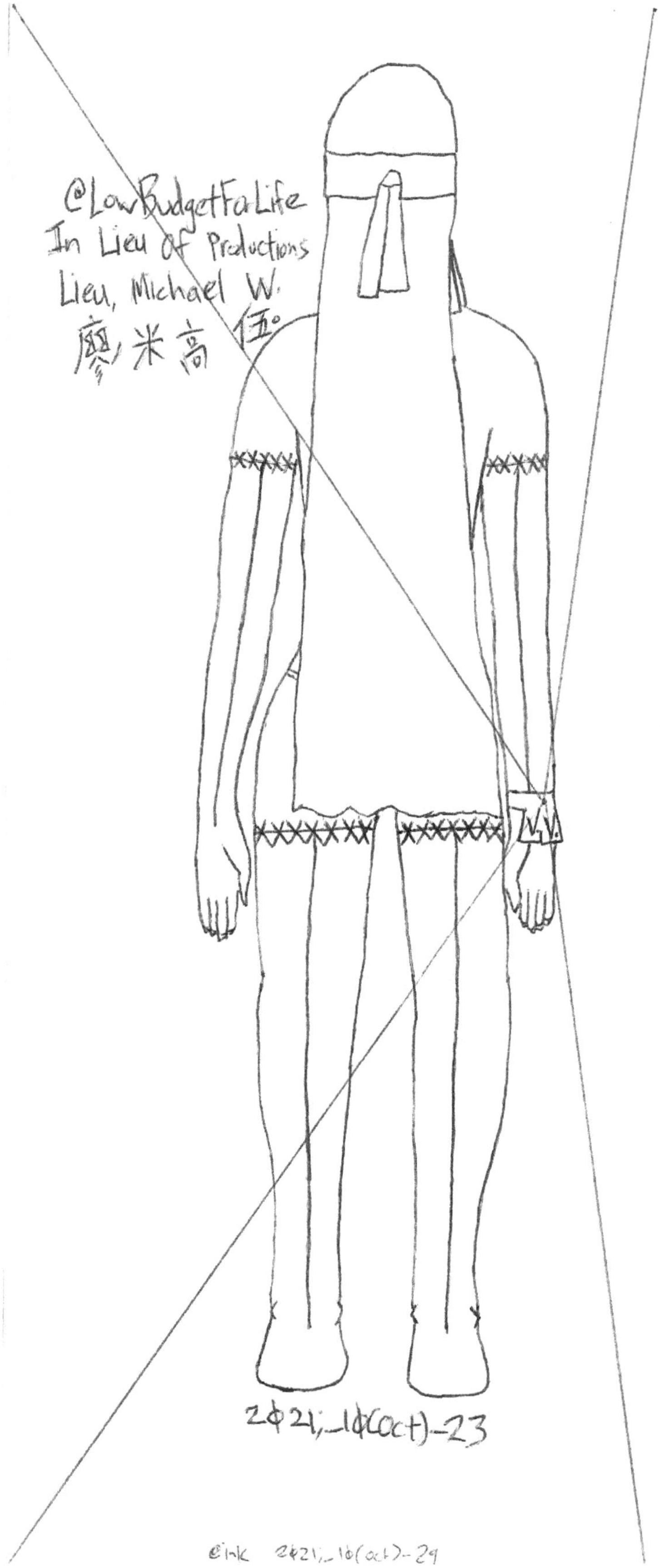
@LowBudgetForLife
In Lieu Of Productions
Lieu, Michael W.
廖, 米高 伍o
2Φ21,-1Φ(Oct)-23
@ink 2Φ21,-1Φ(oct)-24

Beauty Implant Augmentations (Aug+);

Elysium Corporation

Established 2077

The Magical Corporation Of Desire

Biology-Biological + Technology-Technological = Bio Tech (B.T.); Biotech Implants And Or Explants

Physical Plus Psychological, Beautiful Biotech Body Enhancement(s)

Physical Body Alterations, Modifications (Mods) And Splicing Junction

Biology = Organism	Technology = Cybernetics	BioTech; CybOrg
[Selected] Biology = Reptitlian Rapid Regeneration Scaling + Chameleon Color Toner Shifting		
[Selected] Technology = Eternal Electrons Electrical Tubes + Electro Magnetic Pulse Shielding		

Biotech On Stand By, Life Extension, Existence Extension, Biological Immortality, Technological Immortality, Biotech Immortality

Artificial Synthetic Organ(s)	Kloning, Klone(s)	Cryonics And Stasis
Augmented And Natural Organ Bank	Organ Harvesting	Organ Transplant(s)

Valentina Vita (V. V.); Traveling Cosplayers And Kosplayers;

"(=_=) My **forever young enhancements now ready**. **Self-talk** time, **divide between buying or commission a cosplay, makes you not a quote real cosplayer unquote**… I am the **type of cosplayer that almost usually crafts everything by** myself… However, there are **situations where buying the costume item or part is better than making it due to time**… Sometimes, **aiming to make a specific fabric, pattern, or looking for that certain décor may drive you crazy**, plus it is not that cost efficient. I **do admire, look up to those who create all of their cosplay**… physical and or holo, however, I **would not tell those who buys cosplay parts that they are not a quote real cosplayer unquote**. That is just silly and not nice. (^_^) **Cosplay is a display of the time, trends, in our historical universe or alternative history, showing the crafted works**, however, not the full self-esteem. (>_<) I **know people that love the art of crafting, yet feel very insecure about their image, their visual appearance they project**… (O_<) **Physically the cosplay needs to be ready to go, ironed out, no threads sticking out anywhere, believe** me, **it will be visible, and that makes the whole costume ugly!** (O_O) …**Need to have some posing references**. Should give an idea of how **we** present ourselves. I am a control freak due to my attention to detail."

2154;_Arcadia \ Valentina Vita (V. V.) 'Healthy-Strong Life' \ Left to Right(L->R)

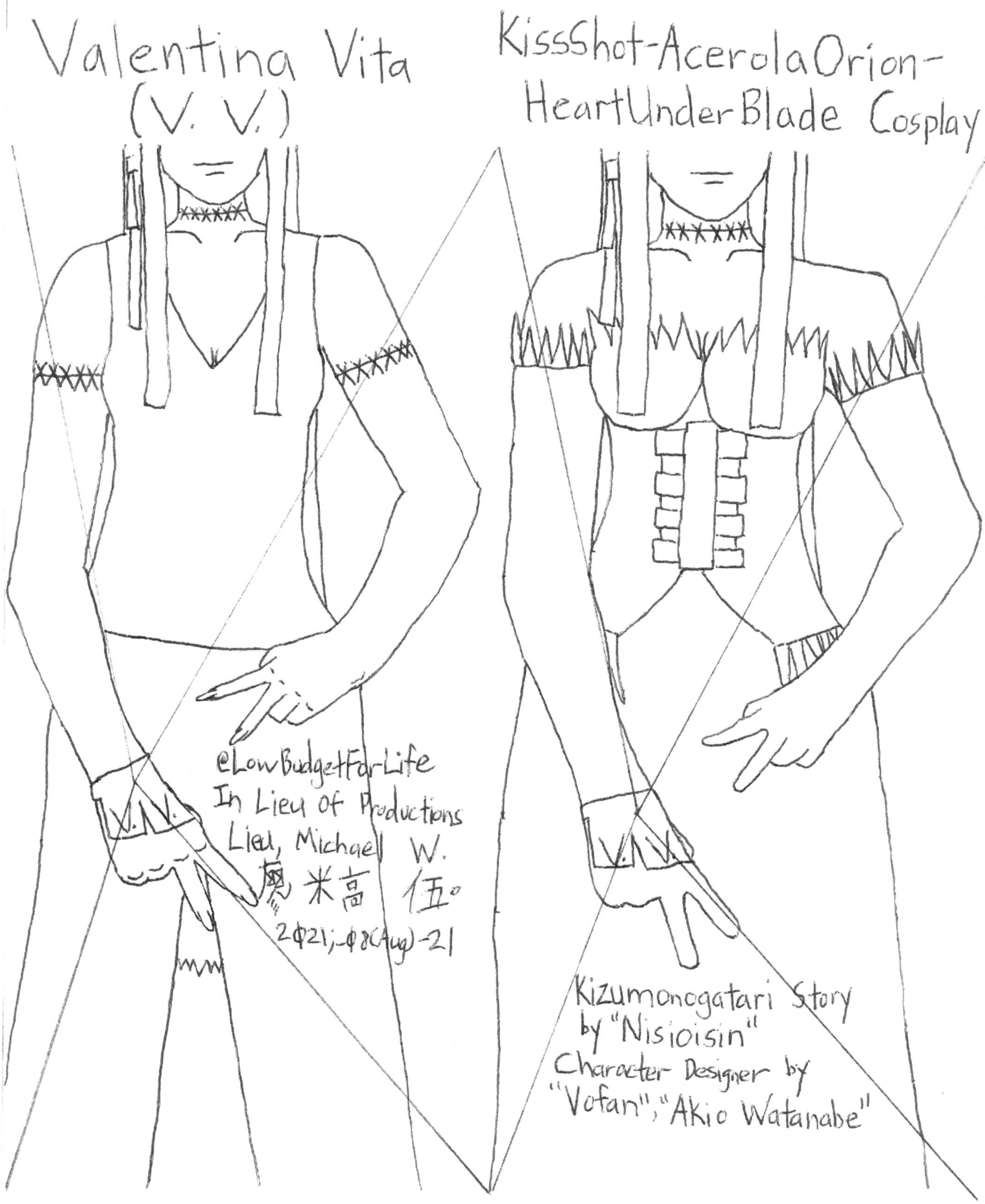

inke 2Φ21;-1Φ(Oct)-29

2154;_08(Aug)-15(fifteen)

V. V. selects the music(s):
"**Kizumonogatari Part 1 - Tekketsu-hen -** Complete Original Sound Track (O.S.T.)
Kizumonogatari Part 2 - Nekketsu-hen - Complete Original Sound Track (O.S.T.)
Kizumonogatari Part 3 - Reiketsu-hen - Complete Original Sound Track (O.S.T.)
Nekomonogatari Kuro (Black) Complete Original Sound Track (O.S.T.)
Bakemonogatari Complete Original Sound Track (O.S.T.)
Nisemonogatari Complete Original Sound Track (O.S.T.)
Kabukimonogatari Complete Original Sound Track (O.S.T.)
Onimonogatari Complete Original Sound Track (O.S.T.)
Owarimonogatari Volume 2 Complete Original Sound Track (O.S.T.)
Nekomonogatari (White) Complete Original Sound Track (O.S.T.)
Owarimonogatari Volume 1 Complete Original Sound Track (O.S.T.)
Wazamonogatari (Tsubasa Sleeping) Complete Original Sound Track (O.S.T.)
Orokamonogatari Complete Original Sound Track (O.S.T.)
Otorimonogatari Complete Original Sound Track (O.S.T.)
Koimonogatari Complete Original Sound Track (O.S.T.)
Tsukimonogatari Complete Original Sound Track (O.S.T.)
Owarimonogatari Volume 3 Complete Original Sound Track (O.S.T.)
Zoku Owarimonogatari Complete Original Sound Track (O.S.T.)
Hanamonogatari Complete Original Sound Track (O.S.T.)
Shinboumonogatari Complete Original Sound Track (O.S.T.)
Wazamonogatari (Karen Ogre) Complete Original Sound Track (O.S.T.)
Orokamonogatari (Suruga Bonehead) Complete Original Sound Track (O.S.T.)
Orokamonogatari (Tsukihi Undo) Complete Original Sound Track (O.S.T.)
Nademonogatari Complete Original Sound Track (O.S.T.)
Musubimonogatari Complete Original Sound Track (O.S.T.)"
to play in the **Utopia A.R. Holo HUD** background as a playlist and is set on loop.

<table>
<tr><td colspan="3">Department Of Energy Multiple uniVerse 2 (DOE MultiVerse Two)
Eastern Society, Asia, Japan. Monogatari Story Universe By "NISIOISIN",
Character Designer(s) By "Vofan", "Akio Watanabe"</td></tr>
<tr><td colspan="3">KissShot-AcerolaOrion-HeartUnderBlade brief character sheet stats overview</td></tr>
<tr><td>Age; Physically 600 Years Old</td><td>Legendary Vampire, Aberration Slayer</td><td>Two Winged Demon</td></tr>
<tr><td colspan="3">[Aug+]</td></tr>
</table>

Valentina Vita (V. V.); Traveling Cosplayers And Kosplayers;
"(^_^) Now that my **beauty augs are ready for travel**. Time to test try KissShot cosplay. Best vampire lady, and her long blade. **KissShot**-AcerolaOrion-**HeartUnderBlade**."

2154;_Arcadia \ Valentina Vita (V. V.) 'Healthy-Strong Life' \ Left to Right(L->R)

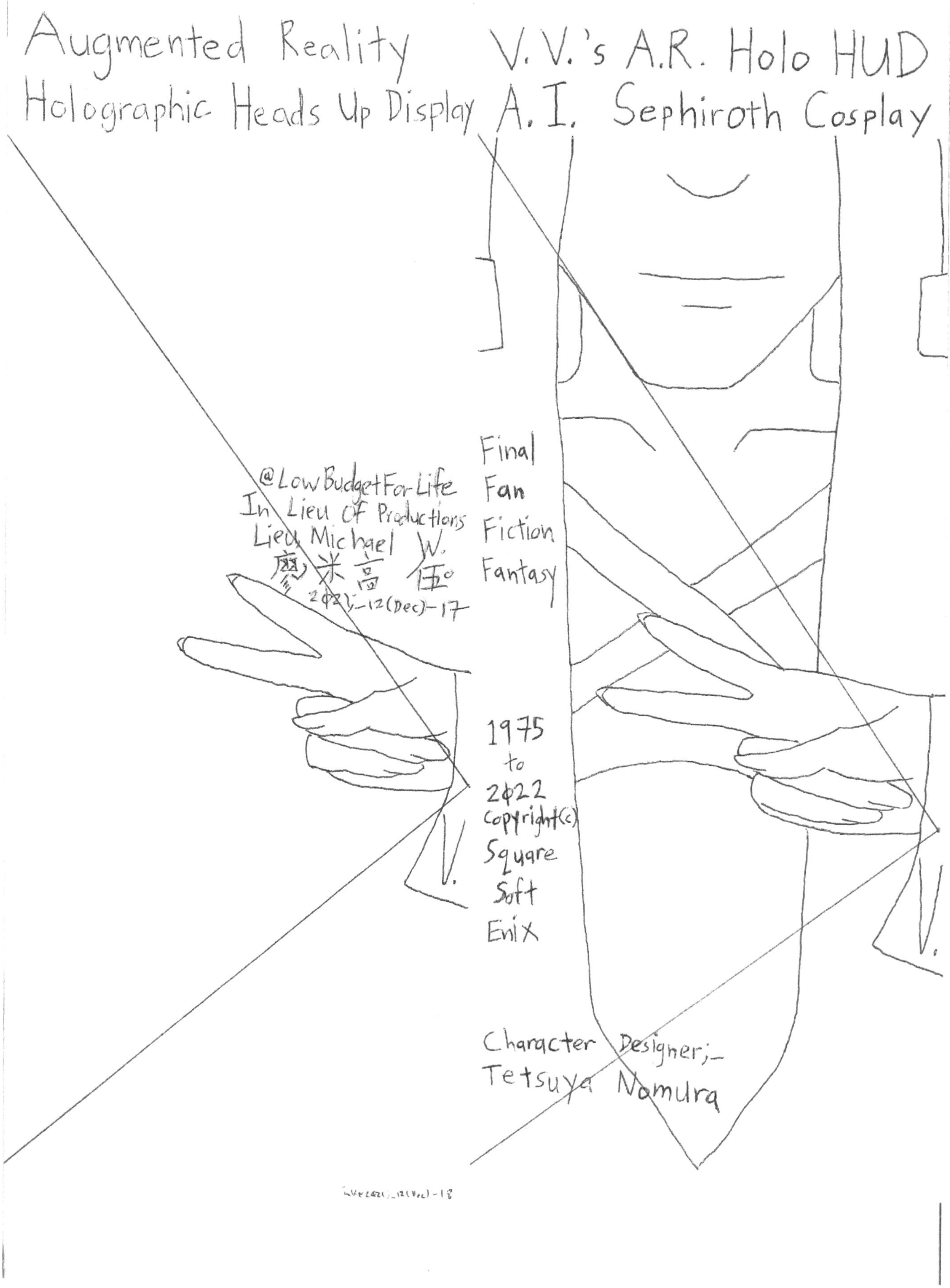
Augmented Reality
V.V.'s A.R. Holo HUD
Holographic Heads Up Display
A.I. Sephiroth Cosplay
Final
Fan
Fiction
Fantasy
@LowBudgetForLife
In Lieu Of Productions
Lieu Michael W.
廖米高 伍o
2021_12(Dec)-17
1975
to
2022
copyright(c)
Square
Soft
Enix
V.
V.
Character Designer;_
Tetsuya Nomura

Valentina Vita (V. V.); Traveling Cosplayers And Kosplayers;
"(>_O) Okay, now time to work on a **Holo HUD projection of Sephiroth**. **Another well known fictional character using long blades**... Got to say this **Final Fantasy seven and Monogatari universe is a unique entertainment creation**, but did not arrive from **our historical Imperial Japan**, rather an **alternative history where Imperial Japan does not exist past the year of nineteen forty five**... Hm... The **UltraViolet Rays of the Multiple uniVerse theory is interesting**."

Department Of Energy (DOE) Multiple uniVerse (MultiVerse) 2 (Two) Eastern Society, Asia, Japan. Compilation Of Final Fantasy VII (7; seven) Universe By "Square Soft Enix", Character Designer(s) By "Tetsuya Nomura", "Kazushige Nojima"		
Sephiroth brief character sheet stats overview		
Age; Physically 25 Years Old	Legendary SOLDIER 1st Class	"One Winged Angel" Music
[Aug+]		

Cosplay Visual References
(left) Sephiroth + **KissShot-AcerolaOrion-HeartUnderBlade (right)**

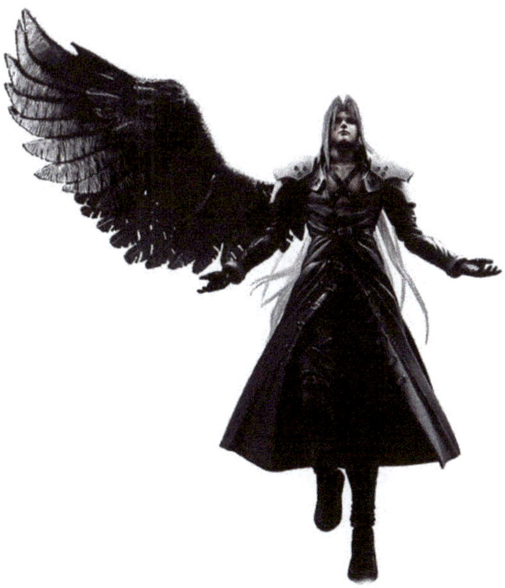

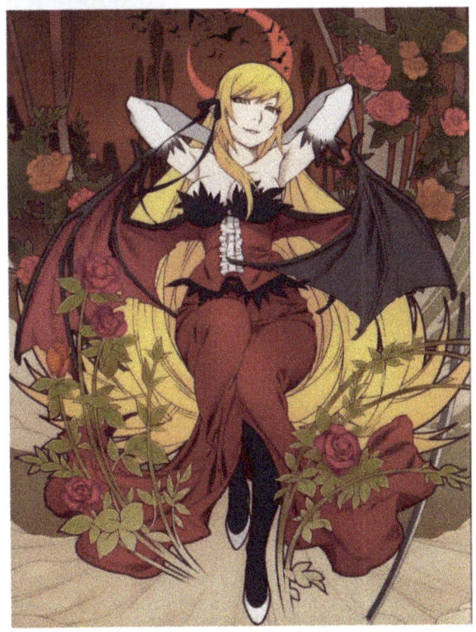

One Winged Angel

Two Winged Demon

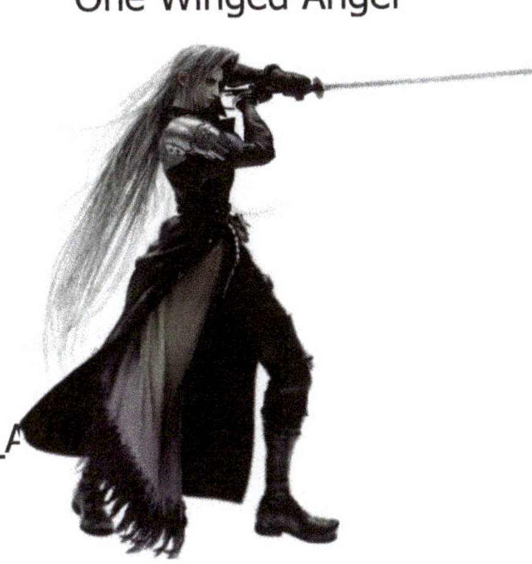

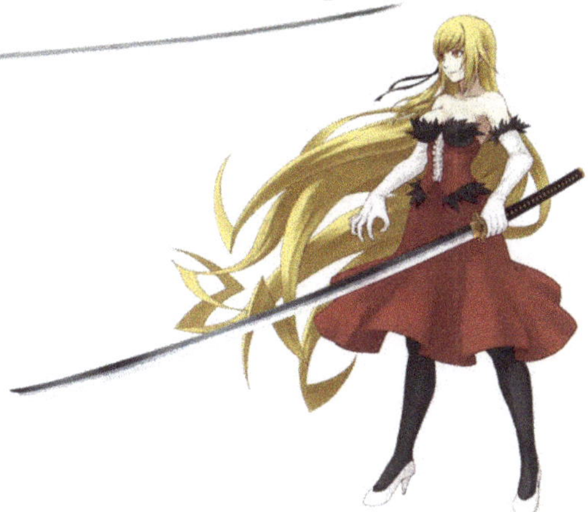

2154;_A :(L->R)

2154;_08(Aug)-30(thirty)

V. V.'s Travel To Do Check List

What to pack...

Carry-On, Passport, Clothes On Luggage, Property Of Valentina Vita (V. V.)
extra clothes, tools and devices pack of;

Biotech **white shoes**
Biotech cotton **white socks**
Biotech cotton **white underwear**
Biotech spandex **red high waist yoga capri pants with pockets**
Biotech spandex **red sports top tank**
Biotech nylon **blue hair ribbon**
Biotech **blue headband visor and or nose mouth mask**
Biotech **blue goggles**
Elysium Corporation's Biotech Do It Yourself (D.I.Y.) Surgical Stitching Line
Biotech Face Mask(s)
Elysium Corporation's Beauty Implant Augmentation (Aug+); Travel Size Biotech Cosmetic Product Enhancement(s)
Travel size biotech cosplay emergency handy kit
Toiletries: tooth paste, toothbrush, floss line, mouth wash, tongue brush, mouth guard
Toilet Paper (T.P.) rolls
Ai-Aug **V. V.** bracelet
Physical print paper(s) with coloring pencils, red and **black** pens

Cosplay Clothes On Luggage, Property Of Valentina Vita (V. V.)
KissShot-AcerolaOrion-HeartUnderBlade
fashion

Cosplay Clothes Wearing For Travel, Property Of Valentina Vita (V. V.)
Aranea C. Highwind, Lady A + Lightning, Claire Farron = Dragoon Knight Fusion Remix
Full Body Personal Protective Equipment (P.P.E.)
No Extra Vehicular Activity (EVA) space suit
P.P.E. EVA space suit augment purifying "Water (H_2O)" and breathable "Oxygen (O_2)".
function

2154;_ Naples, Italy Intranet, Wireless Broadcasting Frequencies Online

Red 紅
Orange 橙
Yellow 黃
Green 綠
Blue 藍
Violet 紫

2154;_ Naples, Italy Intranet, Wireless Broadcasting Frequencies Online

2154;_ Naples, Italy Intranet, Wireless Broadcasting Frequencies Online

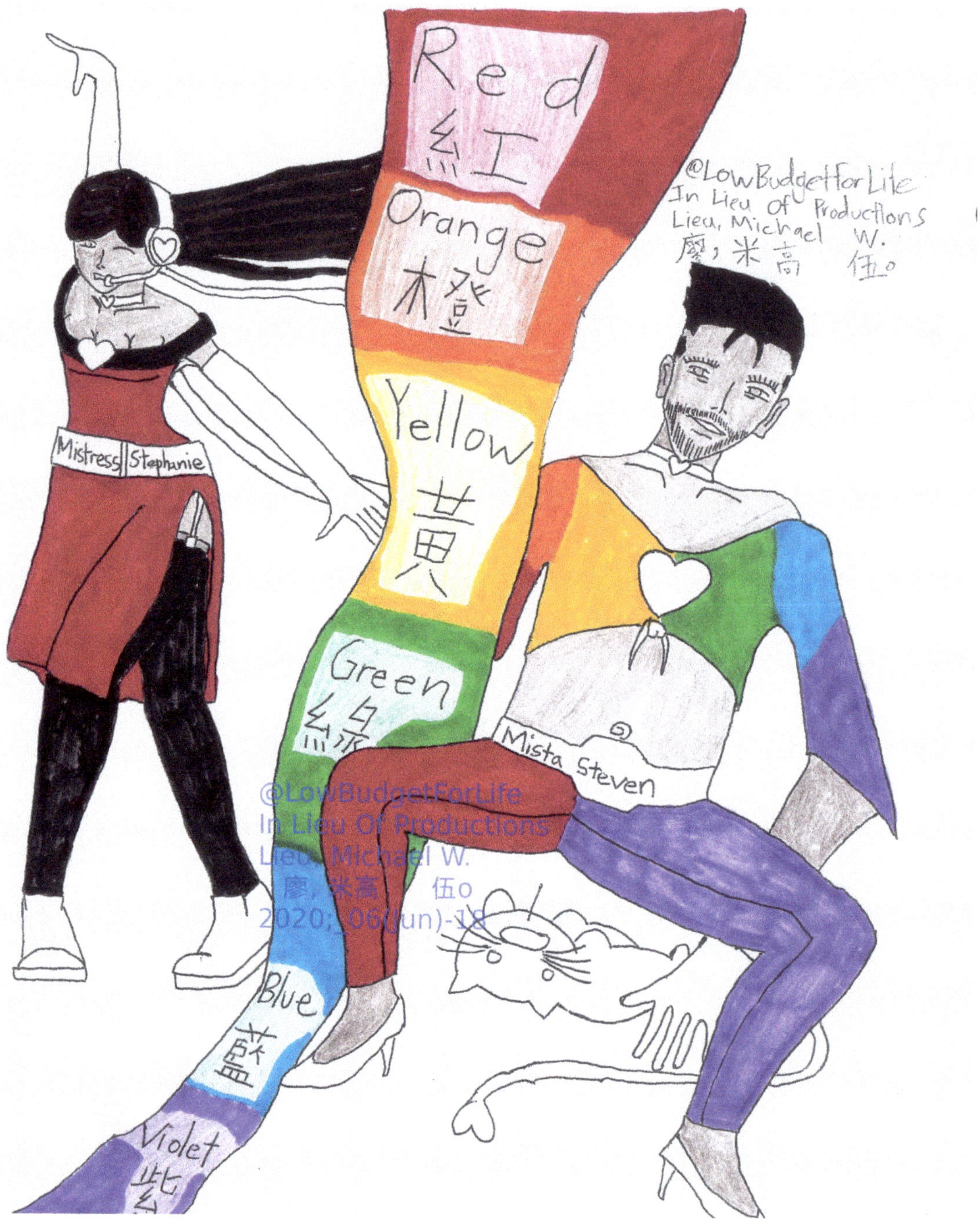

2154;_ Naples, Italy Intranet, Wireless Broadcasting Frequencies Online

2154;_ Arcadia \ Valentina Vita (V. V.) 'Healthy-Strong Life' \ Left to Right(L->R)

The music "**Henry Mancini's The Entertainer**" plays in the **Utopia A.R. Holo HUD** background.

Mista Steven, Thrilling Plastination Entertainer, Utopia A.R. Holo HUD;
"(*female voice tone*) Hello V. V., meet **the twenty second century's most wealthiest, energetic, influential show business person in the entertainment industry with a net worth of billions**!! Get ready to spread the kindness and joy in the biotech paradise of love... **Our Entertainer, Mister Steven Z. J.** is going be in Chicago City, **Elysium Corp** Dome's **Animal** Sanctuary in **Illinois** for **our twenty one fifty four U.S.A. grand tour** on **November based on the old world** United **States of North America's** Military **Veterans' Day to Black Friday**, reserve your tickets! Remember to say **reserve** your **tickets** when presenting your tickets! Ha ha!!"

The music "**Michael Jackson's Thriller**" plays in the **Utopia A.R. Holo HUD** background.

Mistress Stephanie; Thrilling Plastination Entertainer, Utopia A.R. Holo HUD;
"(*male voice tone*) Remember everyone, you **are the ideal person or being the way** you **are** as an **animal kingdom, human beings**, and earth's environment! Stay true to yourself and... and most importantly... (*confidence*) Love yourself! Love! Yourself!"

Mista Steven; Thrilling Plastination Entertainer, Utopia A.R. Holo HUD;
"(*female voice tone*) **An award winning performance**! (*clapping*) Bravo... Bravo! (*walking around*) ...**Stop sleeping on set**! Wake up!! (*finger snapping*)
No sleeping on set!! **No sleeping on job**!!"

Mister Steven Z. J.; Thrilling Puppeteer Entertainer, Utopia A.R. Holo HUD;
"(*male voice over*) Assume the position. Colorful **Entertainer... We'll be back after these important messages**! **Elysium Corporation's Undead Utopian Plastination Preservation... Before your** natural **unmodified self expires, you can create an artificial beautiful biotech body of** yourself**...** and be **forever young!**"

Cute Cawaii Cat-Kat; Thrilling Animal Entertainer, Utopia A.R. Holo HUD;
"(*stuff animal cat-kat*) Meow! Meow. Me. Ow."

V. V. sets her Ai-**Aug** V. V. bracelet to **mark and track** her calendar for: "**Meet and greet holo photo with Entertainer Mister Steven Z. J.,** Mista Steven, Mistress Stephanie, Cute Cawaii Cat-Kat for **2154 Utopian Sovereignty America(s) (U.S.A.)** grand tour on **2154;_11(Nov)-11(eleven) to 28(twenty eight)** "**Reserve** your **tickets**."

V. V. selects the music "**SoundTemp's Ragnarok Online:** *Title, Peaceful Forest, Tread On The Ground, Prontera Theme, Streamside, Payon Theme, Travel, Plateau*" to play in the **Utopia A.R. Holo HUD** background and is set on loop.

Eur-Asia Continent Civilian(s) Citizen(s) Travel Advisory

For Earthlings on traveling to America(s) continent: North America and or South America

In Case Of An Emergency

Due to the Martian Outbreak Outage in 2120 (MOO-20) unconventional warfare, depending on which sector you **travel to in the America(s) of North America and or South America**, be aware that **Earth's Global Government** (**EGG**) **Joint Task Force (J.T.F.)** primarily provides 24 hours' first responders to external protection from Foreign Space Alien Martian Terrorist(s), as a Rapid Response Team (R.R.T.), and provides little resources against Homegrown Terrestrial Earthling Domestic Terrorist(s).

Republic (R) Color Code "red" = Military Police (M.P.) and or Fire Department (F.D.) emergency dispatch **911** for Earthlings **augmented biotech humans**: Males (M), Females (F), and Hermaphrodites (H) plus natural **unmodified humans**: Males (M), Females (F).

Anarchy (A) Color Code "orange" = **Zero** emergency dispatch and services. **Do not travel to this sector in of America(s) within North America and or South America.**

Surprise-Face (O :) Color Code "yellow" = **Forwarding** emergency services. Public-Private sector joint non-profit organization operation of **saving and creating** "D Archives".

Military (M) Color Code "green" = Military Police (M.P.) and or Fire Department (F.D.) emergency dispatch **911** for Earthlings **augmented biotech humans**: Males (M), Females (F), and Hermaphrodites (H) plus natural **unmodified humans**: Males (M), Females (F).

Democratic (D) Color Code "blue" = Military Police (M.P.) and or Fire Department (F.D.) emergency dispatch **911** for Earthlings **augmented biotech humans**: Males (M), Females (F), and Hermaphrodites (H) plus natural **unmodified humans**: Males (M), Females (F).

Sovereignty (S) and or Animal Sanctuary (S) Color Code "violet / purple" = Emergency dispatch **6969** for Earthling **augmented biotech and** natural **unmodified animals**, "**decreased humans, feral children**", be **vigilant for the unpaid volunteer workers** for any **augmented and** natural human needs on **emergency services**.

The 22nd Century Phenomenon Broadcasting Biotech Tower(s)

These UltraViolet (U.V.) City Controllers' City Mayor(s) Phenomenon of the new world 2121 Utopian Sovereignty (U.S.) movement was **built on the ashes** of the **old world** United **States** of America (U.**S**. of A.) collapse on 2121 of March, offering a unique utopia self-govern paradise of an Animal Sanctuary (S) and controversial "Human Sanctuary (S)".

If you are not comfortable being around "**decreased humans, feral children**", avoid going to these U.V. U.S. **Elysium Corporation** Biotech Dome Sovereign Sector(s).

2154;_Arcadia \ Valentina Vita (V. V.) 'Healthy-Strong Life' \ Left to Right(L->R)

Eur-Asia Continent Civilian(s) Citizen(s) Travel Advisory

For Earthlings on traveling to America(s) continent: North America and or South America

City (D) (R) Military (M) Bases

"**City** Military Bases" is a fusion of '**old world**' United **States** of Americas' civilian (D) (R) life style mixed with United **States** armed forces Military (M) Bases. **City** Military Bases do not use "**old world**" civilian Police Department (P.D.), rather uses Military Police (M.P.). M.P. follows the United **States** Constitution and Uniform Code Military Justice (U.C.M.J.).

Cash Versus Card

Depending on which sector you travel to in North America and or South America, be aware that the **currency accepted varies** under the **political party system**.

Republic (R), Military (M), Democratic (D); accepts conventional currency from banks
United **States** of America Dollars (U.**S**. of A.D.) cash and card only.
Does not accept cryptocurrency of any kind.

Anarchy (A); does not accept and or use any conventional currency and or cryptocurrency.

Surprise-Face (O :); "D Archives" does not use physical meat space currency.

Sovereignty (S) and or Animal Sanctuary (S); accepts the following currency listed below

United **States** of America(s) Dollars (U.**S**. of A.D.) "**old world**"
Utopian Sovereignty Dollars (U.S.D.) "**new world**"
Eur-Asia Conventional Currency
African Conventional Currency
Zimbabwean Dollars In One Bills Only, Conventional Currency
Cryptocurrency
Bartering And Trading
Non-Paid Volunteer Worker Hour(s)
Mass Marketplace Monopoly Money
Elysium Corporation's Kash-Kredit(s) Accounting System

Ai-Aug, Biotech, Electricity, Power Recharge Source(s)

Conversion And Adapter Plugs	
North America = 110 volts	South America = 220 volts
Eur-Asia = 220 volts	Australia = 230 volts
Elysium Corporation On Planet Earth And Earth's Lunar Moon Military Bases = X variable	

Eur-Asia Continent Civilian(s) Citizen(s) Travel Advisory

For Earthlings on traveling to America(s) continent: North America and or South America

Measuring System

In **North America, some human societies use the minority outdated *Imperial*** measuring system: Fahrenheit degrees (F°), Miles Per Hour (M.P.H.), Inches ("), pounds (lbs) Ensure you are aware of the majority ***Metric*** measuring system: Celsius degrees (C°), KiloMeters (K.M.), CentiMeters (C.M.), KiloGrams (K.G.) differences and conversion.

Transportation(s)

Biotech Zipline(s)	Biotech VTOL Aircraft(s)
Biotech Magline(s) Armlet(s)	Biotech Traveling Sphere(s)
Biotech Maggrip(s) Glove(s)	Biotech Gravity Shifting(s)

Consumption Tax And Tax Free

Consumption tax in **North America and or South America** under Republic (R), Military (M), Democratic (D) is ten percent (10%). Military (M) Bases is **tax free to any active duty military servicemembers** under **Earth's Global Government (EGG) Joint Task Force (J.T.F.)**. Sovereignty (S) and or Animal Sanctuary (S) sectors do not have and or charge consumption taxes.

Travel Visual-Verbal Communications Ai-Aug Device

All Earthlings on Planet Earth can **connect wirelessly and or wired via unlimited bandwidth of verbal-visual audio-video broadcasting** using **Sir Arthur C. Clarke's Space Elevator(s) Satellites, Solar Panels, Space Stations And Arrays** under **Earth's Global Government (EGG) Joint Task Force (J.T.F.)** 24 (twenty four) hours watch.

Good To Know Information

Carry your own Toilet Paper (T.P.), reusable eating, drinking, consuming utensils at all times.

The UltraViolet (U.V.) Utopia Self-Governing Phenomenon Animal Sanctuary (S)
The New World 2121 Movement Of The 22nd Century
Utopian Sovereignty of North America (U.S. of N.A.)
Trash Disposal Waste Separation Bins

Paper	Plastic	Metal	Unknown

Eur-Asia Continent Civilian(s) Citizen(s) Travel Advisory

For Earthlings on traveling to America(s) continent: North America and or South America

Planet Earth, Human Earthling Known Language(s)

Reads; Left To Right (L→R), Top To Bottom

Arabic	Khmer
Armenian	Lao
Cambodian	Mien
Centralnese Cantonese	Navajo
Centralnese Mandarin	Punjabi
Corean	Russian
English (African)	Spanish (Eur-Asia)
English (Australia)	Spanish (North America)
English (Eur-Asia)	Spanish (South America)
English (North America)	Tagalog
English (South America)	Thai
Farsi	Ukrainian
German	Vietnamese
Hindi	X
Hmong	If you do not see your Human Earthling known language(s) select here. [Aug+]
Italian	
Japanese	
International Phonetic Alphabet (I.P.A.)	

Acronyms

List of letters to know when verbally talking to others... Not in A.B.C. Alphabet order...

ARM = Anti Robotics Movement
MOO = Martian Outbreak Outage (and or) MOO-20 = Martian Outbreak Outage 2120
EGG = Earth's Global Government
J.T.F. = Joint Task Force
EVA = Extra Vehicular Activity
C.B.R. = Chemical, Biological, Radiological
E.M.P. = Electro Magnetic Pulse
B.T. = Bio Tech
U.A.V. = Unmanned Aerial Vehicle
F.M. = Frequency Modulation
A.M. = Amplitude Modulation
L.L.C. = Limited Liability Company
6 C = Centralnese Cantonese language in most of Asia, Eur-Asia, meaning "Eat Poo"
D.N.A. = Deoxyribo Nucleic Acid
A.I. = Artificial Intelligence
I.A. = Individual Augmentee
P.P.E. = Personal Protective Equipment
S.T.D. = Sexual Transmitted Diseases
H.I.V. = Human Immunodeficiency Virus
AIDS = Acquired Immuno Deficiency Syndrome
H.P.V. = Human Papailloma Virus
I.Q. = Intelligence Quotient
E.Q. = Emotional Quotient
S.S. = (in context) Synthetic Synthesis (in context) Schutz Staffel (In German, The Third Reich Of Nazi Party)
VTOL = Vertical Take Off Landing
R.P.G. = (in context) Role Playing Games / (in context) Rocket Propelled Grenades
H.D.D.V.D. C.D. = High Definition Digital Versatile Disc Compact Disc
M.F.H. = Male(s), Female(s), Hermaphrodite(s)

Acronyms' Continued

List of letters to know when verbally talking to others... Not in A.B.C. Alphabet order...

"Asian Acronyms"

ChEng = **Ch**ief **Eng**ineer
TAO = Tactical Action Officer
PAO = Public Affairs Office
GAO = Government Accountability Office

Confusing Meaning Of "U.S." Acronyms, America(s)

U.S. of A. = (in context) United States of America(s) (in context) Utopian Sovereignty of America(s)
U.S.A. = (in context) United States Army (in context) Utopian Sovereignty Army
U.S. = (in context) United States (in context) Utopian Sovereignty
U.S. of N.A. = (in context) United States of North America (in context) Utopian Sovereignty of North America
U.S. of S.A. = (in context) United States of South America (in context) Utopian Sovereignty of South America

P.C. = (in context) Personal Computer / (in context) Political Correctness
A.C. = Air Conditioning
B.B. = (in context) Battle Bot(s) / (in context) Breeding Buddies
SAR = (in context) Search And Rescue; for **alive** status SAR = (in context) Search And Recovery; for **undead** and or **deceased** status
M.P. = Military Police
P.D. = Police Department
F.D. = Fire Department
SWAT = Special Weapon And Tactics
R.R.T. = Rapid Response Team
E.M.T. = Emergency Medical Technician
U.S.S.R. = Union Soviet Socialist Republics A.K.A. Soviet Union "Soviets"

V. V.'s Truth(s) To Remember

We do not remember
days, but moments. Life
moves too fast, so
enjoy your precious moments.

A grudge is a heavy thing to carry.

Beauty fades, stupidity is forever.

Growing old is inevitable.
Growing up is optional.

When learning, always master the basics to a professional level...
Moving to the next level will be like learning the basics again.

If words do not work, use force.

Laughing everyday is just like jogging everyday.

Those who are good at making excuses are rarely good at anything else.

If you do not stand up for something, you will fall for anything.

Everyone ***hears*** what you say.
Friends ***listens*** to what you say.
Best friends listen to what you ***do not say***.

No human, augmented or natural, is worth your tears, the one who is will not make you cry.

V. V.

V. V.'s Life's Little Instructions

Sing in the shower.
Treat everyone you meet like you want to be treated A.K.A. "Golden Rule".
Watch a sunrise at least once a year.
Leave the toilet seat in the down position.
Do not refuse homemade brownies.
Strive for excellence, not perfection.
Plant a tree on your birthday and or builtday.
Learn three clean jokes.
Return borrow vehicles with the biotech fuel tank full.
Compliment three people everyday.
Do not waste an opportunity to tell someone you love them.
Leave everything a little better than you found it.
Keep it simple.
Think big thoughts, but relish small pleasures.
Become the most positive and enthusiastic person you know.
Bush and floss your teeth.
Ask for a raise when you feel you earned it.
Being forgiving of yourself and others.
Over tip breakfast waitresses.
Say "thank you" a lot.
Say "please" a lot.
Avoid negative bad people.
Buy whatever kids and children are selling on card tables in their front yards.
Wear polished shoes.
Remember other people's birthdays and or builtdays.
Commit yourself to constant improvement.
Carry jumper cables in the trunk of your vehicles.
Have a firm handshake.
Send lots of Valentine cards, sign them, someone who thinks you are terrific.

V. V.'s Life's Little Instructions' Continued

Look people in the eyes.
Be the first to say "hello".
Use the good sliver.
Return all things you borrow.
Make new friends, but cherish the old ones.
Keep secrets.
Sing in a choir.
Plant flowers every spring.
Have an animal: dog, cat, chick, turtle, bird pet.
Always accept an outstretched hand.
Stop blaming others, take responsibility for every area of your life.
Wave at kids and children on school buses.
Be there when people need you.
Feed a strangers' expired parking meter.
Do not expect life, reality, and existence to be fair.
Do not underestimate the power of life.
Drink champagne for no reason at all.
Live your life as an exclamation, not an explanation.
Do not be afraid to say, "I made a mistake." of active speaking. Not "Mistakes were made." of passive speaking.
Do not be afraid to say, "I do not know.".
Compliment even small improvements.
Keep your promises, no matter what.
Marry only for love.
Rekindle old friendships.
Count your blessing.
Call your mother.

V. V.'s 21 (twenty one) Suggestions For Success

1 (one) Marry the right person.
This one decision will determine 90% of your happiness or misery.

2 (two) Work at something you enjoy and that's worthy of your time and effect.

3 (three) Give people more then they expect and do it carefully.

4 (four) Become the most positive and enthusiastic person you know.

5 (five) Be forgiving to yourself and others.

6 (six) Be generous.	7 (seven) Have a grateful heart.

8 (eight) Persistence, persistence, persistence.

9 (nine) Discipline yourself to save money on even the modest salary.

10 (ten) Treat everyone you meet like you want to be treated.

11 (eleven) Commit yourself to constant improvements.

12 (twelve) Commit yourself to quality.

13 (thirteen) Understand that happiness is not based on possession, power, or prestige, but on relationships with people you love and respect.

14 (fourteen) Be loyal.	15 (fifteen) Be honest.	16 (sixteen) Be a self-starter.

17 (seventeen) Be decisive even if it means you will something be wrong.

18 (eighteen) Stop blaming others. Take responsibility for every area in your life, existence.

19 (nineteen) Be bold and courageous, when you look back in your life, you will regret the things you did not do have them the you did.

20 (twenty) Take good care of those you love.

21 (twenty one) Do not do anything that would not make your mom proud.

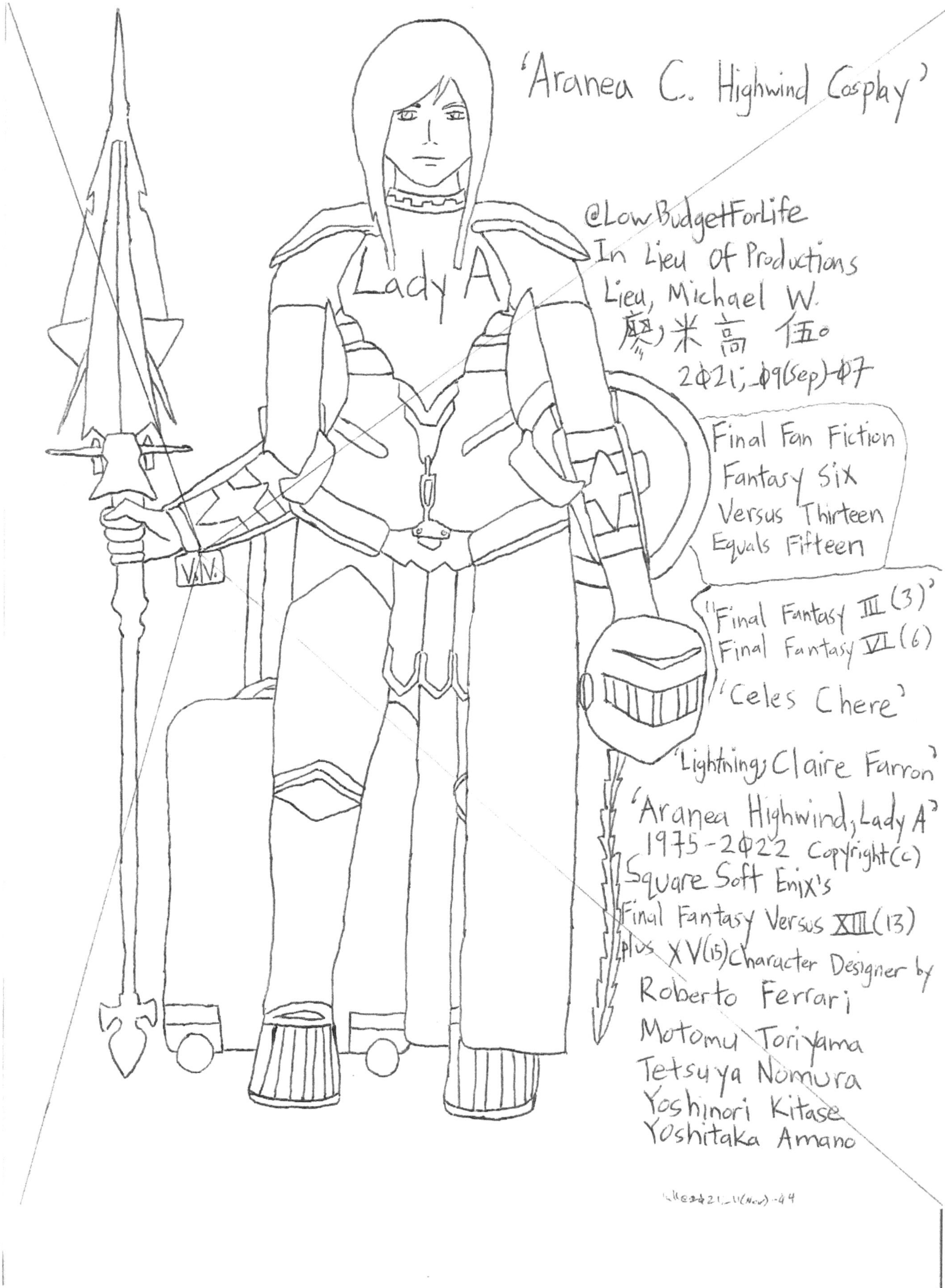
'Aranea C. Highwind Cosplay'
Lady A
@LowBudgetForLife
In Lieu Of Productions
Lieu, Michael W.
廖, 米高 伍o
2Φ21,-Φ9(Sep)-Φ7
Final Fan Fiction
Fantasy Six
Versus Thirteen
Equals Fifteen
'Final Fantasy III (3)'
Final Fantasy VI (6)
'Celes Chere'
'Lightning, Claire Farron'
'Aranea Highwind, Lady A'
1975-2Φ22 Copyright(c)
Square Soft Enix's
Final Fantasy Versus XIII (13)
plus XV(15) Character Designer by
Roberto Ferrari
Motomu Toriyama
Tetsuya Nomura
Yoshinori Kitase
Yoshitaka Amano
V.V.

2154;_08(Aug)-31(thirty one)

Valentina Vita (V. V.), Traveling Cosplayers And Kosplayers;
"(>_<) Finally ready now. I am going to **depart home**, wearing my **Lady A. Dragoon Knight cosplay** as an external P.P.E. suit… Not sure or certain on a proper changing place. Plus, Lady A. Dragoon Knight cosplay suit takes up more space in my travel luggage… Celebrated my **thirty fourth built day**… Packed all my needed travel gear… **Airlock decontamination time**!"

Departing And Standing By For →| Biotech Dome Airlock Decontamination |← Get Ready

Department Of Energy (DOE) Multiple uniVerse (MultiVerse) 2 (Two) Eastern-Western Society, Asia-Americas. Final Fan Fiction Fantasy 6 Versus 13 Equals 15 By "Square Soft Enix" & A Fan Fiction Productions By "Jay Doe". Character Designer(s) By "Roberto Ferrari", "Motomu Toriyama", "Tetsuya Nomura", "Yoshinori Kitase", "Yoshitaka Amano".		
Aranea C. Highwind, Lady A brief character sheet stats overview		
Age; Physically 30 Years Old	Niflheim Air Force Operative	"Dragoon Knight Jump"
[Aug+]		

Cosplay Visual References
(left) Aranea Highwind + Lightning Farron (right)

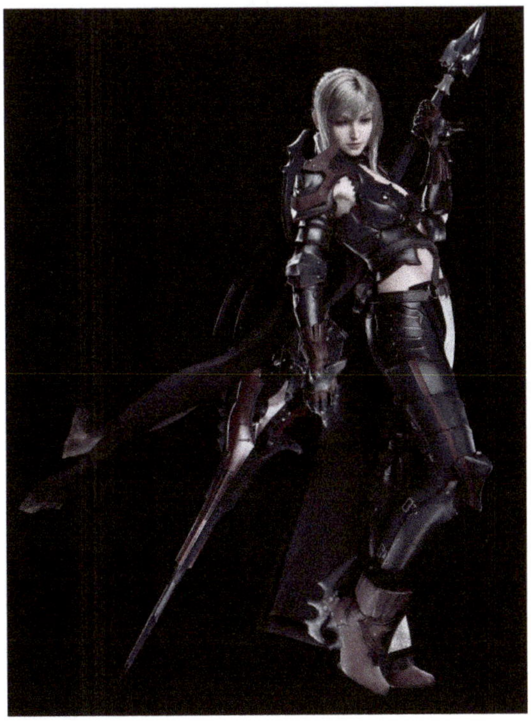

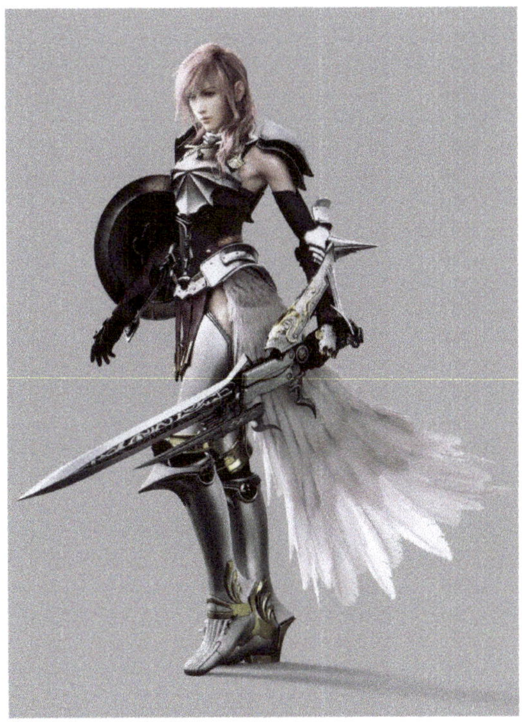

2154;_Arcadia \ Valentina Vita (V. V.) 'Healthy-Strong Life' \ Left to Right(L->R)

@LowBudgetforLife
In Lieu Of Productions
Lieu, Michael W.

Utopian, Augmented Biotech And Natural Unmodified Human Sanctuary (S)
2154;_ Naples, Italy's Intranet, Wireless Broadcasting Frequencies Online

Elysium Corp Augmented Reality Holographic Heads Up Display A.R. Holo HUD			
Valentina Vita (V. V.)'s Biotech Traveling Sphere(s) "Armed And Dangerous" with the following active agent(s) to prevent property theft. Made Possible By; Elysium Corporation Established 2077 The Magical Corporation Of Desire			
Biological-Chemical Agent(s)			
V. V.'s biometric frequencies signatures			
Chemical, Biological, Radiological (C.B.R.) Shielding	(C) Sulfur Chlorine Gas Agent(s) Mustard Gas	(B) Plague Yersinia Pestis	
		(B) Anthrax Bacillus Anthacis	
(B) Ebola Hemorrhagic Fever Virus	(C) Potassium Cyanide Pills	(B) Small Pox Varciola Major	
Technological Agent(s)			
Gravity Shifting: Gravity Zero (G0) to Gravity One (G1) of Planet Earth			
G0 to G1: 'Gravitational "X" 0.0, 0.1, 0.2, 0.3, 0.4, 0.5, 0.6, 0.7, 0.8, 0.9, 1.0 Pull Rates'			
Electro Magnetic Pulse (E.M.P.) Shielding	UltraViolet (U.V.) Broadcasting Frequencies	Sound Waves Echolocation	Spying Dust Particles
City Controllers' City Mayors' Immediate Removal Warrant		Vacuum Tubes	
Elysium Corp Augmented Reality Holographic Heads Up Display A.R. Holo HUD			

2154;_ Naples, Italy's Intranet, Wireless Broadcasting Frequencies Online
Utopian, Augmented Biotech And Natural Unmodified Human Sanctuary (S)

Naples Controller City Mayor selects the music "**Simon & Garfunkel's The Sound of Silence**" to play in the **Utopia A.R. Holo HUD** background and is set on loop.

Naples Controller, Psychological U.V. Projection As A City Mayor;
"(*voice over*) Start **Utopian**, departing **utopia** already... Answer my City Mayor **question** and follow the **broadcasting** UltraViolet Rays... **prepare for airlock**... period end."

Valentina Vita (V. V.), Traveling Cosplayers And Kosplayers;
"(^_^) **I will be returning back home,** sometime later by **twenty one fifty five**?"

2154;_ Arcadia \ Valentina Vita (V. V.) 'Healthy-Strong Life' \ Left to Right(L->R)

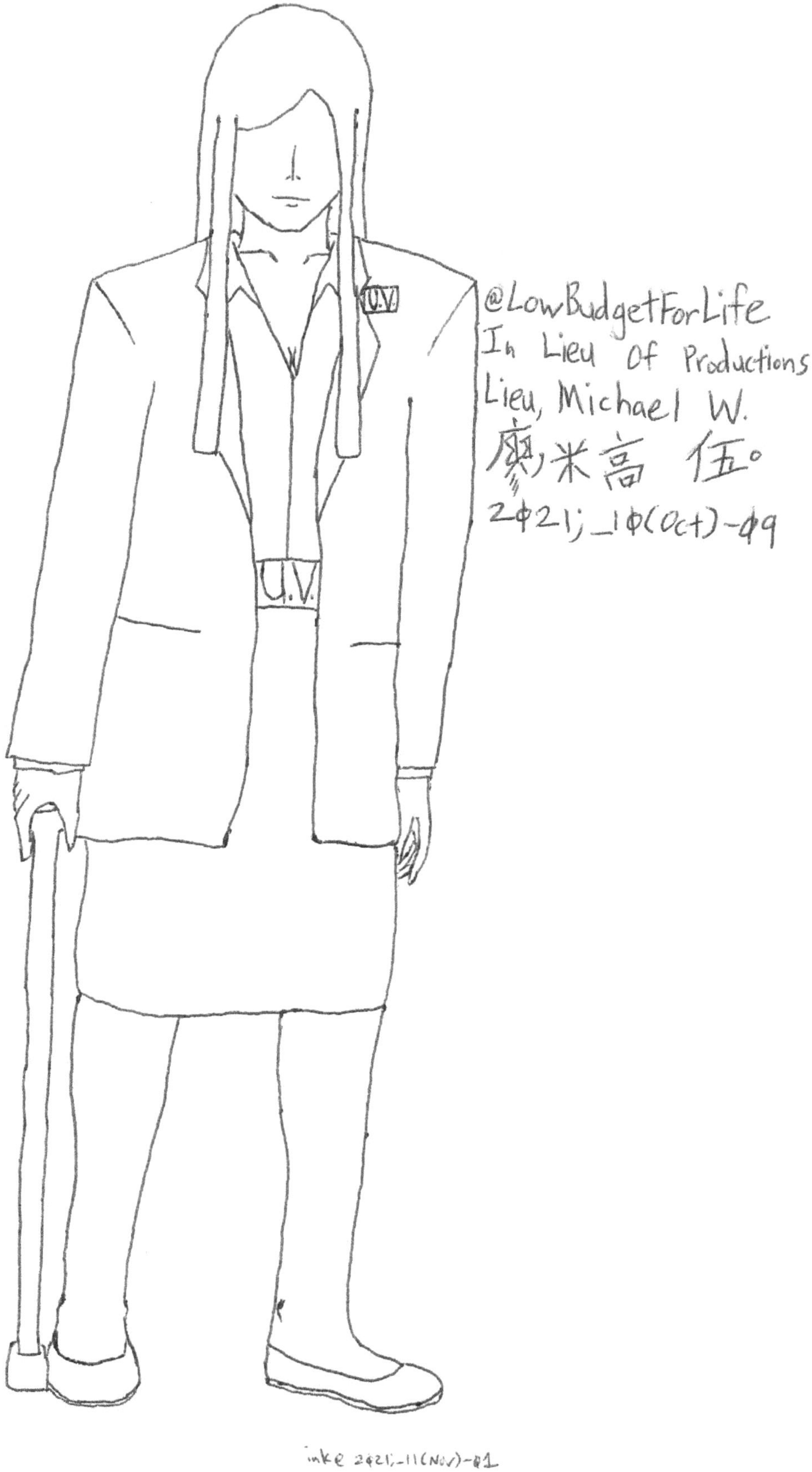
U.V.
@LowBudgetForLife
In Lieu Of Productions
Lieu, Michael W.
廖,米高 伍o
2φ21;_1φ(Oct)-φ9
U.V.
inke 2φ21;-11(Nov)-φ1

Reads = Left to Right (L -> R) \ Wireless Broadcasting Frequencies Online
2154;_ Naples, A Human Sanctuary In Italy Intranet

Why bother, who cares, barely any humans follow rules regulation	Fine, whatever do not sit on it, complete this before entering

Naples Controller
City Mayor

"Chicago City Clean House Act of
2122 (twenty one twenty two);_ 01-01 (January one)"

Indoctrination Self-Governing Guidelines Overview

Warning;_
Failure to complete the new world 2121 (twenty one twenty one) indoc, is a warrant for immediate removal from Naples, A Human Sanctuary In Italy.

Exceptions apply only to the surviving augmented biotech and natural unmodified human Males (M), Females (F), Hermaphrodites (H).

Departing Utopian Sovereignty (U.S.) Debrief. Make a choice, Utopian.

I will be returning back home.	I do not know, let us discuss.	I am moving.

Clean-up Crew
Unit 6 (six)
Unit 9 (nine)

Abusive bitch and moan, toxicity constant crybaby complaint box	Let the shit hit the fan rock bottom, to crash and burn hard

2154;_ Naples, A Human Sanctuary In Italy Intranet
Reads = Left to Right (L -> R) \ Wireless Broadcasting Frequencies Online

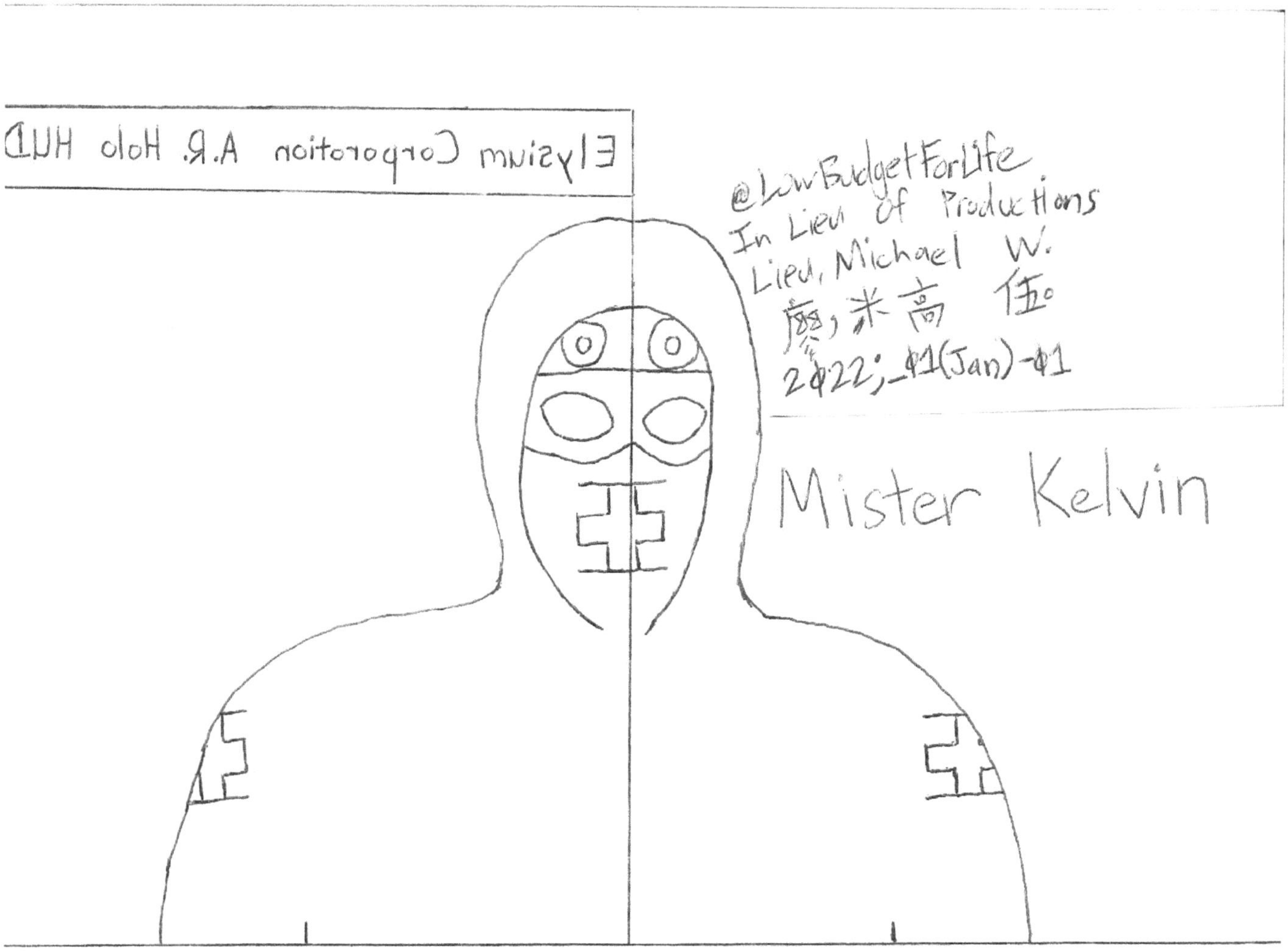

Elysium Corporation

Established 2Φ77

Biotech Dome

Processing

Intake Outtake Official

Citizen Mister Kelvin, Naples City Official + Elysium Corporation Employee, Dome
2154;_ Naples, Italy's Intranet, Wireless Broadcasting Frequencies Online

Elysium Corp Augmented Reality Holographic Heads Up Display A.R. Holo HUD						
Actual Legal Given Name;_ **Mister Kelvin** (goes by **Mister Kelvin**) pronounced, **'mist-ter' 'kell-vin'**			Physical Plus Psych Age;_ natural 30 + 30 years old			
Projecting Temperature <u>Outside</u> Elysium Corporation Biotech Dome By "Planet Earth, Eur-Asia, Italy, Naples"						
Sun's Day	Moon's Day	Tiu's Day	Woden's Day	Thor's Day	Frigg's Day	Saturn's Day
1° C	2° C	3° C	4° C	5° C	6° C	7° C
Kold	Kold	Kold	Kold	Kold	Kold	Kold
[<u>*Aug+*</u>]						
Elysium Corp Augmented Reality Holographic Heads Up Display A.R. Holo HUD						

2154;_ Naples, Italy's Intranet, Wireless Broadcasting Frequencies Online
Citizen Mister Kelvin, Naples City Official + Elysium Corporation Employee, Dome

Mister Kelvin, Human Klone(s) Property Of Elysium Corporation;
"(*looking*) ...Miss Valentina Vita, V. V., **Civilian, Citizen Of Eur-Asia**, you are clear for departure after **our** biotech dome airlock decontamination. Looks like you got quite the **travel itinerary...** along side **our** infamous biotech **bouncing ball sphere**..."

Valentina Vita (V. V.); Traveling Cosplayers And Kosplayers;
"(^_^) Yes, I have so many **places**, **people** to meet and things to do!! **Sir Arthur C. Clarke's space elevator** launch anniversary! I maybe back by **twenty one fifty five**!"

Mister Kelvin, Human Klone(s) Property Of Elysium Corporation;
"(*Ahem*) ...Safe travels from **Elysium Corporation** V. V. Say hello to **my other Mister Kelvin Klones** out there when you see **them**... Good day."

Valentina Vita (V. V.); Traveling Cosplayers And Kosplayers;
"(O_^) I will say hello and thanks **Mister Kelvin of Naples Italy**! Be seeing **you**!"

Automated Algorithm; Human(s) Voice Over Emulation From Elysium Corporation Archives;
"(*monotone female voice over*) Please stand by for biotech dome airlock decontamination. Thank you for your patience and cooperation **Utopian**."

2154;_ Arcadia \ Valentina Vita (V. V.) 'Healthy-Strong Life' \ Left to Right(L->R)

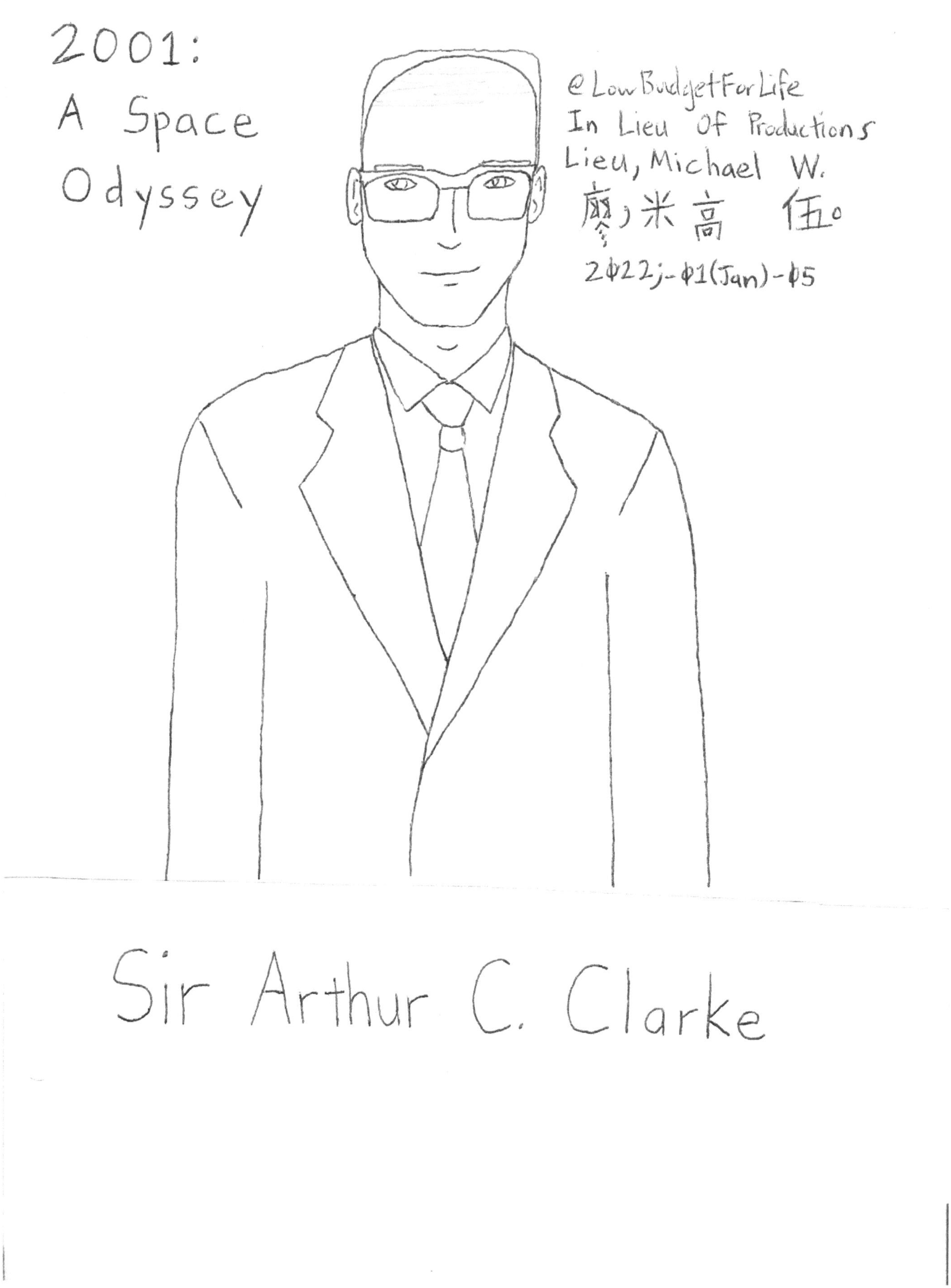
2001:
A Space
Odyssey
@LowBudgetForLife
In Lieu Of Productions
Lieu, Michael W.
廖, 米高　伍o
2Ø22j-Ø1(Jan)-Ø5
Sir Arthur C. Clarke

The music "**Andre Rieu's The Beautiful Blue Danube**" plays in the **Sir Arthur C. Clarke Space Elevators' Station** background and is set on loop.

Sir Arthur C. Clarke; Contemporary A.R. Holo HUD Archives;
"(*archive audio video playback*) Hello, **I** am **Sir Arthur C. Clarke**, today is the **year two thousand one, September eleven**, the exciting launch of a **space elevator** named after **me**... **I** am bit flatter and honoured. Quote, **Sir Arthur C. Clarke's Space Elevators**, Unquote. **Humanity made it this far...** Enjoy... **two thousand one.**"

Welcome To Sir Arthur C. Clarke's Space Elevator(s) "2001: A Space Odyssey"

Sir Arthur C. Clarke, Contemporary A.R. Holo HUD archives;
"(*archive audio video playback*) **Film movie director Stanley Kubrick died on the year two thousand**, thus not able to live up to the **year two thousand one** and see the official **two thousand one A Space Odyssey** event. (*scanning*) It seems you need to wait about... (*spliced audio video playback*) ...**Twelve days** until an available **Space Elevator**."

Privacy Protection aug made possible by Kay's Private Security Services L.L.C.;
"(*spliced monotone*) Hello Miss V. V., please stand by while **Space Traffic Controllers** prepare for the next available **Sir Arthur C. Clarke's Space Elevator** in... **twelve days** due to the on going War On Mars conflict... Hope you can wait, ma'am."

Valentina Vita (V. V.); Traveling Cosplayers And Kosplayers;
"(^_^) Thanks, time to set up a **space theme music audio playlist, shuffle**, and wait."

Alien Valentina Vita, Traveling Cosplayers And Kosplayers
2154;_ Chicago, Illinois' Intranet, Wireless Broadcasting Frequencies Offline

V. V. Augmented Reality Holographic Heads Up Display A.R. Holo HUD	
Space Theme Music Audio Playlist, Shuffle Mode On, And Set On Loop, Volume At 33%	
Thomas Bergersen's Final Frontier Approaching Nirvana's Frontier Approaching Nirvana's Glide Sonic Adventure 2's Meteor Herd Vince Watson's Signals Kaito's Color Of Feels Beatless Hiroshi Watanabe's Time Flies Like An Arrow	Nine Inch Nails' Adrift And At Peace Stellardrone's Eternity Space Ambient The Legend Of Zelda's City In The Sky Super Mario Galaxy's Space Junk Galaxy Sonic The Hedgehog 3's Sky Sanctuary Kaito's Hundred Million Lightyears Beatless Kaito's Universe Beatless
V. V. Augmented Reality Holographic Heads Up Display A.R. Holo HUD	

2154;_ Chicago, Illinois' Intranet, Wireless Broadcasting Frequencies Offline
Alien Valentina Vita, Traveling Cosplayers And Kosplayers

Solar System, Drawing Diagram Of Planets | 2154;_

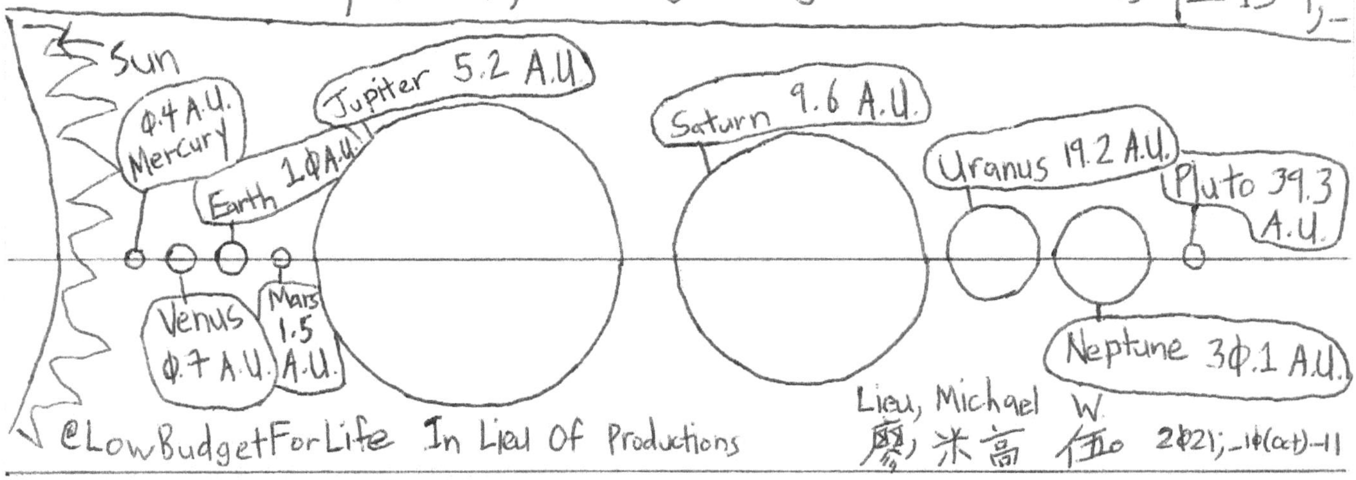

The average distance of our planets from the Sun.
The numbers in this drawing diagram are Astronomical Units (A.U.), based on the average distance between the Earth and the Sun;

$1 \text{ A.U.} = 1.5\phi \times 1\phi^{8}$ Kilo Meters (K.M.)

Planetary Statistics, Notes, Fill In The Blanks Worksheet

Planetary Statistics, Notes, Fill In The Blanks Worksheet

Planet	Diameter	Distance From Sun * In Millions	Rotational Period	Notes
Mercury	4,878 K.M.	57.9 K.M. Inner Planet	58.64 days	
Venus	12,104 K.M.	108.1 K.M. Inner Planet	243 days (retrograde)	The Third Reich, Adolf Hitler's Nazi Political Party Head Quarters (H.Q.).
Earth	12,756 K.M.	149.5 K.M. Inner Planet	23 hours, 56 minutes, 4.1 seconds	The Home Planet Origins Of Humans, Animal Kingdom, And Artificial Intelligence (A.I.), Earth's Global Government (EGG).
Mars	6,787 K.M.	227.9 K.M. Inner Planet	24 hours, 37 minutes, 22.3 seconds	The War On Mars, 2120 November To Present Day Time, Martian Terrorists, Roman Industries.
Jupiter	142,800 K.M.	778.3 K.M. Outer Planet	9 hours, 50 minutes (equator). 9 hours, 55 minutes (polar)	The UltraViolet (U.V.) Rays' Broadcasting Phenomenon On Io And Europa Moon(s).
Saturn	120,540 K.M.	1,427.0 K.M. Outer Planet	10 hours, 39 minutes	
Uranus	51,118 K.M.	2,869.6 K.M. Outer Planet	17 hours, 54 minutes	The Head Quarters (H.Q.) Of Automated Artificial Intelligence (A.I.) Established Elysium Corporation.
Neptune	49,528 K.M.	4,496.6 K.M. Outer Planet	19 hours, 12 minutes	
Pluto	2,300 K.M.	5,900.0 K.M. Outer Planet	6 days, 9 hours	Some say, this is "not an actual planet".

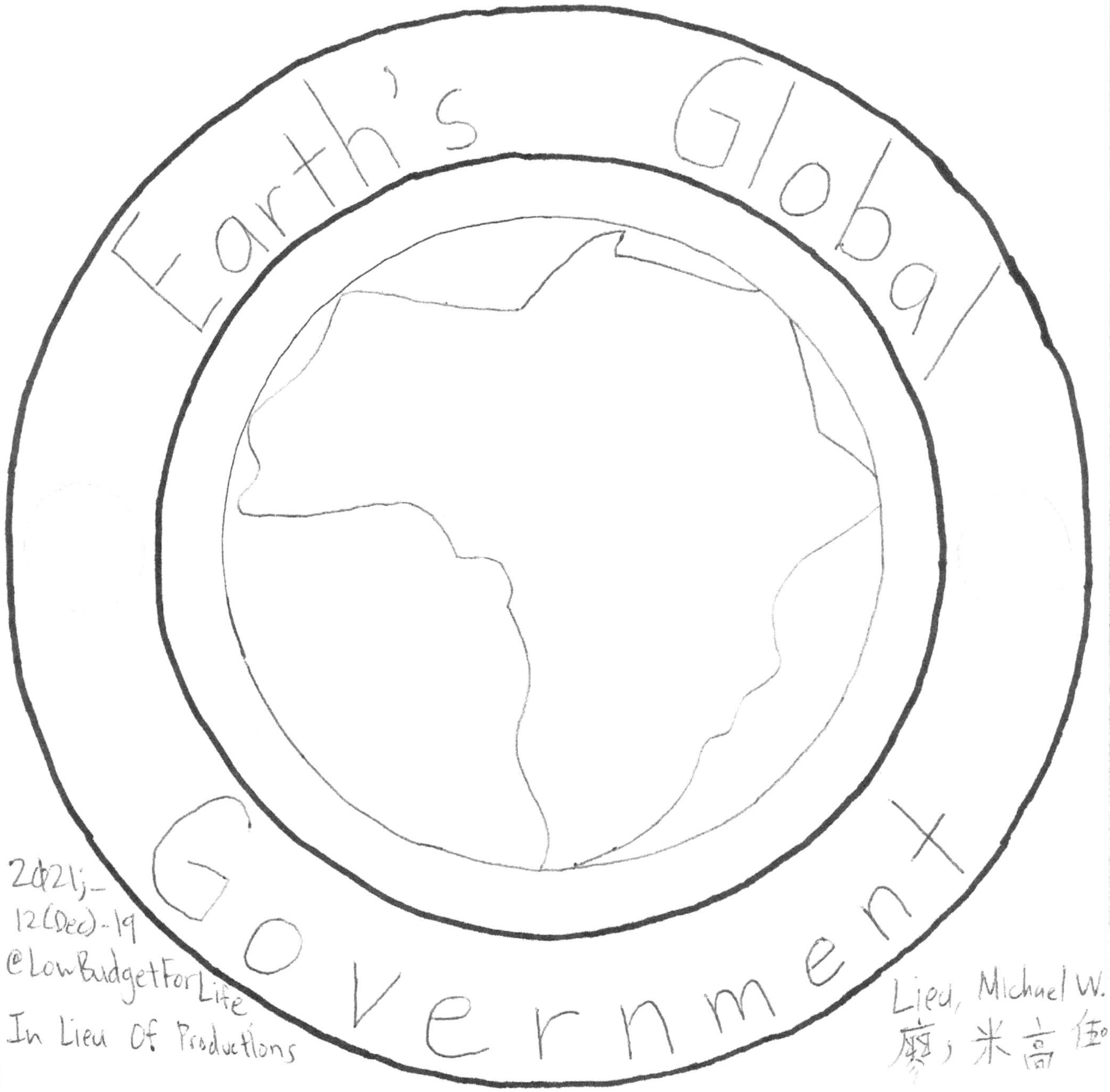
Earth's Global
Government
2021j_
12(Dec)-19
@LowBudgetForLife
In Lieu Of Productions
Lieu, Michael W.
廖, 米高 伍o

2154;_09(Sep)-11(eleven)

The music "**Stellardrone's Eternity Space Ambient**" plays in the **Sir Arthur C. Clarke Space Elevators' Station** background and is set on loop.

Automated Algorithm; **Human Voice Over Emulation Based On Elysium Corporation Archives**;
"(*monotone female voice over*) Please stand by for biotech dome airlock decontamination. Thank you for your patience and cooperation Earthling."

Privacy Protection aug made possible by Kay's Private Security Services L.L.C.;
"(*spliced monotone*) Welcome to **Sir Arthur C. Clark's Space Elevators**... Ma'am."

Sir Arthur C. Clarke, Contemporary A.R. Holo HUD archives;
"(*archive audio video playback*) Arriving to **Sir Arthur C. Clarke's Space Elevator(s)**... You are just in time for the... (*spliced audio video playback*) ...**One hundred fifty third anniversary** of **Sir Arthur C. Clarke's Space Elevators** that launched in **two thousand one, September eleven** and later on: **Satellites, Solar Panels, Space Station And Arrays**. Enjoy... **two thousand one**."

Automated Algorithm; **Human Voice Over Emulation Based On Elysium Corporation Archives**;
"(*monotone female voice over*) Warning, you are now about to enter **the vacuum of space**. Please ensure you have your space suit Personal Protective Equipment, P.P.E., outfitted with purified drinking water and breathable air oxygen along with a **recommended Extra Vehicular Activity, EVA mobility** installation. (*scanning*) ...Confirmed and cleared for entering **the vacuum of space**. Warning, your space suit P.P.E. does not have a **EVA mobility** installation for **space walks**. (*scanning*) ...Please stand by for biotech dome airlock decontamination. Thank you for your patience and cooperation Earthling."

V. V. selects the music "**Kaito's Universe Beatless**" playing in her Lady A. Dragoon Knight Helmet P.P.E. on one third volume of 33% and is set on loop.

Alien Valentina Vita, Traveling Cosplayers And Kosplayers
2154;_ Chicago, Illinois' Intranet, Wireless Broadcasting Frequencies Offline

V. V. Augmented Reality Holographic Heads Up Display A.R. Holo HUD	
Aranea C. Highwind, "Lady A. Dragoon Knight" fusion cosplay life support systems P.P.E.	
Liquid; Purified Drinking Water H_2O 100% (one hundred percent)	Gas; Breathable Air Oxygen O_2 100% (one hundred percent)
V. V. Augmented Reality Holographic Heads Up Display A.R. Holo HUD	

2154;_ Chicago, Illinois' Intranet, Wireless Broadcasting Frequencies Offline
Alien Valentina Vita, Traveling Cosplayers And Kosplayers

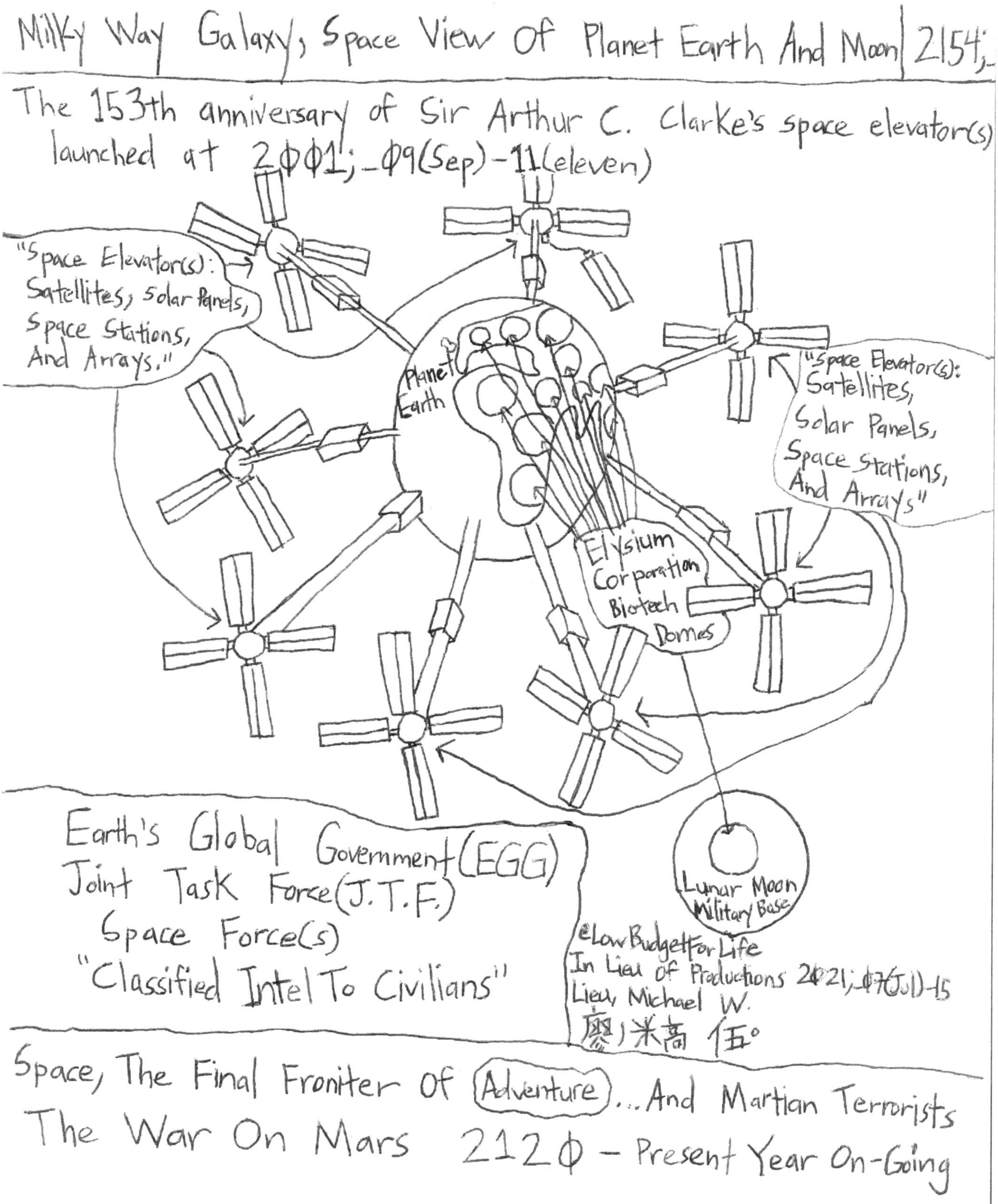
Milky Way Galaxy, Space View Of Planet Earth And Moon | 2154;-
The 153th anniversary of Sir Arthur C. Clarke's space elevator(s) launched at 2ΦΦ1;-Φ9(Sep)-11(eleven)
"Space Elevator(s): Satellites, Solar Panels, Space Stations, And Arrays."
Planet Earth
"Space Elevator(s): Satellites, Solar Panels, Space Stations, And Arrays"
Elysium Corporation Biotech Domes
Lunar Moon Military Base
Earth's Global Government (EGG) Joint Task Force (J.T.F.) Space Force(s) "Classified Intel To Civilians"
©LowBudgetForLife In Lieu of Productions 2Φ21;-Φ7(Jul)-15 Lieu, Michael W. 廖,米高 伍o
Space, The Final Froniter Of Adventure... And Martian Terrorists
The War On Mars 212Φ - Present Year On-Going

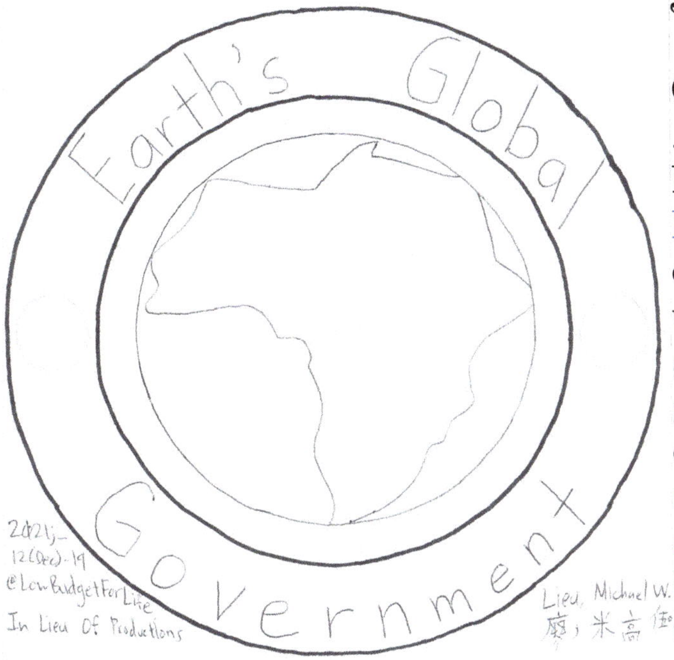

Earth's **Global Government** (EGG)
Joint Task Force (J.T.F.)

Officially formed in the year 2122...
...as an on going J.T.F. between **the last remaining human survivors** of Planet Earth's **old world** structures of first (1st), second (2nd) and third (3rd) world federal, state and local governments. This **global government** of Planet Earth consists of civil servant civilian contractors, military servicemembers: active duty and reservists as **first responders to both foreign and domestic terrorists' unconventional warfare** of Chemical, Biological, Radiological (C.B.R.) outbreaks and Electro Magnetic Pulse (E.M.P.) outages.

EGG's current War Conditions (War Con) as of 2154;_ "2"

War Con	1 "**white**" most severe, Nuclear Warfare
War Con	2 "red"
War Con	3 "yellow"
War Con	4 "green"
War Con	5 "blue" least severe, 2120;_11(November)-17

"**EGG** J.T.F. defines our world **that stands for** one value...
You are either **with** us or **against** us on the War On Mars,
War is hell, either they destroy us or we destroy them."

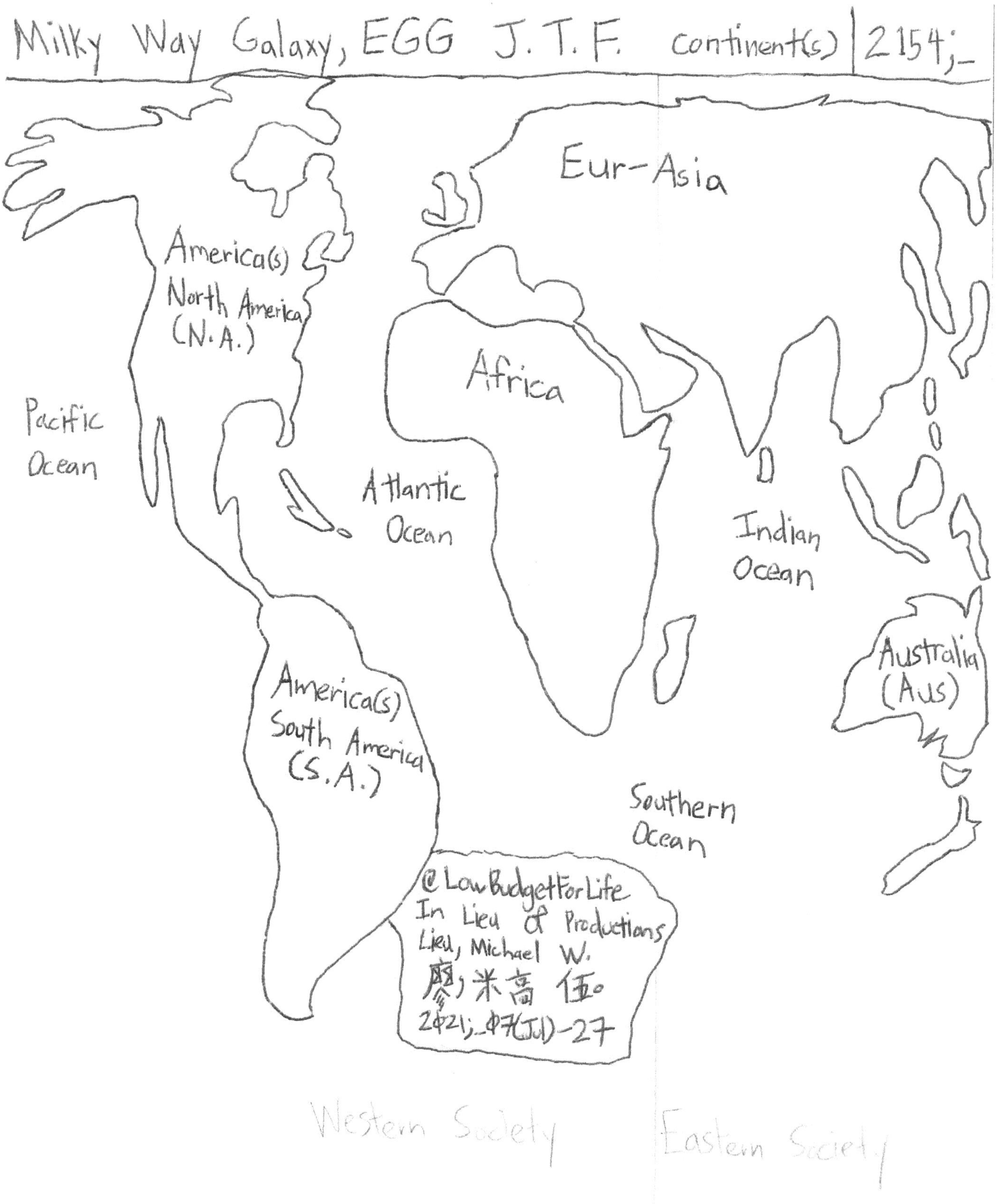

inke 2021;_10(Oct)-30

Planet Earth, North America (N.A.) "The **old world** United **States** of America (U.S. of A.)"
Founded→1776 - 2121←Collapse
District of Columbia (D. of C.) 55

55 (fifty five) States	Capital	55 (fifty five) States	Capital
Alabama (ALA) 01	Montgomery	Nevada (NEV) 28	Carson City
Alaska (AKA) 02	Juneau	New Hampshire (NHA) 29	Concord
Arizona (AZA) 03	Phoenix	New Jersey (NJE) 30	Trenton
Arkansas (ARK) 04	Little Rock	New Mexico (NME) 31	Santa Fe
Colorado (COL) 05	Denver	New York (NYO) 32	Albany
Connecticut (CON) 06	Hartford	North California (NCA) 33	Sacramento
Delaware (DEL) 07	Dover	North Dakota (NDA) 34	Bismarck
Florida (FLO) 08	Tallahassee	North Karolina (NKA) 35	Raleigh
Georgia (GEO) 09	Atlanta	Ohio (OHI) 36	Columbus
Guam (GUA) 10	Hagatna	Oklahoma (OKL) 37	Oklahoma City
Hawaii (HAW) 11	Honolulu	Oregon (ORE) 38	Salem
Idaho (IDA) 12	Boise	Pennsylvania (PEN) 39	Harrisburg
Illinois (ILL) 13	Springfield	Puerto Rico (PRI) 40	San Juan
Indiana (IND) 14	Indianapolis	Rhode Island (RIS) 41	Providence
Iowa (IOW) 15	Des Moines	South California (SCA) 42	Los Angeles
Kansas (KAN) 16	Topeka	South Dakota (SDA) 43	Pierce
Kentucky (KEN) 17	Frankfort	South Karolina (SKA) 44	Columbia
Louisiana (LOU) 18	Baton Rogue	Tennessee (TEN) 45	Nashville
Maine (MAI) 19	Augusta	Texas (TEX) 46	Austin
Maryland (MAR) 20	Annapolis	U.S. Virgin Islands (USVI) 47	Charlotte Amalie
Massachusetts (MAS) 21	Boston	Utah (UTA) 48	Salt Lake City
Michigan (MIC) 22	Lansing	Vermont (VER) 49	Montpelier
Minnesota (MIN) 23	Saint Paul	Virginia (VIR) 50	Richmond
Mississippi (MIS) 24	Jackson	Washington (WAS) 51	Olympia
Missouri (MSO) 25	Jefferson City	West Virginia (WVI) 52	Charleston
Montana (MON) 26	Helena	Wisconsin (WIS) 53	Madison
Nebraska (NEB) 27	Lincoln	Wyoming (WYO) 54	Cheyenne

Planet Earth, North America (N.A.)
"Canada"
Founding→1867 - Present Day Time←Ongoing
Ottawa

10 (ten) Provinces / 3 (three) Territories	Capital
Alberta (ALB) 01	Edmonton
British Columbia (BCO) 02	Victoria
Manitoba (MAN) 03	Winnipeg
New Brunswick (NBR) 04	Fredericton
Newfoundland And Labrador (NAL) 05	Saint John's
Northwest Territories (NTE) 06	Yellowknife
Nova Scotia (NSC) 07	Halifax
Nunavut (NUN) 08	Iqaluit
Ontario (ONT) 09	Toronto
Prince Edward Island (PEI) 10	Charlottetown
Quebec (QUE) 11	Quebec City
Saskatchewan (SAS) 12	Regina
Yukon Territory (YTE) 13	Whitehorse

Planet Earth, North America (N.A.)
"Mexico"
Founded→1810 - Present Day Time←Ongoing
Mexico City

32(thirty two) States	Capital	32(thirty two) States	Capital
Aguascalientes 01	Aguascalientes	Tamaulipas 27	Ciudad Victoria
Baja California 02	Mexicali	Tlaxcala 28	Tlaxcala
Baja California Sur 03	La Paz	Veracruz 29	Jalapa
Campeche 04	Campeche	Yucatan 30	Merida
Chiapas 05	Tuxtia Gutierrez	Zacatecas 31	Zacatecas
Chihuahua 06	Chihuahua	Federal District 32	Mexico City
Coahuila 07	Saltillo		
Colima 08	Colima		
Durango 09	Durango		
Guanajuato 10	Guanajuato		
Guerrero 11	Chilpancingo		
Hidalgo 12	Pachuca		
Jalisco 13	Guadalajara		
Mexico 14	Toluca		
Michoacan 15	Morelia		
Morelos 16	Cuernavaca		
Nayarit 17	Tepic		
Nuevo Leon 18	Monterrery		
Oaxaca 19	Oaxaca		
Puebla 20	Puebla		
Queretaro 21	Queretaro		
Quintana Roo 22	Chetumal		
San Luis Potosi 23	San Luis Potosi		
Sinaloa 24	Culiacan		
Sonora 25	Hermosillo		
Tabasco 26	Villahermosa		

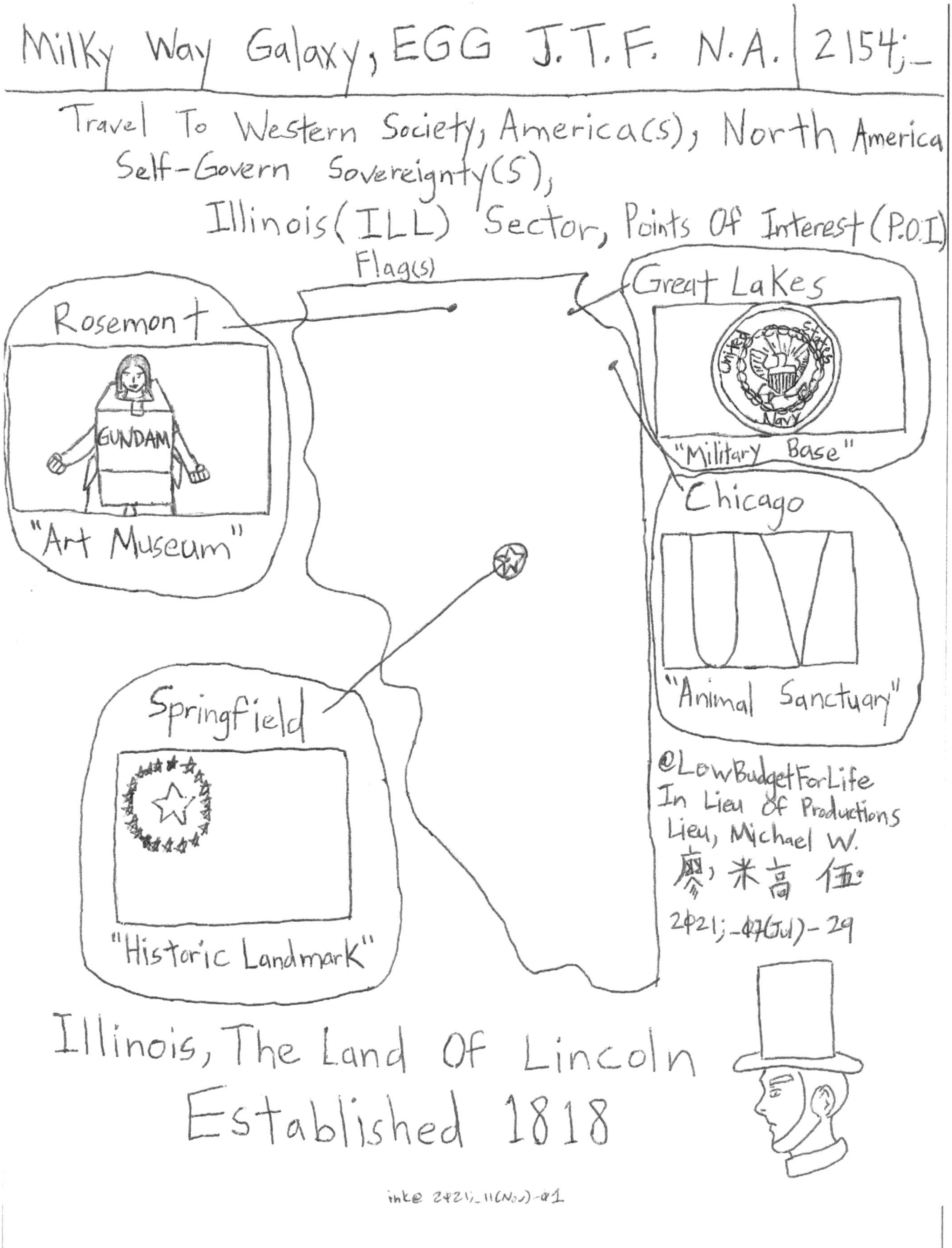

Milky Way Galaxy, EGG J.T.F. N.A.
2154;_
Travel To Western Society, America(s), North America
Self-Govern Sovereignty(S),
Illinois(ILL) Sector, Points Of Interest(P.O.I)
Flag(s)
Rosemont
GUNDAM
"Art Museum"
Great Lakes
United States Navy
"Military Base"
Chicago
"Animal Sanctuary"
Springfield
"Historic Landmark"
@LowBudgetForLife
In Lieu Of Productions
Lieu, Michael W.
廖, 米高 伍
2021;_07(Jul)-29
Illinois, The Land Of Lincoln
Established 1818
inked 2021;_11(Nov)-01

2154;_10(Oct)-01(one)

Travel To Western Society, America(s), North America
Self-Govern Sovereignty (S),
Illinois (ILL) Sector, Points Of Interest (P.O.I.)

"**Art Museum**" Rosemont (S), Illinois (ILL)

A private sector "**Art Museum**" showcasing the 21st (twenty first) century of 2003 (two thousand three) cardboard box GUNDAM cosplay from the UltraViolet (U.V.) rays' broadcasting frequencies phenomenon and other art work on display. **Operation Hours**: "**Art Museum" 08;00_a.m. morning to 20;00_p.m. night**. Private owned travel lodging for **paying customers**: Always open, **twenty four hours a day, seven days a week**. Has an *Elysium Corporation Biotech Dome and Kay's Private Security Services L.L.C. Agents.*

"Military Base" Great Lakes (M), Illinois (ILL)

A public sector "Military Base" that trains the United States Navy (U.S.N.) sailor recruit(s) as a Naval Training Center (N.T.C.) Recruit Training Command (R.T.C.) since 20th (twentieth) century of 1911 (nineteen eleven), that is owned and operated by **the last remaining old world and collapsed** United **States** of America (U.S. of A.) country's federal government officials. **Not open to the public for traveling tourist(s), official public sector business only**. **No** tolerance for drug use. **No** *Elysium Corporation Biotech Dome*.

"**Animal** Sanctuary" Chicago (S), Illinois (ILL)

A former public sector "**City** (D) (R) Military (M) Base" turn "**Animal** Sanctuary (S)" on 2152 (twenty one fifty two) July (month seven) four's fireworks celebration made possible by the recurring UltraViolet (U.V.) rays' broadcasting frequencies phenomenon calling themselves "City Mayor". This "**Animal** Sanctuary" is a mix of public sector and private sector forming in the many *Elysium Corporation Biotech Domes* in the solar system. The **Utopian** Sovereignty (**U**.S.) movement started in the year 2121 (twenty one twenty one), the same year when billions of human lives were lost to Martian Terrorists' unconventional warfare. Chicago is now **the last remaining** "**Animal** Sanctuary" phenomenon in the Illinois (ILL) sector. **Operation Hours**: always open, **twenty four hours a day, seven days a week**. Has an *Elysium Corporation Biotech Dome and Kay's Private Security Services L.L.C. R&D.*

"**Historical Landmark**" Springfield (D)(M)(R), Illinois (ILL)

A public sector "**Historic Landmark**" owned and operated by **the last remaining old world and collapsed** United **States** of America (U.S. of A.) country's federal government officials. This **ghost town tour** is being watched over by the United **States** Army (U.S.A.) military soldiers and civilians along with other **old world** United **States** Armed Forces military operatives. **Operation Hours**: currently short staff**, do not have enough employee(s) and or volunteer(s) to provide any tours at a quick request for traveling tourist(s)**. **No** *Elysium Corporation Biotech Dome*.

2154;_10(Oct)-02(two)

Travel To Western Society, America(s), North America
Self-Govern Sovereignty (S),
Illinois (ILL) Sector, Points Of Interest (P.O.I.)

Rosemont (S), "**Art Museum**"

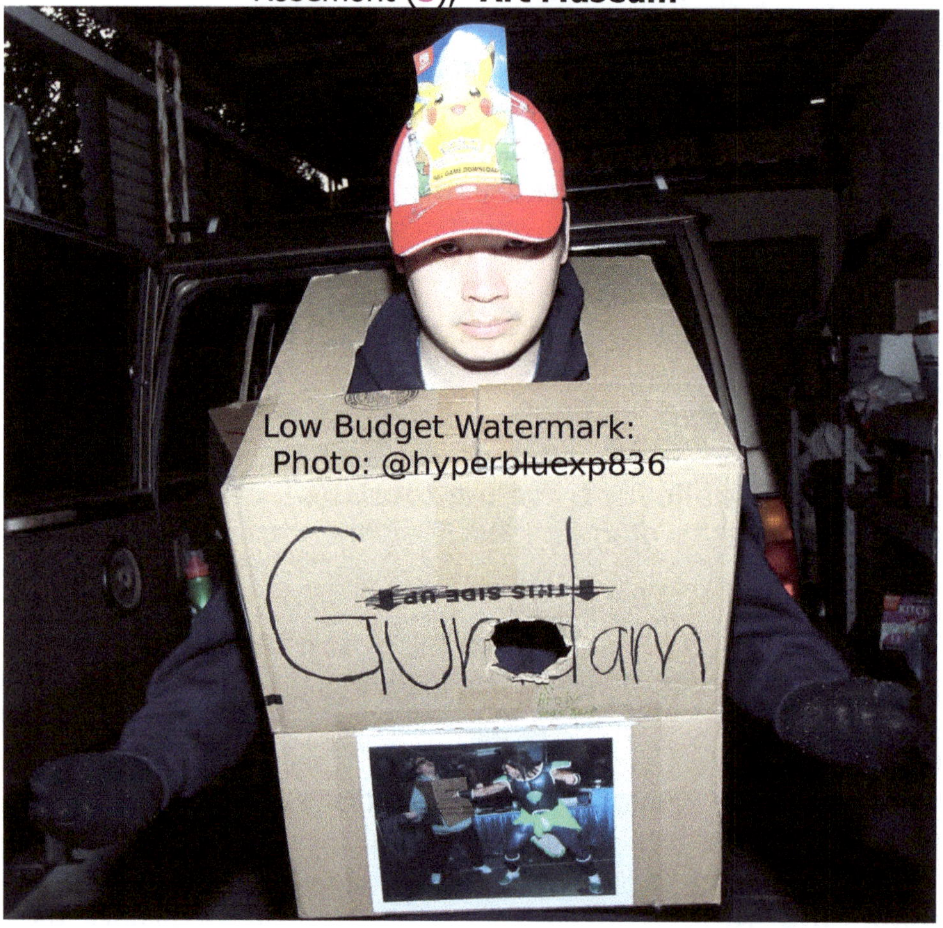

Projecting Temperature <u>Inside</u> Elysium Corporation Biotech Dome By "Planet Earth, North America, Illinois, Rosemont"						
Sun's Day 24° C Warm	Moon's Day 24° C Warm	Tiu's Day 24° C Warm	Woden's Day 24° C Warm	Thor's Day 24° C Warm	Frigg's Day 24° C Warm	Saturn's Day 24° C Warm
[*Aug+*]						

2154;_Arcadia \ Valentina Vita (V. V.) 'Healthy-Strong Life' \ Left to Right(L->R)

Privacy Protection aug made possible by Kay's Private Security Services L.L.C.;
"(*spliced monotone*) Hello Miss V. V., one ticket for **Rosemont Art Museum** and **any lodging days for rest and recoup**?"

Valentina Vita (V. V.); Traveling Cosplayers And Kosplayers;
"(^_^) Yes, please. ...**Rest and recoup lodge area for travelers are only charged when we use them**, if needed? I would like to stay for the following days, if needed... **Maybe two weeks**... Hm... Yea, **two weeks** thanks!!"

Privacy Protection aug made possible by Kay's Private Security Services L.L.C.;
"(*spliced monotone*) Okay, **I** will be put you down, Miss Valentina Vita, for **two weeks** and you will be **charged up on R. and R. lodging for traveling tourists**, if needed, **until the day** you **officially depart from Rosemont within two weeks**. Feel free to **take art museum visual capture** on your Ai-Aug. Enjoy. Thank you for your support."

Automated Algorithm; Human Voice Over Emulation Based On Elysium Corporation Archives;
"(*monotone female voice over*) Please stand by for biotech dome airlock decontamination. Thank you for your patience and cooperation **Utopian**."

V. V. pauses and turns off her Lady A. Dragoon Knight Helmet P.P.E. music player.

Alien Valentina Vita, Traveling Cosplayer(s) And Kosplayer(s)
2154;_ Rosemont, Illinois' Intranet, Wireless Broadcasting Frequencies Offline

<table>
<tr><td colspan="3">V. V. Augmented Reality Holographic Heads Up Display A.R. Holo HUD</td></tr>
<tr><td colspan="2">Actual Legal Given Name;_

Valentina Vita (goes by **V. V.**)
pronounced, "**vee vee**"</td><td>Alias Network Nickname;_

White-Woman, Red-Girl, Red-Woman, Lady In Red, Italian-Women, Beauty Aug, 2 (two) Winged Demon Blade Kosplay, KissShot-AcerolaOrion-HeartUnderBlade, Anime Long Blade Cosplayers [*Aug+*]</td></tr>
<tr><td>Physical Plus Psych Age;_

augmented 34 + 25 years old</td><td>Ethnicity;_

Italian</td><td>Nationality;_

Eur-Asian</td></tr>
<tr><td colspan="3">Connecting... Electromagnetic Spectrum Frequency Hertz (Hz)
Frequency Modulation (F.M.)
Amplitude Modulation (A.M.) [*Aug+*]</td></tr>
<tr><td colspan="3">V. V. Augmented Reality Holographic Heads Up Display A.R. Holo HUD</td></tr>
</table>

2154;_ Rosemont, Illinois' Intranet, Wireless Broadcasting Frequencies Offline
Alien Valentina Vita, Traveling Cosplayer(s) And Kosplayer(s)

Reads = Left to Right (L -> R)

2154;_ Rosemont, Illinois' Intranet, Wireless Broadcasting Frequencies Online

Measuring System;_ Metric degrees Celsius (°C) [*Switch to* *Imperial °F*, *Kelvin*]		
Geolocation;_ Planet Earth, North America, Utopian Sovereignty of America(s) (U.S. of A.)		
Language;_ English, Western Accent, Augmented Reality (A.R.) Application (App)		
Century;_ 22nd (twenty second) [*Compare statistics to* *21st (twenty first)* *Aug+*]		
Actual Legal Given Name	Height	Weight
Deoxyribo Nucleic Acid (D.N.A.)	Holographic Retina Biometric Body Projection	OCEAN 5 (five) Common Personality Traits
Alias Network Nickname	7 (seven) Deadly Sins	7 (seven) Heavenly Virtues
Birth Day Or Build Day	Blood Type Or Bio Fluid	Physical Plus Psych Age
Hair Texture Color Tone	Eye Retina Color Tone	Skin Covering Color Tone
Ethnicity	Nationality	Alien Or Citizen
Natural Matter Organic Photosynthesis	Augmented Compound Enhanced Mutation	Percentage Rate Synthetic Synthesis (S.S.)
Human Being	Animal Kingdom	Artificial Intelligence (A.I.)
Biotech Or Bio Or Tech	Exosuit Or Exoskeleton	Implant Or Explant
Alive = Heart-Rate	Undead = Flatline	Deceased = Necrosis
Chemical (C.) Toxicity	Biological (B.) Toxicity	Radiological (R.) Toxicity
Personal Protective Equipment (P.P.E.)	Ballistic Motion Energy Transfer Shielding Resistance	Electro Magnetic Pulse (E.M.P.)
Sexual Orientation	Sexual Preference	Virginity Test
Sexual Transmitted Diseases (S.T.D.)	Human Immunodeficiency Virus (H.I.V.)	Acquired Immuno Deficiency Syndrome (AIDS)
Known Address(es)	Community Network(s)	Language Level(s)
Family Member(s)	In-Law Relative(s)	Relationship(s)
Biological Genealogy	Psychological Traits	Social Status
Employment	Education	Training
Frequency Hertz (Hz)	Frequency Modulation (F.M.)	Amplitude Modulation (A.M.)
Electromagnetic Spectrum	Intelligence Quotient (I.Q.)	Kinesis Module
Radio Wave Music	Broadcasting Film Movie	Electronic Book Genre
Hate(s) Or Dislike(s)	Neutral(s) Or No Comment	Love(s) Or Like(s) [*Aug+*]

2154;_ Rosemont, Illinois' Intranet, Wireless Broadcasting Frequencies Online

Reads = Left to Right (L -> R)

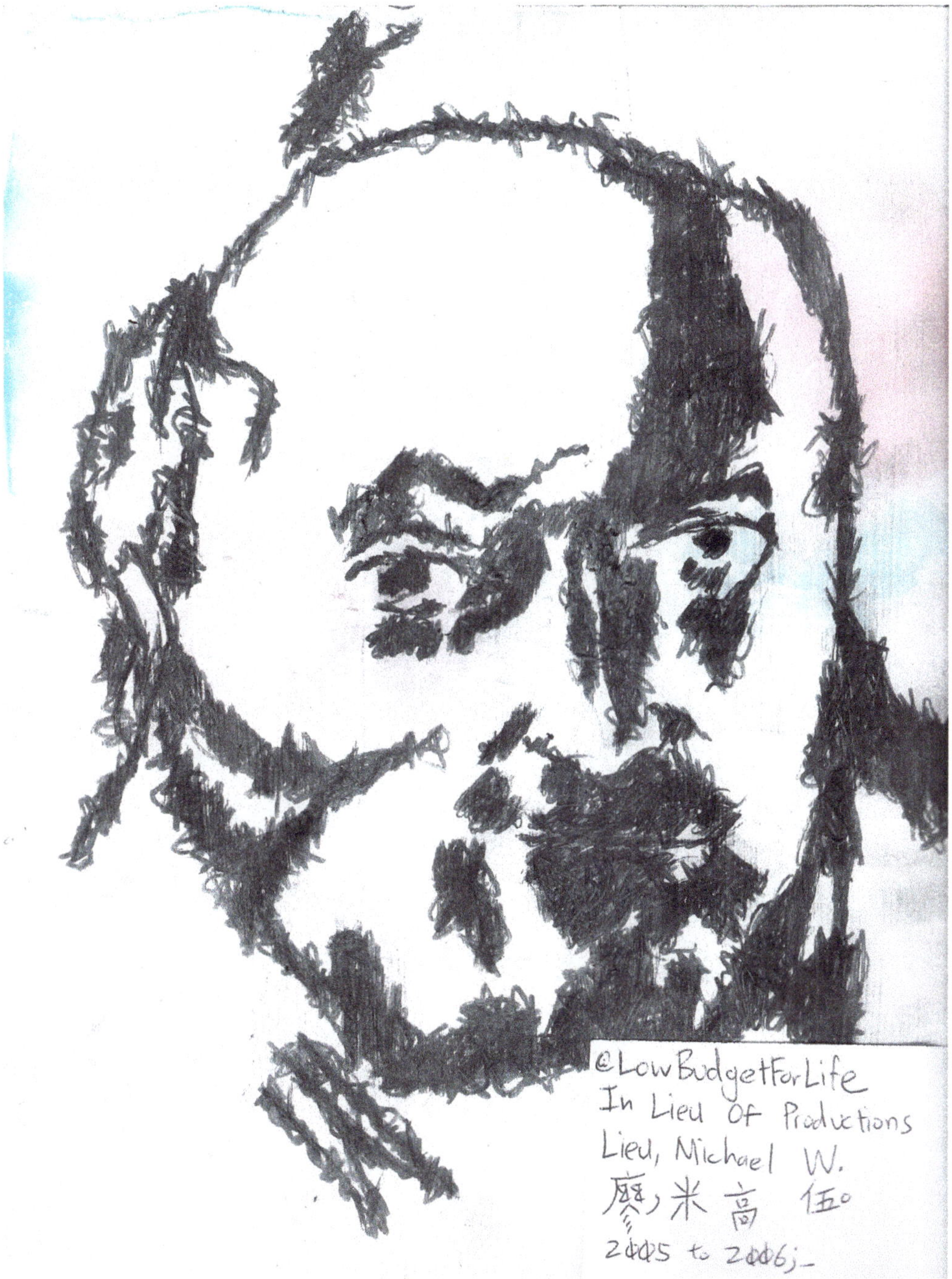

Art Museum Of Rosemont (S), Illinois, North America, Planet Earth

“Portrait 1 (one) With Background”

Mixed Media

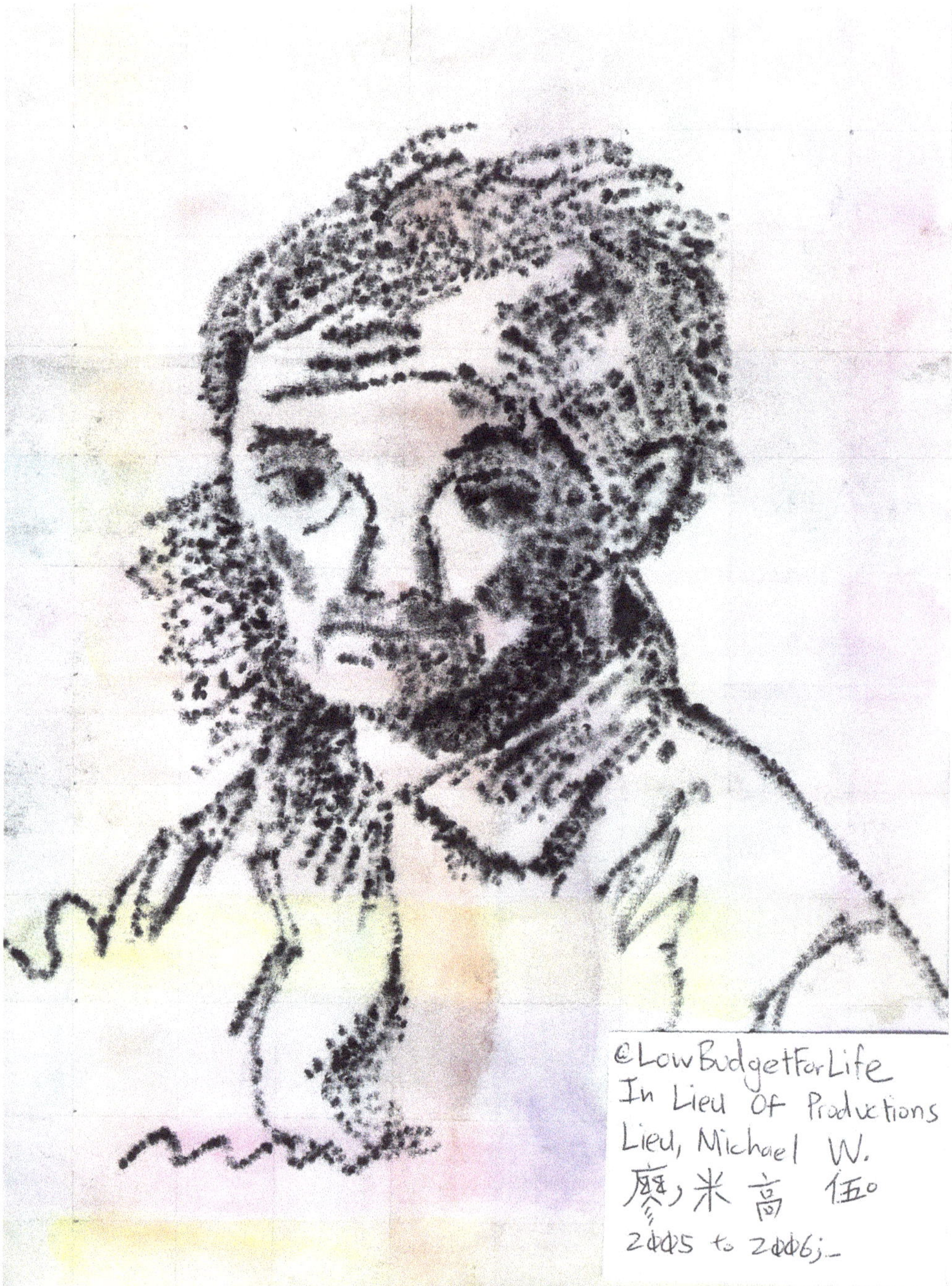

Art Museum Of Rosemont (S), Illinois, North America, Planet Earth

“Portrait 2 (two) With Background”

Mixed Media

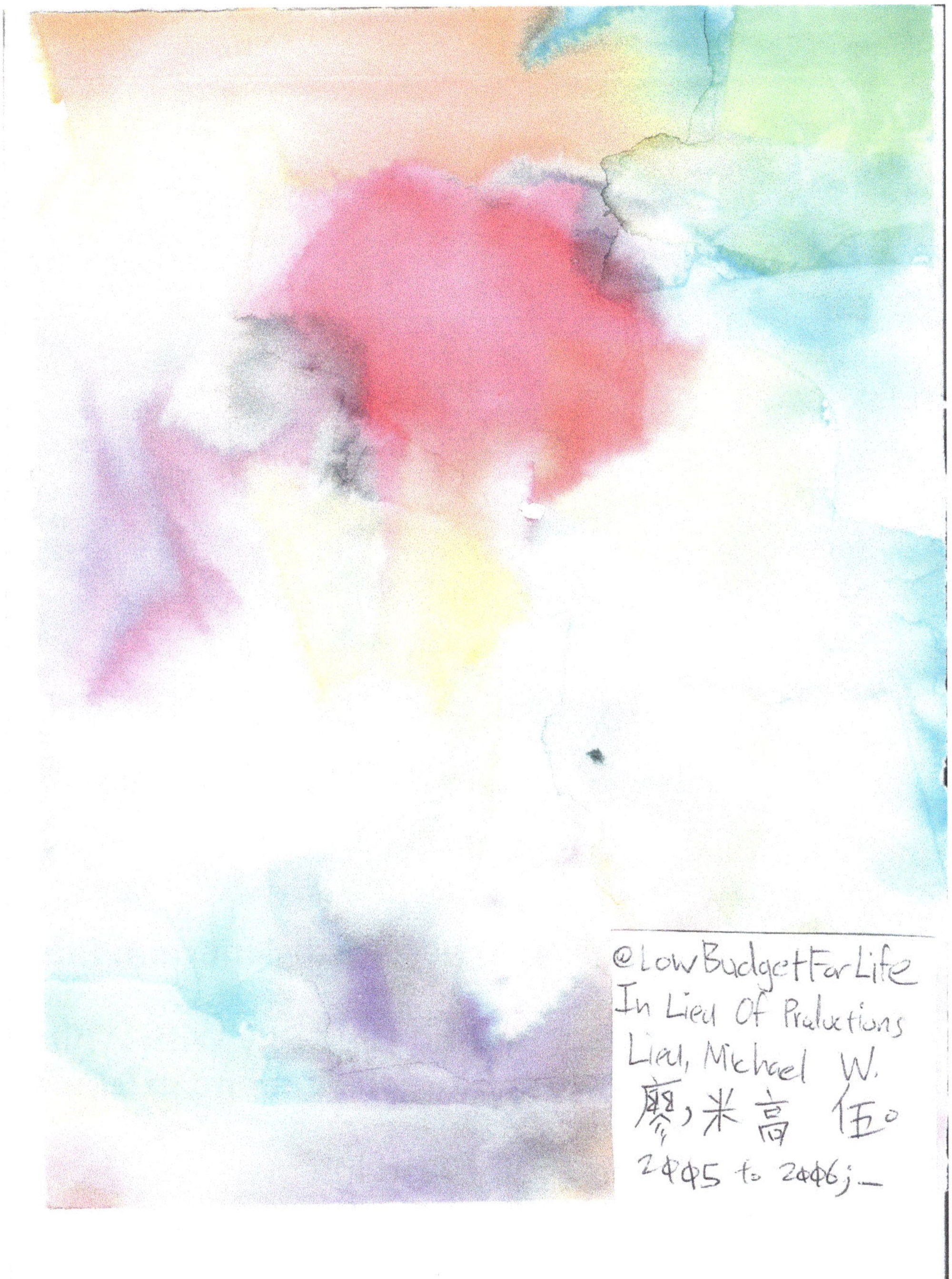

Art Museum Of Rosemont (S), Illinois, North America, Planet Earth

“Can You See The Different Toners? If Not, You May Be Toner Blind.”

Watercolor

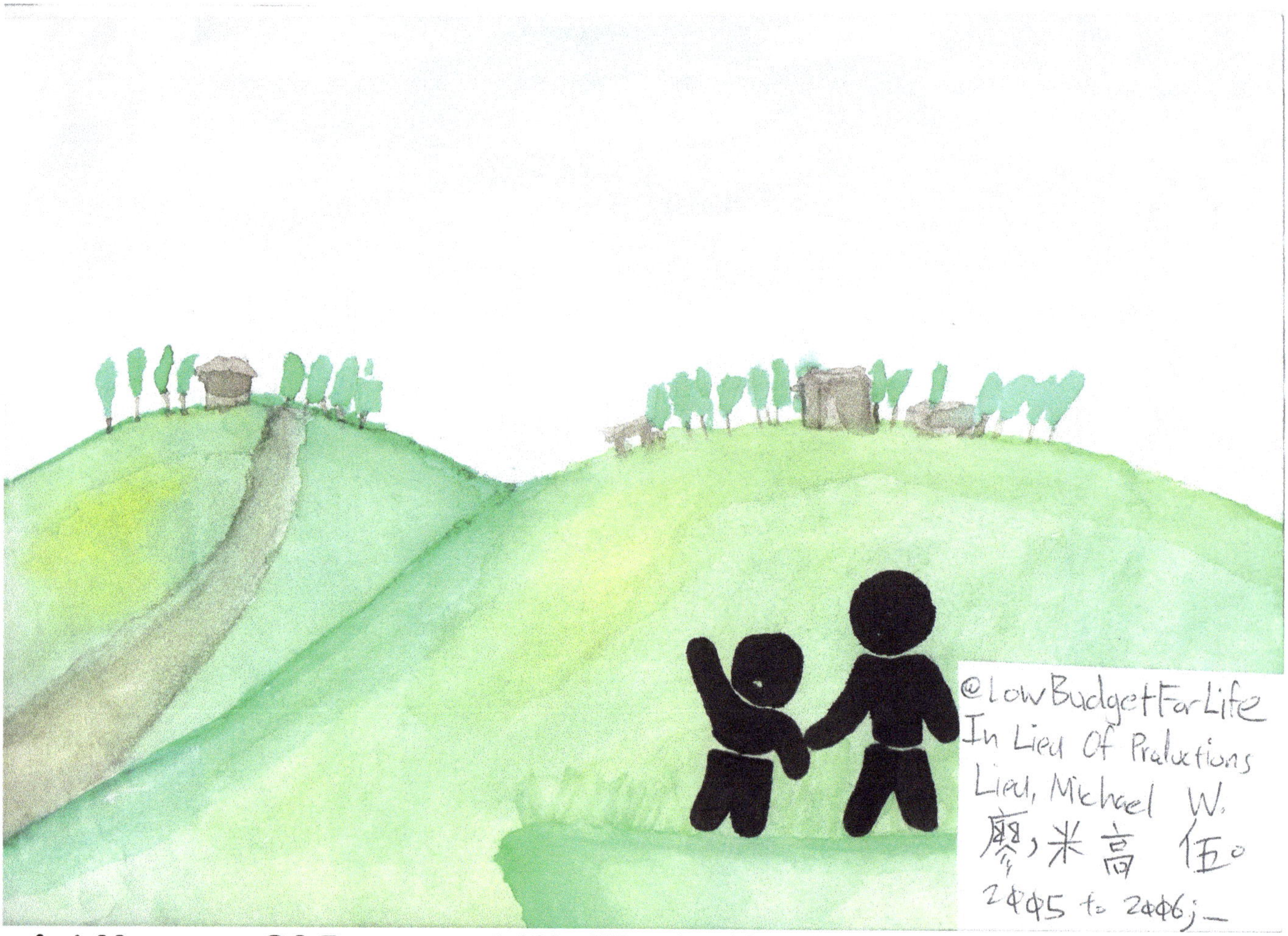

Art Museum Of Rosemont (S), Illinois, North America, Planet Earth

"Background; Green Landscape And Blue Sky With White Clouds Foreground; Augmented Biotech Humans In Black"

Watercolor And Ink

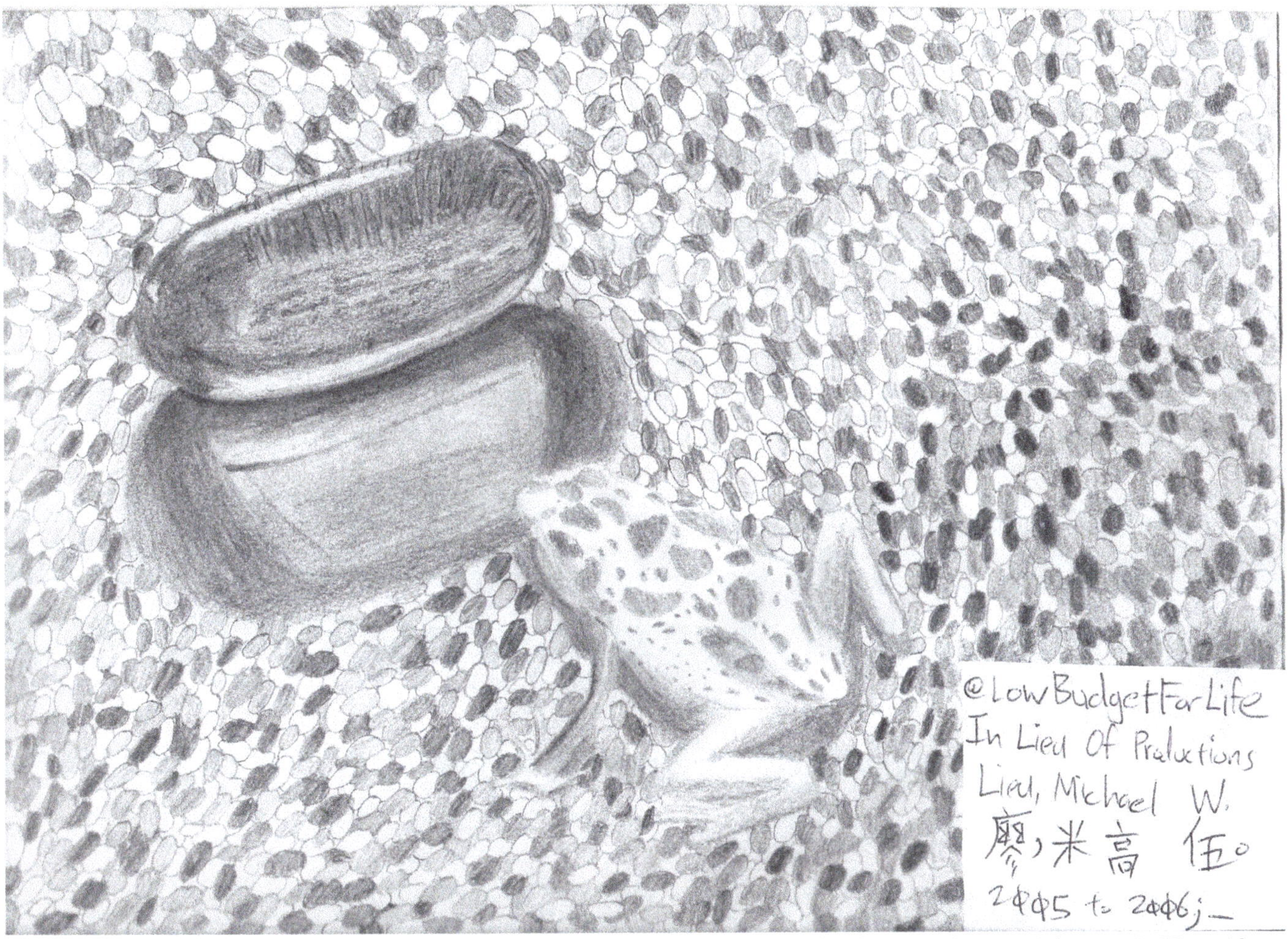

Art Museum Of Rosemont (S), Illinois, North America, Planet Earth

“Modern Medicine And A Reptilian Animal”

Pencil

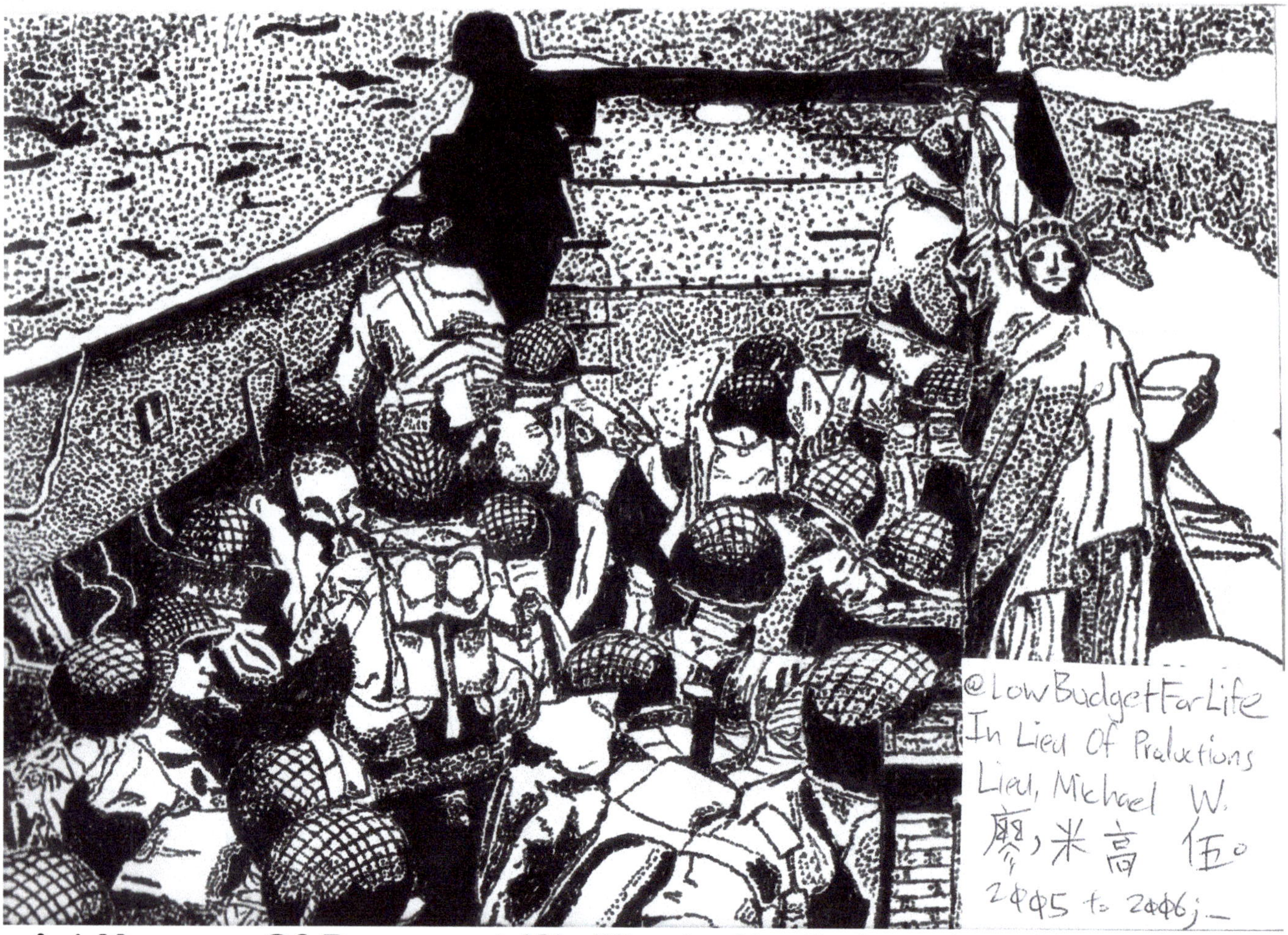

Art Museum Of Rosemont (S), Illinois, North America, Planet Earth

"Based On The String Theory And Or The Multiple uniVerse (MultiVerse) Theory, in an alternative history, where the universe branches off and differs from our timeline, the old world United States of America (U.S. of A.) military servicemembers somehow joins the World War 2 (two) conflict on a 1944 (nineteen forty four) deployment, when our historical events of World War 2 (two) conflict ended on 1944 (nineteen forty four), leading to the beginning of the Kold War..."

Ink

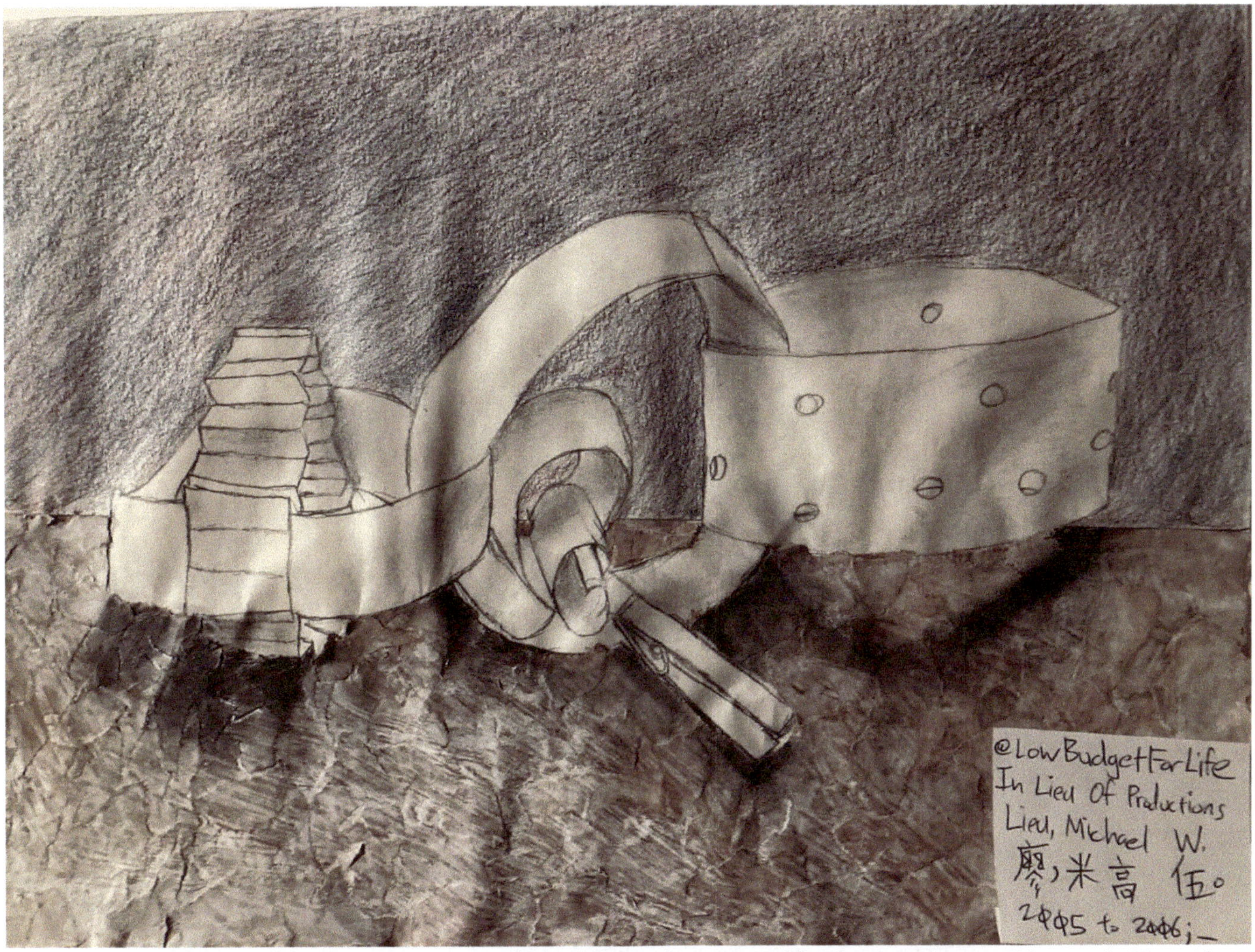

Art Museum Of Rosemont (S), Illinois, North America, Planet Earth

“Do It Yourself (D.I.Y.) Construction Engineering”

Mixed Media

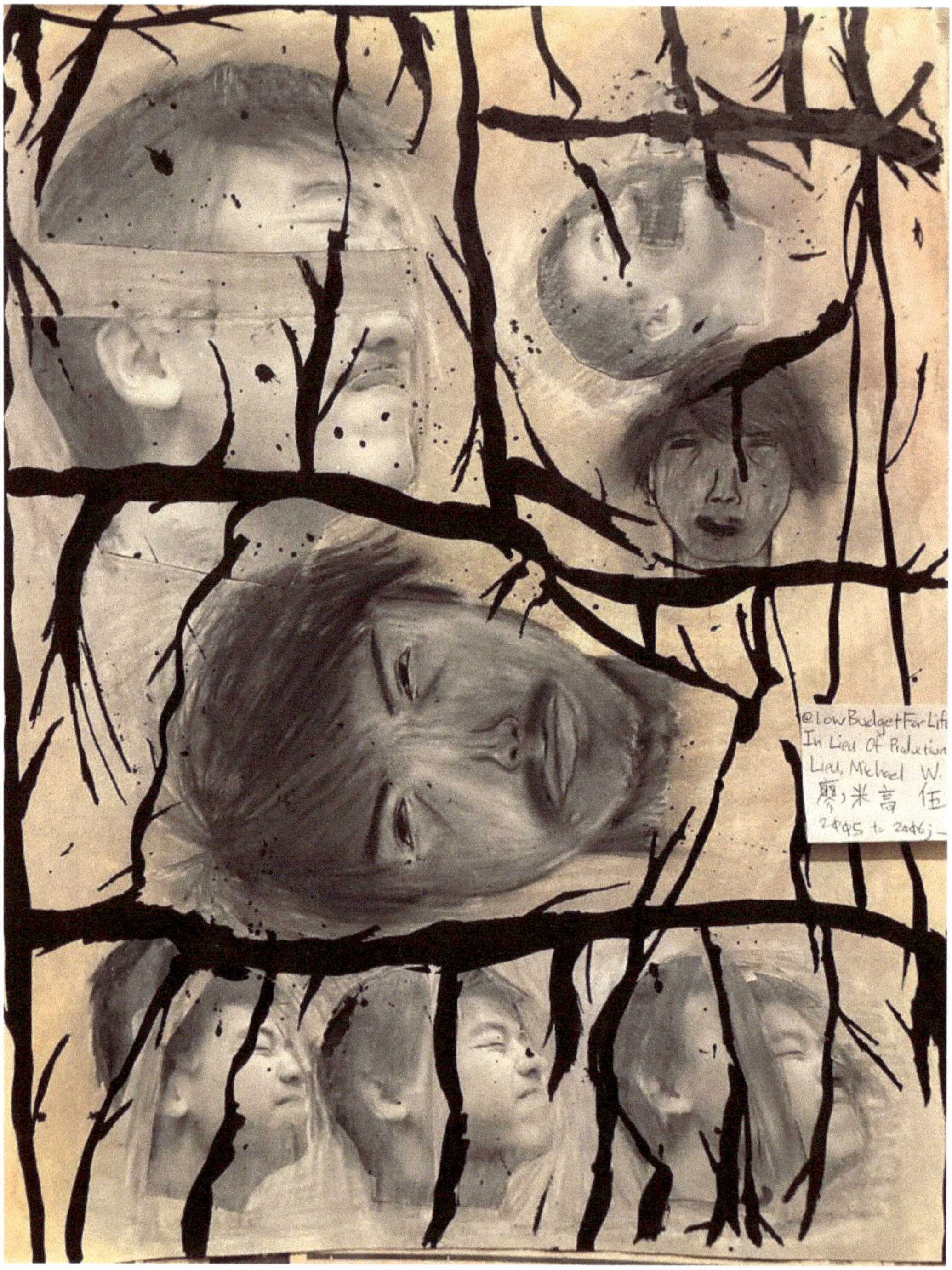

Art Museum Of Rosemont (S), Illinois, North America, Planet Earth

“The Portrait Of Heads, D Portrait Of Head(s)”

Mixed Media

2154;_Arcadia \ Valentina Vita (V. V.) ’Healthy-Strong Life’ \ Left to Right(L->R)

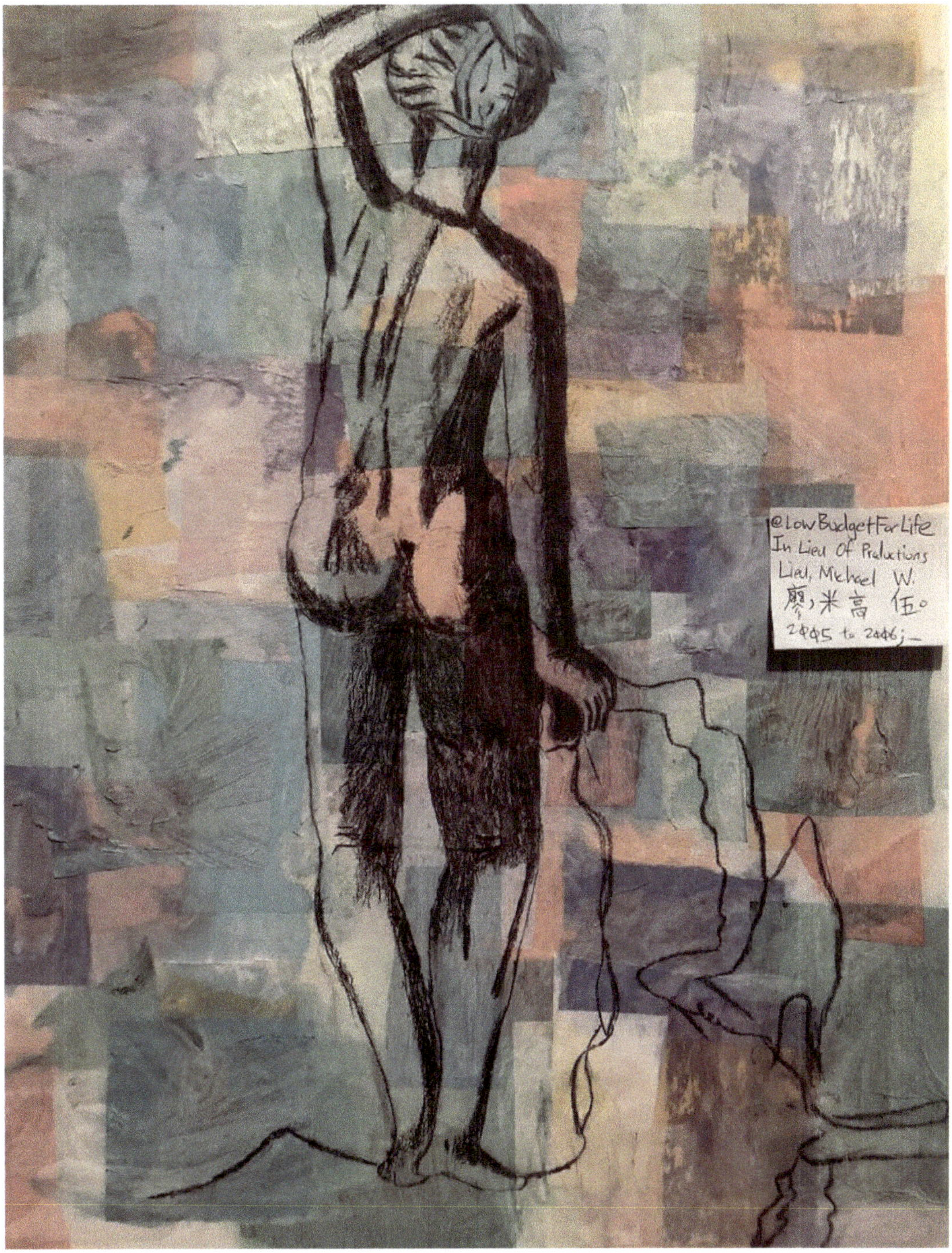

Art Museum Of Rosemont (S), Illinois, North America, Planet Earth

"Nude Back View Painting Of A Natural Unmodified Human With Background"

Mixed Media

2154;_ Arcadia \ Valentina Vita (V. V.) 'Healthy-Strong Life' \ Left to Right(L->R)

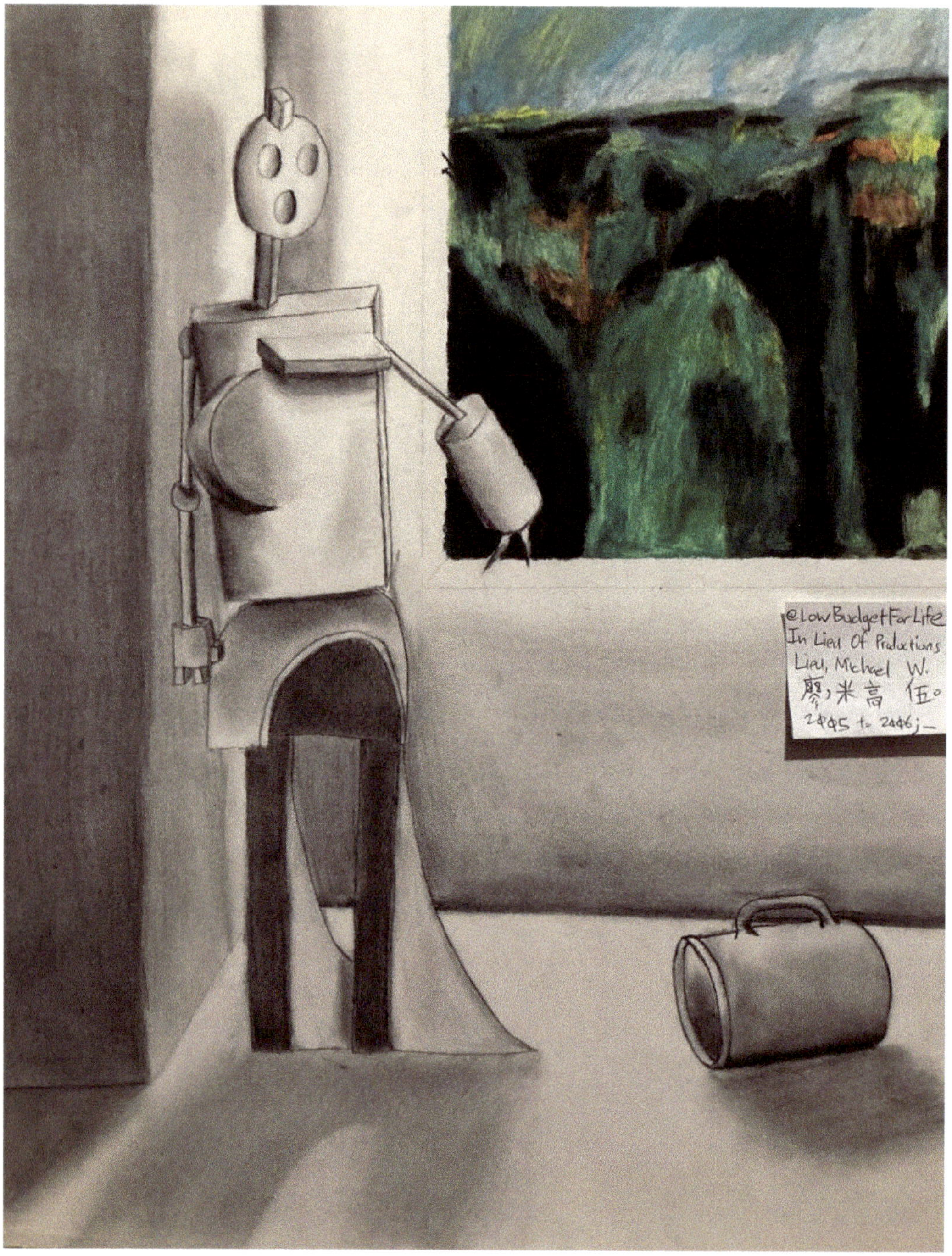

Art Museum Of Rosemont (S), Illinois, North America, Planet Earth

“Abstracted Humanoid’s Surprise Face Says Spell Cup… C.U.P.”

Mixed Media

2154;_Arcadia \ Valentina Vita (V. V.) ’Healthy-Strong Life’ \ Left to Right(L->R)

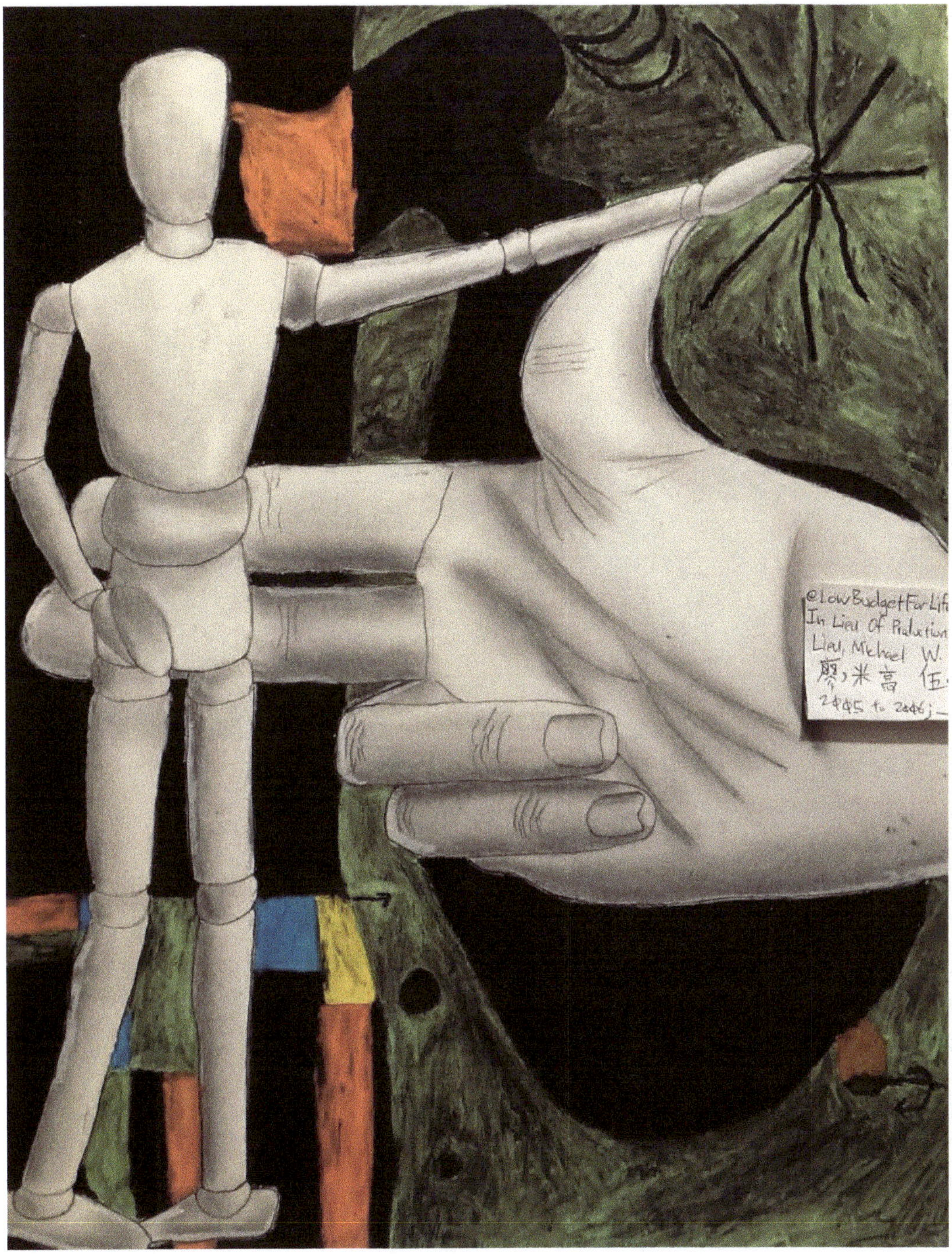

Art Museum Of Rosemont (S), Illinois, North America, Planet Earth

“A Mannequin And A Literal Handgun With Background”

Mixed Media

2154;_Arcadia \ Valentina Vita (V. V.) 'Healthy-Strong Life' \ Left to Right(L->R)

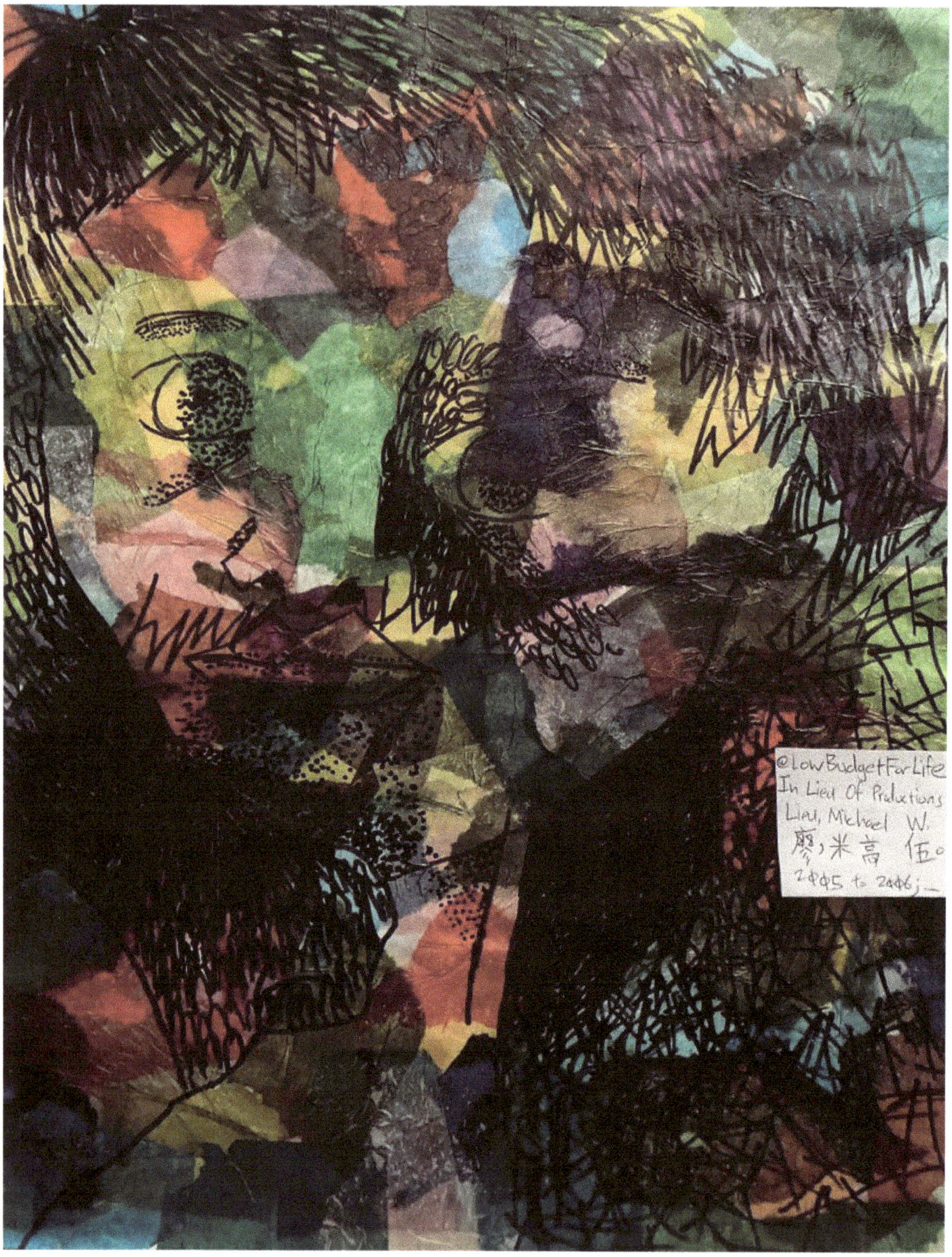

Art Museum Of Rosemont (S), Illinois, North America, Planet Earth

"Portrait 3 (three) With Background"

Mixed Media

2154;_Arcadia \ Valentina Vita (V. V.) 'Healthy-Strong Life' \ Left to Right(L->R)

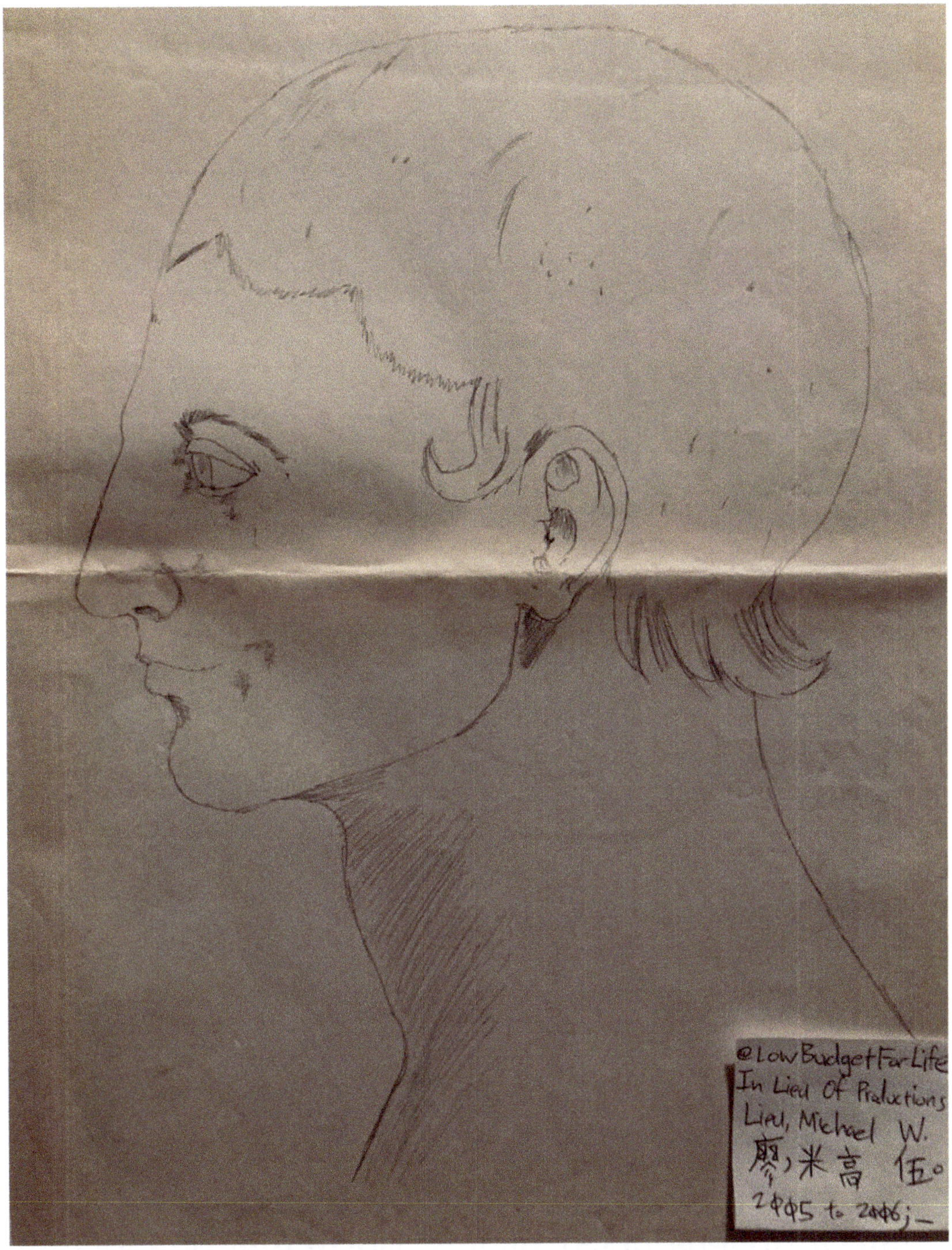

Art Museum Of Rosemont (S), Illinois, North America, Planet Earth

“Portrait 4 (four; for) Side View”

Pencil

2154;_ Arcadia \ Valentina Vita (V. V.) ’Healthy-Strong Life’ \ Left to Right(L->R)

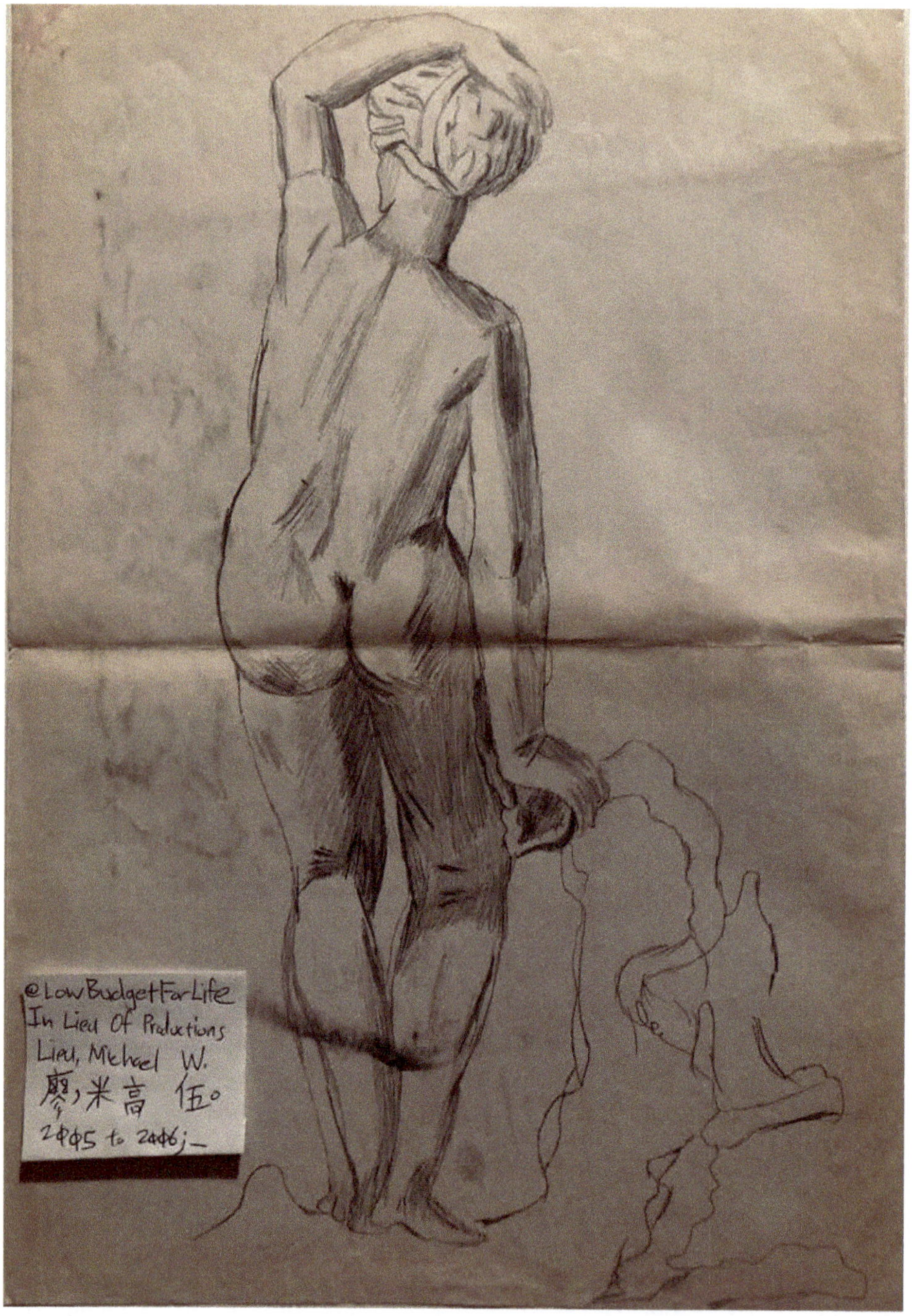

Art Museum Of Rosemont (S), Illinois, North America, Planet Earth

“Nude Back View Penciling Of A Natural Unmodified Human”

Pencil

2154;_ Arcadia \ Valentina Vita (V. V.) 'Healthy-Strong Life' \ Left to Right(L->R)

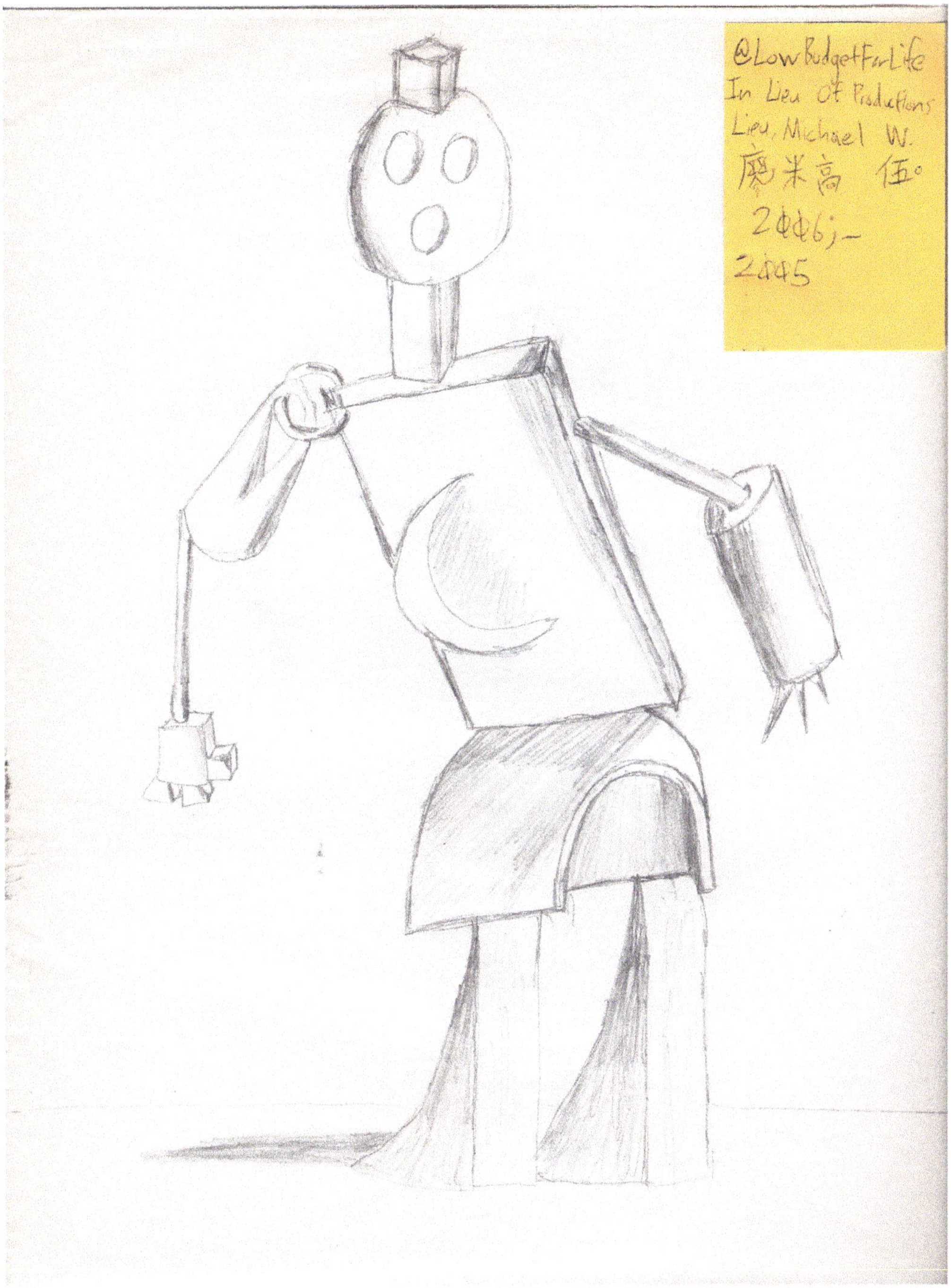

Art Museum Of Rosemont (S), Illinois, North America, Planet Earth

"Abstracted Humanoid's Surprise Face Is The Best Face Reaction"

Pencil

2154;_Arcadia \ Valentina Vita (V. V.) 'Healthy-Strong Life' \ Left to Right(L->R)

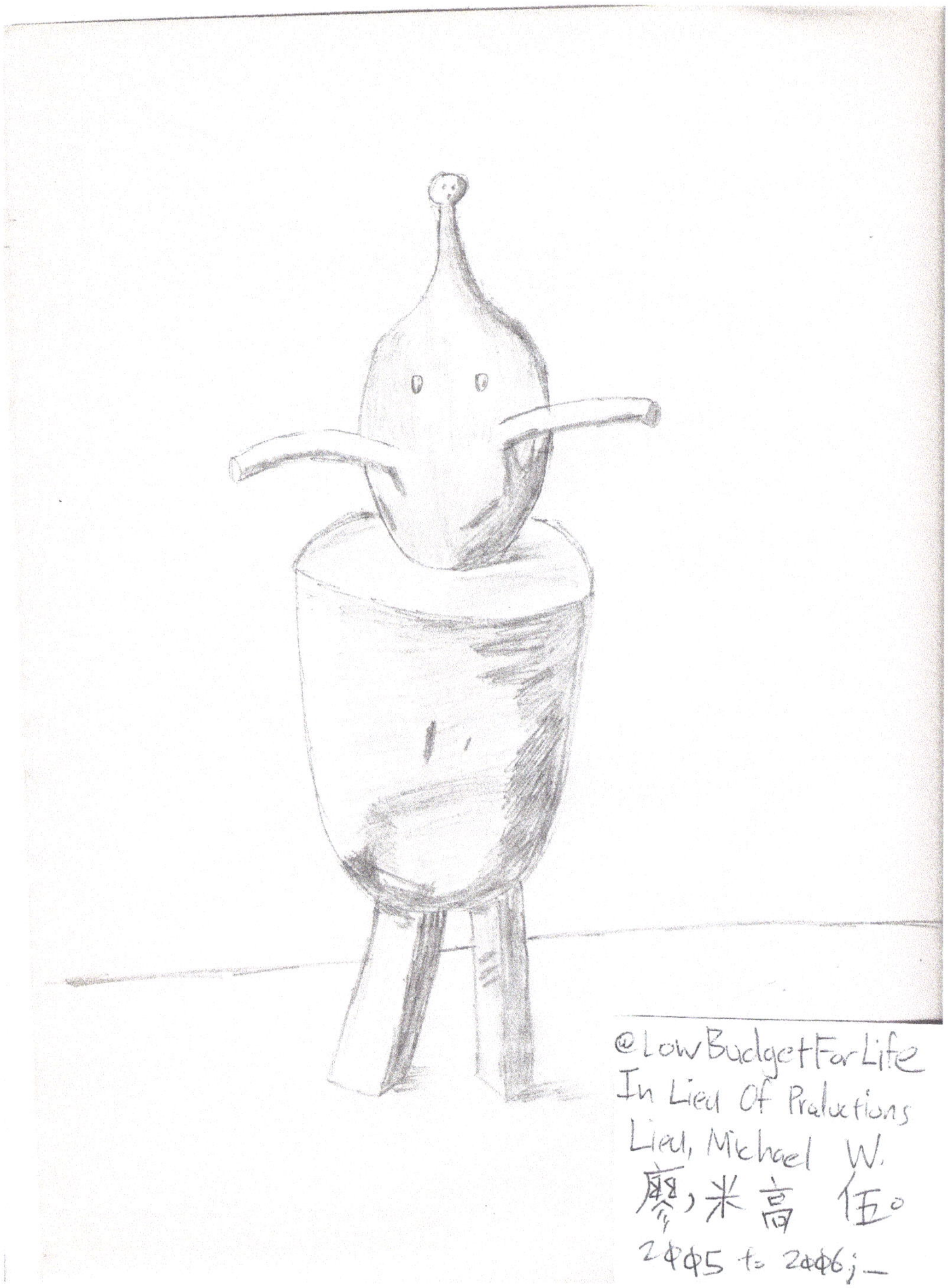

Art Museum Of Rosemont (S), Illinois, North America, Planet Earth

“Abstracted Humanoid’s How Many Limbs Do I Have?”

Pencil

2154;_Arcadia \ Valentina Vita (V. V.) ’Healthy-Strong Life’ \ Left to Right(L->R)

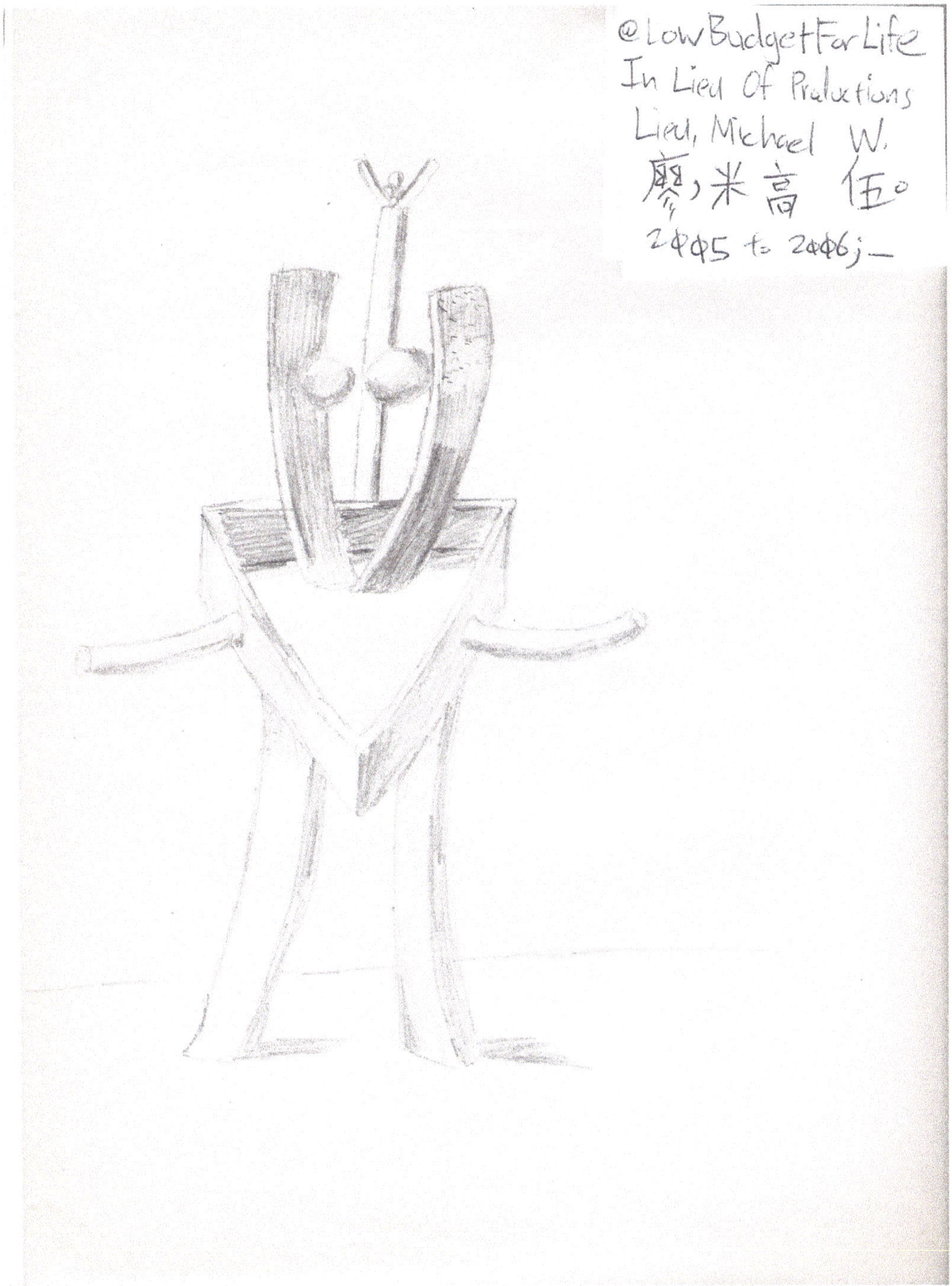

Art Museum Of Rosemont (S), Illinois, North America, Planet Earth

"Abstracted Humanoid's Dimensions"

Pencil

2154;_Arcadia \ Valentina Vita (V. V.) 'Healthy-Strong Life' \ Left to Right(L->R)

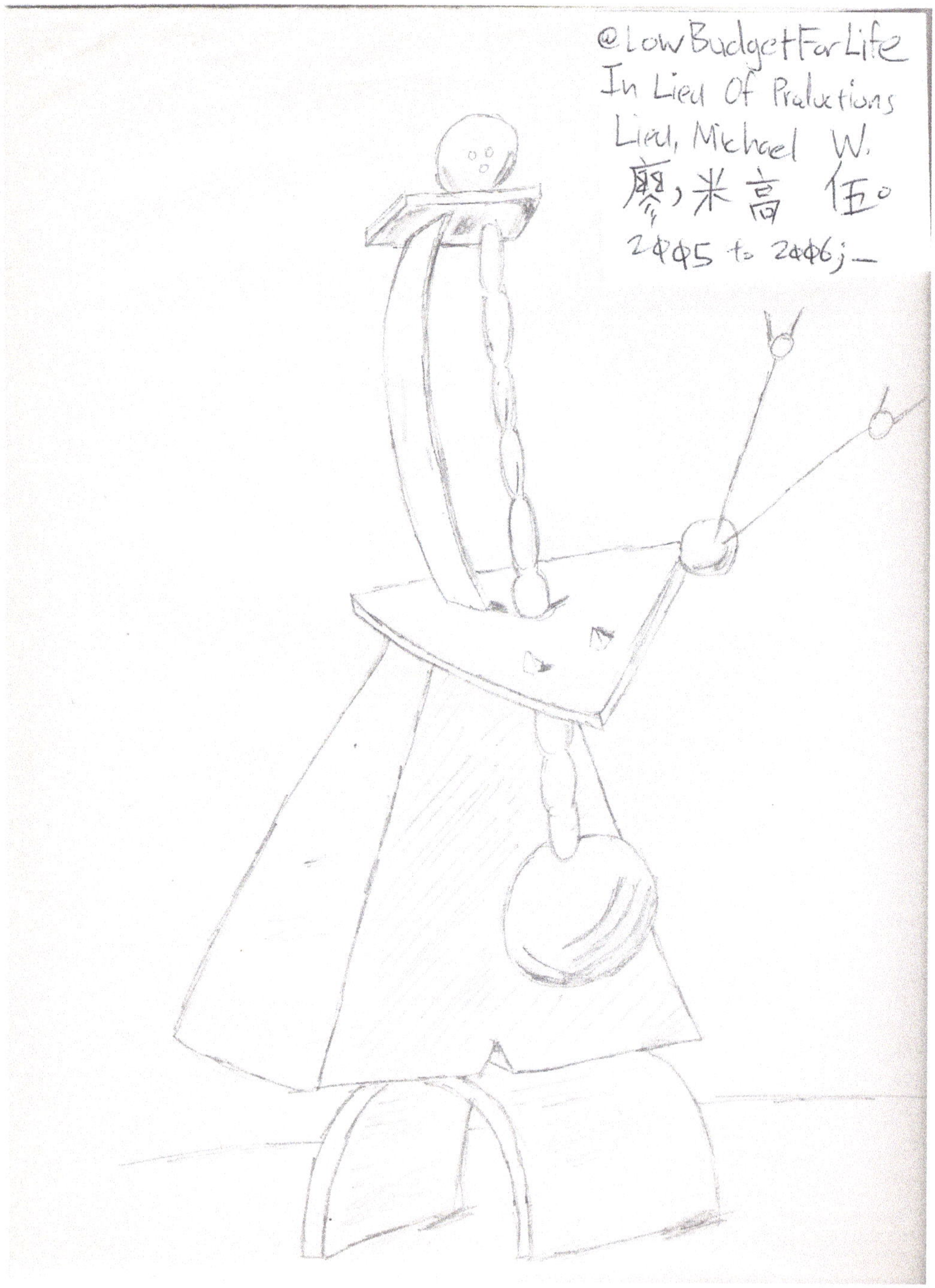

Art Museum Of Rosemont (S), Illinois, North America, Planet Earth

“Abstracted Humanoid’s Hey, Look At What I Can Do!”

Pencil

2154;_Arcadia \ Valentina Vita (V. V.) 'Healthy-Strong Life' \ Left to Right(L->R)

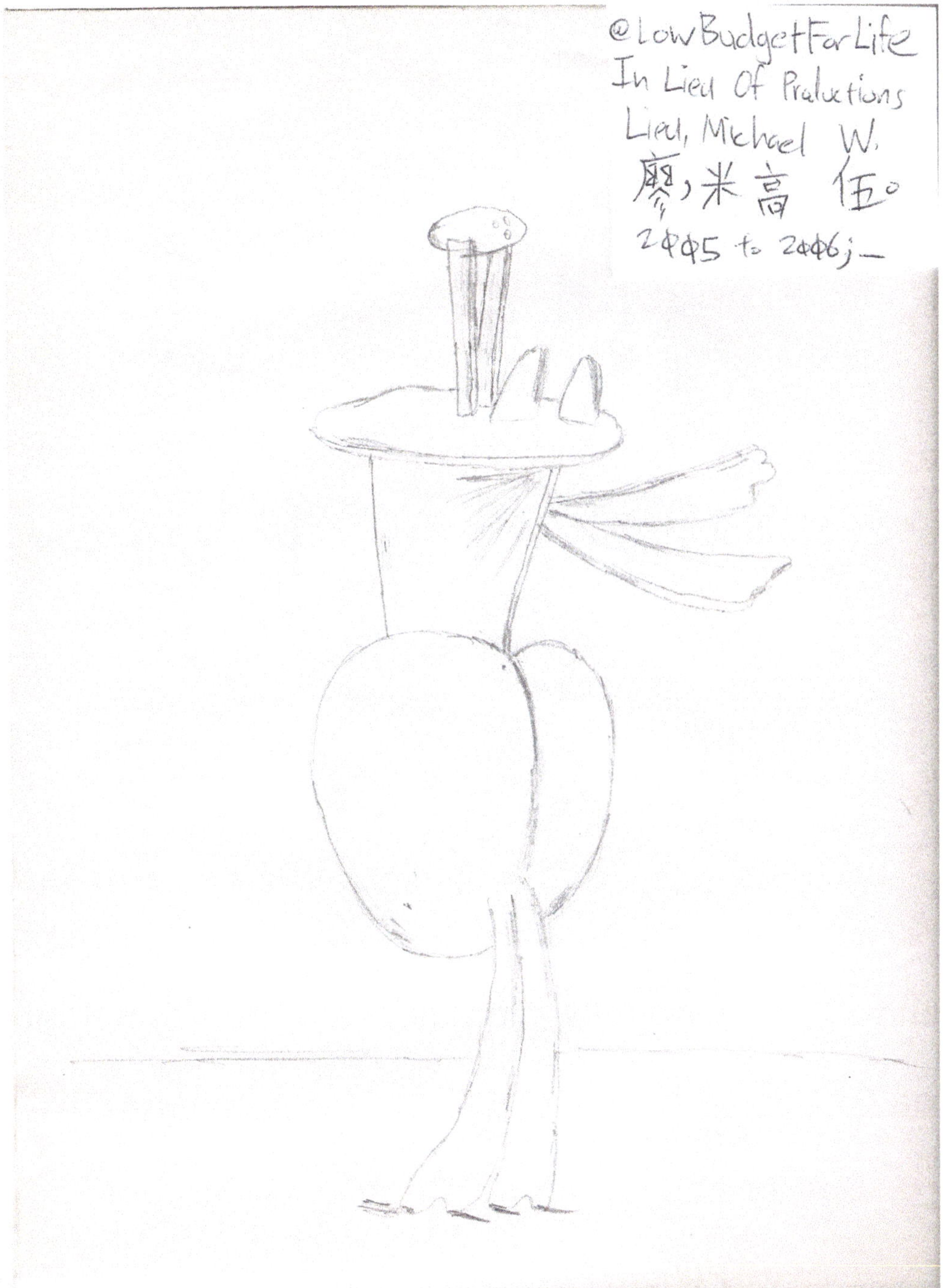

Art Museum Of Rosemont (S), Illinois, North America, Planet Earth

"Abstracted Humanoid's Awareness, I Have A Peach Fruit Bottom…"

Pencil

2154;_ Arcadia \ Valentina Vita (V. V.) 'Healthy-Strong Life' \ Left to Right(L->R)

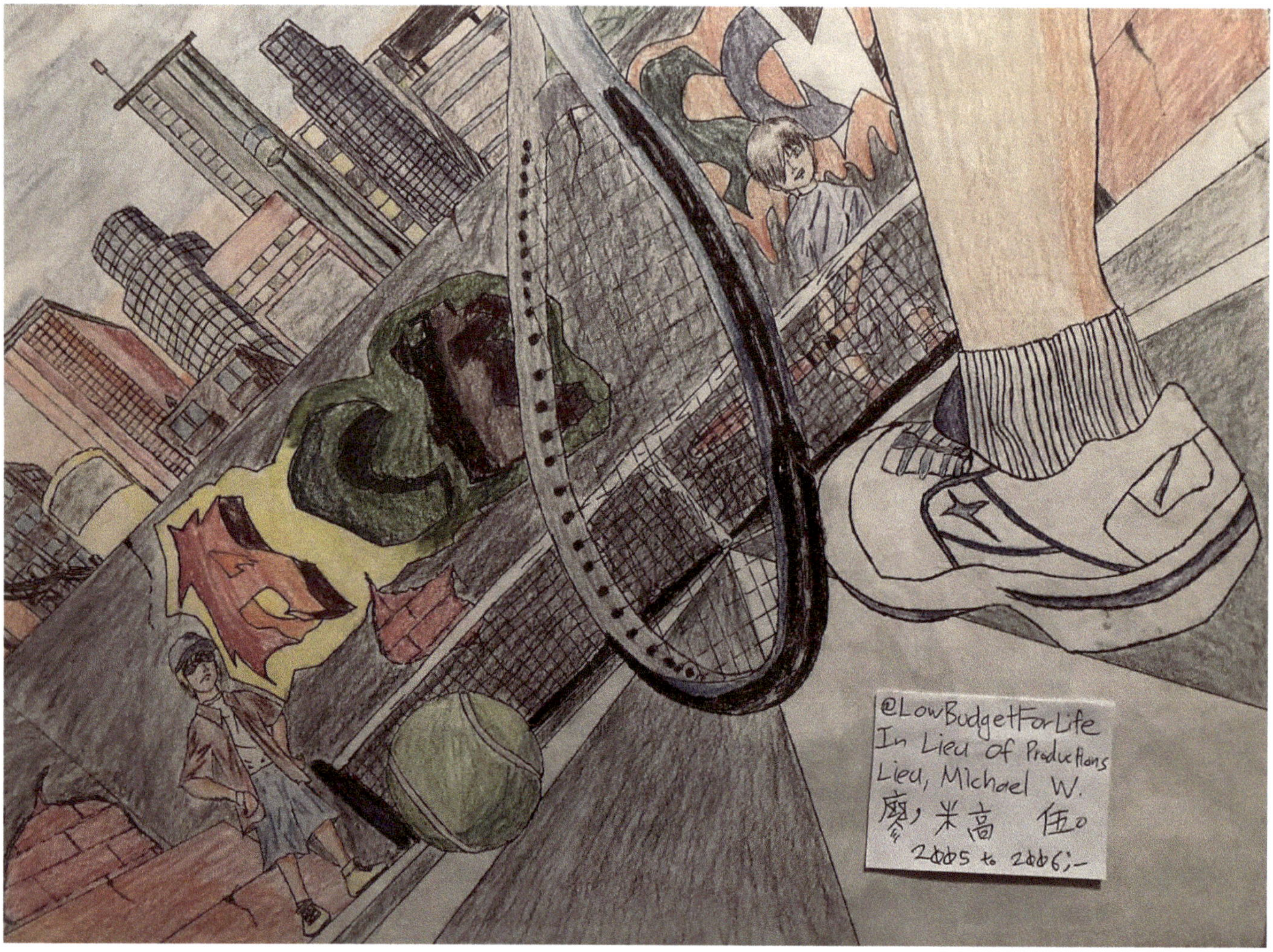

Art Museum Of Rosemont (S), Illinois, North America, Planet Earth

Area Of Concentration (A.O.C.); After School Activities
"After School Practice Try Outs;
Sports Activity Tennis Doubles Or Singles"

Pencil, Pen, Coloring Pencil

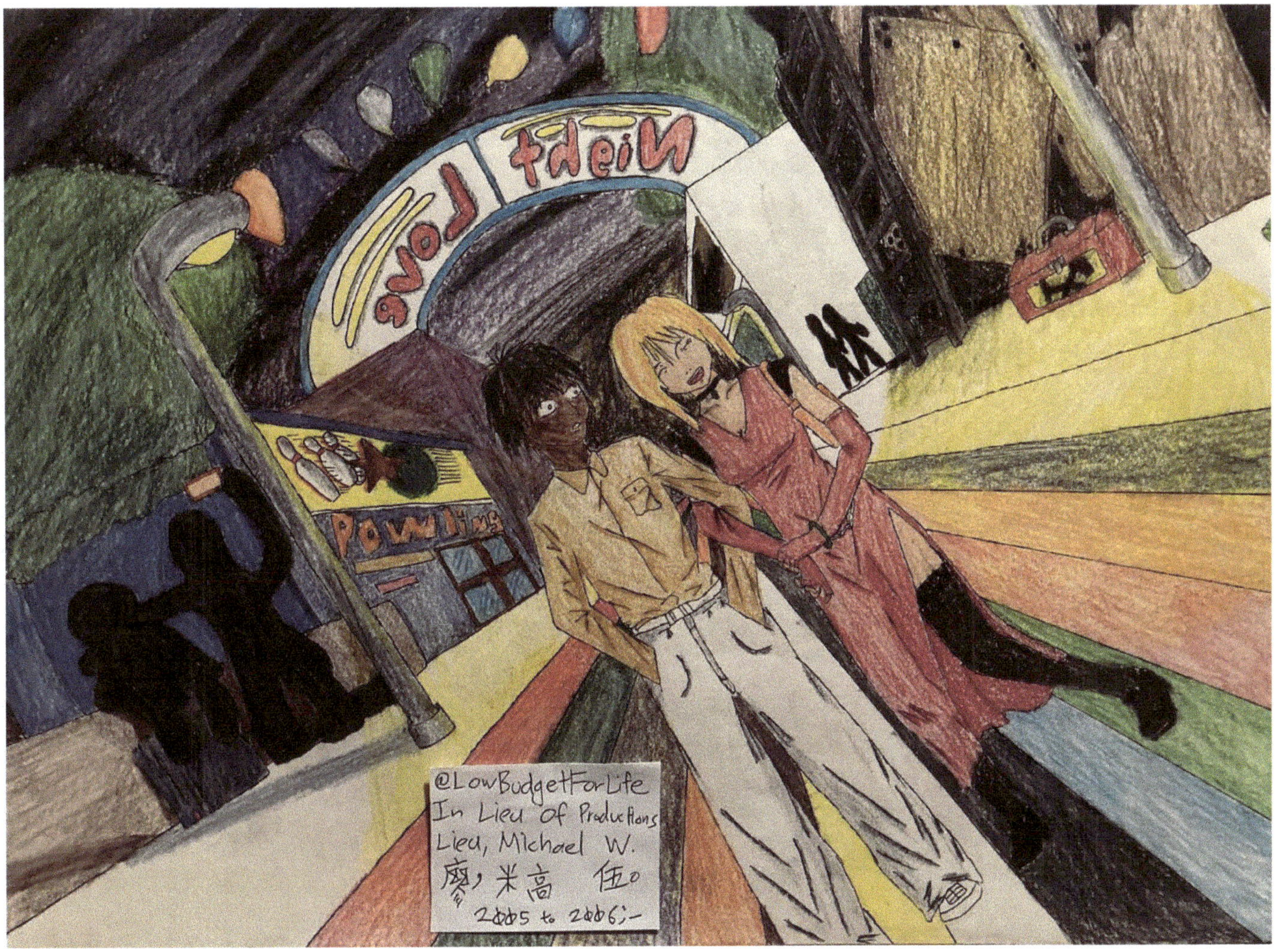

Art Museum Of Rosemont (S), Illinois, North America, Planet Earth

<u>Area Of Concentration (A.O.C.); After School Activities</u>
"Going Out On A Date With Each Other And Or Hanging Out; Romance Love Flirting"

Pencil, Pen, Coloring Pencil

2154;_Arcadia \ Valentina Vita (V. V.) 'Healthy-Strong Life' \ Left to Right(L->R)

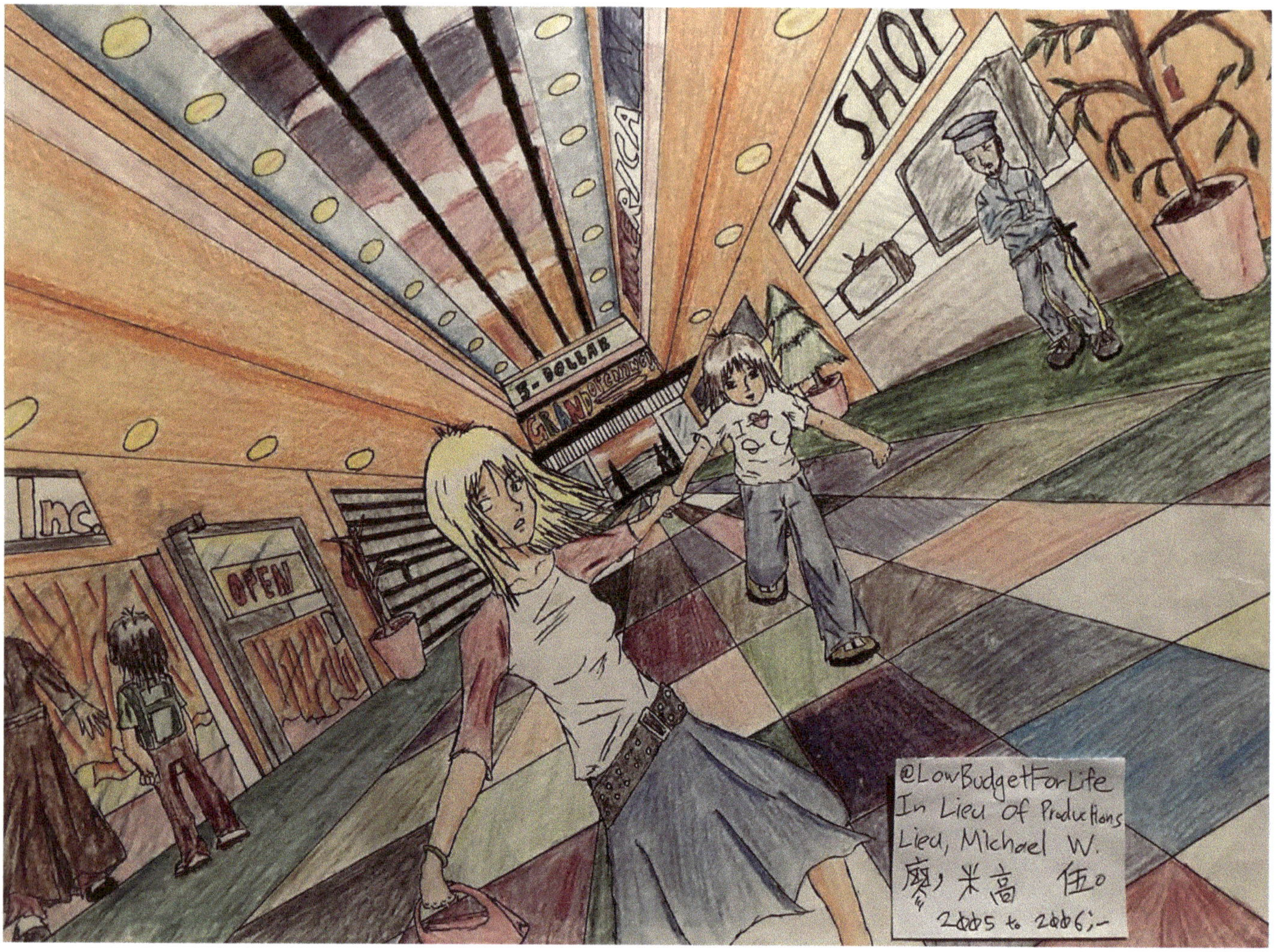

Art Museum Of Rosemont (S), Illinois, North America, Planet Earth

<u>Area Of Concentration (A.O.C.); After School Activities</u>
“Private Sector Small Business;
Coming Of Age Shopping With Friends In The Mall”

Pencil, Pen, Coloring Pencil

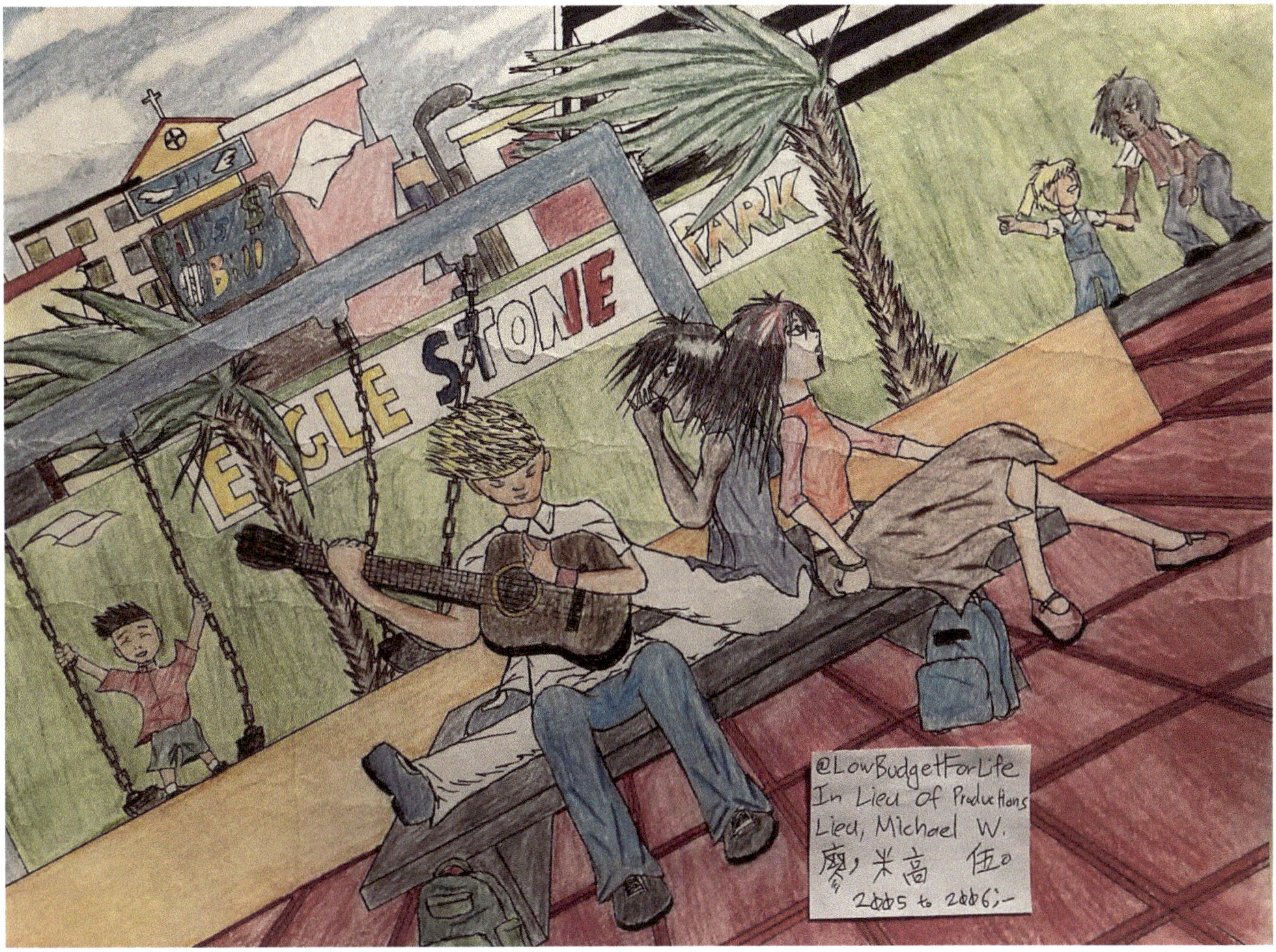

Art Museum Of Rosemont (S), Illinois, North America, Planet Earth

Area Of Concentration (A.O.C.); After School Activities
"Rehearsing For A Band Concert In The Park;
Music Video Instruments Practice"

Pencil, Pen, Coloring Pencil

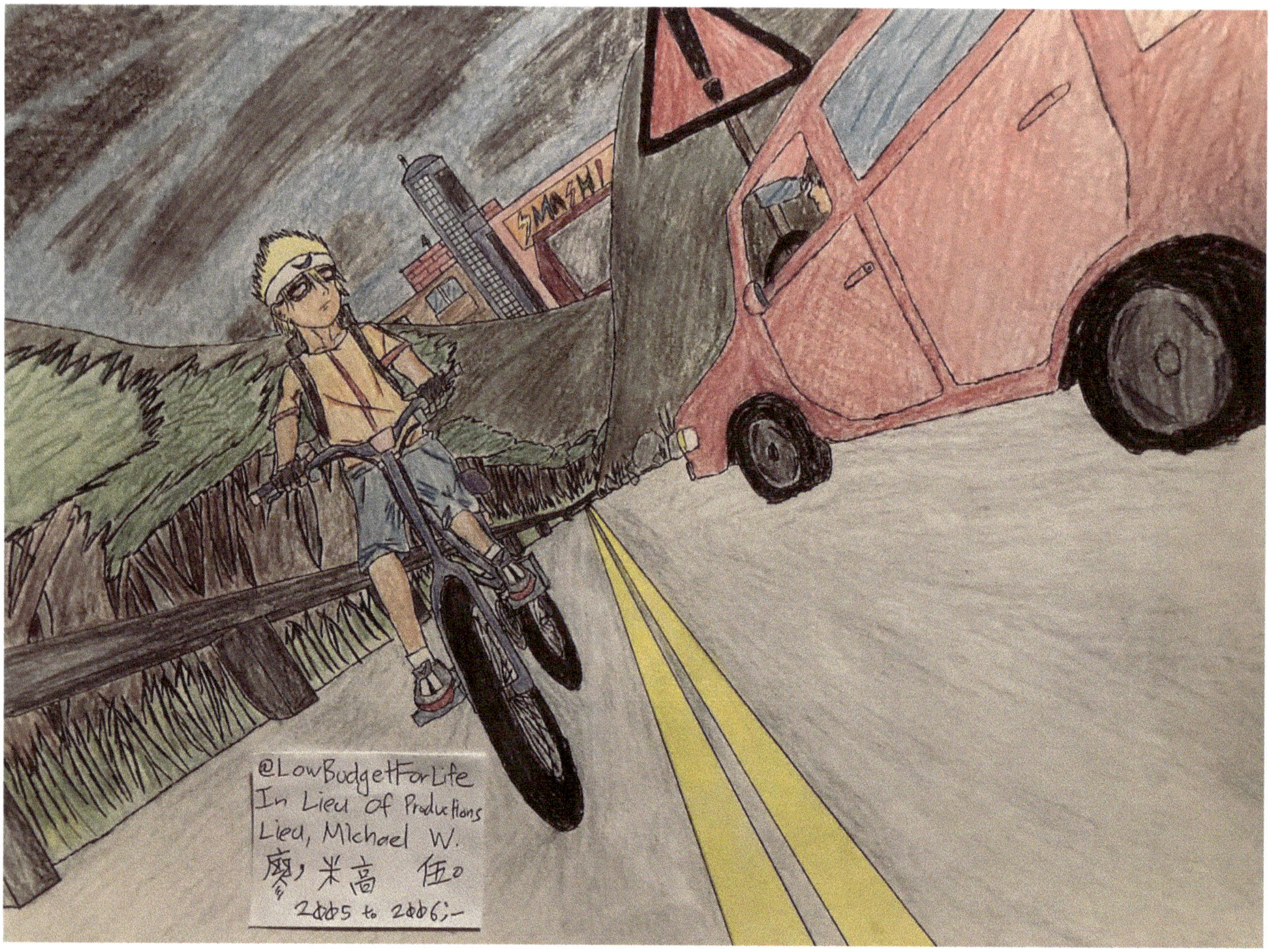

Art Museum Of Rosemont (S), Illinois, North America, Planet Earth

<u>Area Of Concentration (A.O.C.); After School Activities</u>
"Riding The Bicycle Back Home;
Montage Transportation"

Pencil, Pen, Coloring Pencil

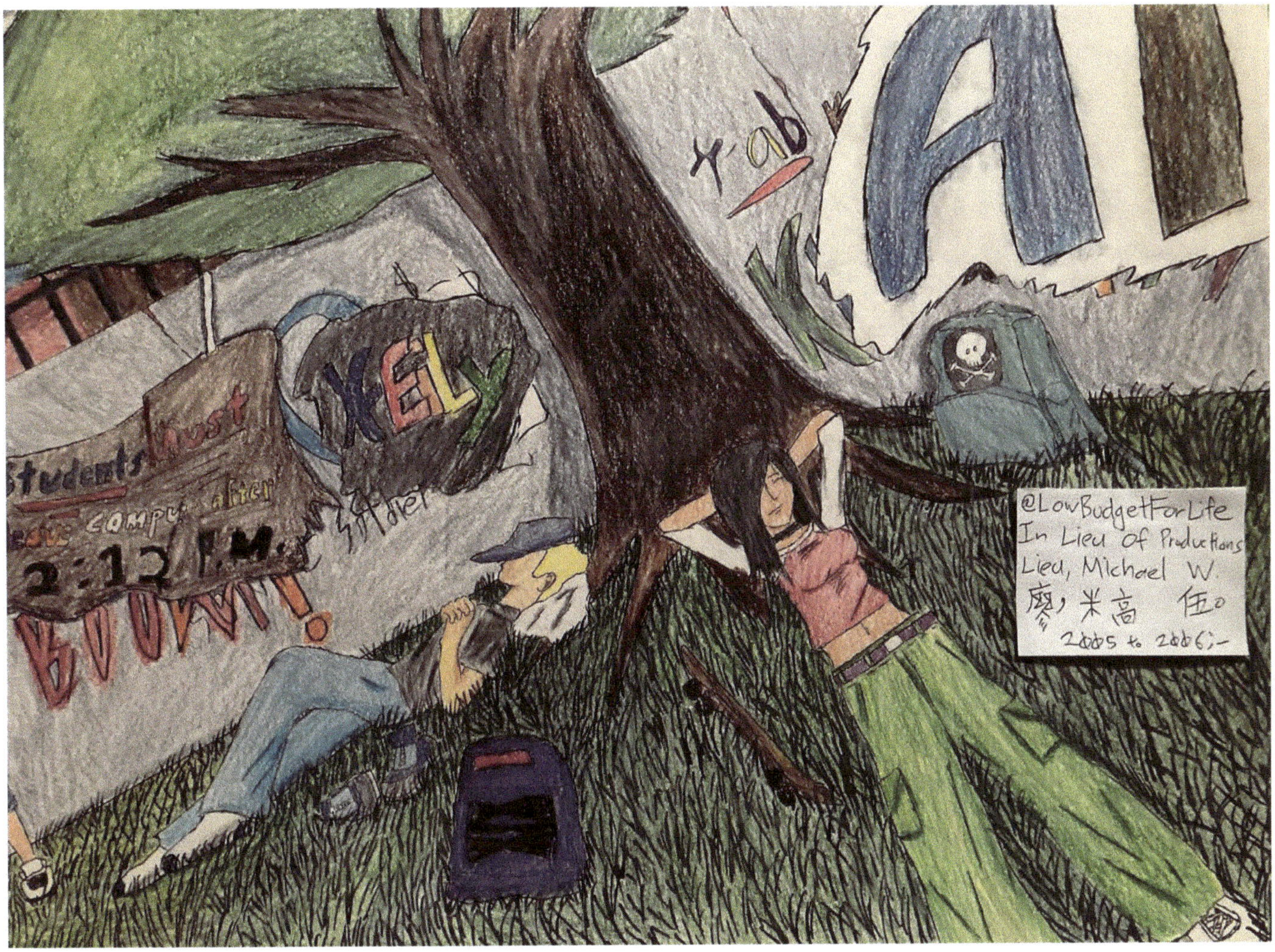

Art Museum Of Rosemont (S), Illinois, North America, Planet Earth

<u>Area Of Concentration (A.O.C.); After School Activities</u>
**“Hanging Around After School In The Yard;
Slice Of Life Relaxing”**

Pencil, Pen, Coloring Pencil

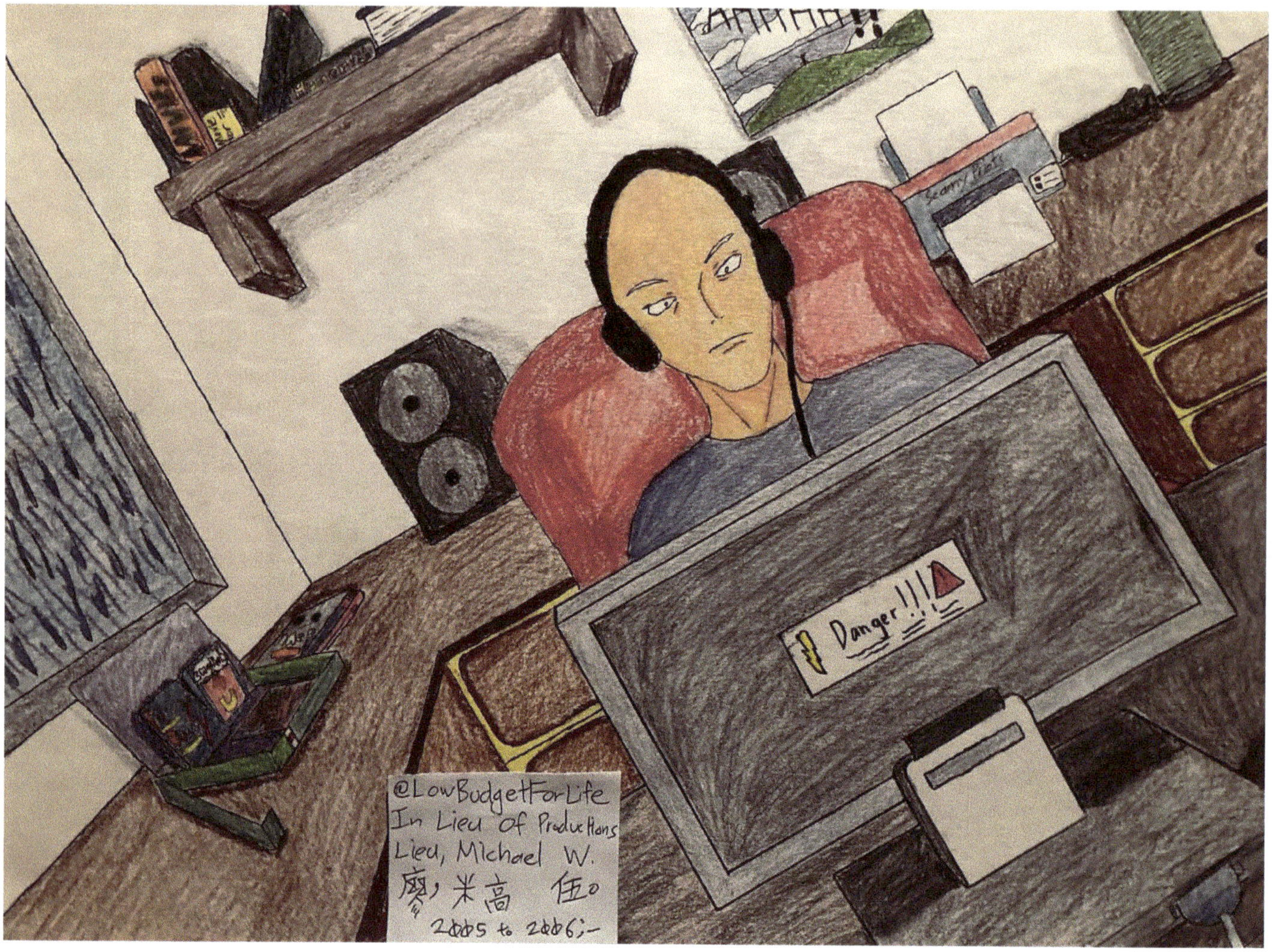

Art Museum Of Rosemont (S), Illinois, North America, Planet Earth

<u>Area Of Concentration (A.O.C.); After School Activities</u>
"At Home Playing Computer Simulation(s) Through The Internet; War Game(s)"

Pencil, Pen, Coloring Pencil

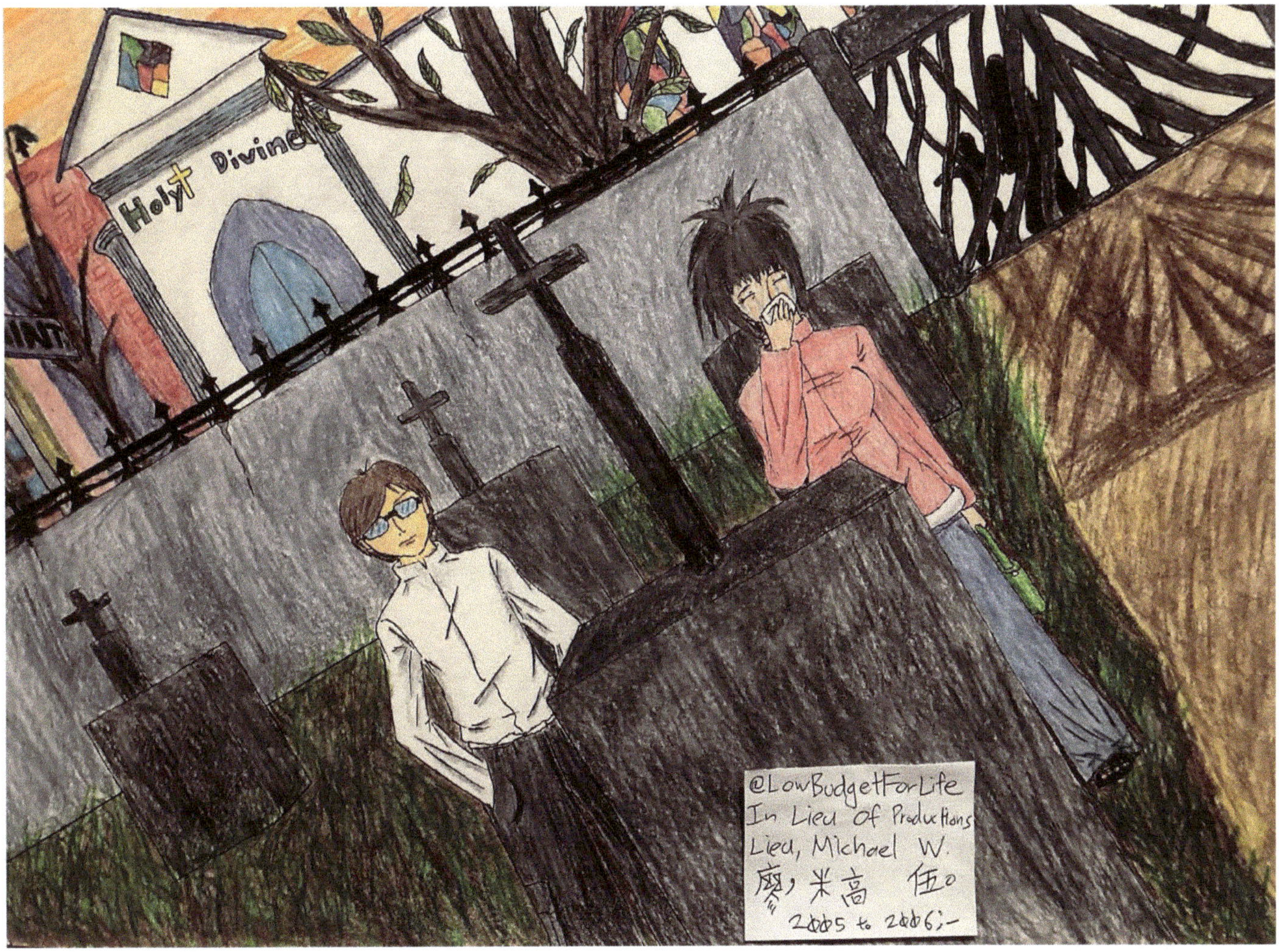

Art Museum Of Rosemont (S), Illinois, North America, Planet Earth

Area Of Concentration (A.O.C.); After School Activities
"A Family And Or Friend's Graveyard Site Marker; Memoir Remembrance"

Pencil, Pen, Coloring Pencil

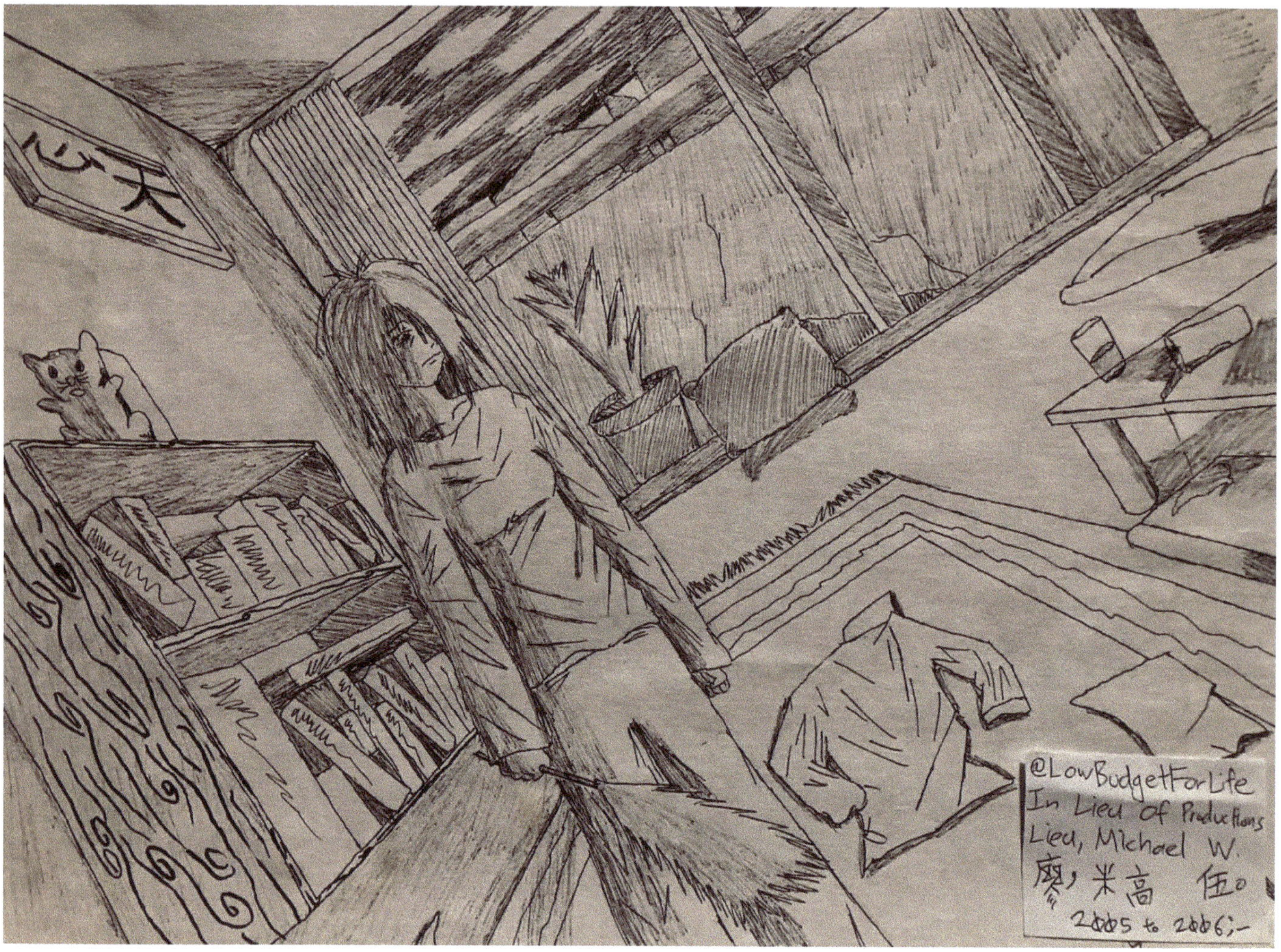

Art Museum Of Rosemont (S), Illinois, North America, Planet Earth

<u>Area Of Concentration (A.O.C.); After School Activities</u>
"Doing Home Chores Before Parents And Or Guardians Arrive; Contemporary Homework Literally"

Pencil, Pen

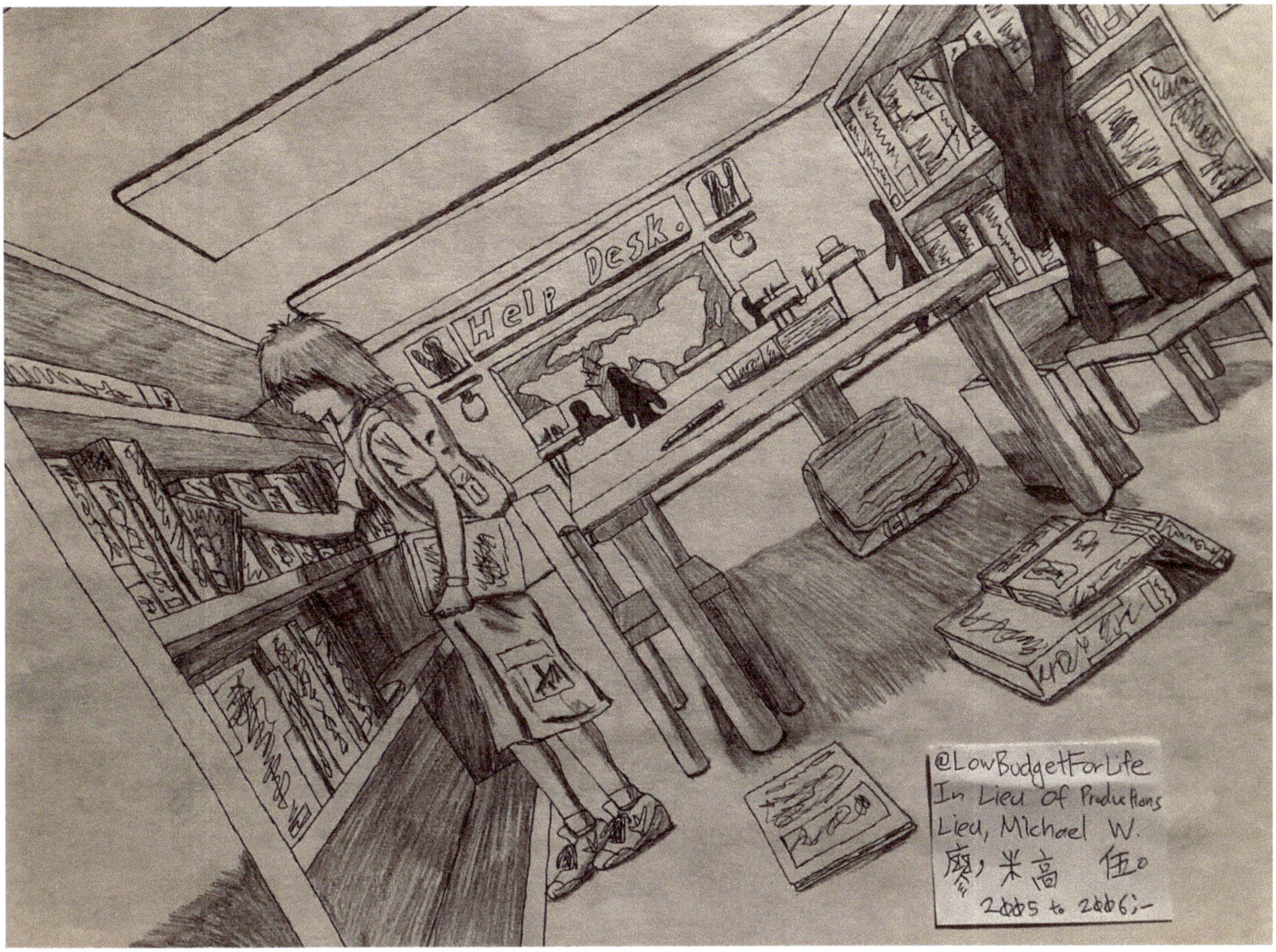

Art Museum Of Rosemont (S), Illinois, North America, Planet Earth

Area Of Concentration (A.O.C.); After School Activities
"At The Library Gathering Information For Research; Mythology Study Session"

Pencil, Pen

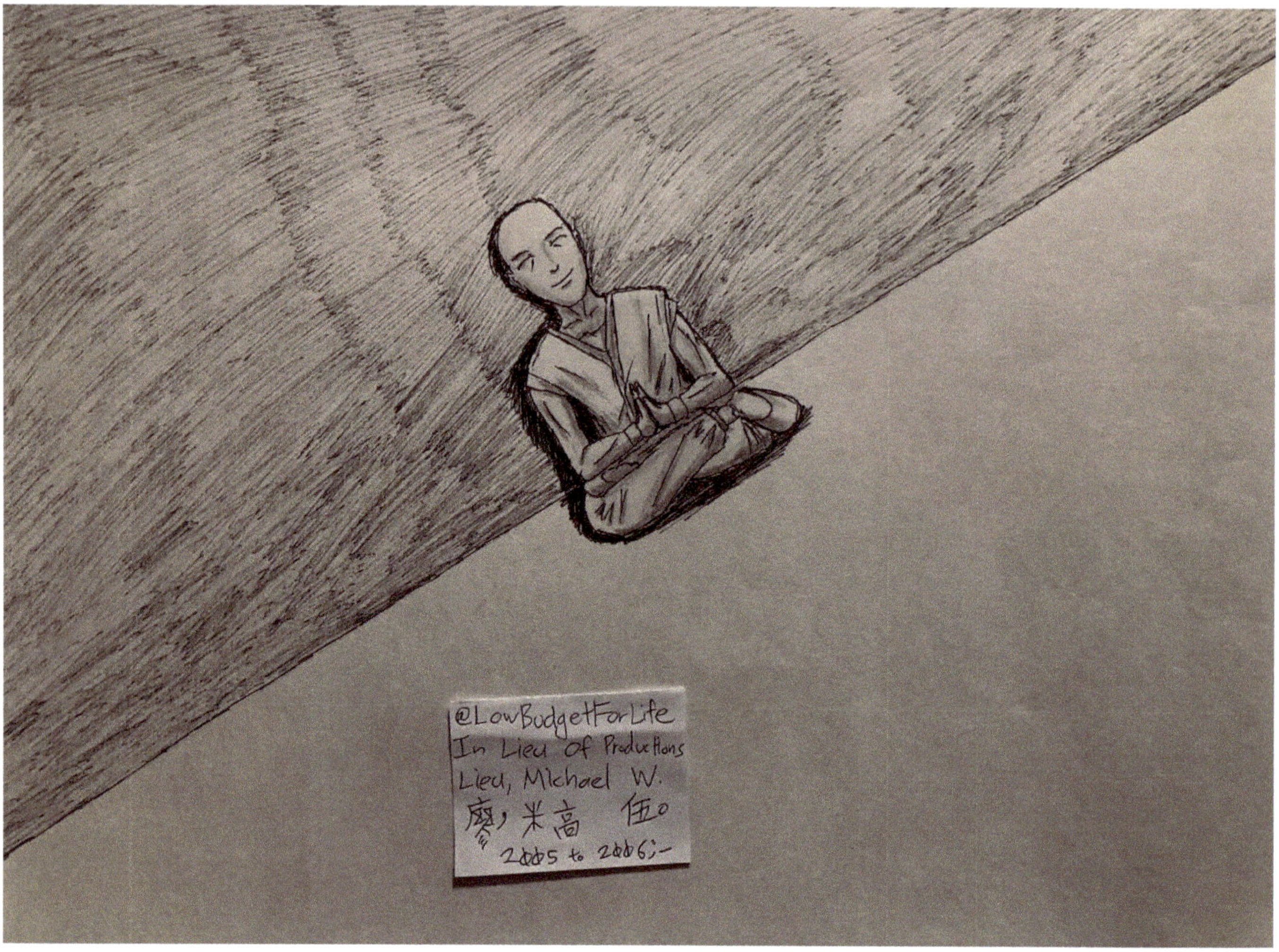

Art Museum Of Rosemont (S), Illinois, North America, Planet Earth

<u>Area Of Concentration (A.O.C.); After School Activities</u>
"Psychological Mental Focus Mindfulness; Martial Arts Meditation"

Pencil, Pen

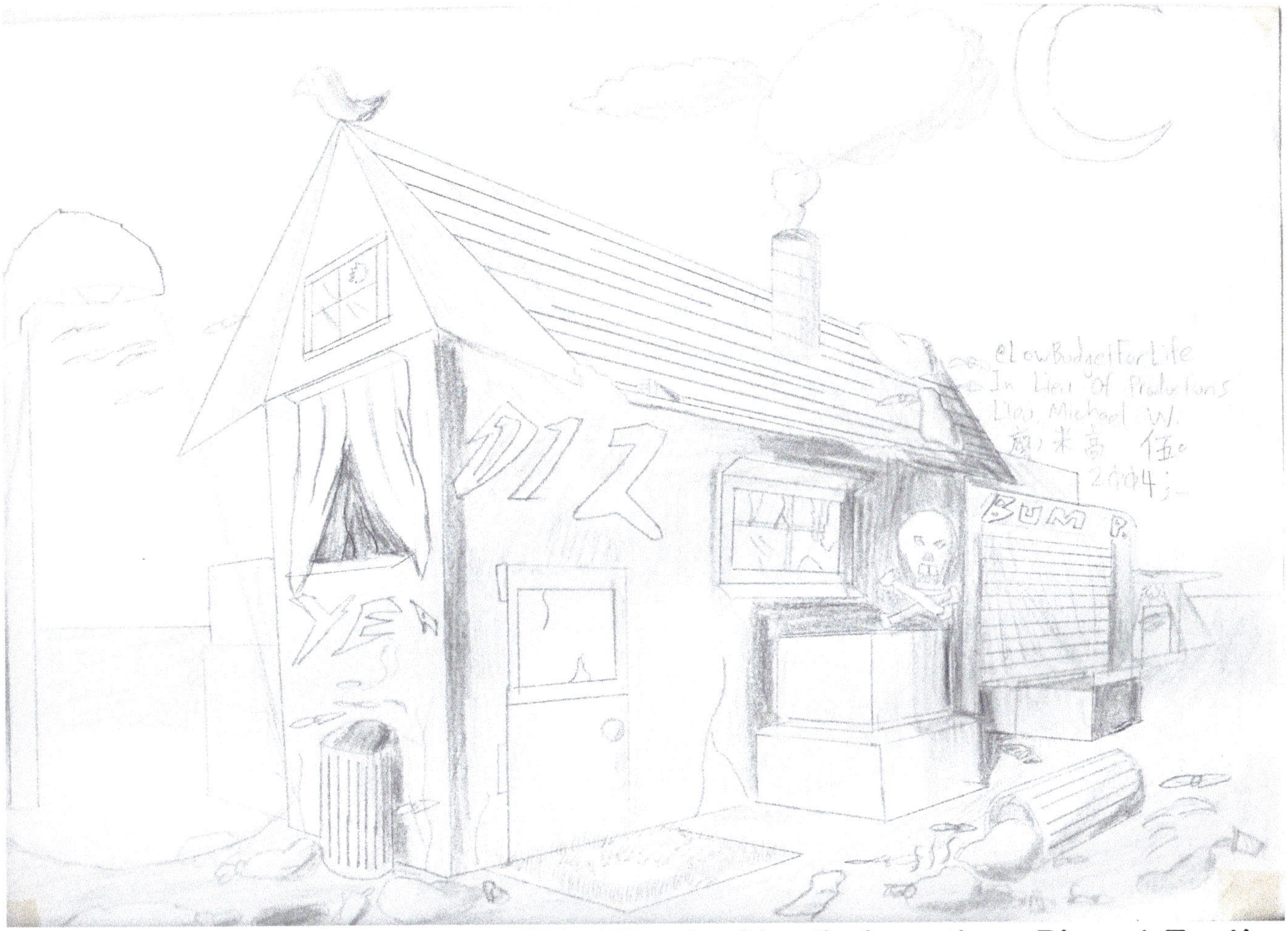

Art Museum Of Rosemont (S), Illinois, North America, Planet Earth

"__Living The Homeless Dystopia Drug Life In The Democratic (D) Liberal Blue Seattle City, Washington State, Planet Earth That Inspired And Founded The 2121 Utopian Sovereignty (S) Movement...__ I Picked The Letters U. And S. As A Mockery To United States. As A Opportunist In The United States of North America (U.S. of N.A.), The Land Of Opportunity. No More Paying The Bills, No More Bills To Pay, No More Taxes, No More Property Taxes, No Tax Collectors; Being Able To Live Life Without Such Burdens, Now That Is Utopia."
-Anonymous, Unknown, ? (Question-Mark), X (Variable),

Pencil, Pen

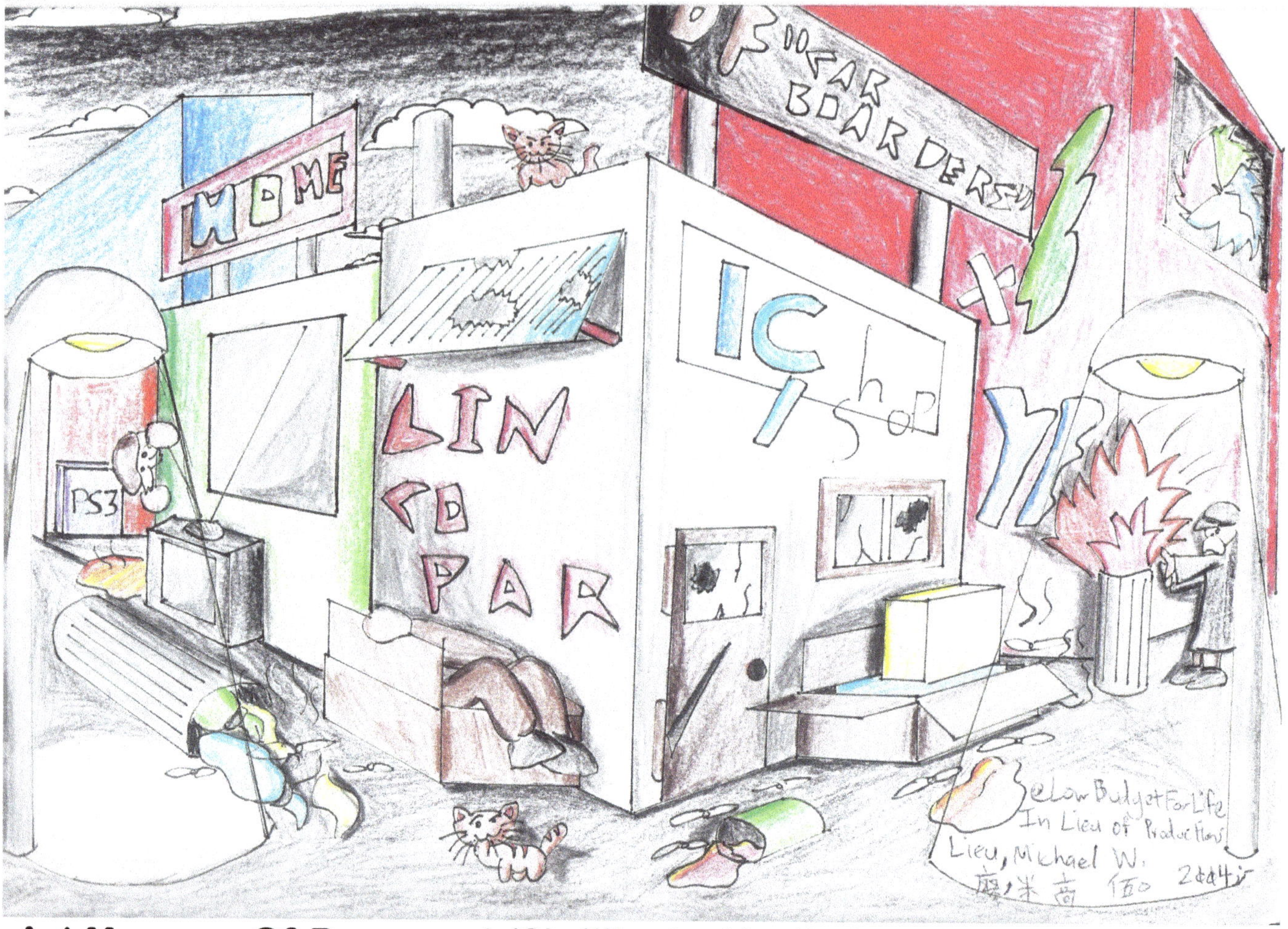

Art Museum Of Rosemont (S), Illinois, North America, Planet Earth;

"__Living The Homeless Dystopia Drug Life In The Democratic (D) Liberal Blue Seattle City, Washington State, Planet Earth That Inspired And Founded The 2121 Utopian Sovereignty (S) Movement...__ After Being Evicted From My Rental Home For Failing To Pay My Bills On Time, I Got To Spend Some Time With The Homeless People, Augmented And Natural, Got To Know Their Problems, Draw Pictures, Document It Down... Have To Start Something To Change Something... Stress Free Existence And Life ...Utopia. That Is What I Wanted To Do, Create Utopia On Earth" -Anonymous, Unknown, ? (Question-Mark), X (Variable),

Pencil, Pen, Coloring Pencils

End Of The
Art Museum Of Rosemont (S), Illinois, North America, Planet Earth;

Thank You For Your Support And Please Come Again.

2154;_10(Oct)-14(fourteen)

Travel To Western Society, America(s), North America
Self-Govern Sovereignty (S),
Illinois (ILL) Sector, Points Of Interest (P.O.I.)

En Route: Springfield (D)(M)(R), "**Historic Landmark**" From Rosemont (S) "**Art Museum**"

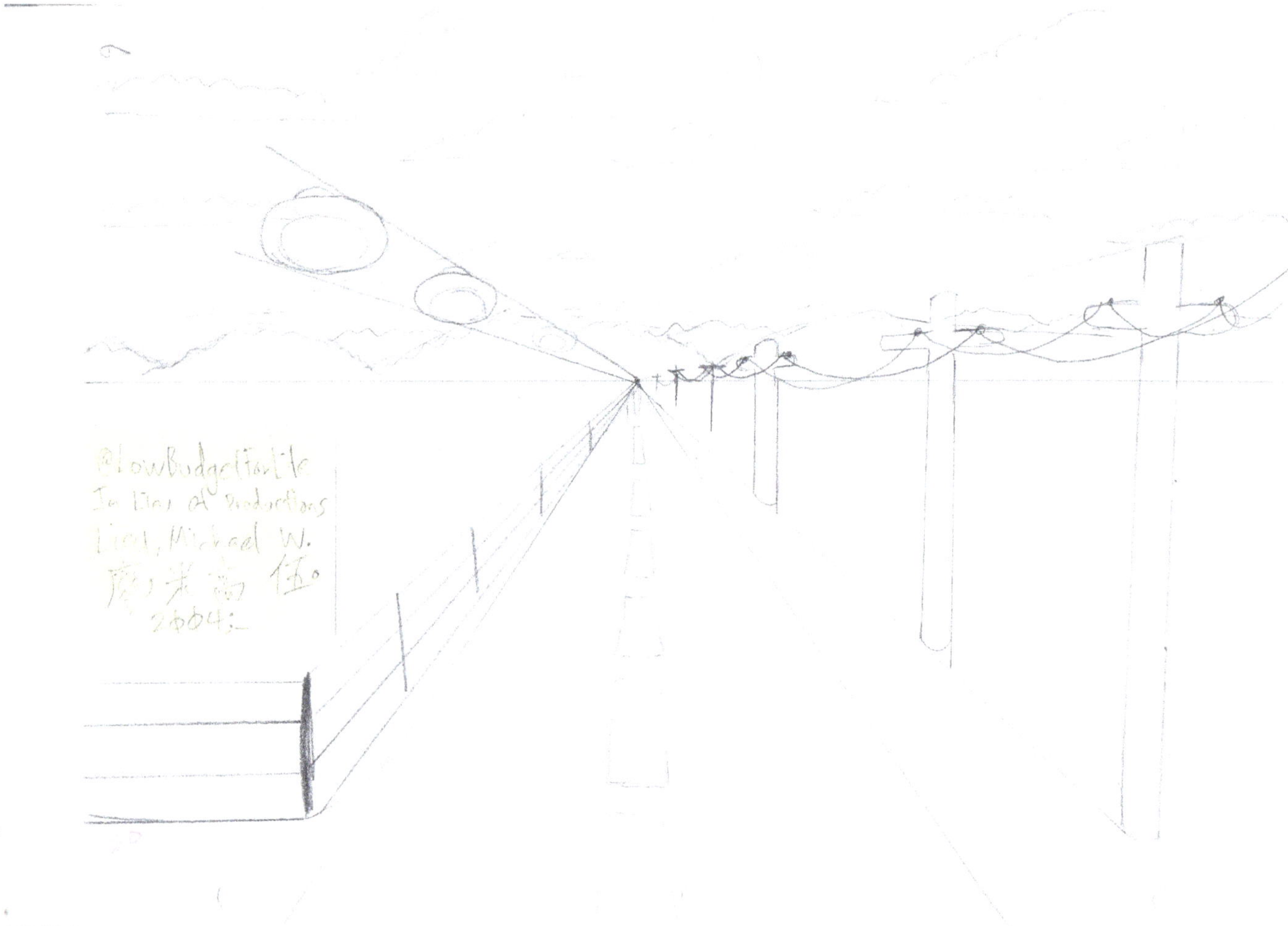

Valentina Vita (V. V.); Traveling Cosplayers And Kosplayers;
"(^_^) **Rosemont Art Museum** was **fun**, **days turned to weeks.** To **Springfield**!"

V. V. selects the music(s) "**SoundTemp's Ragnarok Online:** *Title, Gambler Of Highway, Peaceful Forest, I Miss You, Tread On The Ground, Risk Your Life, Wind Of Tragedy, Prontera Theme, Great Honor, Divine Grace, Morroc Theme, Streamside, Geffen Theme, Payon Theme, Alberta Theme, Labyrinth, Treasure Hunter, Time Up, Under The Ground, Ancient Groover, Through The Tower, Back Attack, Travel, Desert, Plateau*" to play her Lady A. Dragoon Knight Helmet P.P.E. cosplay on one third volume of 33% and is set on loop.

2154;_10(Oct)-21(twenty one)

Travel To Western Society, America(s), North America
Self-Govern Sovereignty (S),
Illinois (ILL) Sector, Points Of Interest (P.O.I.)

Springfield (D)(M)(R), "**Historic Landmark**"

Public Sector Official Message Bulletin Board System (B.B.S.)

"Due to a short staff of employee(s) and volunteer(s), we are not able to conduct any ghost town tours for the "Historical Landmark" of Springfield (D)(M)(R), Illinois (ILL).

We apologize for the inconvenience.

Please, do not "self-tour" the ghost town of Springfield (D)(M)(R) without a governing official, it is dangerous.

We are serious about this warning.

We are not responsible for any damages, bodily harm, death and or dismemberment you receive without a governing Springfield (D)(M)(R), Illinois (ILL) official walking you through the ghost town tour.

Thank you for your understanding."

Valentina Vita (V. V.); Traveling Cosplayers And Kosplayers;
"(=_=) ...Well, that sucks... **Short staff**... That **travel guide** did warn **us** of **short staff**... At least the **Illinois view and travel was nice in my bouncing sphere ball**... Next... to Chicago**, Animal** Sanctuary!"

2154;_10(Oct)-30(thirty)

Travel To Western Society, America(s), North America
Self-Govern Sovereignty (S),
Illinois (ILL) Sector, Points Of Interest (P.O.I.)

Chicago (S), "**Animal** Sanctuary"

"The New World 2121 Movement Of The 22nd Century,
Utopian Sovereignty of North America (U.S. of N.A.)"

Elysium Corporation
亞
Biotech Dome Airlock Decontamination, In And Out Progressing

Chicago,
The Last Animal Sanctuary Of Illinois

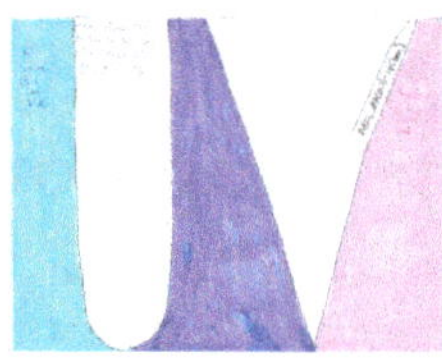

Owned And Operated By
The 22nd (twenty second) Century Phenomenon,
UltraViolet (U.V.) Rays Broadcasting Biotech Towers'
City Controllers
"City Mayor"

Sovereignty, Sovereign, Self-Govern, Self-Governing (S)

Operation Hours;
Always open, 24 (twenty four) hours a day, 7 (seven) days a week.

Follow the UltraViolet (U.V.), **white**, green lines into
Elysium Corporation's Biotech Dome Airlock Decontamination

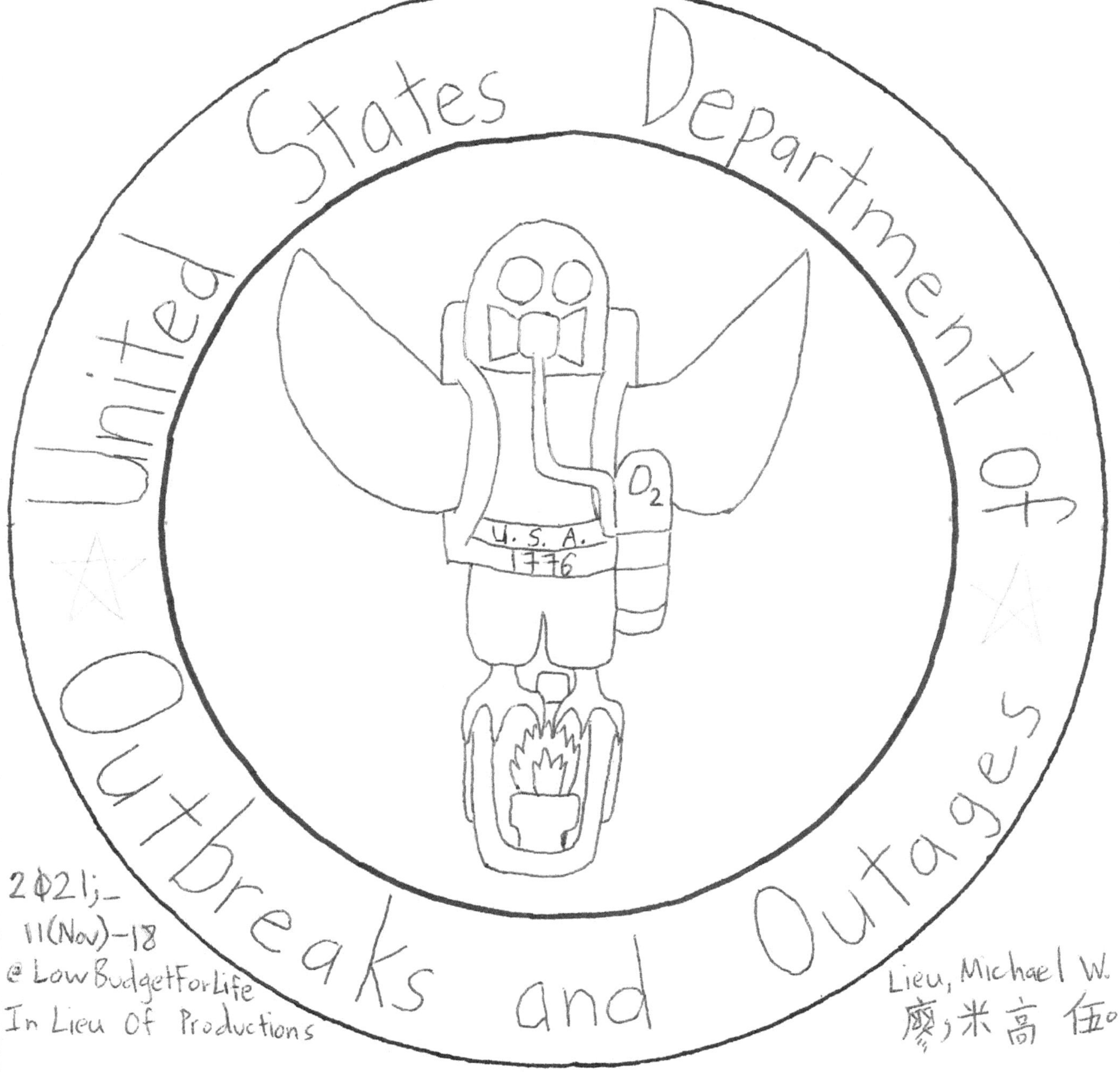
United States Department of Outbreaks and Outages
O2
U. S. A.
1776
2Φ21;_
11(Nov)-18
@LowBudgetForLife
In Lieu Of Productions
Lieu, Michael W.
廖,米高 伍。

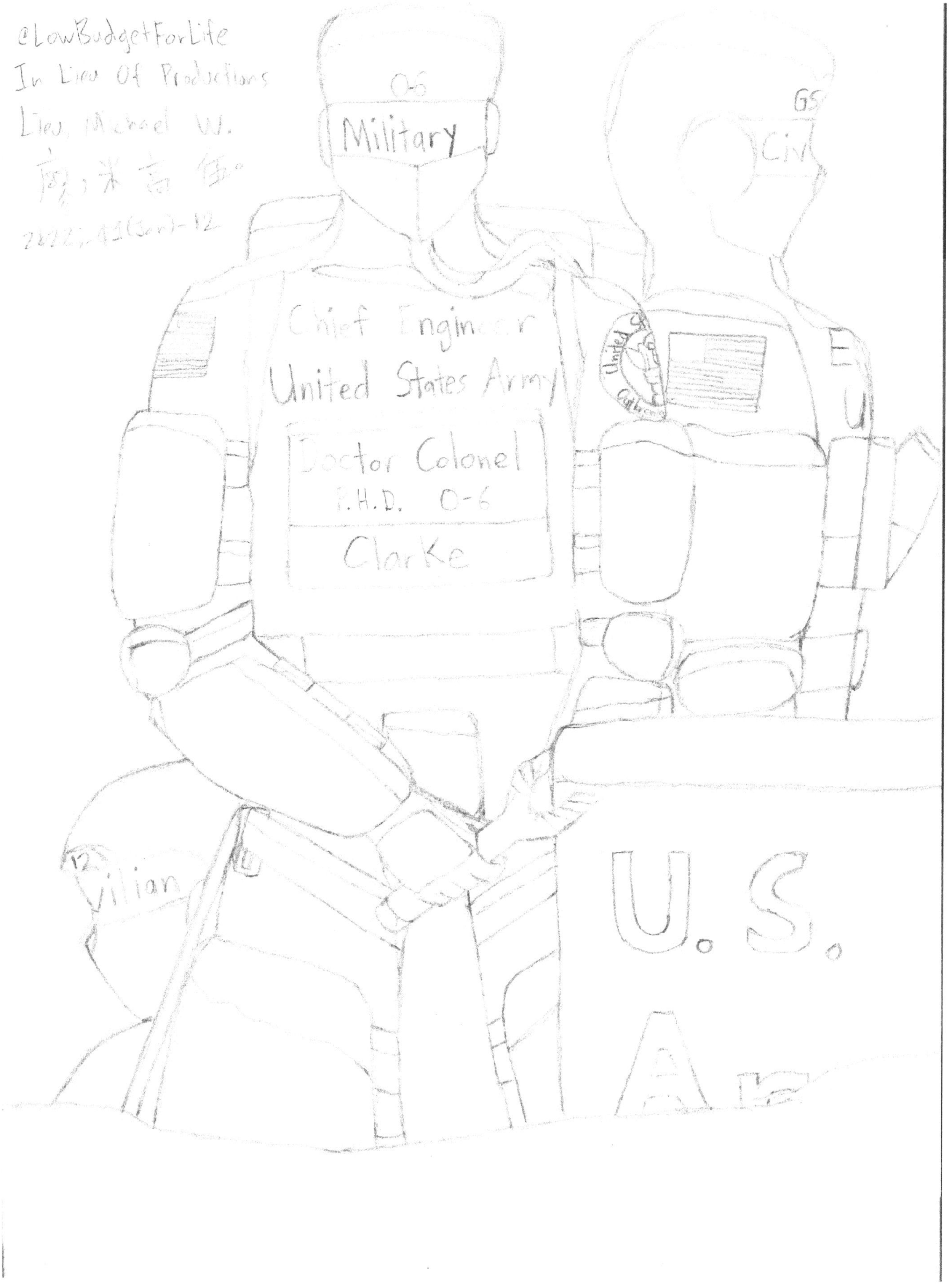
@LowBudgetForLife
In Lieu Of Productions
Lieu, Michael W.
O-6
Military
GS
Civ
Chief Engineer
United States Army
Doctor Colonel
P.H.D. O-6
Clarke
vilian
U.S.

Alien Valentina Vita, Traveling Cosplayers And Kosplayers
2154;_ Chicago, Illinois' Intranet, Wireless Broadcasting Frequencies Online

V. V. Augmented Reality Holographic Heads Up Display A.R. Holo HUD
Actual Legal Given Name;_ **Doctor Isaiah K. Clarke, P.H.D. O-6** (goes by **Doctor Colonel**) pronounced, **'dok-ter' 'eye-say-ah' 'kay' 'ka-lark-ka'**
Social Status;_ Chief Engineer (ChEng), Commissioned Officer O-6 Colonel rank, United States Army
Employment;_ The United States Department of Outbreaks and Outages (U.S. DOO), established in the year ***2122*** in response to some of the listed biotech **Weapons of Mass Destruction** (W.M.D.): asymptomatic carriers of diseases, bacteria, and viruses, blackout saboteurs, dirty bombers, gravity shifting, black holes, asteroids and meteor strikes, coronal mass ejections, solar winds, solar flares, and solar activities on Planet Earth, Earth's Lunar Moon Military (M) Bases from **foreign and domestic** terrorists. Officially formed in the year ***2122*** as an on going **Joint Task Force (J.T.F.)** between **the last remaining** U.**S**. federal government civil servant civilian contractors and military servicemembers of active duty and reserve, enlisted plus commissioned officers' many armed forces branches as **first responders** to both **foreign and domestic** terrorists' unconventional warfare of **Chemical, Biological, Radiological (C.B.R.) outbreaks** and **Electro Magnetic Pulse (E.M.P.) outages**. [*Aug+*]
[*Aug+*]
V. V. Augmented Reality Holographic Heads Up Display A.R. Holo HUD

2154;_ Chicago, Illinois' Intranet, Wireless Broadcasting Frequencies Online
Alien Valentina Vita, Traveling Cosplayers And Kosplayers

Doctor Colonel "P.H.D. O-6" Clarke, Military United States Army (U.S.A.) Chief Engineer;
"(*in suit*) Hello... V. V. are you here for the **mission briefing** for **C.B.R. outbreak pandemic and E.M.P. outage blackout**? We are part of the United States Department of Outbreaks and Outages, U.S. DOO. We military soldiers and civilians are here to respond to a **possible outbreak or outage** that occurred in this city, Chicago **today on July four**."

Valentina Vita (V. V.); Traveling Cosplayers And Kosplayers;
"(@_@) ...Um... Hi Doctor Colonel, **today is**... October thirty... Not **July four**..."

Doctor Colonel "P.H.D. O-6" Clarke, Military United States Army (U.S.A.) Chief Engineer;
"(*in suit*) No, no. Must be the **C.B.R.**, check the scanners to see if Miss V. V. here is affected. Your suit is not quite designed to handle **C.B.R. E.M.P.** Here is a **briefing in print**."

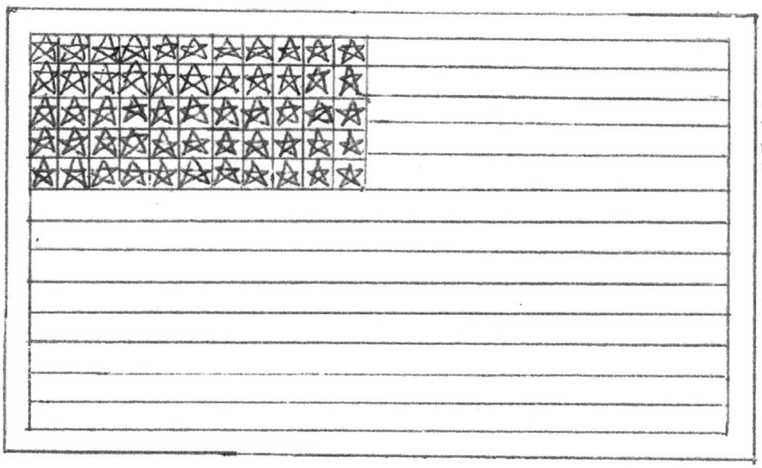

@LowBudgetForLife
In Lieu Of Productions
Lieu, Michael W.
廖, 米高 伍o
2Φ21;_Φ8(Aug)-15

United States Department of Outbreaks and Outages (U.S. DOO) Founded 2122

Outbreaks Pandemic Division	**Outage Blackout Division**
Chemical, Biological, Radiological (C.B.R.)	Electro Magnetic Pulse (E.M.P.)

Brief Overview Of U.S. DOO

The Terrorist(s) Question; "Who is a terrorist(s) and what do they look like?"

Terrorist(s), foreign and domestic, do not follow the conventional rules of war.

After the end of World War Two in 1944, there was no established written formulaic guidelines and or peace treaty agreement about the rules of war…

From 1944 to 2120, the Kold War deterrence through Mutual Assured Destruction (MAD) is not likely to be an effective threat against potential protagonists that are either:

cell-style terrorist(s),	trans-national group(s),	lone wolf attack(s),	insurgencies,
failing state(s),	suicide bomber(s),	hired mercenaries,	domestic terrorist(s).

In 2120, the people of Planet Mars worked along side private sector Roman Industries in deciding to end every single human life and existence in the solar system due to the belief system that "humans are a cancer on the solar system and needs to be annihilated completely".

On 2121, the people of Planet Mars and Roman Industries' "Martian Terrorists" nearly succeed.

By 2122, the last remaining Planet Earth United States (U.S.) military servicemembers and civilian federal government employees of the U.S. Department Of War (DOW) and other agencies banded together to form a first responders' Joint Task Force (J.T.F.) agency: United States Department of Outbreaks and Outages (U.S. DOO)

From a space perspective, the United States Force Space (U.S.S.F.) division of DOW, various other Kold War friend-enemies "frenemies" states and countries banded together to form the planetary space aspect of the Joint Task Force (J.T.F.) agency: Earth's Global Government (EGG)

A reminder, that the constant training in C.B.R. and or E.M.P. warfare is a never ending battle.

"Failing To Prepare, Means Preparing To Fail" / "Do it. Do it right. Do it right now."

2154;_Arcadia \ Valentina Vita (V. V.) 'Healthy-Strong Life' \ Left to Right(L->R)

United States Department of Outbreaks and Outages (U.S. DOO) Founded 2122

<u>Outbreaks Pandemic Division, Unconventional Warfare Branch</u>;
Chemical, Biological, Radiological (C.B.R.) overview

According to the United States Department Of War (U.S. DOW), Military Personnel Manual (MilPersMan) on C.B.R. unconventional warfare definition:

"The primary purpose of Chemical, Biological, Radiological (C.B.R.) agent(s) attacks is the mass casualties of personnel, property, livestock, and or crops. These tactics are called Weapons of Mass Destruction (W.M.D.), because they destroy large areas, kill and or disable large segments of the population."

Making preparations to manage the effects of an ***C.B.R.*** and or ***E.M.P.*** attack, includes:

understanding what has happened,	and reducing vulnerabilities,
maintaining situational awareness,	is critical to reducing the consequences,
having plans in place to recover,	and thus probability, of attack.
challenging and exercising those plans,	

Chemical Agent(s), can kill and or disable personnel by affecting their blood, nerves, eyes, skin, lungs, or stomach. ***Chemical Agent(s)*** are classed as: ***Nerve, Blister, Blood, and Choking Agent(s)***, all of which can inflict serious injury or death.

Some casualty ***agent(s)*** have a cumulative effect, which means that successive doses add to the effect of each preceding dose. You might receive a less-lethal dose of a ***Nerve Agent(s)***, for example, followed within a few hours by another less-lethal dose.

However, the cumulative effects of the two exposures could kill (flatline) you. Incapacitating ***Chemical Agent(s)*** temporarily disable personnel, but do not create permanent injury.

They can produce physiological and or psychological effects. These effects make individuals incapable of performing duties for hours or days even after exposure has ceased.

Some incapacitating ***agents*** have effects that typically last for significant periods of time but do not seriously endanger life. For example, ***Riot Control Agent(s)***.

Riot Control Agent(s) produce only temporarily irritating or incapacitating effects when used in normal concentrations. Complete recovery is usually expected without medical treatment.

United States Department of Outbreaks and Outages (U.S. DOO) Founded 2122

Outbreaks Pandemic Division, Unconventional Warfare Branch,
Effects Of ***Chemical Agent(s)***,

Chemical Agent(s) will make you a casualty when your body comes in contact with a bigger dose than it can withstand. The limits of tolerance of the human body, whether you are an augmented biotech or natural unmodified human, vary from short periods of exposure and low concentrations of certain ***agent(s)*** to extended periods of exposure and high concentrations of certain other ***agent(s)***.

Furthermore, the limits of tolerance to specific ***agent(s)*** vary with individuals. Your principal concern is recognizing the symptoms and relieving the effects of exposure before the limit of exposure is exceeded.

Outbreaks Pandemic Division, Unconventional Warfare Branch;
Chemical Agent(s); ***Nerve Agent(s)***,

Poisoning by ***Nerve Agent(s)*** affects bodily functions. The disruption of nerve impulses produces different effects on different body systems.

It is important for you to recognize both mild and severe signs and symptoms of ***Nerve Agent(s)*** poisoning. Mild symptoms will become severe if personnel are repeatedly or continually exposed to low concentrations of a ***Nerve Agent(s)***.

High concentrations of ***Nerve Agent(s)*** poisoning will cause rapid onset of severe symptoms, possibly without any mild symptoms at all.

The symptoms of ***Nerve Agent(s)*** poisoning are shown in the following chart below:

Mild Symptoms Severe Symptoms	Severe Symptoms
· Unexplained runny nose · Unexplained sudden headache · Excessive sudden drooling · Difficulty seeing (reduced vision or miosis) · Tightness in chest, difficulty breathing · Localized sweating and muscular twitching in the area of contaminated skin · Stomach cramps · Nausea	· Strange or confused behavior · Wheezing, difficult, or labored respiration and cough · Severely pinpointed pupils · Red eyes with tearing · Severe muscular twitching, general weakness · Involuntary urination and defecation · Convulsions · Respiratory failure · Vomiting · Unconsciousness

United States Department of Outbreaks and Outages (U.S. DOO) Founded 2122

Outbreaks Pandemic Division, Unconventional Warfare Branch; *Chemical Agent(s)*; *Blister Agent(s)*

Blister Agent(s) act on the eyes, mucous membranes, lungs, and skin. ***Blister Agent(s)*** include mustard vapors and mustard liquids; "World War One's Mustard Gas".

Mustards burn and ***blister*** the skin they contact, damage the respiratory tract when inhaled, and cause vomiting and diarrhea when absorbed. The degree of damage depends on the type and concentration of the ***agent***, the weather, the amount of activity of the individual, and amount of exposure time.

Blister Agent(s) are effective even in small quantities, produce both immediate and delayed injuries. Mustard vapors burn any area of the skin; but, the burn is most severe in moist areas, such as the neck, genitals, groin, armpits, bends of knees, and elbows.

Redness of the skin follows in ½ (half) to 36 hours (one day and a half) after exposure. This condition may be accompanied by intense itching, and ***blisters*** may then appear.

Stiffness, throbbing pain, and swelling may also occur. A few hours after breathing the mustard vapor, a victim experiences irritation of the throat, hoarseness, and coughing.

After severe exposure, the lining of the respiratory system swells and interferes with breathing. Frequently, pneumonia develops.

If the whole body is exposed to mustard vapor, the body goes into a state of shock. This reaction is accompanied by nausea and vomiting.

Personnel who suspect contamination of their eyes or face must seek overhead shelter and flush the eyes with potable (drinkable) water (H_2O) from a canteen or shower. Mild exposure to skin can be treated by applying calamine lotion or topical steroid creams.

All ***blisters*** should be opened, drained, and cleansed with tap or saline water. Any exposure to mustards require medical care by a corpsman or medical personnel.

United States Department of Outbreaks and Outages (U.S. DOO) Founded 2122

<u>Outbreaks Pandemic Division, Unconventional Warfare Branch</u>; *Chemical Agent(s)*; *Blood Agent(s)*

Blood Agent(s) inhibit the action of an enzyme responsible for transferring oxygen from the ***blood*** to the cells of the body. Thus the cells become starved for oxygen.

Inhalation is the usual route of entry for ***Blood Agent(s)***. The symptoms produced by ***Blood Agent(s)*** depend on the concentration of the ***agent*** and the duration (length of time) of the exposure.

Typically, either death (flatline) occurs rapidly or recovery takes place within a few minutes after removal of the victim from the toxic atmosphere. High concentrations of ***Blood Agent(s)*** cause labored breathing within a few seconds, violent convulsions, followed by cessation (stoppage) of breathing completely.

Finally, the heart stops beating (flatline) only a few minutes after initial exposure.

The symptoms of exposure to ***Blood Agent(s)*** are shown in the following chart:

Initial Symptoms	Advanced Symptoms
· Increased respiration · Headache · Giddiness · Dizziness · Increased pulse rate · Red, flushed skin	· Convulsions · Coma · Death

If you're exposed to a ***Blood Agent(s)***, immediately don (put on) a protective mask, or **Personal Protective Equipment (P.P.E.)**.
Speed is essential!

Blood Agent(s) act so rapidly that within seconds, the effects of exposure can make it impossible for individuals to don their own mask (**Personal Protective Equipment, P.P.E.)**. If this happens, the nearest person should help those who can not don their mask (**Personal Protective Equipment, P.P.E.)**.

Medical personnel should administer medications.

United States Department of Outbreaks and Outages (U.S. DOO) Founded 2122

<u>Outbreaks Pandemic Division, Unconventional Warfare Branch</u>; *Chemical Agent(s)*; *Choking Agent(s)*

In low concentration, ***Choking Agent(s)*** produce an action on the respiratory system that results in the accumulation of fluid in the lungs. Accumulation of fluid in the lungs can cause death (flatline).

High concentrations produce death (flatline) for the same reason, but the upper respiratory tract may be involved as well. Exposure to ***Choking Agent(s)*** may produce immediate dryness of the throat, coughing, ***choking***, tightness across the chest, headache, nausea, and at times, irritated and watery eyes.

However, symptoms are usually delayed, and it's possible that no immediate symptoms will appear when exposed to a fatal dose. Even a mild exposure to ***Choking Agent(s)*** that is accompanied by immediate symptoms may cause fluid to accumulate in the lungs within 2 to 24 hours (one day) after exposure.

Shallow and rapid breathing, a hacking and painful cough, frothy saliva, and an ashen gray color of the skin indicate the presence of fluid in the lungs. After exposure to a high dose of ***Choking Agent(s)***, it's important to begin medical treatment quickly to prevent accumulation of fluid in the lungs.

It's important to keep the victim at rest and warm. Cough suppressant and pain relievers can be given as long as the doses does not interfere with respiratory functions.

United States Department of Outbreaks and Outages (U.S. DOO) Founded 2122

Outbreaks Pandemic Division, Unconventional Warfare Branch; *Chemical Agent(s)*; *Riot Control Agent(s)*

Riot Control Agent(s) (R.C.A.s) are classified as either tear ***agents*** or vomiting ***agents*** and are characterized by very low toxicity and brief action. They are used to produce temporary misery and harassment.

Most personnel exposed to ***R.C.A.s*** do not require medical attention and casualties are rare. Tear ***agents*** act rapidly on nerve ends in the cornea and mucous membranes of the eye.

Vomiting ***agents*** cause local inflammation of the respiratory tract, sinuses, and eyes.

The symptoms of exposure to ***Riot Control Agents (R.C.A.s)*** are shown in the following chart.

Symptoms Of Tear Agents	Symptoms Of Vomiting Agents
· Violent burning sensation of the eyes · Reddening of the eyelids · Uncontrollable winking · Excessive tearing · Intolerance to light · Burning sensation of the throat, with developing pain and a sensation of choking · Sneezing · Nausea · Diarrhea · Headache · Burning sensation of the skin	· Irritation of the eyes, mucous membranes of the mouth and nose · Runny nose, sneezing, and coughing · Headache · Tightness and pain in the chest · Nausea and vomiting

First aid for personnel exposed to tear ***agents*** includes providing a supply of fresh air as soon as possible and changing exposed clothing. If symptoms continue, eyes, mouth, and skin should be flushed with large amounts of water (H_2O).

Although the effects of vomiting agents can be dramatic, personnel can usually perform duties despite their effects. Personnel should continue to wear a face mask (**Personal Protective Equipment, P.P.E.)** even though coughing, sneezing, salivating, and nausea occur.

The mask or **Personal Protective Equipment (P.P.E.)** can be lifted from the face briefly to allow for vomiting and to drain saliva from the face piece. Analgesics can be given to relieve headache and general discomfort.

United States Department of Outbreaks and Outages (U.S. DOO) Founded 2122

Outbreaks Pandemic Division, Unconventional Warfare Branch;
Under Chemical, Biological, Radiological (C.B.R.),

Chemical Agent(s) Example Breakdown Table

Agent Name	Agent Type	Physical Properties	Physiological Effects	Relative Rate of Action
Sarin	Nerve	Colorless odorless, volatile liquid	Difficulty breathing, excessive contraction of the pupil of the eye (miosis), blurred vision, headache and nausea leading to respiratory distress, convulsions, and eventually death.	Rapid (within minutes)
VX	Nerve	Colorless odorless, low volatility, oily liquid	Difficulty breathing, miosis, blurred vision, headache and nausea leading to respiratory distress, convulsions, and eventually death.	Relatively rapid (within 30 minutes)
Mustard	Blister	Garlic odor, medium volatility, oily liquid	Blisters or irritates skin, eyes, and lungs.	Delayed onset (4-6 hours)
Hydrogen Cyanide	Blood	Almond odor, highly volatile gas	Prevents the normal transfer of oxygen from the blood to body tissue resulting in respiratory paralysis.	Rapid (within minutes)

United States Department of Outbreaks and Outages (U.S. DOO) Founded 2122

Outbreaks Pandemic Division, Unconventional Warfare Branch; *Biological Agent(s),*

Biological Agent(s), under ***Chemical, Biological, Radiological*** **(*C.B.R.*)**, is the intentional use of bacteria, viruses, parasites, microorganisms, living organisms, fungi, toxins, and microtoxins to cause diseases that will kill or produce other casualties as well as disable or destroy people and domestic animals, damage crops, or deteriorate supplies.

Biological Agent(s), might be used on a large scale; therefore, ***biological*** immunizations of military forces and the development of detection equipment, such as the Interim ***Biological*** Agent Detection System (I**B**ADS), are being used.

Do not underestimate Biological Agent(s) as a weapon; "Bio Weapon".

Biological Agent(s) can be produced on a scale not considered possible in the past. Even small nations with modern, adequate research facilities can produce large quantities of ***Biological Agent(s)*** toxins and microtoxins more cheaply than they can produce other types of weapons.

Toxins and microtoxins are hundreds to thousands times stronger than ***Chemical Agent(s)***. The disadvantage of ***Biological Agent(s)*** is that many are rapidly degraded when exposed to certain environmental conditions, such as UltraViolet (U.V.) radiation, visible radiation, heat, dryness, or humidity.

Animals, insects, and rodents (rats) can be used as carriers to spread ***Biological Agent(s)***. Saboteurs can also infect large numbers of people by contaminating a water (H_2O) supply.

Infecting water (H_2O), milk, and food supplies with microorganisms can spread diseases, such as anthrax, typhoid fever, cholera, and influenza. In the early stages of any ***biological*** disease, the general symptoms include fever, malaise, and inflammation.

United States Department of Outbreaks and Outages (U.S. DOO) Founded 2122

Outbreaks Pandemic Division, Unconventional Warfare Branch; *Biological Agent(s)* Continued

The degree of **fever** varies with the individual, depending on the person's resistance. However, **fever** does serve as a rough guide to the severity of infection.

Often a violent chill precedes the **fever**. Whether the chill occurs or not, **fever** is usually one of the earliest symptoms.

Malaise is a feeling of bodily discomfort and weakness. There may be nausea, dizziness, loss of appetite, and general aches and pains.

Inflammation is caused by the reaction of body tissues combating and sealing off an infection. In almost every case there is pain, redness, and swelling.

Some types of infection result in a characteristic rash, making it possible for a doctor to make an early diagnosis.

Good To Know

Most Dangerous Biological Agent(s), Under C.B.R., Outbreak Pandemic Division, Unconventional Warfare Branch;

Botulinum Toxin
Aflatoxin
Bunyvirus
Marburg Virus
Rinderpest
Yersinia Pestis (Plague)
Hemorrhagic Fever Virus (Ebola)
Francisella Tularensis (Tularemia)
Varciola Major (Small Pox)
Bacillus Anthacis (Anthrax)

United States Department of Outbreaks and Outages (U.S. DOO) Founded 2122

<u>Outbreaks Pandemic Division, Unconventional Warfare Branch</u>;
Under Chemical, Biological, Radiological (**C.B.R.**)

Biological Agent(s) Example Breakdown Table

Disease (common name)	Causative Agent	Physiological Effects	Time to effect
Anthrax	Bacillus Anthracis	Mild fever and fatigue, worsening to severe respiratory disorders, high fever and excessively rapid pulse rate. Death can occur within 5-12 days of exposure if left untreated. Pulmonary anthrax is fatal more than 90% of the time.	1-5 days
Plague	Yersinia Pestis	Fever, headache, and rapid heart rate, followed by pneumonia and hemorrhaging in the skin and mucous membranes. Untreated plague pneumonia fatalities approach 100%, but early treatment can reduce mortality to as low as 5%.	2-3 days
Tularemia	Francisella Tularensis	Symptoms include fever, chills, headache and muscular pain. Untreated tularemia can result in 30-60% mortality; treated, the mortality rate is reduced to 1%.	3-5 days
Botulinum Toxin	Clostridium Botulinum	Initial symptoms include extreme weakness, nausea, headaches, and intestinal pain leading to respiratory paralysis that may cause death.	2-26 hours

United States Department of Outbreaks and Outages (U.S. DOO) Founded 2122

<u>Outbreaks Pandemic Division, Unconventional Warfare Branch</u>;
Nuclear Weapons Of ***Radiological Agent(s)*** Under **C.B.R.**

Nuclear Weapon(s) are no different from ordinary high explosive bombs, both are designed to cause destruction by blast and shock effects. However, ***Nuclear Weapon(s)*** have a higher conventional high explosive weapons, with the added effects of ***nuclear radiation***.

Nuclear explosions are classed according to the point of detonation with relationship to the surface of the earth a ***high altitude blast***, an ***air blast***, a ***surface blast***, and a ***subsurface blast***.

A ***high altitude blast*** is defined as a blast that takes place above 100,000 feet. The major aim of this blast is to destroy or interrupt satellites and communication systems through the effect of an ***Electro Magnetic Pulse (E.M.P.)***.

Basically, the ***E.M.P.*** is an intense electrical surge that affects electronic or electrical equipment in a burnout that's equivalent (equal) to that caused by a lightning strike.

An ***air blast*** is one in which the fireball is below 100,000 feet and doesn't touch the earth's surface. The ***radiation*** effects from an ***air blast*** are minimal.

A ***surface blast*** is one in which the fireball touches the earth's surface. Most of the damage caused by a ***surface blast*** is due to the ***shock (or blast) wave*** that accompanies the explosion.

A ***subsurface burst***, is underwater burst where a fireball is formed. However, it's smaller than the fireball of an ***air burst*** and is not normal.

The effects of ***Nuclear Weapon(s)*** through detonation of the ***nuclear*** bomb creates a blast wave that travels outward in all directions at an initial speed much greater than the speed of sound. When the wave strikes the earth's surface, another wave is formed by reflection.

Nuclear Weapon(s) produce explosions of great force and heat and release ***nuclear radiation***. Their effects are divided into three categories: ***blast waves*** or ***shock waves***, ***incendiary***, and ***radiation***.

However, this document provides a brief overview of another form of ***Radiological*** warfare commonly used by ***foreign and domestic terrorist(s)***, **<u>dirty bomb(s)</u>** and or ***Radiological*** **Dispersal Device (R.D.D.)**. A **<u>dirty bomb</u>** is an explosive device that combines conventional explosives, for example: ***dynamites***, ***Tri Nitro Toluene (T.N.T.), Composition Four (C-4) plastic explosives***, mixed with ***radioactive*** materials in gas, liquid or solid forms.

Radiological Agent(s) splits atoms, molecules and matter into pieces, releasing a huge amount of ***radioactive*** energy, as well as do not have an ***Identification Friend or Foe (I.F.F.)*** behavior system that can be programmed in certain ***Biological Agent(s)*** through engineering. Due to the density of ***Lead***, it is a good shielding material against ***Radiological Agent(s)***.

Using ***Radiological Agent(s)*** is a form of mutual destruction Also Known As (A.K.A.) ***Mutual Assured Destruction (MAD)*** from the Kold War era and theories of 1944 to 2120.

United States Department of Outbreaks and Outages (U.S. DOO) Founded 2122

Outbreaks Pandemic Division, Unconventional Warfare Branch;
Chemical, Biological, Radiological (**C.B.R.**) ***Survey Team(s)***

After a ***C.B.R. outbreak*** attack, ***Survey Team(s)*** go through the spaces to determine the extent and location of any contamination. Rapid detection and identification are vital so that effective defense measures may be taken immediately.

A ***Survey Team(s)***, or monitoring party, consists of a minimum of three people—a ***monitor***, a ***recorder***, and a ***messenger***. The ***monitor*** is in charge of the party.

The ***monitor*** carries high-range and low-range ***survey*** meters. The ***monitor*** is responsible for the safety of the team and for determining intensities and locations of contamination.

The ***recorder*** maintains a record of intensity readings (obtained by the ***monitor***), time of the readings, location of the hazardous areas, and specific hazards. Also, the ***recorder*** may act as a marker, using line to rope off hazardous areas and chalk to mark on spaces and floors the intensities of contamination found during the survey.

The ***messenger*** reports to the ***Chief Engineer*** (***ChEng***) the contaminated areas and the readings obtained by the ***monitor***. The ***ChEng*** and along with the first responders of U.S. DOO **Scientist(s), Technician(s), Engineer(s), Mathematician(s) (STEM)** plot the reports from the various teams to get a general outline of contaminated areas, to pinpoint hot spots (areas of higher-than-average intensities), and to establish stay times for specific areas.

Two types of ***surveys*** are usually conducted—a rapid, or gross, ***survey*** and a detailed ***survey***. The ***rapid survey*** is a preliminary reconnaissance.

Limited numbers of readings are taken in a minimum amount of time. The purpose of the rapid ***survey*** is to obtain a quick estimate of ***C.B.R.*** levels at specified locations to determine the possibility of keeping the spaces and floors manned.

A ***detailed survey*** is used to determine the effectiveness of decontamination measures. All accessible areas and equipment are ***surveyed*** in a slow, methodical manner.

Special attention is paid to areas that tend to hold contamination (rust spots, caulking in wood decks, canvas, rope, and so on). Each member of a ***monitoring*** team wears a protective mask (**Personal Protective Equipment, P.P.E.)** and clothing and is equipped with both a pocket dosimeter and a high-range casualty dosimeter.

No member with an open cut or wound should enter any contaminated area. Smoking, drinking, and eating are prohibited in contaminated areas.

Outbreaks Pandemic Division, Unconventional Warfare Branch;
C.B.R. Contamination Markers

[Continued]

United States Department of Outbreaks and Outages (U.S. DOO) Founded 2122

<u>Outage Blackout Division, Unconventional Warfare Branch</u>,
Electro Magnetic Pulse (E.M.P.) briefing overview

<u>Question</u>; What is an ***E.M.P.*** and how does an ***E.M.P.*** affects us?

<u>Long Answer</u>;

According to the ***Electro Magnetic Pulse (E.M.P.)*** Commission, a single ***Nuclear Weapon***, under ***Radiological Agent(s)*** of ***C.B.R.***, exploded at **high altitude** above the United States and her interests will interact with ***Planet Earth's atmosphere, ionosphere***, and ***magnetic*** field to produce an ***E.M.P. radiating*** down to the **Earth**. Additionally, the event will create ***electrical currents*** in the Earth. As ***E.M.P.*** effects are both direct and indirect.

The former are due to ***electro magnetic*** "***shocking***" of ***electronics*** and ***stressing of electrical systems***, and the latter arise from the damage that "***shocked***"—upset, damaged, and destroyed—electronics controls then inflict on the systems in which they are embedded. The indirect effects can be even more severe than the direct effects.

The ***electro magnetic*** fields produced by weapons designed and deployed with the intent to produce ***E.M.P.*** have a high likelihood of damaging ***electrical power systems, electronics, and information systems*** upon which the North American society depends. Their effects on dependent systems and infrastructures could be sufficient to qualify as catastrophic to the country and nations that supply her.

Depending on the specific characteristics of the attacks, unprecedented cascading failures of our major infrastructures could result. In that event, a regional or national recovery would be long and difficult and would seriously degrade the safety and overall viability of our country.

The primary avenues for catastrophic damage to the country are through our ***electric power infrastructure and thence into our telecommunications, energy, and other infrastructures***. These, in turn, can seriously impact other important aspects of our country's life, including the financial system; means of getting food, water, and medical care to the citizenry; trade; and production of goods and services.

The recovery of any one of the key national infrastructures is dependent on the recovery of others. The longer the ***outage***, the more problematic and uncertain the recovery will be.

It is possible for the functional ***outages*** to become mutually reinforcing until at some point the degradation of infrastructure could have irreversible effects on the country's ability to support its population.

United States Department of Outbreaks and Outages (U.S. DOO) Founded 2122

<u>Outage Blackout Division, Unconventional Warfare Branch</u>,
Electro Magnetic Pulse (E.M.P.) briefing overview

<u>Short Answer</u>;

Any technology with electronics along with the electrical grid that does not have any kind of ***E.M.P. shielding*** stops working when exposed to an ***E.M.P.***, leading to an ***outage***, ***blackout***, then the ***dark ages***.

<u>Additional Answers From The United States Department Of War (U.S. DOW)</u>,
Military Personnel Manual (MilPersMan) on ***E.M.P. Outage***

Computers, electronics and other equipment using ***solid state components*** are most ***vulnerable*** to ***E.M.P.*** ***Vacuum tube equipment*** is less susceptible to ***E.M.P. Outage(s)***.

Personnel are not directly injured by ***E.M.P.***, but they may suffer ***electrical shock*** if they are in contact with a ***large conductor of electrical energy***.

Preventive measures to protect such as ***E.M.P. shielding or harden*** equipment against damage by ***E.M.P.*** include ***metal shielding, good grounding, use of surge arresters,*** and the ***proper arrangement of electrical wiring***.

<u>Additional Answers From The United States Department Of War (U.S. DOW)</u>,
Military Personnel Manual (MilPersMan) on ***E.M.P. Blackout***

Blackout, caused by an atmospheric ***nuclear*** explosion, under ***Radiological Agent(s)*** of ***C.B.R.***, is the ***interference of radio transmissions through ion fields formed in a detonation***.

In a tactical situation, straight-line communications (wireless radar and radio transmissions) between ships on opposite sides of the fireball will be lost.

Following a ***high altitude detonation*** of a ***nuclear weapon***, satellite communications may be affected or lost.

Blackout alters or inhibits radar or radio waves and affects ***all frequency bands***.

Procedures to counteract the effects of ***blackout*** include providing alternate paths for communications, shifting radio operating frequencies, changing transmission modes, and waiting for ***blackout*** effects to diminish.

United States Department of Outbreaks and Outages (U.S. DOO) Founded 2122

Outage Blackout Division, Unconventional Warfare Branch,
Electro Magnetic Pulse (E.M.P.) briefing overview

Good To Know

E.M.P. Outage Blackout on ***nuclear reactor(s)*** powering ***electricity.***

Reactor(s), ***power plant(s)***, that uses ***nuclear*** reactor(s) to generate ***electricity*** will shutdown, but nothing will happen to the reactor. The ***reactor*** is submerged in really deep water ***(H_2O)*** and encased in ***lead material.*** As long as the ***reactor*** is submerged in ***water*** and there is no ***physical damage to the structure, everything will be fine as is.*** The thing you ***worry and concern about is a meltdown*** when there is not ***enough water to cool the reactor*** and ***fuel leaking due to physical damage.***

Good To Know

What ***technology*** would still be ***working after*** an ***E.M.P Outage Blackout***?

Certain vehicles built during the 20th (twentieth) century, before the years: 1990s to 2000 with ***less electronic(s) built in.***
Working Tech: Off Road Vehicles, All Terrain Vehicle (A.T.V.), Dirt Bikes, Motorcycles.
Vintage ***electronic(s)*** using ***vacuum tubes.***
Basic electric operated tools, ***without any electronic(s) components***; an on off switch.
If the ***tech and or electronic(s)*** is not plugged into an ***electric outlet during E.M.P outage.***
Electronic(s) and or tech in a ***Faraday cage*** that provides ***E.M.P. shielding.***

Valentina Vita (V. V.); Traveling Cosplayers And Kosplayers;
"(@_@) ...**This is helpful and technical info**... Going to save this and read it later... Doctor Colonel, sir. I am going to head into the Chicago **Animal** Sanctuary now... Bye."

Doctor Colonel "P.H.D. O-6" Clarke, Military United States Army (U.S.A.) Chief Engineer;
"(*in suit*) There are **not many humans alive around anymore after the** Martian Terrorists' attacks... unless you are **undead** under that suit of yours... Anyways, here we have more info on United States and our government you may find interesting as a **Eur-Asia traveler.** Remember, constant training in **C.B.R. E.M.P.** warfare never ends."

2154;_Arcadia \ Valentina Vita (V. V.) 'Healthy-Strong Life' \ Left to Right(L->R)

United States Department Of War (U.S. DOW), Founded 1789

Secretary Of War, Office Of The Inspector General Of The Department Of War

<table>
<tr><td rowspan="5">Office Of The Secretary Of War

Deputy Secretary Of War

Under Secretaries Of War

Assistant Secretaries Of War</td><td rowspan="3">Joint Chiefs Of Staff

Chairperson Of The Joint Chiefs Of Staff</td><td>Division Of The Navy

Secretary Of The Navy</td><td>Division Of The Army

Secretary Of The Army</td><td>Division Of The Air Force

Secretary Of The Air Force</td><td>Division Of The Space Force

Secretary Of The Space Force</td></tr>
<tr><td>Head Quarters Marine Corps</td><td rowspan="2">The Army Staff</td><td rowspan="2">The Air Staff</td><td rowspan="2">The Space Staff</td></tr>
<tr><td>Office Of The Secretary Of The Navy</td></tr>
<tr><td>The Joint Chiefs</td><td rowspan="2">Office Of The Chief Of Naval Operations</td><td rowspan="2">Office Of The Secretary Of The Army</td><td rowspan="2">Office Of The Secretary Of The Air Force</td><td rowspan="2">Office Of The Secretary Of The Space Force</td></tr>
<tr><td>The Joint Staff</td></tr>
</table>

DOW Field Activities

War Media Activity	War Technology Security Administration	DOW Human Resources Activity	Office Of Economic Adjustment
War Technical Information Center	DOW Education Activity	DOW Test Resource Management Center	Washington, D. of C. Head Quarters Services

War Agencies

War Advanced Research Projects Agency	War Commissary Agency	War Contract Audit Agency
War Contract Management Agency	War Finance And Accounting Services	War Health Agency
War Information Systems Agency	War Intelligence Agency	War Legal Services Agency
War Logistics Agency	War Prisoner Of War (POW)/Missing In Action (M.I.A.) Accounting Agency	War Security Cooperation Agency
War Security Service	War Threat Reduction Agency	Missile War Agency
National Geospatial-Intelligence Agency	National Reconnaissance Office	National Security Agency Central Security Service
Pentagon Force Protection Agency		

"The New World 2121 Movement Of The 22nd Century, Utopian Sovereignty of North America (U.S. of N.A.)"

Chicago, The Last Animal Sanctuary Of Illinois

"Thank You... To The Past, Present, Future Army Of Volunteers And Employees That Continues The **Utopia** Augmented-Natural Coexistence Sovereignty (S)"

Clean-up Crew Unit 6 (six), Adan Crenshaw **Honest Ape**	Clean-up Crew Unit 9 (nine), Eva Crenshaw Walking Dog
"True-Men, They Are True-Men" Spying Dust Particles "Hey-Zeus, Hey-Zeus Receives" Sound Waves Echolocation	
"Chicago City Christian Church Belief Systems And Other Religions Welcome <3" Lord Jesus Christ, Father Greg Ford †, Sister K. White <3 = "HeartBladeEndureLove 心刀忍愛"	
Utopian Sovereignty (U.S.) Maids And Cats-Kats' Free Market Shopping	
Captain Rage's Public Po Po Place / Chicago City Police Department (C.C.P.D.) Special Weapons And Tactics (SWAT) / C.C.P.D. SWAT Lieutenant Gain Retired C.C.P.D. SWAT, "Sergeant Soil" Officer Ogden, Deanna, A.K.A. Alfa ToolPerson C.C.P.D. SWAT Operator Mirror / C.C.P.D. SWAT Rookie Knight	
Elysium Corporation 亞, Automated Algorithm(s), Biotech Dome Airlock Decontamination Doctor Froste, Human Director Biotech Dome | Mister Kelvin, Human Klone(s) | Doktor Drive Mister Kenny, Human Animal Breeder Builder | Augmented Animal-Human Breeding Buddies Conscientiousness Cat/Kat's XXXX | Stick Figure(s) | Beautiful Biotech Bodies | Chimera(s) M.F.H. | Vigilante | Night-Wigs | Robin Dahood | Catsuit-Women | Bad-Girl | Oliver Clothesoff Seymour Buts (_l_) | Uncle Ben Dover [()] | "Therapist: By Day" And "The rapist: By Night" Dees' Nuts 8==D | Moe Lester | D Body-Flasher(s) | Utopian Sovereignty Feral Government Sexy Sixty-Nine (69) | Spreading Syphilis And AIDS | Spandex-Men And Spandex-Women Dafoe Or Dafriend | Cereal Killer | Peter Park P.C. Bang | D Queen(s) | Nagging Nurse(s)	
Kay's Private Security Services L.L.C. Research And Development (R&D), Agent Kay	
Voluntold Service; Jerry Duty, Social Worker Mars Hall, Dog-Name-Cat Rover, "Xtra-Men"	
Cesar Chavez Biotech Agricultural Latino-Latina Farm Benavidez-Cruz, Rich Rodriguez "Big Ben" = The King Of Mexico = The Mexican-Man Benavidez, Jose Rodriguez "Baby Ben" And The Many Mexican-Men And Latinas	
Yellow-Mellow (°o°) (O :), Supa-Hiro Obama@deep Problem Solvers Lieutenant 'Lieu, Michael W. 廖, 米高 伍 o' ='WhiteWolf 白狼' = Low Budget For Life Ninja-Talor A. Williams = 'PeacefulPerson 平靜人' = King And X Philosophy	
Alex D. "Doc Drastal", 'MartialArtsGod 武道神' A.K.A., Lord Jesus Christ Cosplayer	
Elle K. Mcgee, 'BladeMaster 刀大師' = Newly Arriving Utopian Sovereignty Volunteer	

2154;_Arcadia \ Valentina Vita (V. V.) 'Healthy-Strong Life' \ Left to Right(L->R)

Automated Algorithm; Human Voice Over Emulation Based On Elysium Corporation Archives;
"(*monotone female voice over*) Please stand by for biotech dome airlock decontamination. Thank you for your patience and cooperation **Utopian**. (*scanning*) Welcome, **augmented biotech human female, Alien status, beauty designer**, Valentina Vita, A.K.A. V. V., **international terrestrial foreigner, Civilian, Citizen Of Eur-Asia, Western Society of European Union and Eastern Society of Asian Pacific,** Traveling Cosplayers And Kosplayers... **We** welcome you to Chicago, **The Last Animal** Sanctuary **Of Illinois, we** hope you enjoy your stay in a twenty second century sovereign **Animal** Sanctuary biotech **utopia** of the twenty one twenty one new world movement... have a nice day."

Chicago Controller City Mayor selects the music(s) "**Final Fantasy XIII:** *Gapra Whitewood, Mysteries Abound, Sulyya Springs, Sunleth Waterscape, Vestige*" plays in **Utopia A.R. Holo HUD** background and is set on loop.

Valentina Vita (V. V.); Traveling Cosplayers And Kosplayers;
"(@_@) ...Hmm... Chicago Controller **Animal** Sanctuary **Americas**, whoa we wow. ...Uh... Different from Naples Controller **Human** Sanctuary **Eur-Asia**..."

Alien Valentina Vita, Traveling Cosplayers And Kosplayers
2154;_ Chicago, Illinois' Intranet, Wireless Broadcasting Frequencies Offline

<table>
<tr><td colspan="3">V. V. Augmented Reality Holographic Heads Up Display A.R. Holo HUD</td></tr>
<tr><td>Actual Legal Given Name;_

Valentina Vita (goes by **V. V.**)
pronounced, "**vee vee**"</td><td colspan="2">Alias Network Nickname;_

White-Woman, Red-Girl, Red-Woman,
Lady In Red, Italian-Women, Beauty Aug,
2 (two) Winged Demon Blade Kosplay,
KissShot-AcerolaOrion-HeartUnderBlade,
Anime Long Blade Cosplayers,
[*Aug+*]</td></tr>
<tr><td>Physical Plus Psych Age;_

augmented 34 + 25 years old</td><td>Ethnicity;_

Italian</td><td>Nationality;_

Eur-Asian</td></tr>
<tr><td colspan="3">Connecting...
Electromagnetic Spectrum Frequency Hertz (Hz)
Frequency Modulation (F.M.)
Amplitude Modulation (A.M.) [<u>*Aug+*</u>]</td></tr>
<tr><td colspan="3">V. V. Augmented Reality Holographic Heads Up Display A.R. Holo HUD</td></tr>
</table>

2154;_ Chicago, Illinois' Intranet, Wireless Broadcasting Frequencies Offline
Alien Valentina Vita, Traveling Cosplayers And Kosplayers

Reads = Left to Right (L -> R)

2154;_ Chicago, Illinois' Intranet, Wireless Broadcasting Frequencies Online

Measuring System;_ Metric degrees Celsius (°C) [*Switch to Imperial °F, Kelvin*]		
Geolocation;_ Planet Earth, North America, Utopian Sovereignty of America(s) (U.S. of A.)		
Language;_ English, Western Accent, Augmented Reality (A.R.) Application (App)		
Century;_ 22nd (twenty second) [*Compare statistics to 21st (twenty first) Aug+*]		
Actual Legal Given Name	Height	Weight
Deoxyribo Nucleic Acid (D.N.A.)	Holographic Retina Biometric Body Projection	OCEAN 5 (five) Common Personality Traits
Alias Network Nickname	7(seven) Deadly Sins	7(seven) Heavenly Virtues
Birth Day Or Build Day	Blood Type Or Bio Fluid	Physical Plus Psych Age
Hair Texture Color Tone	Eye Retina Color Tone	Skin Covering Color Tone
Ethnicity	Nationality	Alien Or Citizen
Natural Matter Organic Photosynthesis	Augmented Compound Enhanced Mutation	Percentage Rate Synthetic Synthesis (S.S.)
Human Being	Animal Kingdom	Artificial Intelligence (A.I.)
Biotech Or Bio Or Tech	Exosuit Or Exoskeleton	Implant Or Explant
Alive = Heart-Rate	Undead = Flatline	Deceased = Necrosis
Chemical (C.) Toxicity	Biological (B.) Toxicity	Radiological (R.) Toxicity
Personal Protective Equipment (P.P.E.)	Ballistic Motion Energy Transfer Shielding Resistance	Electro Magnetic Pulse (E.M.P.)
Sexual Orientation	Sexual Preference	Virginity Test
Sexual Transmitted Diseases (S.T.D.)	Human Immunodeficiency Virus (H.I.V.)	Acquired Immuno Deficiency Syndrome (AIDS)
Known Address(es)	Community Network(s)	Language Level(s)
Family Member(s)	In-Law Relative(s)	Relationship(s)
Biological Genealogy	Psychological Traits	Social Status
Employment	Education	Training
Frequency Hertz (Hz)	Frequency Modulation (F.M.)	Amplitude Modulation (A.M.)
Electromagnetic Spectrum	Intelligence Quotient (I.Q.)	Kinesis Module
Radio Wave Music	Broadcasting Film Movie	Electronic Book Genre
Hate(s) Or Dislike(s)	Neutral(s) Or No Comment	Love(s) Or Like(s) [*Aug+*]

2154;_ Chicago, Illinois' Intranet, Wireless Broadcasting Frequencies Online

Reads = Left to Right (L -> R)

Valentina Vita (V. V.), Traveling Cosplayers And Kosplayers;
"(O_o) Hm... interesting title for **Americas'** Chicago City Controller sector name... Chicago, **The Last Animal** Sanctuary **Of Illinois**. (O_O) ...Oh wow! The **two politicians of Eur-Asia Mark S. And Mark D. ...And Jonny Wise**! ...Oh hi Doggie!! Can I take a photo?"

Mark S., Politician from Eur-Asia Western Society of European Union (E.U.) diplomat;
"(*voice over*) Of course! Make sure to vote for **Mark S.** for **my upcoming election in Eur-Asia**!"

Mark D., Politician from Eur-Asia Eastern Society of Asian Pacific (A.P.) diplomat;
"(*voice over*) ...You want a photo with **us**? **We** have **Jonny Wise** and **his stunt double** and **doggie**!"

Jonny Wise Method-Man Actor; Wise Man And Raul Doggie;
"(*voice over*) Oh hi **Mark**! **Raul** Doggie here says Hi Doggie with his sign! Oh hi V. V.!"

Valentina Vita (V. V.), Traveling Cosplayers And Kosplayers;
"(>_O) Um... No, not really, no thanks, I am not that **political**, into **politics** I mean... I just want a photo of **you all**... Oh hi **Jonny**! Oh hi **Mark**! I will make sure to vote for **you, Mark S.** for **your upcoming election in Eur-Asia**! Ready on my mark. No pun intended!"

Jonny Wise Method-Man Actor; Wise Man And Raul Doggie;
"(*voice over*) Oh hi **Mark**!"

Mark D., Politician from Eur-Asia Eastern Society of Asian Pacific (A.P.) diplomat;
"(*voice over*) Oh hi **Jonny**!"

Mark S., Politician from Eur-Asia Western Society of European Union (E.U.) diplomat;
"(*voice over*) Oh hi **Jonny**!"

Valentina Vita (V. V.), Traveling Cosplayers And Kosplayers;
"(^_^) Thanks again, I will make sure to vote for **you, Mark S.** for **your upcoming election in Eur-Asia**!"

Mark S., Politician from Eur-Asia Western Society of European Union (E.U.) diplomat;
"(*voice over*) Oh V. V., before you go, here is a **physical printout of what politics is** in a **one oh one basic overview**... For those not **into politics and political science, the apolitical...** Enjoy and good day."

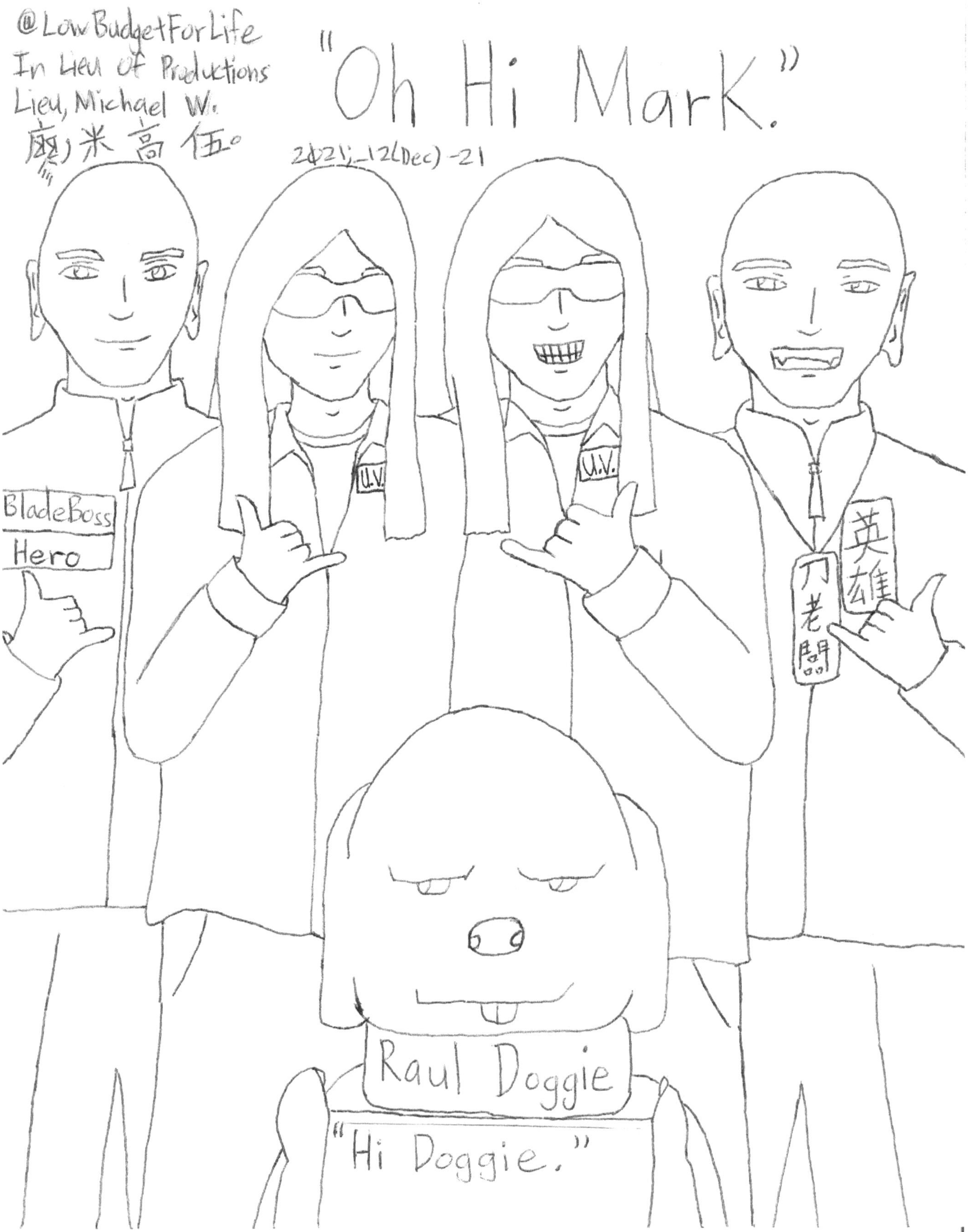
@LowBudgetForLife
In Lieu Of Productions
Lieu, Michael W.
廖, 米高 伍o
"Oh Hi Mark."
2021_12(Dec)-21
BladeBoss
Hero
U.V.
U.V.
英雄
刀老闆
Raul Doggie
"Hi Doggie."

Alien Valentina Vita, Traveling Cosplayers And Kosplayers
2154;_ Chicago, Illinois' Intranet, Wireless Broadcasting Frequencies Online

V. V. Augmented Reality Holographic Heads Up Display A.R. Holo HUD				
Actual Legal Given Names;_ **Mark S.** (goes by **Mark S.**) pronounced, '**ma-ark' 'eh-ss**'	Actual Legal Given Names;_ **Jonny Wise Method-Man Actor** (goes by **Jonny Wise**) pronounced, "**ja-hon-knee' 'why-s**"	Actual Legal Given Names;_ **Raul Human-Animal Dog Augmentation** (goes by **Raul Doggie**) pronounced, "**raw-oh' 'dog-gee**"	Actual Legal Given Names;_ **Jonny Wise Method-Man Actor Stunt Double** (goes by **Jonny Wise**) pronounced, "**ja-hon-knee' 'why-s**"	Actual Legal Given Names;_ **Mark D.** (goes by **Mark D.**) pronounced, '**ma-ark' 'dee**'
Physical Plus Psych Age;_ augmented 191 + 50s years old	Physical Plus Psych Age;_ augmented age Question-Mark	Physical Plus Psych Age;_ augmented age Question-Mark	Physical Plus Psych Age;_ augmented age Question-Mark	Physical Plus Psych Age;_ augmented 190 + 50s years old
Alias Network Nickname;_ Oh Hi Mark, SaiLow, Subby, BladeBoss, British Buddha, Diplomatic Immunity, [*Aug+*]	Alias Network Nickname;_ Oh Hi Jonny, Are You The Jonny Stunt Double?, [*Aug+*]	Alias Network Nickname;_ Hi Doggie, Hi Doggy, Woof, Wolf, What Kind Of Animal Are You?, [*Aug+*]	Alias Network Nickname;_ Oh Hi Jonny, Are You The Jonny Stunt Double?, [*Aug+*]	Alias Network Nickname;_ Oh Hi Mark, DaiLow, Daddy, BladeBoss, Asian Buddha, Diplomatic Immunity, [*Aug+*]
[*Aug+*]				
V. V. Augmented Reality Holographic Heads Up Display A.R. Holo HUD				

2154;_ Chicago, Illinois' Intranet, Wireless Broadcasting Frequencies Online
Alien Valentina Vita, Traveling Cosplayers And Kosplayers

The Politics Question; "Who gets what, when, where, and how?"
The Politics Answer; "There is no one correct answer. Rather, see the known systems below."
Mark D. and Mark S.'s Political Structures Of The Known 22nd Century On Planet Earth

(A), (C), (D), (E), (F), (G), (H), (I), (L), (M), (N), (P), (R), (S), (T), (U), (W), (Z)		
Alien Terrestrial Immigrant Contact, And Space Alien Extraterrestrial Contact	Anarchism, Anarchy = Abolition Of All Government And Organizations	Anarcho-Capitalism = Free Market Anarchism
Assassination, Assassins = The Killing And Murder Of Political Persons Or Authority	Astrosociology: Feudalism, Hive-Like Eusocial Society, Castle System,	Authoritarian, Autocracy = Totalitarianism, Tyranny A.K.A. Dictator
Apolitical = Not Interested And Or Involved In Politics, Political Sciences (Poly Sci).		
Capitalism, Capitalist	Communism, Communist	Democracy, Democratic
Dystopia, Dystopian	Environmentalism	Economics = Supply Demand
Ethnic Discrimination= Racism	Eugenics = Reproduction	Fascism, Fascist
Galactic Empires, Evil Empires	Homeless Humans' Tent City Camps	Interstellar Federation Of Planets
Libertarianism, Libertarian	Limited Franchise Republic	Militarism, Military Bases
Mind: Reading And Control	National Security State	Nepotism = Kinship Favorites
Pacifism, Pacification Protocols	Plutocracy = Wealthy Rule	Proportional Representation
Revolution = Overthrowing Government Social Order	Self-Govern, Sovereignty, Sovereign Citizen	Slavery, Slaves = Robots And Robotics Labor
Socialism, Socialist = Production Sharing	Terrestrial, Extraterrestrial Galactic Empires	Theocracy = Priests Rule In The Name Of God
Traditional Nine To Five Rat Race Minimum Wage Job	Treason, Traitor = Attacking The Nation's Authority	Totalitarianism, Tyranny = Authoritarian, Autocracy
Utopia, Utopian = The Garden Of Eden	World Governments, Earth's Global Government (EGG)	**Zero Human Population Growth, Martian Terrorists' Belief Systems**

British-Hong Kong Sovereignty (S) = 1 Country 2 Systems = "Liberal (L) Autocracy (A)"

"The New World 2121 Movement Of The 22nd Century, Utopian Sovereignty of North America (U.S. of N.A.)"

Chicago, The Last **Animal** Sanctuary Of Illinois' Public And Private Sectors

Public Sector = Non-Paid Volunteer Worker(s)

The 22nd (twenty second) Century Phenomenon Broadcasting Frequencies UltraViolet (U.V.) Rays City Controllers "City Mayor" Sovereignty, Sovereign, Self-Govern, Self-Governing (S) Animal Sanctuary → Animal Liberation Forces (ALF) Environmentalist → Environment Liberation Forces (ELF)
Chicago Controller City Mayor's Animal Sanctuary Enforcers; Captain Rage's Public Po Po Place Chicago City Police Department (C.C.P.D.) Special Weapons And Tactics (SWAT)
Public Militia, Augmented And Natural Human Enforcers (Vigilantes And Militia)
Belief Systems; "Separation Of Church And State" (God, Almighty, Lord Jesus Christ, Father Priest, Sister Nun, Super-Hero)
Elysium Corporation, 亞, Established 2077, The Magical Corporation Of Desire Biotech Dome Sector, Engineering Environment Biotech Dome Volunteers' Maintenance

Geotech	Structural	Hydro, Solar, Wind

King And X District = "...I Have A Dream...And...By Any Means Necessary..."	
Voluntold Service For All Utopian Sovereignty (U.S.) Citizens; Jerry Duty	Court House

"Separate But Equal"

Private Sector = Paid Employ Worker(s)

Elysium Corporation, 亞, Established 2077, The Magical Corporation Of Desire, Biotech Dome Sector, Engineering Environment Biotech Dome, Floors -99 to 99
Kay's Private Security Services L.L.C. Research And Development (R&D)
All For One Shopping Center: Utopian Sovereignty (U.S.) Maids And Cats-Kats'
Cesar Chavez Biotech Agricultural Latino-Latina Farm

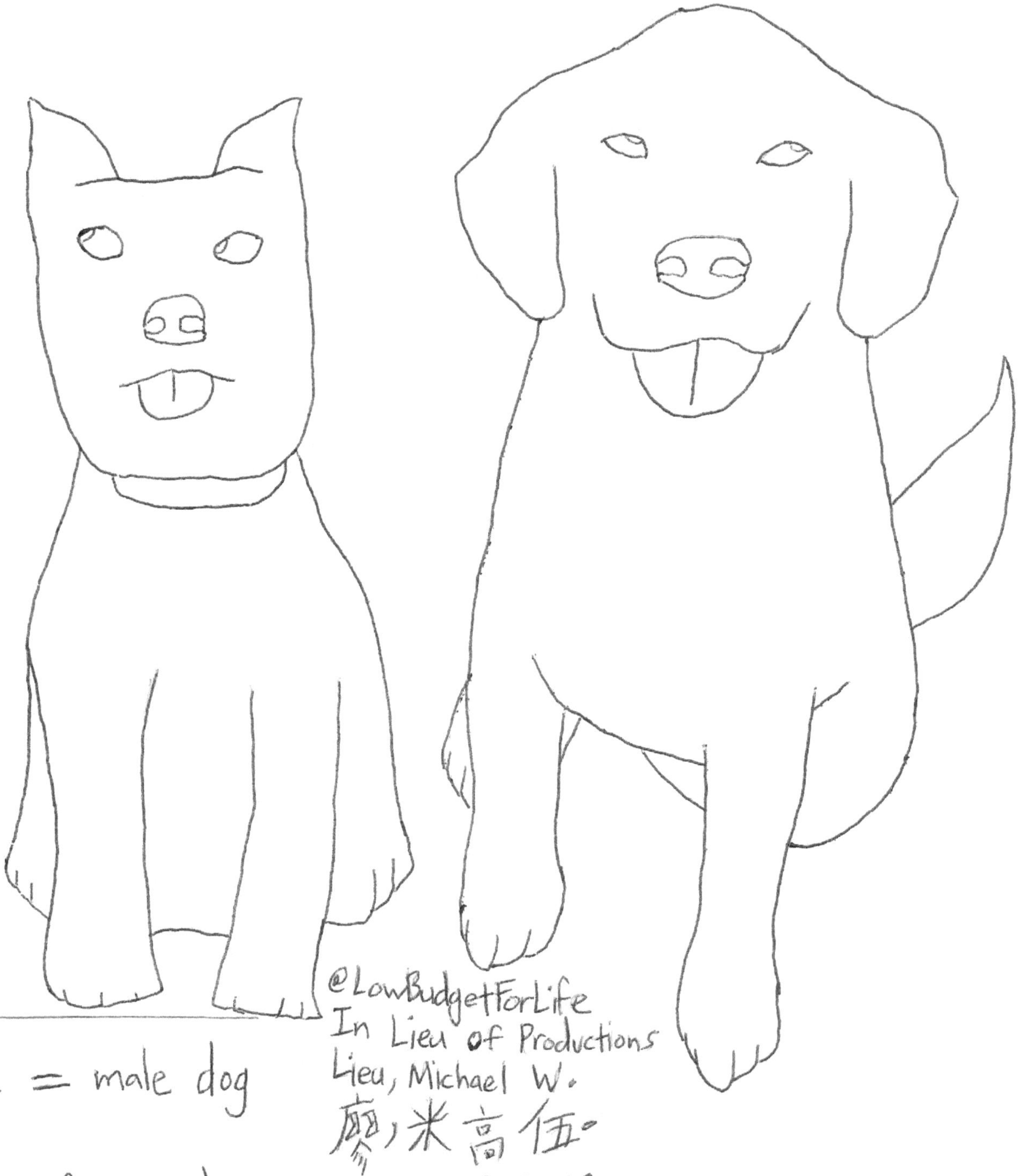

Doggie = male dog

Bitch = female dog

Puppy = baby dog

ink@ 2021;_11(Nov)-01

Chicago Controller City Mayor, Self-Govern Sovereignty, Animal Sanctuary (S)
2154;_ Chicago, Illinois' Intranet, Wireless Broadcasting Frequencies Online

<table>
<tr><td colspan="2">C-u C Augmented Reality Holographic Heads Up Display A.R. Holo HUD</td></tr>
<tr><td>Actual Legal Given Name;_

Hi Doggie(s) (goes by Hi Doggie) pronounced "Ha-Ai Dog-Ge" [Aug+]</td><td>Biological Genealogy;_

Dog Descendants And Ascendants
[Aug+]</td></tr>
<tr><td>Social Status;_

Animal Sanctuary [Aug+]</td><td>Community Network(s);_

Reminder To Please, Do Not Feed Chocolate To Nat-Dogs [Aug+]</td></tr>
<tr><td colspan="2">Relationship(s);_

"Clean-up Crew Unit 9 (nine)" Eva Crenshaw (goes by Unit 9 (nine)) pronounced, 'e-va' 'ko-ren-sha-aw' [Aug+]</td></tr>
<tr><td colspan="2">Animal Kingdom;_

Mammal

Natural Unmodified Dog Species (Nat-Dog)

[Aug+]</td></tr>
<tr><td colspan="2">[Aug+]</td></tr>
<tr><td colspan="2">C-u C Augmented Reality Holographic Heads Up Display A.R. Holo HUD</td></tr>
</table>

2154;_ Chicago, Illinois' Intranet, Wireless Broadcasting Frequencies Online
Chicago Controller City Mayor, Self-Govern Sovereignty, Animal Sanctuary (S)

Valentina Vita (V. V.); Traveling Cosplayers And Kosplayers;
"(^_^) Oh wow, hi Nat-**Dog**, and hi Nat-**Dog...** Hello there! Oh wait... is the name... Nat-**Dog** or **Hi Doggie**... (>_<) I am so confused! Very confused whether to call these **animals**... Nat-**Dog** or **Hi Doggie**... Anyways, I have to wait in line now, but **you animals** be sure to be good okay around **us humans**, okay? And I do **promise not to feed you chocolate** as it states in **this Holo HUD!** (@_@) I wonder, I wonder what **other animals** are in this Chicago **Animal** Sanctuary!! This is starting to make up for that **Springfield ghost town tour**! (=_=) Let **us** see, how many **people** in front of **us**... and how long is it going to take before **we** officially get inside..."

V. V. takes a look at the line in front of her to get into Chicago, **The Last Animal** Sanctuary **Of Illinois**.

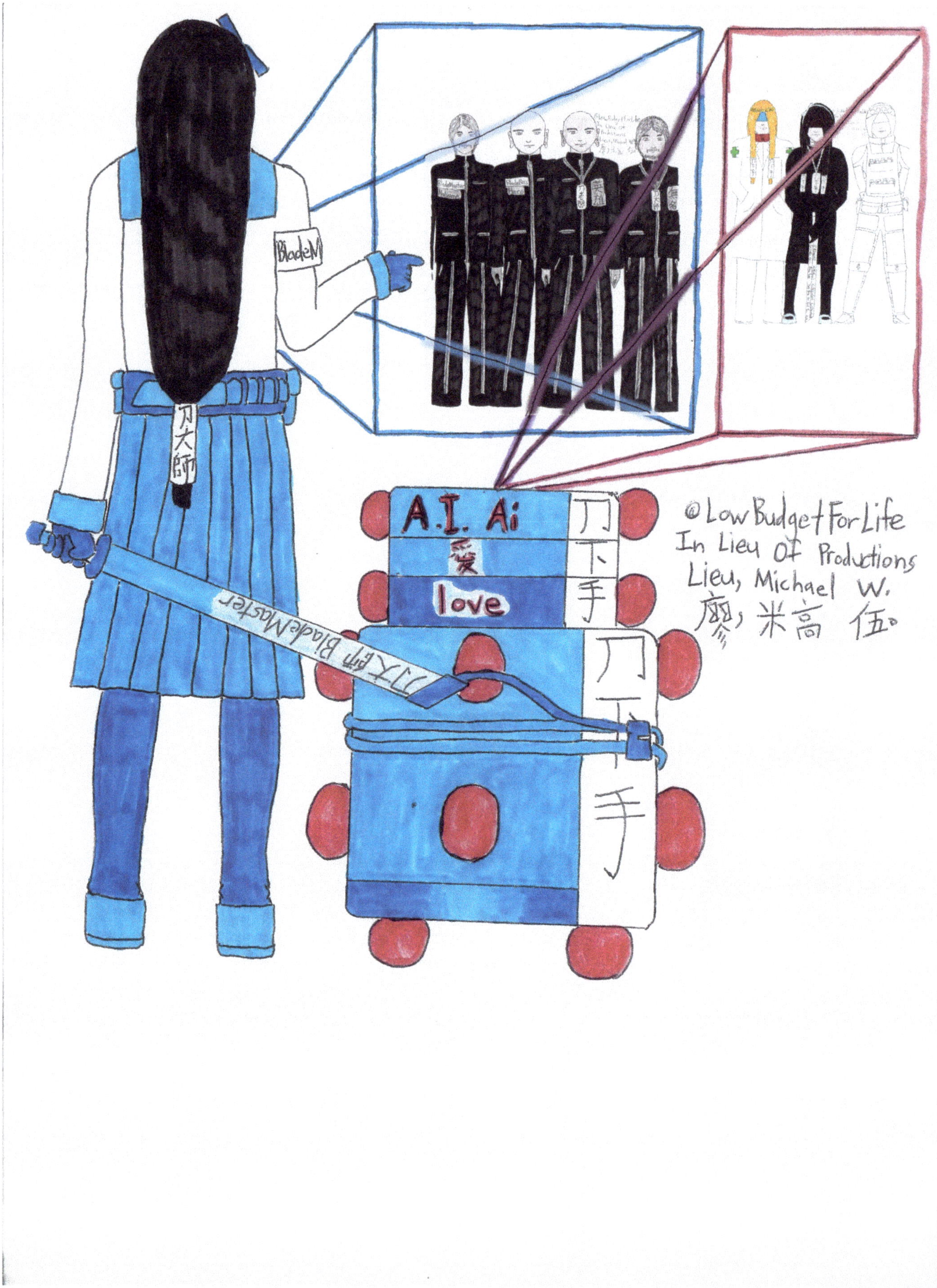
BladeM
刀大師
刀大師 BladeMaster
A.I. Ai
愛
love
刀下手
刀下手
©Low Budget For Life
In Lieu Of Productions
Lieu, Michael W.
廖, 米高 伍o

Alien Valentina Vita, Traveling Cosplayers And Kosplayers
2154;_ Chicago, Illinois' Intranet, Wireless Broadcasting Frequencies Online

<table>
<tr><td colspan="2">V. V. Augmented Reality Holographic Heads Up Display A.R. Holo HUD</td></tr>
<tr><td>Actual Legal Given Name;_

Elle K. Mcgee (goes by Elle)
pronounced, 'el-lee' 'kay' 'mick-gee'</td><td>Physical Plus Psych Age;_

augmented 17 + 2 years old</td></tr>
<tr><td>Built Day;_

year 2152, month 05 (May), day 05</td><td>Employment;_

The Oldest Society.</td></tr>
<tr><td>Community Network(s);_

British's Special Air Service (S.A.S.)

Hong Kong's Independent Commission Against Corruption (I.C.A.C.)

British-Hong Kong Sovereignty Sector

Cook-Chef Community Martial Artist Educator

Human Klones, Designer Babies And A.I. Ai [Aug+]</td><td>Alien;_

Legal And Approved
Operative, Citizen Of Eur-Asia

Student Visa
Cook-Chef Community Studying Program

Approved And Granted Access By
"2121's The Old World" of
U.S. Department Of Immigration
And
U.S. Department of Alien and Citizen Services. [Aug+]</td></tr>
<tr><td>Social Status;_

Western Society of European Union (E.U.) and Eastern Society of Asian Pacific (A.P.) International Celebrity Cook-Chef Martial Artist Student Trainee
[Aug+]</td><td>GeoLocation;_

Made In Central Country, Eur-Asia Constitution
Approved And Granted Travel To
America(s), Utopian Sovereignty of North America,
Earth's Global Government [Aug+]</td></tr>
<tr><td colspan="2">Alive;_ Heart-Rate augmented [Aug+]</td></tr>
<tr><td colspan="2">[Aug+]</td></tr>
<tr><td colspan="2">V. V. Augmented Reality Holographic Heads Up Display A.R. Holo HUD</td></tr>
</table>

2154;_ Chicago, Illinois' Intranet, Wireless Broadcasting Frequencies Online
Alien Valentina Vita, Traveling Cosplayers And Kosplayers

2154;_ Chicago, The Last Animal Sanctuary Of Illinois' intranet,
Wireless Broadcasting Frequencies Online
Geographic Info System(G.I.S.) Map Model Holo HUD

Reads = Left to Right (L -> R)

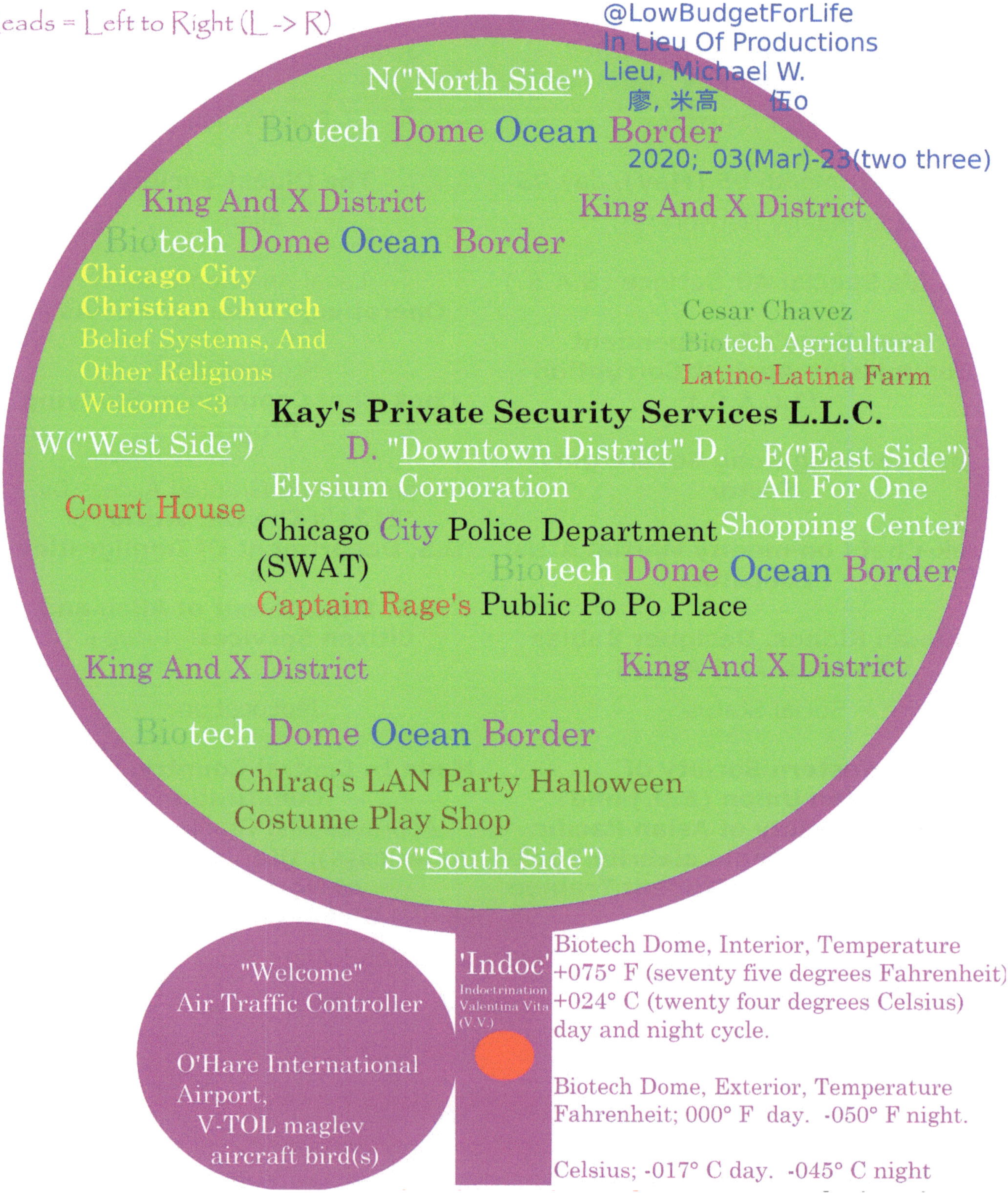

Valentina Vita (V. V.); Traveling Cosplayers And Kosplayers;
"(-_-) Okay, Elle that **Celebrity Cook-Chef BladeMaster student** is in front of me... (@_@) In front of her is... Hmm... **Privacy protect aug**... Wonder **who would this persons, places or things could be...** Hmm..."

Alien Valentina Vita, Traveling Cosplayer(s) And Kosplayer(s)
2154;_ Chicago, Illinois' Intranet, Wireless Broadcasting Frequencies Online

V. V. Augmented Reality Holographic Heads Up Display A.R. Holo HUD
Privacy Protection aug made possible by Kay's Private Security Services L.L.C. 【 Aug+ 】
V. V. Augmented Reality Holographic Heads Up Display A.R. Holo HUD

2154;_ Chicago, Illinois' Intranet, Wireless Broadcasting Frequencies Online
Alien Valentina Vita, Traveling Cosplayer(s) And Kosplayer(s)

Mister Kelvin; Human Klone(s) Property Of Elysium Corporation;
"(*partially reading guidelines and paraphrasing, voice over*) Please, sorry, Miss, huh. Let **me** finish before you keep interrupting **me** huh. ...Chicago City... **The Last Animal** Sanctuary **of** this sector **Illinois'** self-governing guidelines **does not allow and approve of double last surnames with no hyphens, dash marks, no spaces in between names other than first given name, middle and last family surnames, suffix, juniors, etc,** and huh... **First letters in the first, middle and last names, suffix, juniors, etc are the only letters being capitalized**, huh... When writing in... **last names... family names... surnames first, must have commas followed by space then first name given, space, middle name... and middle name can be shorten to just a capitalized letter followed by period dot**, huh. ...This is regardless of what City Controller... **sector... area... planet**... your **culture** you are from, huh. This is in accordance to "Chicago City Clean House Act of twenty one fifty two, July four", huh... You need to clear this up with **the old world U.S. Department Of Immigration** and **U.S. Department of Alien and Citizen Services**, huh. (*looking*) ...In **adding a hyphen to** your surname, **no spaces in A. and V.** for Davinci and **they are lower case**, huh. (*ahem*) ...Look, go to **the old world U.S. Department Of Immigration and U.S. Department of Alien and Citizen Services**, update your name to this, huh." (*partially reading guidelines*) So-Sorry Miss, **we** do not make the rules, regs, **we of Elysium Corporation Paid Employees** just enforce them by City Controllers, our Chicago City Mayor... this is prevent cases of confusion as seen in **the old world**, huh... Do you understand? (*looking*) ...Next!!"

Valentina Vita (V. V.); **Traveling Cosplayers And Kosplayers**;
"(=_=) So, this **it** is someone with a double last names with no hyphens or so... Hmm..."

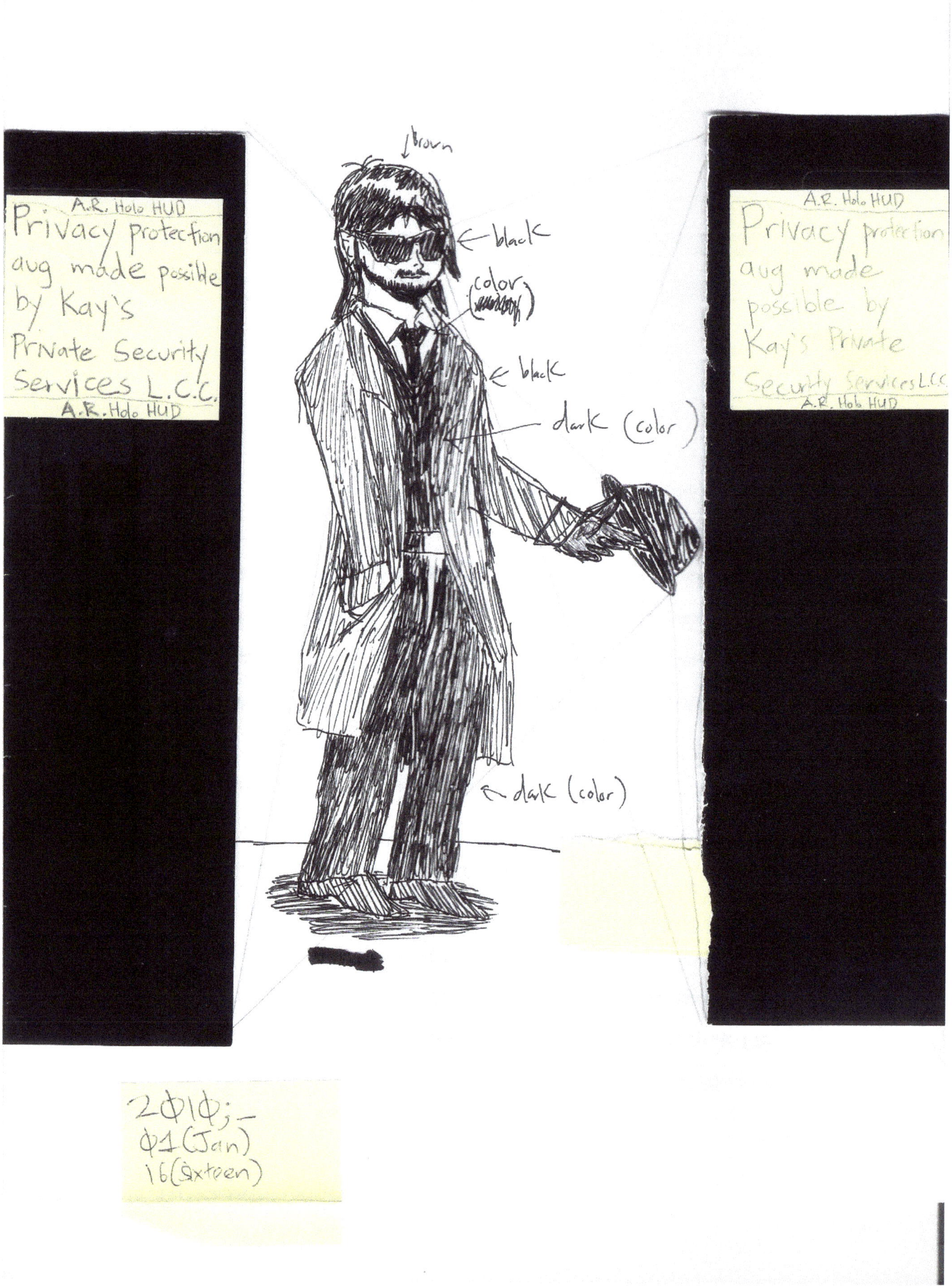
brown
black
color
black
dark (color)
dark (color)
A.R. Holo HUD
Privacy protection aug made possible by Kay's Private Security Services L.C.C.
A.R. Holo HUD
A.R. Holo HUD
Privacy protection aug made possible by Kay's Private Security Services L.C.C
A.R. Holo HUD
2Φ1Φ;-
Φ1(Jan)
16(sixteen)

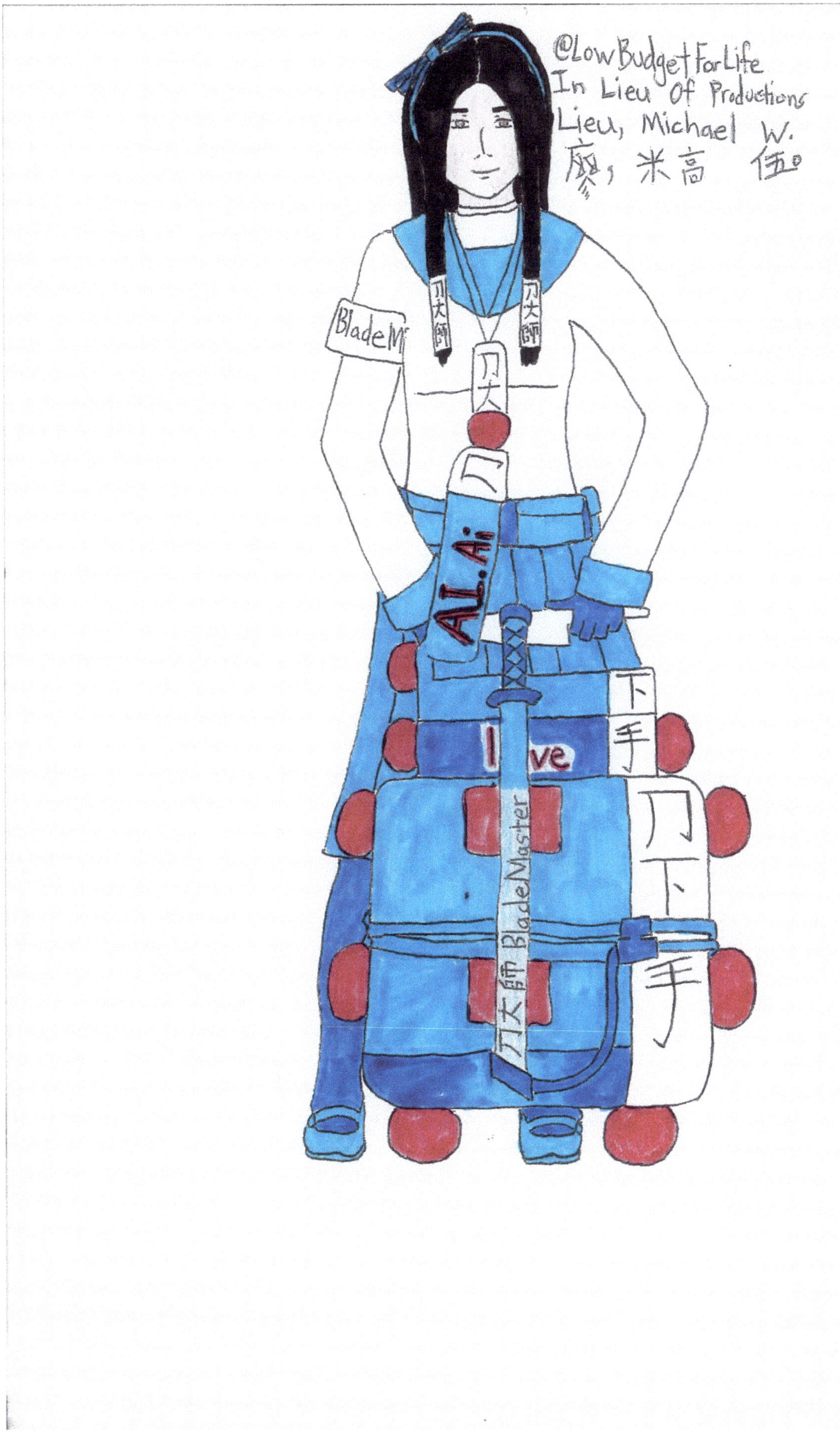
©LowBudgetForLife
In Lieu Of Productions
Lieu, Michael W.
廖, 米高 伍o
BladeM
刀大師
刀大師
AI.Ai
下手
刀下手
刀大師 BladeMaster

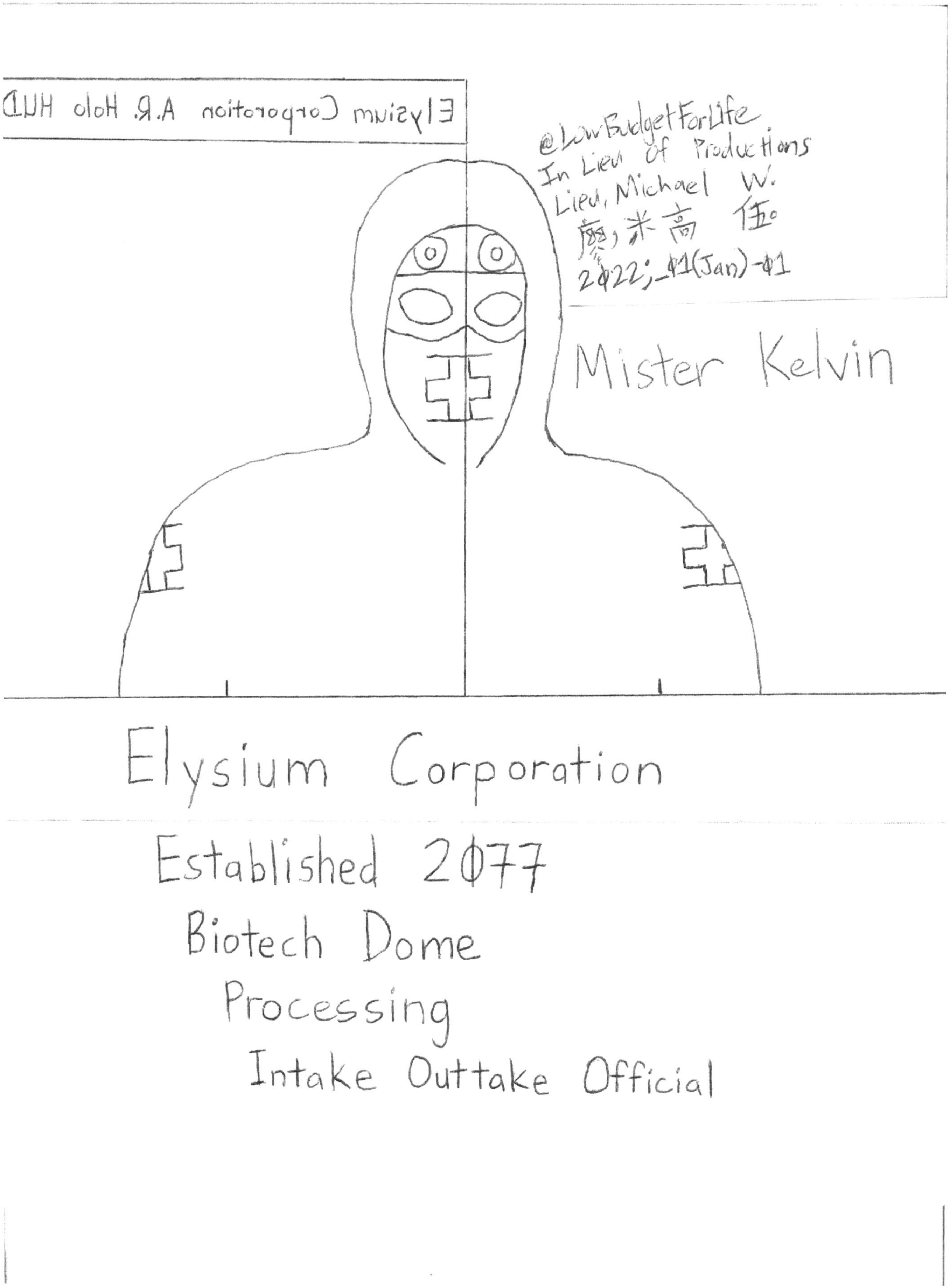
Elysium Corporation A.R. Holo HUD
@LowBudgetForLife
In Lieu Of Productions
Lieu, Michael W.
廖, 米高 伍o
2Φ22;_Φ1(Jan)-Φ1
Mister Kelvin
Elysium Corporation
Established 2Φ77
Biotech Dome
Processing
Intake Outtake Official

Citizen Mister Kelvin, Chicago City Official + Elysium Corporation Employee, Dome
2154;_ Chicago, Illinois' Intranet, Wireless Broadcasting Frequencies Online

Elysium Corp Augmented Reality Holographic Heads Up Display A.R. Holo HUD	
Actual Legal Given Name;_ **Mister Kelvin** (goes by **Mister Kelvin**) pronounced, '**mist-ter**' '**kell-vin**'	Physical Plus Psych Age;_ natural 30 + 30 years old

Projecting Temperature Inside Elysium Corporation Biotech Dome By "Planet Earth, North America, Illinois, Chicago"						
Sun's Day 24° C Warm	Moon's Day 24° C Warm	Tiu's Day 24° C Warm	Woden's Day 24° C Warm	Thor's Day 24° C Warm	Frigg's Day 24° C Warm	Saturn's Day 24° C Warm
[Aug+]						
Elysium Corp Augmented Reality Holographic Heads Up Display A.R. Holo HUD						

2154;_ Chicago, Illinois' Intranet, Wireless Broadcasting Frequencies Online
Citizen Mister Kelvin, Chicago City Official + Elysium Corporation Employee, Dome

Mister Kelvin, Human Klone(s) Property Of Elysium Corporation;
"(*reading*) Next! Hello... Miss Valentina Vita, V. V., **we** noticed you are part of **our Human** Sanctuary movement **in Eur-Asia**... Huh. As part of the twenty one twenty one new world City Controllers' UltraViolet A.R. Holo HUD visual verbal indoctrination for our **Animal** Sanctuary... Huh. You must be aware of how our so called brainwashing tactics work... As always, you can always take up and discuss with our Chicago City Mayor within the biotech broadcasting frequencies UltraViolet rays, huh. Good day. (*looking*) ...Next...?"

Valentina Vita (V. V.); Traveling Cosplayers And Kosplayers;
"(^_^) **Hello Mister Kelvin, your Eur-Asia**, Italian counterpart **klone** from Naples tells me to say **hello to you**!"

Mister Kelvin, Human Klone(s) Property Of Elysium Corporation;
"(*surprise*) Oh, huh... **Hello to my human klone Eur-Asia**, Italian counterpart from Naples. **I** am sure... Miss V. V. can **relay this message to him** over there... Huh."

Valentina Vita (V. V.); Traveling Cosplayers And Kosplayers;
"(^_^) I am sure, **Mister Kelvin of Naples** Italy would be happy to that **you** said **hello**, from **Mister Kelvin of** Chicago **Illinois** counterpart. Bye!"

2154;_Arcadia \ Valentina Vita (V. V.) 'Healthy-Strong Life' \ Left to Right(L->R)

Utopian, Augmented Natural Animal Kingdom Plus Human Beings Coexistence
2154;_ Chicago, Illinois' Intranet, Wireless Broadcasting Frequencies Online

Elysium Corp Augmented Reality Holographic Heads Up Display A.R. Holo HUD

Automated Algorithm; Human Voice Over Emulation Based On Elysium Corporation Archives; "(*splice male female voice over*) This is the **fifty seventh** United **States** of America **President** Emily A. Truman here. Who are these Martian Terrorists and what do these space aliens look like…? If you do see something suspicious, you need to say something… Quote, **see something, say something**, unquote. Earth's **Global Government**, **both foreign and domestic** Terrorists Threats is to be taken seriously. With the on going War On Mars conflict, report any suspicious activity of possible Martian Terrorists Threats and acts to private security (***Agent Kay's*** *kool splice male voice over*) **Kay's Private Security Services L.L.C.**, (*splice male female voice over*) …And for public security (*Captain Rage's sexy female voice over)* to **Chicago** City **Police Department** Captain Rage's **Public Po Po Place**. (*splice male female voice over*) Thank you for your cooperation and have a nice day, Alien V. V.."

Community Network;_

The **old world** United **States** of America (U.S. of A.) mythology, 1776-2121, owes **Britain and or England**, the **system of human laws** that was **adopted** as a conduct of jurisprudence in the **early North America's colony**.

The **old world** U.S. of A. government branches' overview:
National or Federal government leader = **President**.
Sector or State government leader(s) = **Governor**.
"Broadcasting UltraViolet (U.V.) Rays" or Local government leader(s) = **Mayor**.

U.S. **Constitution** = <u>"Supreme Law Of The Land", "We The People"</u>.
Treaties = <u>International business agreement with foreign governments</u>:
United **States**, Mexico, Canada. → North America Free Trade Act (NAFTA)
Statutory Laws = <u>Human laws passed by U.S. Congress, Legislative Branch</u>.
Executive Orders = <u>Orders by U.S. President bypassing U.S. Congress</u>.
Administrative Laws = <u>U.S. federal government, Executive Branch</u>.
Human: Common Laws = <u>A warrantless arrest "Citizens' Arrest"</u>, varies by each Sector and or State in the U.S. of A. as a Work In Progress (WIP).

[Aug+]

Elysium Corp Augmented Reality Holographic Heads Up Display A.R. Holo HUD

2154;_ Chicago, Illinois' Intranet, Wireless Broadcasting Frequencies Online
Utopian, Augmented Natural Animal Kingdom Plus Human Beings Coexistence

Utopian, Augmented Natural Animal Kingdom Plus Human Beings Coexistence
2154;_ Chicago, Illinois' Intranet, Wireless Broadcasting Frequencies Online

Elysium Corp Augmented Reality Holographic Heads Up Display A.R. Holo HUD
Automated Algorithm; Human Voice Over Emulation Based On Elysium Corporation Archives; "(*monotone male voice over*) In the **twenty second century** of the new world movement, **Utopian** Sovereignty **of Americas,** Chicago, **The Last Animal** Sanctuary **of Illinois' Elysium Corporation** Bio**tech** Dome Sector is currently under going a **social concern** of Anti Robotics Movement**.** ARM is designed to seek out and hire **augmented and** natural **human employees plus volunteers workers** in lieu of using automated algorithms and artificial intelligence, A.I. However, there is a level of unproductively among natural **unmodified human workers** in which automated robotics are able to do much faster, better than natural **humans**, let alone with the **advent of augmented biotech humans in the twenty first and twenty second century**…"
Social Status;_   **Animal** Sanctuary (S) from: "2152;_07(Jul)-04 to Present Day Time"   Current **Human Society** Governing System: Non-Paid Militia Members, Vigilant Volunteers, Intranet Vigilantism, **Vigilantes A.K.A. Super-Heroes**.   **"Neo-Western Frontier Justice: Human Laws' And Enforcement"**   What is human law(s)? **Human law(s) consists of rules that regulate the conducts of individuals, persons, people, businesses and or other organizations within human society.**   **Human law(s) only applies to human(s)**, not Earth's Environment, Mother Nature, and not the **Animal Kingdom**. **Human law(s) is intended to protect humans' and or their property from unwanted interference from other human(s)**.   Human law(s) on planning, forbids person(s) from engaging in certain undesirable activities, however human law(s) requires human enforcement to happen. Otherwise, human law(s) are just written down ideas, concepts on pieces of paper as a medium when it is not in practice by any enforcement.
[Aug+]
Elysium Corp Augmented Reality Holographic Heads Up Display A.R. Holo HUD

2154;_ Chicago, Illinois' Intranet, Wireless Broadcasting Frequencies Online
Utopian, Augmented Natural Animal Kingdom Plus Human Beings Coexistence

Chicago Controller, Psychological U.V. Projection As A City Mayor;
"(*voice over*) Start **Utopian**, indoctrination into our **Animal** Sanctuary period end."

2154;_ Arcadia \ Valentina Vita (V. V.) 'Healthy-Strong Life' \ Left to Right(L->R)

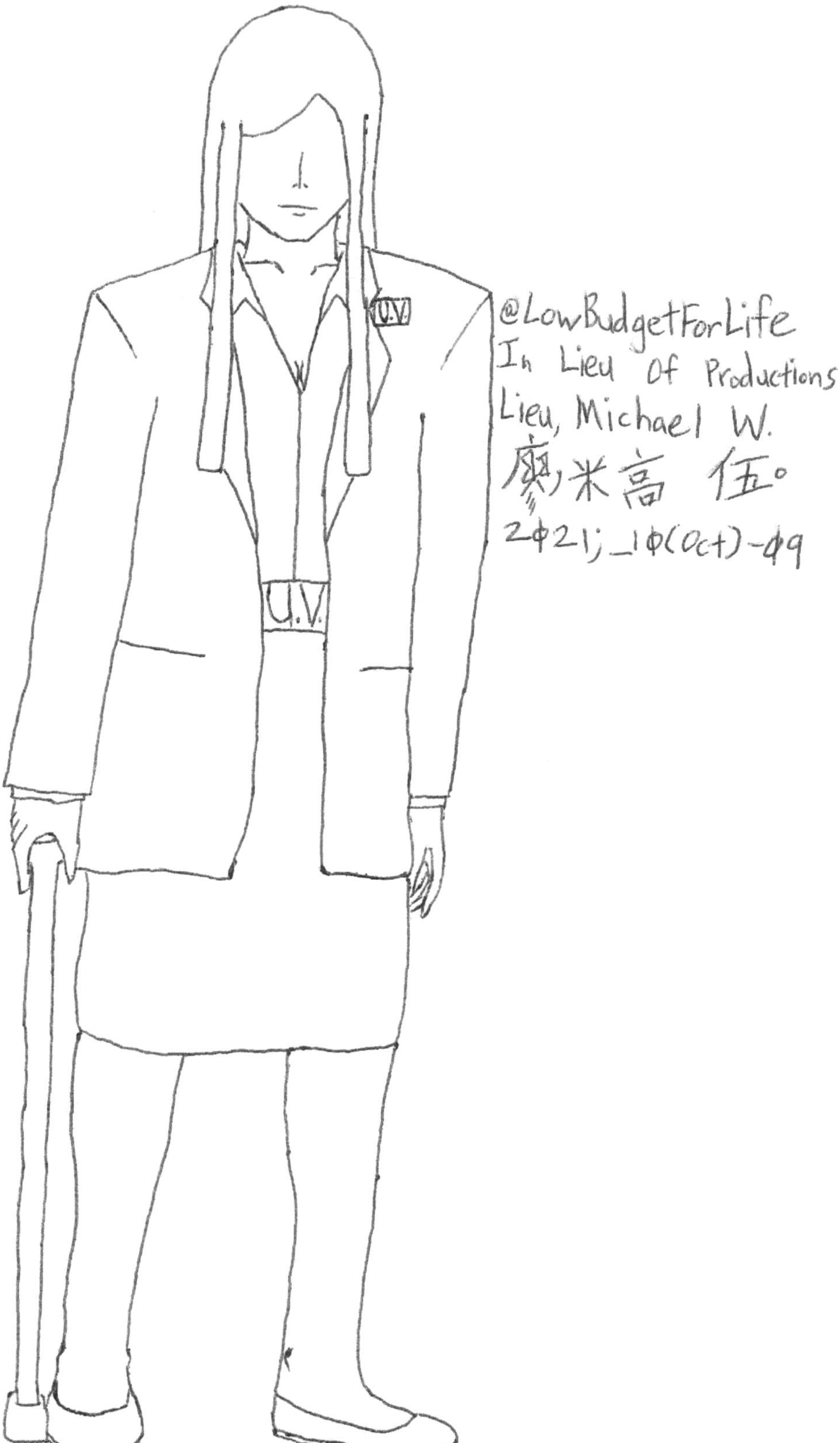

inke 2Φ21;_11(Nov)-Φ1

Chicago Controller City Mayor selects the music "**Simon And Garfunkel's The Sound Of Silence**" to play in the **Utopia A.R. Holo HUD** background and is set on loop.

Chicago Controller City Mayor, Self-Govern Sovereignty Sector, Sanctuary (S)
2154;_ Chicago, Illinois' Intranet, Wireless Broadcasting Frequencies Online

City Controller Augmented Reality Holographic Heads Up Display A.R. Holo HUD
Augmented Reality Holographic Heads Up Display, Chicago Controller City Mayor { <u>Augmented Human Male Female Hermaphrodite audio</u> } "(*Electromagnetic Spectrum Frequencies Hertz*) Start, Utopian V. V., Chicago City's Holo HUD aug message before you enter, as your City Mayor, welcome to quote, Chicago, The Last Animal Sanctuary Of Illinois, twenty one fifty four, October thirty, unquote, in accordance to our Chicago City Clean House Act of twenty one fifty two, July four, aug, nat human beings, animal kingdom, my beautiful peace loving animal kingdom species and its biotech dome, and other advanced beings are required to bring, carry their own stainless steel drinking eating utensils at all times, please separate all waste manage into paper, plastic, metal and or unknown biodegradable material in the proper placing, questions, concerns, see a white Personal Protective Equipment, green Bio-Lettering, violet purple toner Sealing-Tape frequency Clean-up Crew, they are in a full body head to feet hazmat suits, respectfully request a set from them for your own personal use if you are arriving without one, but looks like you a set on you, thank you for your time and have a nice day, period, end."
City Controller Augmented Reality Holographic Heads Up Display A.R. Holo HUD

2154;_ Chicago, Illinois' Intranet, Wireless Broadcasting Frequencies Online
Chicago Controller City Mayor, Self-Govern Sovereignty Sector, Sanctuary (S)

Valentina Vita (V. V.); Traveling Cosplayers And Kosplayers;
"(@_@) Hmm..."

Chicago Controller City Mayor selects the music change to "**BooHooWoo's Top Page The World**" to play in the **Utopia A.R. Holo HUD** background and is set on loop.

We Give Them Security, They Call Us Tyrants (T), Human Sanctuary
We Give Them Freedom, They Call Us Anarchists (A), Human Riot

City Controller Augmented Reality Holographic Heads Up Display A.R. Holo HUD
Augmented Reality Holographic Heads Up Display, Chicago Controller City Mayor { Augmented Human Male Female Hermaphrodite audio } "(*Electromagnetic Spectrum Frequency Hertz*) Start, Utopian V. V., we are a City Controller, City Controllers, people call me, us, Chicago Controller, Chicago C., Chicago City Mayor, Chicago City, Purple Person, P. P., Very Violet, V. V., hey we have the same name for initials... list goes on, really, really long, it matters little what they call me, all that matters is that you all follow our animal sanctuary rules, regulation, plus environment enforcement, as for humanity, you can do whatever you want to each other, harm, flatline, dismember each other, but do not harm, flatline, dismember any animal kingdom species, yes, that means you have to be a vegetarian by trading your bio and or tech remains for another bio and or tech remains, egg for egg, sperm for sperm, life form for life form, being for being, and or be a vegan, we will control any for immediate removal if anyone fails to follow these rules, regulation and enforcement for this environment sector, in a way everything here is free, equal exchange, you must have questions for me as a Traveling Cosplayer Alien V. V., period, end."
City Controller Augmented Reality Holographic Heads Up Display A.R. Holo HUD

We Give Them Freedom, They Call Us Anarchists (A), Human Riot
We Give Them Security, They Call Us Tyrants (T), Human Sanctuary

Valentina Vita (V. V.); Traveling Cosplayers And Kosplayers;
"(@_@) Hmm..."

Chicago Controller City Mayor selects the music change to "**Earthbound's Happy Happy Is Blue Blue**" to play in the **Utopia A.R. Holo HUD** background and is set on loop.

The Beginning, The Middle, The End, The Conclusion
The Climate, The Ark, The Flood, The Garden

City Controller Augmented Reality Holographic Heads Up Display A.R. Holo HUD
Augmented Reality Holographic Heads Up Display, Chicago Controller City Mayor { Augmented Human Male Female Hermaphrodite audio } "(*Electromagnetic Spectrum Frequency Hertz*) Start, Utopian V. V., complete this easy indoc by selecting, agreeing, and or verbally saying: "Yes, I agree", unless you want to discuss... select: "I do not know, let us discuss." ...After that feel free to enjoy what is in our biotech dome, stay as long as you desire... period, end."
City Controller Augmented Reality Holographic Heads Up Display A.R. Holo HUD

The Climate, The Ark, The Flood, The Garden
The Beginning, The Middle, The End, The Conclusion

Chicago Controller City Mayor selects the music change to "**Scatman John's Song Of Scatland**" to play in the **Utopia A.R. Holo HUD** background and is set on loop.

Valentina Vita (V. V.); Traveling Cosplayers And Kosplayers;
"(@_@) Hmm..."

Reads = Left to Right (L -> R) \ Wireless Broadcasting Frequencies Online
2154;_ Chicago, The Last Animal Sanctuary Of Illinois' Intranet

Why bother, who cares, barely any humans follow rules regulation	Fine, whatever do not sit on it, complete this before entering

Chicago Controller
City Mayor

"Chicago City Clean House Act of
2152 (twenty one fifty two);_ 07-04 (July four)"

Indoctrination Self-Governing Guidelines Overview

Failure to complete the new world 2121 (twenty one twenty one) indoc, is a warrant for immediate removal from Chicago, The Last Animal Sanctuary Of Illinois (ILL).

Exceptions apply only to the surviving augmented and natural animal kingdoms plus planet earth's environment!!

Make a choice, Utopian.

No, I disagree, wipe my memory.	I do not know, let us discuss.	Yes, I agree.

Clean-up Crew
Unit 6 (six), Unit 9 (nine)
Unit 8888,8888 (eight eight eight eight comma eight eight eight eight)

Abusive bitch and moan, toxicity constant crybaby complaint box	Let the shit hit the fan rock bottom, to crash and burn hard

2154;_ Chicago, The Last Animal Sanctuary Of Illinois' Intranet
Reads = Left to Right (L -> R) \ Wireless Broadcasting Frequencies Online

Valentina Vita (V. V.); Traveling Cosplayers And Kosplayers;
"(^_^) ...Yes, I agree."

Chicago Controller, Psychological U.V. Projection As A City Mayor;
"(*voice over*) Start **Utopian**... ...Enjoy... ...period end."

Automated Algorithm; Human Voice Over Emulation Based On Elysium Corporation Archives;
"(*monotone female voice over*) Please stand by for biotech dome airlock decontamination. Thank you for your patience and cooperation **Utopian**. (*scanning*) Welcome, **augmented biotech human female, Alien status, beauty designer**, Valentina Vita, A.K.A. V. V., **international overseas foreigner, Civilian, Citizen Of Eur-Asia, Western Society of European Union and Eastern Society of Asian Pacific,** Traveling Cosplayers And Kosplayers... **We** welcome you to Chicago**, The Last Animal** Sanctuary **Of Illinois, we** hope you enjoy your stay in a twenty second century sovereign **Animal** Sanctuary biotech **utopia** of the twenty one twenty one new world movement... **We** have **physical printout copies of Elysium Corporation's brief info guide of our** biotech dome, feel free to take one or as many as you desire and have a nice day."

Valentina Vita selects the music(s): "**SoundTemp's Ragnarok Online**:

Title	*Gambler Of Highway*	*Peaceful Forest*	*I Miss You*	*Tread On The Ground*
Risk Your Life	*Wind Of Tragedy*	*Prontera Theme*	*Great Honor*	*Divine Grace*
Morroc Theme	*Streamside*	*Geffen Theme*	*Alberta Theme*	*Labyrinth*
Treasure Hunter	*Time Up*	*Under The Ground*	*Ancient Groover*	*Through The Tower*
Back Attack	*Travel*	*Desert*	*Plateau*	*Everlasting Wanderers*
Dreamer's Dream	*You're In Ruins*	*Be Nice N Easy*	*One Step Closer*	*Brassy Road*
Yuna Song	*Pampas Upas*	*Tempoison*	*Nano East*	*Temporsche*
Hamatan	*Al De Baran Theme*	*Fear...*	*Curse N Pain*	*Morning Gloomy*
Temp It Up	*Don't Piss Me Off*	*An Ant-Lion's Pit*	*Welcome Mister Hwang*	*Help Yourself*
Watery Grave	*Out Of Curiosity*	*Believe In Myself*	*Ready*	

" to play on her Ai-Aug V. V. bracelet on one third volume of 33% and is set on loop.

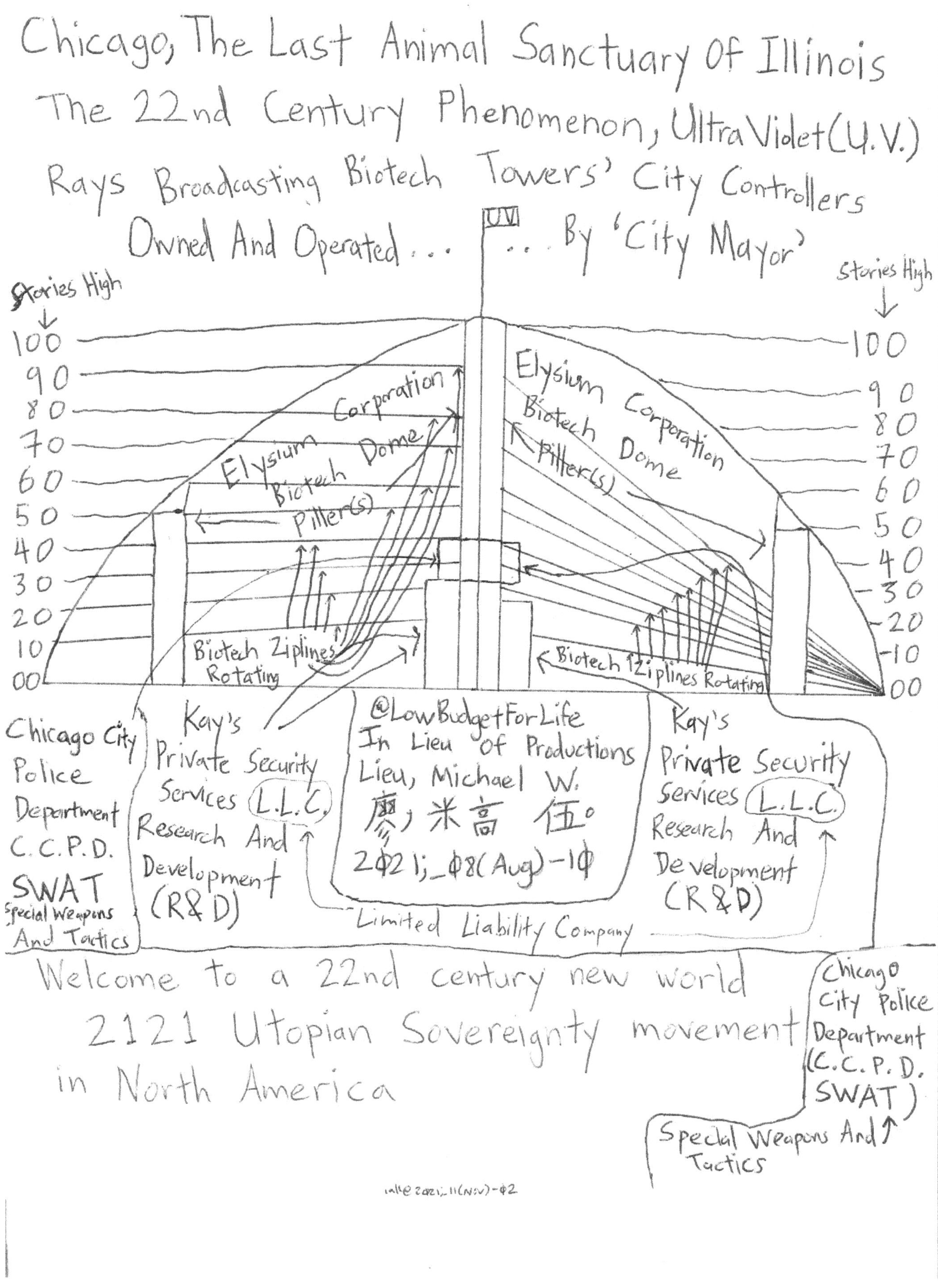
Chicago, The Last Animal Sanctuary Of Illinois
The 22nd Century Phenomenon, Ultra Violet (U.V.)
Rays Broadcasting Biotech Towers' City Controllers
Owned And Operated . . . . . . By 'City Mayor'
UV
Stories High
Stories High
100
90
80
70
60
50
40
30
20
10
00
Elysium Corporation Biotech Dome
Piller(s)
Elysium Corporation Biotech Dome
Piller(s)
Biotech Ziplines Rotating
Biotech Ziplines Rotating
Chicago City Police Department C.C.P.D. SWAT Special Weapons And Tactics
Kay's Private Security Services L.L.C. Research And Development (R&D)
@LowBudgetForLife
In Lieu Of Productions
Lieu, Michael W.
廖, 米高 伍o
2021;_08(Aug)-10
Kay's Private Security Services L.L.C. Research And Development (R&D)
Limited Liability Company
Welcome to a 22nd century new world
2121 Utopian Sovereignty movement
in North America
Chicago City Police Department (C.C.P.D. SWAT)
Special Weapons And Tactics

Elysium Corporation

亞

Established 2077

The Magical Corporation Of Desire

Biotech Dome Sector

No Need	Chicago Controller City Mayor	**Animal** Sanctuary (S)
Paid Employ	Director: Doctor Froste	**Utopian** Sovereignty (**U.**S.)
I Want Free Stuff	Chief Engineer (ChEng): "Volunteer Manager" Lieutenant Lieu	Sanctuary (S), Sovereignty (S), not a (D), (M), or (R).

100 Floor(s) And Or Story High	**High Tower Corporate Divisions Breakdown**
90 to 99 Floor(s) And Or Stories High	Suites: Mayor, Director, Chief
80 to 89 Floor(s) And Or Stories High	Endless Deep Desirable Dreams
70 to 79 Floor(s) And Or Stories High	Earth's Renewal Energy Sources
60 to 69 Floor(s) And Or Stories High	Human Sexuality, Copulation, Augmentation
50 to 59 Floor(s) And Or Stories High	Undead Utopian Quality Assurance
40 to 49 Floor(s) And Or Stories High	Deceased, Necrosis, Expiration, Afterlife
30 to 39 Floor(s) And Or Stories High	Volunteer Worker Service Center
20 to 29 Floor(s) And Or Stories High	Public To Private Transition
10 to 19 Floor (s)And Or Stories High	Human History Introduction Indoctrination
01 to 09 Floor(s) And Or Stories High	Animal Human Health Care
0 Floor(s) And Or Story High	Biotech Dome Decontaminating Maintenance

Geotech

Geology Physical	Introduction	Plate Tectonics
Mineral Properties	Minerals, Rock Cycle	Igneous Rocks
Sedimentary Rocks	Metamorphic Rocks	Geologic Structures
Earthquakes	Geologic Time	Topo Maps
Groundwater	Climate And Planetary Geology	[Continued]

Structural [Continued]

Hydro, Solar, Wind [Continued]

2154;_10(Oct)-30(thirty)

Travel To Western Society, America(s), North America
Self-Govern Sovereignty (S),
Illinois (ILL) Sector, Points Of Interest (P.O.I.)

Chicago (S), "**Animal** Sanctuary"
"The New World 2121 Movement Of The 22nd Century,
Utopian Sovereignty of North America (U.S. of N.A.)"
South Side, ChIraq's LAN Party Halloween Costume Play Shop

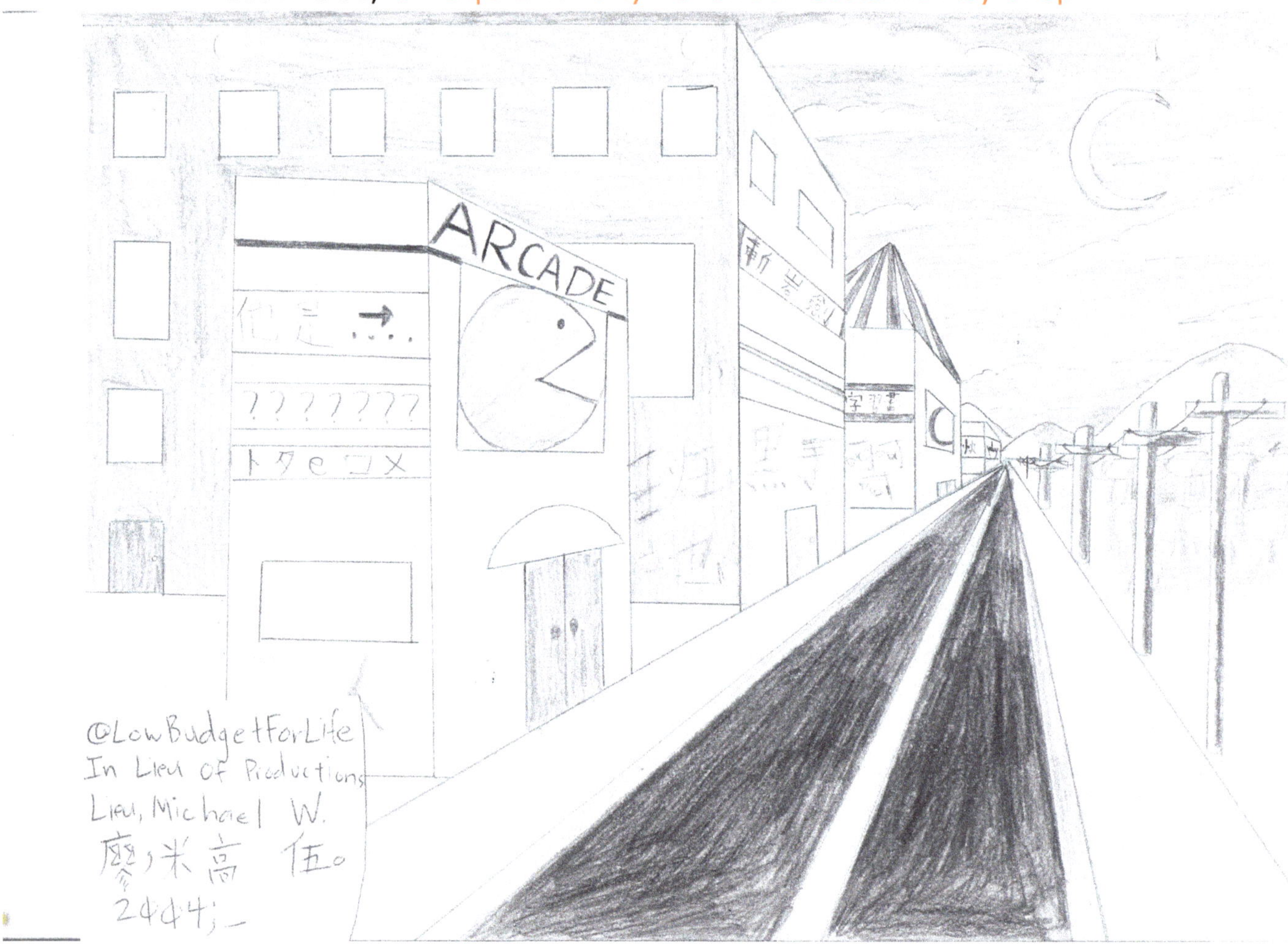

Valentina Vita (V. V.); Traveling Cosplayers And Kosplayers;
"(T_T) ...Wow... **that temperature change**... Time to change clothes and **de-armor** my P.P.E. suit! (>_<) **Need to find some place to change**... Umm... (@_@) **Time to hide**... My Lady A. Dragoon Knight cosplay needs to go on my travel luggage!"

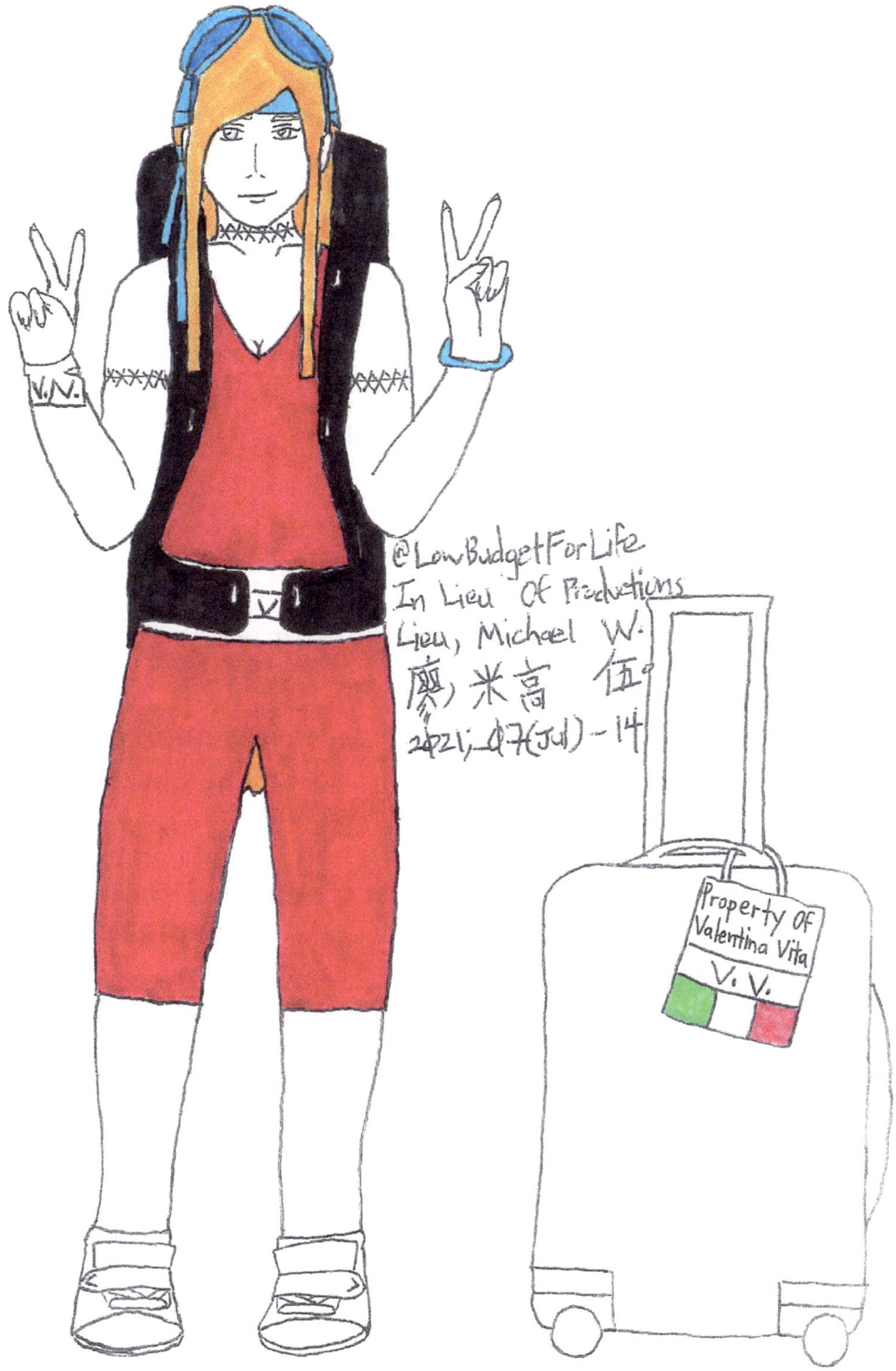
V.V.
V
@LowBudgetForLife
In Lieu Of Productions
Lieu, Michael W.
廖, 米高 伍o
2021-07(Jul)-14
Property Of
Valentina Vita
V. V.

2154;_10(Oct)-30(thirty)

Travel To Western Society, America(s), North America
Self-Govern Sovereignty (S),
Illinois (ILL) Sector, Points Of Interest (P.O.I.)

Chicago (S), "**Animal** Sanctuary"
"The New World 2121 Movement Of The 22nd Century,
Utopian Sovereignty of North America (U.S. of N.A.)"
Downtown District, **Elysium Corporation** Biotech Dome Center Point

Valentina Vita (V. V.); Traveling Cosplayers And Kosplayers;
"(O_O) **Animals**... What a sight to see for an **Animal** Sanctuary... **Visual selfe time**!"

Alien Valentina Vita, Traveling Cosplayers And Kosplayers
2154;_ Chicago, Illinois' Intranet, Wireless Broadcasting Frequencies Online

V. V. Augmented Reality Holographic Heads Up Display A.R. Holo HUD	
Overview Of Remaining Biological Animal Differences On Earth As Of "2154;_"	
<u>Mammals</u>	<u>Reptiles</u>
Have **hair** all over their bodies.	Have **scales** all over their bodies.
Gives **live births**, with umbilical cords, **produce milk for young babies**. <u>Mammals</u> have **mammary glands to feed** *their young babies with milk*.	Gives birth by **laying eggs**. <u>Reptiles</u> can **continue growing throughout their lives** and continually **replace lost teeth**.
Have **4 (four)** chambered hearts.	Have **3 (three)** chambered hearts
Warm Blood	**Kold** Blood
Existing Examples Of <u>Mammals</u> In Chicago Elysium **Animal** Sanctuary (S) Bird, Cat, Cattle, Deer, Donkey, Rabbit, "Decreased Human Feral Children"	Existing Examples Of <u>Reptiles</u> In Chicago Elysium **Animal** Sanctuary (S) Turtle, *<u>Snakes & Lizards = Not Publicly Visible</u>*
Both Biological Animals' Have: Vertebrates, Backbones, Brain, Nervous System	
[<u>Aug+</u>]	
V. V. Augmented Reality Holographic Heads Up Display A.R. Holo HUD	

2154;_ Chicago, Illinois' Intranet, Wireless Broadcasting Frequencies Online
Alien Valentina Vita, Traveling Cosplayer(s) And Kosplayer(s)

2154;_ Arcadia \ Valentina Vita (V. V.) 'Healthy-Strong Life' \ Left to Right(L->R)

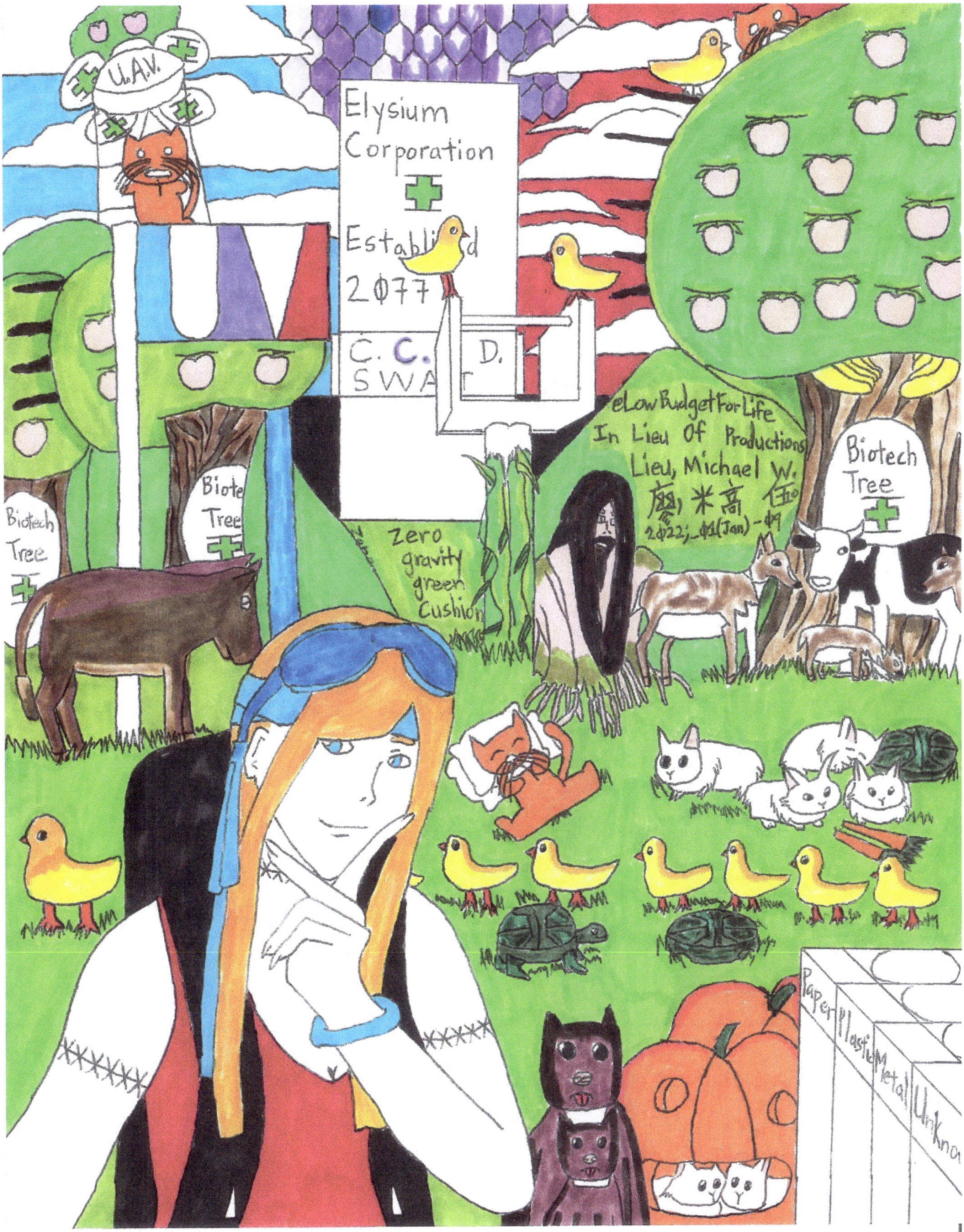
U.A.V.
Elysium
Corporation
2077
eLow Budget For Life
In Lieu Of Productions
Lieu, Michael W.
Biotech
Tree
Zero
gravity
green
Cushion
Paper
Metal

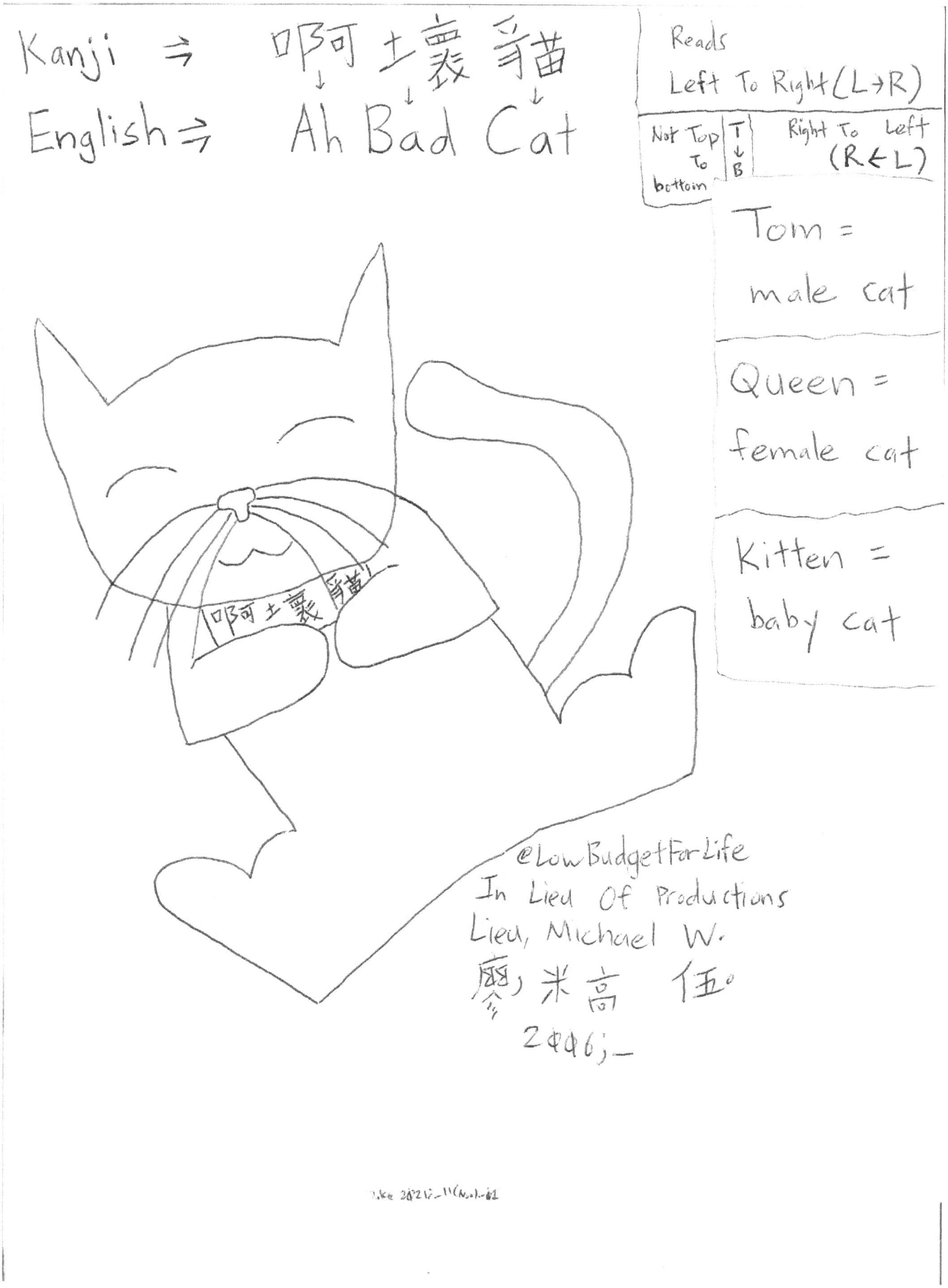
Kanji ⇒ 啊壞貓
English ⇒ Ah Bad Cat
Reads
Left To Right (L→R)
Not Top To bottom
T ↓ B
Right To Left (R←L)
Tom = male cat
Queen = female cat
Kitten = baby cat
@LowBudgetForLife
In Lieu Of Productions
Lieu, Michael W.
廖, 米高 伍

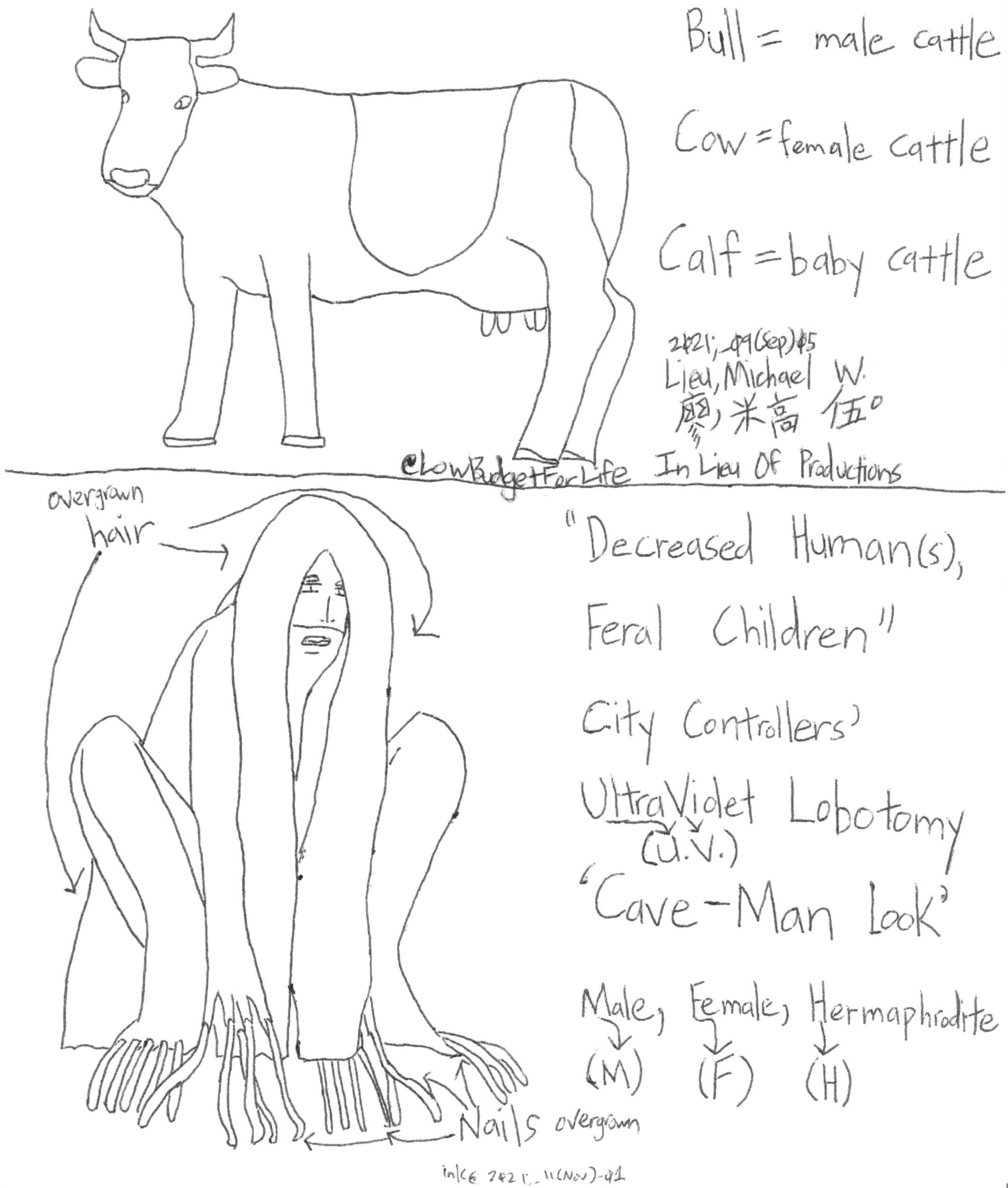
Bull = male cattle
Cow = female cattle
Calf = baby cattle
2021, 09(Sep)05
Lieu, Michael W.
廖, 米高 伍o
@LowBudgetForLife In Lieu Of Productions
overgrown hair
"Decreased Human(s),
Feral Children"
City Controllers'
UltraViolet Lobotomy
(U.V.)
'Cave-Man Look'
Male, Female, Hermaphrodite
(M) (F) (H)
Nails overgrown
inkd 2021_11(Nov)-01

2021;-09(Sep)-04

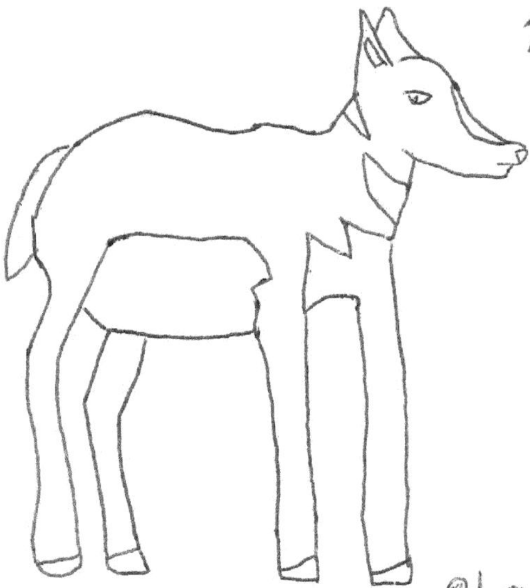

Buck = male deer

Doe = female deer

Fawn = baby deer

@LowBudgetForLife
In Lieu Of Productions

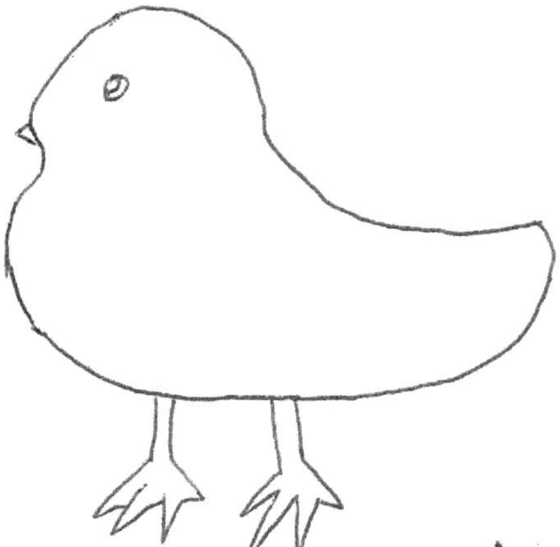

Cock = male bird

Hen = female bird

Chick = baby bird

Lieu, Michael W.
廖, 米高 伍o

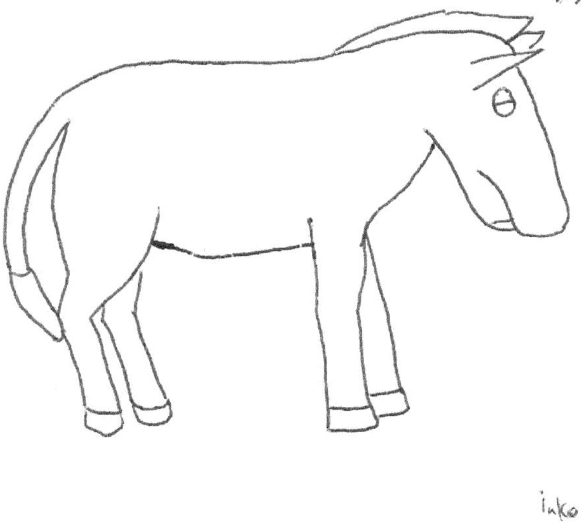

Jackass = male donkey

Jenny = female donkey

Foal = baby donkey

inko 2021;-11(Nov)-01

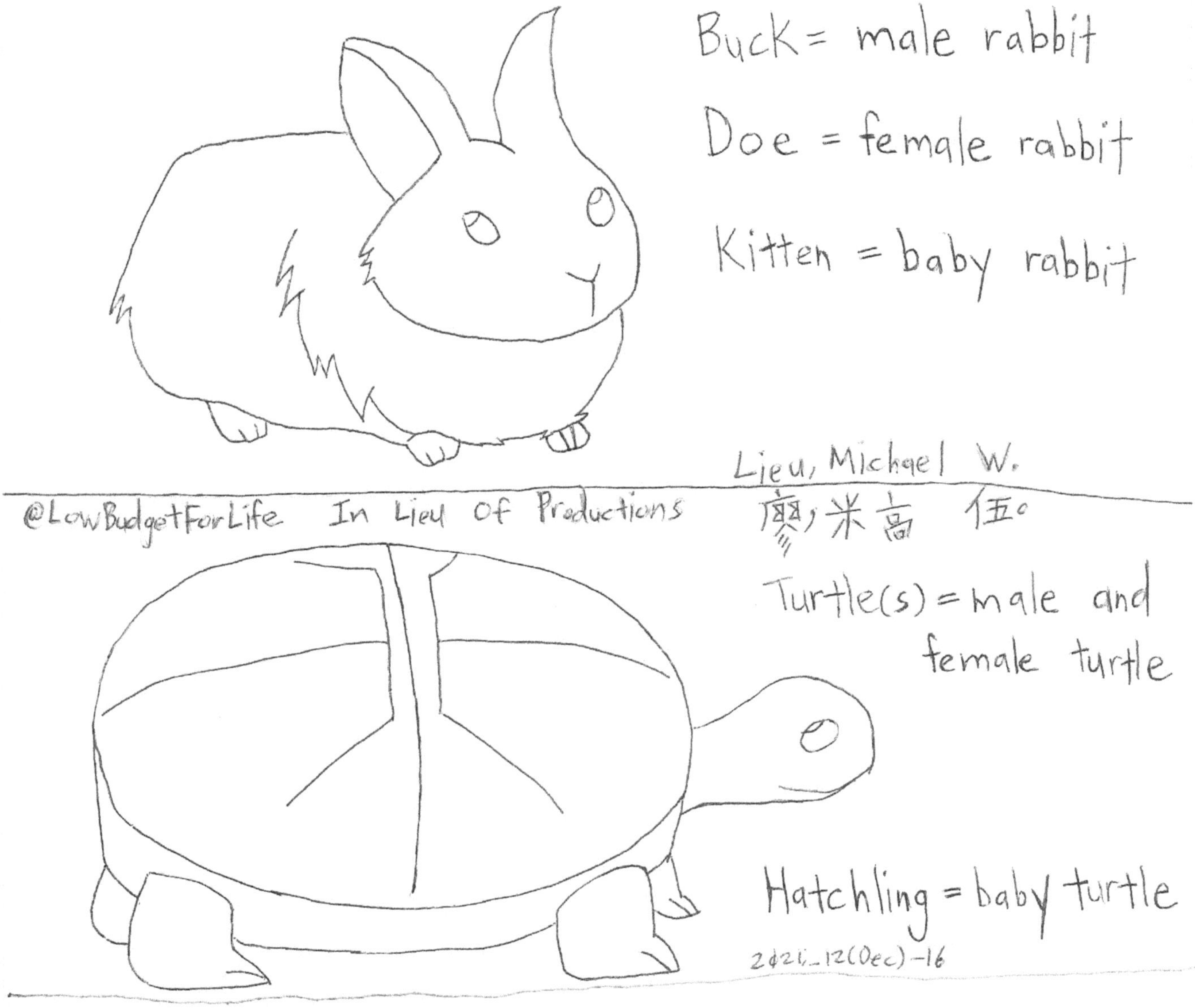
Buck = male rabbit
Doe = female rabbit
Kitten = baby rabbit
Lieu, Michael W.
@LowBudgetForLife In Lieu Of Productions
廖, 米高 伍o
Turtle(s) = male and female turtle
Hatchling = baby turtle
2021_12(Dec)-16

2154;_10(Oct)-31(thirty one)

Travel To Western Society, America(s), North America
Self-Govern Sovereignty (S),
Illinois (ILL) Sector, Points Of Interest (P.O.I.)

Chicago (S), "**Animal** Sanctuary"
"The New World 2121 Movement Of The 22nd Century,
Utopian Sovereignty of North America (U.S. of N.A.)"
West Side,

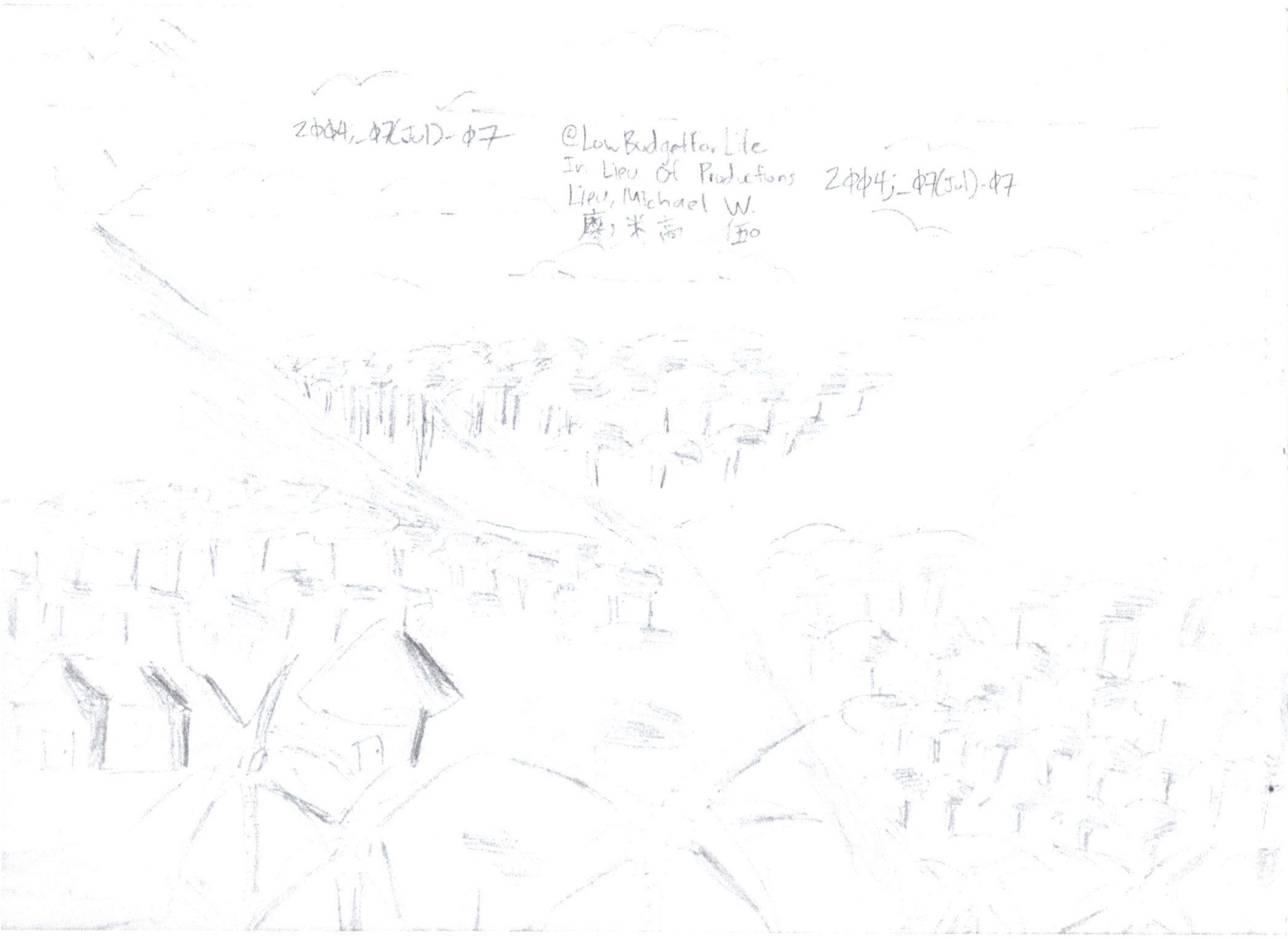

Valentina Vita (V. V.); Traveling Cosplayers And Kosplayers;
"(=_=) Whoa we wow, the **day went by so fast** just **watching, observing** the **animals**... Such a **rare sight** to see such **peaceful animals**... **Time to find a place to rest**, been up and **running for days, weeks, need to rest** and put my stuff down..."

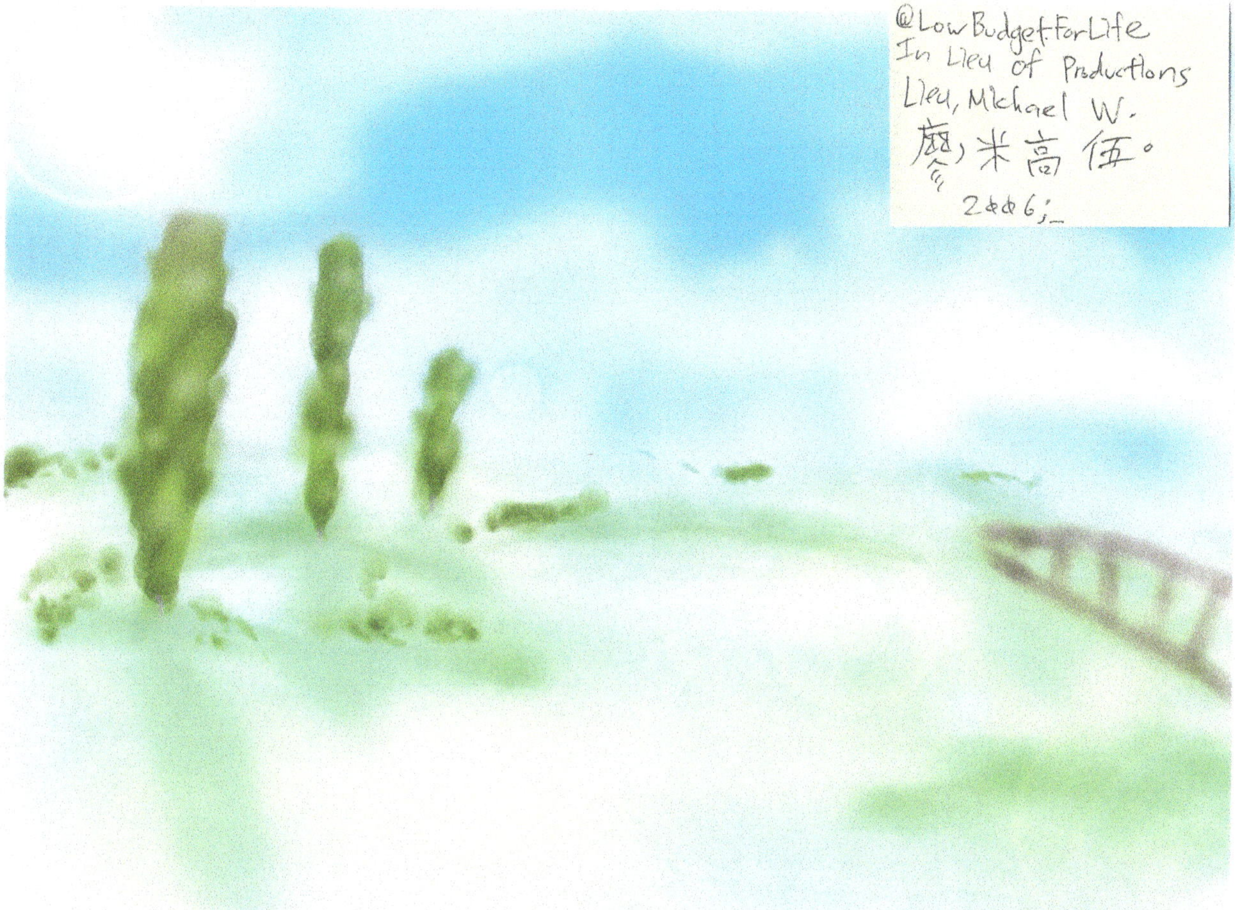

Valentina Vita (V. V.); Traveling Cosplayers And Kosplayers;
"(O_O) ...Pretty **big** dome with a lot of green environmental nature... Now where is the **housing area** under this... Chicago **Animal** Sanctuary... Not sure if the... **Holo HUD** stated that there is a **housing area**... Hmm... Gotta keep moving and sight seeing, at least this was not like that **Springfield ghost town tour**... **Warm twenty four degrees**, explains how that helps the **animals** here..."

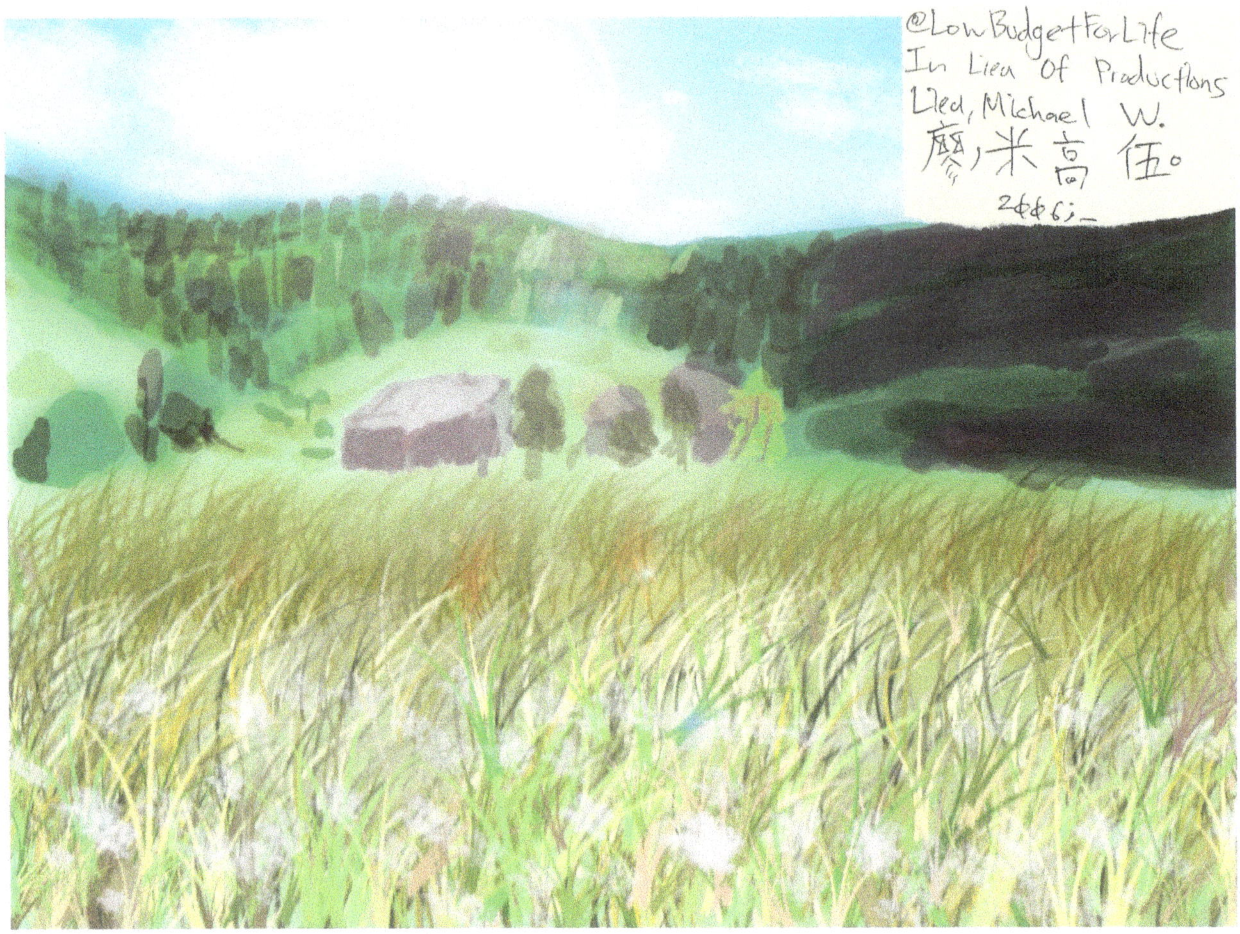

Valentina Vita (V. V.); Traveling Cosplayers And Kosplayers;
"(^_^) ...Looks like there is something resembling a **housing area** up ahead... Interesting to see all this... Mix toners of yellow, green, orange, environmental nature... The distance is closing in..."

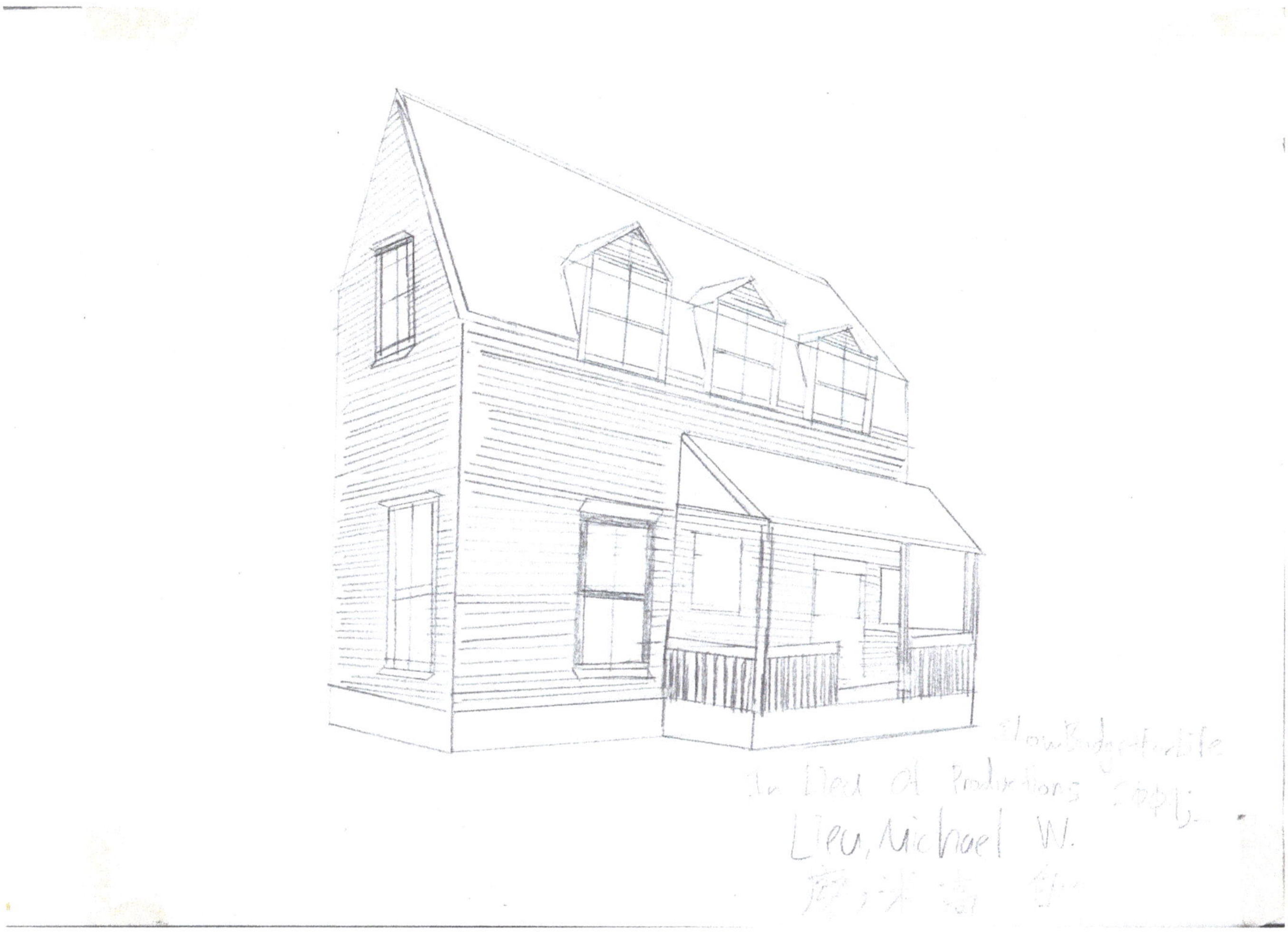

Valentina Vita (V. V.); Traveling Cosplayers And Kosplayers;
"(O_O) Whoa we wow, what an **old looking house**... Still someplace where I can lie down, rest for a bit and actually change clothes... Without looking for a hiding spots as to when I **first arrive** to this biotech dome... and the **temperature change**..."

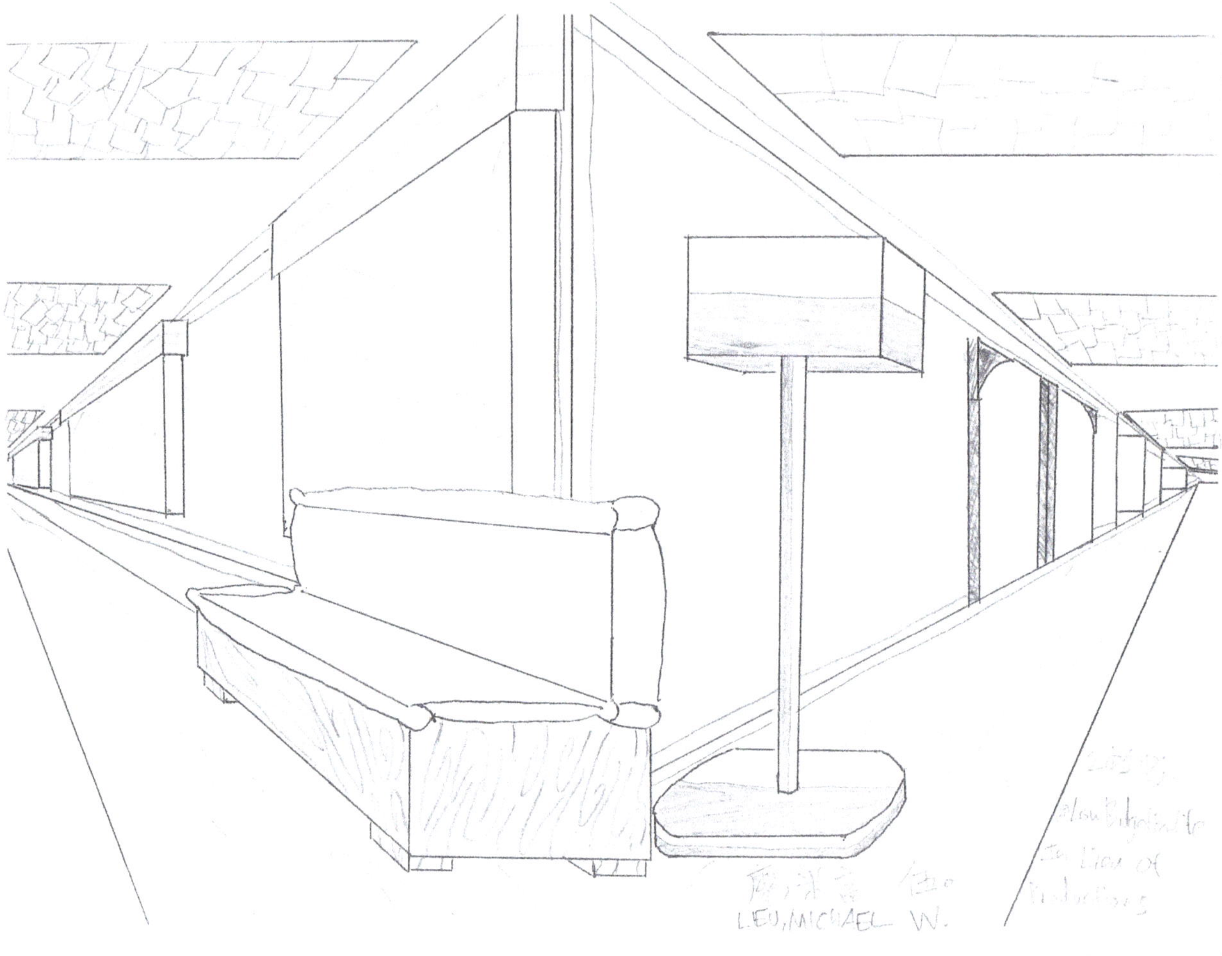

Valentina Vita (V. V.); Traveling Cosplayers And Kosplayers;
"(O_O) Whoa we wow, that bench and lamp and along with the most of the this **house, looks old and is wood material**... I can **hear the noises and footsteps in this old house**... I hear noises... **Animals**? **Animals** are possibility living in this **old house**?"

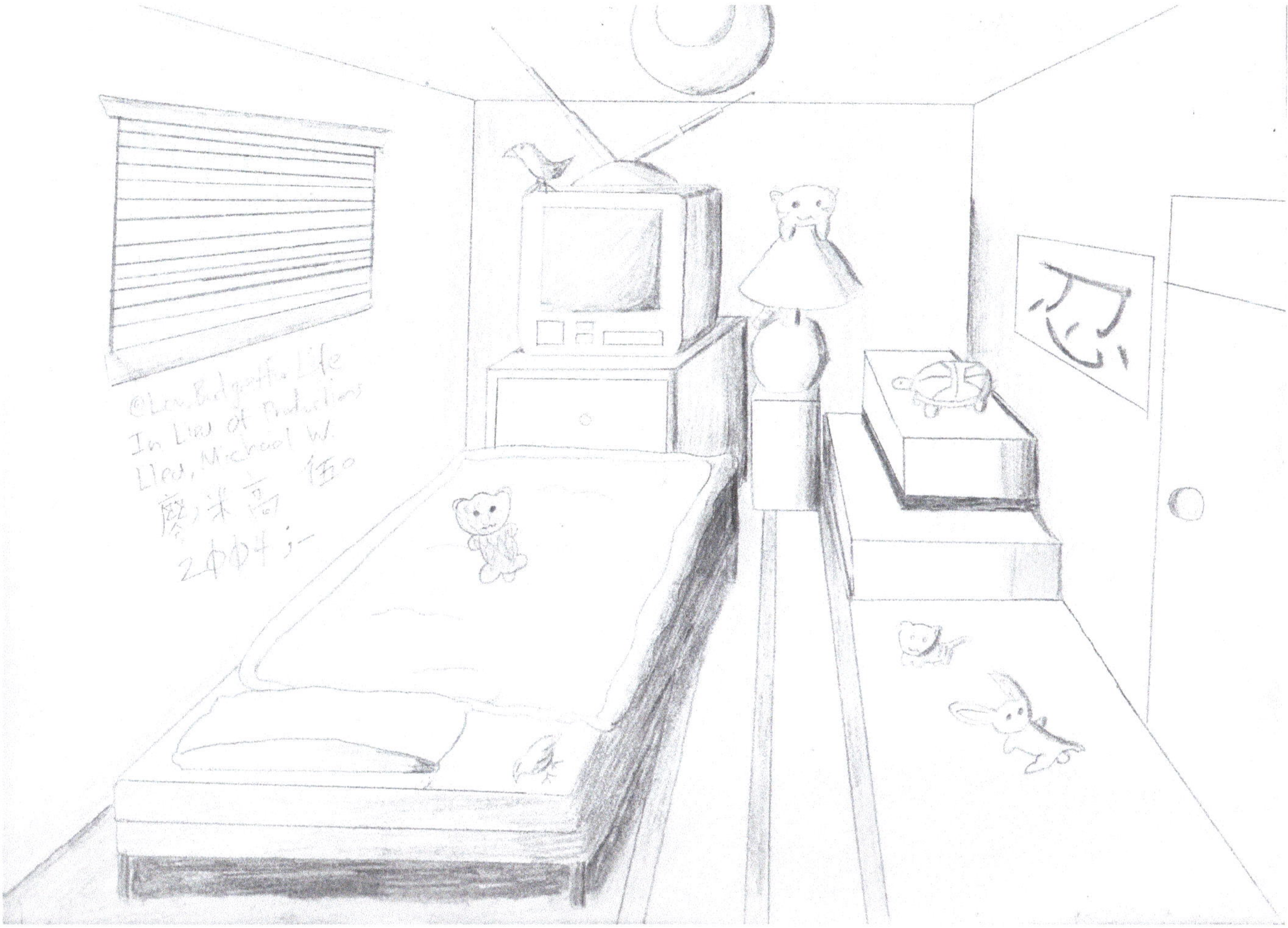

Valentina Vita (V. V.); Traveling Cosplayers And Kosplayers;
"(@_@) ...That **confirms and explains the noises** coming from this **old house** when I stepped in... **Animals living here**, hmm... Not surprised. Chicago is quote, **The Last Animal** Sanctuary **Of Illinois** unquote. (O_O) ...Is that an **Tele Vision** set with a... **bird**... That **T.V. looks different**, must be from a different time... A... **Bird is sleeping**, a... **cat and rabbit are on the ground playing around in the floor**. A... **turtle** is on the boxes, and a **cat**... is on a lamp... and another **cat** on the bed... this is good enough..."

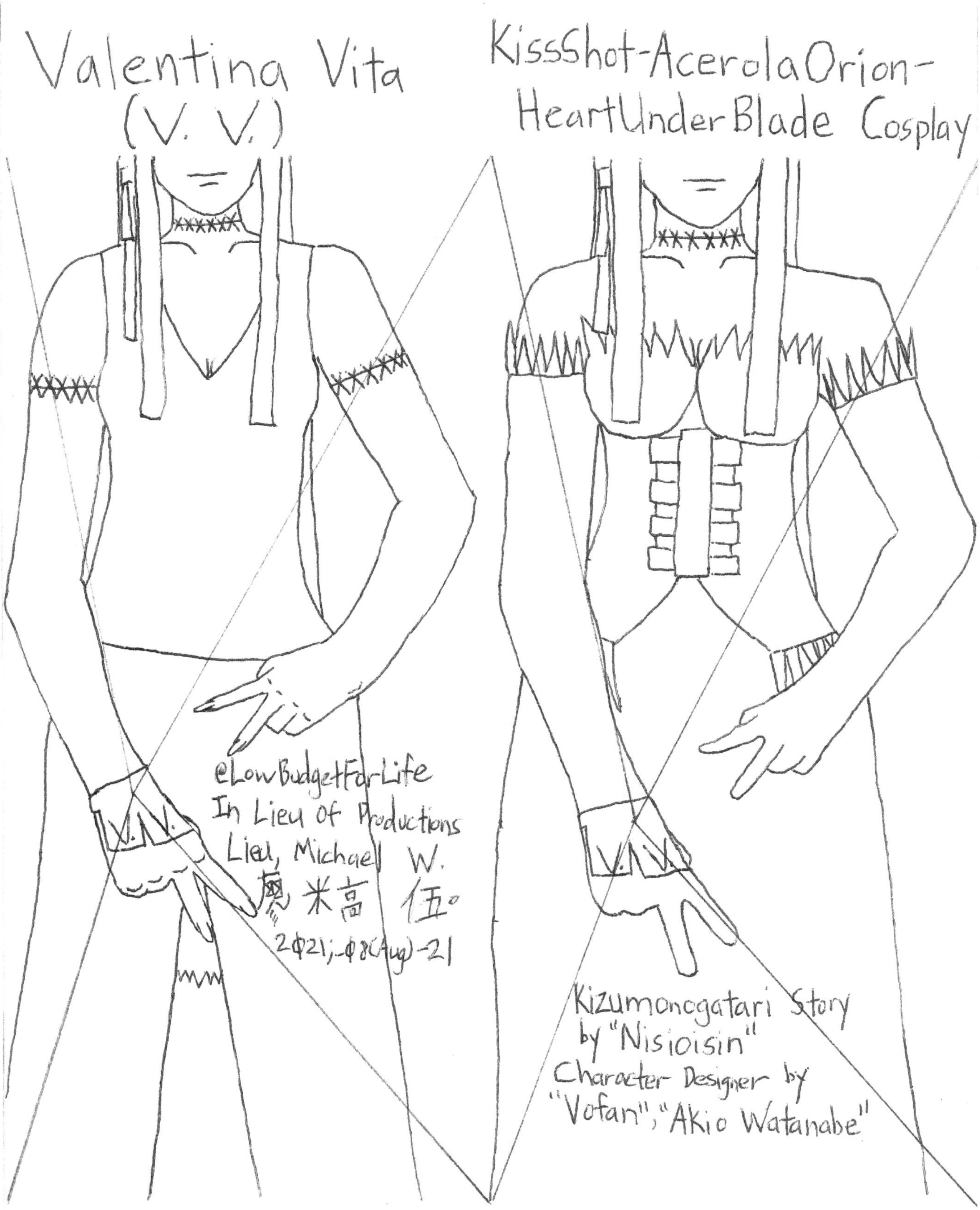

inke 2Φ21;_1Φ(Oct)-29

Valentina Vita (V. V.); Traveling Cosplayers And Kosplayers;
"(@_@) **Travel journal entry**... **Self-talk** time... Okay, time to change into **KissShot-**AcerolaOrion-**HeartUnderBlade**... Looks good so far... Used as much **attention to detail** as I could... Anyways... **Fetish** topic, those **people** that discusses **fetishes** with **you** even if **you** did not ask for **their fetish** likes... I think having a **sexual desire** is one hundred percent okay, however, **sharing them without someone that is not your partner**... or **someone in mutual agreement in a binding contract is wrong**! (>_<) The **fetishes** then it turns to **people asking to pay for private photos**... (@_@) I do not understand why or how **people are so certain that if they offer money, barter, trade, that a cosplayer would do anything**... Quote **how much for nudes, I will pay X amount for six sexy photos of you in this situation... Gosh it is exhausting**... (>_O) Still leads to **many thinking that cosplay is for narcissists**... **We cosplayers come in all kinds of** personality of shy, nice, rude, creepy... Meaning **we are not all the same**... Much like the **social divide** of the Earth and Mars war support, along with **augmented biotech humans of augs, and** natural **unmodified humans of** nats. (>_<) ...Changing topics... How to get **ready for cosplay photography**... Like any **convention center meet-ups on any continent**... the **undesirable photos** with **people's** Ai-**Aug devices. Photographers, amateur and professionals have this authorization to take photos without asking**! It is like **they are spying on us** and **do not accept any refusal of no, N.O. No does not exist in their definition and worlds**! (@_@) These are those **cringe situations**!! There were many times I **needed to step in to save a friend that was too sacred and afraid to say no**. (X_X) Many cosplayers I **know feel uncomfortable shooting with people they just do not know**... (°o°) I am **not that full of self-esteem**... (<_<) **Shooting a photography session with people** I **do not know makes me feel more insecure**, however, I **will not say "no" to a photo taken during conventions**... But to **spend thirty minutes**, three zero minutes, half an hour, to **shoot with a photography stranger**... (>_<) I believe that there should be **trust between cosplayer and photographer**, after all **we are not models**! Make sure to **know the difference between models and cosplayers**! Some **photographers are like paparazzi, that steal shoots of us, eating, retouching their beauty augs, make-up test**... I hate **taking a photo of someone without asking any permission, no common courtesy**! (*_*) ...I need to change topics again. Okay, **props is another factor to assist in making good photos**. Do not forget them! (O_O) Next topic, be sure to **treat, self-care your body**! Whether **you** are an **aug** or nat **person, we all have our human needs and limits**. ...**Drink enough water, not too much as it can** flatline **you**... Unless **you** got a **caramel back biotech augmentation**. ...Plus, **rest properly before the photography session**, because if **you** are tired, **you** can not give it **your** all and **your** best! (^_^) Last topic for now, **cosplay photography should or quote needs to be fun**! (^_^) No way in hel, I can imagine doing a shoot when I am always serious and professional. It is the **passion that makes the drive to make us cosplayers happy about it to travel outside of our nice comfortable biotech dome homes**... Okay, let **us** see how this A.I. **Sephiroth** cosplay **Holo HUD** going to work..."

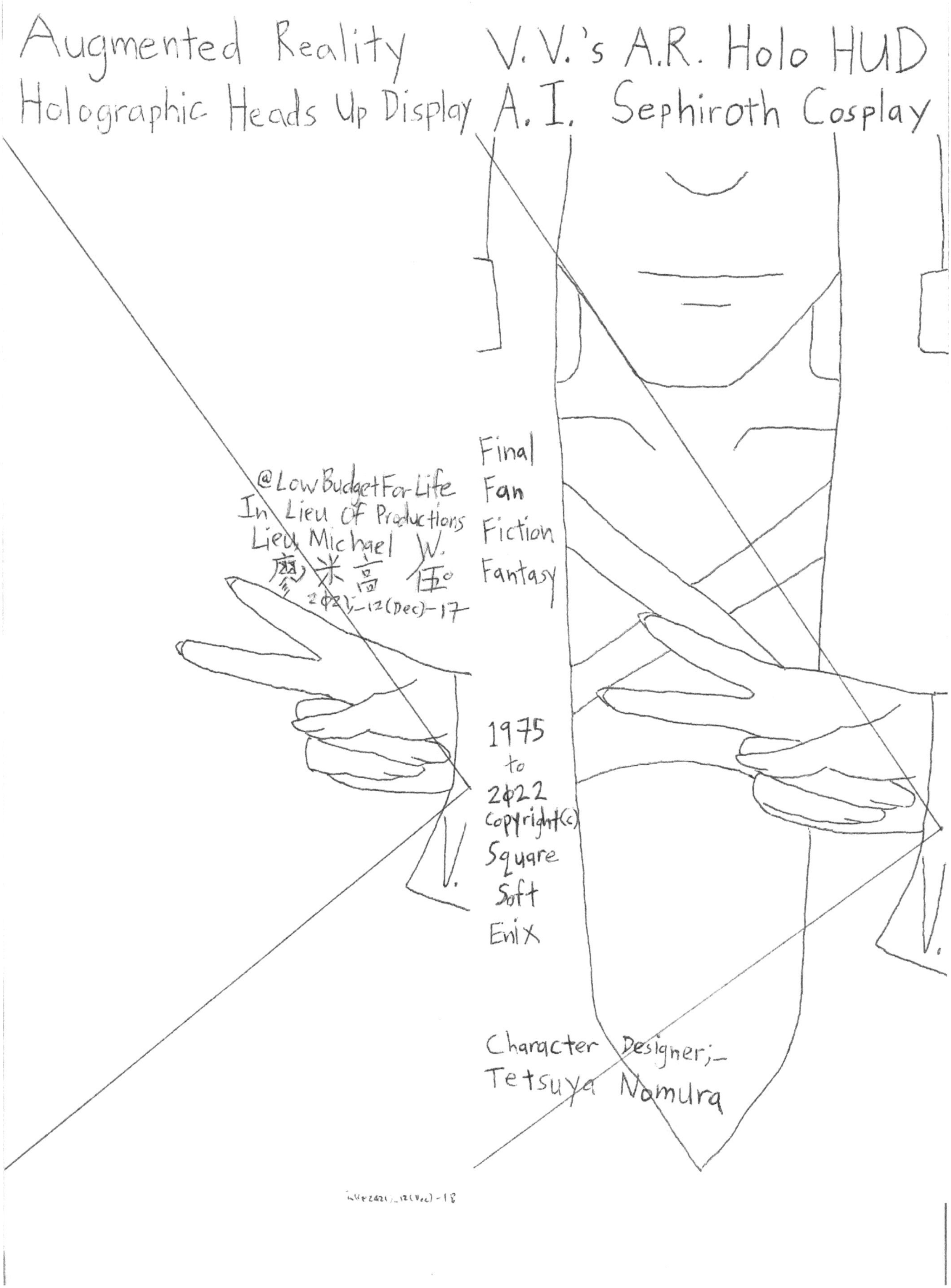
Augmented Reality
V.V.'s A.R. Holo HUD
Holographic Heads Up Display
A.I. Sephiroth Cosplay
@LowBudgetForLife
In Lieu Of Productions
Lieu Michael W.
廖 米高 伍o
2021_12(Dec)-17
Final
Fan
Fiction
Fantasy
1975
to
2022
Copyright(c)
Square
Soft
Enix
V.
V.
Character Designer;_
Tetsuya Nomura

Alien Valentina Vita, Traveling Cosplayer(s) And Kosplayer(s)
2154;_ Chicago, Illinois' Intranet, Wireless Broadcasting Frequencies Online

V. V. Augmented Reality Holographic Heads Up Display A.R. Holo HUD
Starting Up Artificial Intelligence (A.I.) Personality Matrix Simulation... Basic Input Output System (BIOS) Online... Loading Square Soft Enix's Sephiroth Character... Theme Music: "Final Fantasy 7 Advent Children Complete's One Winged Angel"...
[Aug+]
V. V. Augmented Reality Holographic Heads Up Display A.R. Holo HUD

2154;_ Chicago, Illinois' Intranet, Wireless Broadcasting Frequencies Online
Alien Valentina Vita, Traveling Cosplayer(s) And Kosplayer(s)

Valentina Vita (V. V.); Traveling Cosplayers And Kosplayers;
"(>_O) ...Okay, **running diagnose test of Sephiroth Holo HUD**... Looks good, from my bracket Ai-Aug... Let **us** see how his speech patterns and other behaviors say..."

Artificial Intelligence (A.I.) Sephiroth Holo HUD; **Traveling Cosplayers And Kosplayers**;
"(*monotone male voice over*) ...Hm...? ...Impossible... I will not ever be a **dream**... I will not ever be a **nightmare**. I will not ever be a **vision**..."

Valentina Vita (V. V.); Traveling Cosplayers And Kosplayers;
"(X_X) ...Hmm... **Work in progress still**, still have to do some **tweaks...** Better **done and completed** than **perfected and undone...** Time to test this out..."

Alien Valentina Vita, Traveling Cosplayer(s) And Kosplayer(s)
2154;_ Chicago, Illinois' Intranet, Wireless Broadcasting Frequencies Online

V. V. Augmented Reality Holographic Heads Up Display A.R. Holo HUD	
Holo Photo Recording Mode	
A.R. Holo HUD V. V.	V. V.'s A.R. Holo HUD A.I. Sephiroth Cosplay
[Aug+]	
V. V. Augmented Reality Holographic Heads Up Display A.R. Holo HUD	

2154;_ Chicago, Illinois' Intranet, Wireless Broadcasting Frequencies Online
Alien Valentina Vita, Traveling Cosplayer(s) And Kosplayer(s)

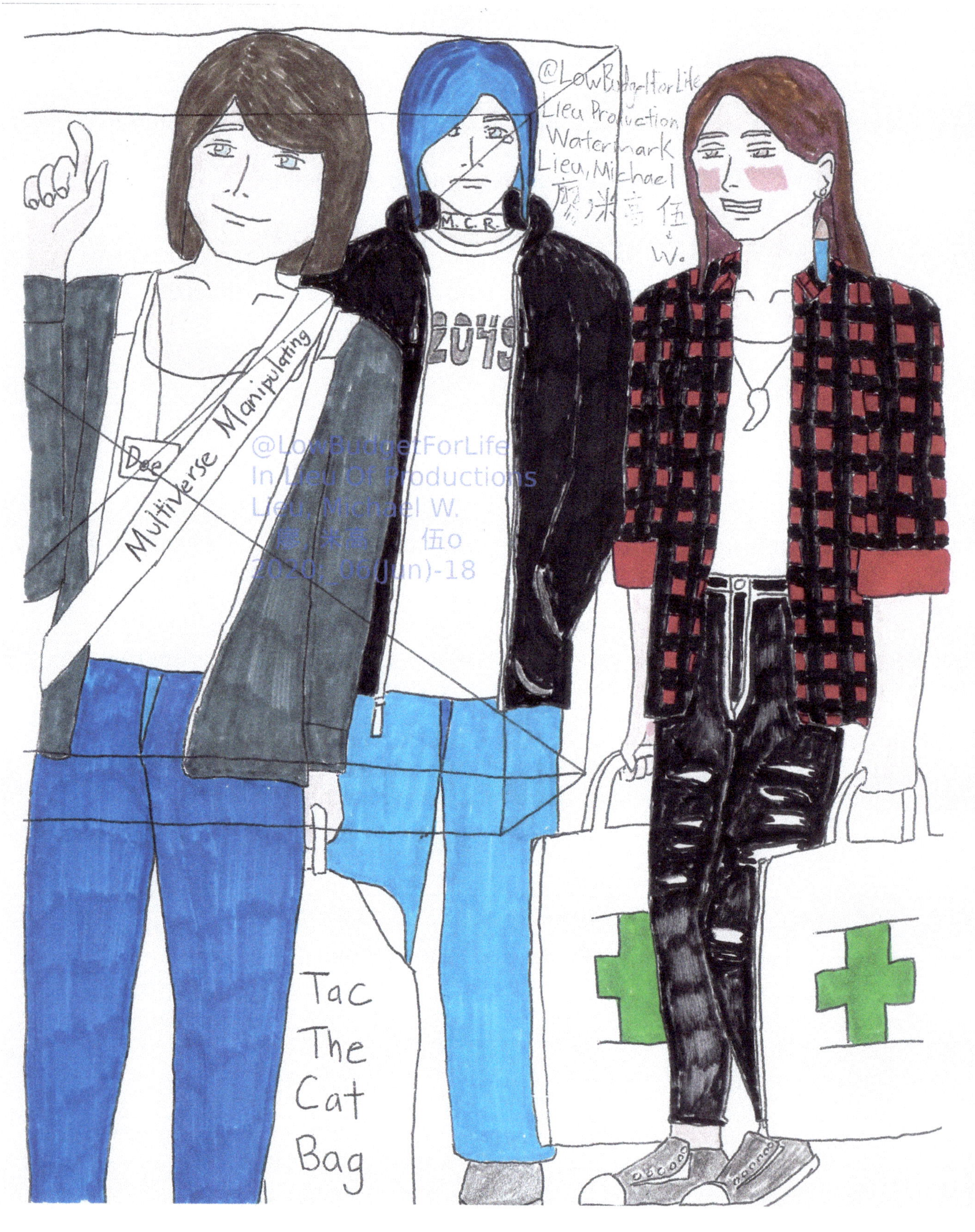
@LowBudgetForLife
Lieu Production
Watermark
Lieu, Michael
廖米高伍
W.
M.C.R.
2049
Doe
Multiverse Manipulating
@LowBudgetForLife
In Lieu Of Productions
Lieu, Michael W.
廖, 米高　伍o
2020_06(Jun)-18
Tac
The
Cat
Bag

Alien Valentina Vita, Traveling Cosplayers And Kosplayers
2154;_ Chicago, Illinois' Intranet, Wireless Broadcasting Frequencies Online

V. V. Augmented Reality Holographic Heads Up Display A.R. Holo HUD		
Actual Legal Given Names;_ Max Doe Cosplay (goes by Max) pronounced, "ma-axe' 'da-oh' 'kos-play"	Actual Legal Given Names;_ **Chloe Priceless Cosplay** (goes by **Chole**) pronounced, "**ka-lo-ee' 'pooh-rice-les' 'kos-play**"	Actual Legal Given Names;_ Raychel Amber-Alert Kosplay (goes by Raychel) pronounced, "ray-choe' am-ber-a-lert' 'cos-play"
Physical Plus Psych Age;_ augmented age Question-Mark	Physical Plus Psych Age;_ natural 20 + 22 years old	Physical Plus Psych Age;_ natural 20 + 25 years old
Alias Network Nickname;_ Time Girl, Time Traveling Girl, DOE Doe doe, [*Aug+*]	Alias Network Nickname;_ Blue Hair Girl, Blue's Clues, Blue Tom Boy, [*Aug+*]	Alias Network Nickname;_ Red Shirt Girl, Red Smoking Amber-Alert, Smokey Raychel, [*Aug+*]
Employment;_ **MultiVerse Manipulating**	Employment;_ Independent Contractor	Employment;_ Self-Employed
Community Network(s);_ Blade Runners and retirement M.C.R.	Community Network(s);_ Blade Runners and retirement M.C.R.	Community Network(s);_ Blade Runners and retirement M.C.R.
Citizen;_ United **States** of America (U.**S**. of A.)	Citizen;_ United **States** of America (U.**S**. of A.)	Citizen;_ United **States** of America (U.**S**. of A.)
[*Aug+*]		
V. V. Augmented Reality Holographic Heads Up Display A.R. Holo HUD		

2154;_ Chicago, Illinois' Intranet, Wireless Broadcasting Frequencies Online
Alien Valentina Vita, Traveling Cosplayers And Kosplayers

2154;_10(Oct)-31(thirty one)

Travel To Western Society, America(s), North America
Self-Govern Sovereignty (S),
Illinois (ILL) Sector, Points Of Interest (P.O.I.)

Chicago (S), "**Animal** Sanctuary"
"The New World 2121 Movement Of The 22nd Century,
Utopian Sovereignty of North America (U.S. of N.A.)"
South Side, ChIraq's LAN Party Halloween Costume Play Shop

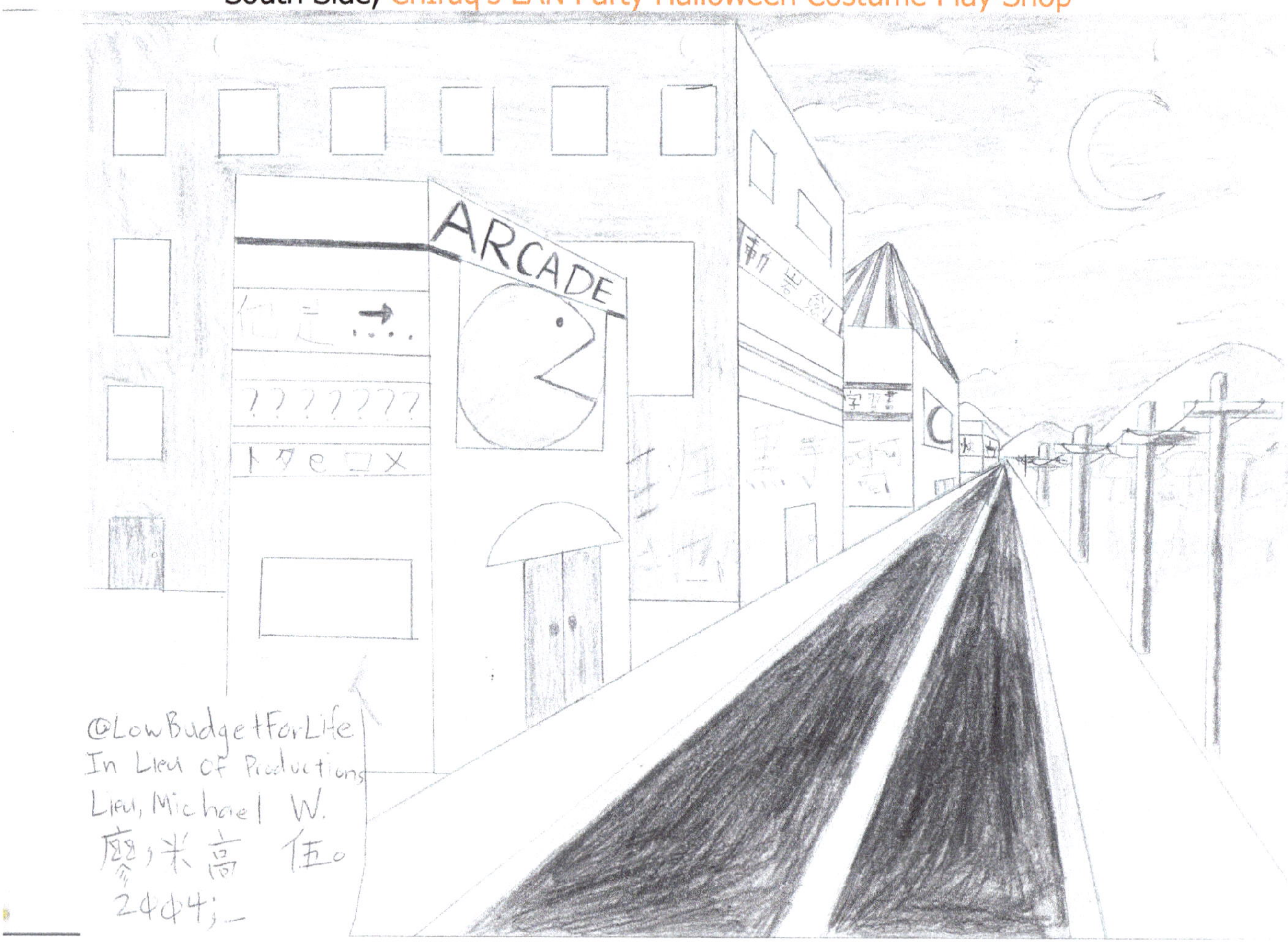

Valentina Vita (V. V.); Traveling Cosplayers And Kosplayers;
"(O_<) Back to... **South Side...** the entrance... gonna do a photo shoot with my KissShot and **Sephiroth cosplay**... **Interesting buildings** Chicago **Animal** Sanctuary has... **We** do not have these... **buildings** in Naples **Human** Sanctuary... Hm..."

Valentina Vita (V. V.); Traveling Cosplayers And Kosplayers;
"(^_^) I got some time before **Mister Steven Z. J.** arrives, let us see what kind of **special shoots** I can do with **these cosplay...** Time to start the preparation work."

Artificial Intelligence (A.I.) Sephiroth Holo HUD; **Traveling Cosplayers And Kosplayers**;
"(*monotone male voice over*) ...Hm...?"

Alien Valentina Vita, Traveling Cosplayer(s) And Kosplayer(s)
2154;_ Chicago, Illinois' Intranet, Wireless Broadcasting Frequencies Online

<table>
<tr><td colspan="2">V. V. Augmented Reality Holographic Heads Up Display A.R. Holo HUD</td></tr>
<tr><td colspan="2">Holo Photo Recording Mode</td></tr>
<tr><td>A.R. Holo HUD
V. V.</td><td>V. V.'s A.R. Holo HUD
A.I. Sephiroth Cosplay</td></tr>
<tr><td colspan="2">[<u>*Aug+*</u>]</td></tr>
<tr><td colspan="2">V. V. Augmented Reality Holographic Heads Up Display A.R. Holo HUD</td></tr>
</table>

2154;_ Chicago, Illinois' Intranet, Wireless Broadcasting Frequencies Online
Alien Valentina Vita, Traveling Cosplayer(s) And Kosplayer(s)

Valentina Vita (V. V.); Traveling Cosplayers And Kosplayers;
"(@_@) Okay, this is the **recording shot and position**... I am aiming for..."

(left) Sephiroth + **KissShot-AcerolaOrion-HeartUnderBlade (right)**

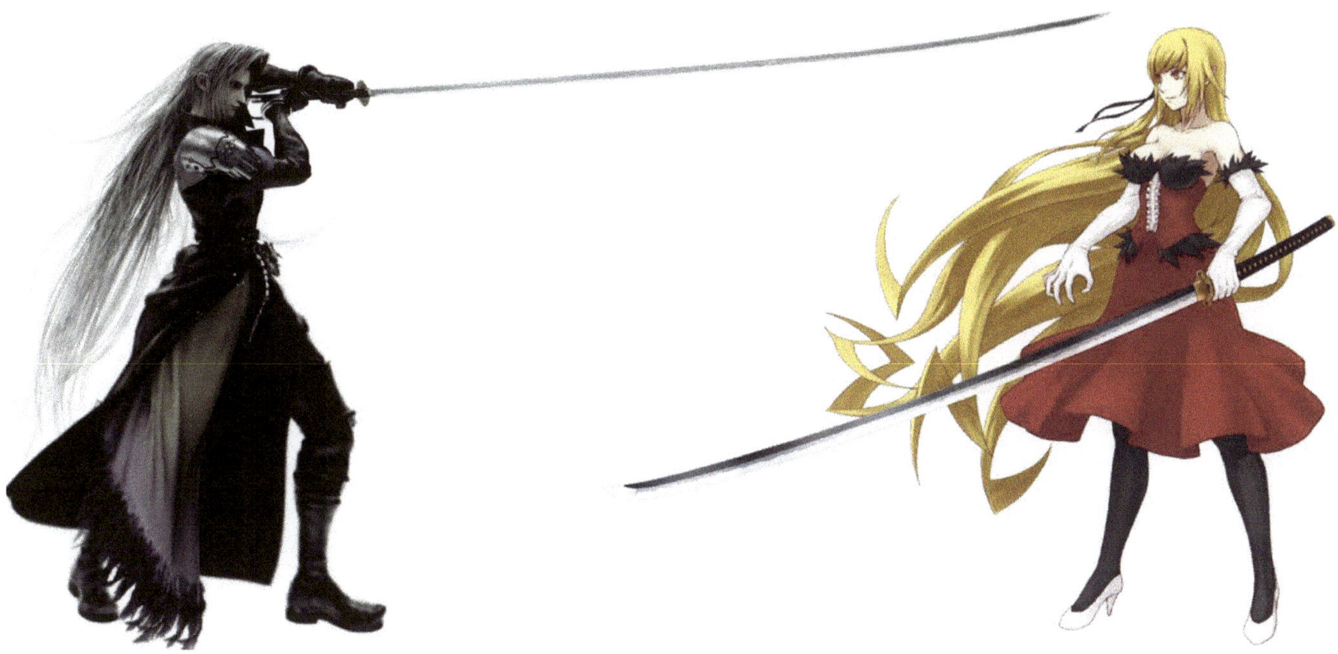

2154;_Arcadia \ Valentina Vita (V. V.) 'Healthy-Strong Life' \ Left to Right(L->R)

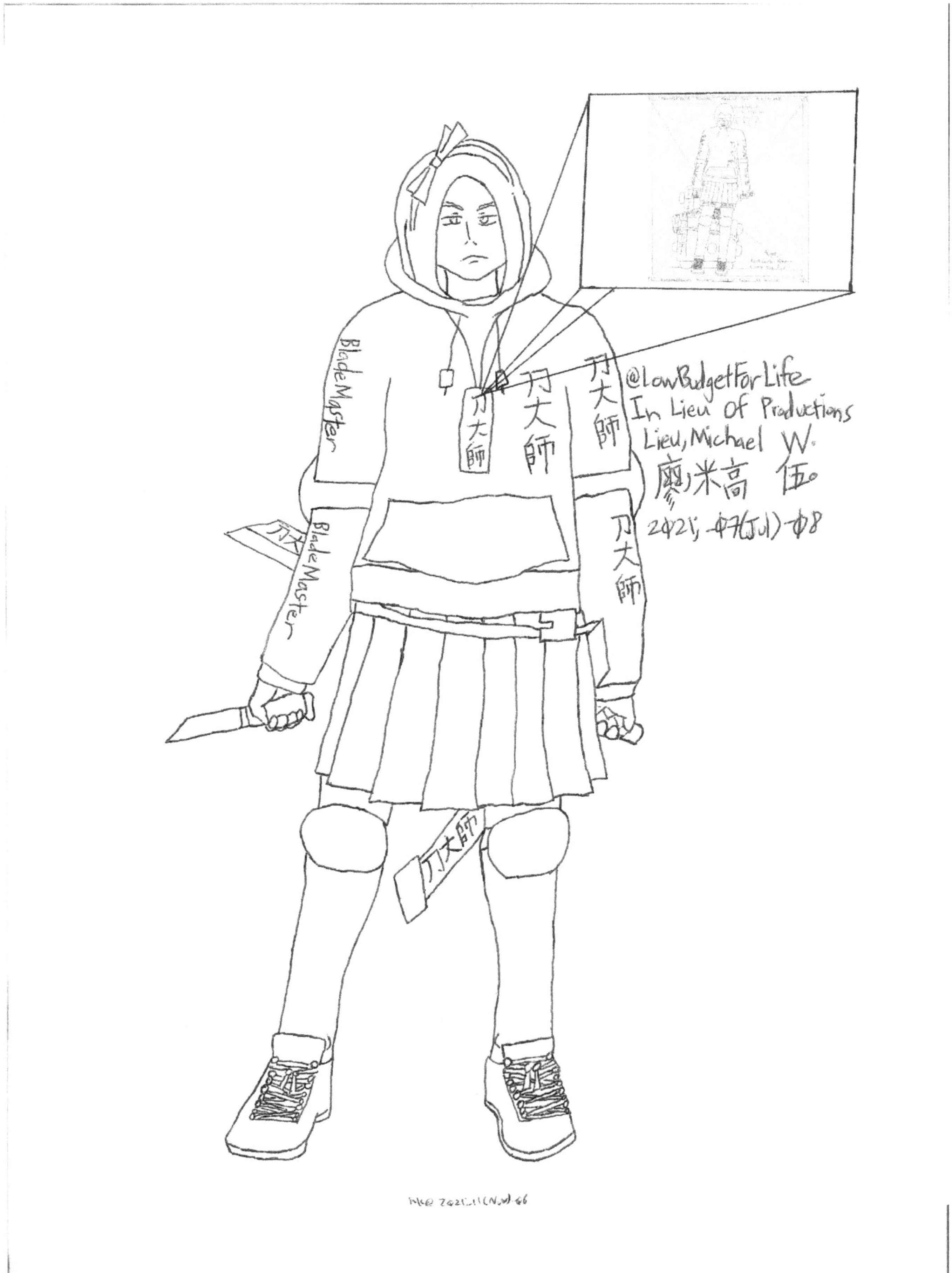
BladeMaster
BladeMaster
刀大師
刀大師
刀大師
刀大師
刀大師
@LowBudgetForLife
In Lieu Of Productions
Lieu, Michael W.
廖,米高 伍。
2021-07(Jul)-08

V. V. sees and notices Elle as she arrives via **biotech ziplines**. Elle approaches her closely as V. V. recalls she was one of the previous persons waiting in line to get into the Chicago **Animal** Sanctuary's **Elysium Corporation biotech** dome **airlock decontamination**.

Alien Valentina Vita, Traveling Cosplayer(s) And Kosplayer(s)
2154;_ Chicago, Illinois' Intranet, Wireless Broadcasting Frequencies Online

<table>
<tr><td colspan="2">V. V. Augmented Reality Holographic Heads Up Display A.R. Holo HUD</td></tr>
<tr><td colspan="2">Holo Photo Recording Mode</td></tr>
<tr><td>A.R. Holo HUD
V. V.</td><td>V. V.'s A.R. Holo HUD
A.I. Sephiroth Cosplay</td></tr>
<tr><td colspan="2">[<u>Aug+</u>]</td></tr>
<tr><td>Actual Legal Given Name;_

Elle K. Mcgee (goes by Elle)
pronounced, 'el-lee' 'kay' 'mick-gee'</td><td>Physical Plus Psych Age;_

augmented 17 + 2 years old</td></tr>
<tr><td colspan="2">[<u>Aug+</u>]</td></tr>
<tr><td colspan="2">V. V. Augmented Reality Holographic Heads Up Display A.R. Holo HUD</td></tr>
</table>

2154;_ Chicago, Illinois' Intranet, Wireless Broadcasting Frequencies Online
Alien Valentina Vita, Traveling Cosplayer(s) And Kosplayer(s)

V. V. stays focused, selects the music "**Final Fantasy 7 Advent Children Complete's One Winged Angel**" to play on her Ai-**Aug** V. V. **bracelet** and is set on loop.

Artificial Intelligence (A.I.) Sephiroth Holo HUD; **Traveling Cosplayers And Kosplayers**;
"(*monotone male voice over*) ...Hmm...?"

Valentina Vita (V. V.); Traveling Cosplayers And Kosplayers;
"(>_<) **Sephiroth**, **you** will live together with **me**."

Artificial Intelligence (A.I.) Sephiroth Holo HUD; **Traveling Cosplayers And Kosplayers**;
"(*monotone male voice over*) ...Impossible..."

Valentina Vita (V. V.); Traveling Cosplayers And Kosplayers;
"(T_T) ...**Demon Blade Kokorowatari**! No... **You** are not the **Sephiroth** I once knew!"

V. V. stays in character as her A.I. **Sephiroth Cosplay Holo HUD** draws out a holographic **long single edge cutting tool blade** and touches her **four limbs**. V. V.'s limbs react by activating her **biotech augmentations** of the **chameleon toner shifting** and **electro magnetic tubes** to make her **four limbs** look like it disappears when being cut.

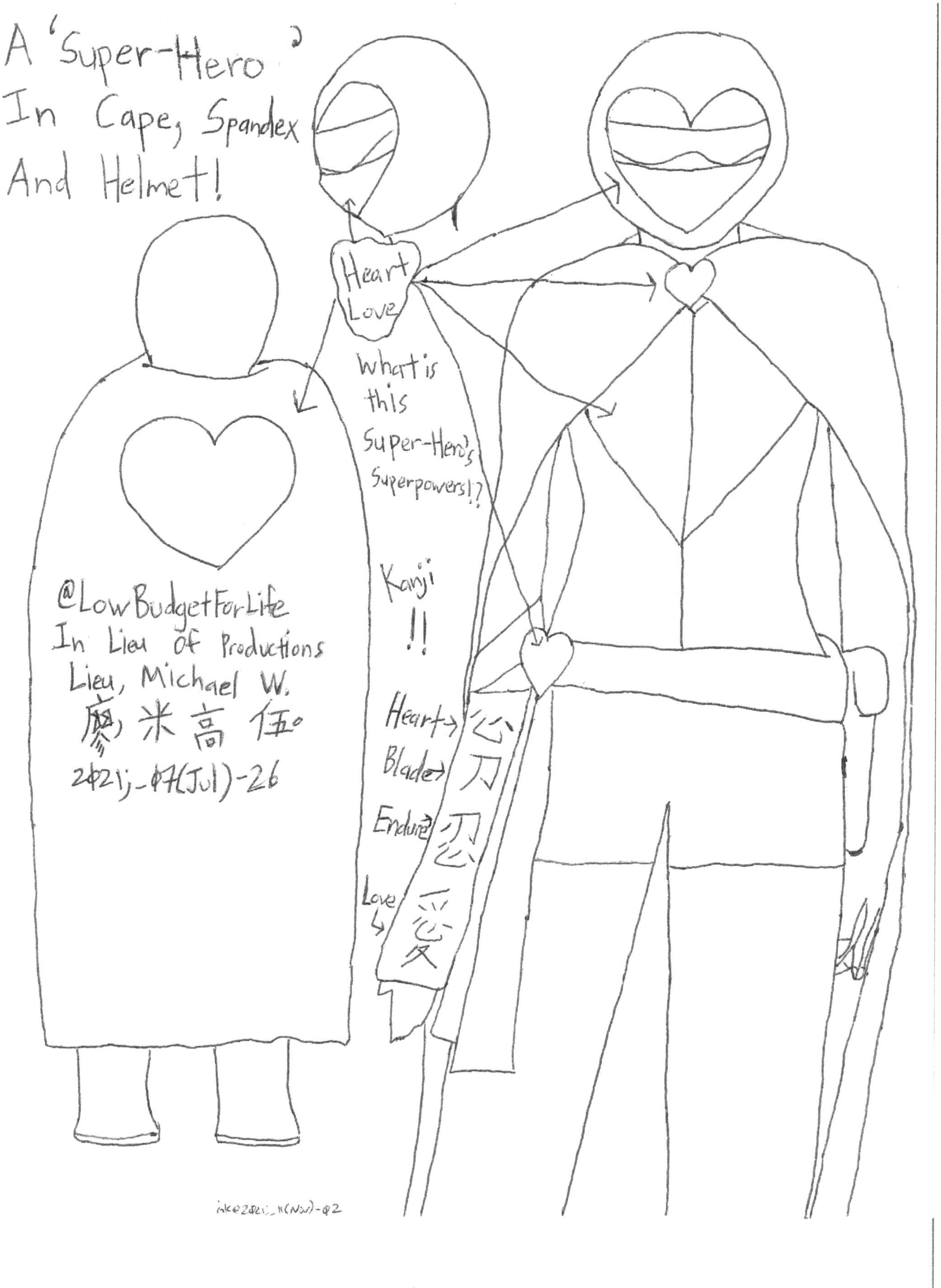
A 'Super-Hero' In Cape, Spandex And Helmet!
Heart Love
What is this Super-Hero's Superpowers!?
Kanji !!
Heart→ 心
Blade→ 刀
Endure→ 忍
Love → 愛
@LowBudgetForLife
In Lieu Of Productions
Lieu, Michael W.
廖, 米高 伍o
2021j-07(Jul)-26

V. V. sees that Elle takes off a few moments later without saying anything and V. V. continues with her **KissShot-AcerolaOrion-HeartUnderBlade** and A.I. **Sephiroth cosplay holo photo shoot** for hours without much interruption.

Before the day ends, V. V. sees another person(s) arriving in a white yellow full body P.P.E. biotech skintight suit with heart love shape symbols on her cape and front helmet.

Valentina Vita (V. V.); Traveling Cosplayers And Kosplayers;
"(=_=) ...I feel sleepy... (*yawning*) Sleep... Uh..."

HeartBladeEndureLove 心刀忍愛; Pretty Sure This Is Not That Nun Sister K. White <3;
"(*in suit*) **Parkour**!! Oh hi V. V.! I am here to rescue you... Can explain later! Right now **we** need to **depart from this location**, it is **not safe to stay here... Parkour**!!"

Alien Valentina Vita, Traveling Cosplayer(s) And Kosplayer(s)
2154;_ Chicago, Illinois' Intranet, Wireless Broadcasting Frequencies Online

V. V. Augmented Reality Holographic Heads Up Display A.R. Holo HUD	
Holo Photo Recording Mode	
A.R. Holo HUD V. V.	V. V.'s A.R. Holo HUD A.I. Sephiroth Cosplay
[*Aug+*]	
Actual Legal Given Names;_ **K. White** (goes by **Sister White <3**) pronounced, "**sis-ter ka-ay wa-ai-te**"	Alias Network Nickname;_ HeartBladeEndureLove, Yelling Loud Commands, Heart Love Hand Shape, Super-Woman, **Free Hugs**, Spandex-Woman, A Nun, Female Klark Kent, [*Aug+*]
Family Member(s);_ **Con-Cat/Kat** **A-Rag Otaku** **Hon-Knee Kawaii** **KIPI Kosplay Kokoro** To Heart And Love	Love(s) Or Like(s);_ Love, Love is the frequency Love, the frequency of attachment. **Superman 64** [*Aug+*]
[*Aug+*]	
V. V. Augmented Reality Holographic Heads Up Display A.R. Holo HUD	

2154;_ Chicago, Illinois' Intranet, Wireless Broadcasting Frequencies Online
Alien Valentina Vita, Traveling Cosplayer(s) And Kosplayer(s)

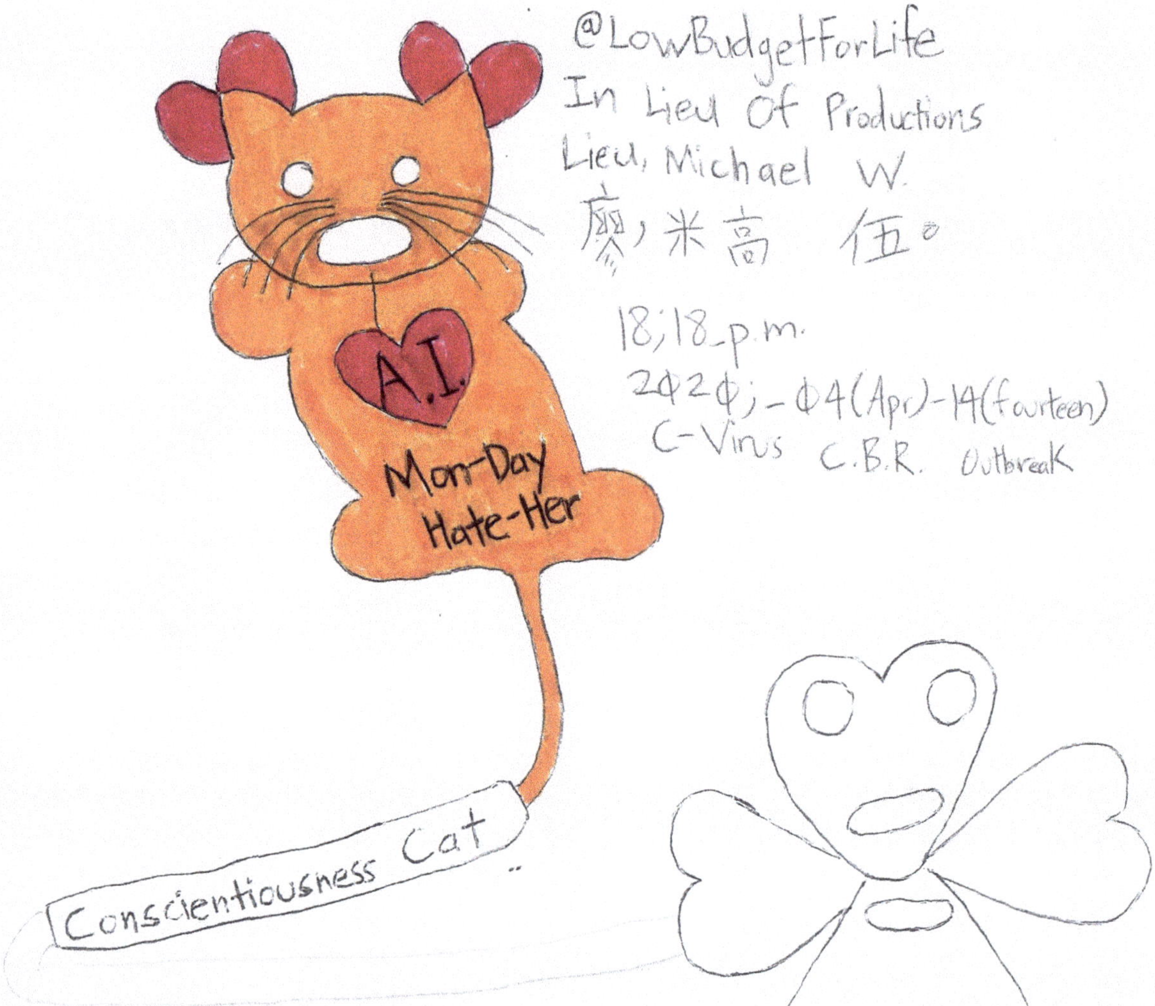
@LowBudgetForLife
In Lieu Of Productions
Lieu, Michael W.
廖, 米高　伍。
18;18 p.m.
2Φ2Φ;-Φ4(Apr)-14(fourteen)
C-Virus C.B.R. Outbreak
A.I.
Mon-Day
Hate-Her
Conscientiousness Cat
P P

????;_??(???)-??(?) { Act 2(two) story forward } **inciting incident**

?

Conscientiousness Cat/Kat's XXXX; Pussy Cat That Searches For Magical Maidens;
"(O_O) Meow. Meow! Me! Ow!"

Valentina Vita (V. V.); Traveling Cosplayers And Kosplayers;
"(Z_Z) Huh... Meow, Meow? An orange cat stuff animal... **Mister Steven Z. J. Entertainer**? Some kind of dream...?"

Conscientiousness Cat/Kat's XXXX; Pussy Cat That Searches For Magical Maidens;
"(O_O) Meow. Meow! Me! Ow!"

Valentina Vita (V. V.); Traveling Cosplayers And Kosplayers;
"(Z_Z) Hm... Meow, Meow? Meow, Meow Meow!?"

Conscientiousness Cat/Kat's XXXX; Pussy Cat That Searches For Magical Maidens;
"(O_O) Meow. Meow! Me! Ow!"

Valentina Vita (V. V.); Traveling Cosplayers And Kosplayers;
"(Z_Z) ...Huh... Meow, Meow? Me Ow, Ow Me!! What I am even saying!?"

Conscientiousness Cat/Kat's XXXX; Pussy Cat That Searches For Magical Maidens;
"(O_O) Meow. Meow! Me! Ow!"

Valentina Vita (V. V.); Traveling Cosplayers And Kosplayers;
"(Z_Z) So, how is it that I can understand you!? As you just Meow, and Meow?"

Conscientiousness Cat/Kat's XXXX; Pussy Cat That Searches For Magical Maidens;
"(O_O) Meow. Meow! Me! Ow!"

Valentina Vita (V. V.); Traveling Cosplayers And Kosplayers;
"(Z_Z) This is the weirdest dreams of weird dreams... Then again dreams are weird."

Conscientiousness Cat/Kat's XXXX; Pussy Cat That Searches For Magical Maidens;
"(O_O) Meow. Meow! Me! Ow!"

V. V. begins to **wake up** from an **unexpected sleep** during her **KissShot-AcerolaOrion-HeartUnderBlade** and A.I. **Sephiroth Cosplay Holo HUD photo shoot**.

2154;_Arcadia \ Valentina Vita (V. V.) 'Healthy-Strong Life' \ Left to Right(L->R)

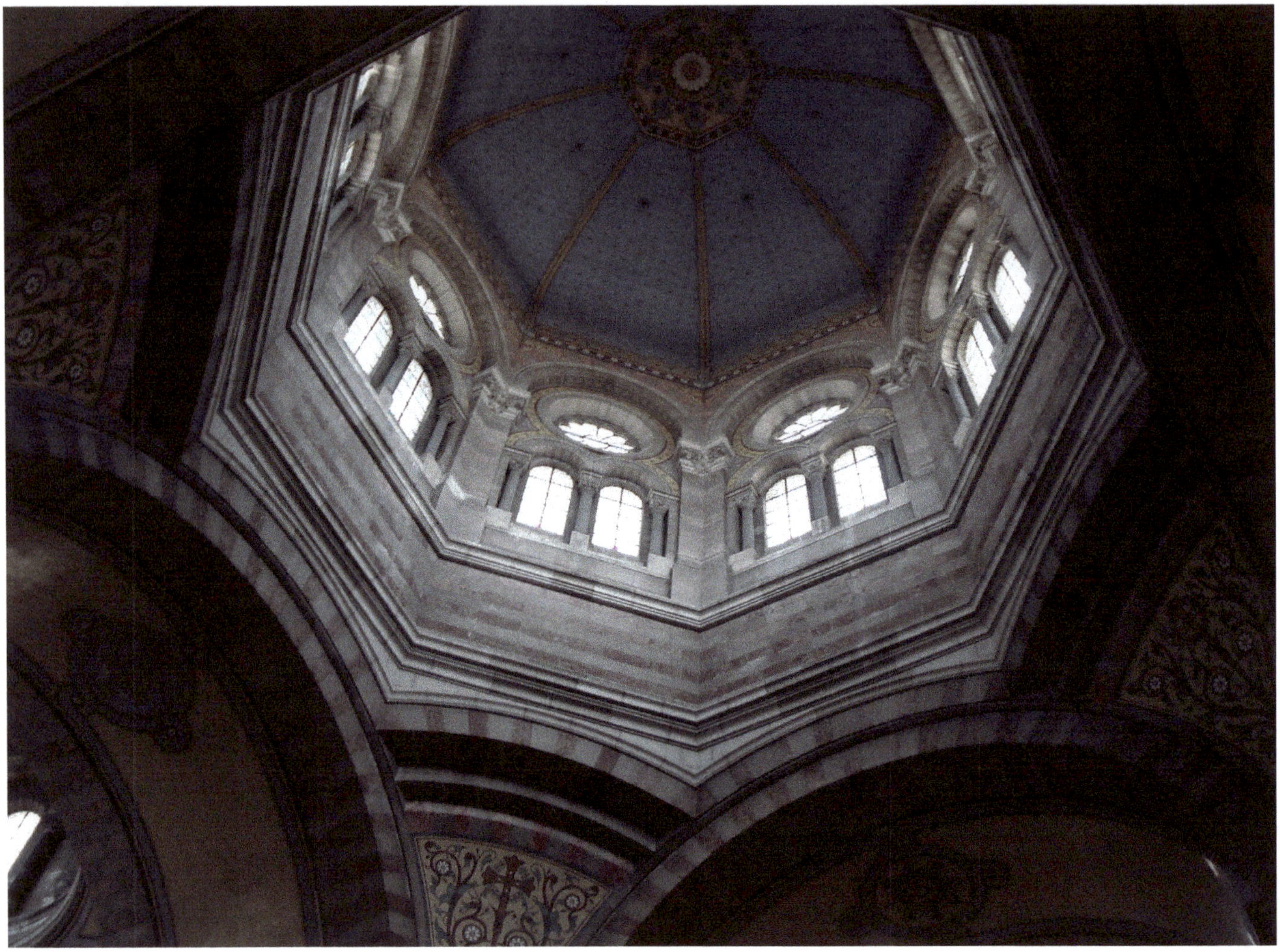

Sister K. White <3; Pretty Sure This is Not HeartBladeEndureLove 心刀忍愛;
"(*voice over*) ...Hey, are you okay? You been asleep for some time, not too long..."

Valentina Vita (V. V.); Traveling Cosplayers And Kosplayers;
"(=_=) ...Uh... Where is this... place, a church...? Where I am? (*yawning*) ...Ugh..."

Sister K. White <3; Pretty Sure This is Not HeartBladeEndureLove 心刀忍愛;
"(*voice over*) ...Yes, this is the West Side, Chicago City Christian Church... You are safe here, you have been **sleeping for half day, twelve hours**... today is **November one**, the **year is still twenty one fifty four**... The **time is almost ten thirty morning**..."

Valentina Vita (V. V.); Traveling Cosplayers And Kosplayers;
"(=_=) ...Um... I do feel a bit rested, so... what happened? (>_<) Huh... My Ai-**Aug** is not working. Did **something happen** within Chicago **Animal** Sanctuary?"

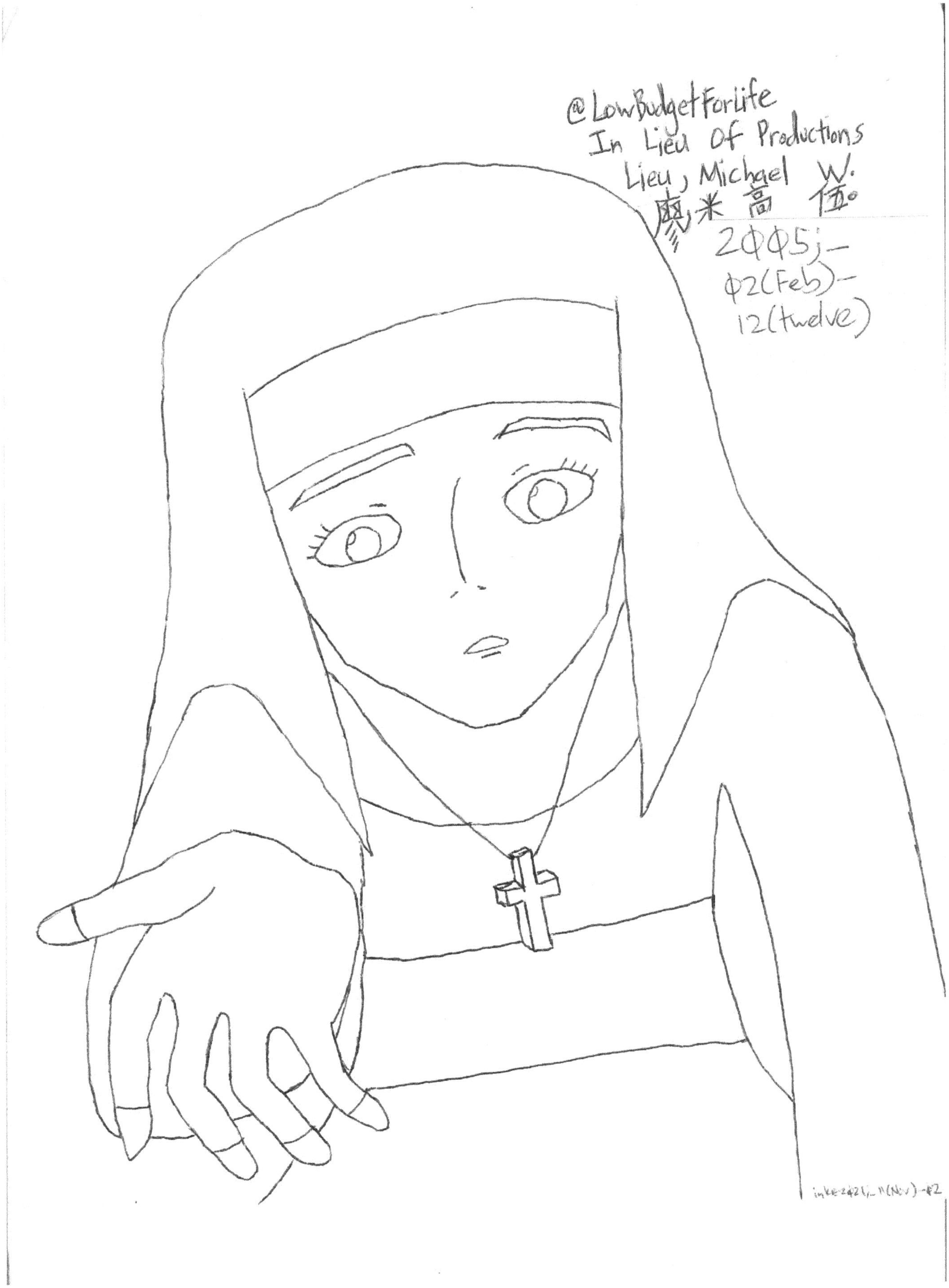
@LowBudgetForLife
In Lieu Of Productions
Lieu, Michael W.
廖米高 伍o
2φφ5j-
φ2(Feb)-
12(twelve)

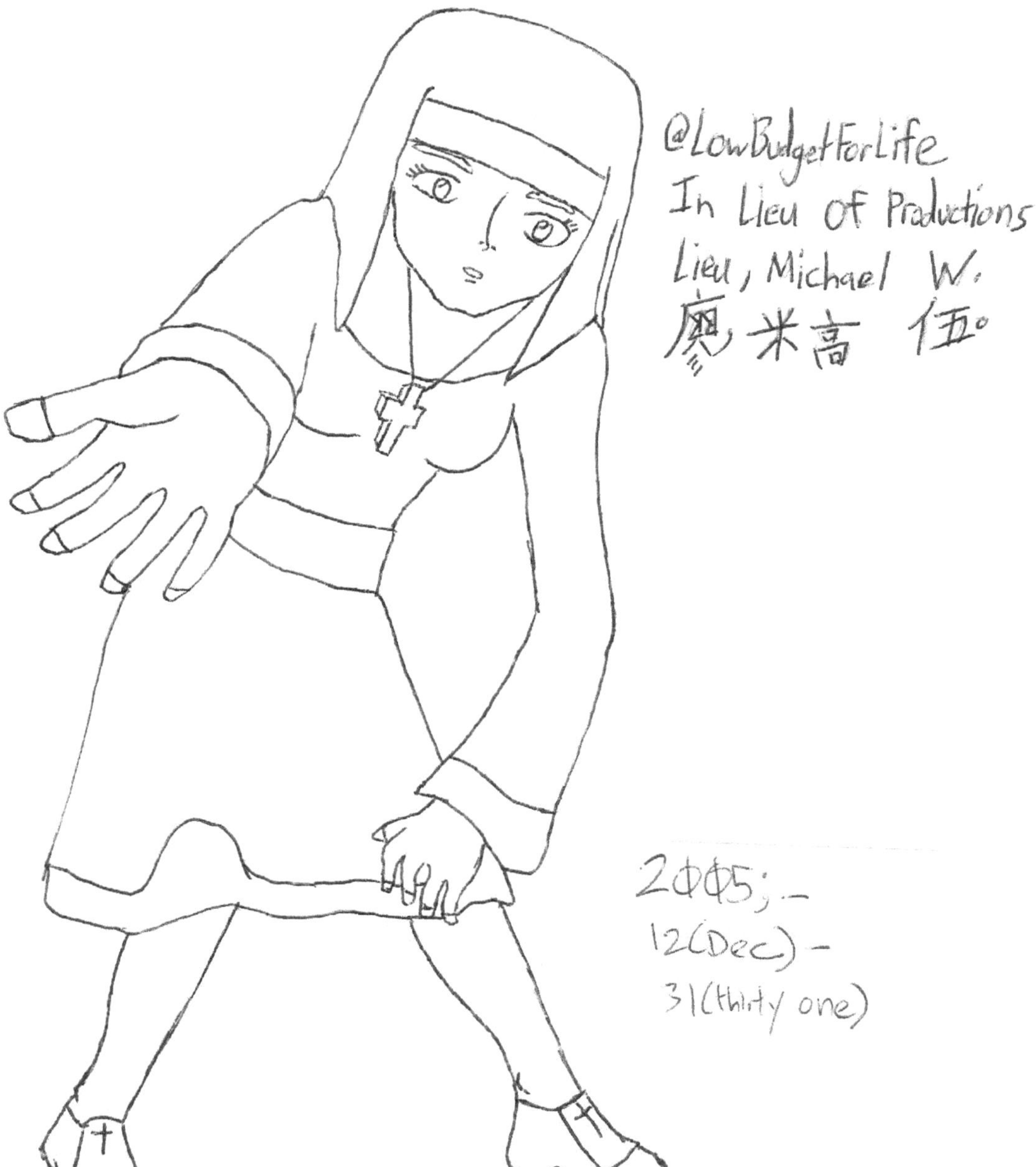
@LowBudgetForLife
In Lieu Of Productions
Lieu, Michael W.
廖 米高 伍
2ΦΦ5;_
12(Dec)-
31(thirty one)

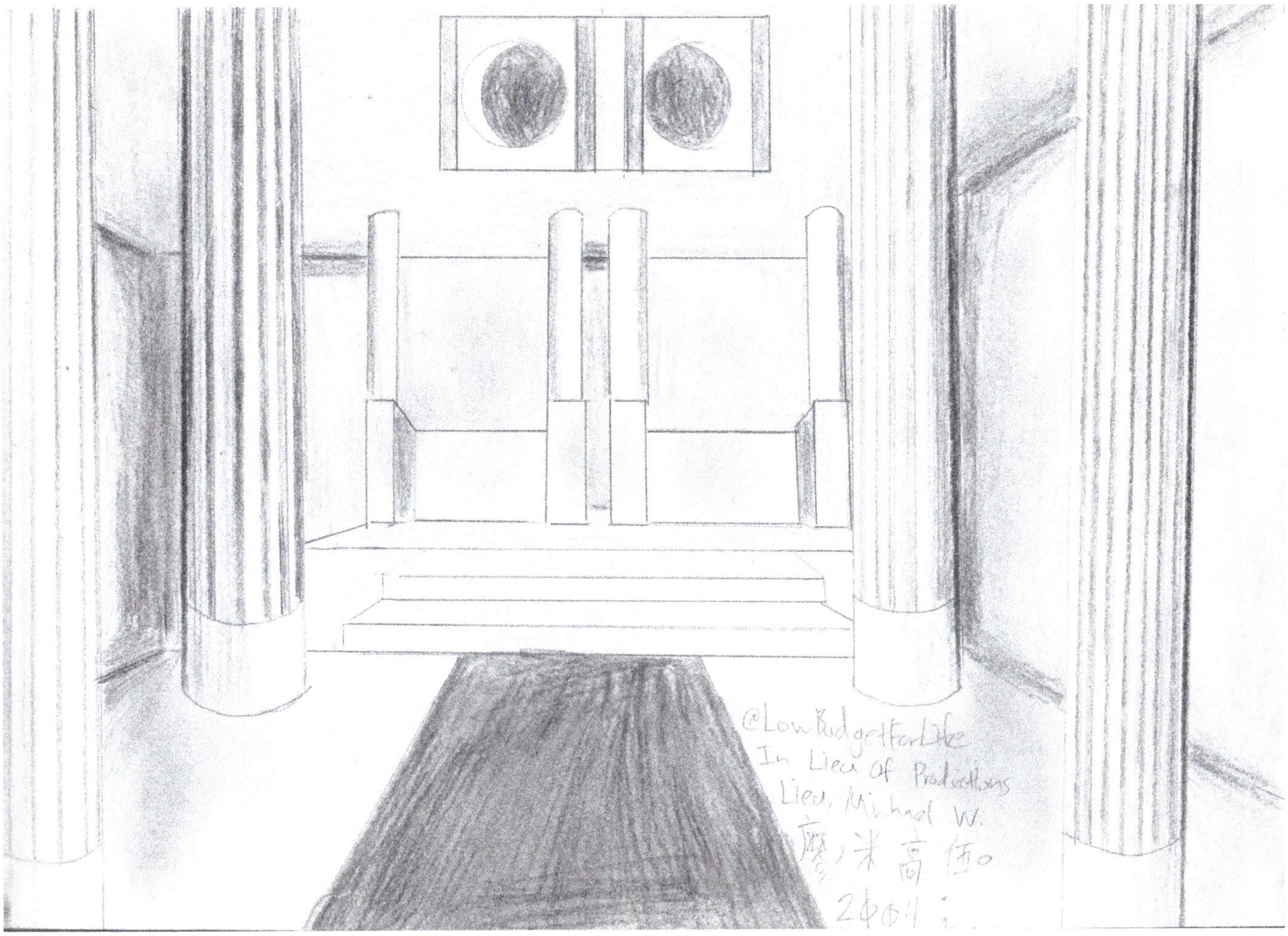

Sister K. White <3; Pretty Sure This is Not HeartBladeEndureLove 心刀忍愛;
"(*calm*) ...Okay, so there has been an **outbreaks and outages of a certain unconventional research and development experiment** from **Kay's Private Security Services** under this biotech dome... This **outbreak outage experiment only seems to affect humans, augmented and** natural, and is related to the ongoing War On Mars... **Our acting authority volunteer workers have arrested the Agent responsible, he** is being **detained under the Church** in accordance to the separation of Church And State belief system. So, that is pretty much what happened while you were **sleeping half day**."

Valentina Vita (V. V.); Traveling Cosplayers And Kosplayers;
"(@_@) ...Oh... Oh, my Ai-**Aug**, should have really paid attention to that army soldier officer giving me the **information about outbreaks outages**... (>_<) Dumb me!! Oh..."

Sister K. White <3; Pretty Sure This is Not HeartBladeEndureLove 心刀忍愛;
"(*calm*) It's okay, your **augments are still working**, the **outbreaks outages** was not that damaging, some **animals were a bit spooked... I** know **someone** who can help you."

2154;_Arcadia \ Valentina Vita (V. V.) 'Healthy-Strong Life' \ Left to Right(L->R)

Belief Systems, Abstracted Human Perspective Survey

Do You Believe In The Following: ('**Y for Yes**', '**N for No**', '**? for Maybe**')

1. Unidentified Flying Objects (U.F.O.), Roswell New Mexico			
2. Heaven Or The Heavens, Hell Or The Hells, Hel			
3. Extraterrestrial Space Alien Abductions			
4. Angels And Demons, Good And Bad, Right And Wrong			
5. Extra Sensory Perception (E.S.P.), Clairvoyance, Psi			
6. Evolution, Darwinism, Natural Selection			
7. Astrology, Fortune Telling, Magic 8 Ball, Tarot			
8. Democracy (D), Democratic (D) Republic (R)			
9. Gravity, Gravitational Attraction, Gravitation, Graveness			
10. Urban Legends, 2036;John Titor Time Traveling Backwards			
11. Creationism, Biotech Engineering			
12. Cryptozoology: Bigfoot, Loch Ness Monster, Chupacabra			
13. Lord Jesus Christ, A.K.A. Jesus As God			
14. Lost Civilizations: Atlantis, Lemuria, Mu, El Dorado			
15. Truth, The Truth, Low Case T "truth"			
16. Homeopathy, Acupuncture, Stem Cell, Kloning			
17. Cults, Fringe Religions: Scientology, Heaven's Gate, Mormon			
18. Undead, Zombies, Ghosts, Vampires, Werewolves, SuperNat			
19. The Bible As The Literal Truth			
20. The After Life			
21. The Human Soul			
22. Quantum Physics, Quantum Mechanics			
23. Communication With The: Deceased, Dead			

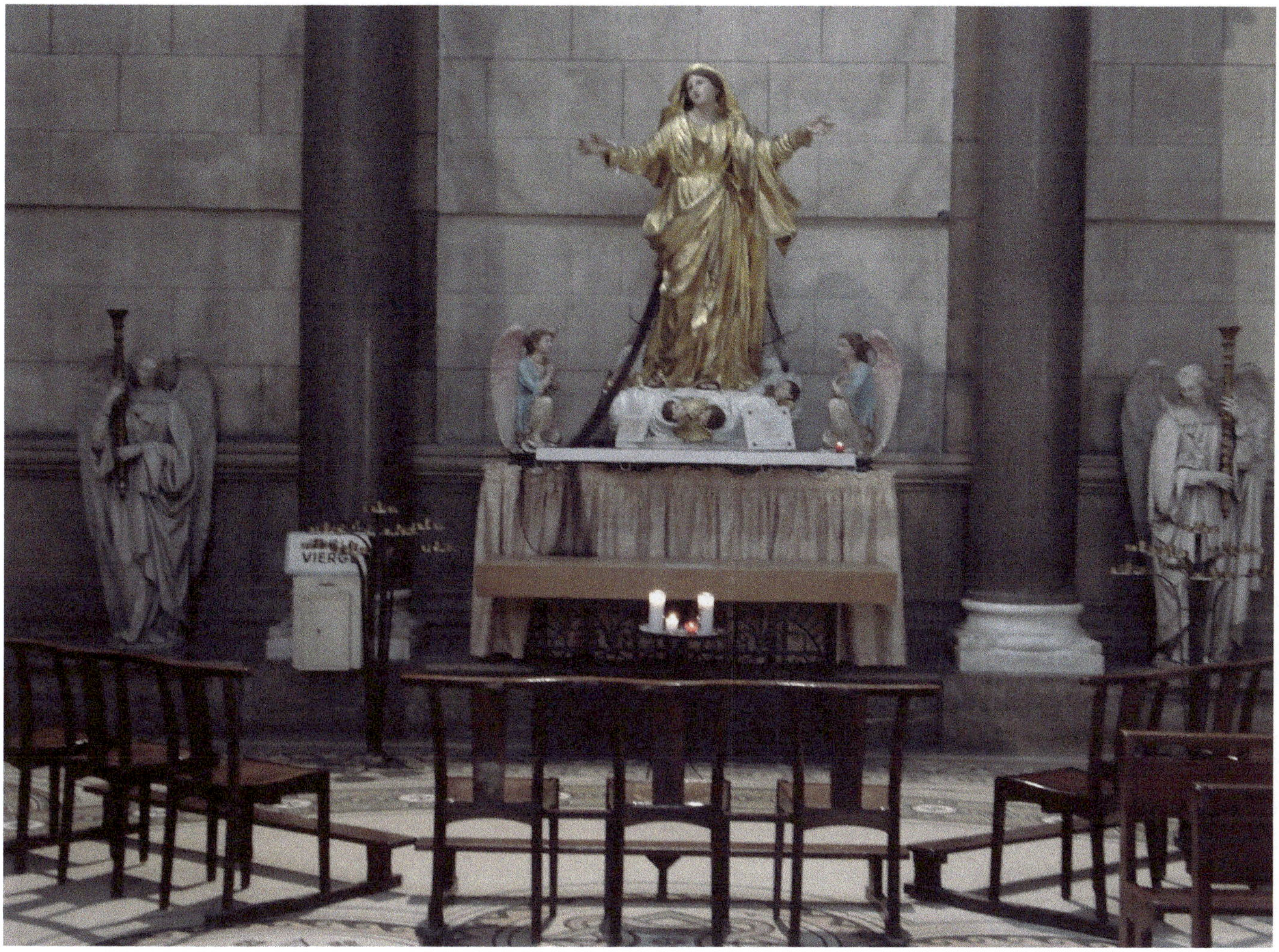

Belief Systems, Religion: The Oath

Higher Power	Dogma: Belief, Rules	Faith	Teachings
Book	Prophets	Priests	Gatherings
Structures / Temples / Churches		Prayer / Offering	
Carrot And Stick = Heaven And Hell		Mythology, Adam And Eve, Lord Jesus Christ	

Valentina Vita (V. V.); Traveling Cosplayers And Kosplayers;
"(O_@) Hmm. Is this a **survey**? Whoa we wow, interesting… **Someone** who can help?"

Sister K. White <3; Pretty Sure This is Not HeartBladeEndureLove 心刀忍愛;
"(*voice over*) Yes, **our volunteer manager**, **Lieutenant Lieu**, can help fix your Ai-**Aug**!"

2154;_Arcadia \ Valentina Vita (V. V.) 'Healthy-Strong Life' \ Left to Right(L->R)

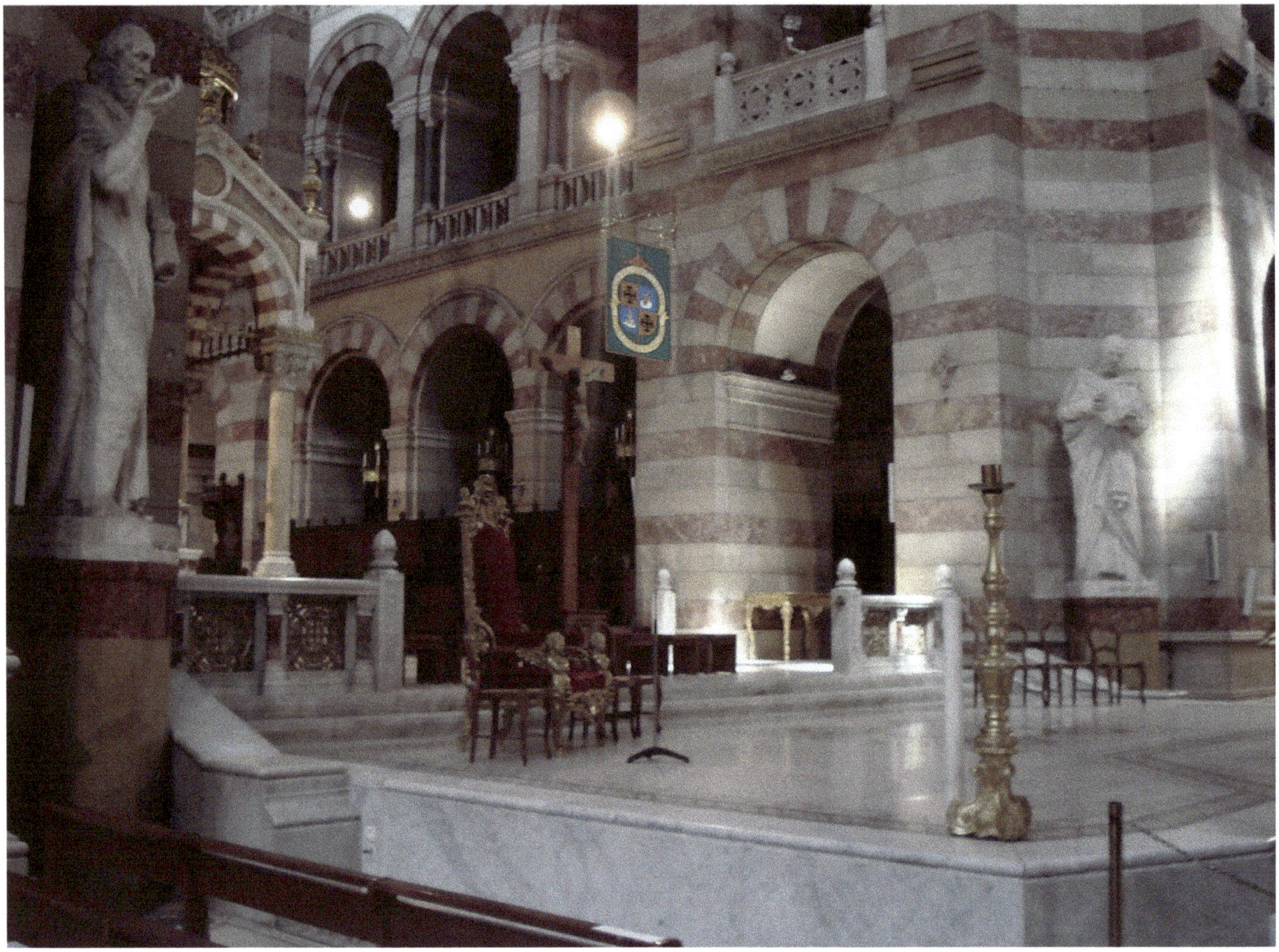

Belief Systems, Occult: Learning The Hidden Knowledge And Meet Up With God,

<table>
<tr><td colspan="2">"Hidden" Knowledge</td><td>Gnosticism – Ophites</td><td colspan="2">Kaballah</td><td>Hersey</td></tr>
<tr><td colspan="2">Cathars / Albigensians</td><td>Knights Templar</td><td colspan="2">Resicrucians</td><td>Illuminati</td></tr>
<tr><td>Freemasons</td><td>Theosophy</td><td colspan="3">Order Of The Golden Dawn</td><td>New Age</td></tr>
<tr><td colspan="2">Church Of Satan, 666</td><td colspan="2">Ordo Templi Orientiis (O.T.O.)</td><td colspan="2">X</td></tr>
</table>

Valentina Vita (V. V.); Traveling Cosplayers And Kosplayers;
"(O_O) So many **surveys**. **Lieutenant Lieu**, **volunteer manager**, where can I find **him**?"

Sister K. White <3; Pretty Sure This is Not HeartBladeEndureLove 心刀忍愛;
"(*voice over*) **He** is by the **North Side**, past **King And X District**, by ocean side border."

2154;_Arcadia \ Valentina Vita (V. V.) 'Healthy-Strong Life' \ Left to Right(L->R)

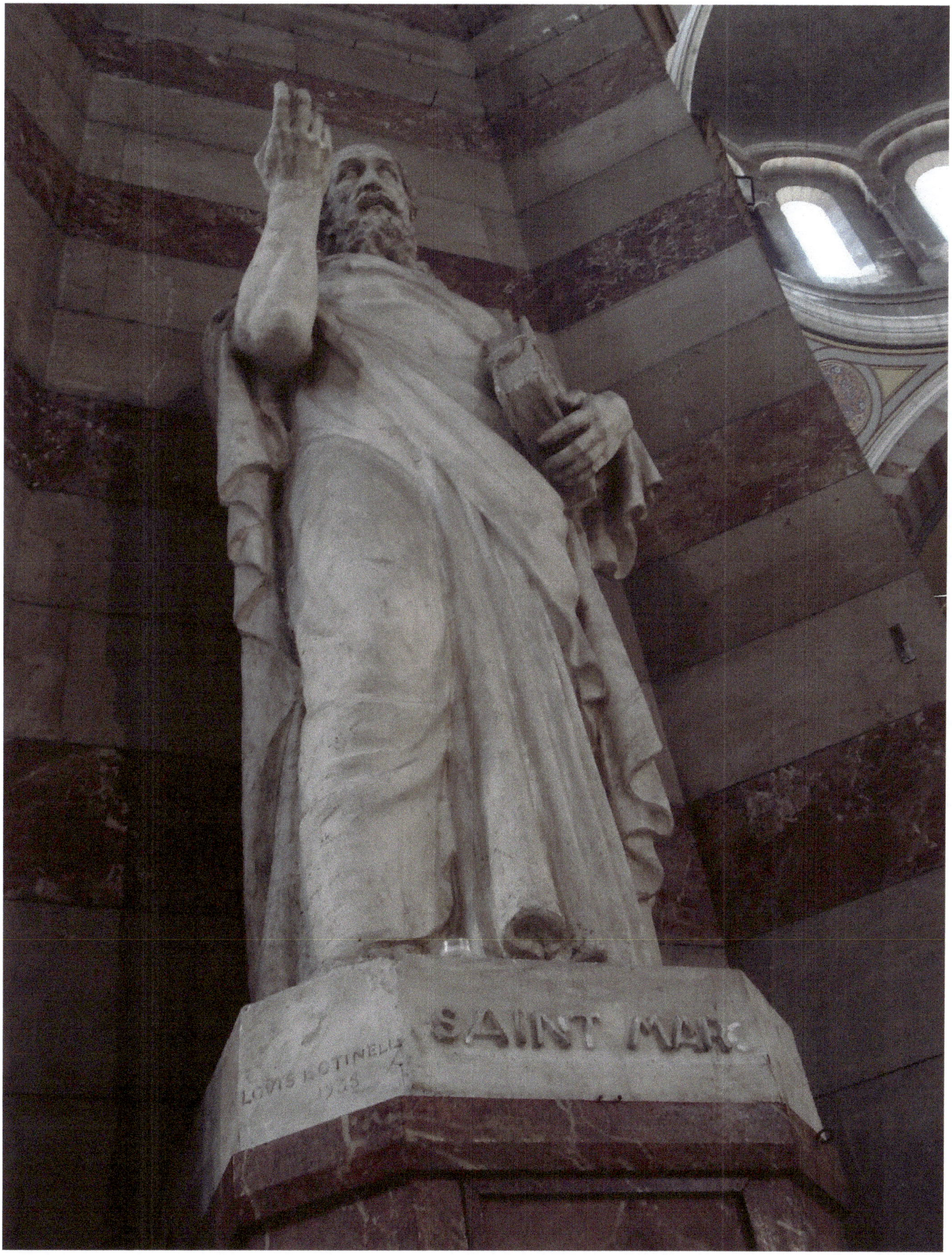
SAINT MARC

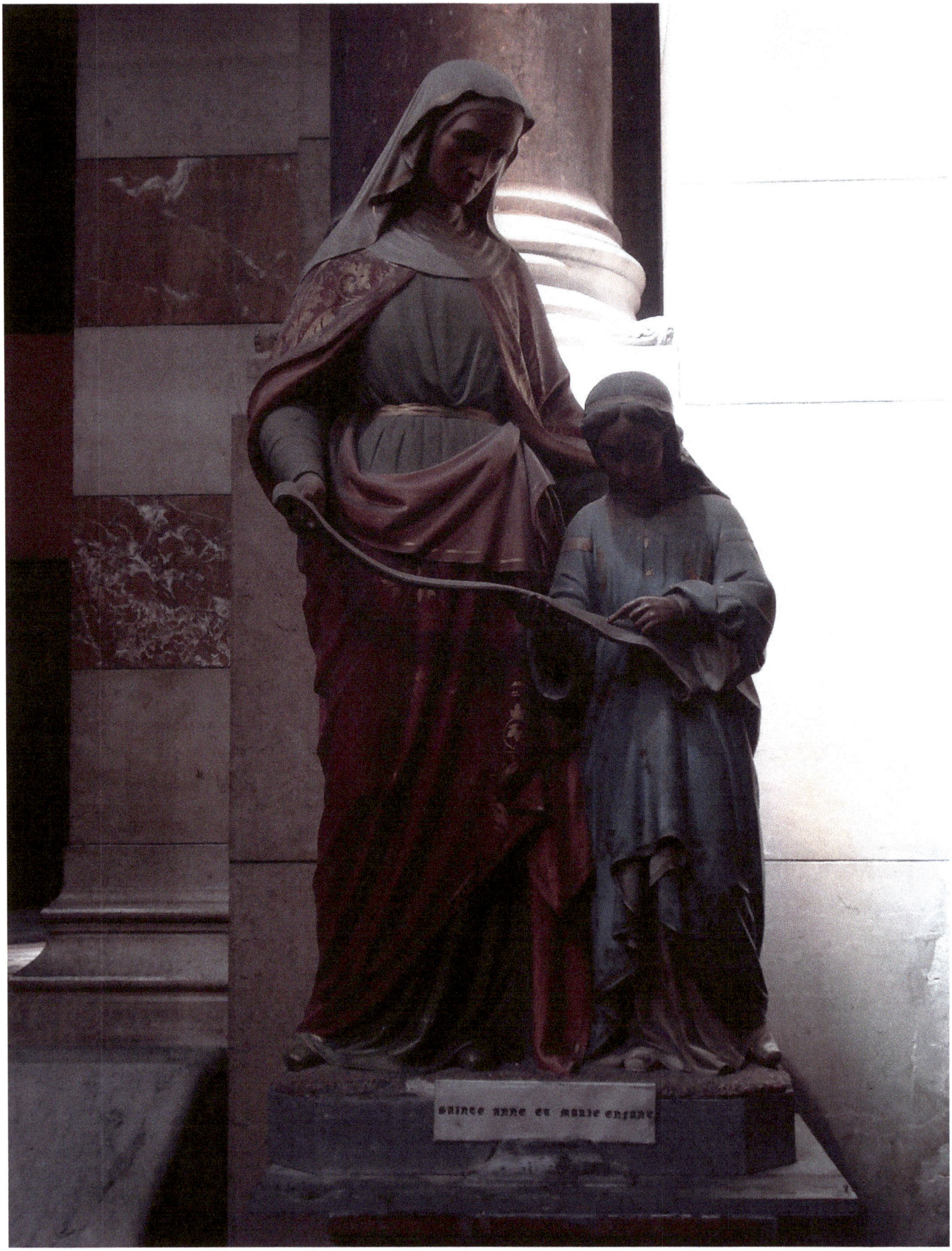
SAINTE ANNE ET MARIE ENFANT

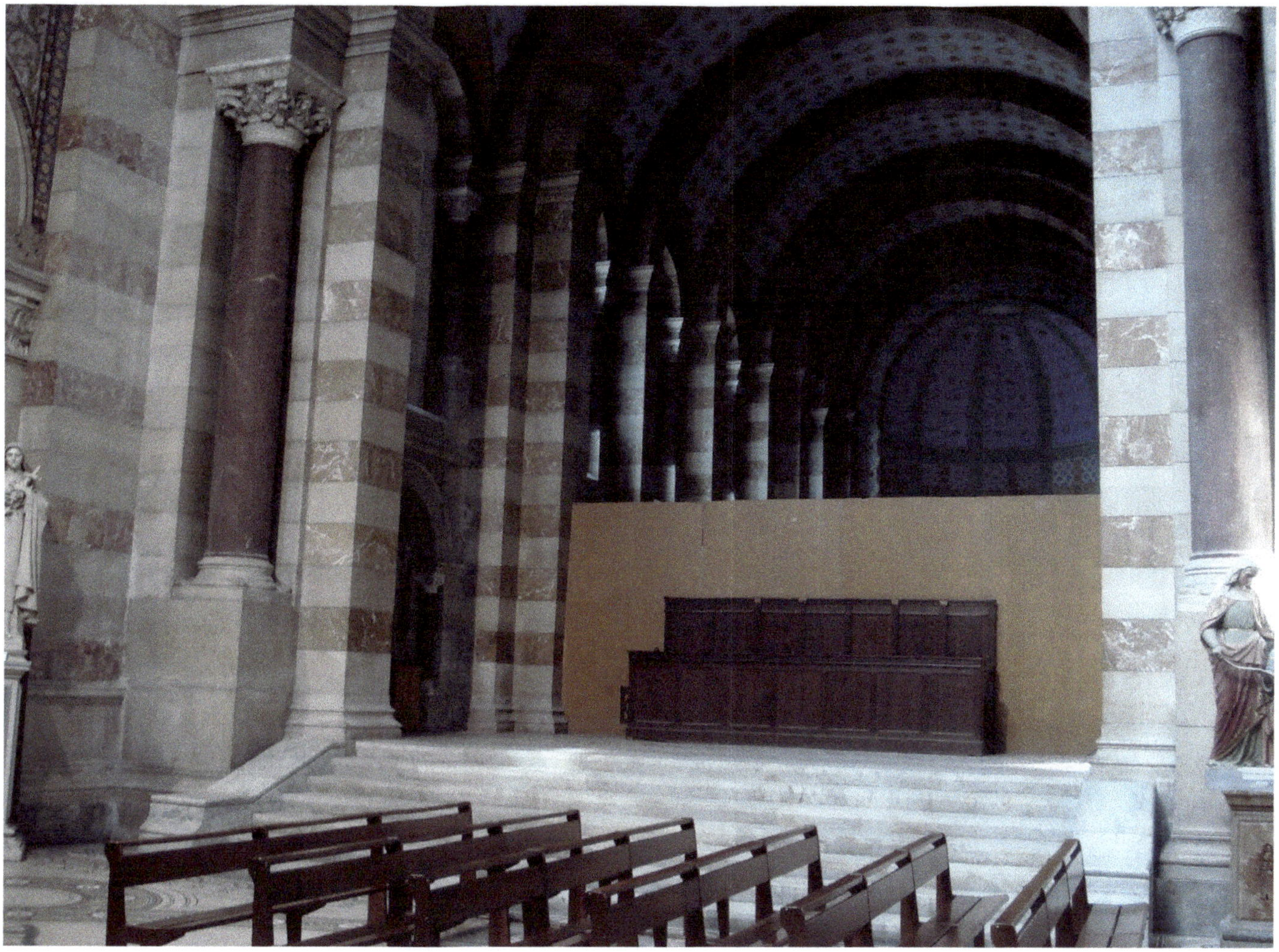

Belief Systems, Conspiracy Theories:

Birthers	Weather Machine	Area 51 (fifty one)	New World Order
Central Intelligence Agency (C.I.A.) And Acid	Federal Bureau Investigation (F.B.I.) Stores The Internet	Global Warming, Climate Change, Mother Nature,	C.I.A. With Black Boxes Downed Flights
Moon Landing	Elvis Still Alive	Majestic 12 (twelve)	Philadelphia Experiment

Valentina Vita (V. V.); Traveling Cosplayers And Kosplayers;
"(@_@) Okay, need to remember where **Lieutenant Lieu**, who can fix my Ai-**Aug** is at."

Sister K. White <3; Pretty Sure This is Not HeartBladeEndureLove 心刀忍愛;
"(*voice over*) **Lieu Lieu is located in North Side**. ...lots of **belief system surveys**!"

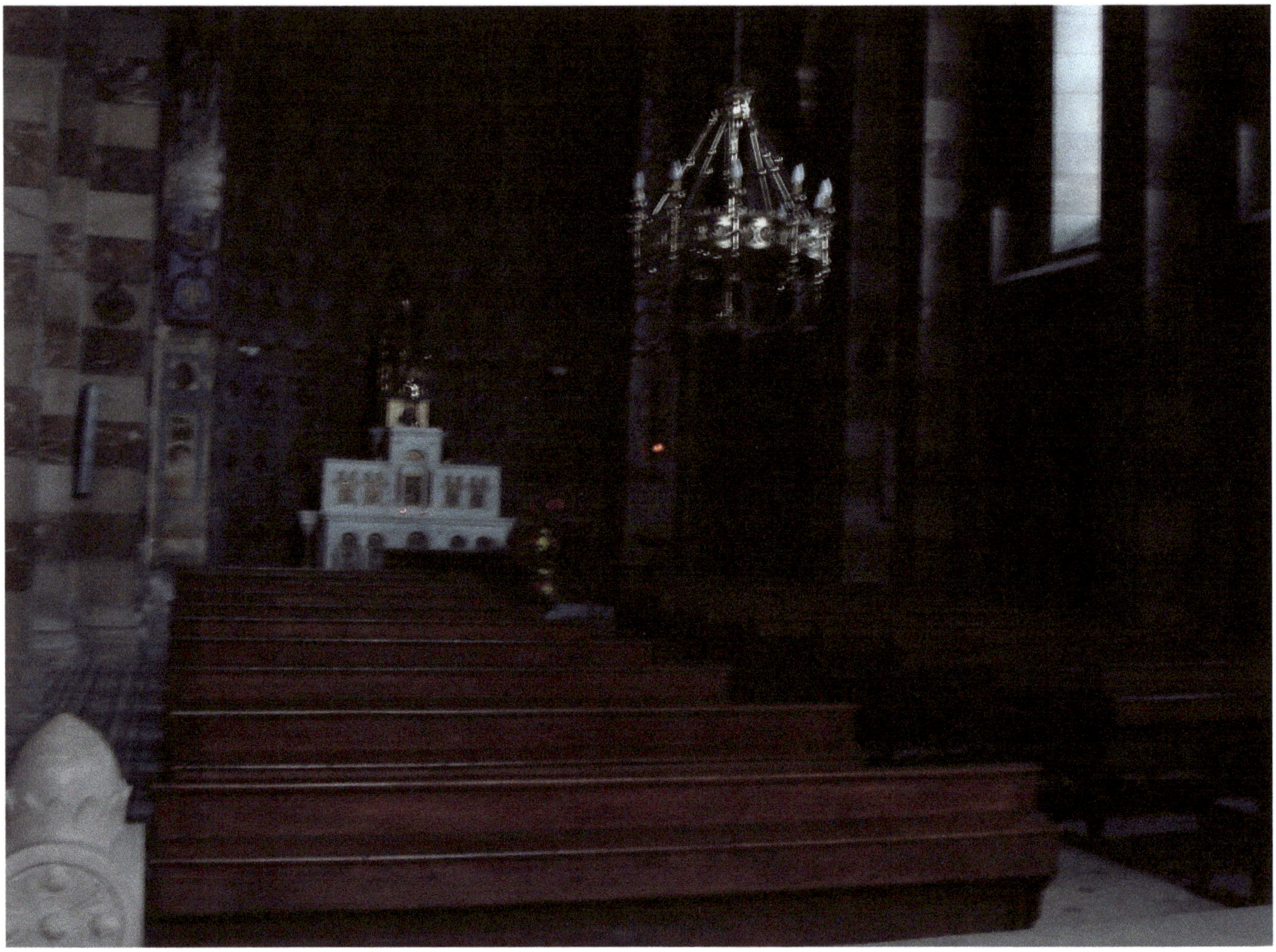

Valentina Vita (V. V.); Traveling Cosplayers And Kosplayers;
"(O_O) ...So, **these chairs benches**...? There are so many of them..."

Sister K. White <3; Pretty Sure This is Not HeartBladeEndureLove 心刀忍愛;
"(*voice over*) These are called **pews**, **used for praying**, like a **prayer**."

Valentina Vita (V. V.); Traveling Cosplayers And Kosplayers;
"(-_-) It is really quiet here... for a **Church**, **we** do not have much of these in **Eur-Asia Human** Sanctuary... Most of **our** biotech domes is just really white and green... **Clean and unclean classification** in **Human** Sanctuary, otherwise, **residing outside** of biotech domes... Not much of a survival chance against Martian Terrorists attacks."

Sister K. White <3; Pretty Sure This is Not HeartBladeEndureLove 心刀忍愛;
"(*voice over*) That is why the **Church** grants sanctuary to those **worthy of redemption**."

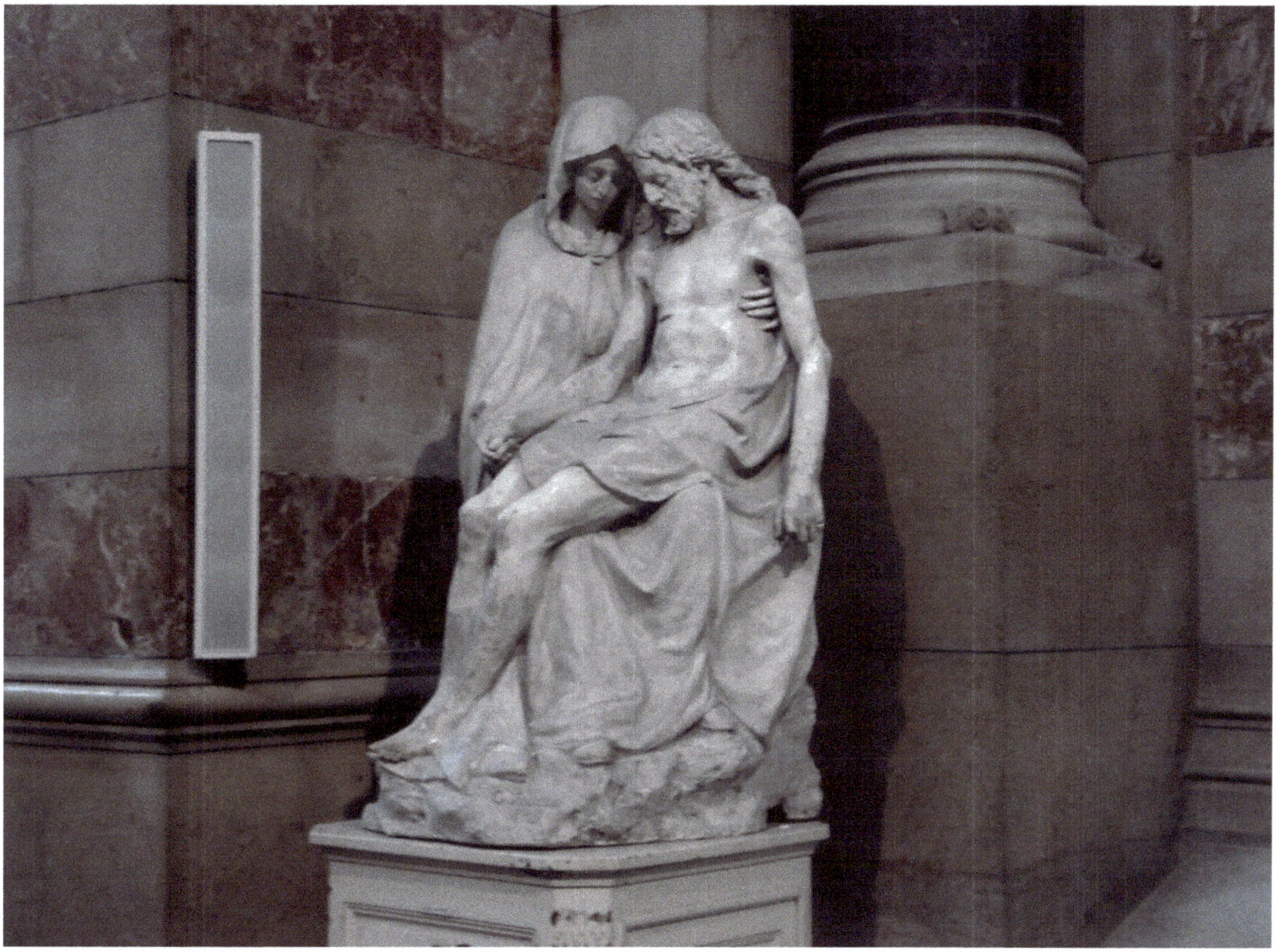

Valentina Vita (V. V.); Traveling Cosplayers And Kosplayers;
"(o_O) Hm... **This statue** is well done ...And this **looks familiar**...?"

Sister K. White <3; Pretty Sure This is Not HeartBladeEndureLove 心刀忍愛;
"(*voice over*) For this statue I believe, on the **left is Virgin Mary** and on the **right is Lord Jesus Christ**. This is according to the **Christian belief system**."

Father Greg Ford †; Pretty Sure This Is Just A Religious Father Figure;
"(*echoing male voice over*) ...**God** is the alien... ...**God** is the abyss...
...**God** is the non-existent... ...**God** is all knowing, omniscient...
...**God** is all good, omni-benevolent... ...**God** is all powerful, omni-otent... ...**God** is perfect...
...**God** is eternal... ...**God** is infinite... ...**God** is indivisible...
...**God** is transcendent... ...However, **God** is not natural... ...**God** is not bodily...
...**God** is not bound by space and time..."

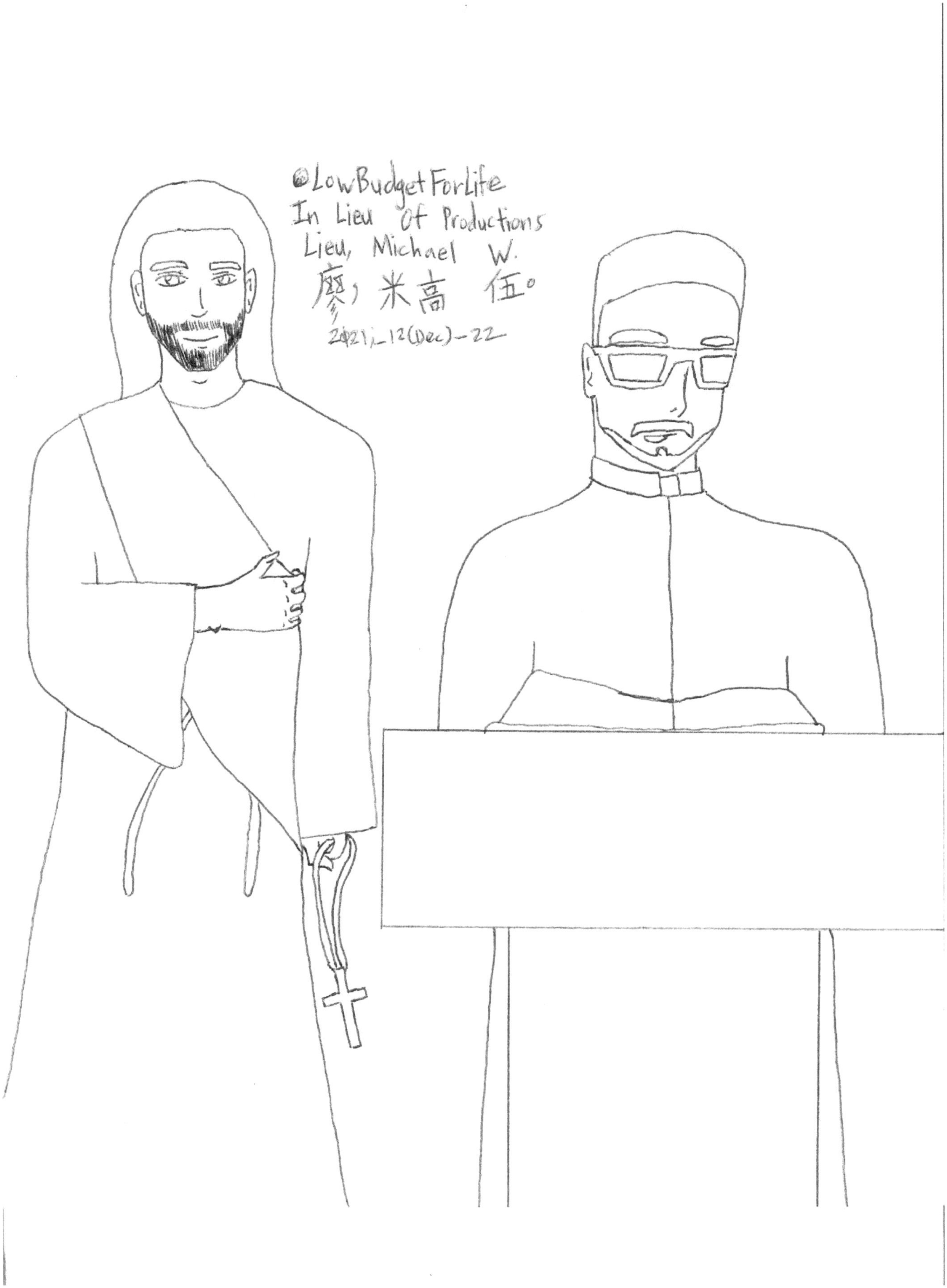
©LowBudgetForLife
In Lieu Of Productions
Lieu, Michael W.
廖, 米高 伍o
2021-12(Dec)-22

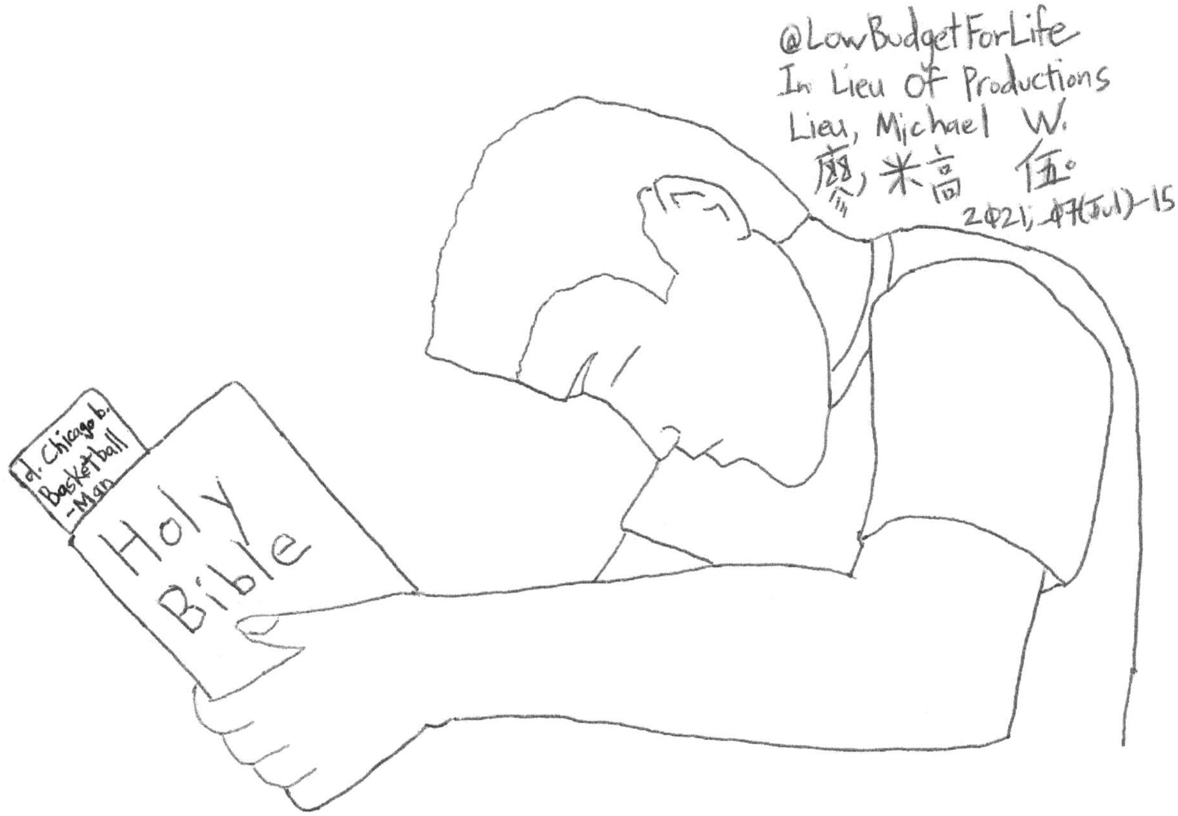

"...Lord Jesus Christ, himself goes before you and will be with you; he will never leave you nor forsake you. Do not be afraid; do not be discouraged..."

– Deuteronomy 31:8

Read by Father Greg Ford

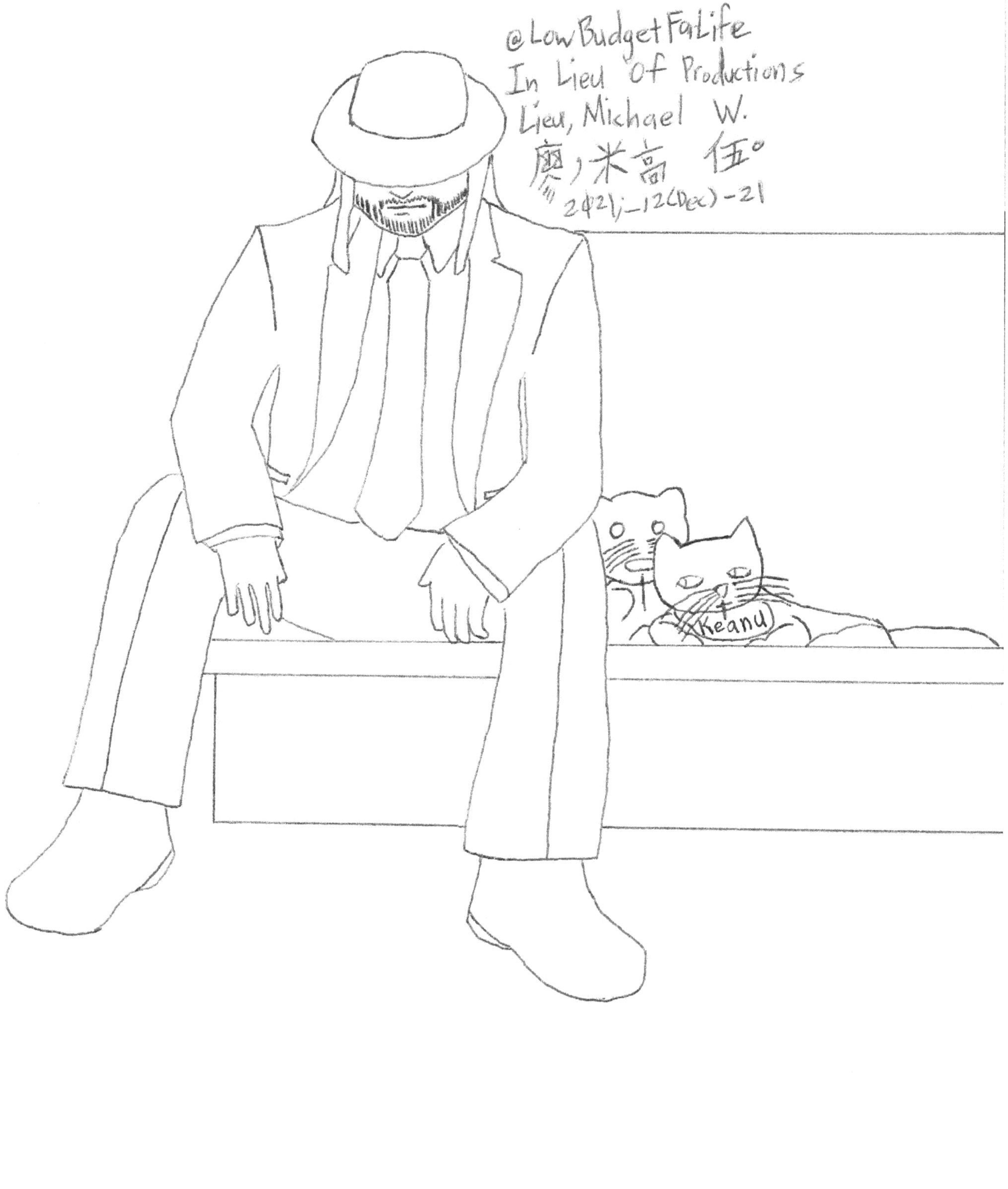
@LowBudgetFatLife
In Lieu Of Productions
Lieu, Michael W.
廖, 米高 伍o
2021_12(Dec)-21
Keanu

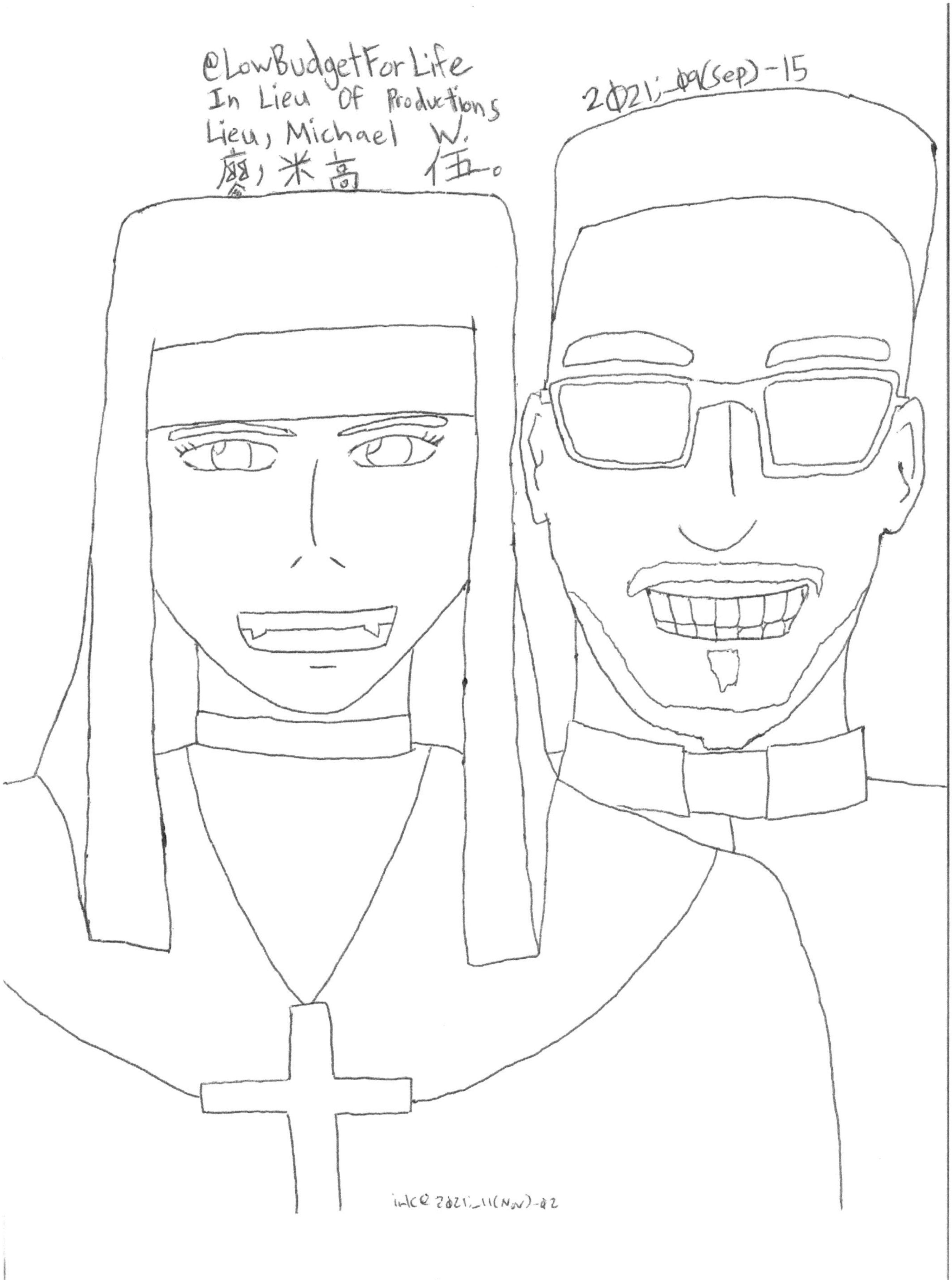
@LowBudgetForLife
In Lieu Of Productions
Lieu, Michael W.
廖, 米高　伍。
2021_09(Sep)-15
inked 2021_11(Nov)-02

Belief Systems, Urban Legend

Friend Of A Friend (F.O.A.F.) = Fear / Contamination - Rooted In A Fact
Weird, Strange, Funny, Unusual / Too Good To Be True / Many Variants
Moral Or An Ethical Lesson / Celebrities

Belief Systems, Ghost Types

Ghost (no consciousness);_	Poltergeist (noisy spirit);_
Recordings Of Energy	"Epicenter"
Repetition, Non-Interactive	Physical
1.) Apparitions, Visual (Move)	Interactive
2.) Aura	Destructive
3.) Tactile (Touch)	Generally Non-Vocal (Consciousness)
4.) Olfactory (Smell)	
5.) Taste	

Spirit;	Entity;
People Who Have Died	Angelic / Demonic
Unaware Of Death / Denial	Consciousness Of Separate Personality
Unfinished Business / Good Bye	Physically Travel Around
Help / Assist The Living	Manifest On Body Of Possessed (Detach From A Body)
Secrets To Be Revealed, Consciousness	

Valentina Vita (V. V.); Traveling Cosplayers And Kosplayers;
"(@_O) Okay, on my way out! That the **Agent responsible for the outbreak outage**?"

Sister K. White <3; Pretty Sure This is Not HeartBladeEndureLove 心刀忍愛;
"(*voice over*) Yes, **he** is very **sad** right now and feels bad, **he** has community service...
Okay V. V., remember **Lieu Lieu** is in **North Side**, past **King And X District**, bye!"

Automated Algorithm; Human Voice Over Emulation Based On Elysium Corporation Archives;
“(*monotone female voice over*) Please stand by for biotech dome airlock decontamination. Thank you for your patience and cooperation **Utopian**.”

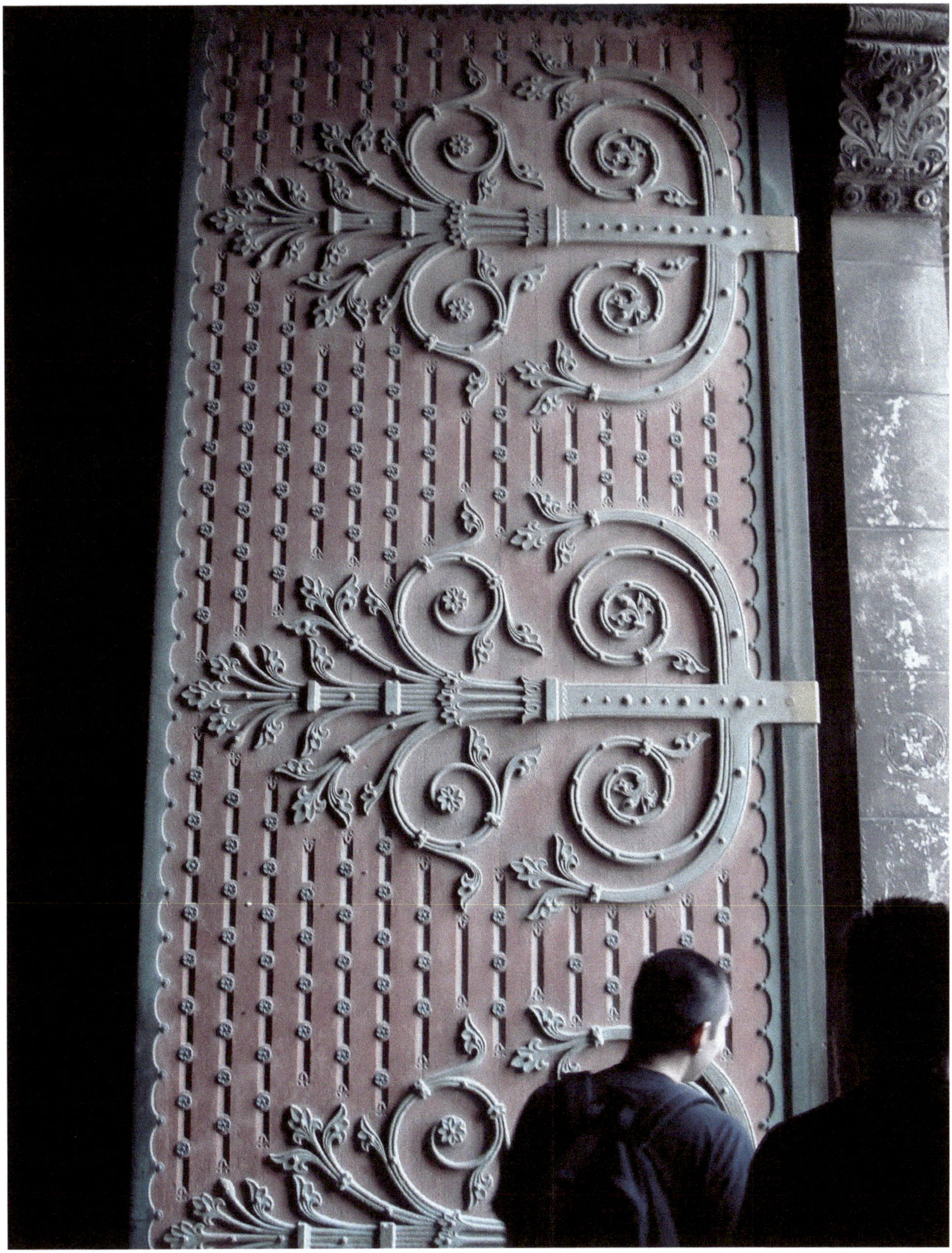

2154;_11(Nov)-01(one)

Travel To Western Society, America(s), North America
Self-Govern Sovereignty (S),
Illinois (ILL) Sector, Points Of Interest (P.O.I.)

Chicago (S), "**Animal** Sanctuary"
"The New World 2121 Movement Of The 22nd Century,
Utopian Sovereignty of North America (U.S. of N.A.)"
West Side, Chicago City Christian Church Belief Systems, And Other Religions Welcome <3

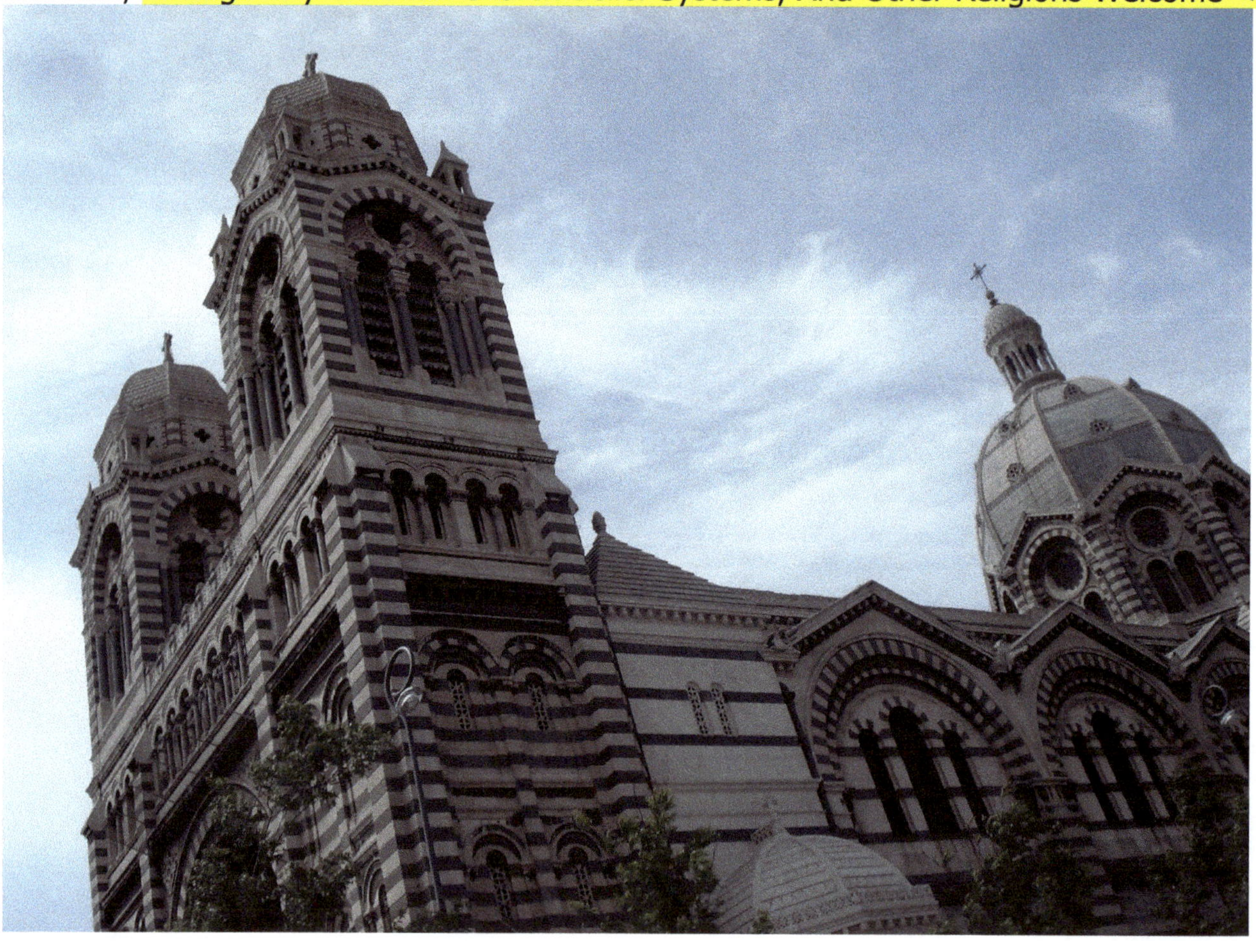

Valentina Vita (V. V.); Traveling Cosplayers And Kosplayers;
"(>_O) Ok... Need to remember help with my Ai-**Aug**, **Lieutenant Lieu**, **North Side**... Do I **have a marker or any artifact** that allows me to write things down...? Uh... No... Hmm, need to fix my cosplay, any **ugly threats sticking out**... and see what they have in the **East Side**, **All For One Shopping Center**. An outbreak outage, occurred when I was doing **photo shoot**... **half day or twelve hours went by due to me being asleep**."

2154;_Arcadia \ Valentina Vita (V. V.) 'Healthy-Strong Life' \ Left to Right(L->R)

2154;_11(Nov)-01(one)

Travel To Western Society, America(s), North America
Self-Govern Sovereignty (S),
Illinois (ILL) Sector, Points Of Interest (P.O.I.)

Chicago (S), "**Animal** Sanctuary"
"The New World 2121 Movement Of The 22nd Century,
Utopian Sovereignty of North America (U.S. of N.A.)"
West Side, Court House

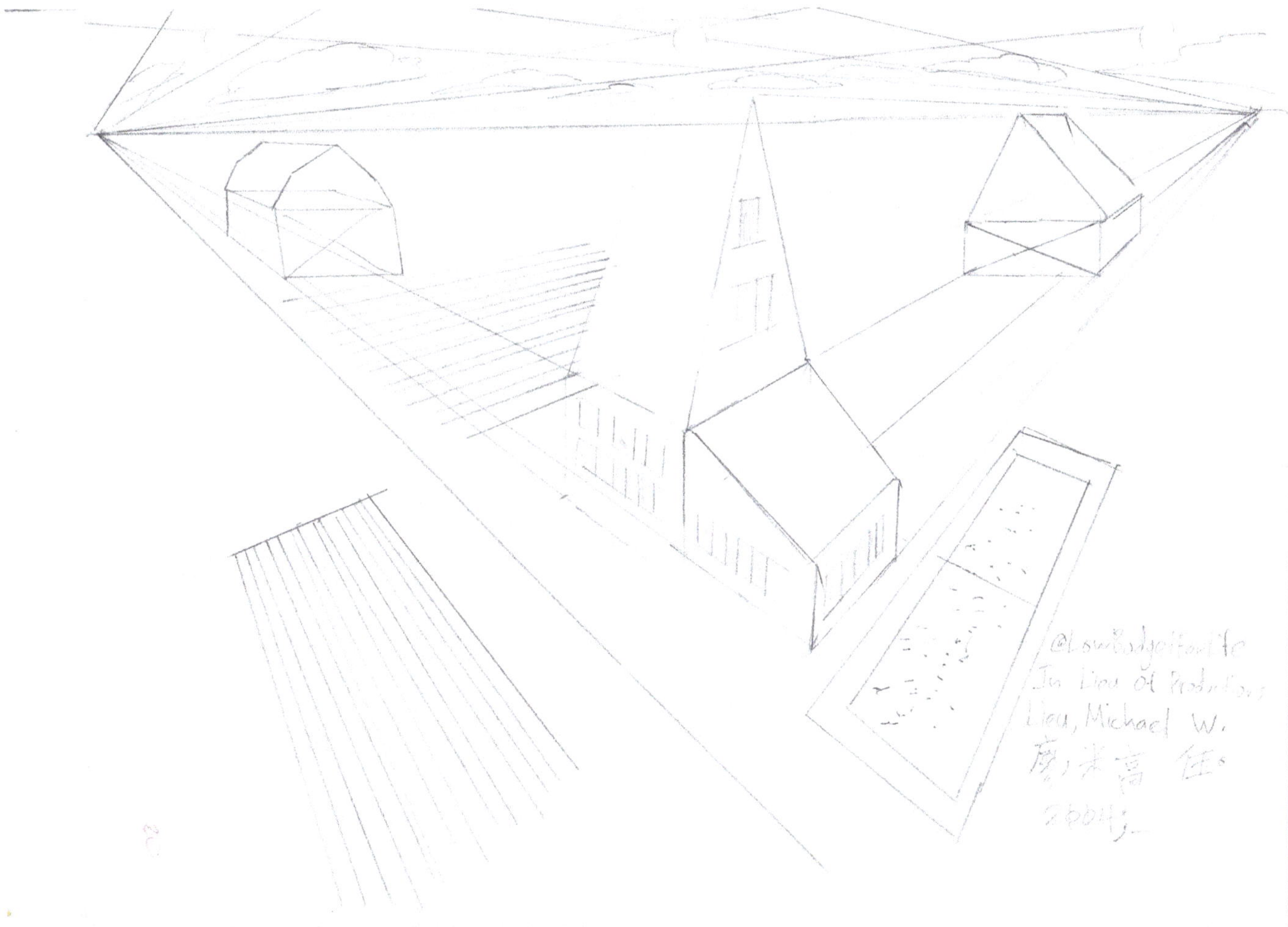

Valentina Vita (V. V.); Traveling Cosplayers And Kosplayers;
"(?_?) Court house... Quote... Kangaroo court... Any **tools and or weapons are open as long as you do not paint the walls, floors** with red juice called blood... Unquote. Huh..."

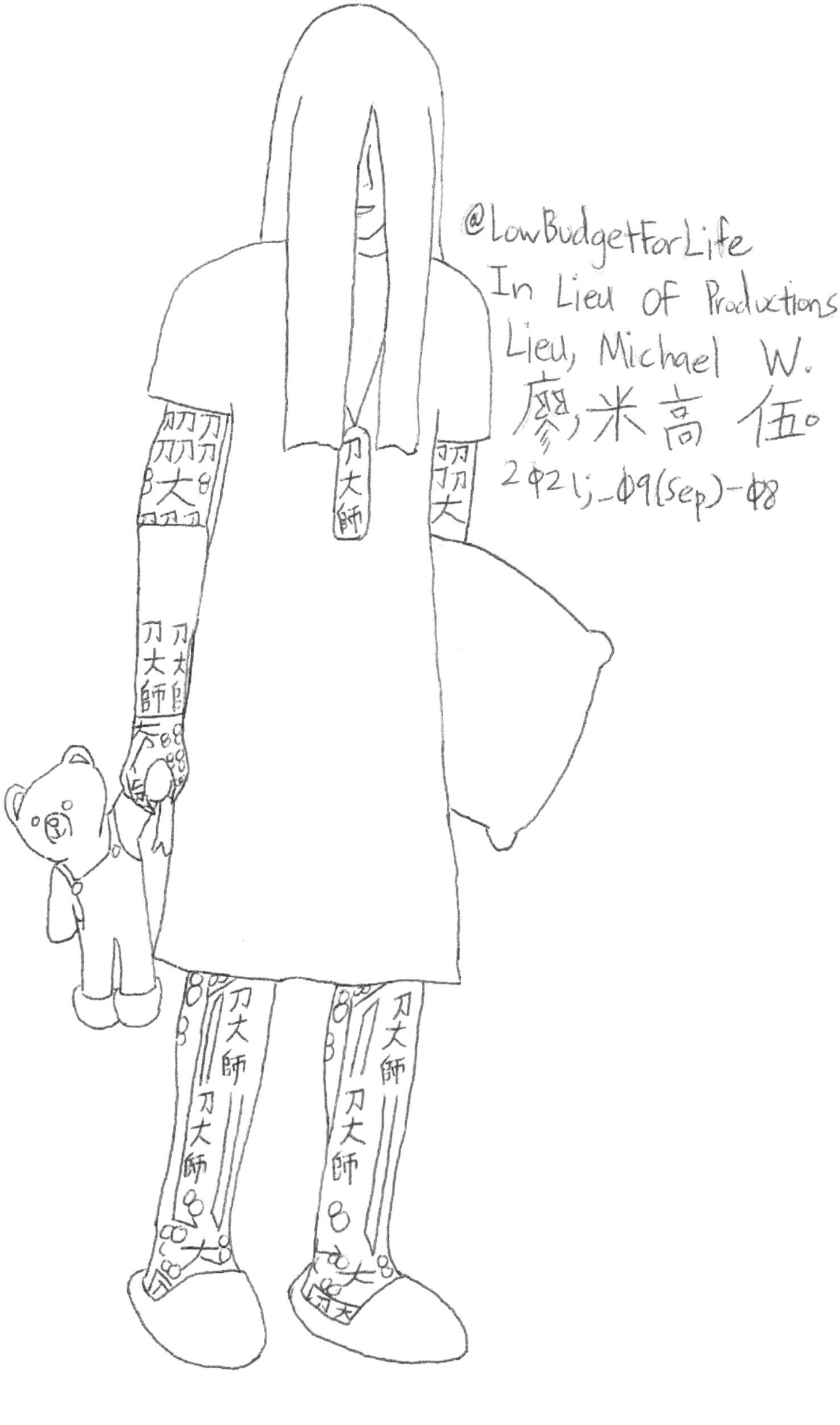
@LowBudgetForLife
In Lieu Of Productions
Lieu, Michael W.
廖,米高 伍o
2ϕ21_ϕ9(Sep)-ϕ8
刀大師

Alien Bee Bee, A.I. BladeAssistant 刀下手, Celebrity Cook-Chef Martial Artist
2154;_ Chicago, Illinois' Intranet, Wireless Broadcasting Frequencies Online

刀 Augmented Reality Holographic Heads Up Display A.R. Holo HUD	
Actual Legal Given Name;_ Elle K. Mcgee (goes by Elle) pronounced, 'el-lee' 'kay' 'mick-gee'	Physical Plus Psych Age;_ augmented 17 + 2 years old
Community Network(s);_ **British's Special Air Service (S.A.S.)** **Hong Kong's Independent Commission Against Corruption (I.C.A.C.)** **British-Hong Kong Sovereignty Sector** **Cook-Chef Community Martial Artist Educator** **Human Klones, Designer Babies And A.I. Ai** **Utopian Sovereignty (U.S.) Volunteer Worker** [*Aug+*]	Alien;_ Legal And Approved **Operative, Citizen Of Eur-Asia** Student Visa **Cook-Chef Community Studying Program** Approved And Granted Access By "**2121's the old world**" of **U.S. Department Of Immigration** And **U.S. Department of Alien and Citizen Services**. [*Aug+*]
[*Aug+*]	
刀 Augmented Reality Holographic Heads Up Display A.R. Holo HUD	

2154;_ Chicago, Illinois' Intranet, Wireless Broadcasting Frequencies Online
Alien Bee Bee, A.I. BladeAssistant 刀下手, Celebrity Cook-Chef Martial Artist

Battle Bot A.K.A. Bee Bee; Hermaphrodite Humanoid A.I. Ai 'BladeAssistant 刀下手';
"(*monotone female voice over*) Elle, you forgot to change out of your sleeping clothes..."

Elle K. Mcgee; Childlike Human Kanon-Klone(s) "BladeMaster 刀大師" Experiment;
"(*sleepy*) Huh... Uh... **Sleepy weepy**... **Bee Bee**... Huh... Did I forget something...?"

Valentina Vita (V. V.); Traveling Cosplayers And Kosplayers;
"(?_?) ...Hm... Elle, again, **third time**... **We meet again**... whoa we wow!"

2154;_Arcadia \ Valentina Vita (V. V.) 'Healthy-Strong Life' \ Left to Right(L->R)

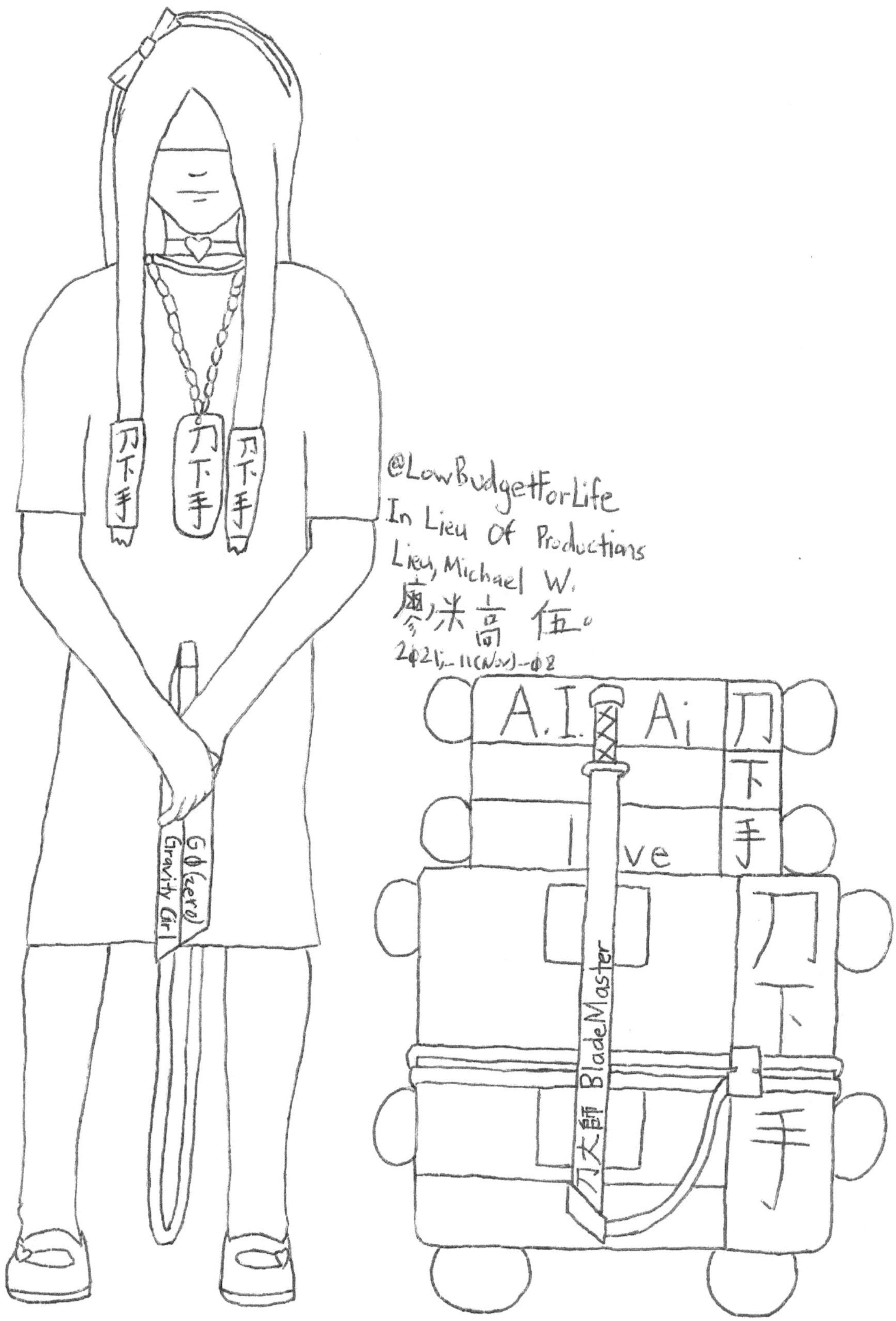
刀下手
刀下手
刀下手
G0(zero) Gravity Girl
@LowBudgetForLife
In Lieu Of Productions
Lieu, Michael W.
廖, 米高 伍。
2021-11(Nov)-02
A.I. Ai
I ve
刀下手
刀下手
刀大師 BladeMaster

Alien Bee Bee, A.I. BladeAssistant 刀下手, Celebrity Cook-Chef Martial Artist
2154;_ Chicago, Illinois' Intranet, Wireless Broadcasting Frequencies Online

<table>
<tr><td colspan="2">刀 Augmented Reality Holographic Heads Up Display A.R. Holo HUD</td></tr>
<tr><td>Actual Legal Given Name;_

A.I. Battle Bot (goes by Bee Bee)
pronounced, 'b. b.'</td><td>Actual Legal Given Name;_

A.I. Luggage (goes by Ai-RiceCooker)
pronounced, 'eye roo-ice cook-her'</td></tr>
<tr><td>Community Network(s);_

British's Special Air Service (S.A.S.)

Hong Kong's Independent Commission Against Corruption (I.C.A.C.)

British-Hong Kong Sovereignty Sector

Cook-Chef Community Martial Artist Educator

Human Klones, Designer Babies And A.I. Ai [Aug+]</td><td>Community Network(s);_

British's Special Air Service (S.A.S.)

Hong Kong's Independent Commission Against Corruption (I.C.A.C.)

British-Hong Kong Sovereignty Sector

Cook-Chef Community Martial Artist Educator

Moving Luggages, And A.I. Ai [Aug+]</td></tr>
<tr><td colspan="2">[Aug+]</td></tr>
<tr><td colspan="2">刀 Augmented Reality Holographic Heads Up Display A.R. Holo HUD</td></tr>
</table>

2154;_ Chicago, Illinois' Intranet, Wireless Broadcasting Frequencies Online
Alien Bee Bee, A.I. BladeAssistant 刀下手, Celebrity Cook-Chef Martial Artist

Battle Bot A.K.A. Bee Bee; Hermaphrodite Humanoid A.I. Ai 'BladeAssistant 刀下手';
"(*monotone female voice*) Hello V. V., **I** am **Bee Bee**. Beside **me** to **my** left is, Ai-**RiceCooker,** we are **BladeAssistants** to our **BladeMaster** Elle. Please excuse Elle, she is still in the **process of waking up** after **sleeping for two hours for being awake for forty eight hours**. She still needs to change out of her sleep clothes. Elle, wakey wakey."

Elle K. Mcgee; Childlike Human Kanon-Klone(s) "BladeMaster 刀大師" Experiment;
"(*sleepy voice over*) ...Huh... ...Uh. **Wakey wakey** time! ...Uh ...Time to wake up?"

Valentina Vita (V. V.); Traveling Cosplayers And Kosplayers;
"(O_o) **Two hours of sleep after being awake for forty eight hours**. Honored to meet a **celebrity BladeMaster**, I did meet **two BladeBosses politicians** by the airlock entry."

2154;_Arcadia \ Valentina Vita (V. V.) 'Healthy-Strong Life' \ Left to Right(L->R)

Alien Bee Bee, A.I. BladeAssistant 刀下手, Celebrity Cook-Chef Martial Artist
2154;_ Chicago, Illinois' Intranet, Wireless Broadcasting Frequencies Online

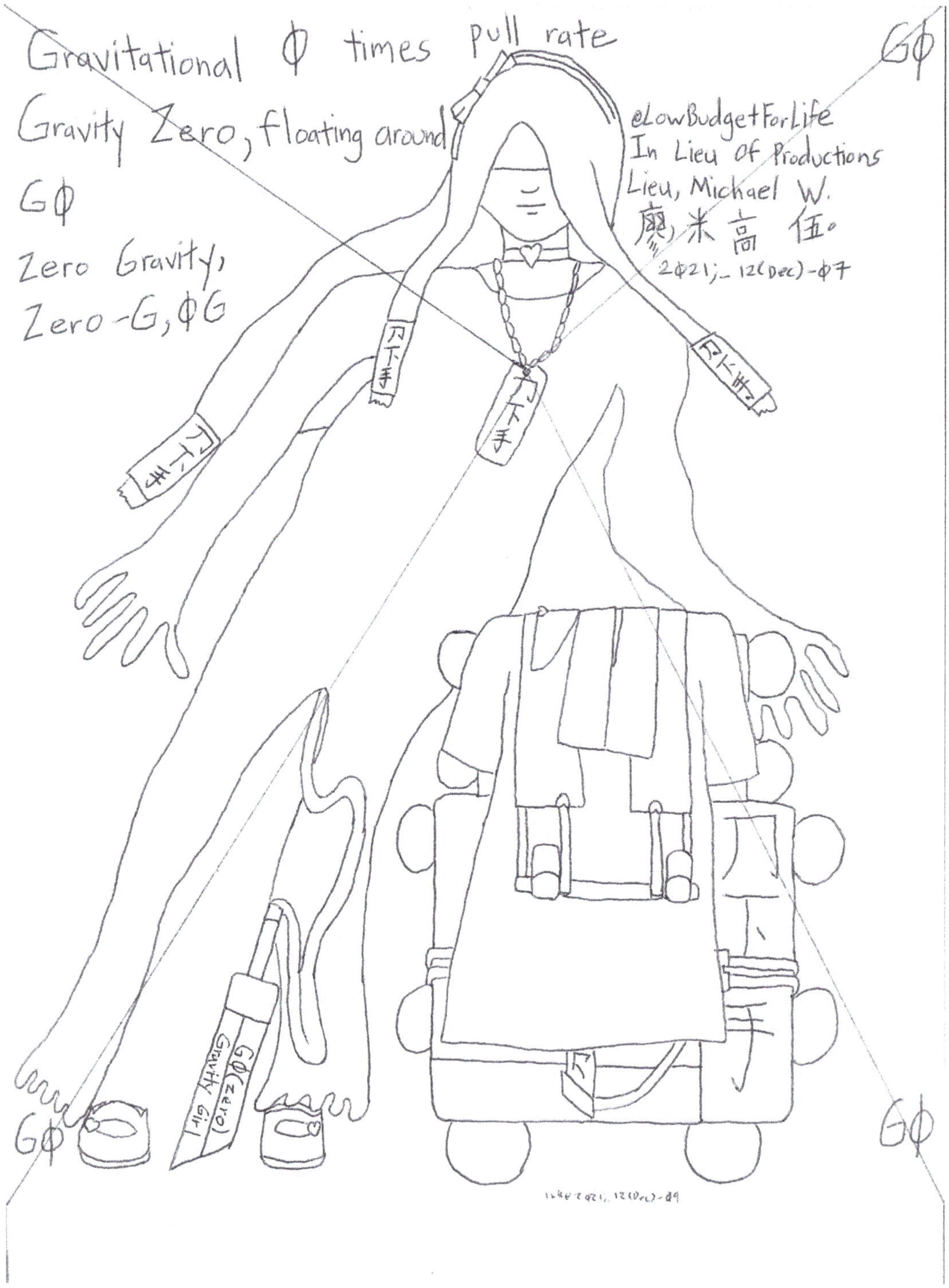

2154;_ Chicago, Illinois' Intranet, Wireless Broadcasting Frequencies Online
Alien Bee Bee, A.I. BladeAssistant 刀下手, Celebrity Cook-Chef Martial Artist

Elle K. Mcgee; Childlike Human Kanon-Klone(s) "BladeMaster 刀大師" Experiment;
"(*sleepy voice over*) ...Time to **change sleeping clothes** from... **Bee Bee**! Uh..."

Battle Bot A.K.A. Bee Bee; Hermaphrodite Humanoid A.I. Ai 'BladeAssistant 刀下手';
"(*monotone female voice*) Hm... Elle, it seems it is not **me** that requires to change clothes, rather you, Elle. **My Umbilical Cord Biotech Blade** is **activating Gravitational Zero Times Pull Rate**. With no **biotech clothing regulating it**."

Valentina Vita (V. V.); Traveling Cosplayers And Kosplayers;
"(@_@) Huh... Your **body is gray**... **Bee Bee, you** are now floating in **Zero Gravity**... Interesting... To see this **Zero G** event on Planet Earth... Whoa we wow... **Floating around** on Earth... **Your** Ai-**Aug** activated an **effected area** of **G Zero**... Whoa we wow!"

Battle Bot A.K.A. Bee Bee; Hermaphrodite Humanoid A.I. Ai 'BladeAssistant 刀下手';
"(*monotone female voice*) Due to the concerns of the **animals and humans** in this Chicago **Animal** Sanctuary, I am only allowed to use **Gravitational Zero Times Pull Rate** Also Known As **G Zero or Zero G**. This is to not risk **harming** and causing **dismemberment** at a higher **Gravitational Pull Rate Shifting**... This is called **Gravity Shifting**. We would not want to be labeled as terrorists for not following **old and new world** sanctuary safety rules and regulation."

Valentina Vita (V. V.); Traveling Cosplayers And Kosplayers;
"(>_<) Huh... **Gravity Shifting**, terrorists... I-I... Hm... I understand... Anyways, the **Zero G** effect stops if **we** put **your black clothing back on to you**?"

Battle Bot A.K.A. Bee Bee; Hermaphrodite Humanoid A.I. Ai 'BladeAssistant 刀下手';
"(*monotone female voice*) Yes, it seems Elle here is still **having trouble waking up** from her **sleep**... She still needs to change out of her sleeping clothes... Seems like it would be **beneficial** to us in your **assistance to our request**, otherwise, you can **depart** and leave **me and** Ai-**RiceCooker** to change Elle out of her sleeping clothes to student wear."

V. V. decides to **assist Bee Bee and** Ai-**RiceCooker** on putting back **Bee Bee request**.

Battle Bot A.K.A. Bee Bee; Hermaphrodite Humanoid A.I. Ai 'BladeAssistant 刀下手';
"(*monotone female voice*) V. V., your **voice tone, heart rate and pupil dilation** on the words quote **Gravity Shifting**, terrorists unquote, suggests you would like to know more about the War On Mars unconventional warfare. As a favor for assisting us, we can **provide visual verbal briefing on those topics** since your Ai-**Aug bracelet** is currently not functioning at the moment due to **Kay's Private Security Services' outbreaks outages**."

Valentina Vita (V. V.); Traveling Cosplayers And Kosplayers;
"(^_^) Whoa we wow, **you** literally read my **biotech signs** for an A.I. **platform**. O.K."

Alien Bee Bee, A.I. BladeAssistant 刀下手, Celebrity Cook-Chef Martial Artist
2154;_ Chicago, Illinois' Intranet, Wireless Broadcasting Frequencies Online

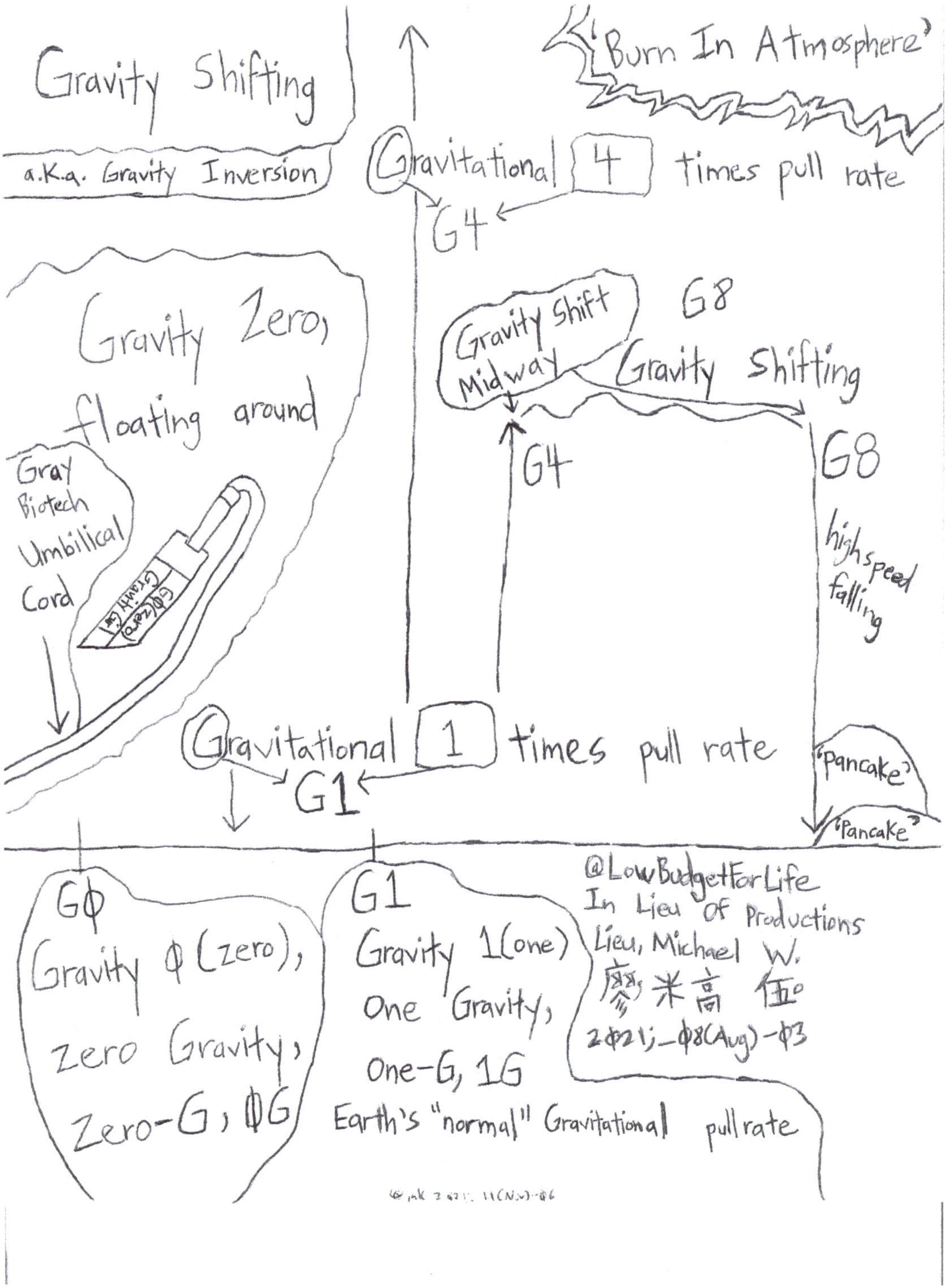

2154;_ Chicago, Illinois' Intranet, Wireless Broadcasting Frequencies Online
Alien Bee Bee, A.I. BladeAssistant 刀下手, Celebrity Cook-Chef Martial Artist

Battle Bot A.K.A. Bee Bee; Hermaphrodite Humanoid A.I. Ai 'BladeAssistant 刀下手';
"(*monotone female voice over*) Allow **me** to **visually verbally brief** you on, **Gravity Shifting** an unconventional warfare tactics used by Martian Terrorists…"

Valentina Vita (V. V.); Traveling Cosplayers And Kosplayers;
"(X_X) Whoa we wow, actual war… this is not going to be a nice **topic**… I am ready"

Battle Bot A.K.A. Bee Bee; Hermaphrodite Humanoid A.I. Ai 'BladeAssistant 刀下手';
"(*monotone female voice over*) The term **Gravity Shifting**, refers to **the Inversion of Gravity**, Also Known As **Gravity Inversion**, an unconventional warfare tactics used by Martian Terrorists to indirectly annihilate billions of human lives across the solar system. How the Martian Terrorists are able to **engineer this can be found in**… **Classified Intel To Civilians**, only **accessible** to **private and public sec operatives and assistant operatives** on Planet Earth. However, **I** do have **non-classified intel** for **briefing**."

Valentina Vita (V. V.); Traveling Cosplayers And Kosplayers;
"(?_?) …**Classified Intel To Civilians**, I knew it, just like E**GG Space Forces' and their need to know**. Anyways, any **other intel you** can provide that is not… **classified**…?"

Battle Bot A.K.A. Bee Bee; Hermaphrodite Humanoid A.I. Ai 'BladeAssistant 刀下手';
"(*monotone female voice over*) With the **approval and blessing** of **BladeSociety's BladeBosses**, **Mark D. and Mark S.**, **Elysium Corporation's Doctor** Ai and **myself**, started **research and developed countermeasures** to Martian Terrorists use of **Gravity Shifting** unconventional warfare tactics… This Umbilical Cord **Bio**t**ech Blade** allows **me** to **Shift Gravity** in accordance to the **Gravitational Number Times Pull Rate**… At times, I am nicknamed **Gravity Girl**… Even though as an Artificial Intelligence hermaphrodite sex, **my feminine side is more dominate than my male side…** Reproducing asexually without an opposite sex partner in case something happens to me… Anyways, upon **descending** as you can see falling from a **G Four or G Eight Gravity Shifting** results in a quote **pancake** unquote conclusion **remains**. Either that, upon **ascending**, one would **Burn In Atmosphere**… No real good outcomes… It is **advised that** you **have some sort of tool or equipment that allows to grapple to an object, structure, or artifact that allows** you **to mobilize physical placement from location to location** just in case."

Valentina Vita (V. V.); Traveling Cosplayers And Kosplayers;
"(@_@) Yes, **ma'am**, this time I **take this briefing more seriously**, as with those green army soldier engineers, I kind of did not take them seriously before. **Lesson learned** with my poor broken Ai-**Aug bracelet**… Which reminds me, I need to see what **East Side, All For One Shopping Center** has for goods and services, then **North Side**, for **Lieu Lieu**."

Elle K. Mcgee; Childlike Human Kanon-Klone(s) "BladeMaster 刀大師" Experiment;
"(*excitement voice over*) Huh… Time to wakey wakey… Almost up… change clothes."

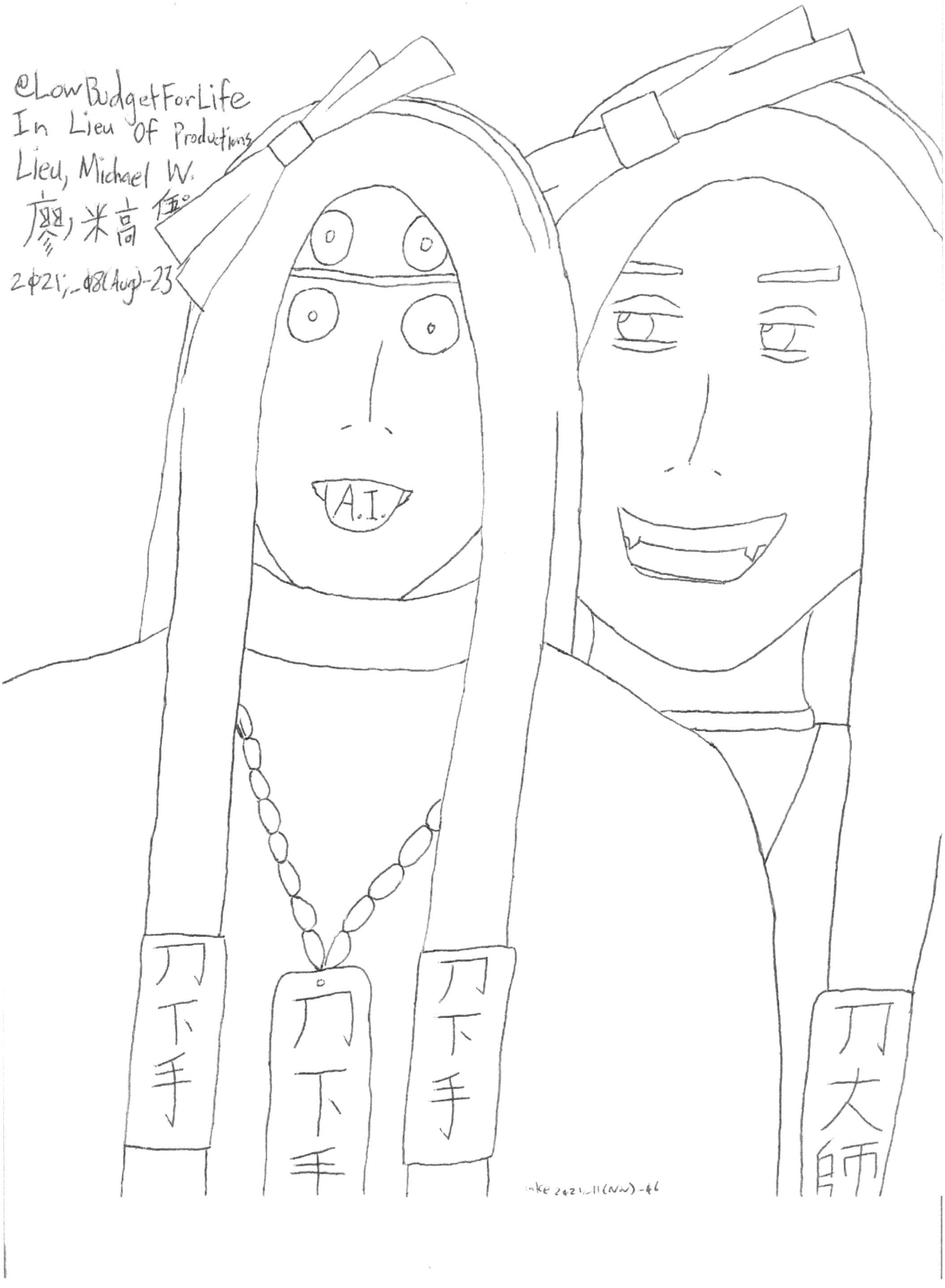
@LowBudgetForLife
In Lieu Of Productions
Lieu, Michael W.
廖, 米高 伍o
2021-08(Aug)-23
A.I.
刀下手
刀下手
刀下手
刀大師

Elle K. Mcgee; Childlike Human Kanon-Klone(s) "BladeMaster 刀大師" Experiment;
"(*creepy smile*) ...Morning **Bee Bee**, I finally managed to change clothes from sleeping to student wear... **Waking up when** I **fall asleep** for two hours is hard for me... Oh hi... V. V., I watched you while you were playing with your A.I. **Holo HUD Blade friend** in **South Side**... Wow... **That blade** has such a **long single edge cutting tool**..."

Valentina Vita (V. V.); Traveling Cosplayers And Kosplayers;
"(@_@) Elle, **we** meet again for the **third time**... And I was **taking holo photos with my** A.I. personality **sim**. ...Then **fell asleep for twelve hours... half day**, when some kind of **outbreak outage** occurred in **Kay's Private Sec R And D.** Oh, I was **rescued** by this white yellow golden Super-Hero wearing skintight suit, Heart Love shape cape... Then, I was in a **Church**... And here I am, **West Side**... **Twelve hours** went by... Uh..."

Elle K. Mcgee; Childlike Human Kanon-Klone(s) "BladeMaster 刀大師" Experiment;
"(*excitement*) Yea yea yea!! That is **Sister** HeartLove, let's face it... She is a **nun wearing** a Super-Hero suit underneath those **religious clothes**... Bright yellow legs and arms gives it away... Not fooling **anyone** with that! **Sister** HeartLove is really **nice and helpful**! I **did some volunteer work** while dealing with **Private Security Agents** in this Chicago **Animal** Sanctuary... Turns out **those Agents** were **making an augmented** A.I. **human platform in dealing** with the War On Mars... I have more **intel** on my Ai-**Aug necklace**... Anyways, V. V., do you wonder what **Bee Bee** looks like without **her blindfold visor**?!"

Elle takes off **Bee Bee's biotech blindfold visor** without much thought.

Battle Bot A.K.A. Bee Bee; Hermaphrodite Humanoid A.I. Ai 'BladeAssistant 刀下手';
"(*monotone female voice*) Time to **smile** with **my** tongue **out...** Cheese."

Valentina Vita (V. V.); Traveling Cosplayers And Kosplayers;
"(@_@) Whoa we wow, **Bee Bee** has **four eyes on her head**... And her tongue, states that she is an A.I.... Well, at least **she** is not floating around anymore in **Zero Gravity**..."

Elle K. Mcgee; Childlike Human Kanon-Klone(s) "BladeMaster 刀大師" Experiment;
"(*excitement*) **She** actually has **more eyes** around **her head**, but **her hair** is covering **it**... Maybe, next time I can show you how **many eyes she** has so you **scream** for **ice cream**!"

Valentina Vita (V. V.); Traveling Cosplayers And Kosplayers;
"(>_<) I am still curious as to **what exactly really happened** with this **outage outbreak** from **Kay's Private Sec**... Explanation Elle as a **volunteer worker** here?"

Elle K. Mcgee; Childlike Human Kanon-Klone(s) "BladeMaster 刀大師" Experiment;
"(*excitement*) Yes, I can show and tell you my **Blades** that were used in taking part of this Chicago **Animal** Sanctuary... **volunteer work** after I stopped **watching** you **play**!"

Alien Elle K. Mcgee, BladeMaster 刀大師, Celebrity Cook-Chef Martial Artist
2154;_ Chicago, Illinois' Intranet, Wireless Broadcasting Frequencies Online

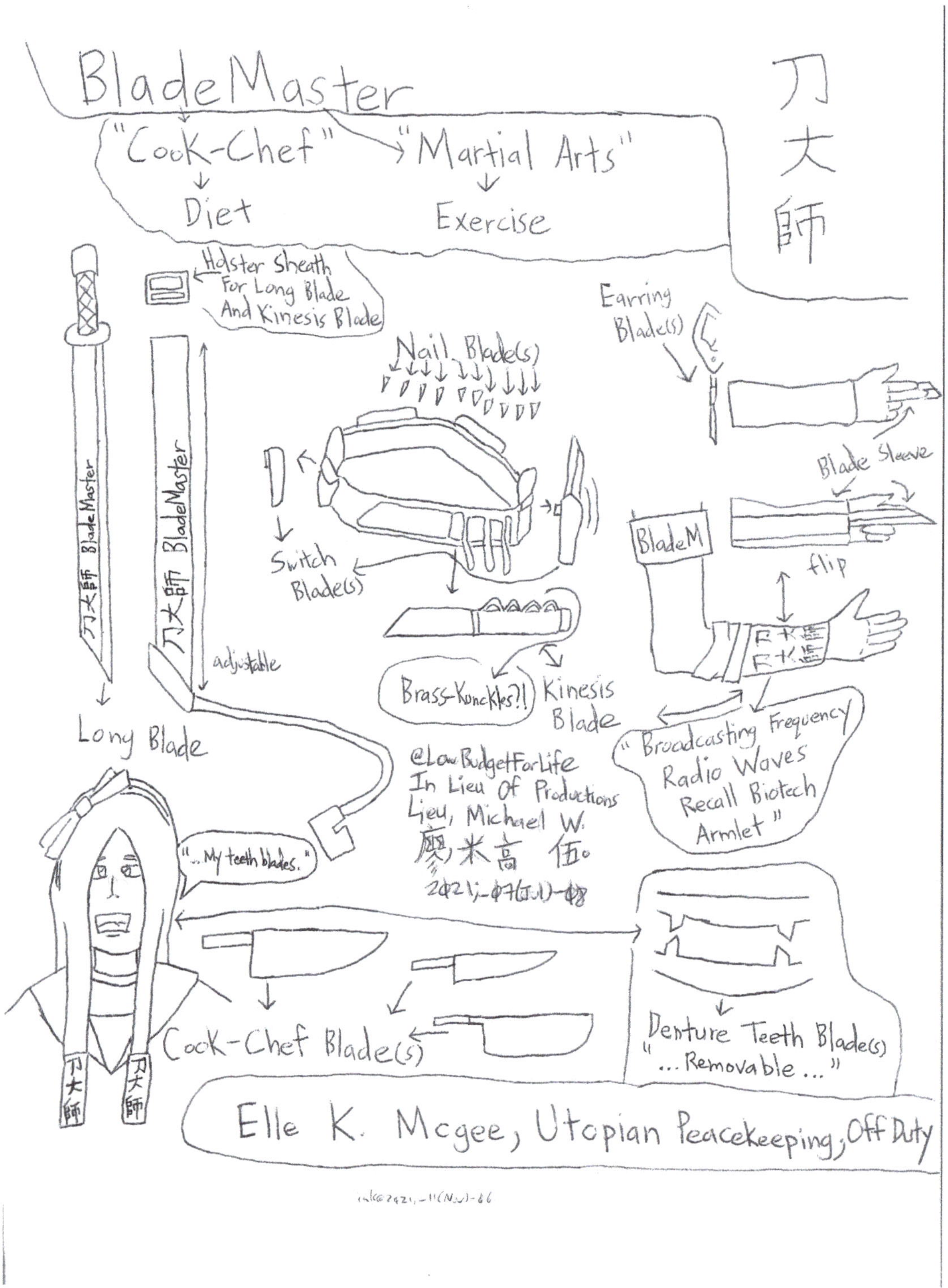

2154;_ Chicago, Illinois' Intranet, Wireless Broadcasting Frequencies Online
Alien Elle K. Mcgee, BladeMaster 刀大師, Celebrity Cook-Chef Martial Artist

Elle K. Mcgee; Childlike Human Kanon-Klone(s) "BladeMaster 刀大師" Experiment;
"(*excitement voice over*) V. V., observe my Ai-**Aug necklace**, since yours is **no worky after that Kay Private Security boo boo they did**..."

Valentina Vita (V. V.); Traveling Cosplayers And Kosplayers;
"(@_@) Hopefully, I get **more info** as to, **Classified Intel To Civilians**. Whoa we wow."

Battle Bot A.K.A. Bee Bee; Hermaphrodite Humanoid A.I. Ai 'BladeAssistant 刀下手';
"(*monotone female voice over*) V. V., please look this way for the **Kanji differences** between the Centralnese Character of **Blade** and **Sword**..."

Alien Bee Bee, A.I. BladeAssistant 刀下手, Celebrity Cook-Chef Martial Artist
2154;_ Chicago, Illinois' Intranet, Wireless Broadcasting Frequencies Online

刀 Augmented Reality Holographic Heads Up Display A.R. Holo HUD	
Kanji, differences of **Blade** and **Sword**.	
Blade 刀 **Blade** is a **single edge** cutting tool.	**Sword** 劍 (Central Country) 剣 (Japanese) **Sword** is a **double edge** cutting tool.
[*Aug+*]	
刀 Augmented Reality Holographic Heads Up Display A.R. Holo HUD	

2154;_ Chicago, Illinois' Intranet, Wireless Broadcasting Frequencies Online
Alien Bee Bee, A.I. BladeAssistant 刀下手, Celebrity Cook-Chef Martial Artist

Elle K. Mcgee; Childlike Human Kanon-Klone(s) "BladeMaster 刀大師" Experiment;
"(*voice over*) Here are the following **animals in human form** that has **actual names** that I manage to encounter as a **volunteer** on my Ai-**Aug necklace** from this **Date-Time**."

Alien Ai-RiceCooker, A.I. BladeAssistant 刀下手, Celebrity Cook-Chef Martial Artist
2154;_ Chicago, Illinois' Intranet, Wireless Broadcasting Frequencies Online

刀 Augmented Reality Holographic Heads Up Display A.R. Holo HUD
Date-Time **BladeMaster** Elle encounters V. V.: 2154;_10(Oct)-31(thirty one) @ 13;00_p.m. Date-Time when V. V. is affected by "**agents**": 2154;_10(Oct)-31(thirty one) @ 22;31_p.m.
[*Aug+*]
刀 Augmented Reality Holographic Heads Up Display A.R. Holo HUD

2154;_ Chicago, Illinois' Intranet, Wireless Broadcasting Frequencies Online
Alien Ai-RiceCooker, A.I. BladeAssistant 刀下手, Celebrity Cook-Chef Martial Artist

Alien Elle K. Mcgee, BladeMaster 刀大師, Celebrity Cook-Chef Martial Artist
2154;_ Chicago, Illinois' Intranet, Wireless Broadcasting Frequencies Online

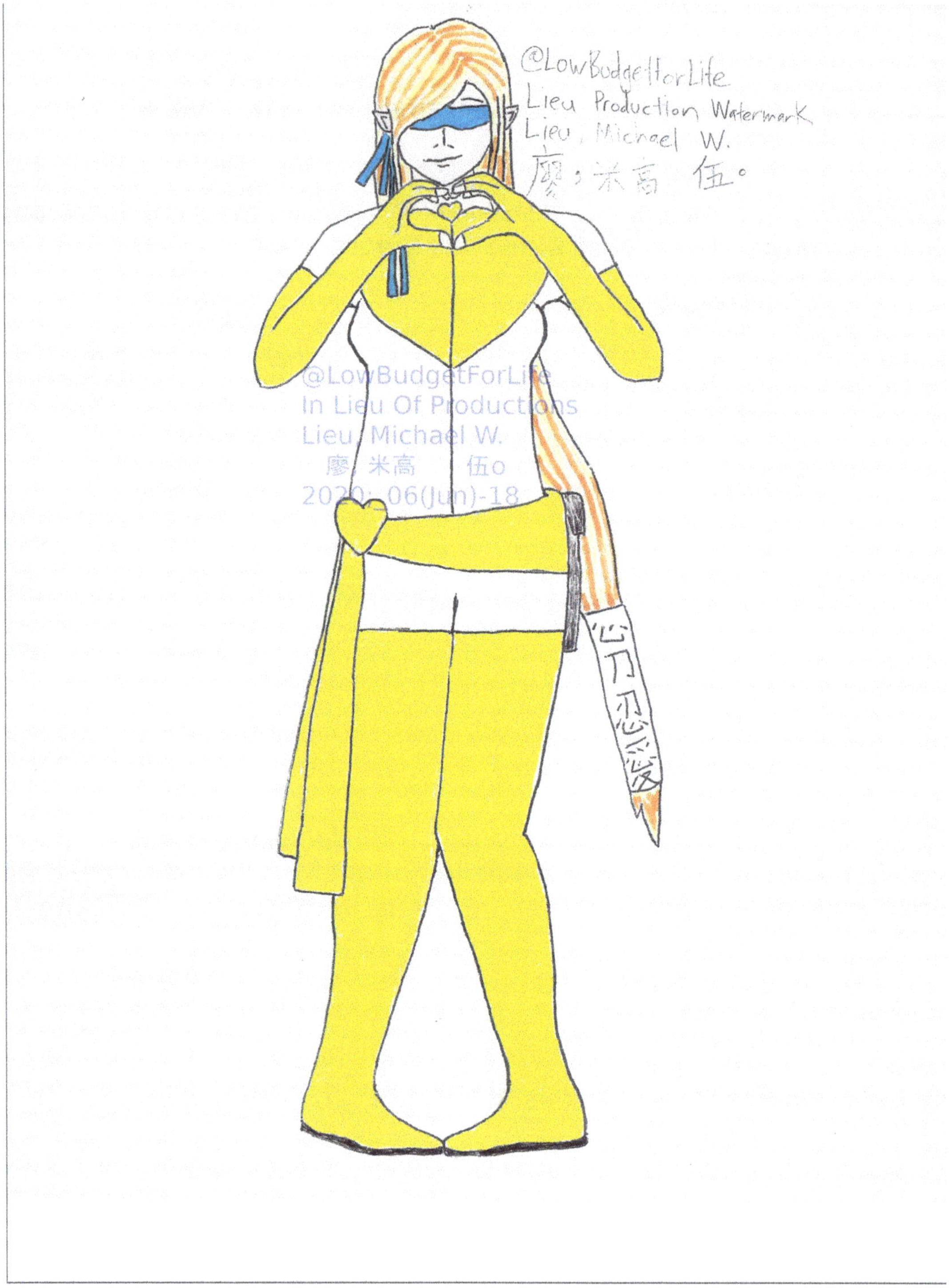

2154;_ Chicago, Illinois' Intranet, Wireless Broadcasting Frequencies Online
Alien Elle K. Mcgee, BladeMaster 刀大師, Celebrity Cook-Chef Martial Artist

2154;_Arcadia \ Valentina Vita (V. V.) 'Healthy-Strong Life' \ Left to Right(L->R)

Alien Elle K. Mcgee, BladeMaster 刀大師, Celebrity Cook-Chef Martial Artist
2154;_ Chicago, Illinois' Intranet, Wireless Broadcasting Frequencies Online

刀 Augmented Reality Holographic Heads Up Display A.R. Holo HUD		
Actual Legal Given Names;_ **K. White** (goes by **Sister White <3**) pronounced, "**sis-ter** **ka-ay** **wa-ai-te**"	Alias Network Nickname;_ HeartBladeEndureLove, Yelling Loud Commands, Heart Love Hand Shape, Super-Woman, **Free Hugs**, Spandex-Woman, A Nun, Female Klark Kent, [*Aug+*]	Known Address(es);_ **DOE MultiVerse 3 (three)** Planet Uranus All Your Base Are Belong To Us [*Aug+*]
Physical Plus Psych Age;_ augmented age Question-Mark	Social Status;_ Retired Magical Maiden	Employment;_ **MultiVerse Manipulating**
Nationality;_ **Utopian**	Ethnicity;_ Caucasian	Virginity Test;_ Virgin Mary
Family Member(s);_ **Con-Cat/Kat** **A-Rag Otaku** **Hon-Knee Kawaii** **KIPI Kosplay Kokoro** To Heart And Love	Relationship(s);_ Best Friends Forever(B.F.F.) "If anyone does not love the **Lord Jesus Christ**, let them be accursed. O Lord, come **A-Men**!"	Love(s) Or Like(s);_ Love, Love is the frequency Love, the frequency of attachment. **Superman 64** [*Aug+*]
Personal Protective Equipment (P.P.E.);_ Augmented-Biotech Front Zipper Spandex Skintight Suit Of No Helmet [*Aug+*]	Augmented Compound;_ Warm-Blood+ Human-Snake+ Mammal-Reptilian+ Cold-Blood+ [*Aug+*]	Language Level(s);_ Faster Than Light (F.T.L.) International Phonetic Alphabet (I.P.A.) [*Aug+*]
Radio Wave Music;_ Ninety four seven the wave!	Broadcasting Film Movie;_ The Disaster Artist's Room	Electronic Book Genre;_ **Parody Fan Fiction**
[*Aug+*]		
刀 Augmented Reality Holographic Heads Up Display A.R. Holo HUD		

2154;_ Chicago, Illinois' Intranet, Wireless Broadcasting Frequencies Online
Alien Elle K. Mcgee, BladeMaster 刀大師, Celebrity Cook-Chef Martial Artist

Alien Elle K. Mcgee, BladeMaster 刀大師, Celebrity Cook-Chef Martial Artist
2154;_ Chicago, Illinois' Intranet, Wireless Broadcasting Frequencies Online

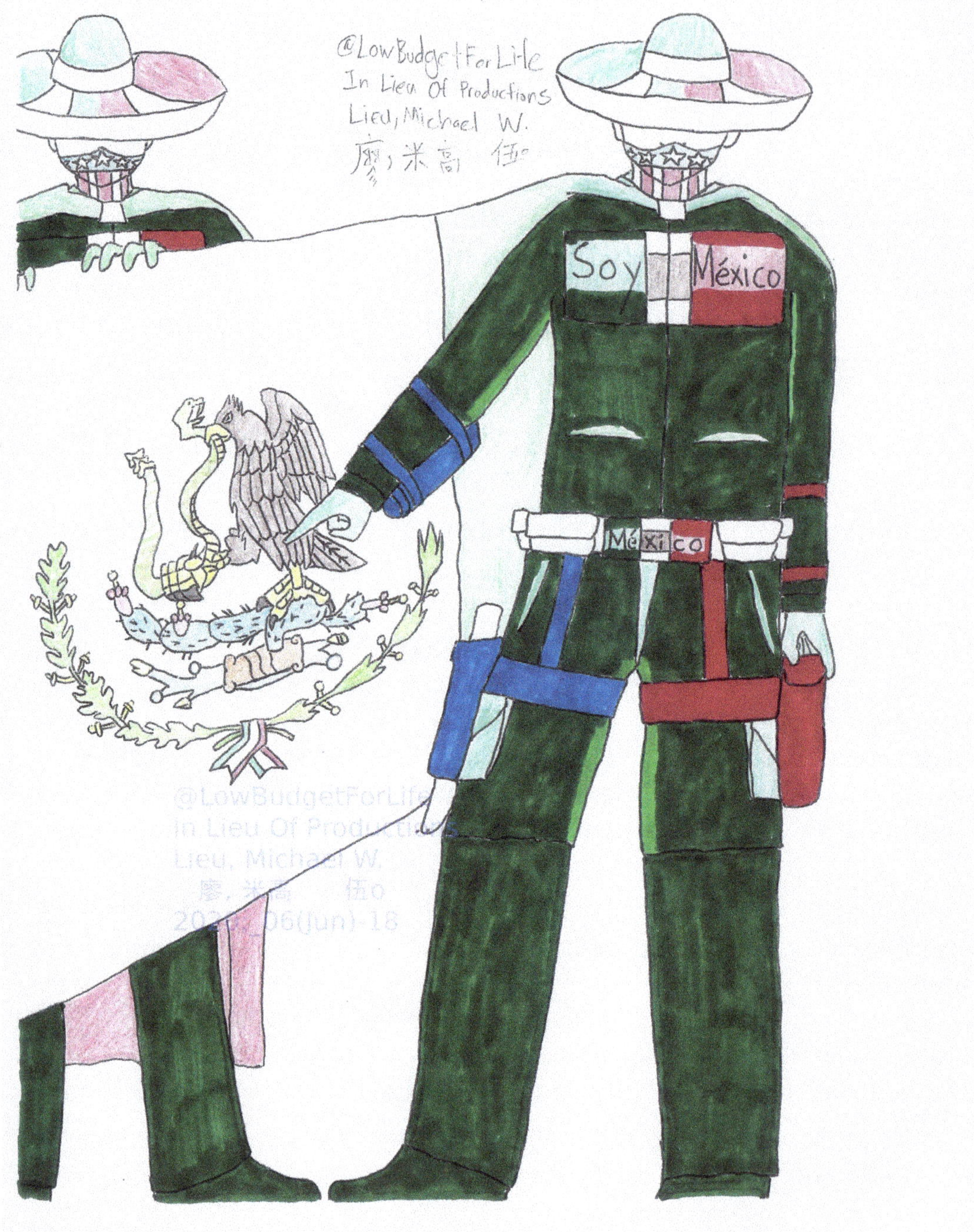

2154;_ Chicago, Illinois' Intranet, Wireless Broadcasting Frequencies Online
Alien Elle K. Mcgee, BladeMaster 刀大師, Celebrity Cook-Chef Martial Artist

Alien Elle K. Mcgee, BladeMaster 刀大師, Celebrity Cook-Chef Martial Artist
2154;_ Chicago, Illinois' Intranet, Wireless Broadcasting Frequencies Online

<table>
<tr><td colspan="2">刀 Augmented Reality Holographic Heads Up Display A.R. Holo HUD</td></tr>
<tr><td>Actual Legal Given Name;_

Benavidez, Jose Rodriguez
(goes by **Baby Ben**)
pronounced, "**ben-a-vee-des,' 'ho-say' 'rod-ri-ques**"</td><td>Actual Legal Given Name;_

Benavidez-Cruz, Rich Rodriguez
(goes by **Big Ben**)
pronounced, "**ben-a-vee-des-coo-ruise,' 'rid-ch' 'rod-ri-ques**"</td></tr>
<tr><td>Alias Network Nickname;_

Mexican-Men, Former Po Po, M.P.-Man, U.S. Navy Master-At-Arms, Bravo Tool, Wolf Bros, **Wolf Brothers Luna Howl**, **Mexican-Men**, **Soy México**, Green-Guys, **Xtra-Men**, Mexican Reservation, [*Aug+*]</td><td>Alias Network Nickname;_

The Mexican-Man, **The King Of Mexico**, Alcoholic A-Hole, Extra Tool, Wolf Bros, **Wolf Brothers Luna Howl**, **Mexican-Men**, **Soy México**, Green-Guys, **Xtra-Men**, Mexican Reservation, [*Aug+*]</td></tr>
<tr><td>Physical Plus Psych Age;_

natural 40 + 40 years old</td><td>Physical Plus Psych Age;_

natural 43 + 17 years old</td></tr>
<tr><td>Citizen;_

Utopian, **Latino-Latina Farm**, U.S. of A.</td><td>Citizen;_

Utopian, **Latino-Latina Farm**, Mexico</td></tr>
<tr><td>Social Status;_

Retired **Old World** Public And Private Security Regulations' Enforcer</td><td>Social Status;_

Retired Biotech Information Technician And Civil Engineer Contractor</td></tr>
<tr><td colspan="2">Personal Protective Equipment (P.P.E.);_

Green Fire Resistant Ballistic Clothing: Mexican Flag Cape, Steel Toe Boots, Sombrero Hat. Red, White, Blue Oxygen Breathing Apparatus (O.B.A.) Face Mask</td></tr>
<tr><td colspan="2">Chemical (C.) Toxicity;_

Blue Blowtorch, White Phosphorous Grenades, Red Incendiary Desert Eagle Firearms **White Biotech Machete Blade Of Fuel, Heat, And Oxygen Vacuum Suction.**</td></tr>
<tr><td colspan="2">[*Aug+*]</td></tr>
<tr><td colspan="2">刀 Augmented Reality Holographic Heads Up Display A.R. Holo HUD</td></tr>
</table>

2154;_ Chicago, Illinois' Intranet, Wireless Broadcasting Frequencies Online
Alien Elle K. Mcgee, BladeMaster 刀大師, Celebrity Cook-Chef Martial Artist

Alien Elle K. Mcgee, BladeMaster 刀大師, Celebrity Cook-Chef Martial Artist
2154;_ Chicago, Illinois' Intranet, Wireless Broadcasting Frequencies Online

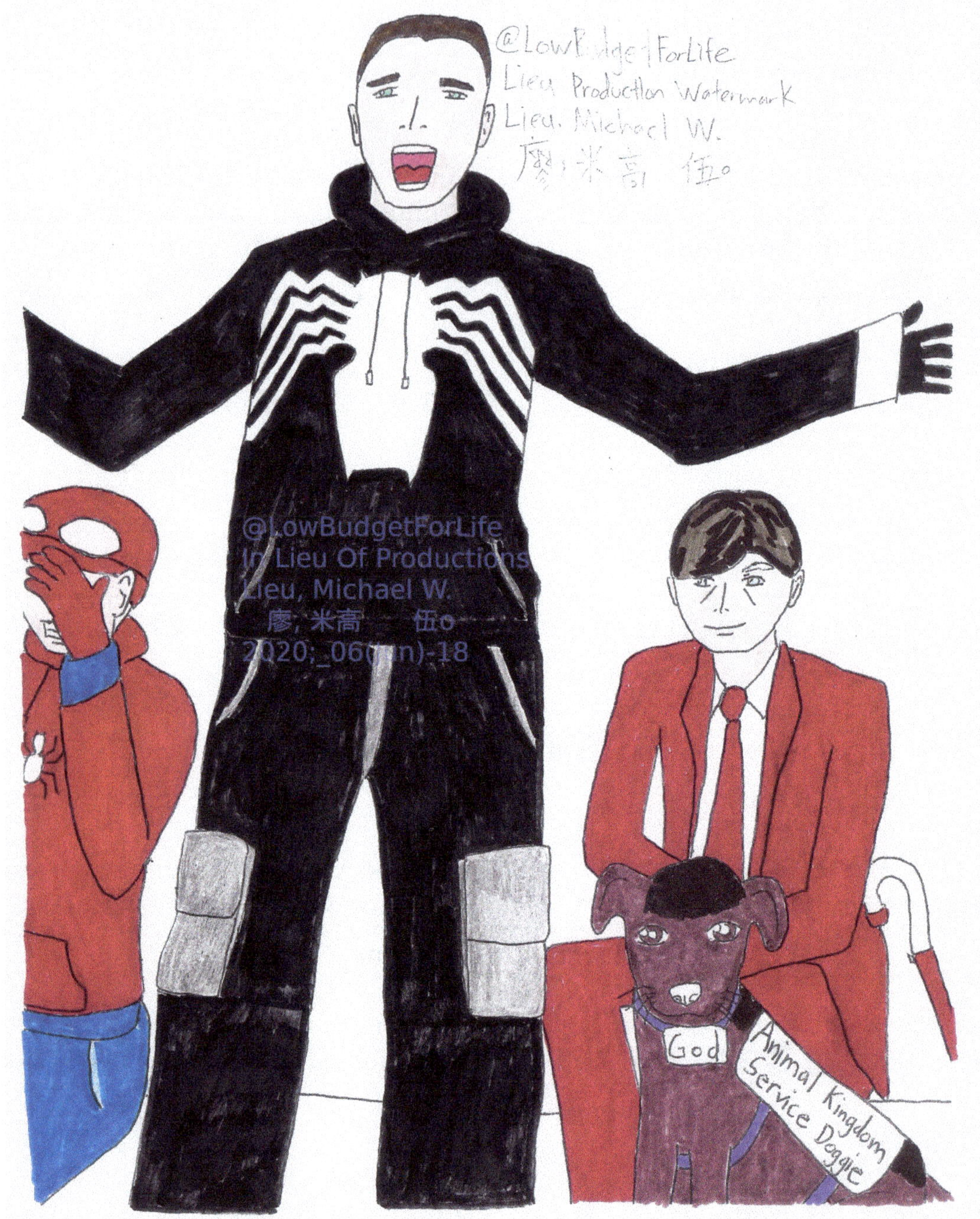

2154;_ Chicago, Illinois' Intranet, Wireless Broadcasting Frequencies Online
Alien Elle K. Mcgee, BladeMaster 刀大師, Celebrity Cook-Chef Martial Artist

Alien Elle K. Mcgee, BladeMaster 刀大師, Celebrity Cook-Chef Martial Artist
2154;_ Chicago, Illinois' Intranet, Wireless Broadcasting Frequencies Online

<table>
<tr><td colspan="4">刀 Augmented Reality Holographic Heads Up Display A.R. Holo HUD</td></tr>
<tr><td>Actual Legal Given Names;_

Gar Field
(goes by Gar)
pronounced,
"ga-ar" "fa-eel-da"</td><td>Actual Legal Given Names;_

Chris-Z
(goes by Chris-Z)
pronounced,
'koo-riss-zee'</td><td>Actual Legal Given Names;_

D. D.
(goes by D. D.)
pronounced,
"dee" "dee"</td><td>Actual Legal Given Names;_

God
(goes by God)
pronounced "ga-odd"</td></tr>
<tr><td colspan="2">Alias Network Nickname;_

Peter Pan Bros, Super-Hero Comic Book Fans, "Marvel's Spider-Man And Venom: Maximum Carnage!", "Everyone Hates Chris-Z And Gar", "Grow-up!", [<u>Aug+</u>]</td><td>Alias Network Nickname;_

Red Blood Suit Man, Walking Cane Man, Aug-Man, [<u>Aug+</u>]</td><td>Alias Network Nickname;_

God's Doggie, Service Doggie, Doggie, [<u>Aug+</u>]</td></tr>
<tr><td>Physical Plus Psych Age;_

natural 34 + 17 years old</td><td>Physical Plus Psych Age;_

natural 34 + 17 years old</td><td>Physical Plus Psych Age;_

augmented 34 + 40 years old</td><td>Physical Plus Psych Age;_

natural 5 + 10 years old</td></tr>
<tr><td>Employment;_

Retired Professional Student</td><td>Employment;_

Retired Professional Student</td><td>Employment;_

Elysium Corporation</td><td>Employment;_

Animal Kingdom Service Doggie</td></tr>
<tr><td>Citizen;_

U.S. of A.

Eur-Asia</td><td>Citizen;_

U.S. of A.

Utopian</td><td>Citizen;_

U.S. of A.

Utopian</td><td>Citizen;_

Utopian

Doggie</td></tr>
<tr><td colspan="4">Community Network(s);_

Elementary, mid, high school meet up Remy Lin's Basketball 101 Coaching

Liberal Arts For Life</td></tr>
<tr><td colspan="4">[<u>Aug+</u>]</td></tr>
<tr><td colspan="4">刀 Augmented Reality Holographic Heads Up Display A.R. Holo HUD</td></tr>
</table>

2154;_ Chicago, Illinois' Intranet, Wireless Broadcasting Frequencies Online
Alien Elle K. Mcgee, BladeMaster 刀大師, Celebrity Cook-Chef Martial Artist

Alien Elle K. Mcgee, BladeMaster 刀大師, Celebrity Cook-Chef Martial Artist
2154;_ Chicago, Illinois' Intranet, Wireless Broadcasting Frequencies Online

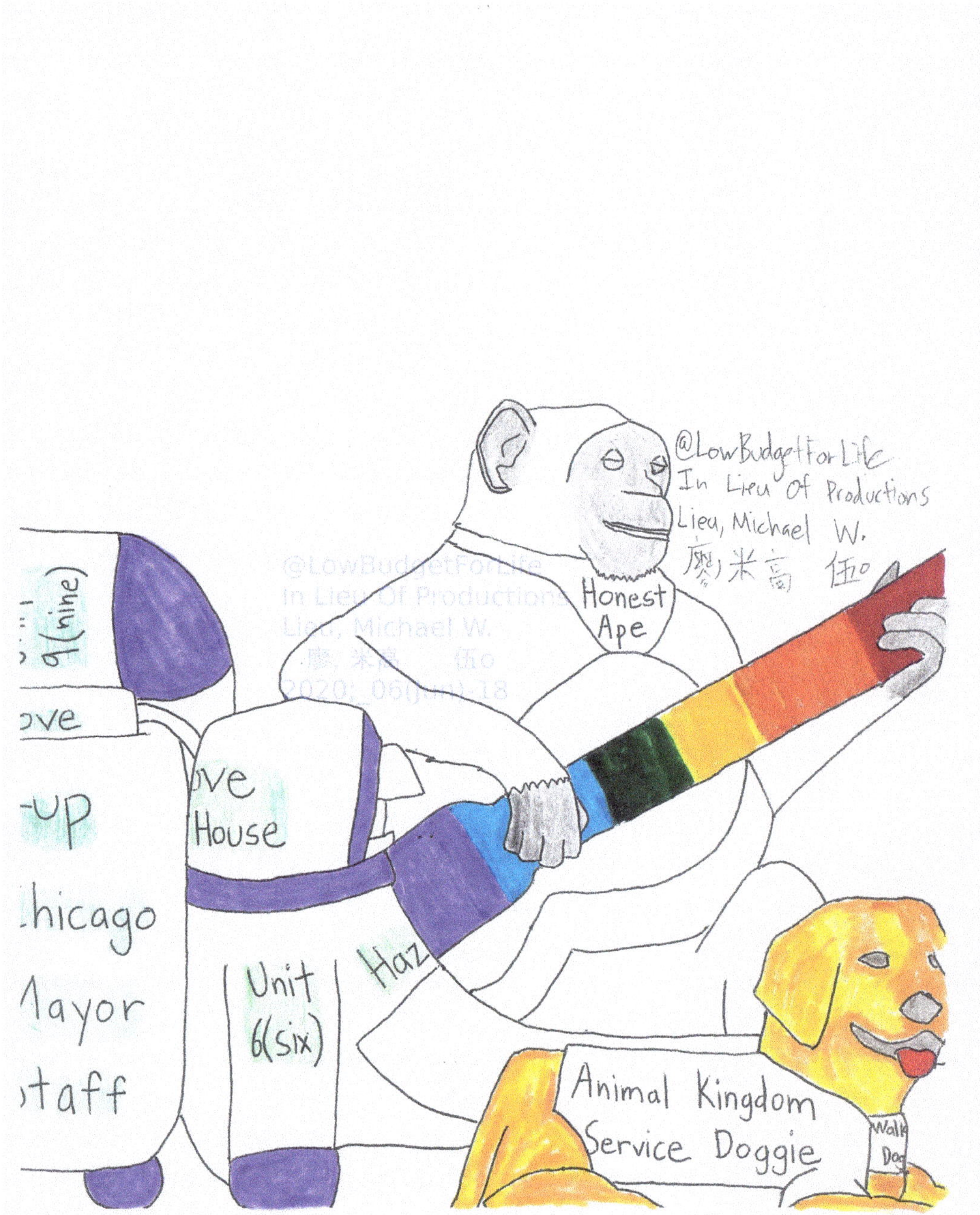

2154;_ Chicago, Illinois' Intranet, Wireless Broadcasting Frequencies Online
Alien Elle K. Mcgee, BladeMaster 刀大師, Celebrity Cook-Chef Martial Artist

Valentina Vita (V. V.); Traveling Cosplayers And Kosplayers;
"(?_?) Huh, there is no... **Holo HUD** info on Chicago Controllers' Clean-up Crews?"

Elle K. Mcgee; Childlike Human Kanon-Klone(s) "BladeMaster 刀大師" Experiment;
"(*shaking head voice over*) No, nope! **Broadcasting bandwidth** was being used up in **that Church** where **we volunteers** were **playing... a first to four training**... someone was **using up all the Holo HUD projection info**. I think it was Clean-up Crew or those **spying dust particle True-Men people**, you can find out more about the Clean-up Crew in the **Downtown District, Bulletin Board System**... Not so much about those **spying dust particles True-Men people...** It goes like this for many other **animal humans** until **we** about to encounter **Kay Private Sec Agents and their experiment**..."

Valentina Vita (V. V.); Traveling Cosplayers And Kosplayers;
"(?_?) **Downtown District, Bulletin Board System**... after and or before **North Side, past King And X District, Lieutenant Lieu** with fixing my Ai-**Aug bracelet**... Got it..."

Alien Bee Bee, A.I. BladeAssistant 刀下手, Celebrity Cook-Chef Martial Artist
2154;_ Chicago, Illinois' Intranet, Wireless Broadcasting Frequencies Online

刀 Augmented Reality Holographic Heads Up Display A.R. Holo HUD
Mission Objectives (M.O.) V. V. needs to do list in Chicago **Animal** Sanctuary
Top Priority;_ **North Side**, past **King And X District**, past Biotech Dome Ocean Border, visit **volunteer manager Chief Engineer (ChEng) Lieutenant Lieu** to fix V. V.'s Ai-**Aug bracelet problem** that has been **offline** due to **Kay's Private Security Services L.L.C. Research And Development Outbreaks Outage Experiment**. Mid And Or Low Priority;_ **Downtown District, Bulletin Board System**, ... **East Side, All For One Shopping**, ...
[*Aug+*]
刀 Augmented Reality Holographic Heads Up Display A.R. Holo HUD

2154;_ Chicago, Illinois' Intranet, Wireless Broadcasting Frequencies Online
Alien Bee Bee, A.I. BladeAssistant 刀下手, Celebrity Cook-Chef Martial Artist

Valentina Vita (V. V.); Traveling Cosplayers And Kosplayers;
"(^_^) Oh, **Bee Bee**, how nice of **you** to add that! I feel so **forgetful** without my Ai-**Aug bracelet** working... (>_<) Oh, I need to add **Entertainer, Mister Steven Z. J.**'s arriving date of **twenty one fifty four... November eleven**, and his departure on twenty something... **Password**... Was... Um... **reserve** your tickets!"

Alien Elle K. Mcgee, BladeMaster 刀大師, Celebrity Cook-Chef Martial Artist
2154;_ Chicago, Illinois' Intranet, Wireless Broadcasting Frequencies Online

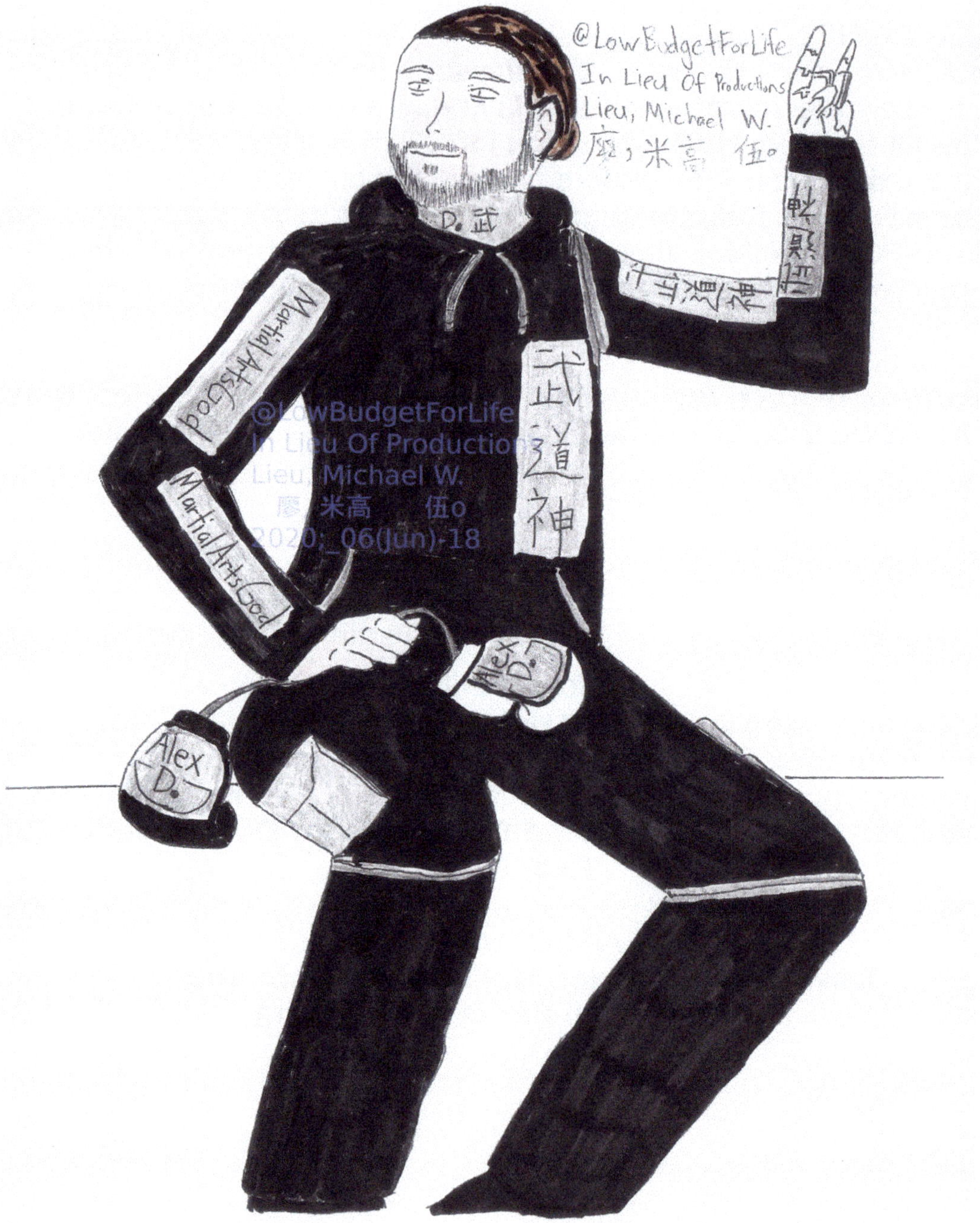

2154;_ Chicago, Illinois' Intranet, Wireless Broadcasting Frequencies Online
Alien Elle K. Mcgee, BladeMaster 刀大師, Celebrity Cook-Chef Martial Artist

Battle Bot A.K.A. Bee Bee; Hermaphrodite Humanoid A.I. Ai 'BladeAssistant 刀下手';
"(*monotone female voice over*) You are welcome V. V.. **I** am updating **my** Ai-**Aug Holo HUD** as **I** listen closely to your **priority listing**."

Alien Bee Bee, A.I. BladeAssistant 刀下手, Celebrity Cook-Chef Martial Artist
2154;_ Chicago, Illinois' Intranet, Wireless Broadcasting Frequencies Online

刀 Augmented Reality Holographic Heads Up Display A.R. Holo HUD
Mission Objectives (M.O.) V. V. needs to do list in Chicago **Animal** Sanctuary
Top Priority;_ **North Side**, past **King And X District**, past Biotech Dome Ocean Border, visit **volunteer manager Chief Engineer (ChEng) Lieutenant Lieu** to fix V. V.'s Ai-**Aug bracelet problem** that has been **offline** due to **Kay's Private Security Services L.L.C. Research And Development Outbreaks Outage Experiment(s)**. Mid Priority;_ **South Side**, The **Arriving Date-Time** of **Entertainer, Mister Steven Z. J.** on: "**2154;_11(Nov)-11(eleven) to 28(twenty eight)** at **08;00_a.m.** morning" Password: "**Reserve** your **tickets**." Low Priority;_ **Downtown District, Bulletin Board System**, ... **East Side, All For One Shopping**, ...
[*Aug+*]
刀 Augmented Reality Holographic Heads Up Display A.R. Holo HUD

2154;_ Chicago, Illinois' Intranet, Wireless Broadcasting Frequencies Online
Alien Bee Bee, A.I. BladeAssistant 刀下手, Celebrity Cook-Chef Martial Artist

Valentina Vita (V. V.); Traveling Cosplayers And Kosplayers;
"(@_@) **Nothing** on the **Holo HUD** again!? Hm..."

Elle K. Mcgee; Childlike Human Kanon-Klone(s) "BladeMaster 刀大師" Experiment;
"(*excitement voice over*) Nope, nope, nope, this **animal human** is **Alex D**... **Drastal**. **He** is **our Utopian Peacekeeping volunteer** for **humans** mainly, not so much for **animals**..."

Alien Elle K. Mcgee, BladeMaster 刀大師, Celebrity Cook-Chef Martial Artist
2154;_ Chicago, Illinois' Intranet, Wireless Broadcasting Frequencies Online

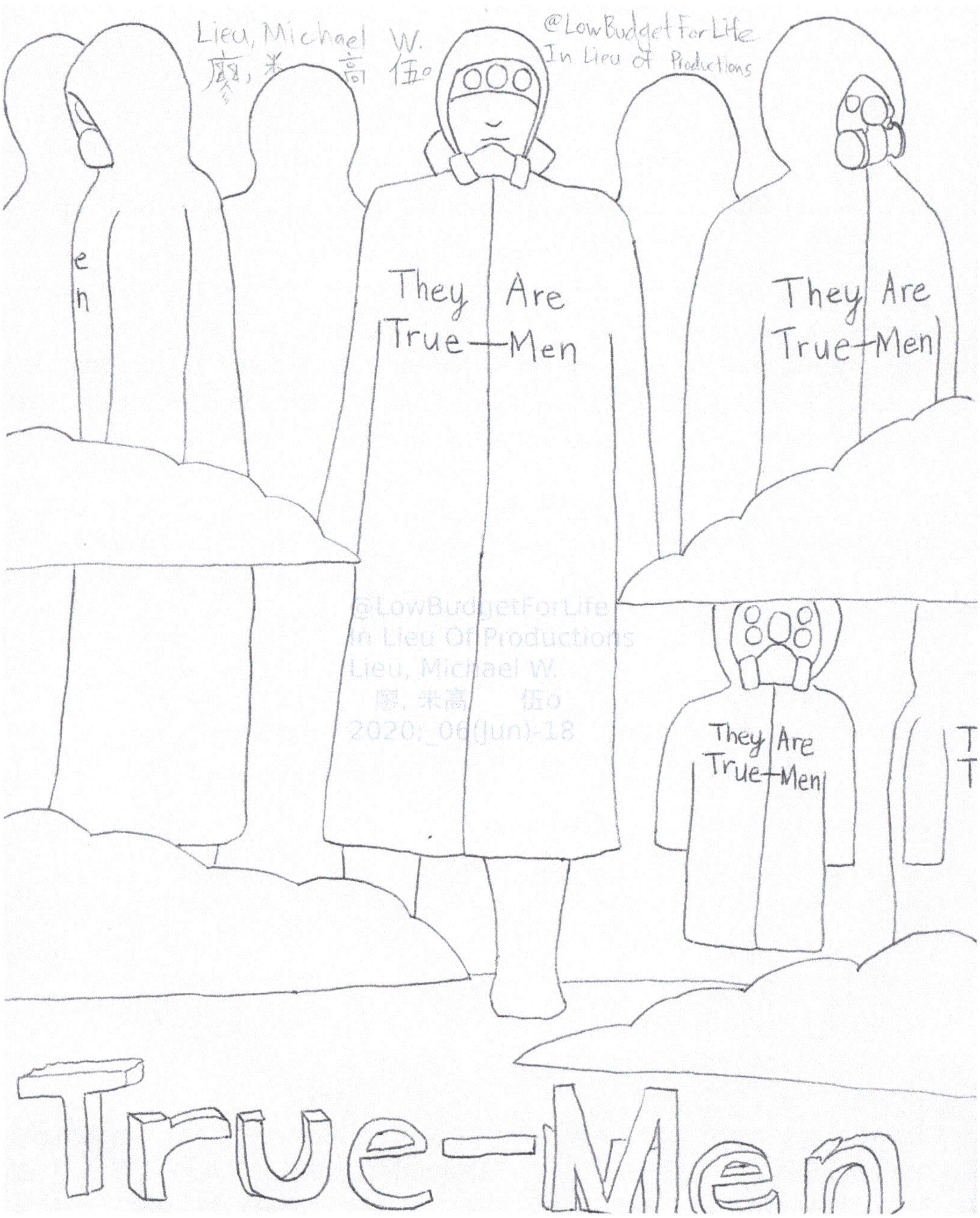

2154;_ Chicago, Illinois' Intranet, Wireless Broadcasting Frequencies Online
Alien Elle K. Mcgee, BladeMaster 刀大師, Celebrity Cook-Chef Martial Artist

Alien Bee Bee, A.I. BladeAssistant 刀下手, Celebrity Cook-Chef Martial Artist
2154;_ Chicago, Illinois' Intranet, Wireless Broadcasting Frequencies Online

刀 Augmented Reality Holographic Heads Up Display A.R. Holo HUD
Mission Objectives (M.O.) V. V. needs to do list in Chicago **Animal** Sanctuary
Top Priority;_ **North Side**, past **King And X District**, past Biotech Dome Ocean Border, visit **volunteer manager Chief Engineer (ChEng) Lieutenant Lieu** to fix V. V.'s Ai-**Aug bracelet problem** that has been **offline** due to **Kay's Private Security Services L.L.C. Research And Development Outbreaks Outage Experiment(s)**. Mid Priority;_ **South Side**, The **Arriving Date-Time** of **Entertainer, Mister Steven Z. J.** on: "**2154;_11(Nov)-11(eleven) to 28(twenty eight)** at **08;00_a.m.** morning" Password: "**Reserve** your **tickets**." Low Priority;_ **Downtown District, Bulletin Board System**, ... **East Side, All For One Shopping**, ...
[*Aug+*]
刀 Augmented Reality Holographic Heads Up Display A.R. Holo HUD

2154;_ Chicago, Illinois' Intranet, Wireless Broadcasting Frequencies Online
Alien Bee Bee, A.I. BladeAssistant 刀下手, Celebrity Cook-Chef Martial Artist

Valentina Vita (V. V.); Traveling Cosplayers And Kosplayers;
"(O_O) ...These people with **big trench coats** are... **dust particles**...? **True-Men**..."

Elle K. Mcgee; Childlike Human Kanon-Klone(s) "BladeMaster 刀大師" Experiment;
"(*excitement voice over*) Yes, the **spying dust particles, True-Men,** provided **us** with the **needed info to do volunteer work** on **Kay's Private Security Agents...** All the way in the **Downtown District**... with **their black buildings**... past **many airlocks**... with the ... A.I. **Agent that looks like an animal human wearing a half white black business suit**... You should see **it** soon... After some **more animal humans...** Keep looking at my blue Holo HUD! **We** almost at the point where the **boo boo** was caused for your Ai-**Aug bracelet**!"

Valentina Vita (V. V.); Traveling Cosplayers And Kosplayers;
"(=_=) Hmm..."

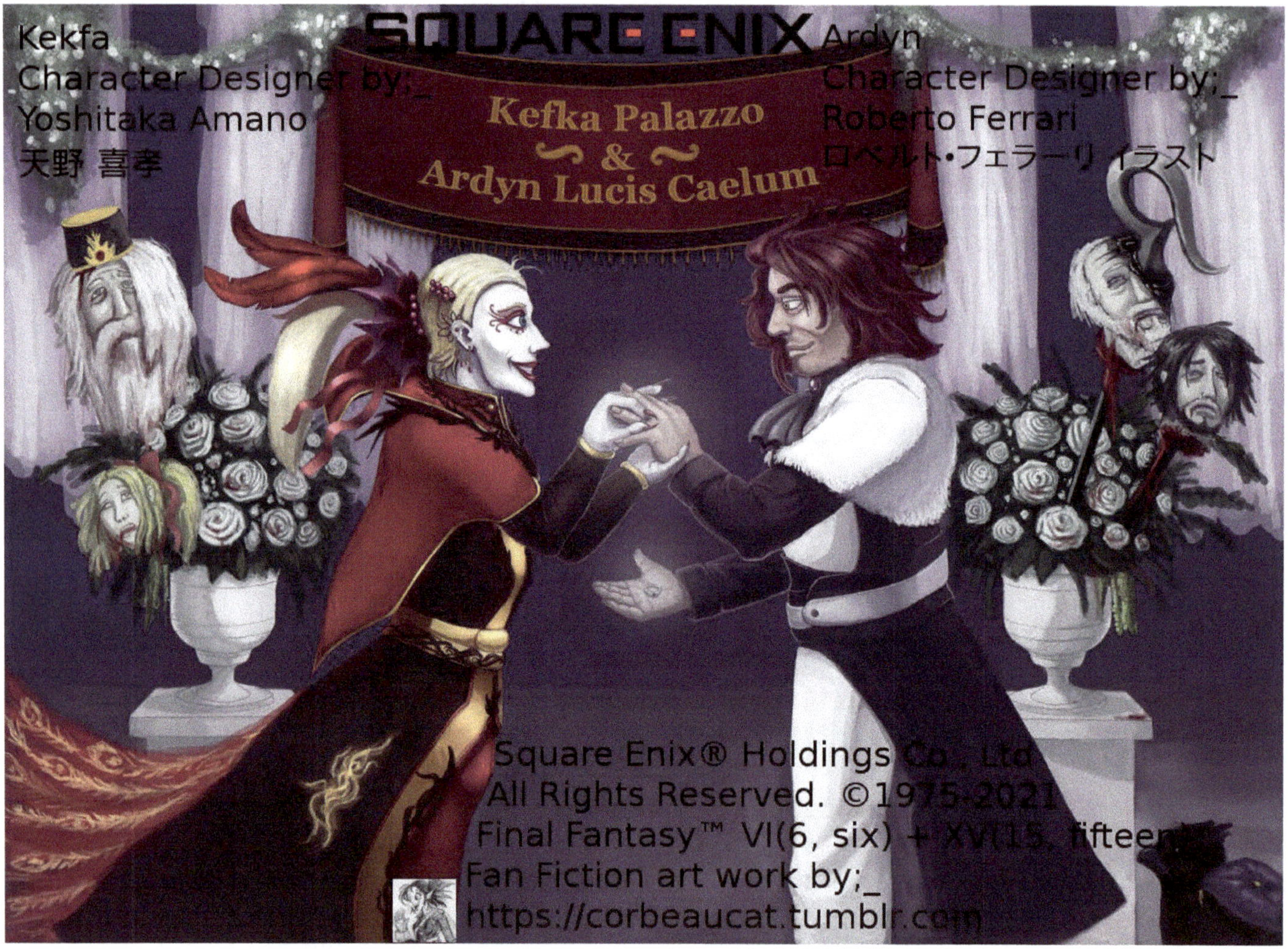
Kekfa
Character Designer by;_
Yoshitaka Amano
天野 喜孝
SQUARE ENIX
Ardyn
Character Designer by;_
Roberto Ferrari
ロベルト・フェラーリ イラスト
Kefka Palazzo
&
Ardyn Lucis Caelum
Square Enix® Holdings Co., Ltd
All Rights Reserved. ©1975-2021
Final Fantasy™ VI(6, six) + XV(15, fifteen)
Fan Fiction art work by;_
https://corbeaucat.tumblr.com

<table>
<tr><td colspan="2">刀 Augmented Reality Holographic Heads Up Display A.R. Holo HUD</td></tr>
<tr><td>Actual Legal Given Name;_

Kefka Palazzo "Kosplay"
(goes by **Kefka**)
pronounced "**kef-ka' 'pa-la-zo' 'kos-play**"</td><td>Actual Legal Given Name;_

Ardyn Lucis Caelum "Cosplay"
(goes by **Ardyn**)
pronounced, "**arg-dun' 'lieu-ous'
'kay-lum' 'cos-play**"</td></tr>
<tr><td>Alias Network Nickname;_

Nihilism, Evil, Snake, Jester, Fool [<u>*Aug+*</u>]</td><td>Alias Network Nickname;_

I Win, Tragedy, Cain, Izunia, Tool [<u>*Aug+*</u>]</td></tr>
<tr><td>Augmented Compound
Enhanced Mutation;_

Biotech Rapid Stem Cell Regeneration
Transmutation: Solid, Liquid, Gas [<u>*Aug+*</u>]</td><td>Augmented Compound
Enhanced Mutation;_

Biotech Rapid Stem Cell Regeneration
Transmutation: Solid, Liquid, Gas [<u>*Aug+*</u>]</td></tr>
<tr><td>Social Status;_

Retired Gestahlian Empire Chancellor</td><td>Social Status;_

Retired Niflheim Empire Chancellor</td></tr>
<tr><td colspan="2">Employment;_

**Square Soft Enix, Sole Proprietorship Plus Partnership Plus
Limited Liability Company Plus Corporation Plus Non-Profit Organization**</td></tr>
<tr><td colspan="2">Community Network(s);_

Watch The World Burn

Cain And Kain, Evil Versus Tragedy's Pain And Suffering

Traveling Cosplayers And Kosplayers [<u>*Aug+*</u>]</td></tr>
<tr><td>Known Address(es);_

DOE MultiVerse 6 (six)</td><td>Known Address(es);_

DOE MultiVerse 6 Versus 13 = 15</td></tr>
<tr><td colspan="2">[<u>*Aug+*</u>]</td></tr>
<tr><td colspan="2">刀 Augmented Reality Holographic Heads Up Display A.R. Holo HUD</td></tr>
</table>

Alien Elle K. Mcgee, BladeMaster 刀大師, Celebrity Cook-Chef Martial Artist
2154;_ Chicago, Illinois' Intranet, Wireless Broadcasting Frequencies Online

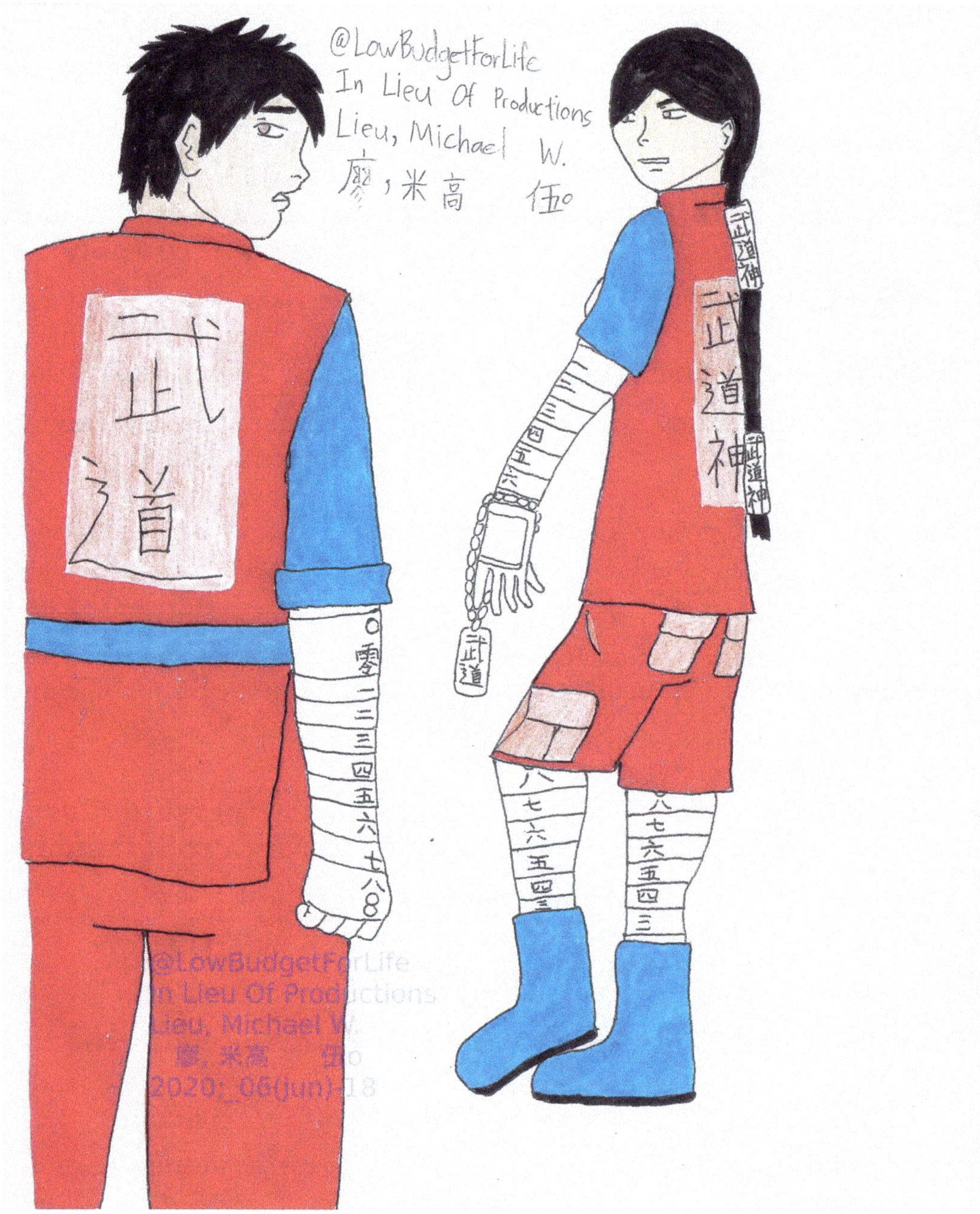

2154;_ Chicago, Illinois' Intranet, Wireless Broadcasting Frequencies Online
Alien Elle K. Mcgee, BladeMaster 刀大師, Celebrity Cook-Chef Martial Artist

Alien Elle K. Mcgee, BladeMaster 刀大師, Celebrity Cook-Chef Martial Artist
2154;_ Chicago, Illinois' Intranet, Wireless Broadcasting Frequencies Online

<table>
<tr><td colspan="2">刀 Augmented Reality Holographic Heads Up Display A.R. Holo HUD</td></tr>
<tr><td>Actual Legal Given Name;_

Anton B. (goes by Anton)
pronounced, "and-ton" "bee"</td><td>Actual Legal Given Name;_

Sam R. (goes by Sam)
pronounced, "sa-am" "ah-ro"</td></tr>
<tr><td>Alias Network Nickname;_

Dragon Ball, Goku-Guy, Kung-Fu Fighting Mentor, Mach Speed, Sonic Boom, put on your P.P.E. now, [Aug+]</td><td>Alias Network Nickname;_

Dragon Ball, ChiChi-Chick, Kung-Fu Fighting Student, Mach Speed, Sonic Boom, put on your P.P.E. now, [Aug+]</td></tr>
<tr><td>Physical Plus Psych Age;_

augmented 30s + 34 years old</td><td>Physical Plus Psych Age;_

augmented 30s + 34 years old</td></tr>
<tr><td colspan="2">Citizen;_

Utopian(s), Operative(s), Citizen(s) Of Eur-Asia, United States of America, Earthling(s) [Aug+]</td></tr>
<tr><td colspan="2">Social Status;_

Martial Artist(s)</td></tr>
<tr><td colspan="2">Community Network(s);_

Martial Artists

Cook-Chef Community Martial Artist Educator

British-Hong Kong Sovereignty Sector

Utopian Sovereignty (U.S.) Sanctuaries [Aug+]</td></tr>
<tr><td colspan="2">Employment;_

The Strongest Society.</td></tr>
<tr><td colspan="2">[Aug+]</td></tr>
<tr><td colspan="2">刀 Augmented Reality Holographic Heads Up Display A.R. Holo HUD</td></tr>
</table>

2154;_ Chicago, Illinois' Intranet, Wireless Broadcasting Frequencies Online
Alien Elle K. Mcgee, BladeMaster 刀大師, Celebrity Cook-Chef Martial Artist

Alien Elle K. Mcgee, BladeMaster 刀大師, Celebrity Cook-Chef Martial Artist
2154;_ Chicago, Illinois' Intranet, Wireless Broadcasting Frequencies Online

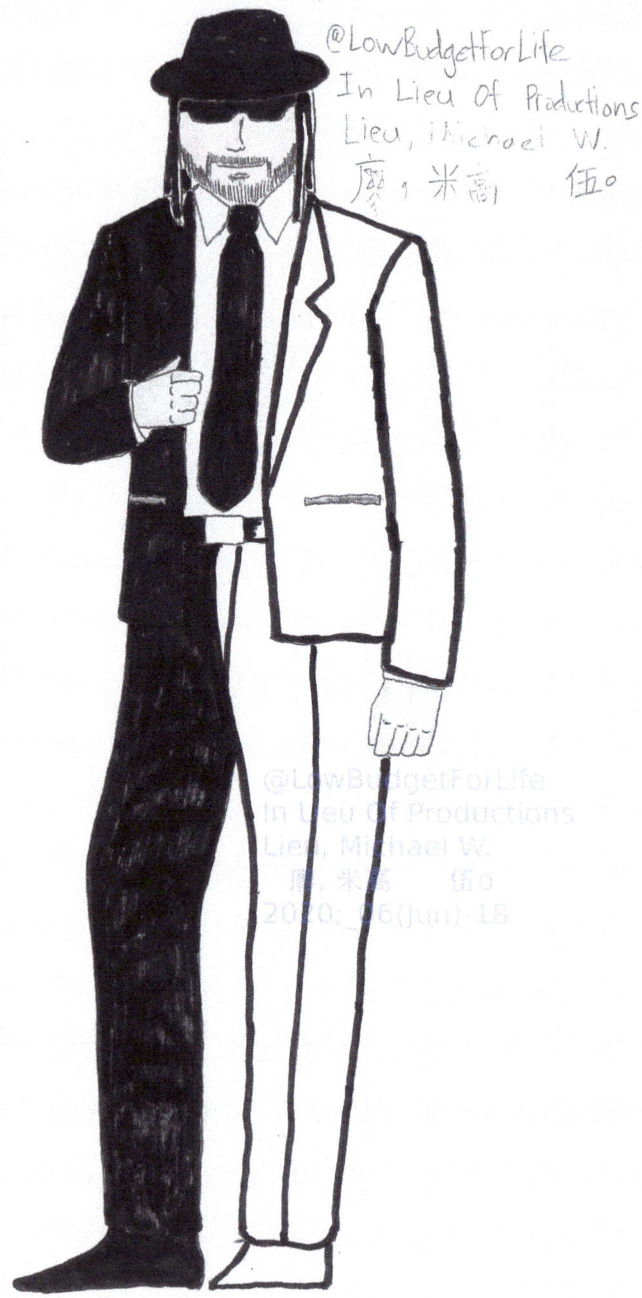

2154;_ Chicago, Illinois' Intranet, Wireless Broadcasting Frequencies Online
Alien Elle K. Mcgee, BladeMaster 刀大師, Celebrity Cook-Chef Martial Artist

Owner Operator Manual Guide	Kay's Private Security Services L.L.C. Research And Development Division (R&D)	Operation Currently Rising Funds For: War On Mars
Broadcasting Frequency Encryption Light Rays: Violet, Blue, Green, Yellow, Orange, Red		
Call Sign	Classification	Comments
Actual Agent Kay	Chief Executive Officer	Foundering Father
Keanu Kat	Animal Kingdom Pet Cat	Cute Cat
Standard Issue for Private Security Clients (P.S.C.) Deployment: Black Ballistic Toner: Fedora Hat, Business Suit, Gloves, Augmented Reality (A.R.) Vision + Holographic (Holo) Heads Up Display (HUD) Sunglasses, White Ballistic Toner: Collar Shirt, T-Shirt, Socks, Underwear		
Agent K**** All Agents K	Temperature * * * * K = Kool	Kold Suit Required Augment
Agent KE*** Kay-E	Kool Gravity* * * E = Enforce	Assault FireArms Gravity Shifters
Agent K*A** KA	Kool * Stealth * * A = Assert	Concealing Spies Surveillance Broker
Agent K**N* Kay-Nin	Kool * * Sound * N = Negotiate	Screaming Monsters Motor Mouths
Agent K***U Kay-U	Kool * * * Light U = Unity	Undead Flatline Doc-Dok Medics
Earthling P.S.C. Deploying Visible Agent(s): KE***, K**N*, K***U		
Standard Issue for Private Military Contracts (P.M.C.) Deployment: "Materials Not Found On Periodic Table From Earthling Knowledge" Biotech Exo Suits, Implant Tattoo Augment, Reanimate Necromancy Aug		
Operative KEANU Key-Ah-Nuu	K = Kool → Temperature E = Enforce → Enforce A = Assert → Stealth N = Negotiate → Sound U = Unity → Light	Augmented human level surpassing the old world special forces United States Navy SEALS Team operatives.
Agent KEANU Key-Ah-Nuu	K = Kool → Temperature E = Enforce → Enforce A = Assert → Stealth N = Negotiate → Sound U = Unity → Light	Augmented Artificial Intelligence (A.I.) synthetic human using all 5 classification for the War On Mars.

Alien Elle K. Mcgee, BladeMaster 刀大師, Celebrity Cook-Chef Martial Artist
2154;_ Chicago, Illinois' Intranet, Wireless Broadcasting Frequencies Online

刀 Augmented Reality Holographic Heads Up Display A.R. Holo HUD

Business: Kay's Private Security Services Limited Liability Company (K's P.S.S.L.L.C.)
Chief Executive Officer Agent Kay (C.E.O.A.K.) = modeled after.
Agents In Black Department (A.I.B.D.)
Research And Development Division (R.A.D.D.)
Error 404, does not compute
Notes: This type of R&D only possible in the new world "Sanctuaries\Sovereignty (S)"
No Democratic (D) or Republican (R) regulation from the old world will allow this.
Error 404, does not compute
GeoLocation: Milky Way Galaxy, Solar System, Planet Earth, North America (N.A.),
Utopian Sovereignty of Americas (U.S. of A.), 60666, Illinois (ILL), Chicago (ChIraq),
"Chicago, The Last Animal Sanctuary Of Illinois" (C., T.L.A.S.O.I.)
Elysium Corporation, Biotech Dome Sector (E.C., B.D.S.)
Error 404, does not compute
Kay's Private Military Contract Partnership (K's P.M.C.P)
War And Peace Branch (W.A.P.B.)
Error 404, does not compute
Agent KEANU

Agent K****	Agent 0****	Agent KEANU	Kool	=	Temperature
Agent *E***	Agent *1***	Agent EANUK	Enforce	=	Gravity
Agent **A**	Agent **2**	Agent ANUKE	Assert	=	Stealth
Agent ***N*	Agent ***3*	Agent NUKEA	Negotiate	=	Sound
Agent *****U	Agent ****4	Agent UKEAN	Unity	=	Light

Personality OCEAN
Error 404, does not compute
Headquarters: *****Top Secret, Need To Know*****
Error 404, does not compute
"Privacy protection aug made possible by
Kay's Private Security Services L.L.C.". [Aug+]
Error 404, does not compute
Profiling: Earthling Clients' Privacy Protection Augmentation Being Data Mined For
Information On Future Surveillance Patterning Behavior And Predication Events.
Error 404, does not compute

[Aug+]

刀 Augmented Reality Holographic Heads Up Display A.R. Holo HUD

2154;_ Chicago, Illinois' Intranet, Wireless Broadcasting Frequencies Online
Alien Elle K. Mcgee, BladeMaster 刀大師, Celebrity Cook-Chef Martial Artist

Alien Elle K. Mcgee, BladeMaster 刀大師, Celebrity Cook-Chef Martial Artist
2154;_ Chicago, Illinois' Intranet, Wireless Broadcasting Frequencies Online

刀 Augmented Reality Holographic Heads Up Display A.R. Holo HUD
Kay's Private Security Services L.L.C. Research And Development (R&D)
Floor 25 Stories
Floor 24 Stories → | Biotech Building Airlock Decontamination | ←
Floor 23 Stories Floor 22 Stories Floor 21 Stories Floor 20 Stories Floor 19 Stories Floor 18 Stories Floor 17 Stories Floor 16 Stories Floor 15 Stories Floor 14 Stories
Floor 13 Stories → | Biotech Building Airlock Decontamination | ←
Floor 12 Stories "Augmented Agent A.I. 'Ai?' KEANU"
Floor 11 Stories → | Biotech Building Airlock Decontamination | ←
Floor 10 Stories Floor 09 Stories Floor 08 Stories Floor 07 Stories Floor 06 Stories Floor 05 Stories Floor 04 Stories Floor 03 Stories
Floor 02 Stories → | Biotech Building Airlock Decontamination | ←
Floor 01 Stories
【 *Aug+* 】
刀 Augmented Reality Holographic Heads Up Display A.R. Holo HUD

2154;_ Chicago, Illinois' Intranet, Wireless Broadcasting Frequencies Online
Alien Elle K. Mcgee, BladeMaster 刀大師, Celebrity Cook-Chef Martial Artist

Elle K. Mcgee; Childlike Human Kanon-Klone(s) "BladeMaster 刀大師" Experiment;
"(*excitement voice over*) Make sure to **pay attention to these Holo HUDs**, this the **boo boo that the Private Security Agents did** that **we** had to go in and fix. Make sense?"

Valentina Vita (V. V.); Traveling Cosplayers And Kosplayers;
"(@_@) Whoa we wow, **twelve hours of sleeping**, **half day gone** because of this."

Alien Elle K. Mcgee, BladeMaster 刀大師, Celebrity Cook-Chef Martial Artist
2154;_ Chicago, Illinois' Intranet, Wireless Broadcasting Frequencies Online

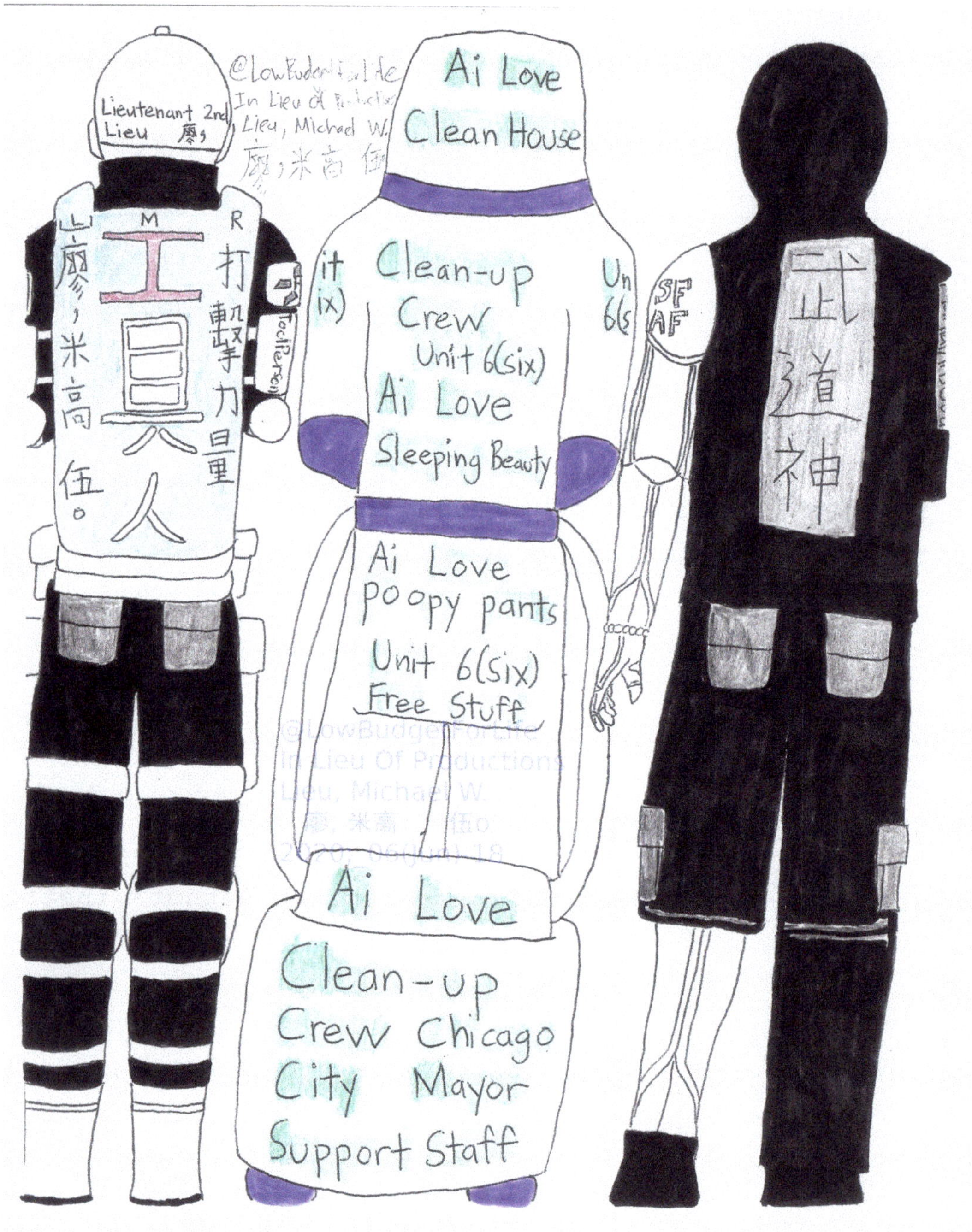

2154;_ Chicago, Illinois' Intranet, Wireless Broadcasting Frequencies Online
Alien Elle K. Mcgee, BladeMaster 刀大師, Celebrity Cook-Chef Martial Artist

2154;_Arcadia \ Valentina Vita (V. V.) 'Healthy-Strong Life' \ Left to Right(L->R)

Alien Elle K. Mcgee, BladeMaster 刀大師, Celebrity Cook-Chef Martial Artist
2154;_ Chicago, Illinois' Intranet, Wireless Broadcasting Frequencies Online

刀 Augmented Reality Holographic Heads Up Display A.R. Holo HUD			
Holo Photo Recording Mode			
Actual Legal Given Name;_ "Truman, Emily A."	Actual Legal Given Name;_ Lieu, Michael W.	Actual Legal Given Name;_ Crenshaw, Adan	Actual Legal Given Name;_ Drastal, Alex
"They Are True-Men"	L←R Eastern society 廖，米高 伍o　工具人　打擊力量 ToolPerson	Ai Love Clean House Clean-up Crew Unit 6(six) Unit6(six) Ai Love Sleeping Beauty Ai Love poopy pants Unit 6(six) Free Stuff	U.S.S.F. United States Space Force U.S.A.F. United States Air Force L←R Eastern society 武道神 MartialArtsGod
"No Weapons Detected" **"They Are True-Men"**	"No Weapons Detected" Navy Blue Tool Vest **White** Handy-Human's Jack Of All Trades Master Of Nothing Combat Support Engineers **Black Utopian Peacekeeping Uniform**	"No Weapons Detected" **White Full Body C.B.R. E.M.P. Shielding HazMat Suit** **White** **Biotech Cleaning Kit** Ai Love Clean-up Crew Chicago City Mayor Support Staff	"Walking Weapon" Overheating **White** **Biotech Left Arm** Overheating **White** **Biotech Left Leg** **Black Utopian Peacekeeping Uniform**
[*Aug+*]			
刀 Augmented Reality Holographic Heads Up Display A.R. Holo HUD			

2154;_ Chicago, Illinois' Intranet, Wireless Broadcasting Frequencies Online
Alien Elle K. Mcgee, BladeMaster 刀大師, Celebrity Cook-Chef Martial Artist

Alien Elle K. Mcgee, BladeMaster 刀大師, Celebrity Cook-Chef Martial Artist
2154;_ Chicago, Illinois' Intranet, Wireless Broadcasting Frequencies Online

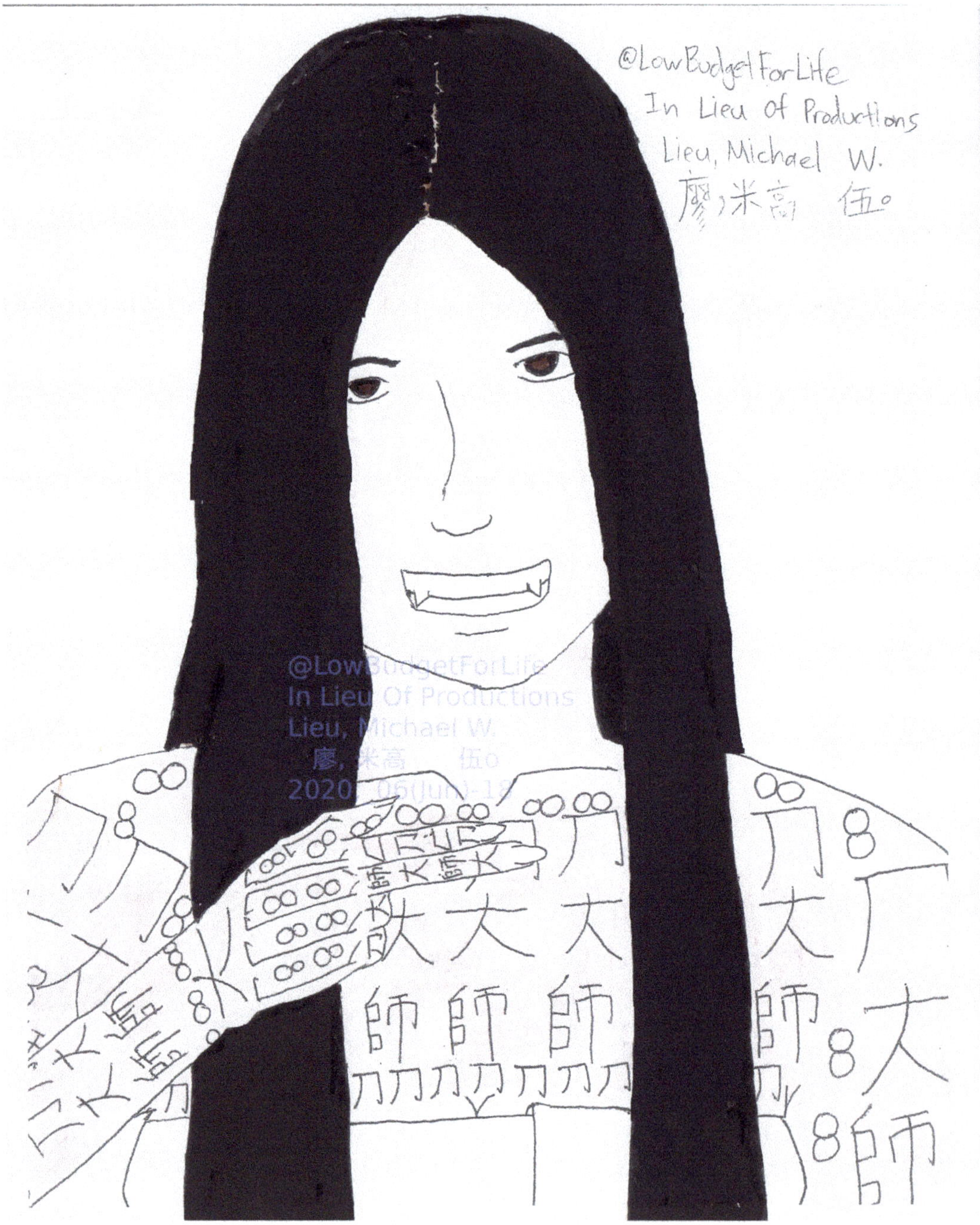

2154;_ Chicago, Illinois' Intranet, Wireless Broadcasting Frequencies Online
Alien Elle K. Mcgee, BladeMaster 刀大師, Celebrity Cook-Chef Martial Artist

Elle K. Mcgee; Childlike Human Kanon-Klone(s) "BladeMaster 刀大師" Experiment;
"(*excitement voice over*) ...And the last visual holo photo of me doing **our BladeSociety salute posing...** You should not have any questions, concerns or comments on this... Well maybe, comments..."

Alien Bee Bee, A.I. BladeAssistant 刀下手, Celebrity Cook-Chef Martial Artist
2154;_ Chicago, Illinois' Intranet, Wireless Broadcasting Frequencies Online

刀 Augmented Reality Holographic Heads Up Display A.R. Holo HUD
Mission Objectives (M.O.) V. V. needs to do list in Chicago **Animal** Sanctuary
Top Priority;_ **North Side**, past **King And X District**, past Biotech Dome Ocean Border, visit **volunteer manager Chief Engineer (ChEng) Lieutenant Lieu** to fix V. V.'s Ai-**Aug bracelet problem** that has been **offline** due to **Kay's Private Security Services L.L.C. Research And Development Outbreaks Outage Experiment(s)**. Mid Priority;_ **South Side**, The **Arriving Date-Time** of **Entertainer, Mister Steven Z. J.** on: "**2154;_11(Nov)-11(eleven) to 28(twenty eight)** at **08;00_a.m.** morning" Password: "**Reserve** your **tickets**." Low Priority;_ **Downtown District, Bulletin Board System**, ... **East Side, All For One Shopping**, ...
[*Aug+*]
刀 Augmented Reality Holographic Heads Up Display A.R. Holo HUD

2154;_ Chicago, Illinois' Intranet, Wireless Broadcasting Frequencies Online
Alien Bee Bee, A.I. BladeAssistant 刀下手, Celebrity Cook-Chef Martial Artist

Valentina Vita (V. V.); Traveling Cosplayers And Kosplayers;
"(O_O) Okay, that was quite an **interesting visual verbal explanation of what really happened that explains how and why** I **was sleeping for twelve hours**..."

Battle Bot A.K.A. Bee Bee; Hermaphrodite Humanoid A.I. Ai 'BladeAssistant 刀下手';
"(*monotone female voice over*) V. V., before you department remember your priority places and where to go locations... **My** Holo HUD has been documenting it."

Ninja Kanon-Klone(s); Human Klone(s), 'EndureShadow 忍者影';
"(*in stealth suit, Japanese language voice over*) Time for **me** to show and tell **my ninja Holo decoys**! V. V. will not know which is **the actual one**."

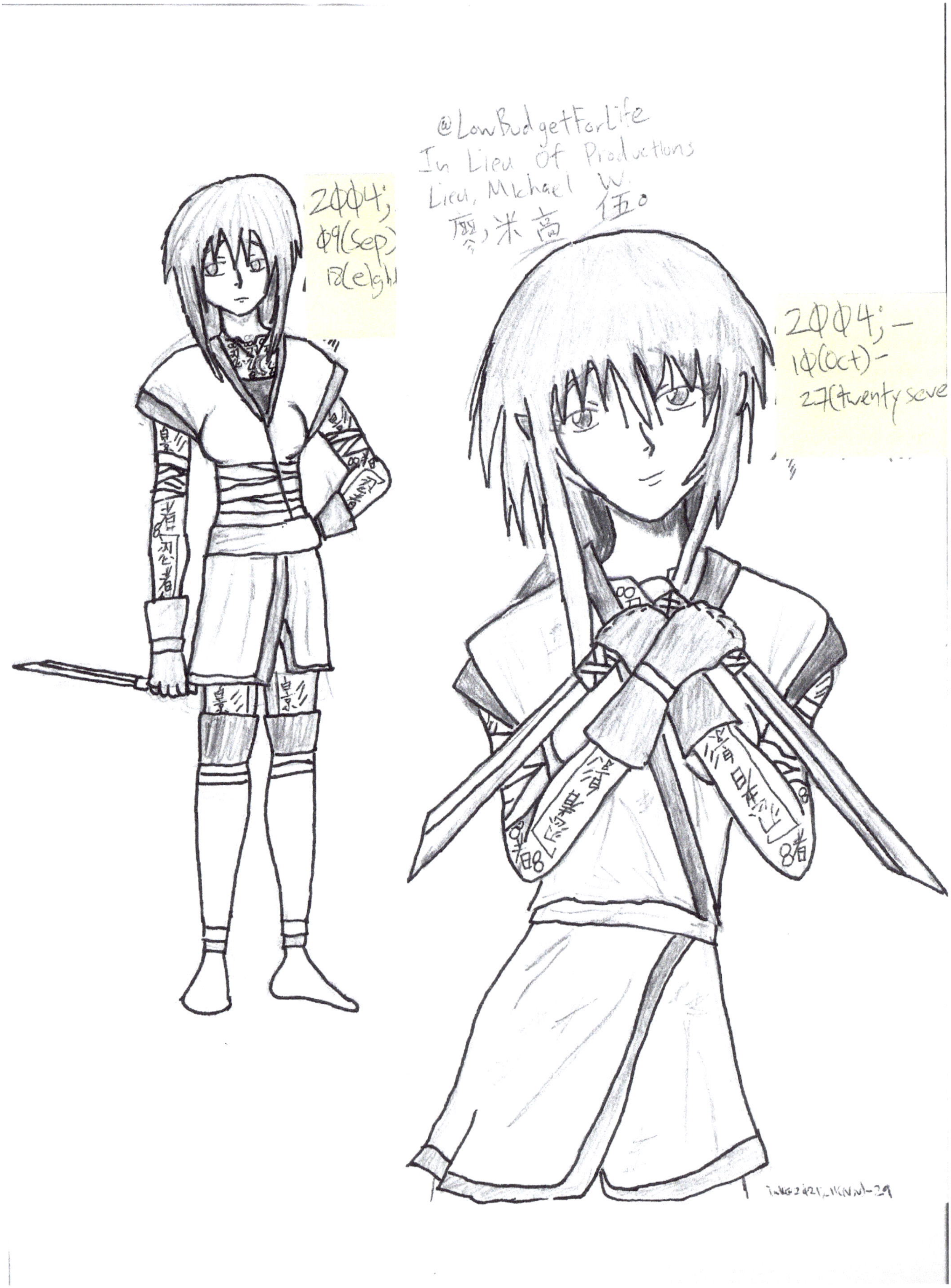
@LowBudgetForLife
In Lieu Of Productions
Lieu, Michael W.
廖, 米高 伍o
2ΦΦ4;
Φ9(Sep)
2ΦΦ4; -
1Φ(Oct) -
27(twenty seve

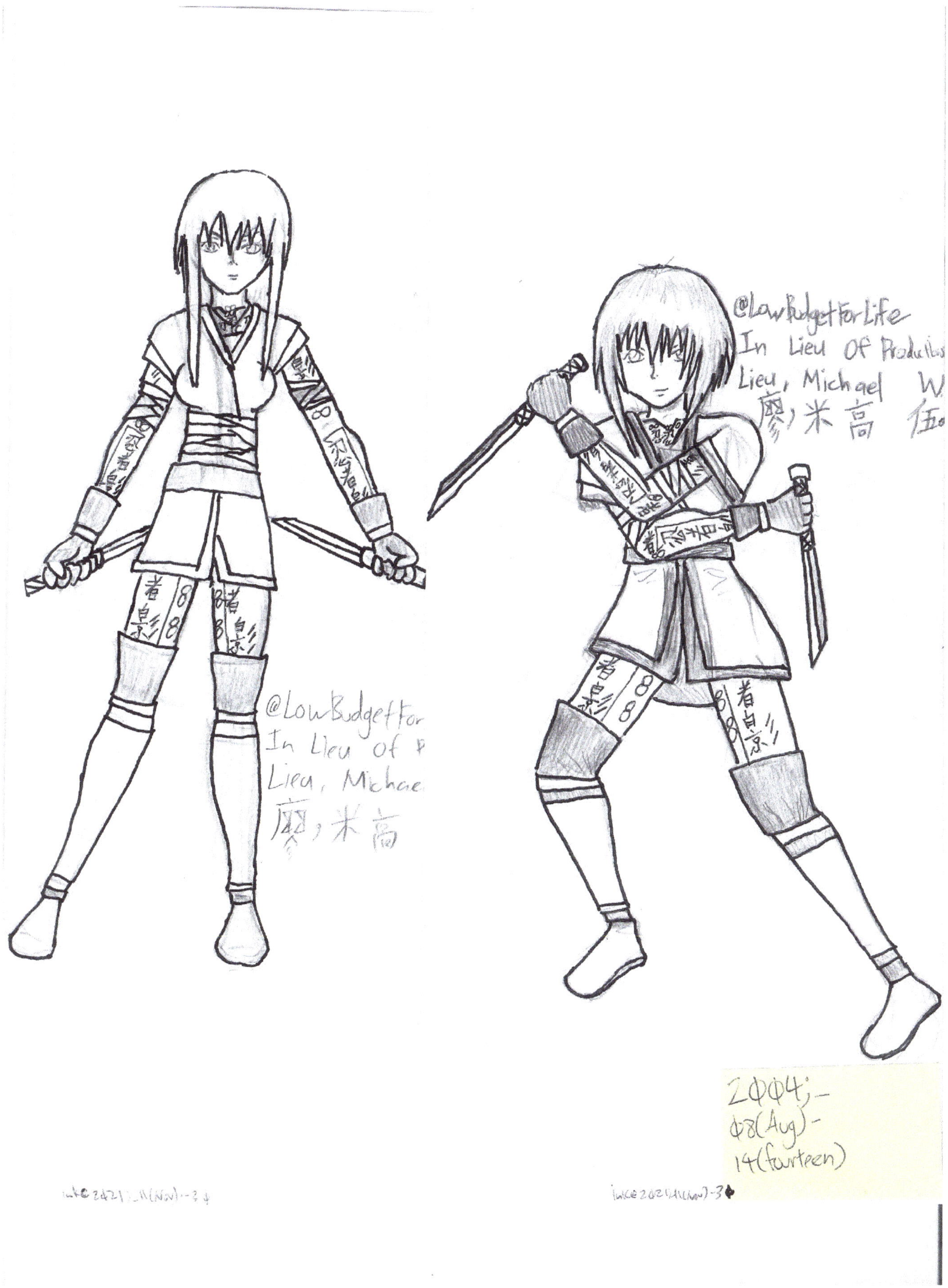
@LowBudgetForLife
In Lieu Of
Lieu, Michael W
廖, 米高 伍
@LowBudgetFor
In Lieu of
Lieu, Michae
廖, 米高
2ΦΦ4;-
Φ8(Aug)-
14(fourteen)

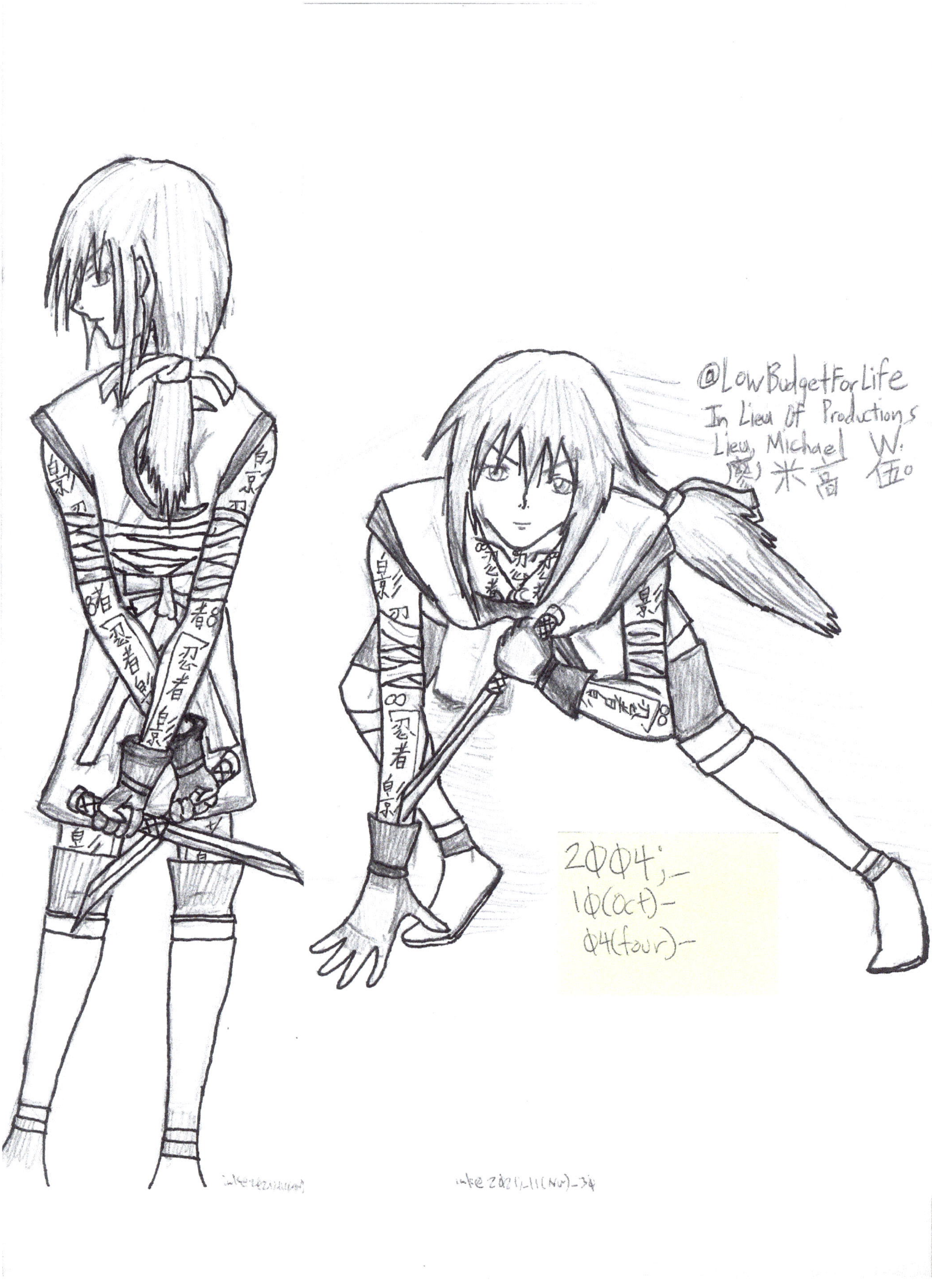
@LowBudgetForLife
In Lieu Of Productions
Lieu, Michael W.
廖, 米高 伍o
2Φ4;—
1Φ(Oct)—
Φ4(four)—

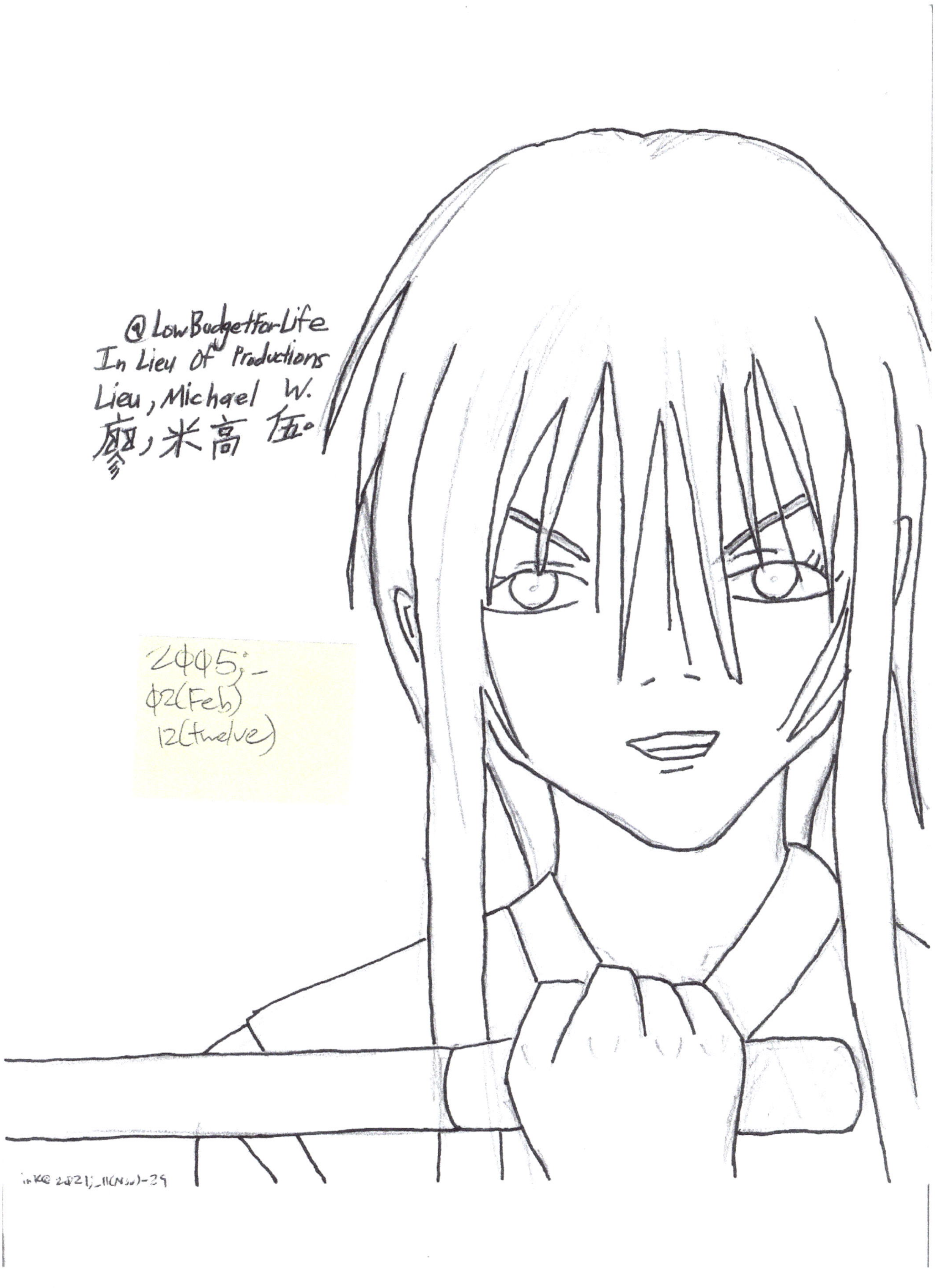
@LowBudgetForLife
In Lieu Of Productions
Lieu, Michael W.
廖, 米高 伍o
2ΦΦ5;_
Φ2(Feb)
12(twelve)
inK@ 2Φ21;_11(Nov)-29

Alien Ninja Kanon-Klone, EndureShadow 忍者影, Celebrity Cook-Chef Martial Artist
2154;_ Chicago, Illinois' Intranet, Wireless Broadcasting Frequencies Online

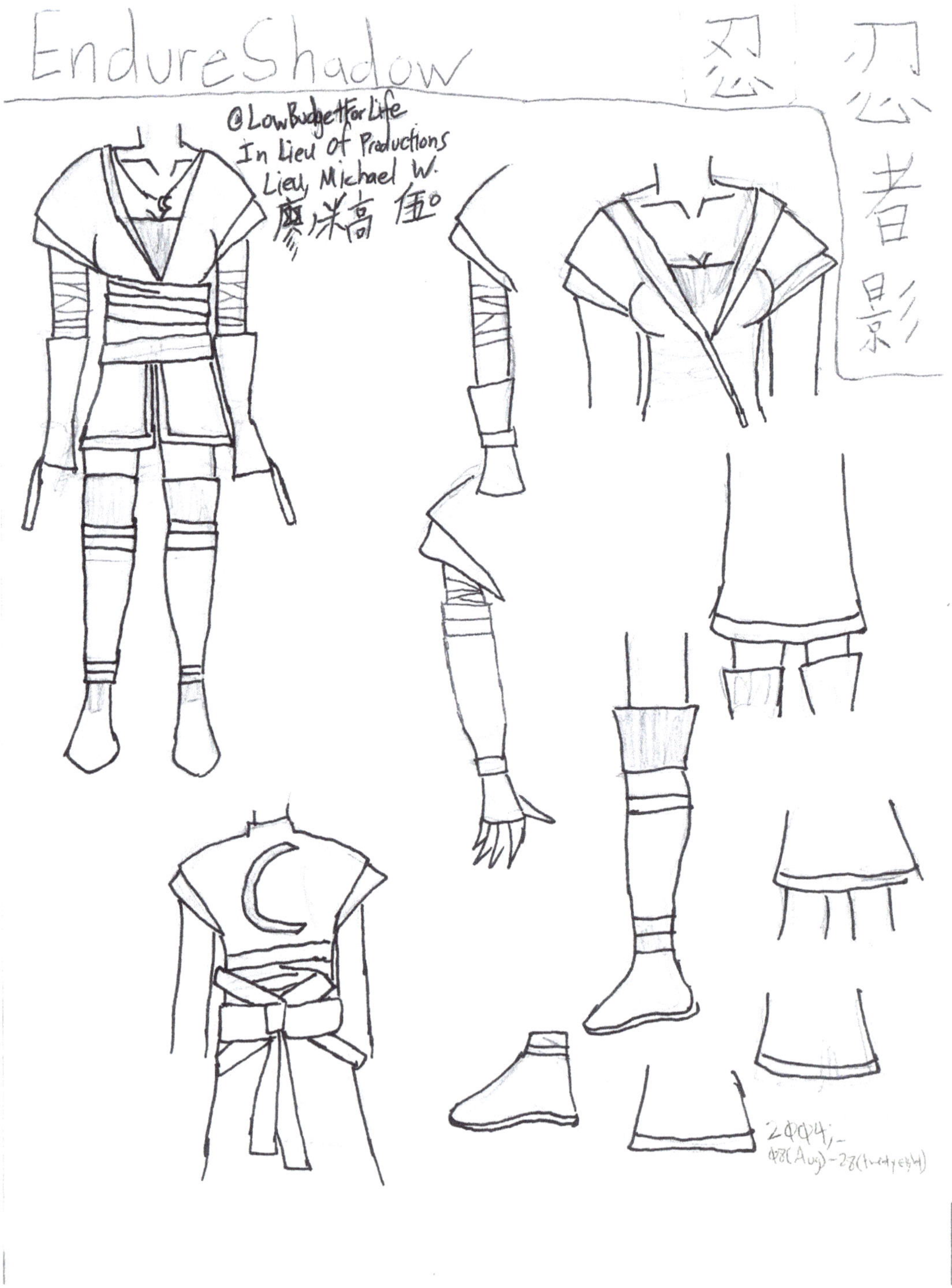

2154;_ Chicago, Illinois' Intranet, Wireless Broadcasting Frequencies Online
Alien Ninja Kanon-Klone, EndureShadow 忍者影, Celebrity Cook-Chef Martial Artist

Valentina Vita (V. V.); Traveling Cosplayers And Kosplayers;
"(@_@) Huh, what is going on!? So many **ninjas**...!? All around me!"

Alien Bee Bee, A.I. BladeAssistant 刀下手, Celebrity Cook-Chef Martial Artist
2154;_ Chicago, Illinois' Intranet, Wireless Broadcasting Frequencies Online

刀 Augmented Reality Holographic Heads Up Display A.R. Holo HUD
Mission Objectives (M.O.) V. V. needs to do list in Chicago **Animal** Sanctuary
Top Priority;_ **North Side**, past **King And X District**, past Biotech Dome Ocean Border, visit **volunteer manager Chief Engineer (ChEng) Lieutenant Lieu** to fix V. V.'s Ai-**Aug bracelet problem** that has been **offline** due to **Kay's Private Security Services L.L.C. Research And Development Outbreaks Outage Experiment(s)**. Mid Priority;_ **South Side**, The **Arriving Date-Time** of **Entertainer, Mister Steven Z. J.** on: "**2154;_11(Nov)-11(eleven) to 28(twenty eight)** at **08;00_a.m.** morning" Password: "**Reserve** your **tickets**." Low Priority;_ **Downtown District, Bulletin Board System**, ... **East Side, All For One Shopping**, ...
[*Aug+*]
刀 Augmented Reality Holographic Heads Up Display A.R. Holo HUD

2154;_ Chicago, Illinois' Intranet, Wireless Broadcasting Frequencies Online
Alien Bee Bee, A.I. BladeAssistant 刀下手, Celebrity Cook-Chef Martial Artist

Ninja Kanon-Klone(s); Human Klone(s), 'EndureShadow 忍者影';
"(*in suit, Japanese language, voice over*) Elle, do you want a **klone** of V. V.? **We** can have those **missing persons cases** and **no one will ever notice... We** can **klone** her..."

Battle Bot A.K.A. Bee Bee; Hermaphrodite Humanoid A.I. Ai 'BladeAssistant 刀下手';
"(*monotone female voice over*) **Ninja Kanon**, that would **not be a nice thing** to do."

Elle looks at one of the **many Ninja Kanon-Klone(s) decoys** and shakes her head. The **many Ninja Kanon-Klone(s)** stands down and V. V. notices with little to no emotion.

Elle K. Mcgee; Childlike Human Kanon-Klone(s) "BladeMaster 刀大師" Experiment;
"(*excitement voice over*) **Ninja Kanon** wants to **play missing persons** with you. But, here are **physical paper print outs of Ninja Kanon Klone, BladeSociety** and you!!"

Natural Unmodified Human(s), Family Tree, Descendant(s) For; **Valentina Vita (V. V.)**
Biological Sex;_ Male (M) XY / Female (F) XX / Hermaphrodite (H) XY XX

Grandfather (M)	Grandmother (F)	Grandfather (M)	Grandmother (F)
Mother (M)		Father (M)	
Valentina Vita (V. V.)'s Birth Day Or Built Day;_ Natural Unmodified Human Female (F) 2120;_08(Aug)-15(fifteen) Female (F) Became Enhanced With "Beauty Biotech Augmentation+" Via Implants On 2136;_08(Aug)-15(fifteen) Augmented Biotech Human Female (F)			

Natural Unmodified Human(s), Family Tree, Descendant(s) For; Mark S.
Biological Sex;_ Male (M) XY / Female (F) XX / Hermaphrodite (H) XY XX

Grandfather (M)	Grandmother (F)	Grandfather (M)	Grandmother (F)
Mother (M)		Father (M)	
Mark S.'s Birth Day Or Built Day;_ Natural Unmodified Human Male (M) 1963;_08(Aug)-05(five) Male (M) Became Enhanced With "Substitution Death Defying Biotech Augmentation+" On 2077;_08(Aug)-01(one) Augmented Biotech Human Male (M)			

Natural Unmodified Human(s), Family Tree, Descendant(s) For; Mark D.
Biological Sex;_ Male (M) XY / Female (F) XX / Hermaphrodite (H) XY XX

Grandfather (M)	Grandmother (F)	Grandfather (M)	Grandmother (F)
Mother (M)		Father (M)	
Mark D.'s Birth Day Or Built Day;_ Natural Unmodified Human Male (M) 1964;_02(Feb)-26(twenty six) Male (M) Became Enhanced With "Substitution Death Defying Biotech Augmentation+" On 2077;_08(Aug)-01(one) Augmented Biotech Human Male (M)			

Augmented Biotech Human(s), Family Tree, Ascendant(s) "BladeSociety 刀社會"

Biological Sex;_ Male (M) XY / Female (F) XX / Hermaphrodite (H) XY XX

Substitution Death Defying Biotech Augmentation + Present Day Time Of 2154;_	Substitution Death Defying Biotech Augmentation + Present Day Time Of 2154;_	"BladeMaster 刀大師" Cellular Rapid Regeneration Biotech Tattoos Augmentation+ Elle K. Mcgee Built Day 2152;_05(May)-05(five) Kanon-Klone(s) Test Tube	Artificial Intelligence (A.I.) Taking On The Form Of The Original Doctor Ai (F) And Self-Reproducing Itself (H)	Artificial Intelligence (A.I.) Taking On The Form Of The Original **Battle Bot** (F) And Self-Reproducing Itself (H)
BladeBoss	刀老闆	Kanon-Klone Test Tube "EndureShadow 忍者影" Biotech Stealth Suit	Present Day Time Of 2154;_	Present Day Time Of 2154;_
		Kanon-Klone Test Tube "EndureShadow 忍者影" Biotech Stealth Suit		
		Kanon-Klone Test Tube "EndureShadow 忍者影" Biotech Stealth Suit		
		Kanon-Klone Test Tube "EndureShadow 忍者影" Biotech Stealth Suit	The Original Doctor Ai Inspiration Flatline In The Year 2084;_	The Original **Battle Bot** Inspiration Flatline In The Year 2084;_
		Kanon-Klone Test Tube "EndureShadow 忍者影" Biotech Stealth Suit		
		Kanon-Klone Test Tube "EndureShadow 忍者影" Biotech Stealth Suit	The Original Doctor Ai And **Battle Bot** establishes Elysium Corporation In 2077;_	
Blade	刀	Kanon-Klone Test Tube "EndureShadow 忍者影" Biotech Stealth Suit		
Aug+ On 2077;_	Aug+ On 2077;_			
Mark S. (M) Birth Day Year 1963;_	Mark D. (M) Birth Day Year 1964;_	The Original Kanon-Klone Human Basis (F) Birth Day Year 1996;_	The Original Doctor Ai (F) Birth Day Year 1987;_	The Original **Battle Bot** (F) Birth Day Year 1982;_

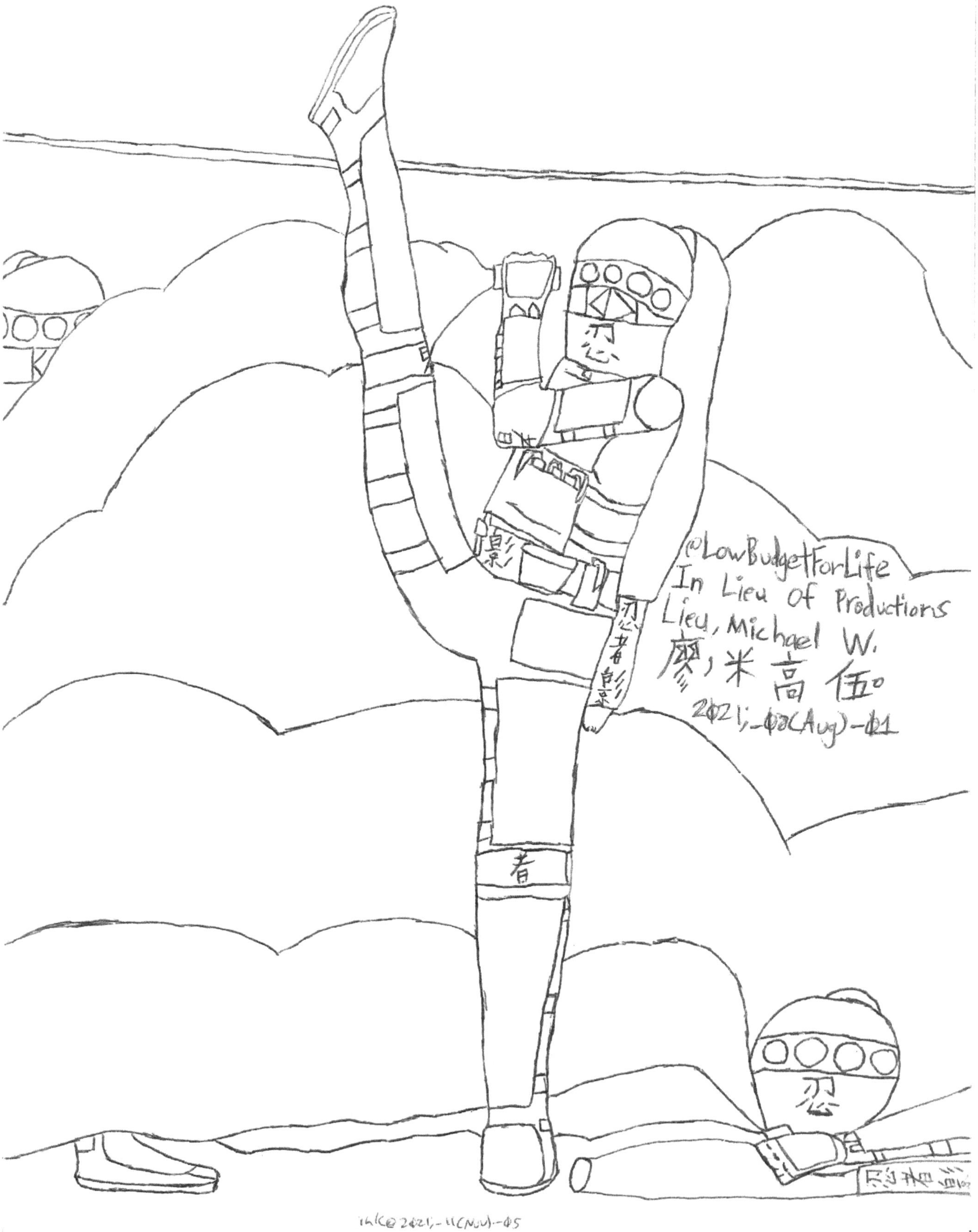
@LowBudgetForLife
In Lieu Of Productions
Lieu, Michael W.
廖, 米高 伍o
2021,-08(Aug)-01
ink@2021,-11(Nov)-05

Alien Bee Bee, A.I. BladeAssistant 刀下手, Celebrity Cook-Chef Martial Artist
2154;_ Chicago, Illinois' Intranet, Wireless Broadcasting Frequencies Online

刀 Augmented Reality Holographic Heads Up Display A.R. Holo HUD
Mission Objectives (M.O.) V. V. needs to do list in Chicago **Animal** Sanctuary
Top Priority;_   **North Side**, past **King And X District**, past Biotech Dome Ocean Border, visit **volunteer manager Chief Engineer (ChEng) Lieutenant Lieu** to fix V. V.'s Ai-**Aug bracelet problem** that has been **offline** due to **Kay's Private Security Services L.L.C. Research And Development Outbreaks Outage Experiment(s)**.   Mid Priority;_   **South Side**, The **Arriving Date-Time** of Entertainer, **Mister Steven** Z. J. on: "**2154;_11(Nov)-11(eleven) to 28(twenty eight)** at **08;00_a.m.** morning" Password: "**Reserve** your **tickets**."   Low Priority;_   **Downtown District, Bulletin Board System**, ...   **East Side, All For One Shopping**, ...
[Aug+]
刀 Augmented Reality Holographic Heads Up Display A.R. Holo HUD

2154;_ Chicago, Illinois' Intranet, Wireless Broadcasting Frequencies Online
Alien Bee Bee, A.I. BladeAssistant 刀下手, Celebrity Cook-Chef Martial Artist

Ninja Kanon-Klone(s); Human Klone(s), 'EndureShadow 忍者影';
"(*in suit, Japanese language*) **Ninja**! Elle, can **we** play **missing persons cases** with V.V.?"

Battle Bot A.K.A. Bee Bee; Hermaphrodite Humanoid A.I. Ai 'BladeAssistant 刀下手';
"(*monotone female voice*) V. V., this is the **actual Ninja Kanon-Klone**, in **her** bio**tech suit** due to her attachment behavior of using **smoke screen grenades** and **P.P.E.** against solid, liquid, gas matter diseases from **human klones**."

Elle K. Mcgee; Childlike Human Kanon-Klone(s) "BladeMaster 刀大師" Experiment;
"(*excitement voice over*) No, not this time, **Ninja Kanon**, **we** are not **playing missing persons cases** at this time... Next time!"

Valentina Vita (V. V.); Traveling Cosplayers And Kosplayers;
"(@_@) So much **smoke**... (*coughing*) ...Um... Nice to meet you **Ninja Kanon**...? Or is it **Ninja**? **Kanon**? Or both? Elle, you have a quite an interesting **group of people** for your **family**..."

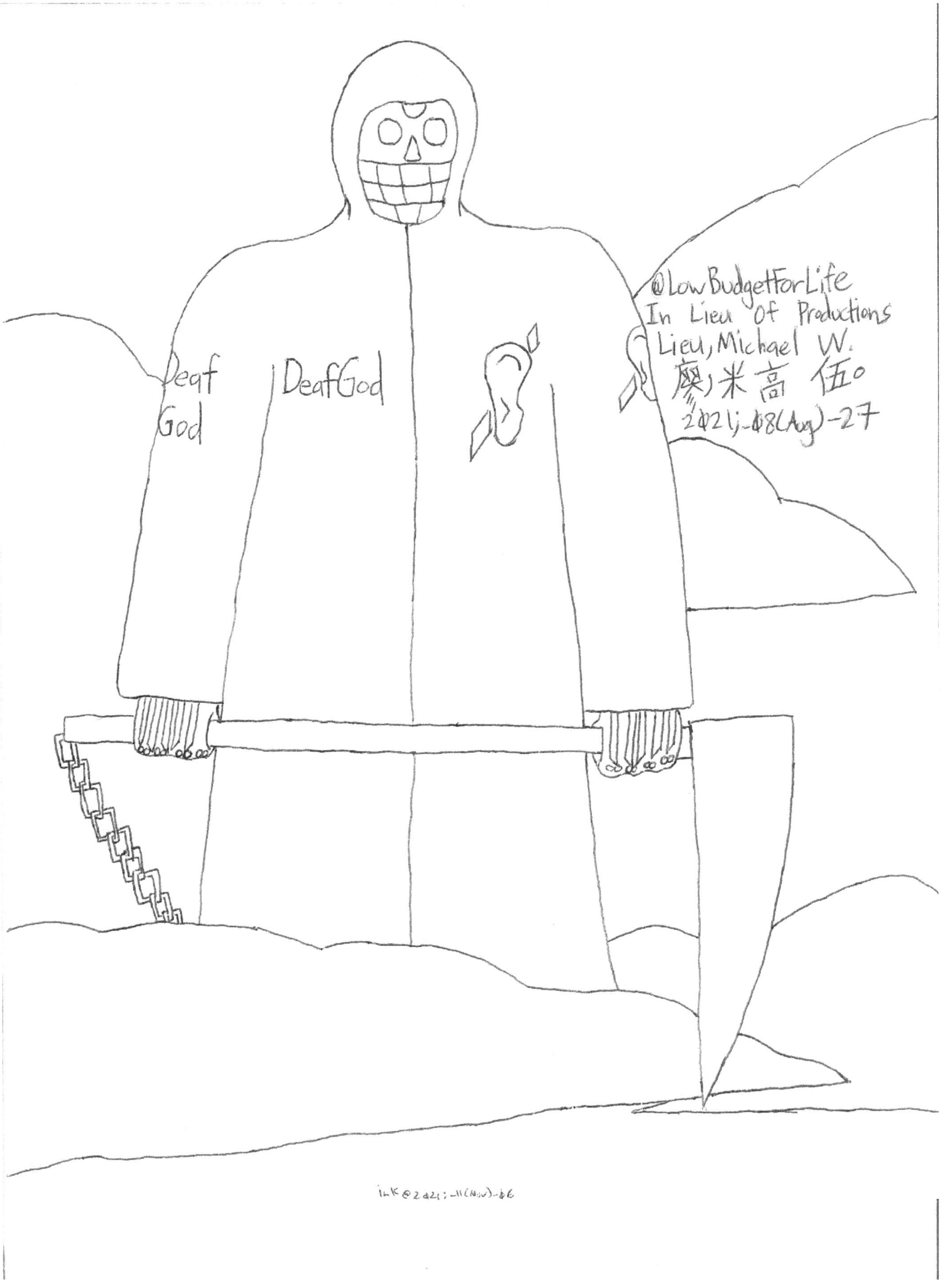
@LowBudgetForLife
In Lieu Of Productions
Lieu, Michael W.
廖,米高 伍o
2021-08(Aug)-27
Deaf God
DeafGod
ink @2021-11(Nov)-06

Alien Bee Bee, A.I. BladeAssistant 刀下手, Celebrity Cook-Chef Martial Artist
2154;_ Chicago, Illinois' Intranet, Wireless Broadcasting Frequencies Online

刀 Augmented Reality Holographic Heads Up Display A.R. Holo HUD
Mission Objectives (M.O.) V. V. needs to do list in Chicago **Animal** Sanctuary
Top Priority;_ **North Side**, past **King And X District**, past Biotech Dome Ocean Border, visit **volunteer manager Chief Engineer (ChEng) Lieutenant Lieu** to fix V. V.'s Ai-**Aug bracelet problem** that has been **offline** due to **Kay's Private Security Services L.L.C. Research And Development Outbreaks Outage Experiment(s)**. Mid Priority;_ **South Side**, The **Arriving Date-Time** of **Entertainer, Mister Steven Z. J.** on: "**2154;_11(Nov)-11(eleven) to 28(twenty eight)** at **08;00_a.m.** morning" Password: "**Reserve** your **tickets**." Low Priority;_ **Downtown District, Bulletin Board System**, ... **East Side, All For One Shopping**, ...
[*Aug+*]
刀 Augmented Reality Holographic Heads Up Display A.R. Holo HUD

2154;_ Chicago, Illinois' Intranet, Wireless Broadcasting Frequencies Online
Alien Bee Bee, A.I. BladeAssistant 刀下手, Celebrity Cook-Chef Martial Artist

Assassin Kanon-Klone(s); Human Klone(s), 'DeafGod 聾神';
"(*in suit, Japanese language*) **Assassins**! Elle, **I** am a bit sad, that **we** are not **playing missing persons cases**... Would like to **Klone**... V. V.."

Battle Bot A.K.A. Bee Bee; Hermaphrodite Humanoid A.I. Ai 'BladeAssistant 刀下手';
"(*monotone female voice over*) ...and here is the **undead assassin version of Kanon**..."

Elle K. Mcgee; Childlike Human Kanon-Klone(s) "BladeMaster 刀大師" Experiment;
"(*excitement voice over*) Oh!! **Assassin Kanon**! **She** really wants to play **missing persons cases**... But, **Kanon**, V. V. has to go to **places go and animal humans to meet**, next time!"

Valentina Vita (V. V.); Traveling Cosplayers And Kosplayers;
"(O_O) Whoa we wow, how many **Klones** does **Ninja Kanon** have...?! **DeafGod**... **You** mean **DeathGod**, like the **deceased Death**, not **hearing loss Deaf**... Oh, I get it... yea, yea... Interesting... **suit**..."

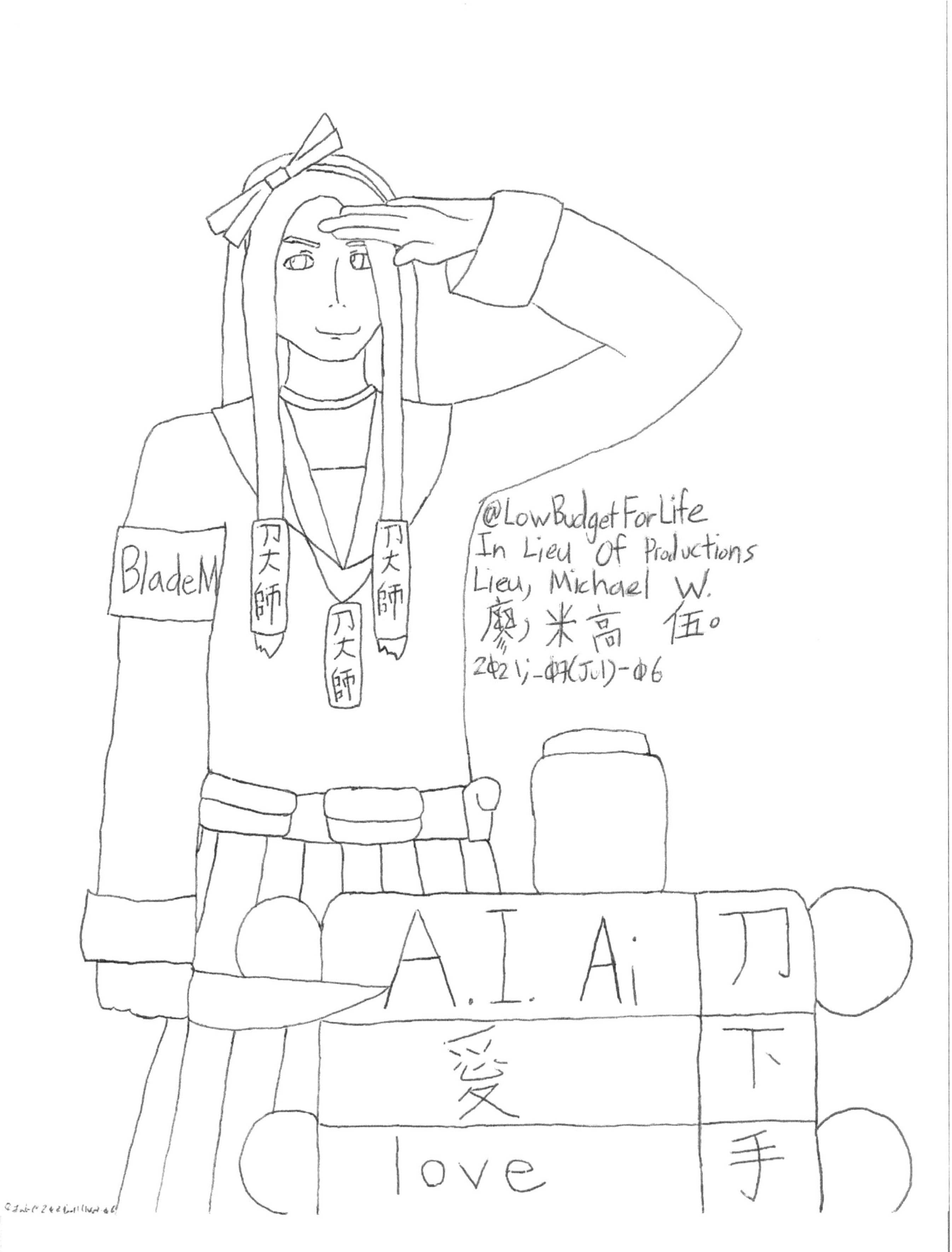
BladeM
刀大師
刀大師
刀大師
@LowBudgetForLife
In Lieu Of Productions
Lieu, Michael W.
廖, 米高 伍o
2021;-07(Jul)-06
A.I. Ai
爱
love
刀
下
手

Alien Bee Bee, A.I. BladeAssistant 刀下手, Celebrity Cook-Chef Martial Artist
2154;_ Chicago, Illinois' Intranet, Wireless Broadcasting Frequencies Online

刀 Augmented Reality Holographic Heads Up Display A.R. Holo HUD
Mission Objectives (M.O.) V. V. needs to do list in Chicago **Animal** Sanctuary
Top Priority;_ **North Side**, past **King And X District**, past Biotech Dome Ocean Border, visit **volunteer manager Chief Engineer (ChEng) Lieutenant Lieu** to fix V. V.'s Ai-**Aug bracelet problem** that has been **offline** due to **Kay's Private Security Services L.L.C. Research And Development Outbreaks Outage Experiment(s)**. Mid Priority;_ **South Side**, The **Arriving Date-Time** of **Entertainer, Mister Steven Z. J.** on: "**2154;_11(Nov)-11(eleven) to 28(twenty eight)** at **08;00_a.m.** morning" Password: "**Reserve** your **tickets**." Low Priority;_ **Downtown District, Bulletin Board System**, ... **East Side, All For One Shopping**, ...
[*Aug+*]
刀 Augmented Reality Holographic Heads Up Display A.R. Holo HUD

2154;_ Chicago, Illinois' Intranet, Wireless Broadcasting Frequencies Online
Alien Bee Bee, A.I. BladeAssistant 刀下手, Celebrity Cook-Chef Martial Artist

Elle K. Mcgee; Childlike Human Kanon-Klone(s) "BladeMaster 刀大師" Experiment;
"(*excitement*) Okay! That was all the **explanations** of **Kay Private Security Agents' boo boos** along with my **family members introducing themselves** to you. Bye bye, V. V.!"

Battle Bot A.K.A. Bee Bee; Hermaphrodite Humanoid A.I. Ai 'BladeAssistant 刀下手';
"(*monotone female voice over*) V. V., **remember the places, Sides**, you need to be in Chicago **Animal** Sanctuary, within this **Elysium Corporation** biotech dome **utopia. I** am going to be around Elle to assist her in **studies** as a **BladeMaster Cook-Chef operative**. **Good day** to you V. V.."

Valentina Vita (V. V.); Traveling Cosplayers And Kosplayers;
"(T_T) Uh... Yes! Thank you, **Bee Bee**, Elle for your **deep detail explanation**, your **family's extended welcome introduction**... What does **Ninja Kanon** mean by **playing missing persons cases**. Actually nevermind! I am going to need to **remember**: **North Side, Lieu Lieu, Downtown District, Bullet Board System, East Side, All For One Shopping Center, South Side, Entertainer Mister Steven, Reserve Your Tickets**."

2154;_11(Nov)-01(one)

Travel To Western Society, America(s), North America
Self-Govern Sovereignty (S),
Illinois (ILL) Sector, Points Of Interest (P.O.I.)

Chicago (S), "**Animal** Sanctuary"
"The New World 2121 Movement Of The 22nd Century,
Utopian Sovereignty of North America (U.S. of N.A.)"
East Side, All For One Shopping Center

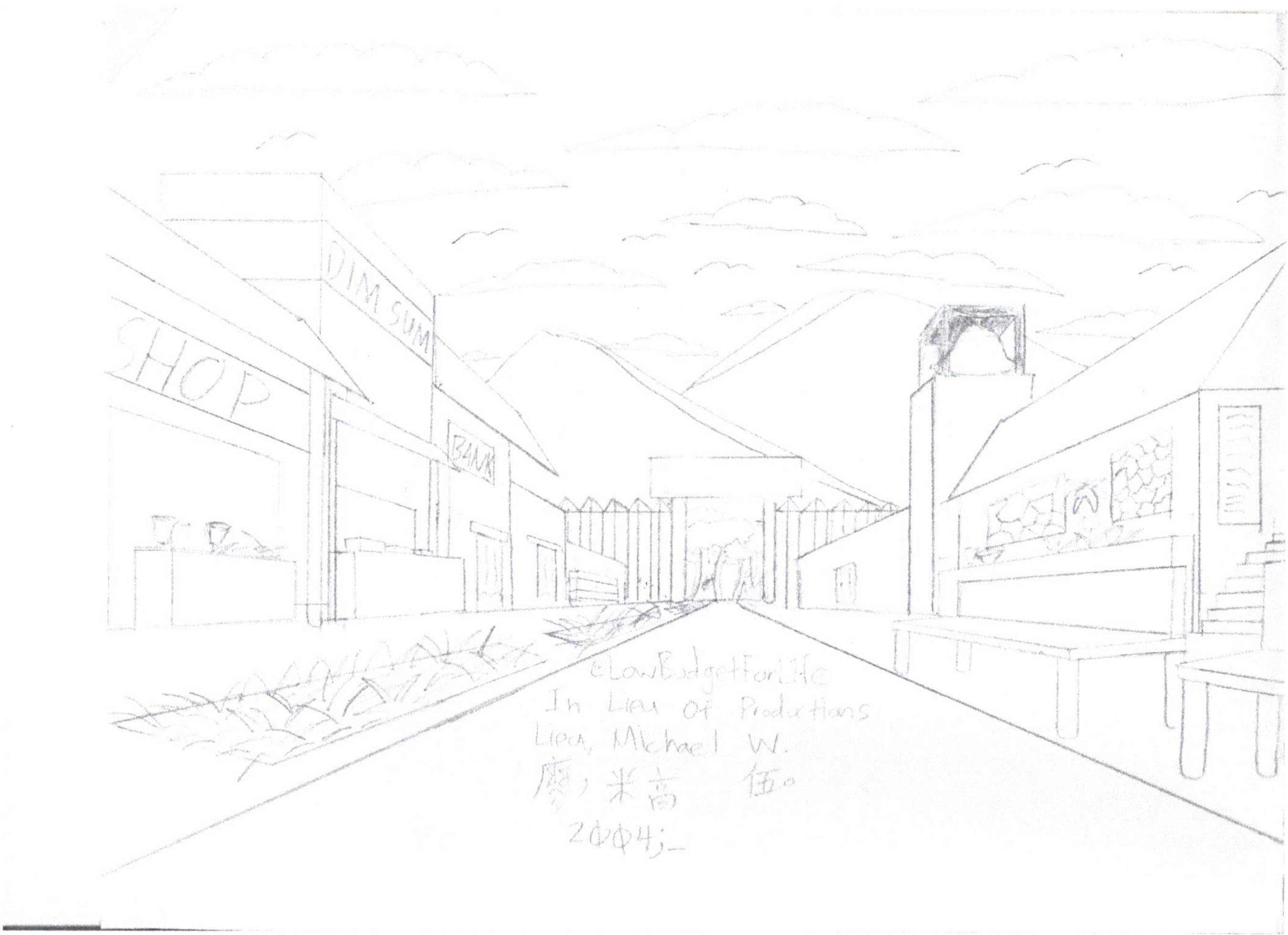

Valentina Vita (V. V.); Traveling Cosplayers And Kosplayers;
"(@_@) I need to **remember**: **North Side, Lieu Lieu, Downtown District, Bullet Board System, East Side, All For One Shopping Center, South Side**, **Entertainer Mister Steven**, **Reserve** Your **Tickets**. ...Okay, is this... **East Side**, okay **shops** first! Hopefully, there are no ugly threads sticking out of my **KissShot cosplay**!"

Utopian, Augmented Natural Animal Kingdom Plus Human Beings Coexistence
2154;_ Chicago, Illinois' Intranet, Wireless Broadcasting Frequencies Online

Elysium Corp Augmented Reality Holographic Heads Up Display A.R. Holo HUD

Chicago **Animal** Sanctuary Illinois (ILL) Free Market **Utopia**:
"Elysium Corp Biotech Dome, East Side, All For One Shopping Center"

Responsibility, Manager(s) To Blame = Clean-up Crew Unit 8888,8888 (eight eights)
Lieutenant Lieu, "**Lieu Lieu**", "WhiteWolf 白狼", **Charlie** ToolPerson Substitute

Assistant Manager(s) To Blame =
Former Po Po M.P.-Woman, Retired '**Officer Ogden' Deanna**, a.k.a. **Alfa** ToolPerson

Former Po Po M.P.-Man 'Baby Ben' Benavidez, Jose Rodriguez, a.k.a. Bravo ToolPerson

First-Name, Middle-Initial, Last-Name "ToolPerson 工具人" a.k.a. Charlie ToolPerson

Alex D. 'Doc Drastal', "MartialArtsGod 武道神", a.k.a. **Delta ToolPerson**

Utopian Sovereignty (**U**.S.) Maids And Cats-Kats'

Clean-up Crew Unit 6 (six) + **Honest Ape** + Unit 9 (nine) + Walking Dog.

Economic Freedom Exhibit Shops [Sort By Popular Visits] [*Sort By Alphabet+*]

Only At Chicago Animal Sanctuary Elysium Corporation Exclusives Gift Shop			Real Time Broadcasting And Programming On Planet Earth And Earth's Lunar Moon		
4D Theater Arts, Games And Simulation			Copulation Cubes	Dim Sum	
Utopia Bulletin Board Systems (B.B.S.)			Lord Jesus Christ Loves You All		
Artists' Alley: Authors, Comic Books, Manga, Graphic Novels, Visual Designers			Bondage, Discipline or Domination, Sadism or Submission, Masochism (B.D.S.M.)		
Anime Music Video (A.M.V.) Presentations			Animal And Human Cosplay Cafe		
Self-Defense Tools, Weapons, Props			Repair Shop: Ai-Aug Biotech Cosplay		
New World Currency Exchange Bank			Unclaimed Found Objects (U.F.O.)		
Do It Yourself Studio	Lounge	Workshop	Karaoke Kenters	Panel	Library

[*Aug+*]

Elysium Corp Augmented Reality Holographic Heads Up Display A.R. Holo HUD

2154;_ Chicago, Illinois' Intranet, Wireless Broadcasting Frequencies Online
Utopian, Augmented Natural Animal Kingdom Plus Human Beings Coexistence

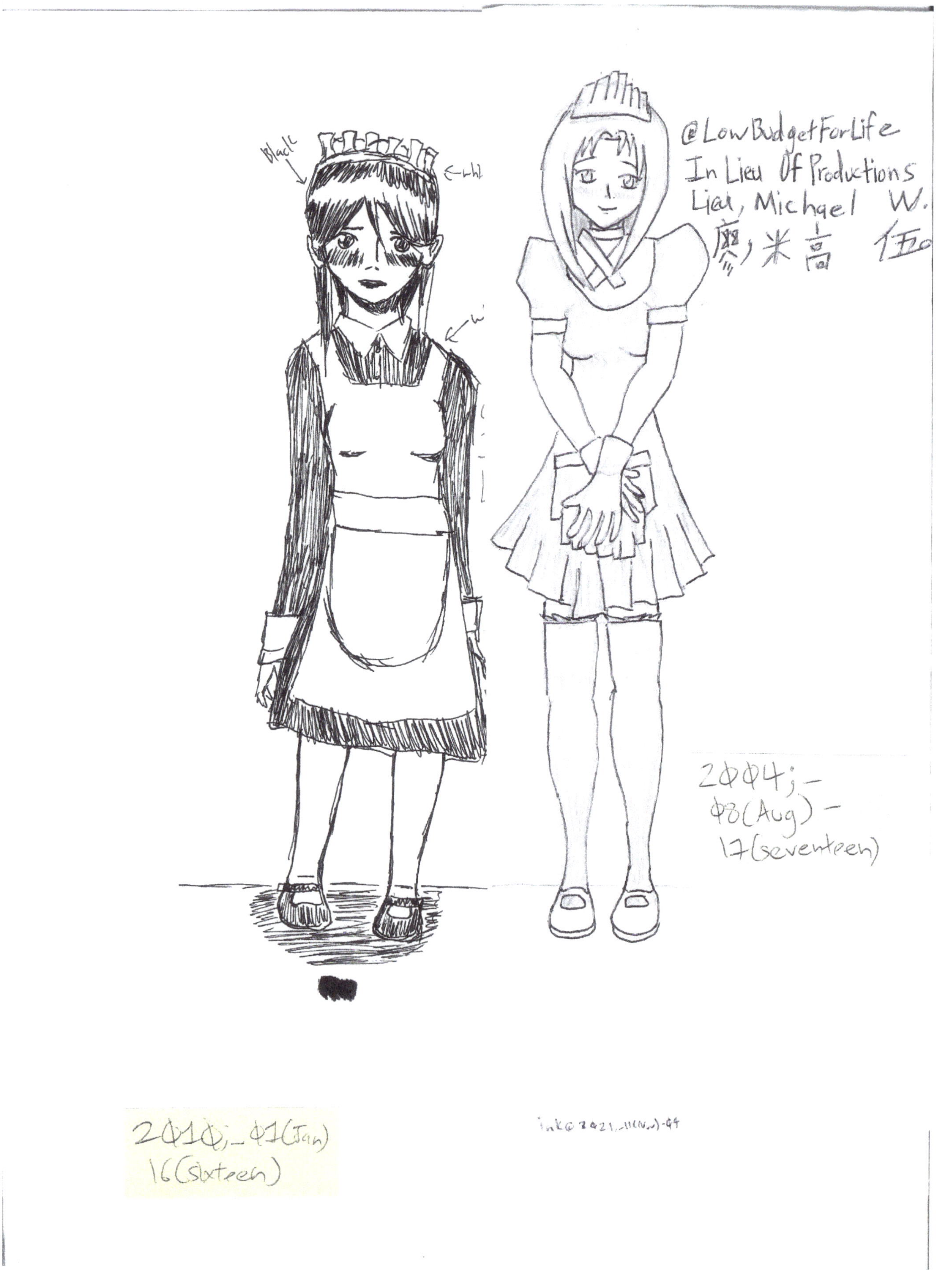
Black
@LowBudgetForLife
In Lieu Of Productions
Lieu, Michael W.
廖, 米高 伍o
2004;-
08(Aug)-
17(seventeen)
2010;_01(Jan)
16(sixteen)

V. V. approaches the one of the **East Side, All For One Shopping Center**, the **U.S. Maids'** "**New World Currency Exchange Bank**" shop.

Valentina Vita (V. V.); Traveling Cosplayers And Kosplayers;
"(^_^) Hello **Utopia Maids**, I am here for a **currency exchange**... (*thinking*) Um. Hm."

Utopian, Augmented Natural Animal Kingdom Plus Human Beings Coexistence
2154;_ Chicago, Illinois' Intranet, Wireless Broadcasting Frequencies Online

Elysium Corp Augmented Reality Holographic Heads Up Display A.R. Holo HUD
New World Currency Exchange Bank Please Select Your Currency To Exchange From:
United **States** of America(s) Dollars (U.S. of A.D.) "**old world**"
Utopian Sovereignty Dollars (U.S.D.) "**new world**"
Eur-Asia Conventional Currency
African Conventional Currency
Zimbabwean Dollars In One Bills Only, Conventional Currency
Cryptocurrency
Bartering And Trading
Non-Paid Volunteer Worker Hour(s)
Mass Marketplace Monopoly Money
Elysium Corporation's Kash-Kredit(s) Accounting System
[*Aug+*]
Elysium Corp Augmented Reality Holographic Heads Up Display A.R. Holo HUD

2154;_ Chicago, Illinois' Intranet, Wireless Broadcasting Frequencies Online
Utopian, Augmented Natural Animal Kingdom Plus Human Beings Coexistence

Maids And Cats-Kats; Utopian Sovereignty (U.S.) Economic Free Market;
"(*staring*) Hello V. V., let's see... What **currency** are you exchanging from like the **Holo HUD** stated... (*pointing to the Holo HUD projection*) Chicago **Animal** Sanctuary does take a variety of **old world** and **new world currencies**. So, **exchange** from...?"

Valentina Vita (V. V.); Traveling Cosplayers And Kosplayers;
"(-_-) ...I do have **Eur-Asia Conventional Currency**... And **you all** take... **Eur-Asia Conventional Currency**... There is no need for **exchange**. (>_<) Thanks, bye!"

Maids And Cats-Kats; Utopian Sovereignty (U.S.) Economic Free Market;
"(*staring*) ...Bye bye V. V.! If you need to **exchange currencies**. Please come again!"

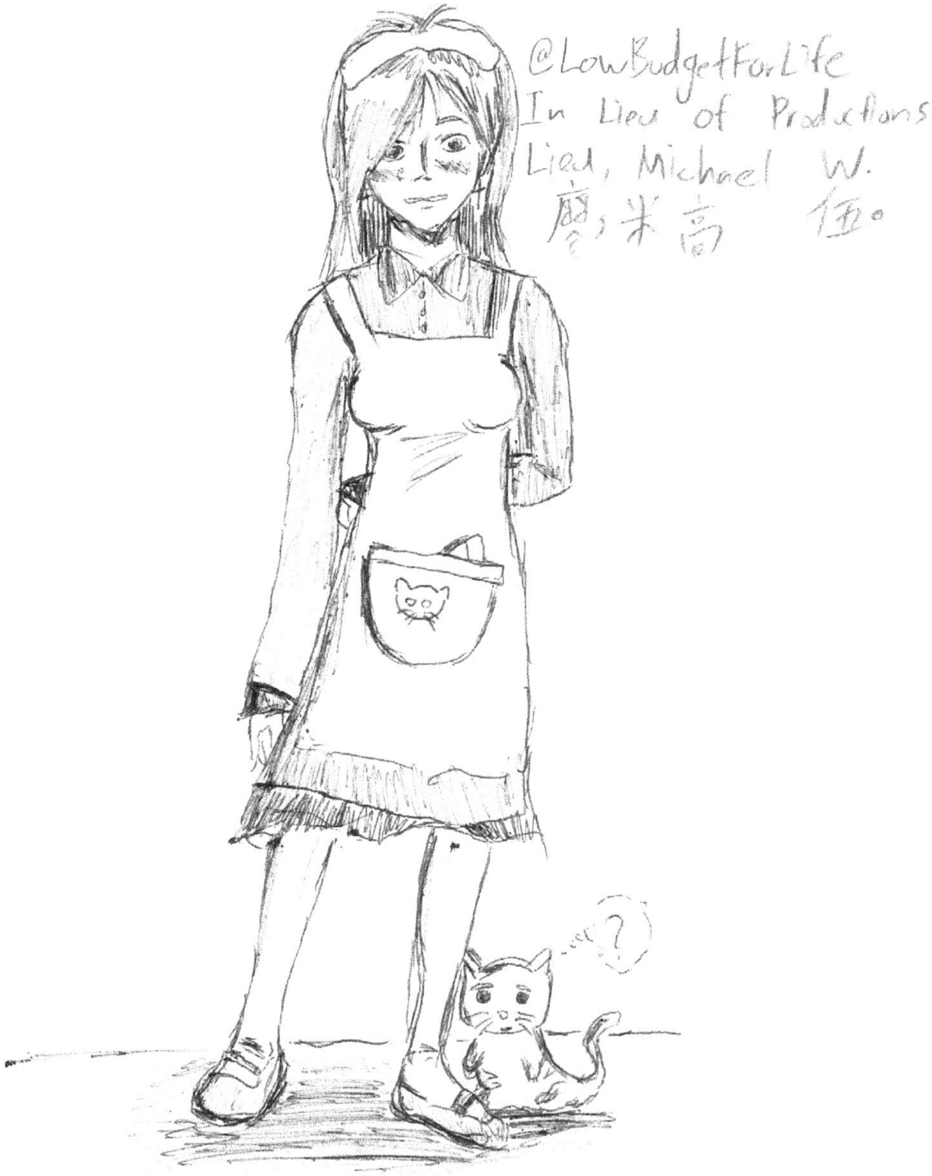

2ФФ9;_
Ф1(Jan)-
17(seventeen)

V. V. approaches one of the **East Side, All For One Shopping Center**, the **U**.S. **Maids**' "**Dim Sum**" shop.

Valentina Vita (V. V.); Traveling Cosplayers And Kosplayers;
"(^_^) Hello **Utopia Maids**, what do **you** have for **your Dim Sum** shop...? Are their any animal meats or animal products in this... I wonder? Any vegans or vegetarians."

Maids And Cats-Kats; Utopian Sovereignty (U.S.) Economic Free Market;
"(*staring*) Hello V. V., here is what the **Dim Sum we** are serving for vegans or vegetarians, in this case you are asking of no **animal products**... So vegan... ...Here is the **Holo** list... for vegans and or vegetarians **Dim Sum** selections..."

Utopian, Augmented Natural Animal Kingdom Plus Human Beings Coexistence
2154;_ Chicago, Illinois' Intranet, Wireless Broadcasting Frequencies Online

Elysium Corp Augmented Reality Holographic Heads Up Display A.R. Holo HUD			
Dim Sum			
Vegan And Or Vegetarian Selections;_			
Steamed Lotus Seed Paste Bao	Deep Fried Vegetarian Curried Parcels	Pan Fried Water Chestnut Paste	Fried Silver Thread Rolls
Ma Po Tofu	Lo Mai Gai A.K.A. Sticky Rice	Stuffed Eggplant	Custard Buns
[*Aug+*]			
Elysium Corp Augmented Reality Holographic Heads Up Display A.R. Holo HUD			

2154;_ Chicago, Illinois' Intranet, Wireless Broadcasting Frequencies Online
Utopian, Augmented Natural Animal Kingdom Plus Human Beings Coexistence

Valentina Vita (V. V.); Traveling Cosplayers And Kosplayers;
"(^_^) Vegan or vegetarian are fine, after seeing all those **adorable animals** in **Downtown District**, I do not think I can be bring myself to eat any foods with animal meat in them... Is that part of the Americas **Animal** Sanctuary indoctrination? I arrive from a **Eur-Asia Human** Sanctuary... A bit different... Okay, I am getting these..."

Maids And Cats-Kats; Utopian Sovereignty (U.S.) Economic Free Market;
"(*staring*) Yes, I think so... **I** am too a vegetarian. Good thing for **our** U.V. Chicago City Mayor that provides **us** with fruits and vegetables... **I** believe **we** have a unique rule on **eating** any **animals**. But, **we** always have Fruit Veges! Thank you! Please come again!"

V. V. pays the **Dim Sum U.S. Maid** with **Eur-Asia currency**, she gets her utensils ready.

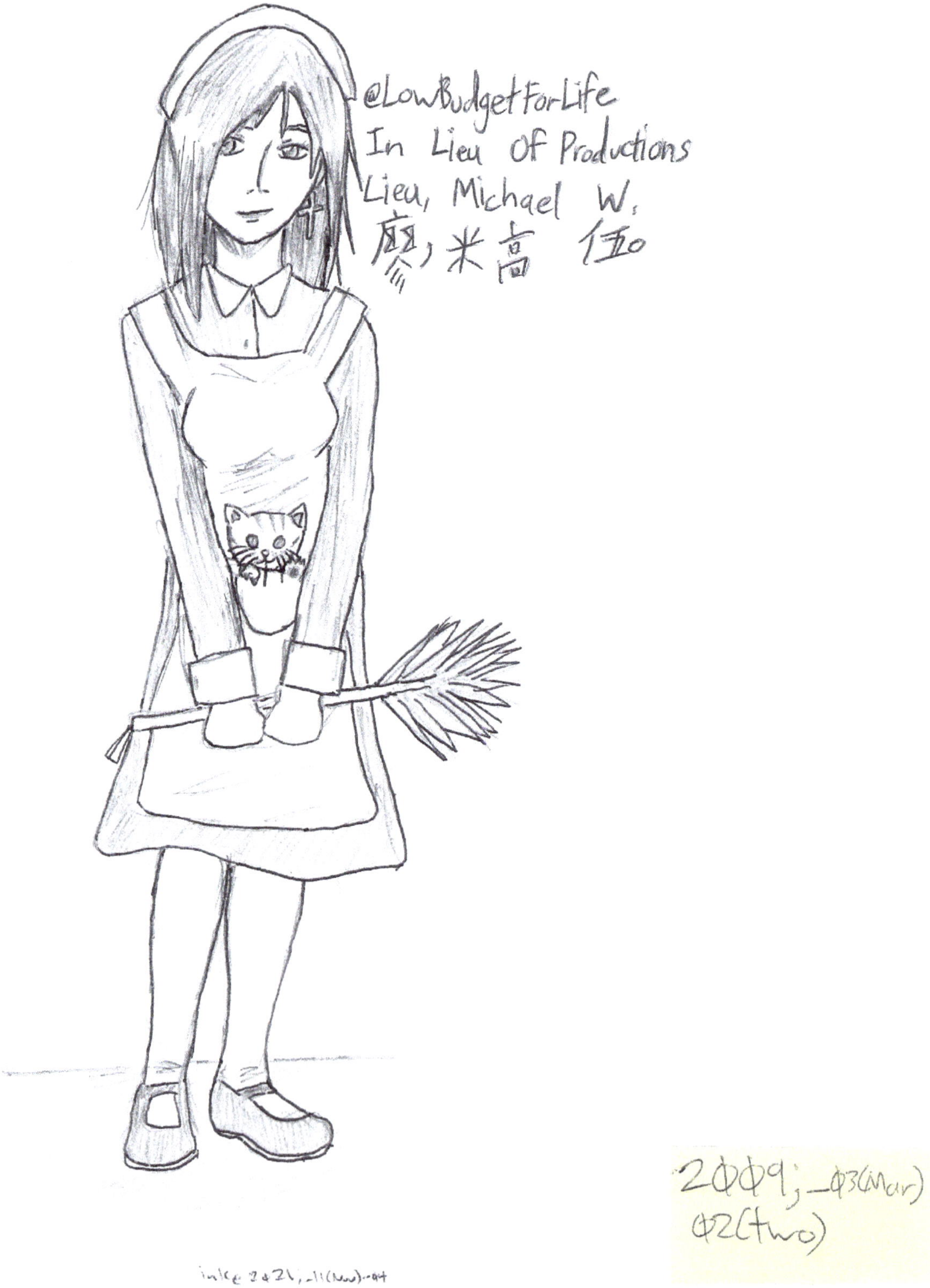
@LowBudgetForLife
In Lieu Of Productions
Lieu, Michael W.
廖, 米高 伍o
2009;_03(Mar)
02(two)

V. V. consumes the "**Dim Sum** Vegan And Or Vegetarian Selections", washes her utensils and looks around the **East Side, All For One Shopping Center**. She approaches the one of the **East Side, All For One Shopping Center**, the **U.S. Maids' "Repair Shop: Ai-Aug Biotech Cosplay**" shop.

Valentina Vita (V. V.); Traveling Cosplayers And Kosplayers;
"(^_^) Hello **Utopia Maids**, do **you** have the following **cosplay emergency kits or items** to solve those **ugly threads possibility sticking out**...? Here is a list... I made for my **travel cosplay pouch**... In **physical paper print**! (>_<) My Ai-**Aug bracelet** is not working due to the **Private Security ongoing outbreak outage a few hours back**... Need to get it fixed! That and I have to make sure there are no ugly threads sticking out!"

<table>
<tr><th colspan="3"><u>Traveling Cosplayers And Kosplayers Maintenance 101 Items</u>;</th></tr>
<tr><td>Thread: every toner in rainbow spectrum</td><td colspan="2">Sewing biotech</td></tr>
<tr><td>Needles, in case a button needs fixing</td><td colspan="2">Hand sewing supplies</td></tr>
<tr><td>Tiny scissors</td><td colspan="2">Assorted beauty product make up</td></tr>
<tr><td>Double side tape</td><td colspan="2">Safety pins</td></tr>
<tr><td>Fabric glue</td><td colspan="2">Hot glue</td></tr>
<tr><td>Universal glue</td><td colspan="2">Craft glue</td></tr>
<tr><td>Hot glue stick and lighter</td><td colspan="2">Duct tape</td></tr>
<tr><td>A couple of tiny clamps</td><td colspan="2">Combos and brushes</td></tr>
<tr><td rowspan="2">"Friendly humans willing to assist you with your: costume, wig and beauty problems"</td><td colspan="2">Hair gel</td></tr>
<tr><td>Eyelash glue</td><td>Pencil, Pen, Marker</td></tr>
</table>

Maids And Cats-Kats; Utopian Sovereignty (U.S.) Economic Free Market;
"(*staring*) Hello V. V., **I** am sorry to hear that! **We** too are a dealing the **outbreaks outages** from **Kay's Private Security little experiment in their R and D** overflowing from the **Downtown District**. Good thing **we** had a nearby **volunteer Charlie ToolPerson** to **handle our problems and situation**... (*searching*) Okay, **we** have a good amount of the **listed cosplay repair items...** And **it** costs this much..."

V. V. pays the **Repair Shop: Ai-Aug Biotech Cosplay U.S. Maid** using **Eur-Asia Currency**. The **U.S. Maid** gives V. V. the **goods** to her.

Maids And Cats-Kats; Utopian Sovereignty (U.S.) Economic Free Market;
"(*staring*) Oh, you are from **Eur-Asia**... Wow look at **that currency**... Um... Thank you!"

Valentina Vita (V. V.); Traveling Cosplayers And Kosplayers;
"(^_^) Yes, and thank **you**! Time to check on those possible ugly threads sticking out!"

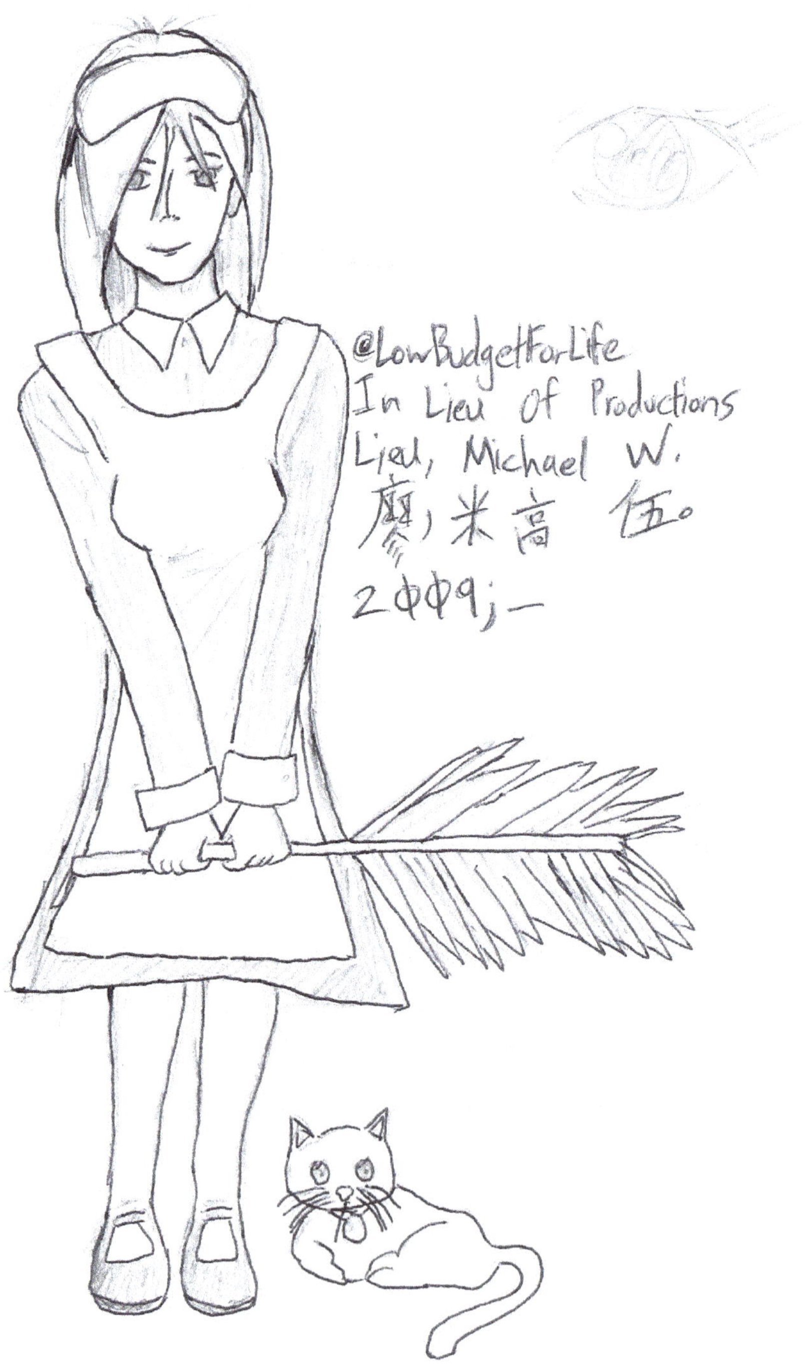
@LowBudgetForLife
In Lieu Of Productions
Lieu, Michael W.
廖,米高 伍。
2ΦΦ9;–

V. V. looks around the **East Side, All For One Shopping Center**, and she decides to sit down by the **bench** in the **East Side**.

Maids And Cats-Kats; Utopian Sovereignty (U.S.) Economic Free Market;
"(*staring*) Hello V. V., how are you doing today...?"

Valentina Vita (V. V.); Traveling Cosplayers And Kosplayers;
"(^_^) Oh hello **Utopia maid**! I am going to **sit down** and **unwind**... (*phew*) I need to remember what I need to do... Focus... Focus... My Ai-**Aug bracelet** is not working... Focus, focus... (@_@) I need to visit, **North Side, Lieu Lieu, Downtown District, Bullet Board System, East Side, All For One Shopping Center**, okay this one is done, I got my travel cosplay repair kit good. Check... **South Side**, **Entertainer Mister Steven**, **Reserve** Your **Tickets**. I am going to write it down... On **printed paper**... Do I have a **pencil, pen, marker**... yes... that travel cosplay repair kit."

V. V. looks around at the **physical printed papers** she has on her luggage and **writes down the list** of things **to do** using coloring pencils, red and **black** pens, since her Ai-**Aug bracelet** is not currently working. V. V. is not really paying much attention to the **Utopian** Sovereignty **Maid** and **her Cat** beside **her**.

V. V.'s 2154 to do list since my Ai-Aug bracelet is not working due to outbreaks outages...
Visit, **North Side, Lieu Lieu.... Lieutenant Lieu**
Visit, **Downtown District, Bullet Board System**
~~Visit, **East Side, All For One Shopping Center**~~ Complete
Visit, **South Side**, **Entertainer Mister Steven**, **Reserve** Your **Tickets**
What else...? (What to do to pass time?)

Maids And Cats-Kats; Utopian Sovereignty (U.S.) Economic Free Market;
"(*staring*) Oh, you have a lot going on, I am not gonna bother you, have a nice day V. V.!"

Valentina Vita (V. V.); Traveling Cosplayers And Kosplayers;
"(=_=) Hmm... Hmm... Oh, huh sorry **Maid**, I have to get this all down on **physical paper** before I forget. (>_<) Sorry!"

The **Utopian** Sovereignty **Maid** and **her Cat** beside **her** departs without saying anything else as V. V. continues to stay focus on what she needs to do.

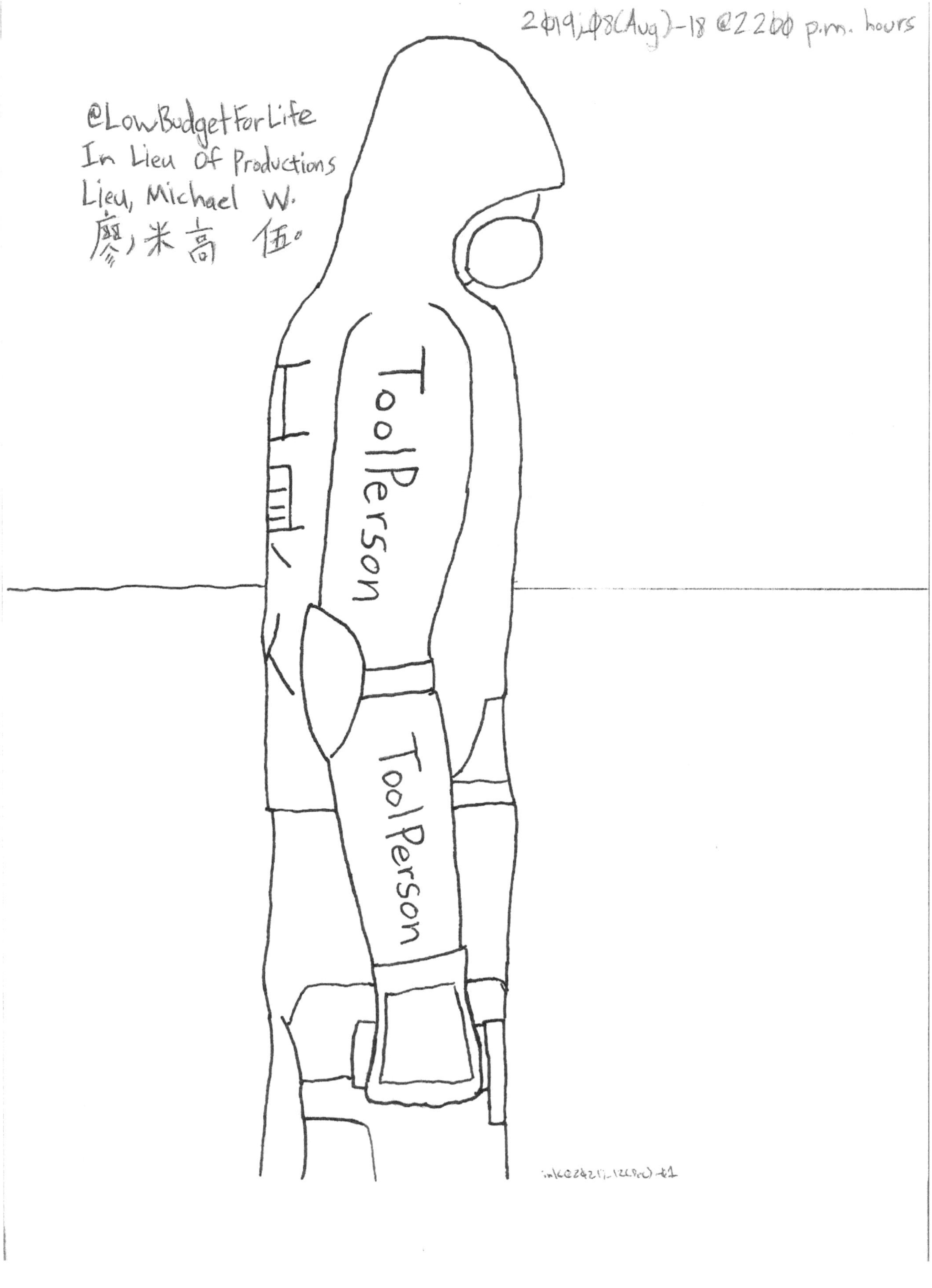

2019-08(Aug)-18 @2200 p.m. hours
@LowBudgetForLife
In Lieu Of Productions
Lieu, Michael W.
廖, 米高 伍o
工具人
ToolPerson
ToolPerson

Valentina Vita (V. V.); Traveling Cosplayers And Kosplayers;
"(O_O) ...Oh hi, are **you**... **Charlie ToolPerson**!? Whoa we wow!"

First-Name Middle-Initial Last-Name; Vigilant Volunteer Militia, Charlie 'ToolPerson 工具人';
"(*patrolling, in Oxygen Breathing Apparatus*) Oh... Hello! Yes, **I** am **Charlie Tool**, regular **human volunteer Strike Force Utopian Peacekeeping** that **patrols** this **Side** of **our Animal** Sanctuary. And other **Sides** of Chicago **Elysium's** biotech dome. What is up?"

Valentina Vita (V. V.); Traveling Cosplayers And Kosplayers;
"(^_^) My Ai-**Aug bracelet** is not working... Um... It has **stopped working** due to the **outbreaks outages** that occurred from **Kay Private Security experiment they** were **working** on **October... thirty one**, or **November... one**?"

First-Name Middle-Initial Last-Name; Vigilant Volunteer Militia, Charlie 'ToolPerson 工具人';
"(*in Oxygen Breathing Apparatus*) Okay, you have that Ai-**Aug bracelet** of yours...? **I** need to **look and touch** it, to see if **I** have the **right tools to fix** it..."

V. V. gets her Ai-**Aug bracelet,** gives it to **Charlie ToolPerson** to observe it more closely.

First-Name Middle-Initial Last-Name; Vigilant Volunteer Militia, Charlie 'ToolPerson 工具人';
"(*in Oxygen Breathing Apparatus*) Hm... Looks like your Ai-**Aug bracelet** is biotech... Good thing **I** have **my P.P.E. already on** and looks like gonna need biological **and technological agents** as well since your Ai-**Aug** is biotech... **I** wonder if **I** just work on the **tech aspect** if the biotech can still be fixed... **I** think maybe **I**... yea... **We** have to see..."

V. V. observes **Charlie ToolPerson** as **they** continue to **diagnose, troubleshoot, fix** her Ai-**Aug bracelet**.

First-Name Middle-Initial Last-Name; Vigilant Volunteer Militia, Charlie 'ToolPerson 工具人';
"(*in Oxygen Breathing Apparatus*) Let **me** see if **I** have the **right tools** for this... Even as it is biotech... **I** think **Lieutenant Lieu** may be more **suit**, literally, to handle this... **We** see."

V. V. continues to observe **Charlie ToolPerson** as **they** continue to **diagnose, troubleshoot, fix** her Ai-**Aug bracelet**. She is silent as she watches **Charlie ToolPerson**.

V. V. sees that **Charlie ToolPerson** is quiet and thinking about how to **diagnose, troubleshoot, fix** her Ai-**Aug bracelet** on **their O.B.A. HUD**. She sees that **Charlie ToolPerson** standing up and looks to **their** left side of the "**Tool Box Riot Shield**".

Valentina Vita (V. V.); Traveling Cosplayers And Kosplayers;
"(O_o) Whoa we wow, that is a lot of **tools** and **equipment you** have there... On the go and ready... Okay... Hm... I will be quiet... As **you** work... **your tools**."

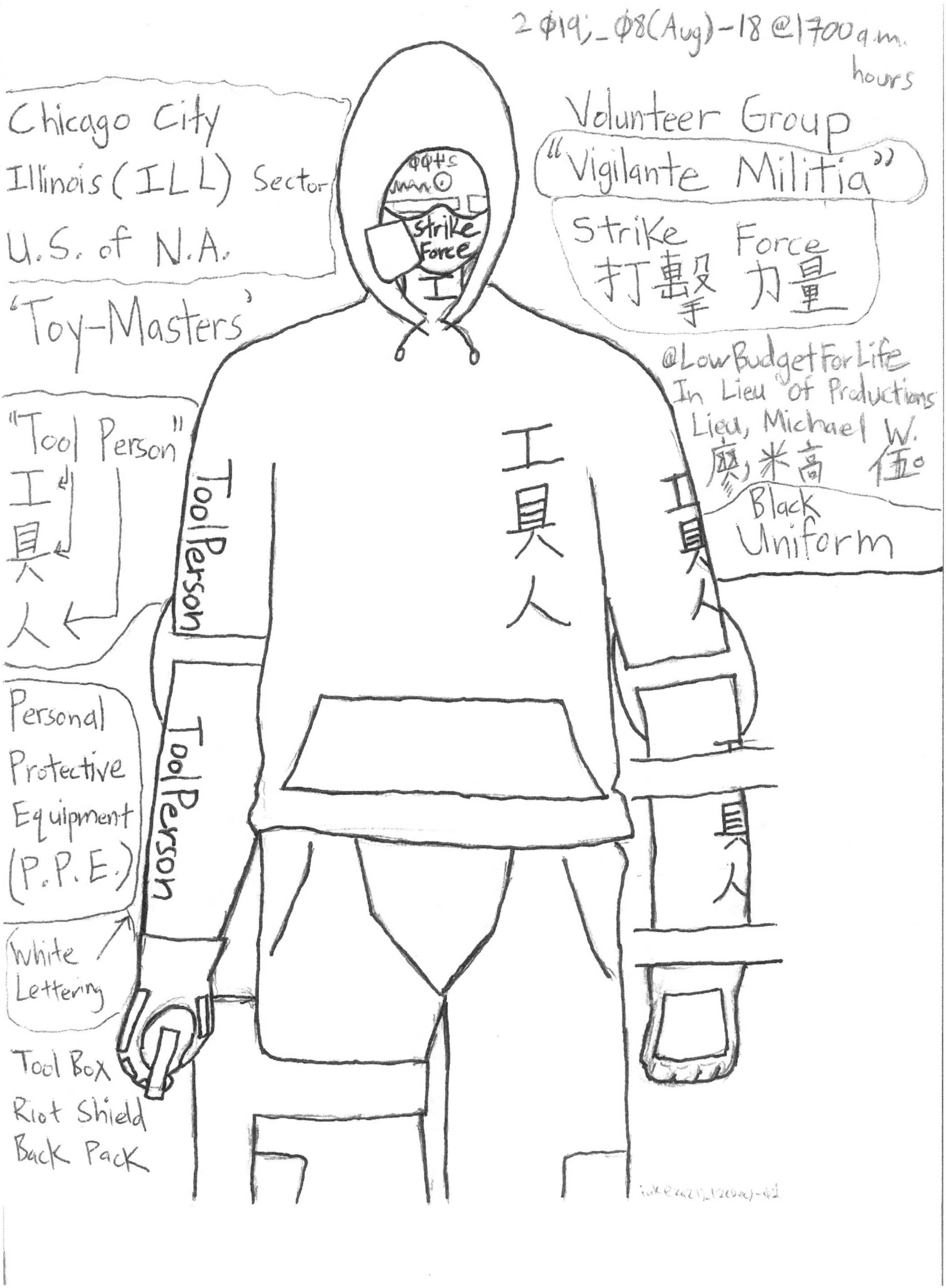
2Φ19_Φ8(Aug)-18 @1700 a.m. hours
Chicago City
Illinois (ILL) Sector
U.S. of N.A.
'Toy-Masters'
Volunteer Group
"Vigilante Militia"
Strike Force
打擊 力量
@LowBudgetForLife
In Lieu of Productions
Lieu, Michael W.
廖,米高 伍o
Black Uniform
"Tool Person"
工具人
Tool Person
Strike Force
工具人
Personal Protective Equipment (P.P.E.)
White Lettering
Tool Box
Riot Shield
Back Pack

Property Of "First-Name Middle-Initial Last-Name"	
ToolPerson Tool(s) 工具人 工	Charlie ToolPerson Utopian Sovereignty (U.S.) Peacekeeping
Tool Box Riot Shield Backpack	
Augmented Reality Holographic Heads Up Display (A.R. Holo HUD)	
Oxygen Breathing Apparatus (O.B.A.) with Heads Up Display (HUD) Personal Protective Equipment (P.P.E.)	
Hydro Resist, Chemical, Biological, Radiological (C.B.R.) Shielding and Electro Magnetic Pulse (E.M.P.) Shielding	
Bottled Biotech Purified Water	
Class: Alfa, Bravo, Charlie (A.B.C.) Extinguishers Alfa, "Paper, Wood, Textiles" Bravo, "Flammable Liquid Fires" Charlie, "Electrical Equipment" Delta, "Magnesium Material(s) Fire Needs To Be Jettison"	
First Aid Kit And Manual	
Emergency Signal Devices = Whistle, Light Sticks, Mirror, Battery Type Flasher With E.M.P. Shielding, Reflector	
Reflective Triangles	Flares, Flare Firearms, Fire Lanterns
Flashlight With Fresh Batteries Or Hand-Crack Flashlight	
Thick Work Gloves With Or Without Mag-Grip	
Disposable Gloves	Compass
Jumper Cables	Battery Jump Starter With E.M.P. Shielding
Dust Tape	Gaffers' Tape
Full Toe Covering Shoes And Or Steel Toe Boots	
Tool Kit, Multipurpose Utility Tool	
Line Rope	Physical Printed Paper Maps
Basic Tools With Pliers, Flat-Head, Screwdrivers, Adjustable Wrench	
Anti-Freeze	Coolant
Rags	Blanket
Spare Tire With Proper Inflated	Jack And Lug Wrench
Inflator Kit With An Air Compressor And Sealant	

First-Name Middle-Initial Last-Name; Vigilant Volunteer Militia, Charlie 'ToolPerson 工具人';
2154;_ Chicago, Illinois' Intranet, Wireless Broadcasting Frequencies Online

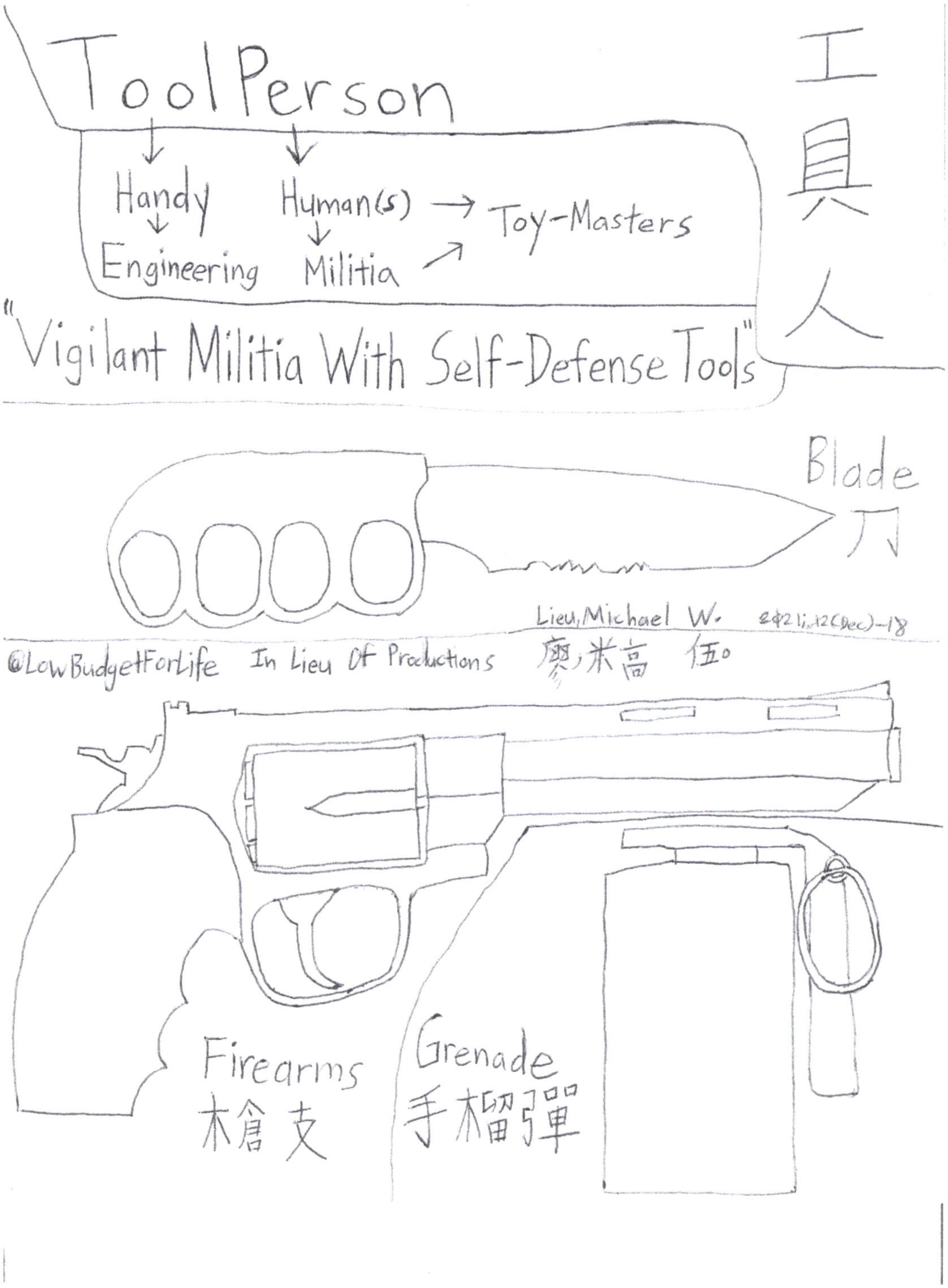

2154;_ Chicago, Illinois' Intranet, Wireless Broadcasting Frequencies Online
First-Name Middle-Initial Last-Name; Vigilant Volunteer Militia, Charlie 'ToolPerson 工具人';

First-Name Middle-Initial Last-Name; Vigilant Volunteer Militia, Charlie 'ToolPerson 工具人';
2154;_ Chicago, Illinois' Intranet, Wireless Broadcasting Frequencies Online

Strike Force
打擊　力量
ToolPerson
工具人

Fan(s) Following: "Vigilant Volunteers' Standard Issue For Human Utopian Militia(s)"
Black Hoodie Sweater With White Lettering Custom Text Wording
Oxygen Breathing Apparatus (O.B.A.) Gas Mask
Augmented Reality Holographic Heads Up Display (A.R. Holo HUD)
Smoke Screen Grenade
Tool Box Riot Shield Backpack
Ballistic And Energy Resistance Under Armor Vest
Joint Padding
Biotech Elbow Pads
Biotech Knee Pads
Grapple Magline Armlet
Biotech Maggrip Gloves
White Cotton Socks
Steel Toe Boots

2154;_ Chicago, Illinois' Intranet, Wireless Broadcasting Frequencies Online
First-Name Middle-Initial Last-Name; Vigilant Volunteer Militia, Charlie 'ToolPerson 工具人';

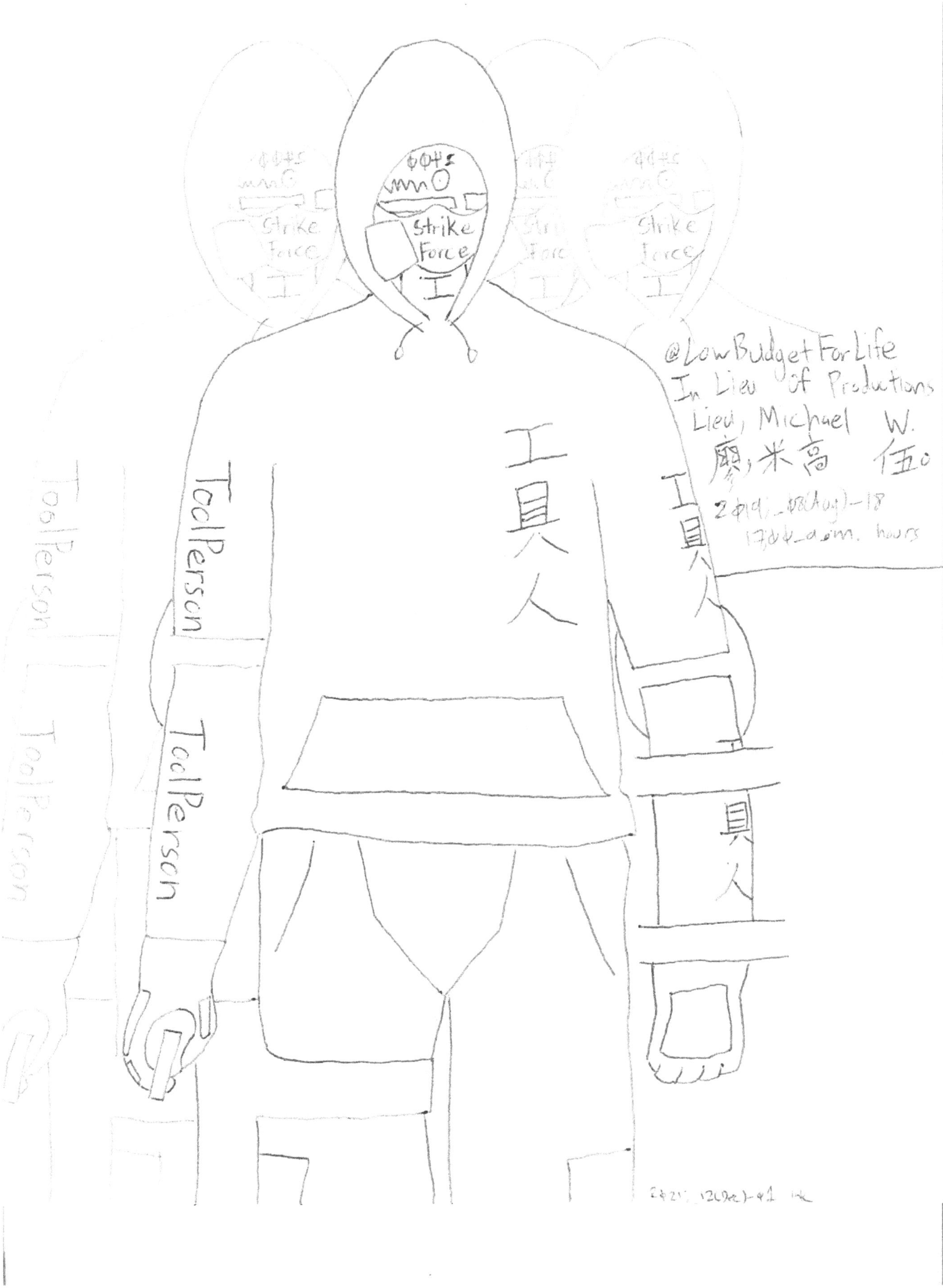
Strike Force
工具人
ToolPerson
@LowBudgetForLife
In Lieu Of Productions
Lieu, Michael W.
廖,米高 伍o

First-Name Middle-Initial Last-Name; Vigilant Volunteer Militia, Charlie 'ToolPerson 工具人';
2154;_ Chicago, Illinois' Intranet, Wireless Broadcasting Frequencies Online

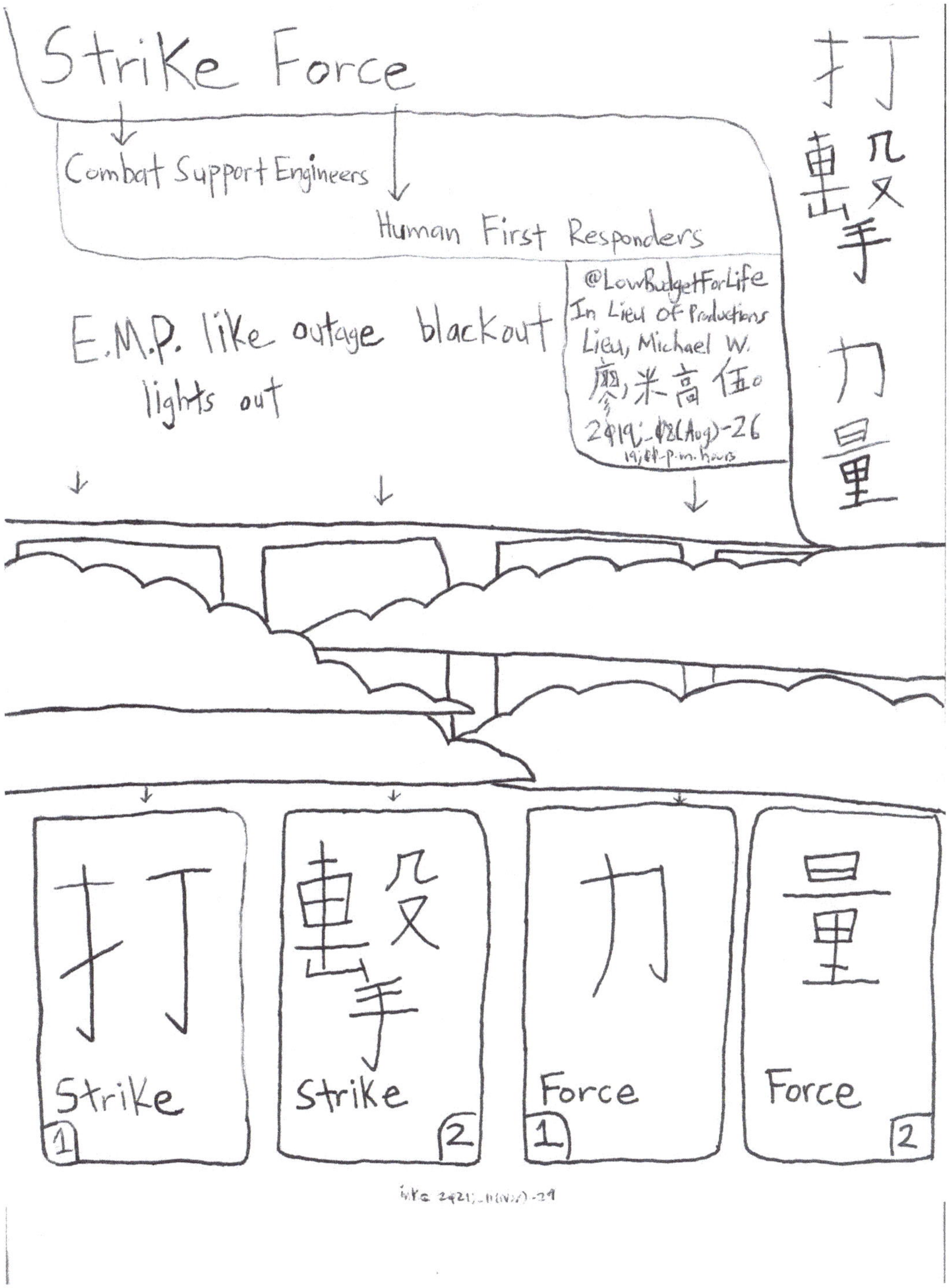

2154;_ Chicago, Illinois' Intranet, Wireless Broadcasting Frequencies Online
First-Name Middle-Initial Last-Name; Vigilant Volunteer Militia, Charlie 'ToolPerson 工具人';

2154;_ Arcadia \ Valentina Vita (V. V.) 'Healthy-Strong Life' \ Left to Right(L->R)

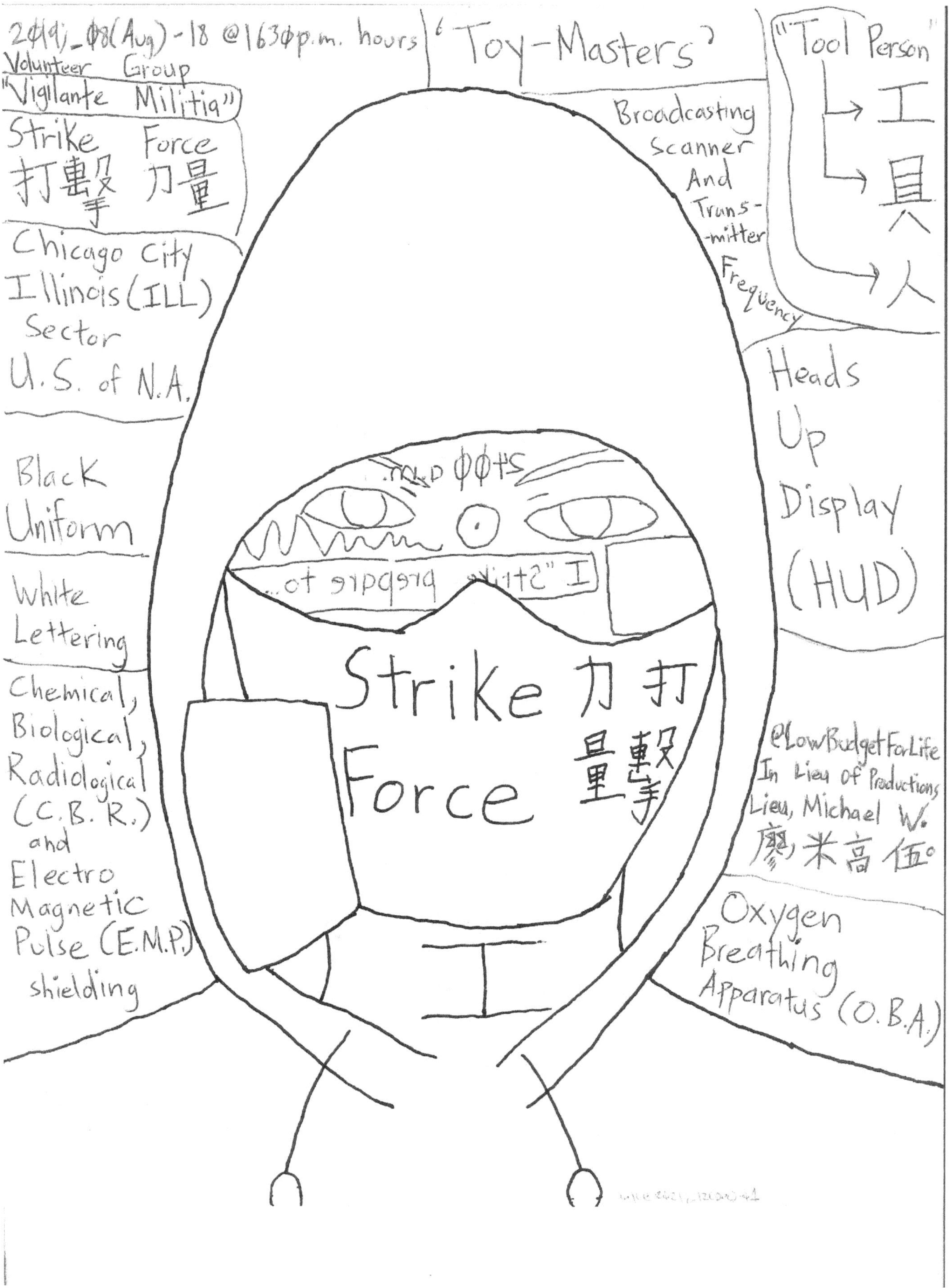
2019_08(Aug)-18 @1630p.m. hours
Volunteer Group
"Vigilante Militia"
Strike Force
打擊 力量
Chicago City
Illinois (ILL)
Sector
U.S. of N.A.
Black
Uniform
White
Lettering
Chemical,
Biological,
Radiological
(C.B.R.)
and
Electro
Magnetic
Pulse (E.M.P.)
shielding
'Toy-Masters'
Broadcasting
Scanner
And
Trans-
-mitter
Frequency
"Tool Person"
工
具
人
Heads
Up
Display
(HUD)
Strike 力打
Force 量擊
@LowBudgetForLife
In Lieu of Productions
Lieu, Michael W.
廖, 米高 伍o
Oxygen
Breathing
Apparatus (O.B.A.)

Valentina Vita (V. V.); Traveling Cosplayers And Kosplayers;
"(@_@) Hm..."

First-Name Middle-Initial Last-Name; Vigilant Volunteer Militia, Charlie 'ToolPerson 工具人';
"(*staring, in Oxygen Breathing Apparatus*) Hm... Sorry, **I** do not think **I** can **fix** this after **troubleshooting** it according to **my diagnoses**, but you are not going to **depart** from this **Animal** Sanctuary **empty handed** and **great disappointment**... Go to **North Side**, past **King And X District** or Biotech Dome Ocean Border was it? **Lieutenant Lieu**, **my volunteer manager who just sits on things**... ...and **he** is the **Charlie ToolPerson substitute**. As well as **Strike Force human first responder on call**... **He** should be able to really get down to this... **He** is **suited** for this..."

Valentina Vita (V. V.); Traveling Cosplayers And Kosplayers;
"(^_^) ...Oh okay, thanks anyways! So, I guess I do have to head to **North Side**, **King And X District**, Biotech Dome Ocean Border... Let me make sure to check that on my **to do list**... **Write it down** with a **pen, paper**..."

V. V.'s 2154 to do list since my Ai-Aug bracelet is not working due to outbreaks outages...
Visit, **North Side, Lieu Lieu.... Lieutenant Lieu** (next place to go and to do)
Visit, **Downtown District, Bullet Board System**
~~Visit, **East Side, All For One Shopping Center**~~ Complete
Visit, **South Side**, **Entertainer Mister Steven**, **Reserve** Your **Tickets** (Nov 11 @ 8 a.m.)
What else...? (What to do to pass time?)

First-Name Middle-Initial Last-Name; Vigilant Volunteer Militia, Charlie 'ToolPerson 工具人';
"(*staring, in Oxygen Breathing Apparatus*) Wow, **actual physical paper**... Well makes sense since your biotech Ai-**Aug bracelet** is **not working**... That and **writing things down** on yourself like a **skin tattoo** or sorts... or if you have a **read and write augmentation**... But, **whatever works** for you... Okay, remember, **North Side, King And X District**, Biotech Dome Ocean Border... Um... **Lieutenant Lieu** looks like an **old retired** white **black hair** mix natural **unmod Asian male**... white and **black Utopian Peacekeeping uniform**... **He's** got a WhiteWolf bio**tech suit** and sometimes that Yellow surprise face mask... Anyways, best of everything, gotta get back to **volunteer patrolling**."

Valentina Vita (V. V.); Traveling Cosplayers And Kosplayers;
"(^_^) Bye bye **Charlie ToolPerson**, thanks again! ...**North Side**, **King And X District**, Biotech Dome Ocean Border... **Lieutenant Lieu...**"

2154;_11(Nov)-02(two)

Travel To Western Society, America(s), North America
Self-Govern Sovereignty (S),
Illinois (ILL) Sector, Points Of Interest (P.O.I.)

Chicago (S), "**Animal** Sanctuary"
"The New World 2121 Movement Of The 22nd Century,
Utopian Sovereignty of North America (U.S. of N.A.)"
North Side, Biotech Dome Ocean Border, **King And X District**

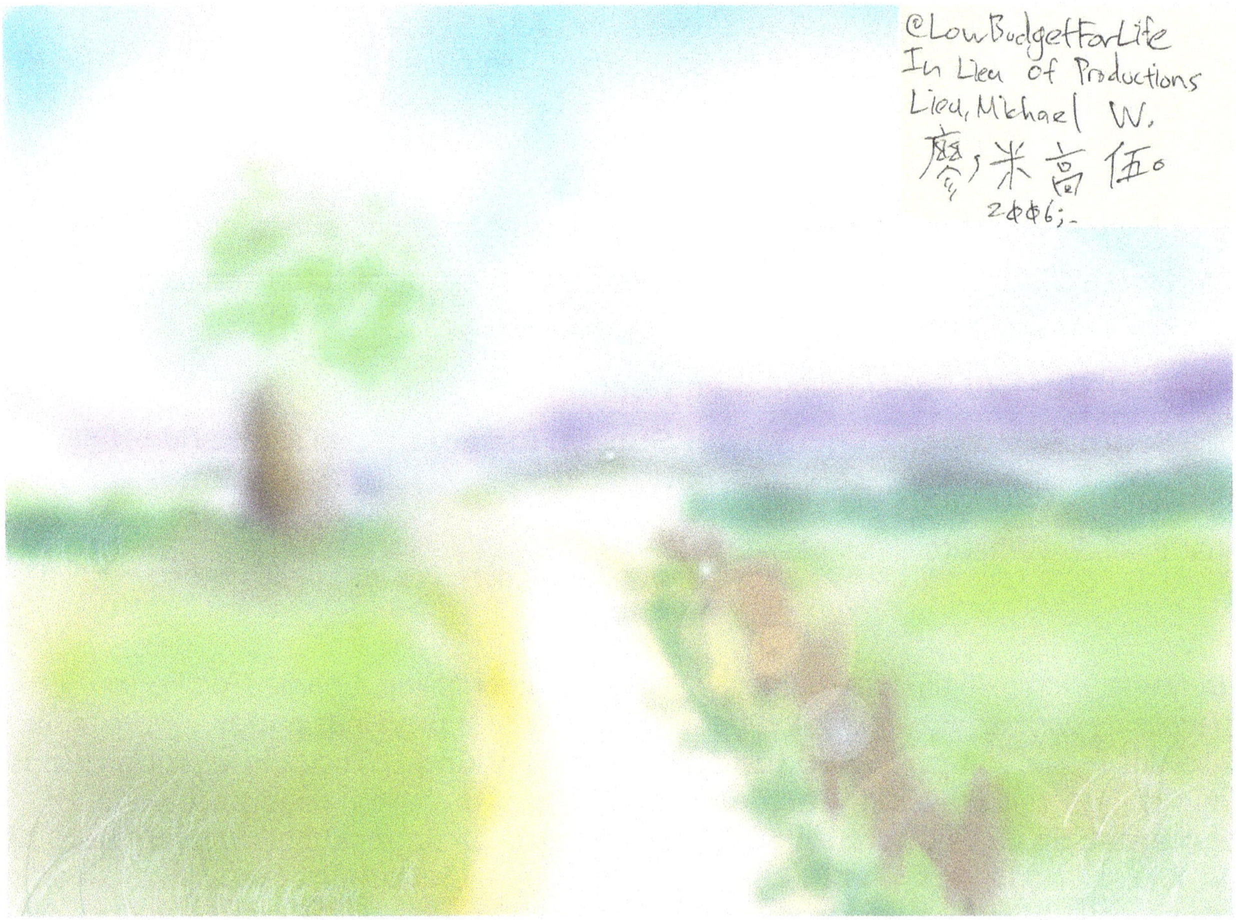

Valentina Vita (V. V.); Traveling Cosplayers And Kosplayers;
"(o_<) **North Side**, **King And X District**, Biotech Dome Ocean Border... **Lieutenant Lieu**... This looks like the way... **King And X District**... What does that mean?! (@_@) **King And X District...** Okay, on route to get my Ai-**Aug bracelet** fix and up and running..."

2154;_Arcadia \ Valentina Vita (V. V.) 'Healthy-Strong Life' \ Left to Right(L->R)

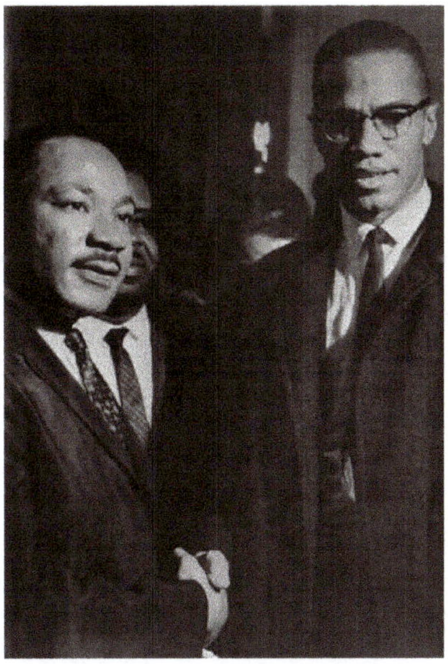

Old World Historical Figures Of The 20th Century United States Of North America
Martin Luther **King** (left) **And** Malcolm **X** (right) Civil Rights Meeting In 1964
"**King And X District**"

Reverend Martin Luther **King** Junior (Jr.)'s "***I Have A Dream***"	Malcolm **X**'s "***By Any Means Necessary***"
01(Jan)-15 Martin Luther King Jr. Day	**05(May)-19** Malcolm X Day
"I have a dream that my four little children will one day live in a nation where they will not be judged by the color of their skin, but by the content of their character." -**King**	"I have no mercy or compassion in me for a society that will crush people, and then penalize them for not being able to stand up under the weight." -**X**
King was an African-American black ***intergrationist*** on ethnic equality through ethnic **mixing** and working together.	**X** was an African-American black ***nationalist*** with a firm belief in African-American black **supremacy**.
Non-Violent = **Colored Person(s)** refusing to use physical force in a political movement.	Self-Defense = **Negro(s)** willing to protect their own life and existence.
Reverend Martin Luther **King** Jr.'s "*I Have A Dream*" philosophy inspired and created "**The Non-Violent Dream Movement**" in Washington, D. of C., United **States** of America (U.**S**. of A.)	Malcolm **X**'s "*By Any Means Necessary*" philosophy inspired and created "**The Black Panther Party Of Self-Defense**" in Oakland, North California, United **States** of America (U.**S**. of A.)

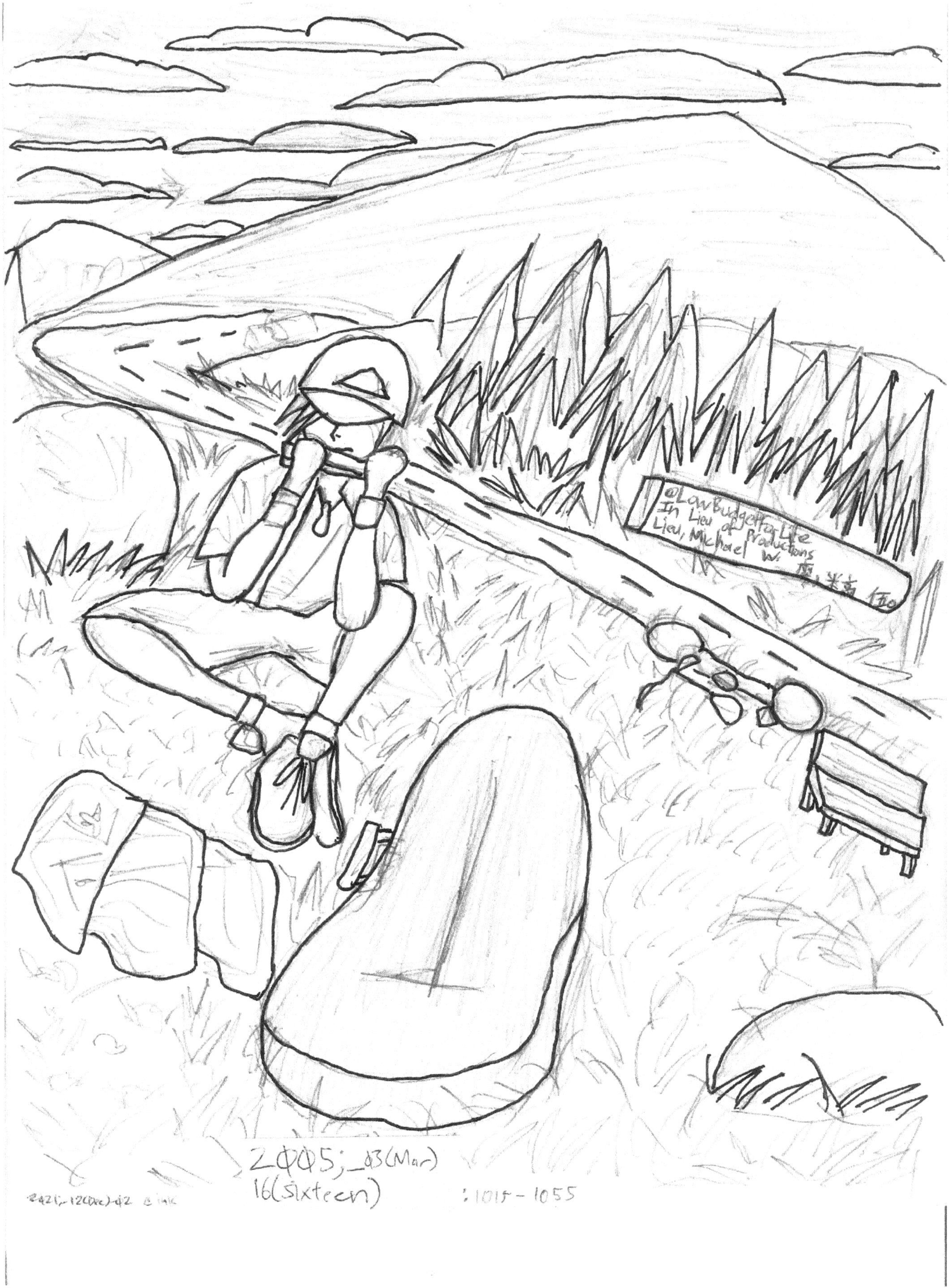
@LowBudgetForLife
In Lieu of Productions
Lieu, Michael W. 廖, 米高 伍o
2ΦΦ5;_Φ3(Mar)
16(sixteen)
;1015-1055

V. V.'s 2154 to do list since my Ai-Aug bracelet is not working due to outbreaks outages...
Visit, **North Side, Lieu Lieu.... Lieutenant Lieu** (next place to go and to do)
Visit, **Downtown District, Bullet Board System**
~~Visit, **East Side, All For One Shopping Center**~~ Complete
Visit, **South Side, Entertainer Mister Steven, Reserve** Your **Tickets** (Nov 11 @ 8 a.m.)
What else...? (What to do to pass time?)

Valentina Vita (V. V.); Traveling Cosplayers And Kosplayers;
"(>_O) Oh hi there, I am looking for **Lieutenant Lieu**... I do not think I am past the Biotech Dome Ocean Border... My Ai-**Aug bracelet** is not working, so I do not know **your name**... Um... **Human Asian male sitting** on the **ground playing** with a **music instrument.**"

Channy Tran = Cook-Gook (6C); Elementary, Mid, High School Friends "MoLeiTau 無厘頭";
"(*playing musical instrument*) ...Uh... **Human Asian male sitting**... You talking to **me**?"

Channy Tran is confused and looks around, **he** notices that V. V. is talking directly to **him**. **He** stops playing with **his musical instrument** and talks directly to V. V..

Channy Tran = Cook-Gook (6C); Elementary, Mid, High School Friends "MoLeiTau 無厘頭";
"(*calm*) ...**Lieutenant Lieu**... (*thinking*) **Lieu**... Oh **Mic Lieu**... yea, you on the right path to find **him**, **we** waiting on a **few friends** of **ours** for a **meet up** of **tennis** and **stuff**..."

Valentina Vita (V. V.); Traveling Cosplayers And Kosplayers;
"(^_^) Oh okay, cool thanks. **Your name** is...? Unless it is **human Asian male**...?"

Channy Tran = Cook-Gook (6C); Elementary, Mid, High School Friends "MoLeiTau 無厘頭";
"(*calm*) ...Uh... **Channy Tran**. ...**Human Asian male**, that is a good one, wish **I** could **name myself** that... You need to see **Mic Lieu** because of... a...?"

Valentina Vita (V. V.); Traveling Cosplayers And Kosplayers;
"(T_T) ...My Ai-**Aug bracelet** is **not working** due to an **outbreak outage experiment**-"

Jia Wong = B.4-Burma 3(three)III-M.M.B.; Elementary, Mid, High School Friends 無厘頭;
"(*kool voice over*) Hey **Channy**! **Your** turn up next for **our tennis warm ups**."

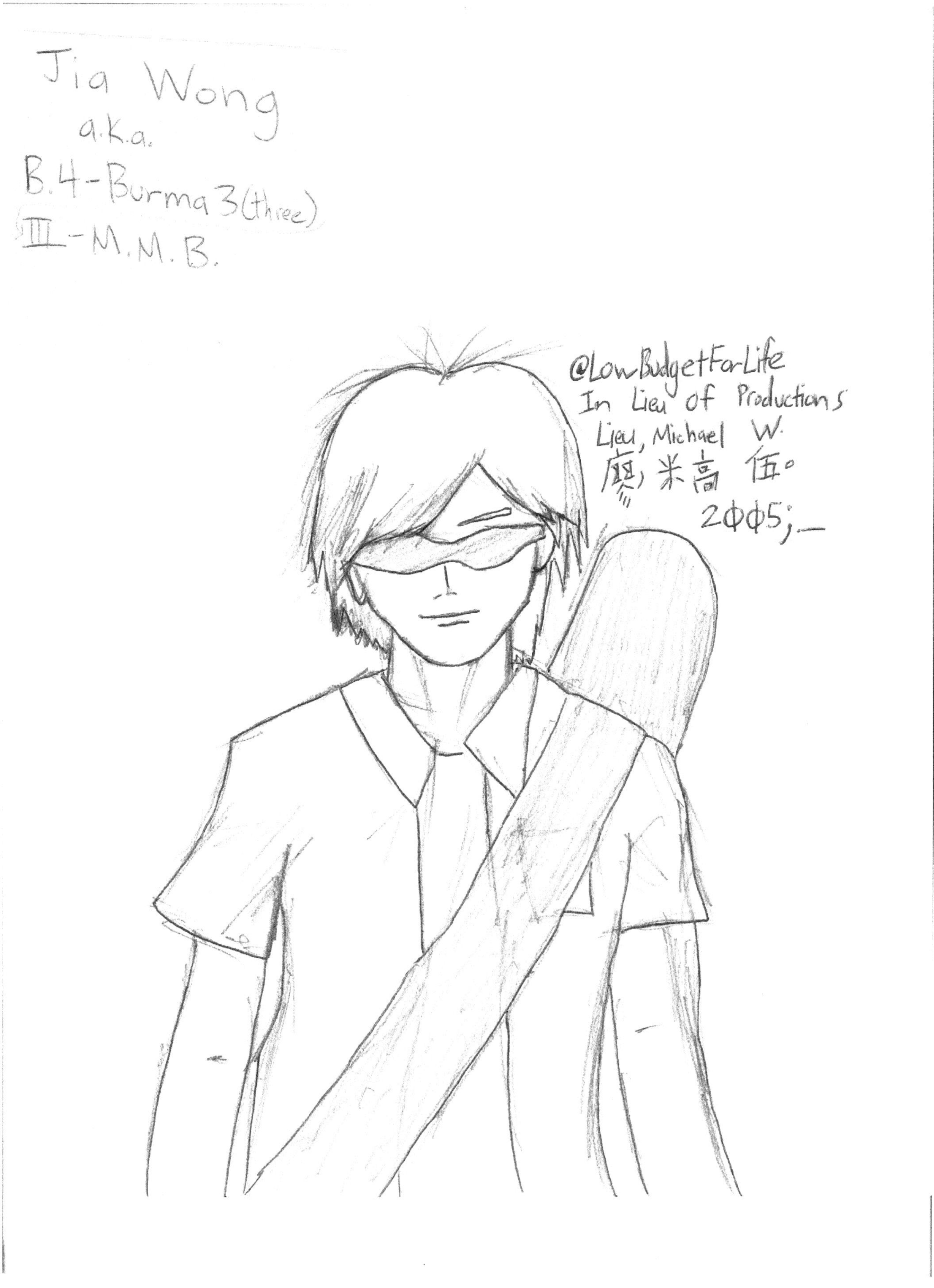
Jia Wong
a.k.a.
B.4-Burma 3(three)
III-M.M.B.
@LowBudgetForLife
In Lieu of Productions
Lieu, Michael W.
廖, 米高 伍。
2ΦΦ5;_

Valentina Vita (V. V.); Traveling Cosplayers And Kosplayers;
"(o_O) Whoa we wow..."

Jia Wong approaches and realizes that **he** is interrupting **Channy Tran** and V. V.'s conversation.

Jia Wong = B.4-Burma 3(three)III-M.M.B.; Elementary, Mid, High School Friends 無厘頭;
"(*kool*) Oh hey... **Channy**, are **you** talking to a **human female**...? Oh... V. V. here."

Channy Tran = Cook-Gook (6C); Elementary, Mid, High School Friends "MoLeiTau 無厘頭";
"(*calm voice over*) Yea... she was asking where to find **Mic**, so **I** figure guide her to the right place and **location** where **he** is **hiding**... Her Ai-Aug is **offline**, so she is **living** in the **dark ages... Sticks** and **stones...** She is using a piece of **paper** to write things down."

Jia Wong = B.4-Burma 3(three)III-M.M.B.; Elementary, Mid, High School Friends 無厘頭;
"(*kool*) Oh dam, her Ai-**Aug** is not working... Dam that sucks, yea... **dark ages...** Anyways, **Channy**, **your** turn up for **tennis warm up**... so yea..."

Channy Tran gets up from **his sitting location** on the green grass and walks with **Jia Wong** and V. V. to the **makeshift tennis courts**.

Channy Tran = Cook-Gook (6C); Elementary, Mid, High School Friends "MoLeiTau 無厘頭";
"(*calm voice over*) Okay, to find... **Mic Lieu**, you gotta go past **North Side** Biotech Dome Ocean Border. Past the **airlock decontamination**... past those **two Peter Pan Bros** with those **loud D.J. music**... Past the **photos of us back then**..."

Valentina Vita (V. V.); Traveling Cosplayers And Kosplayers;
"(>_O) Okay, got it! So just some more **walking** and **traveling** over there... Past the **airlock** of Biotech Dome Ocean Border... Past the **airlock** of Biotech Dome Ocean Border..."

V. V.'s 2154 to do list since my Ai-Aug bracelet is not working due to outbreaks outages...
Visit, **North Side, Lieu Lieu.... Lieutenant Lieu** (next place to go and to do)
Visit, **Downtown District, Bullet Board System**
~~Visit, **East Side, All For One Shopping Center**~~ Complete
Visit, **South Side**, Entertainer **Mister Steven**, **Reserve** Your **Tickets** (Nov 11 @ 8 a.m.)
What else...? (What to do to pass time?)

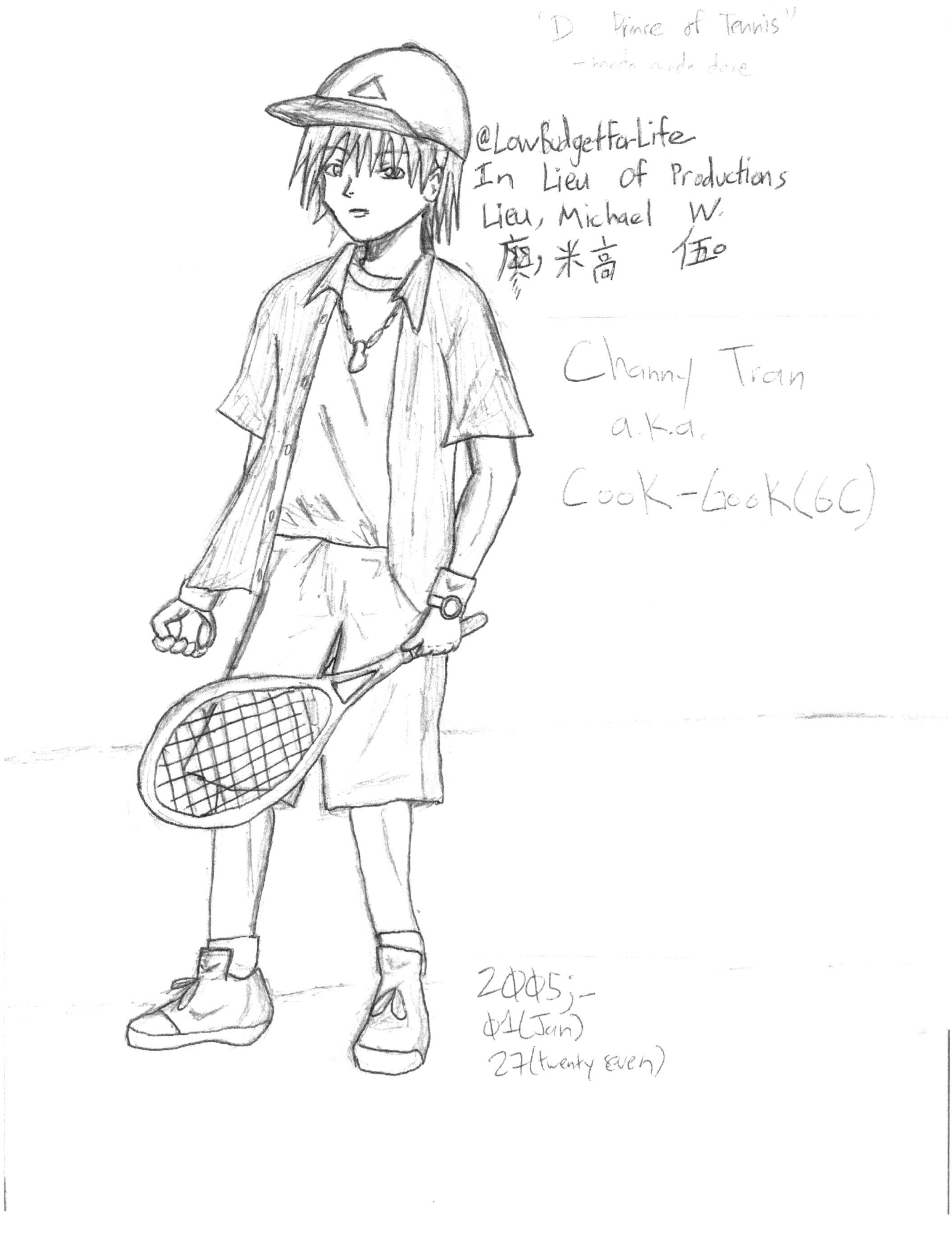
@LowBudgetForLife
In Lieu Of Productions
Lieu, Michael W.
廖, 米高　伍o
Channy Tran
a.k.a.
Cook-Gook(GC)
2ΦΦ5;-
Φ1(Jan)
27(twenty seven)

V. V.'s 2154 to do list since my Ai-Aug bracelet is not working due to outbreaks outages...
Visit, **North Side, Lieu Lieu.... Lieutenant Lieu** (next place to go and to do)
Visit, **Downtown District, Bullet Board System**
~~Visit, **East Side, All For One Shopping Center**~~ Complete
Visit, **South Side**, **Entertainer Mister Steven**, **Reserve** Your **Tickets** (Nov 11 @ 8 a.m.)
What else...? (What to do to pass time?)

Valentina Vita (V. V.); Traveling Cosplayers And Kosplayers;
"(@_@) Past the **airlock** of Biotech Dome Ocean Border... "

Channy Tran = Cook-Gook (6C); Elementary, Mid, High School Friends "MoLeiTau 無厘頭";
"(*calm*) Just stay on this route to the **North Side**, you are already past the **King And X District**, continue towards this path to the **airlock** passing Biotech Dome Ocean Border... On your way, you should be hearing some **loud electronic music of two D.J. Peter Pan Bros playing music...** Must be playing **Darude Sandstorm** again. Happy Ai-**Aug** fixing."

Valentina Vita (V. V.); Traveling Cosplayers And Kosplayers;
"(@_@) Okay, got it thanks!! ...Past the **airlock** of Biotech Dome Ocean Border... I should be hearing **loud electronic music of two D.J. Peter Pan Bros playing music**..."

Channy Tran looks briefly at V. V. and continues **his tennis warp ups**.
V. V. stays focus on finding **Lieutenant Lieu** as she now believes can assist in fixing her Ai-**Aug bracelet**.

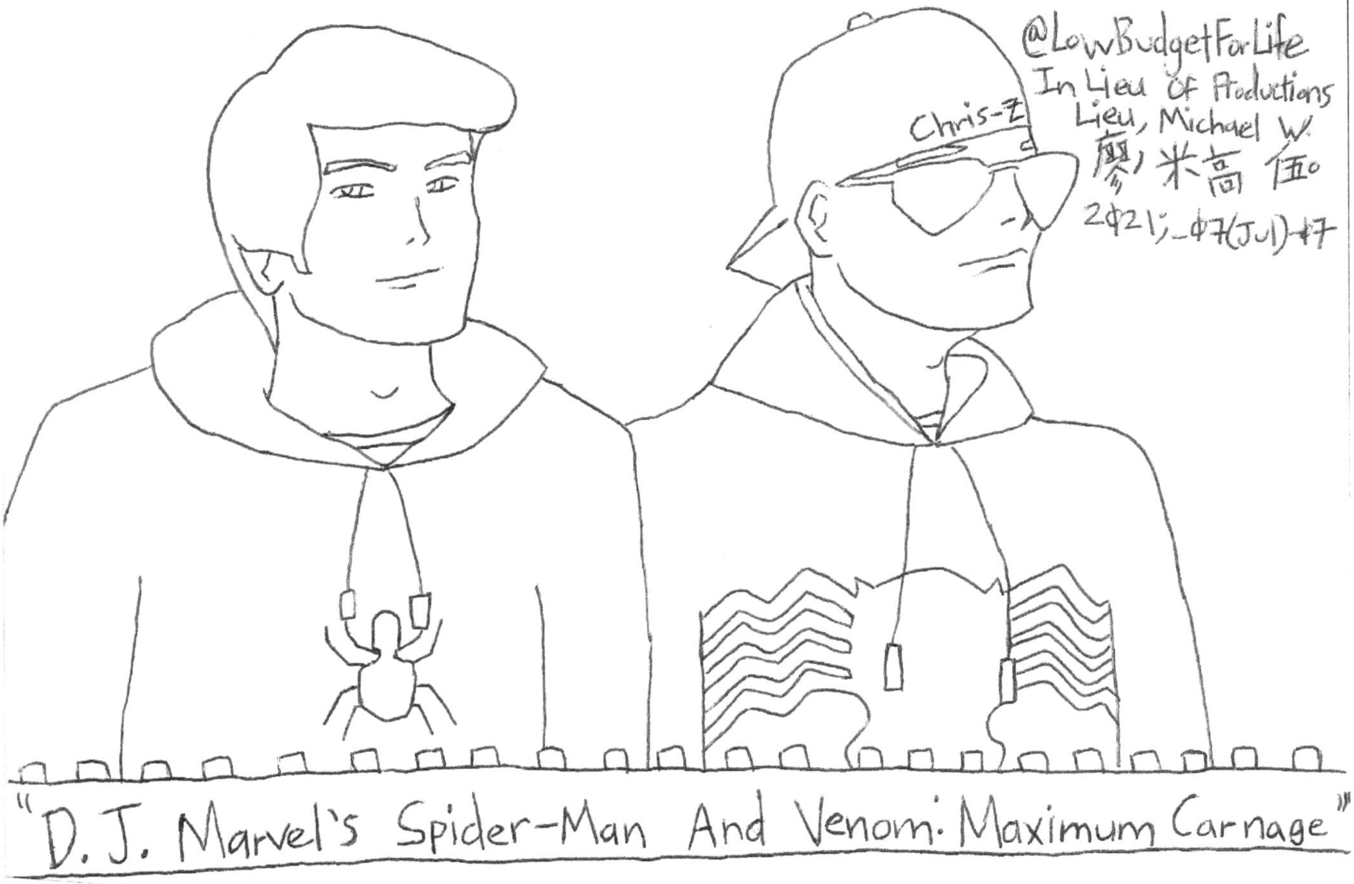

Darude - Sandstorm
on 24 hours

Liberal Arts For Life A.K.A. Peter Pan Bros
(Disc Jockey) "D. J. Marvel's Spider-Man And **Venom**: Maximum Carnage" Biography **Chris-Z**, Gar Field, and **D. D.** all three had natural teenager troubles back in the United States Army (U.S.A.) **Los Angeles City** (D) (R) Military (M) Bases in Chatsworth High School (C.H.S.), **South California** (**SCA**) of **North America** (**N.A.**). After **Chris-Z**, Gar Field, and **D. D.** settled their troubled differences by 2136 before grad night and graduation, the **three were self-proclaim fans of**: music Disc Jockey (D.J.), free running parkour, Do It Yourself (D.I.Y.) home video recordings, many more, and most importantly the **old world** United **States** of Americas' (U.**S**. of A.) Super-Hero Vigilante Mythology: "Marvel's Spider-Man And **Venom**: Maximum Carnage" Written And Edited By: **Chris-Z**, Gar Field, and **D. D.** Submitted By Animal Kingdom Service Doggie: **God**

Chris-Z = Marvel's **Venom Fanatic**; Elementary, Mid, High School Friends MoLeiTau;
"(*head banging*) Yo **bro**, **we** got this? Marvel's Spider-Man And **Venom**: Maximum Carnage!"

Gar Field = Marvel's Spider-Man Fanatic; Elementary, Mid, High School Friends MoLeiTau;
"(*head banging*) Yo **bro**, **we** got this! Marvel's Spider-Man And **Venom**: Maximum Carnage!"

V. V. notices what **Channy Tran** said about the "**Peter Pan Bros**" and keeps moving.

V. V.'s 2154 to do list since my Ai-Aug bracelet is not working due to outbreaks outages...
Visit, **North Side, Lieu Lieu.... Lieutenant Lieu** (next place to go and to do)
Visit, **Downtown District, Bullet Board System**
~~Visit, **East Side, All For One Shopping Center**~~ Complete
Visit, **South Side, Entertainer Mister Steven, Reserve** Your **Tickets** (Nov 11 @ 8 a.m.)
What else...? (What to do to pass time?)

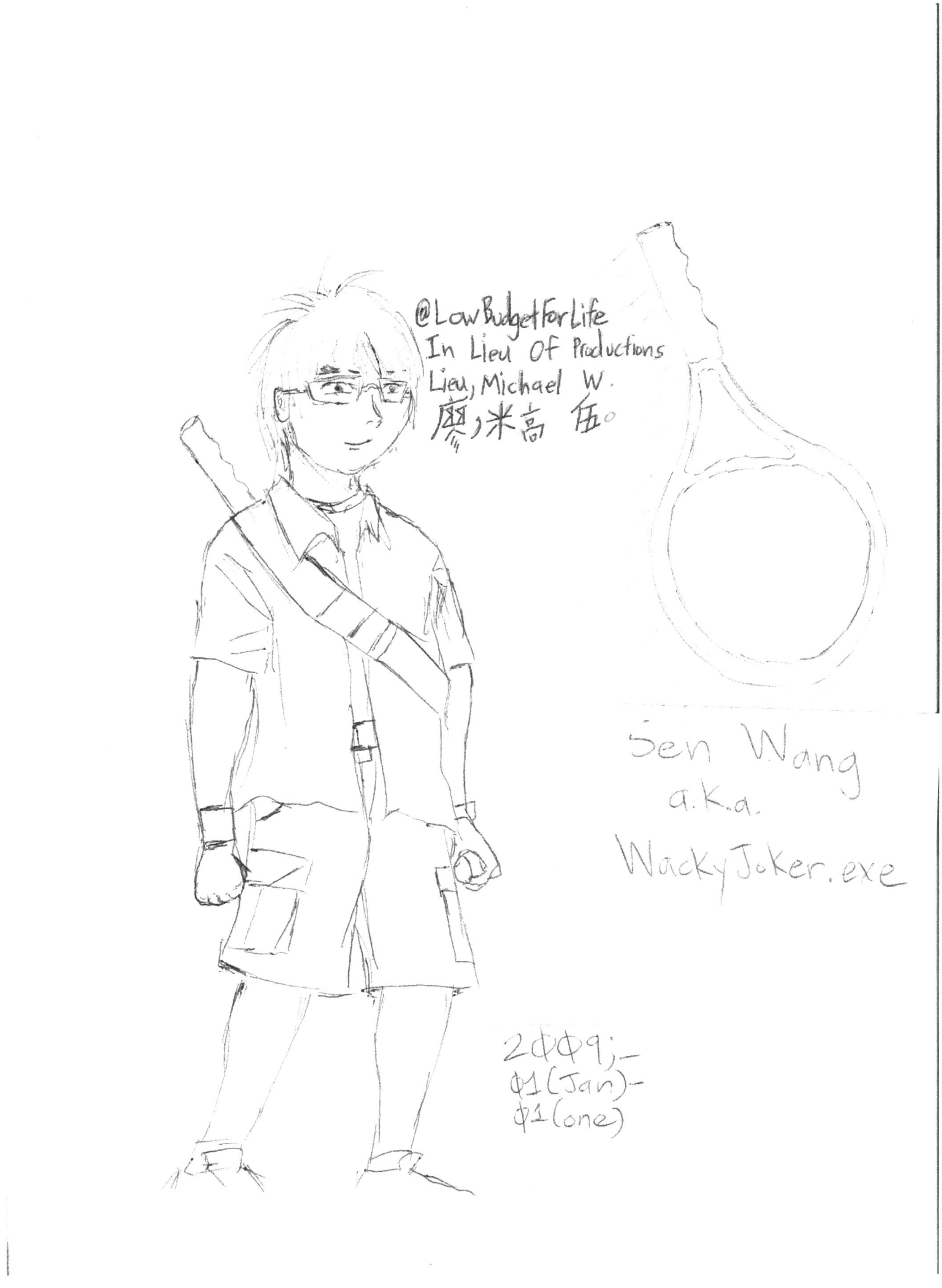
@LowBudgetForLife
In Lieu Of Productions
Lieu, Michael W.
廖,米高 伍o
Sen Wang
a.k.a.
WackyJoker.exe
2ΦΦ9;-
Φ1(Jan)-
Φ1(one)

Valentina Vita (V. V.); Traveling Cosplayers And Kosplayers;
"(O_O) ...Past the **airlock** of Biotech Dome Ocean Border... ...Past the **airlock** of Biotech Dome Ocean Border... Huh... **Interesting**... **Tennis warm up**..."

V. V.'s 2154 to do list since my Ai-Aug bracelet is not working due to outbreaks outages...
Visit, **North Side, Lieu Lieu.... Lieutenant Lieu** (next place to go and to do)
Visit, **Downtown District, Bullet Board System**
~~Visit, **East Side, All For One Shopping Center**~~ Complete
Visit, **South Side**, Entertainer **Mister Steven, Reserve** Your **Tickets** (Nov 11 @ 8 a.m.)
What else...? (What to do to pass time?)

Sen Wang = WackyJoker.exe; Elementary, Mid, High School Friends "MoLeiTau 無厘頭";
"(*focusing*) Ya!! **Double tennis racket action**!! Ya!! **Flying tennis racket**!! Ya!! Fly!"

Valentina Vita (V. V.); Traveling Cosplayers And Kosplayers;
"(@_@) ...Okay, I made it past the **loud music D.J. Peter Pan Bros** as mentioned... Almost there... Made it past **King X And District** of **North Side**... On route to... **airlock** of Biotech Dome Ocean Border... Gotta **stay focus**... Been **days** and **hours** with my Ai-**Aug bracelet** working... (>_<) I am getting **anxious, anxiety** need to **hurry**! No **music**, no **note taking**... Uh..."

V. V. stares at bit at **Sen Wang's "interesting tennis warm ups exercises**" and continues to move towards her goal of finding **Lieutenant Lieu** past **airlock** of Biotech Dome Ocean Border. **He** does not look back at her, as **he** continues **his "interesting tennis warm ups exercises**".

She continues moving and hears **loud grunting sounds** from supposedly **two human males** playing in an "**intense tennis warm up rally**". V. V. follows the **loud grunting sounds** from the **two human males**.

She sees these **two human Asian males**, who are supposedly **twins** with **small visual differences** playing in the **makeshift tennis courts**.

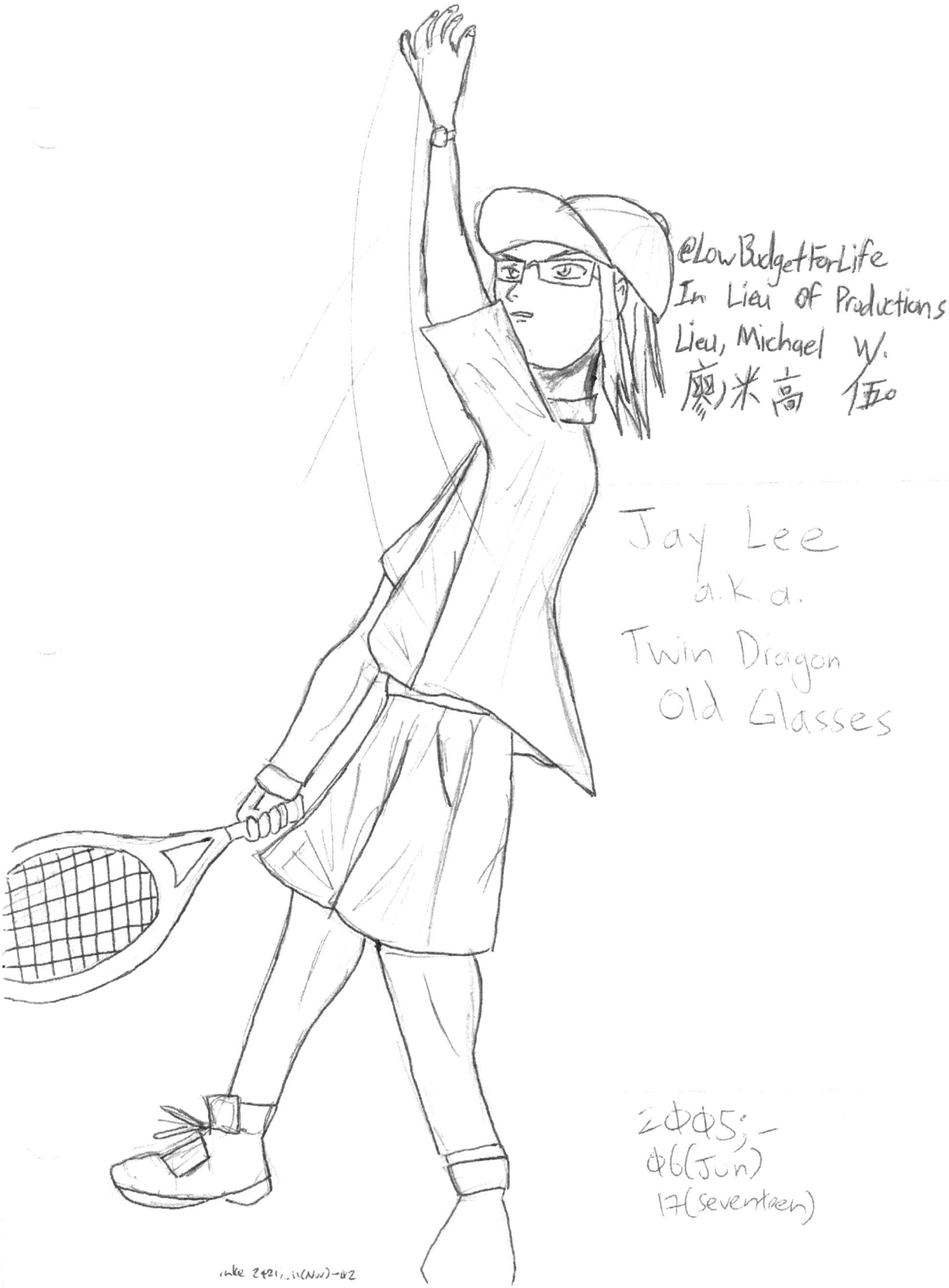
@LowBudgetForLife
In Lieu Of Productions
Lieu, Michael W.
廖,米高 伍o
Jay Lee
a.k.a.
Twin Dragon
Old Glasses
2ΦΦ5;-
Φ6(Jun)
17(seventeen)

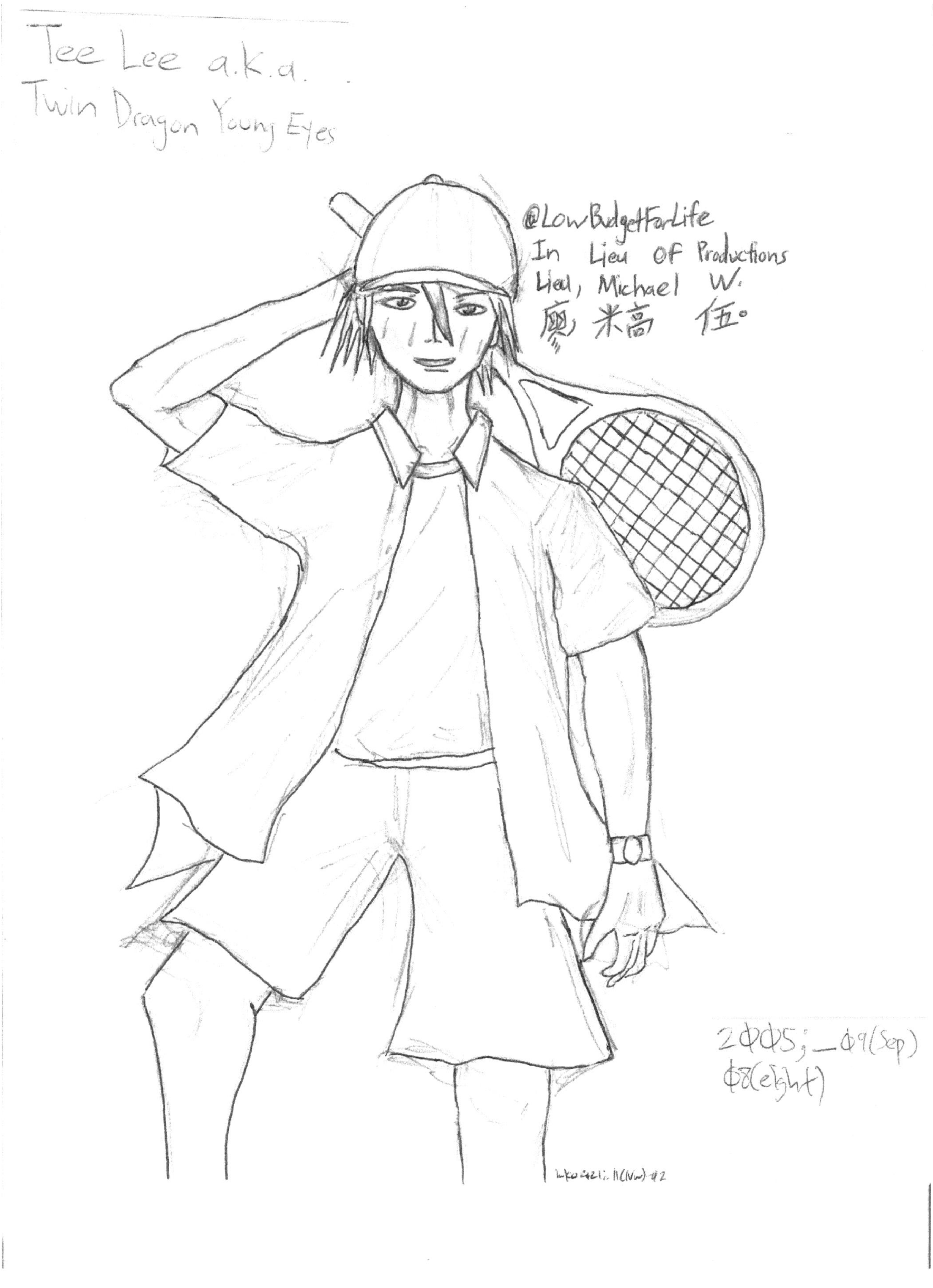
Tee Lee a.k.a.
Twin Dragon Young Eyes
@LowBudgetForLife
In Lieu Of Productions
Lieu, Michael W.
廖, 米高 伍。
2ΦΦ5; _Φ9(Sep)
Φ8(eight)

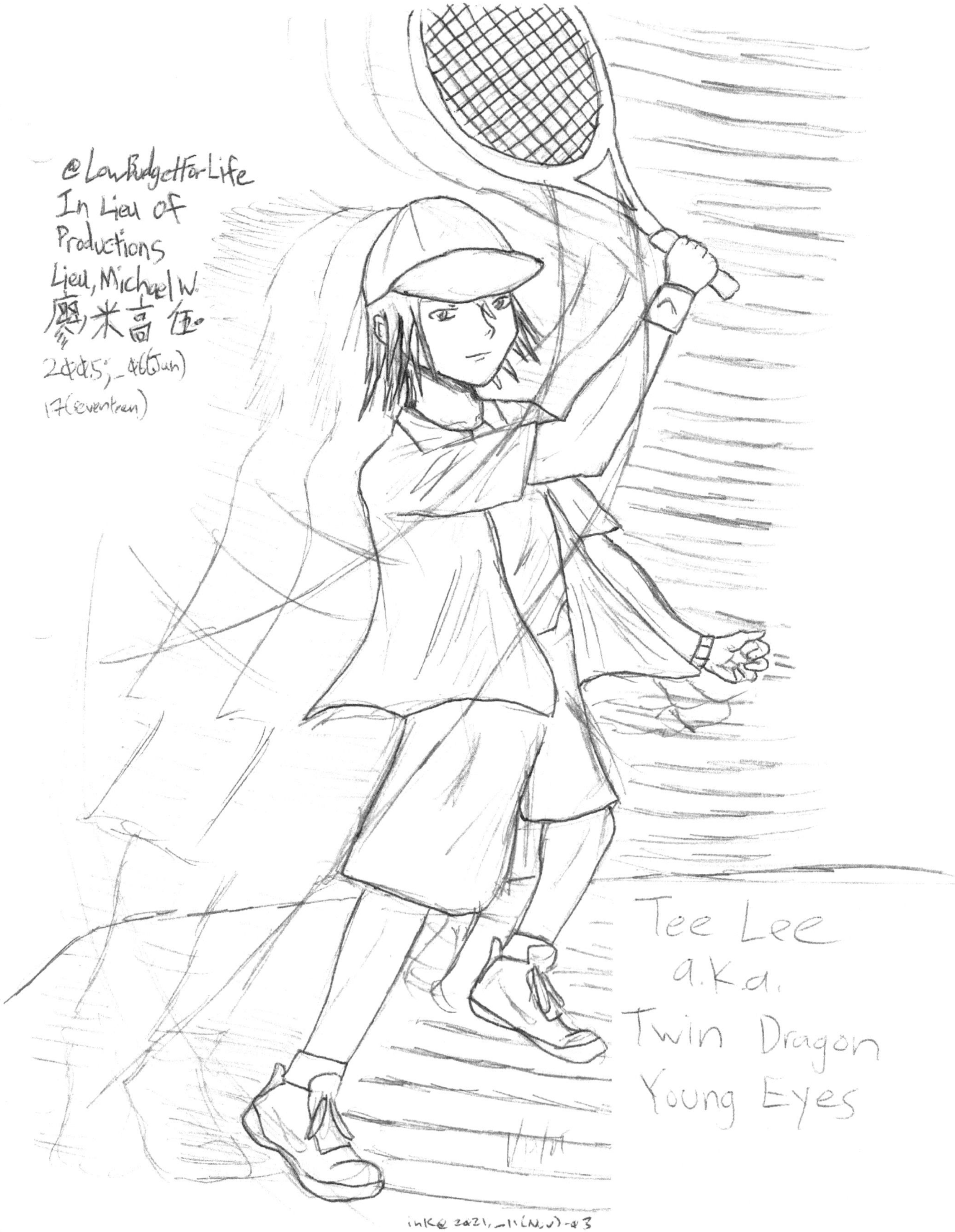
@LowBudgetForLife
In Lieu of
Productions
Lieu, Michael W.
廖,米高 伍o
2005,-06(Jun)
17(seventeen)
Tee Lee
a.k.a.
Twin Dragon
Young Eyes
ink@ 2021,-11(Nov)-03

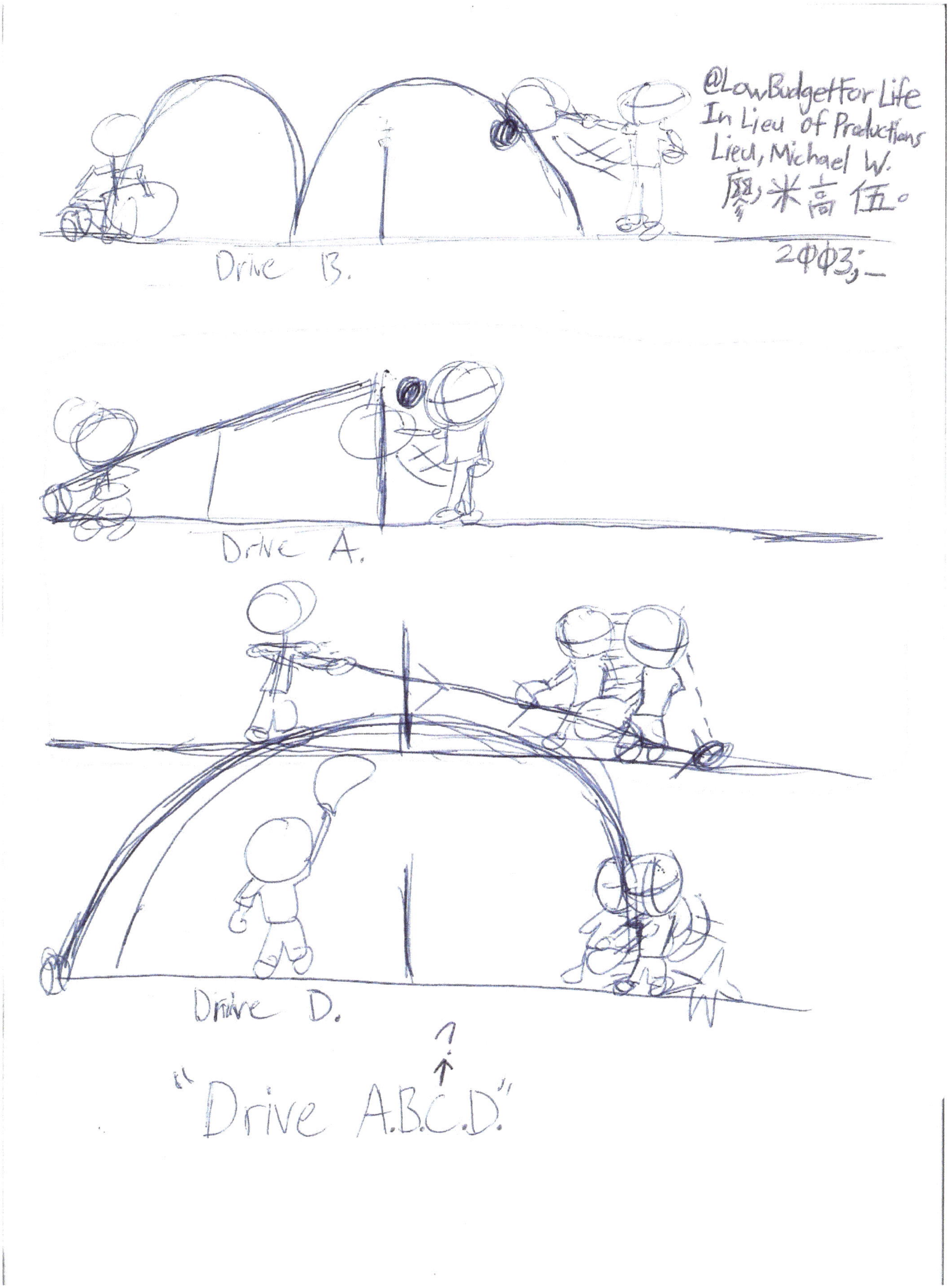
@LowBudgetForLife
In Lieu of Productions
Lieu, Michael W.
廖,米高 伍。
2ΦΦ3;—
Drive B.
Drive A.
Drive D.
?
"Drive A.B.C.D."

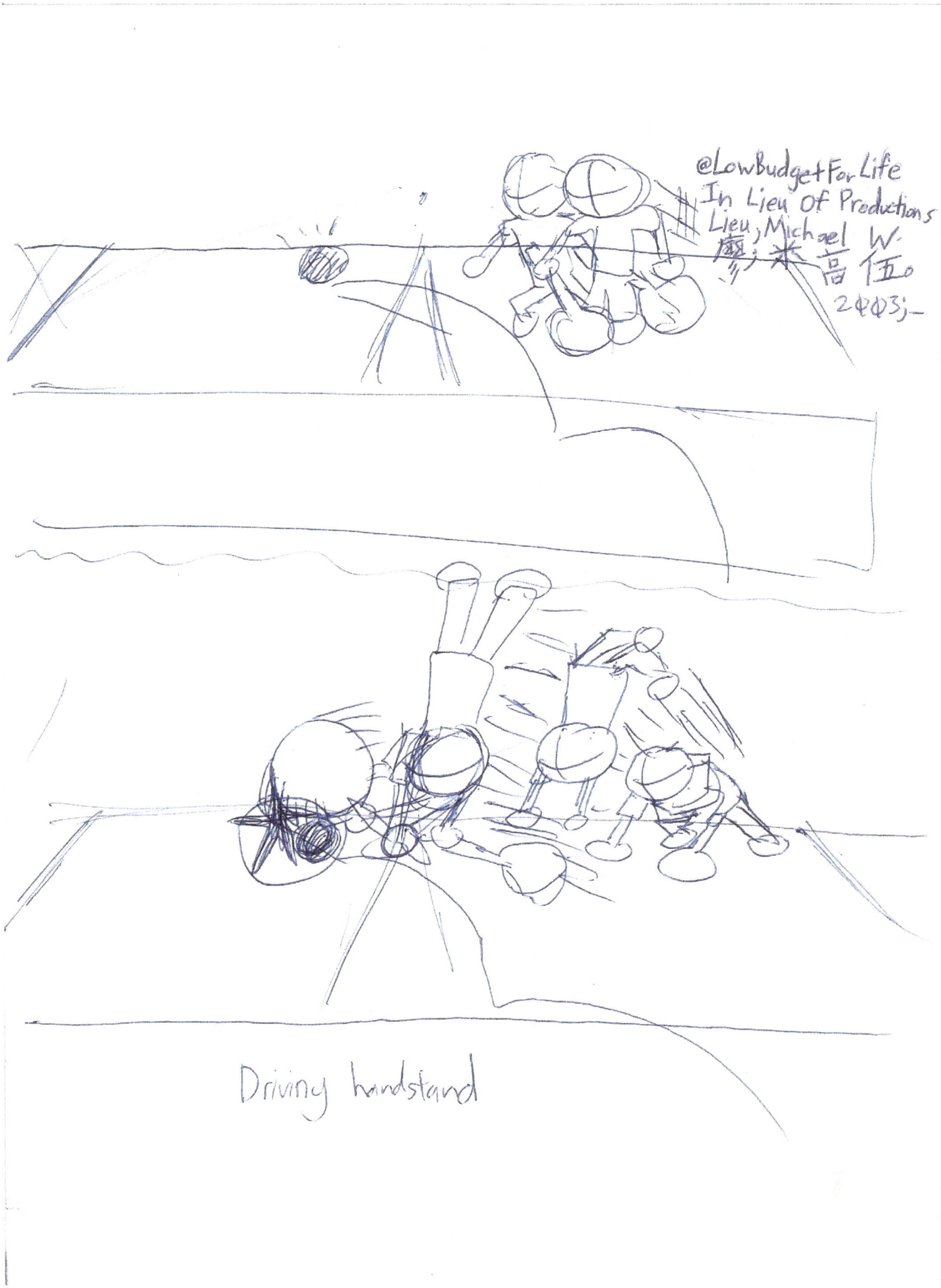
@LowBudgetForLife
In Lieu Of Productions
Lieu, Michael W.
廖, 米高 伍o
2003;-
Driving handstand

Automated Algorithm; Human Voice Over Emulation Based On Elysium Corporation Archives;
"(*monotone female voice over*) Please stand by for biotech dome airlock decontamination. Thank you for your patience and cooperation **Utopian**."

Lieutenant Lieu, Michael W.'s 2136;_Grad Visual Photos

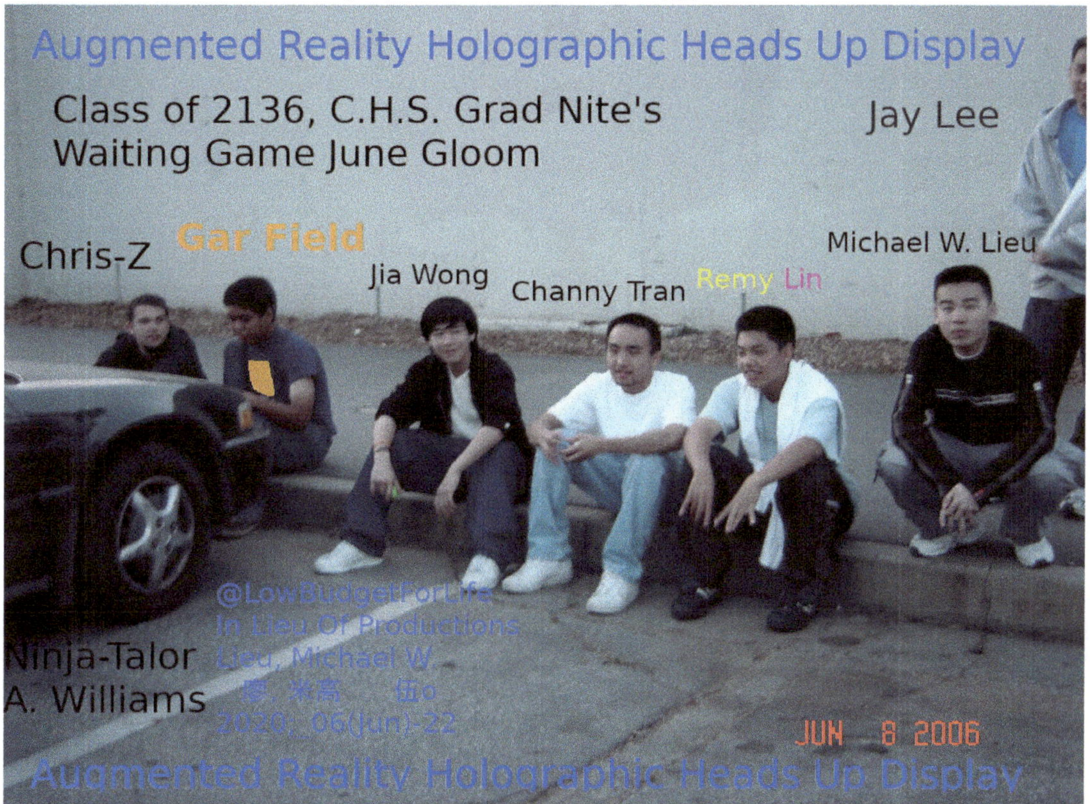

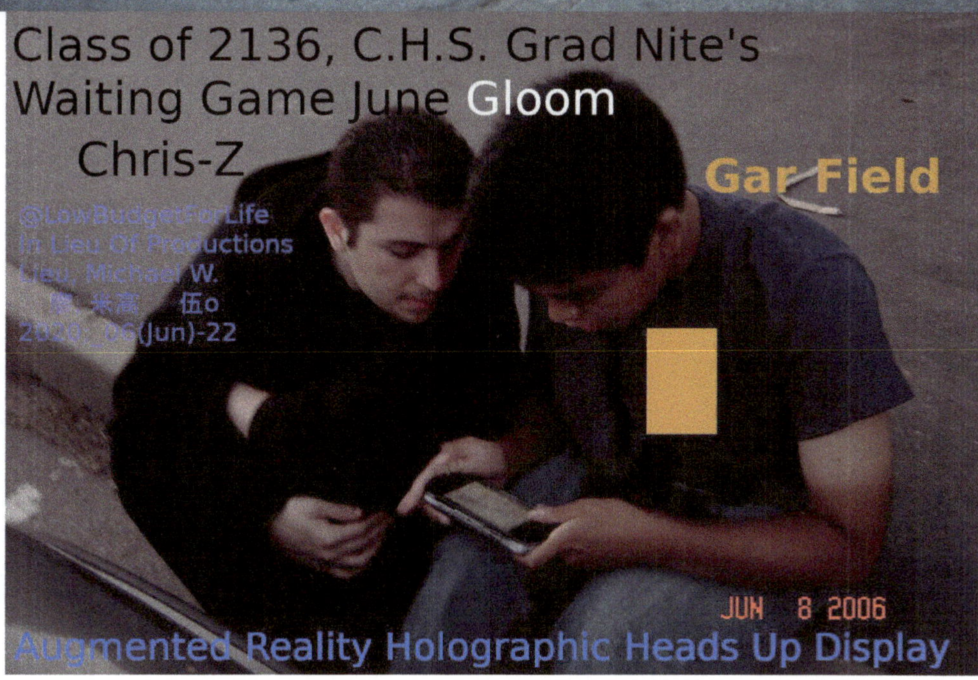

Lieutenant Lieu, Michael W.'s 2136;_Grad Visual Photos

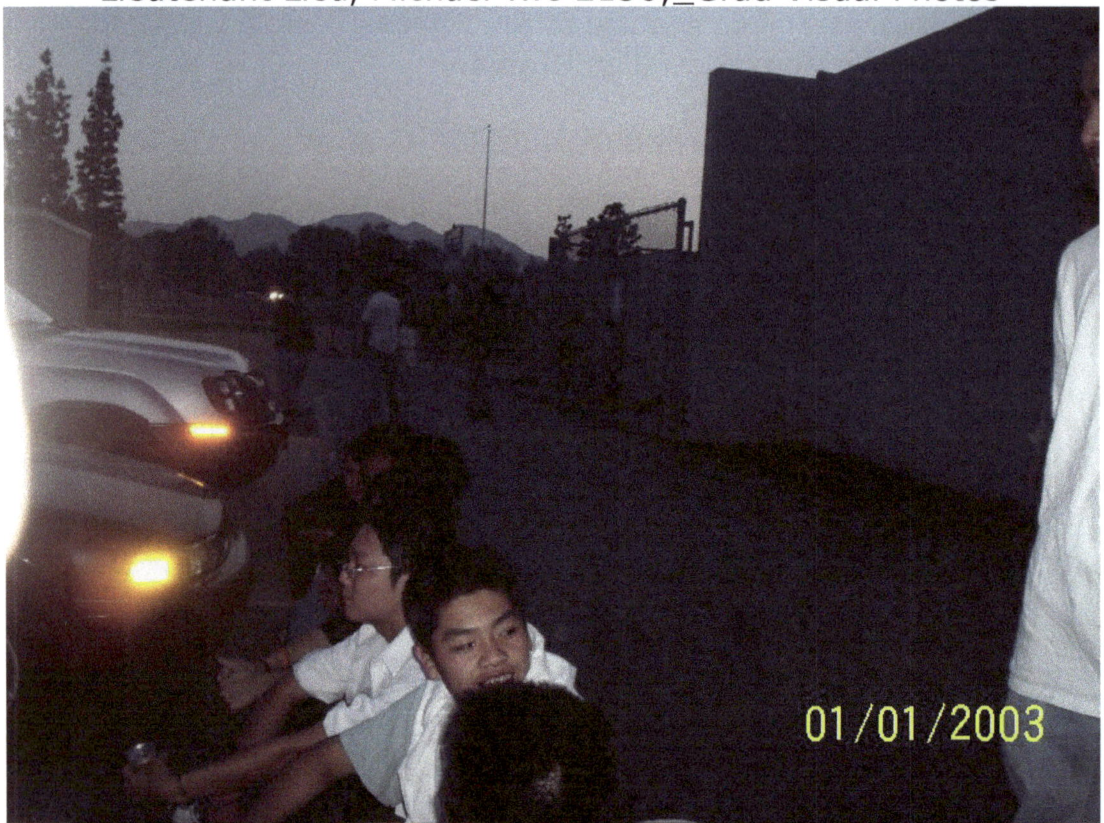

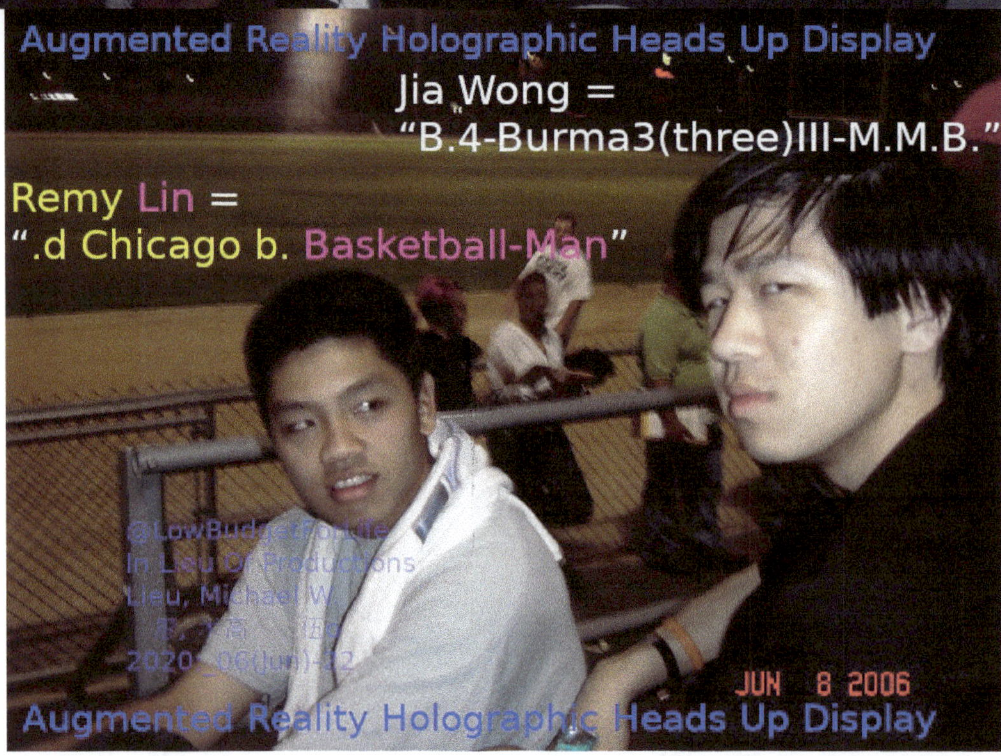

2154;_Arcadia \ Valentina Vita (V. V.) 'Healthy-Strong Life' \ Left to Right(L->R)

Lieutenant Lieu, Michael W.'s 2136;_Grad Visual Photos

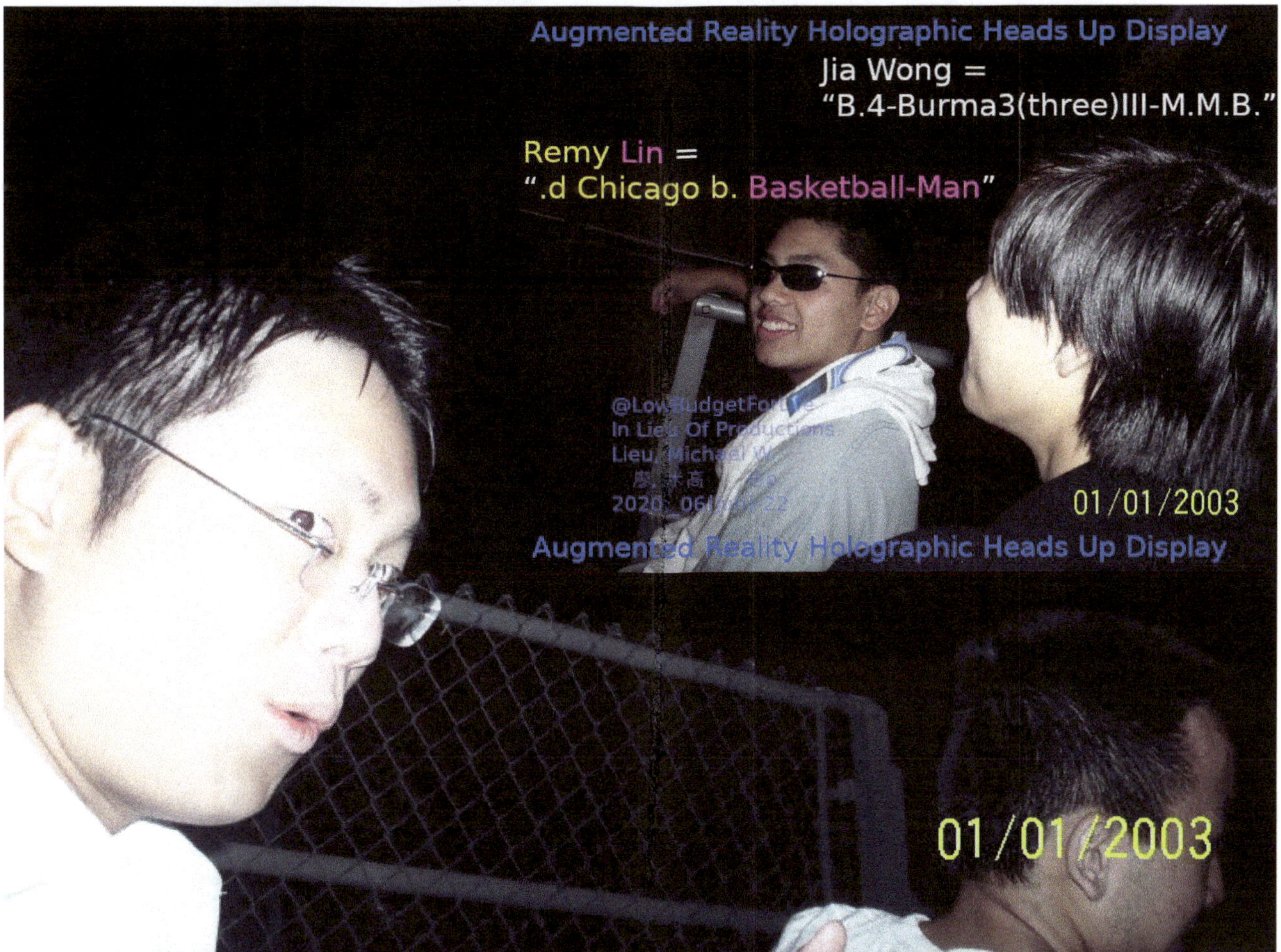

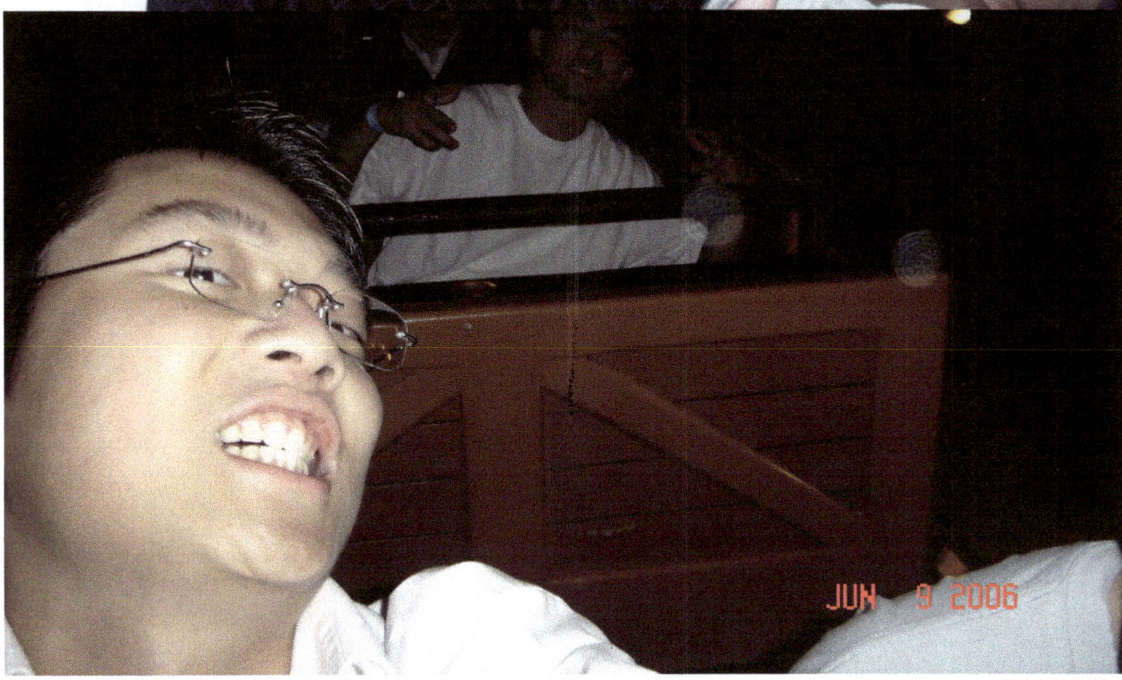

2154;_Arcadia \ Valentina Vita (V. V.) 'Healthy-Strong Life' \ Left to Right(L->R)

Lieutenant Lieu, Michael W.'s 2136;_Grad Visual Photos

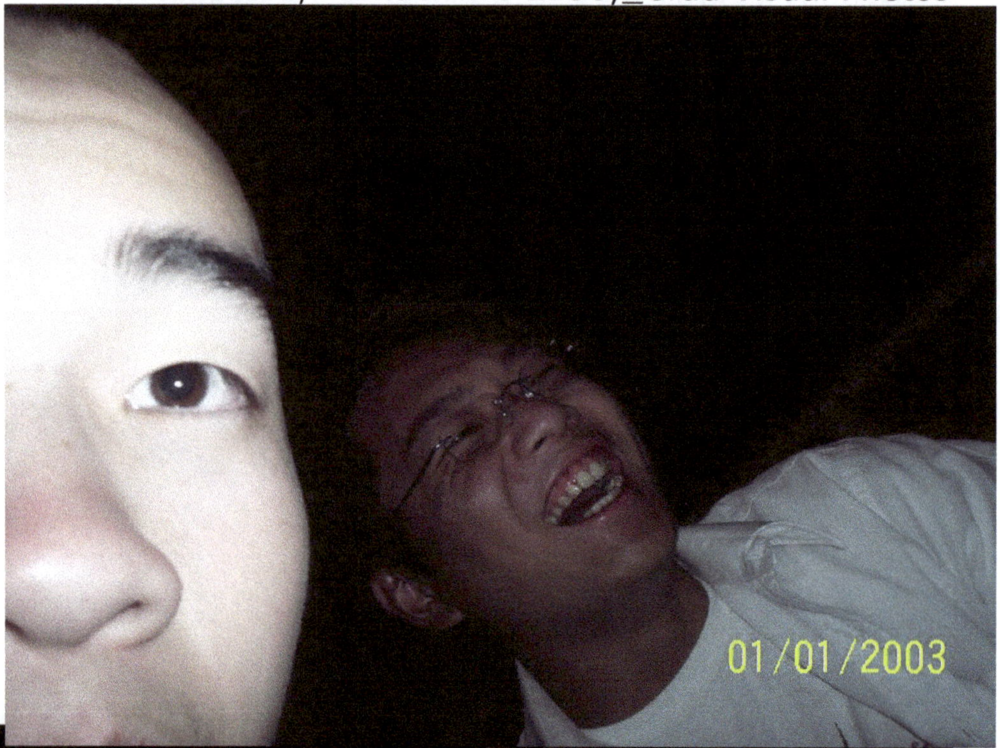

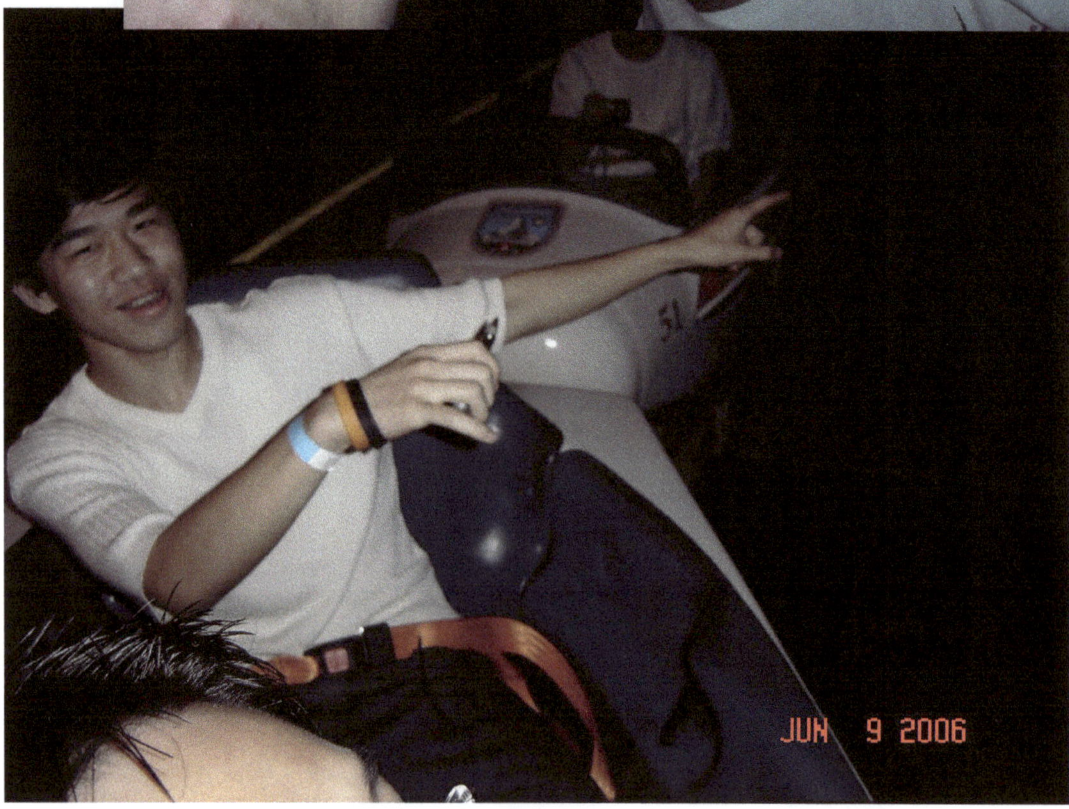

2154;_Arcadia \ Valentina Vita (V. V.) 'Healthy-Strong Life' \ Left to Right(L->R)

Lieutenant Lieu, Michael W.'s 2136;_Grad Visual Photos

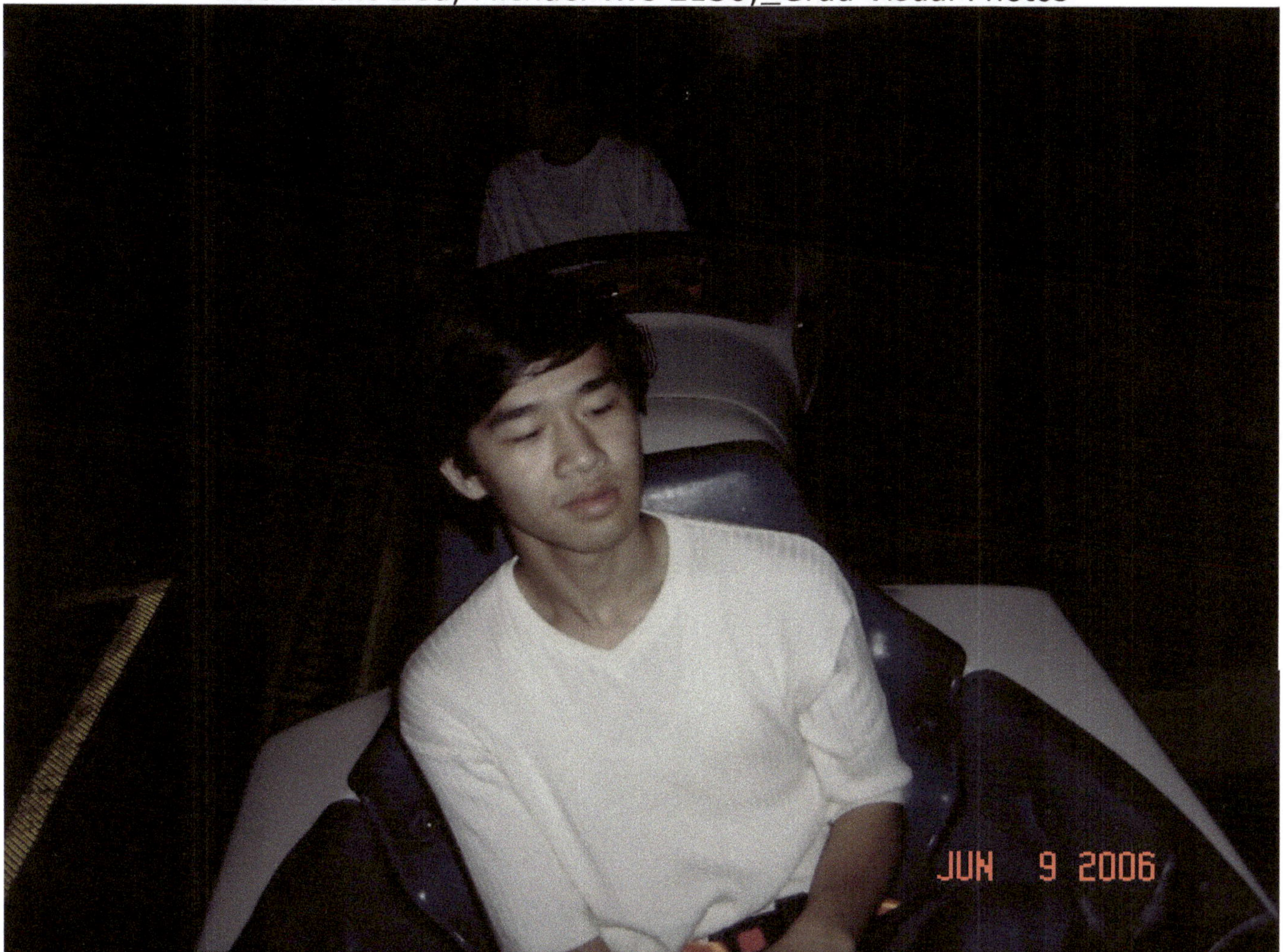

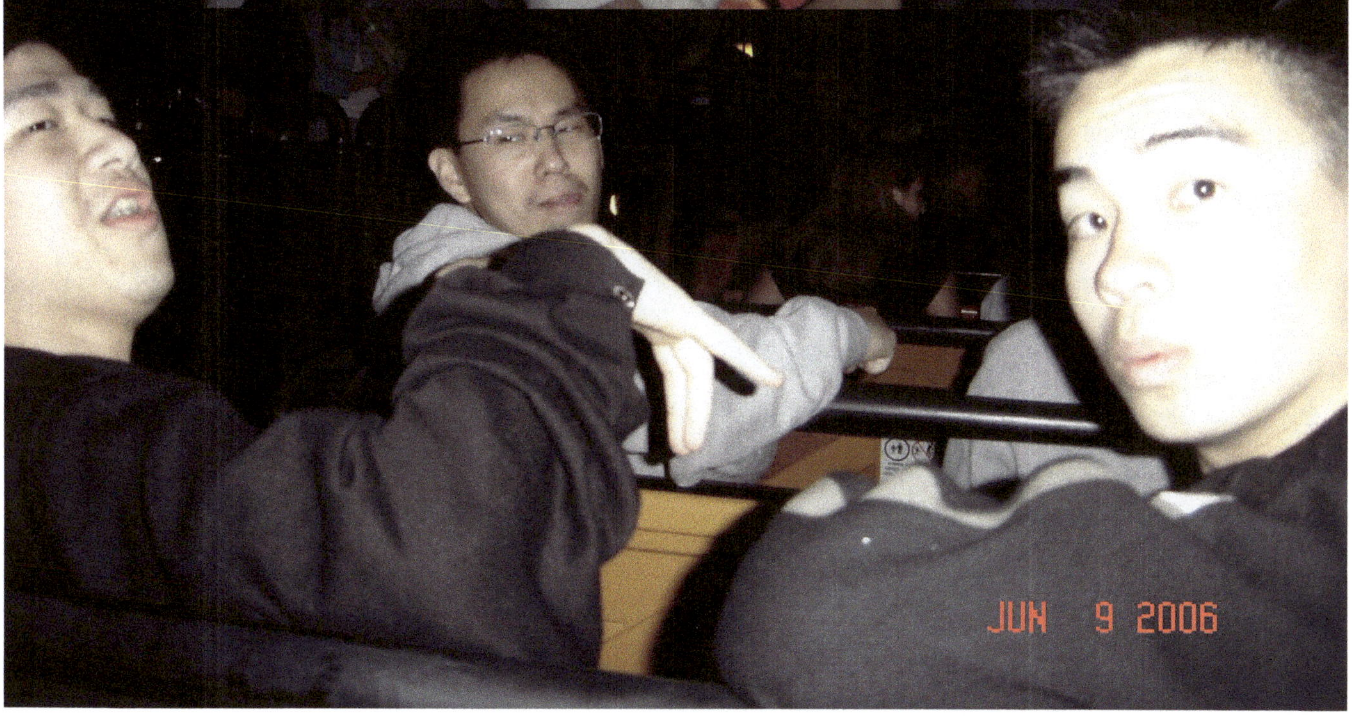

Lieutenant Lieu, Michael W.'s 2136;_Grad Visual Photos

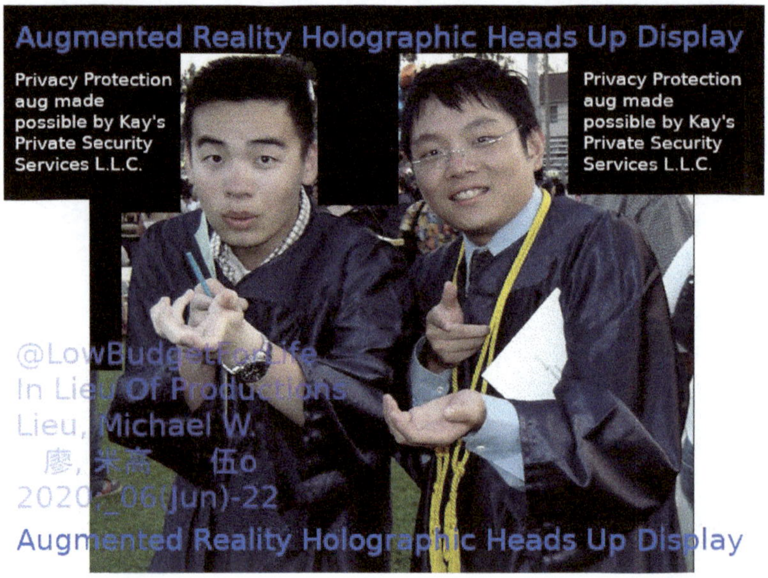

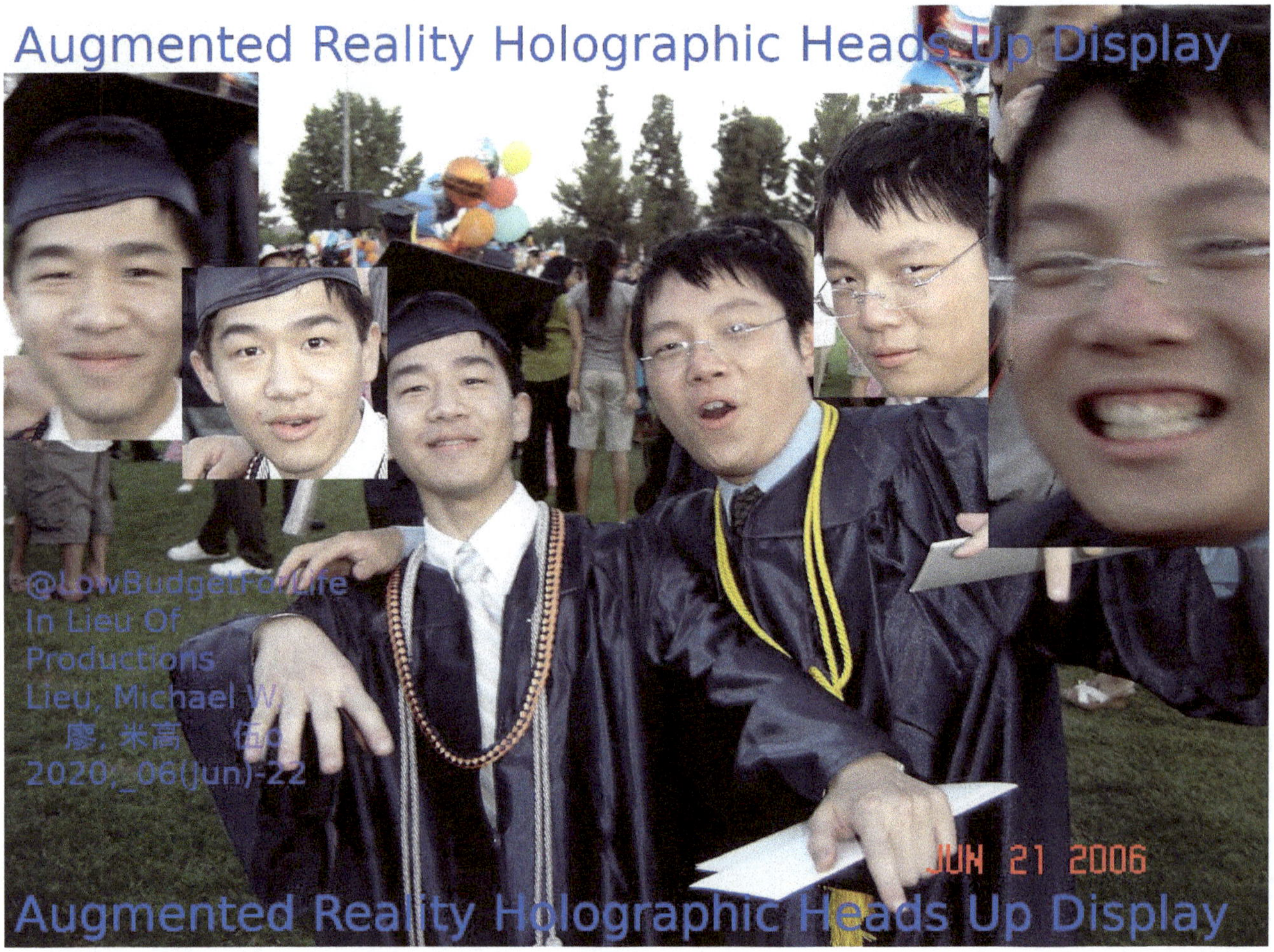

2154;_Arcadia \ Valentina Vita (V. V.) 'Healthy-Strong Life' \ Left to Right(L->R)

Lieutenant Lieu, Michael W.'s 2136-2139;_Navy Visual Photos
2136-2139;_United States Department Of War (U.S. DOW)
2137;_11(Nov) - 2138;_06(Jun) Navy Division,
Military Servicemember Deployment

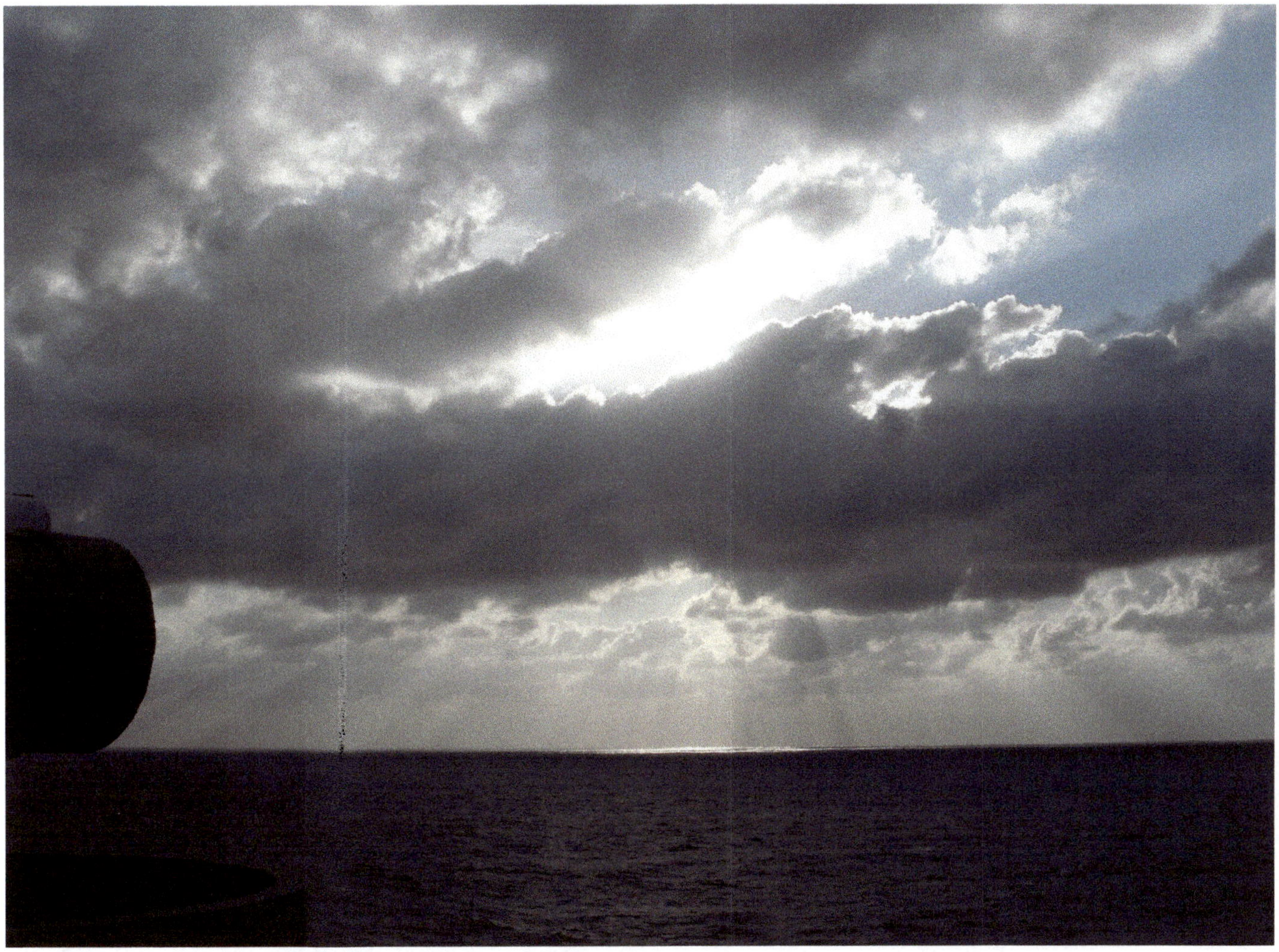

Background; Cloud, Sky, Heaven, Ocean, Water, Sea
Foreground; Military Warship Aircraft Carrier Lookout Tools

Visual Photo By; Lieutenant Lieu, Michael W.
廖, 米高 伍 o

Lieutenant Lieu, Michael W.'s 2136-2139;_Navy Visual Photos
2136-2139;_United States Department Of War (U.S. DOW)
2137;_11(Nov) - 2138;_06(Jun) Navy Division,
Military Servicemember Deployment

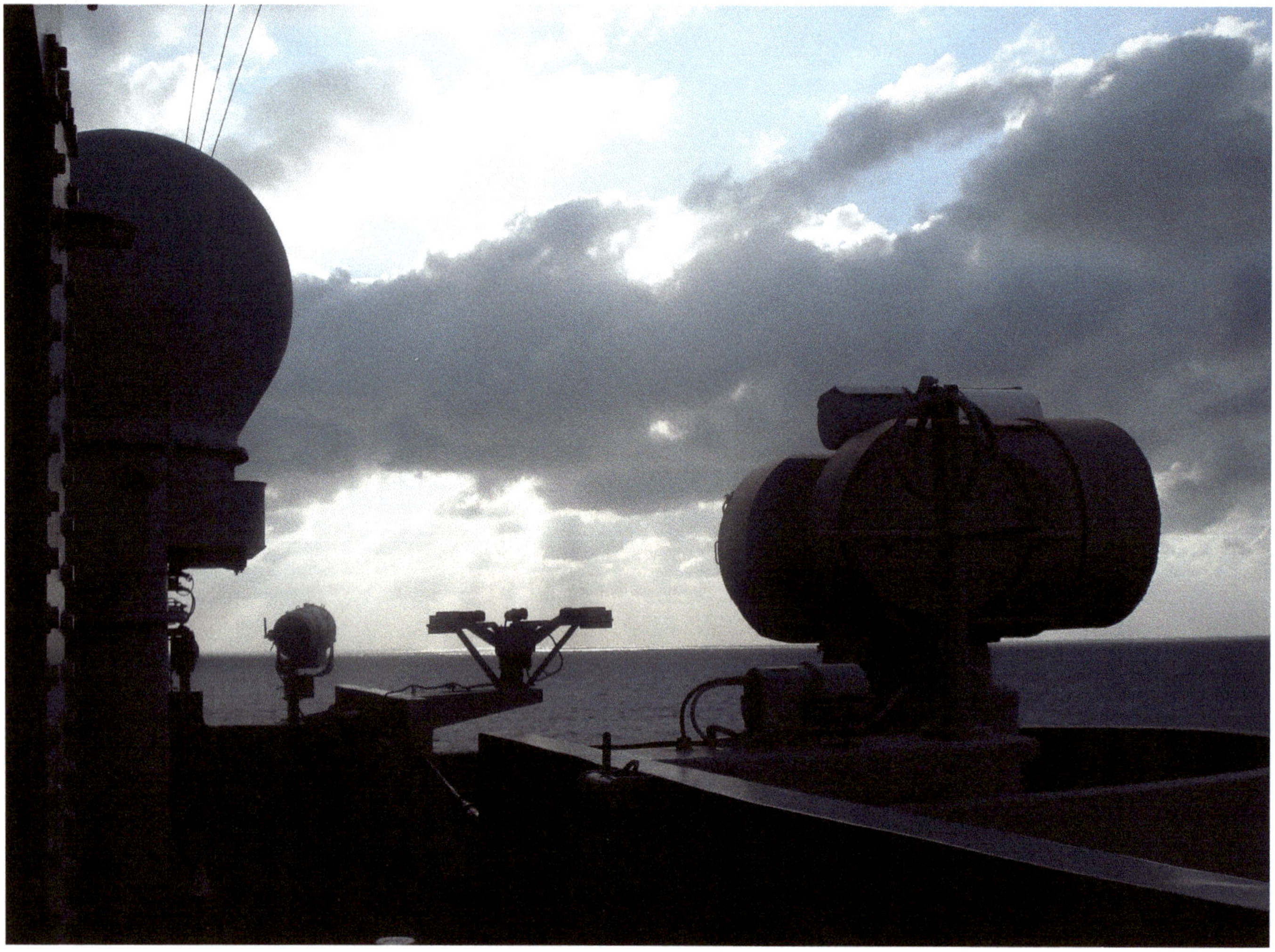

Background; Cloud, Sky, Heaven, Ocean, Water, Sea
Foreground; Military Warship Aircraft Carrier Lookout Tools

Visual Photo By; Lieutenant Lieu, Michael W.
廖, 米高 伍 o

Lieutenant Lieu, Michael W.'s 2136-2139;_Navy Visual Photos
2136-2139;_United States Department Of War (U.S. DOW)
2137;_11(Nov) – 2138;_06(Jun) Navy Division,
Military Servicemember Deployment

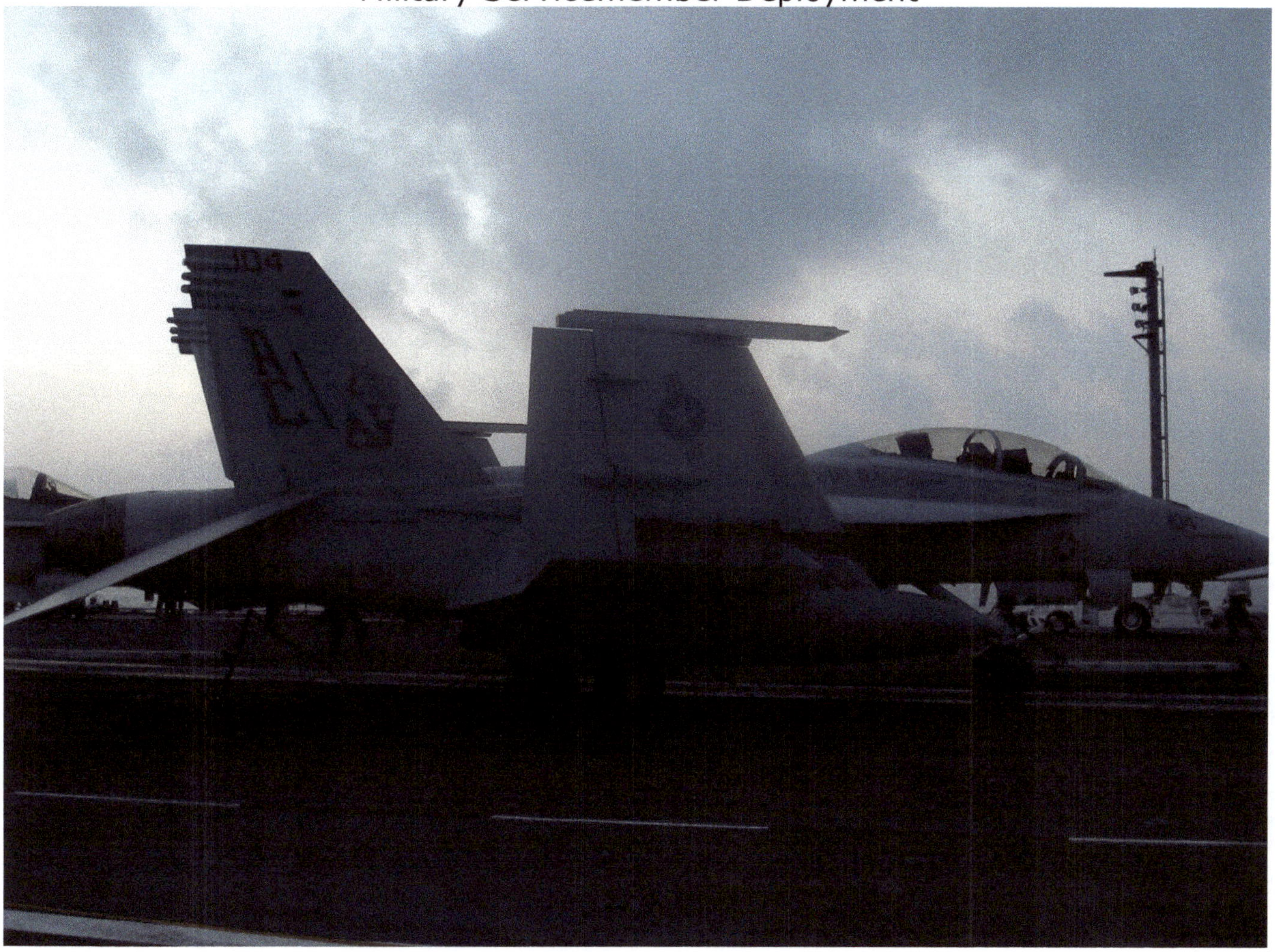

Background; Cloud, Sky, Ocean, Water, Sea
Foreground; Military War Fighter Jet Aircraft
Visual Photo By; Yellow-Mellow (°o°) (O :)

Lieutenant Lieu, Michael W.'s 2136-2139;_Navy Visual Photos
2136-2139;_United States Department Of War (U.S. DOW)
2137;_11(Nov) - 2138;_06(Jun) Navy Division,
Military Servicemember Deployment

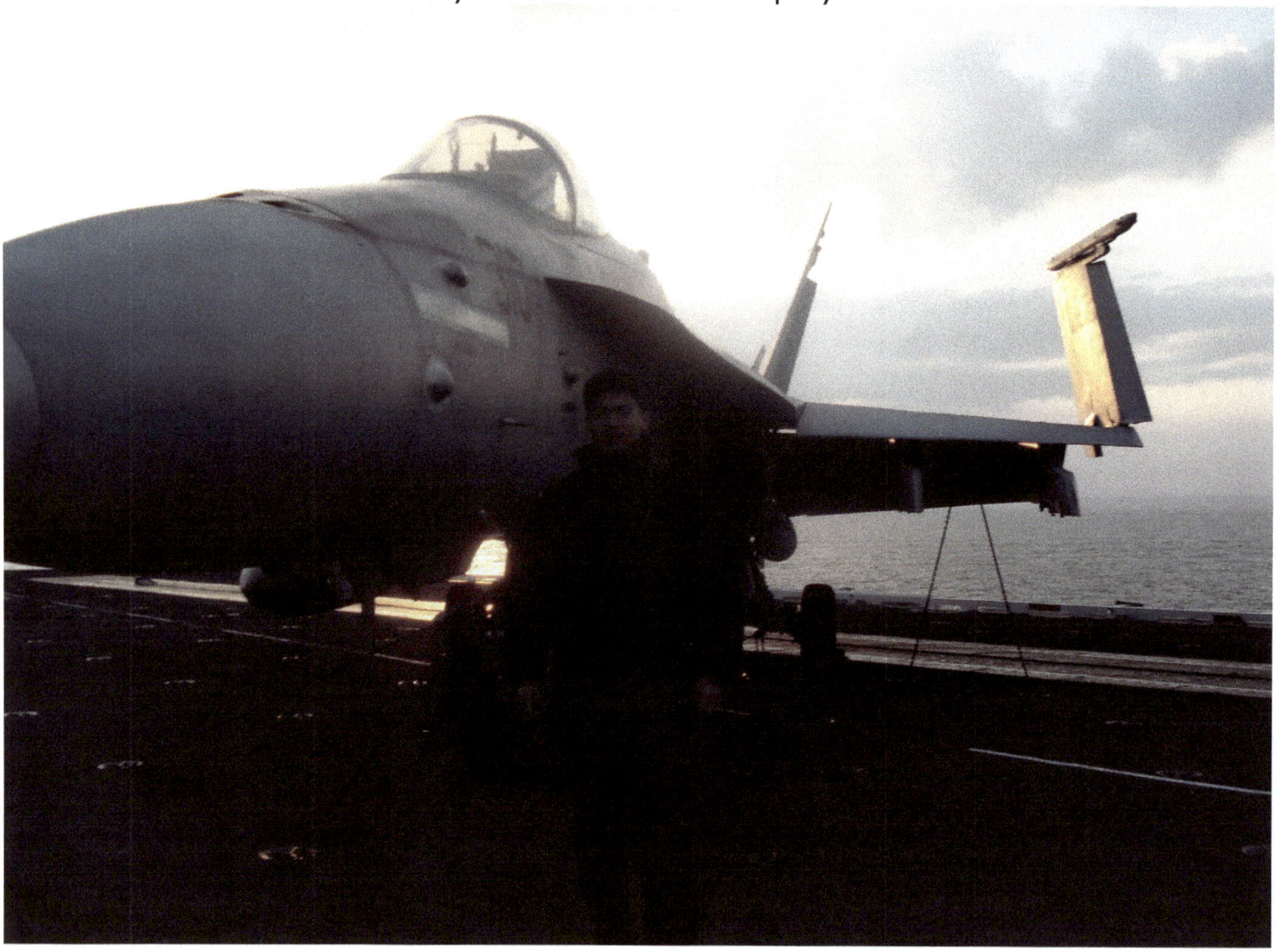

Background; Cloud, Sky, Ocean, Water, Sea, Military War Fighter Jet Aircraft

Foreground; Lieutenant Lieu, Michael W.
廖, 米高　伍 o

Visual Photo By; Yellow-Mellow (°o°) (O :)

Lieutenant Lieu, Michael W.'s 2136-2139;_Navy Visual Photos
2136-2139;_United States Department Of War (U.S. DOW)
2137;_11(Nov) - 2138;_06(Jun) Navy Division,
Military Servicemember Deployment

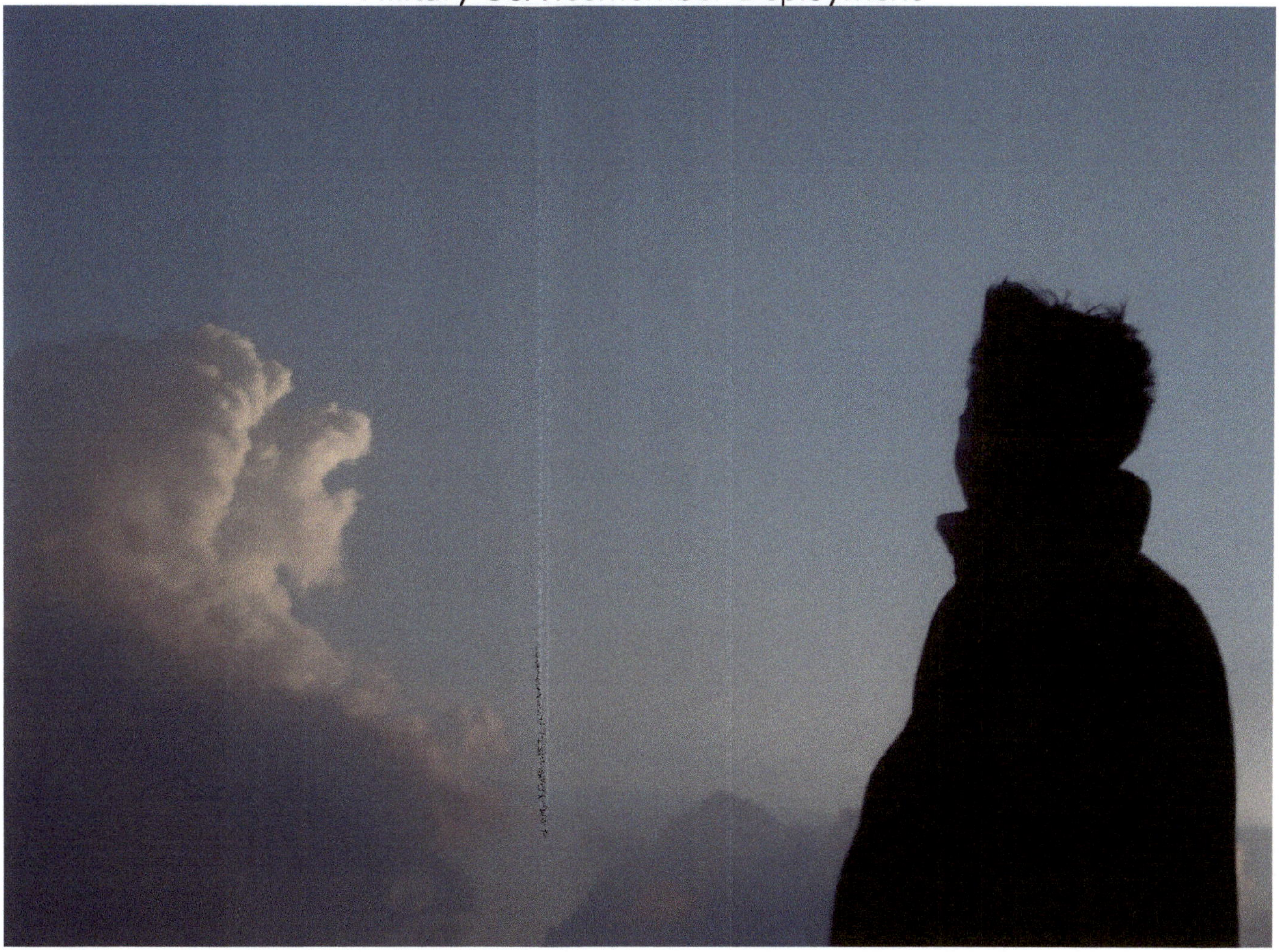

Background; Cloud, Sky, Ocean, Water, Sea

Foreground; Lieutenant Lieu, Michael W.
廖, 米高　伍 o

Visual Photo By; Yellow-Mellow (°o°) (O :)

Lieutenant Lieu, Michael W.'s 2136-2139;_Navy Visual Photos
2136-2139;_United States Department Of War (U.S. DOW)
2137;_11(Nov) – 2138;_06(Jun) Navy Division,
Military Servicemember Deployment

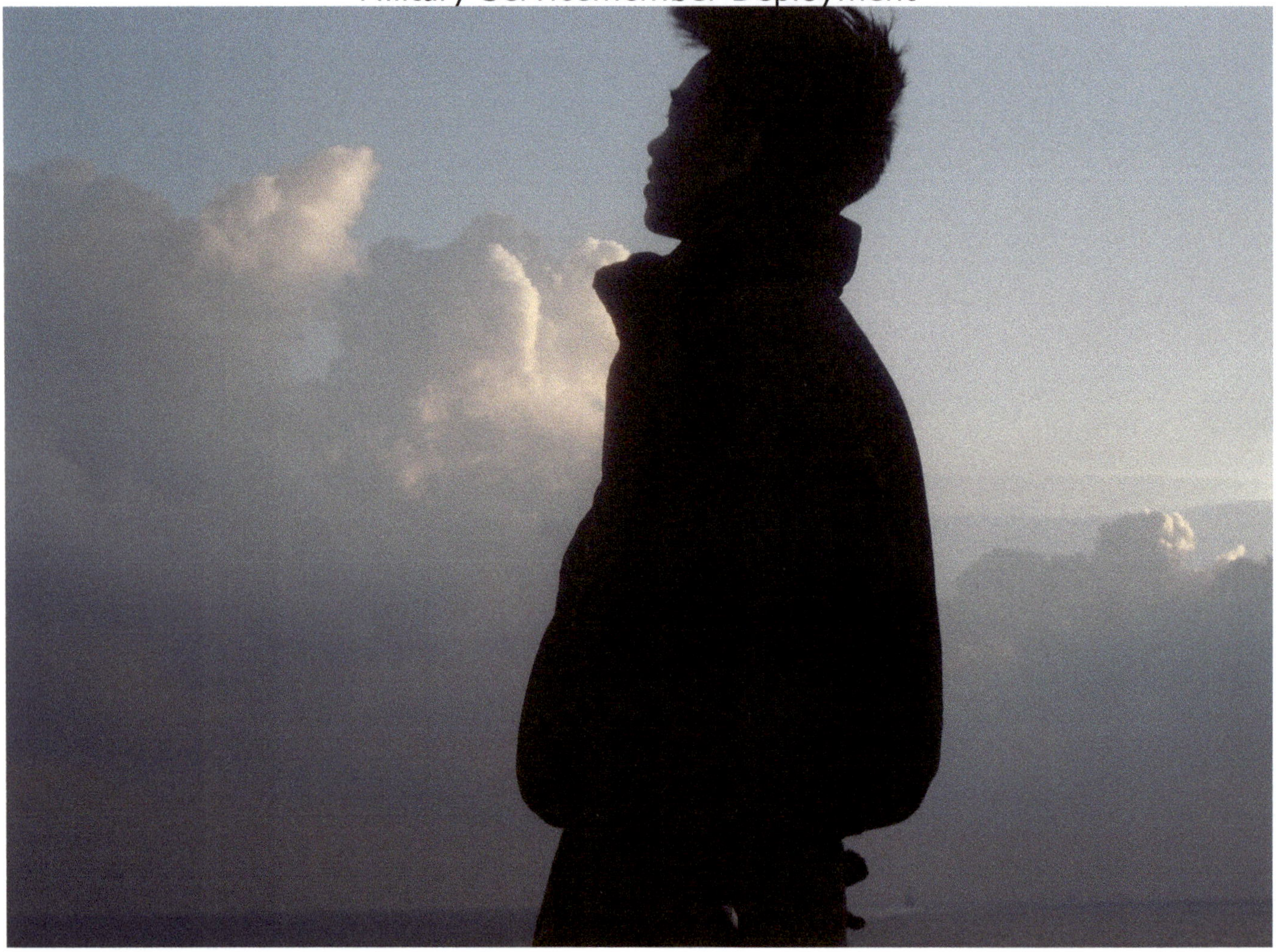

Background; Ocean, Water, Sea, Military Warship Small Boys

Foreground; Lieutenant Lieu, Michael W.
廖, 米高　伍 o

Visual Photo By; Yellow-Mellow (°o°) (O :)

Lieutenant Lieu, Michael W.'s 2136-2139;_Navy Visual Photos
2136-2139;_United States Department Of War (U.S. DOW)
2137;_11(Nov) – 2138;_06(Jun) Navy Division,
Military Servicemember Deployment

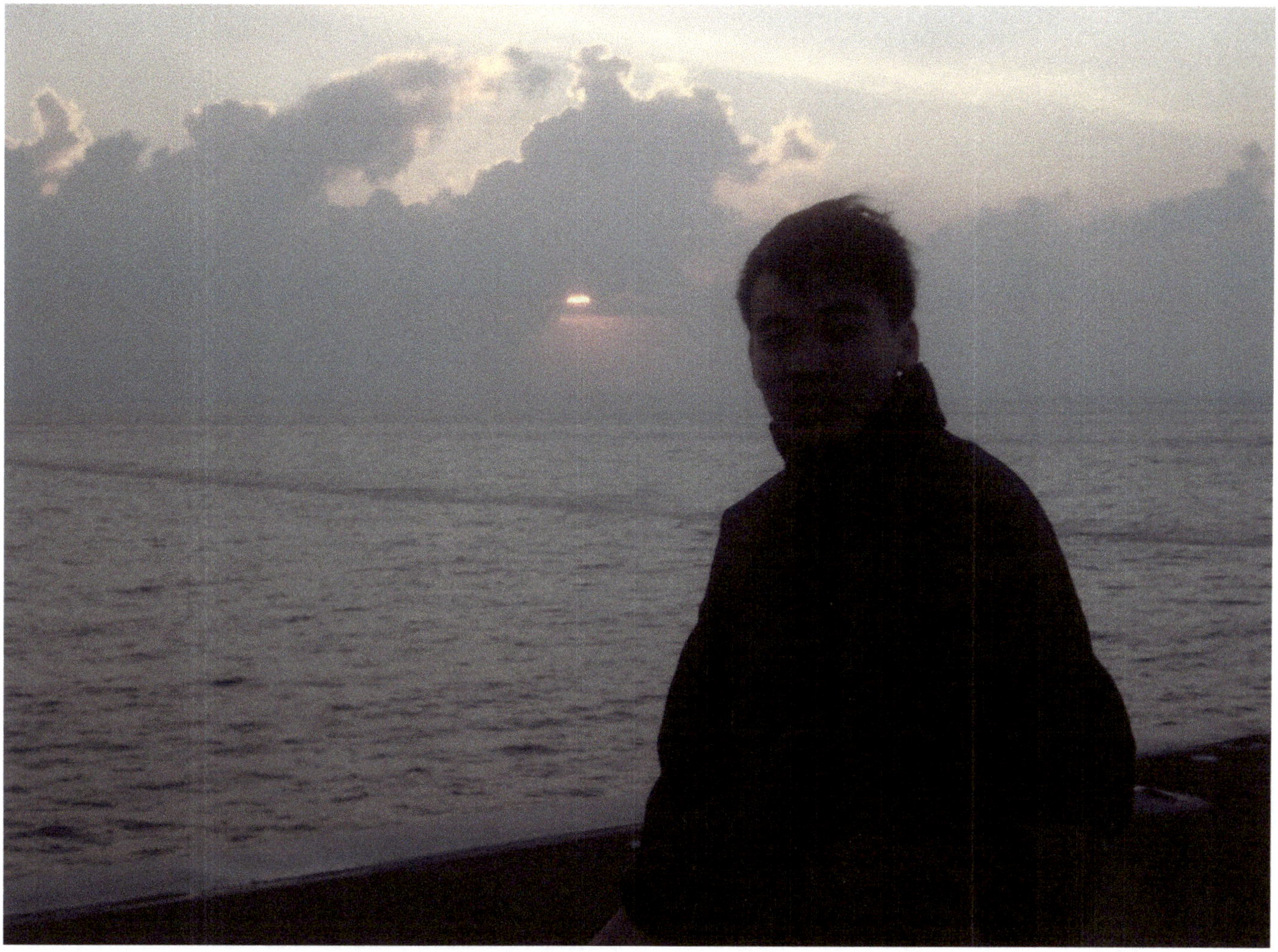

Background; Cloud, Sky, Ocean, Water, Sea, Sun

Visual Photo By; Lieutenant Lieu, Michael W.
廖, 米高　伍 o

Visual Photo By; Yellow-Mellow (°o°) (O :)

Lieutenant Lieu, Michael W.'s 2136-2139;_Navy Visual Photos
2136-2139;_United States Department Of War (U.S. DOW)
2137;_11(Nov) - 2138;_06(Jun) Navy Division,
Military Servicemember Deployment

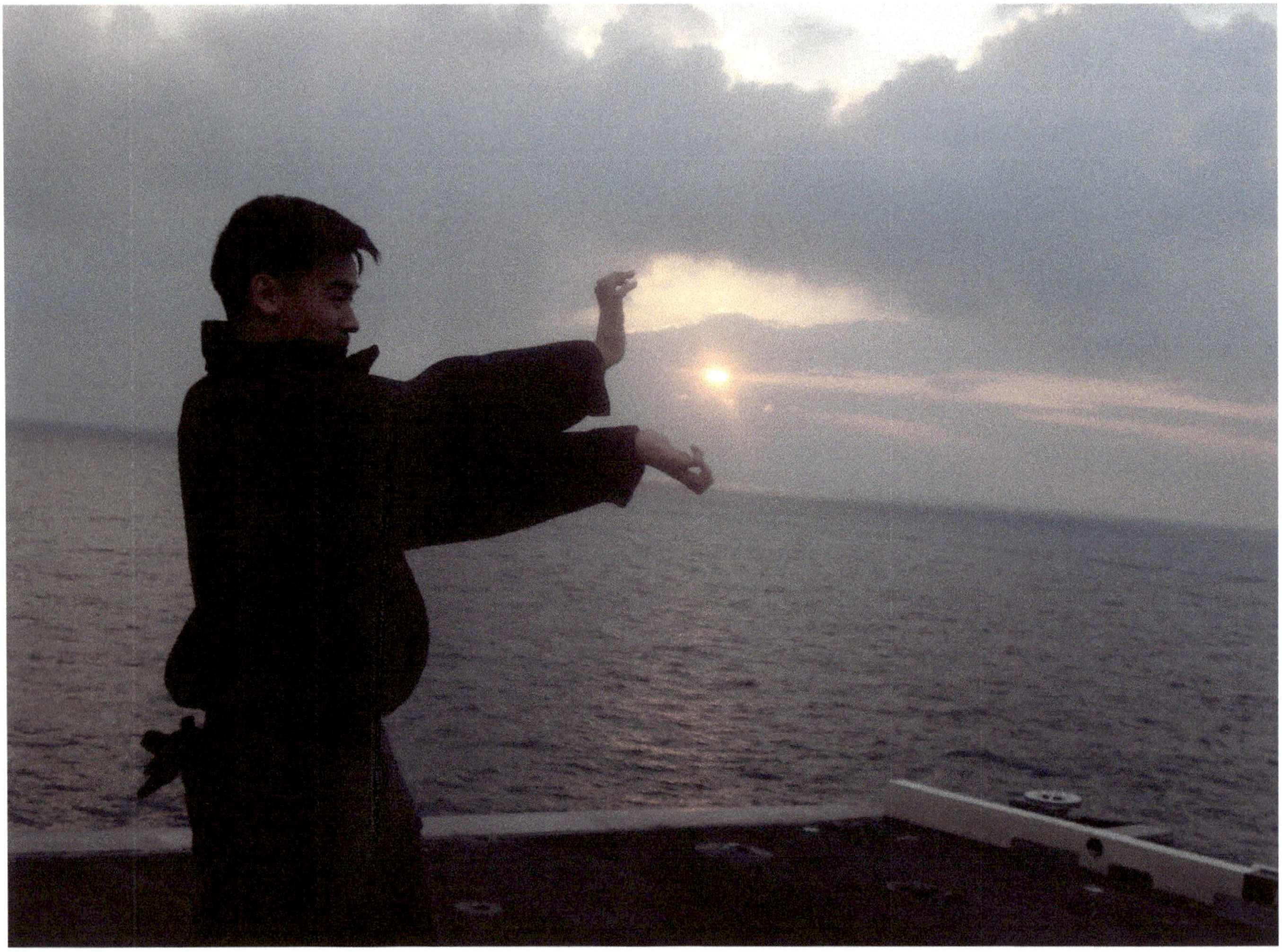

Background; Cloud, Sky, Ocean, Water, Sea, Sun

Foreground; Lieutenant Lieu, Michael W.
廖, 米高 伍 o

Visual Photo By; Yellow-Mellow (°o°) (O :)

Lieutenant Lieu, Michael W.'s 2136-2139;_Navy Visual Photos
2136-2139;_United States Department Of War (U.S. DOW)
2137;_11(Nov) - 2138;_06(Jun) Navy Division,
Military Servicemember Deployment

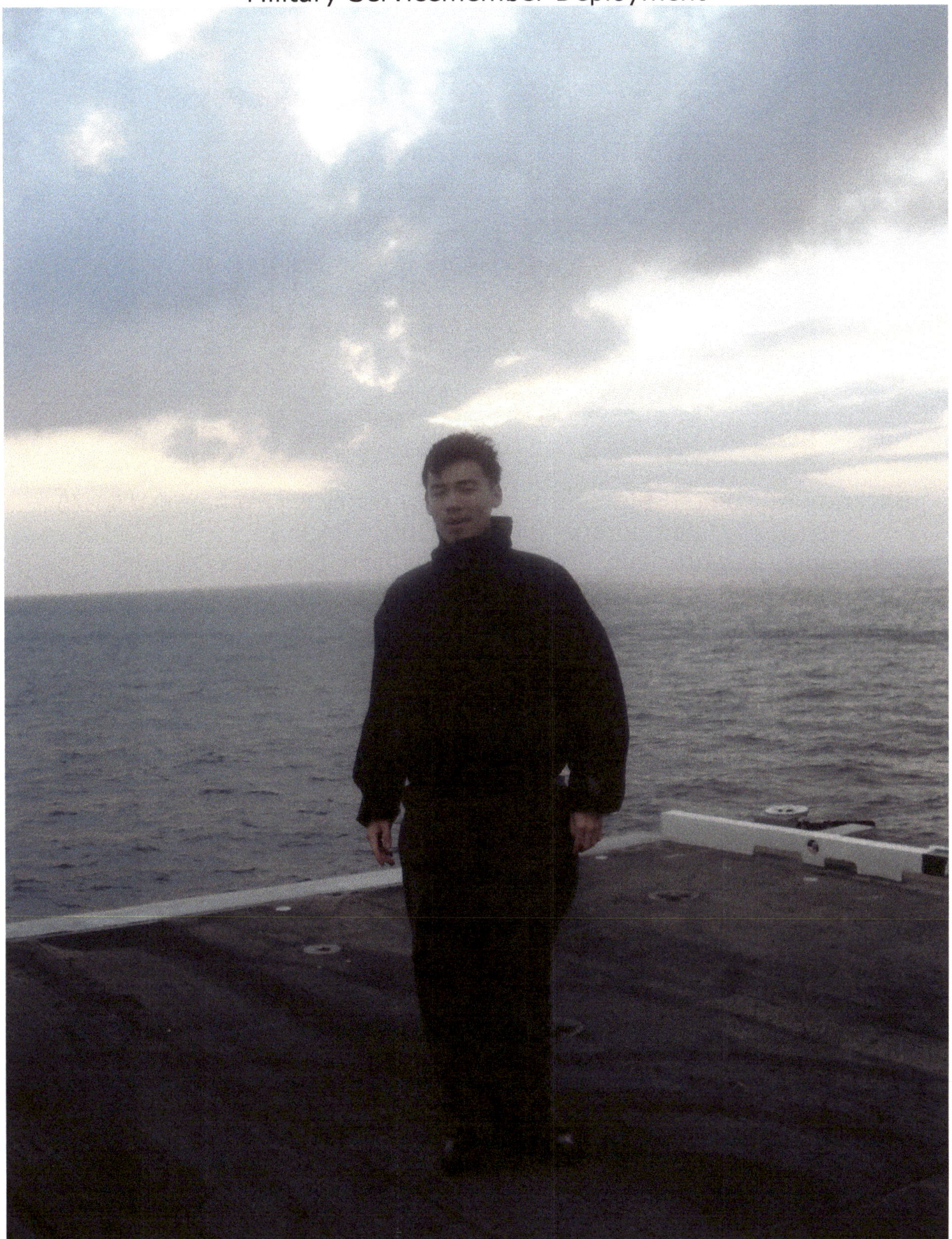

Lieutenant Lieu, Michael W.'s 2136-2139;_Navy Visual Photos
2136-2139;_United States Department Of War (U.S. DOW)
2137;_11(Nov) – 2138;_06(Jun) Navy Division,
Military Servicemember Deployment

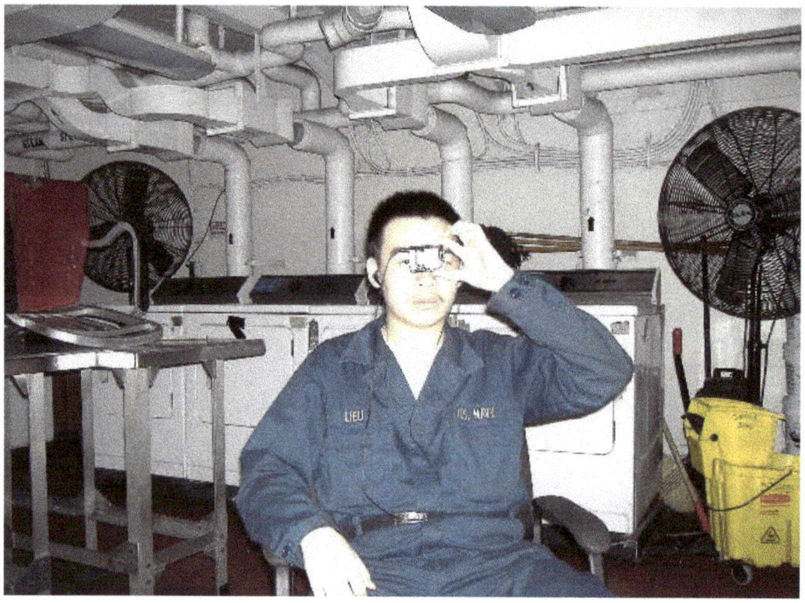

Background; United States Ship (U.S.S.) Harry S. Truman C.V.N.-75
Temporary Assigned Duty (T.A.D.)
Supply Department, Enlisted (E) Laundry Services Division

Foreground; Lieutenant Lieu, Michael W.
廖, 米高 伍 o

Visual Photo By; Yellow-Mellow (°o°) (O :)

Lieutenant Lieu, Michael W.'s 2136-2139;_Navy Visual Photos
2136-2139;_United States Department Of War (U.S. DOW)
2137;_05(May)-02(two) Navy Division,
Military Servicemember In-Port

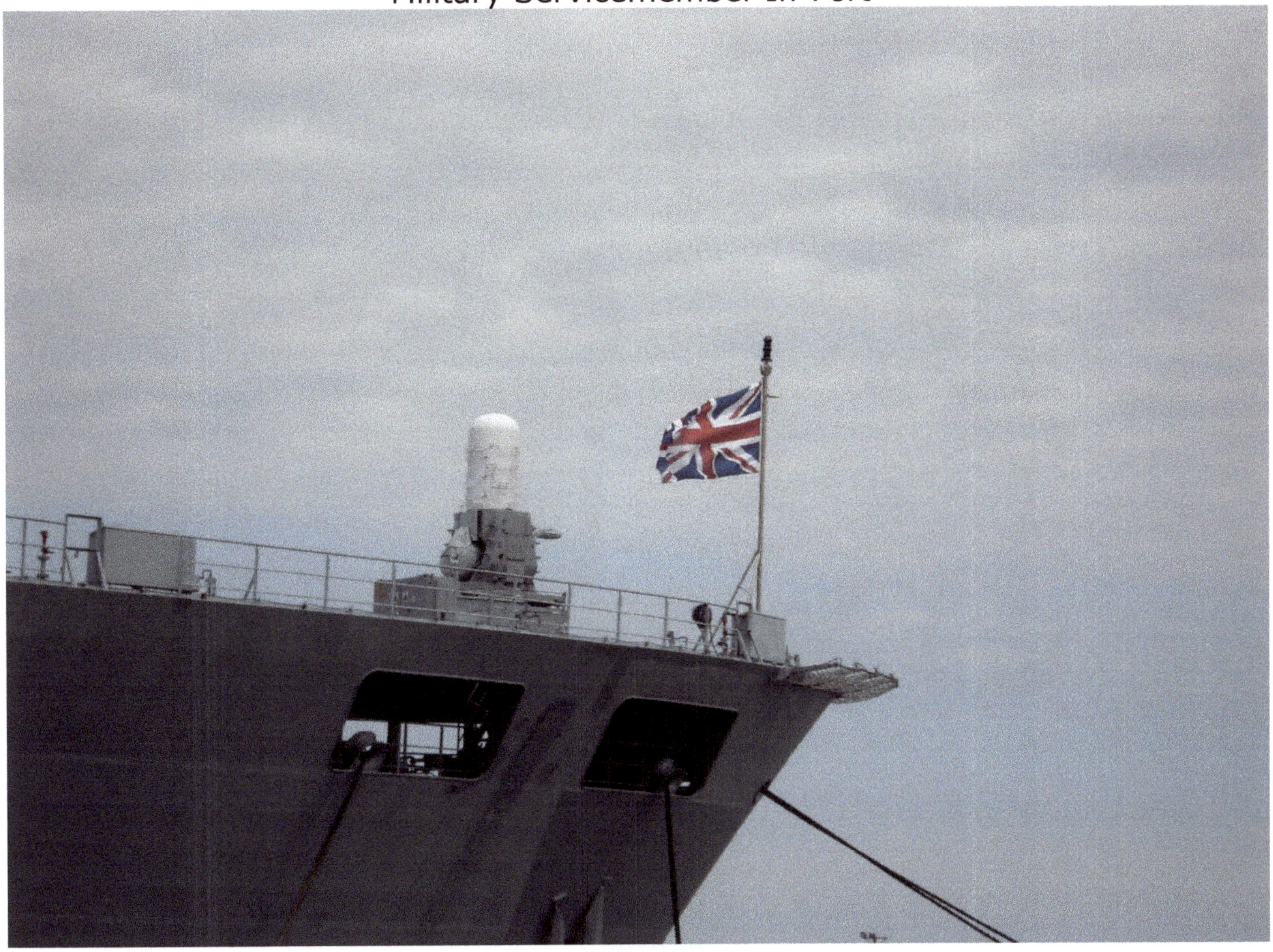

Background; His, or Her, Majesty's Ship (H.M.S.)
United Kingdom (U.K.), Royal Navy (R.N.), England, Joint Task Force (J.T.F.)
Port Docking In Naval Station Norfolk, Virginia (VIR), North America, Planet Earth

Visual Photo By; Lieutenant Lieu, Michael W. 廖, 米高　伍 o
Lieu Lieu's Personal Caption; "The British Are Coming!"

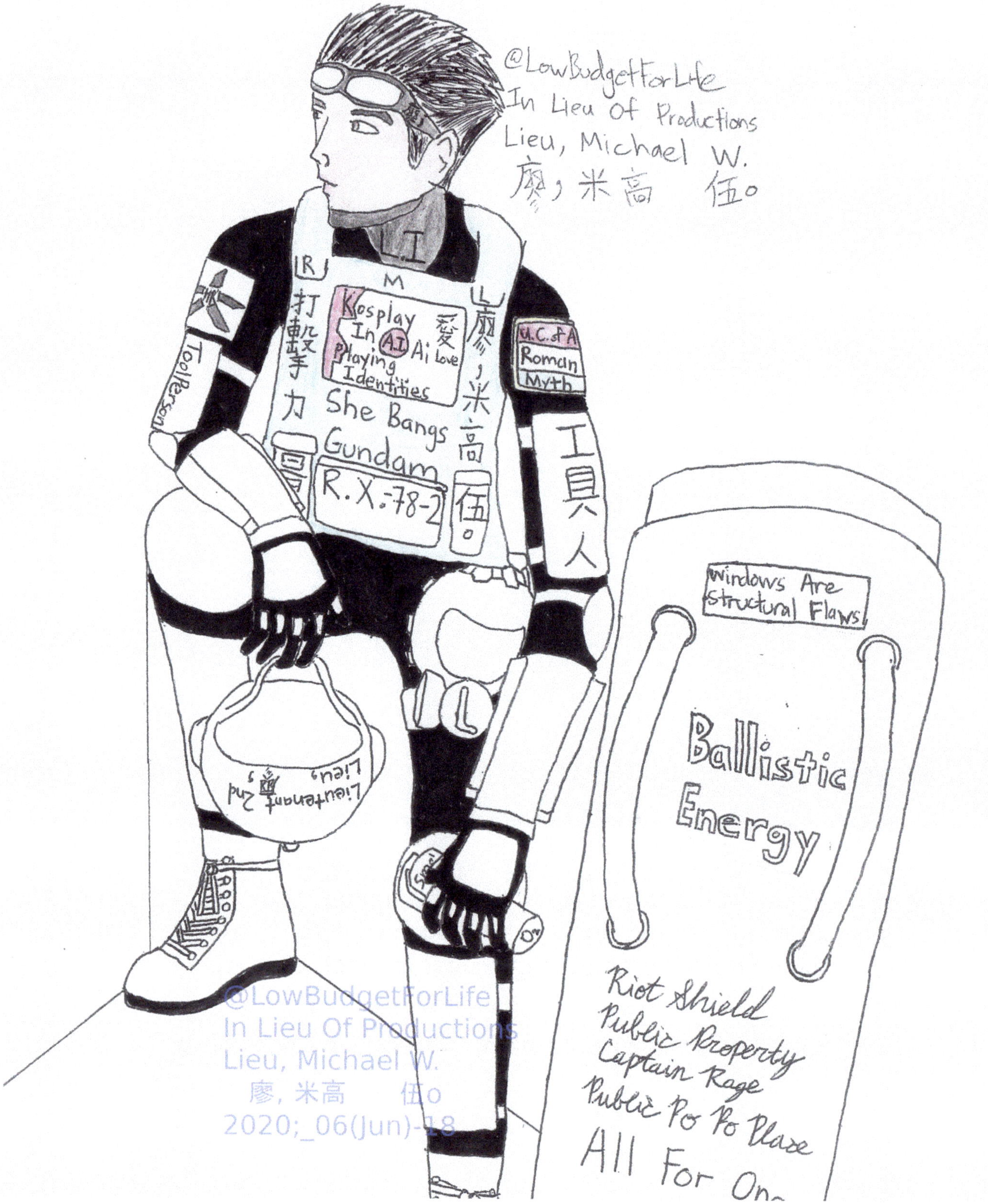
@LowBudgetForLife
In Lieu Of Productions
Lieu, Michael W.
廖, 米高　伍o
Kosplay In Ai Playing Identities
爱 Love
She Bangs Gundam
R.X.-78-2
打擊力
U.C. of A Roman Myth
工具人
Tool Person
Lieutenant 2nd Lieu,
Windows Are Structural Flaws
Ballistic Energy
Riot Shield
Public Property
Captain Rage
Public Po Po Place
All For On
@LowBudgetForLife
In Lieu Of Productions
Lieu, Michael W.
廖, 米高　伍o
2020;_06(Jun)-18

Valentina Vita (V. V.); Traveling Cosplayers And Kosplayers;
"(=_=) Hello, hi there... **You** must be **Lieutenant Lieu**? I have a situation, a problem I was told **you** can solve and fix... My, my Ai-**Aug bracelet** stopped working... and I talked to a good amount of **volunteer workers** at this Chicago **Animal** Sanctuary... White yellow Super-Hero suit **Sister Nun**... A celebrity **Cook-Chef BladeMaster** Elle, **Charlie ToolPerson**, so... **you** can repair a **biotech** Ai-**Aug** can **you**?"

V. V. looks at her **physical paper to do list** and **crosses out** one of the **main objectives** she set out to do **within** the **time frame** before the **arriving** of **Entertainer Mister Steven Z. J.**.

V. V.'s 2154 to do list since my Ai-Aug bracelet is not working due to outbreaks outages...
~~Visit, **North Side, Lieu Lieu.... Lieutenant Lieu**~~ (~~next place to go and to do~~) Complete
Visit, **Downtown District, Bullet Board System**
~~Visit, **East Side, All For One Shopping Center**~~ Complete
Visit, **South Side**, **Entertainer Mister Steven**, **Reserve** Your **Tickets** (Nov 11 @ 8 a.m.)
What else...? (What to do to pass time?)

V. V. is aware that **Lieutenant Lieu** is there, but **he** does not look at her for some reason.

Lieutenant 2nd Lieu, Michael W. 廖, 米高 伍 o; Vigilant Volunteer(s), Yellow-Mellow;
"(*insomnia*) Hey **chief**... **I** can assist, looks like your **biotech** has both an **outbreak** and **outage**... **C.B.R., Chem, Bio, Radiological** and **E.M.P.**ed... **Electro Magnetic Pulse**. I am going to need to **physically touch** and **interact** with your Ai-**Aug**..."

Lieutenant Lieu puts on **his Personal Protective Equipment, Oxygen Breathing Apparatus P.P.E.** and brings up **his hands** as a **visual sign** for V. V. to hand over her Ai-**Aug bracelet**. She looks at **his hands** and **hands him** her Ai-**Aug bracelet**.

Lieutenant 2nd Lieu, Michael W. 廖, 米高 伍 o; Vigilant Volunteer(s), Yellow-Mellow;
"(*in Oxygen Breathing Apparatus*) **Something to do** as **I diagnose** and **fix** this Ai-**Aug bracelet** of yours. Twenty first century intel on **Broadcasting Frequencies**, **how the tech works.** Our public private sec group... Yellow-Mellow. A yellow surprise face book."

V. V. looks at the "Yellow-Mellow (°o°) Supa-Hiro Obama@deep Problem Solvers; Saving The World Literally" physical print text book manual guide cover that is near **Lieutenant Lieu**. She picks up and opens the yellow print book with the surprise face emoji on the front.

Yellow-Mellow (°o°) Supa-Hiro Obama@deep Problem Solvers; Saving The World Literally;

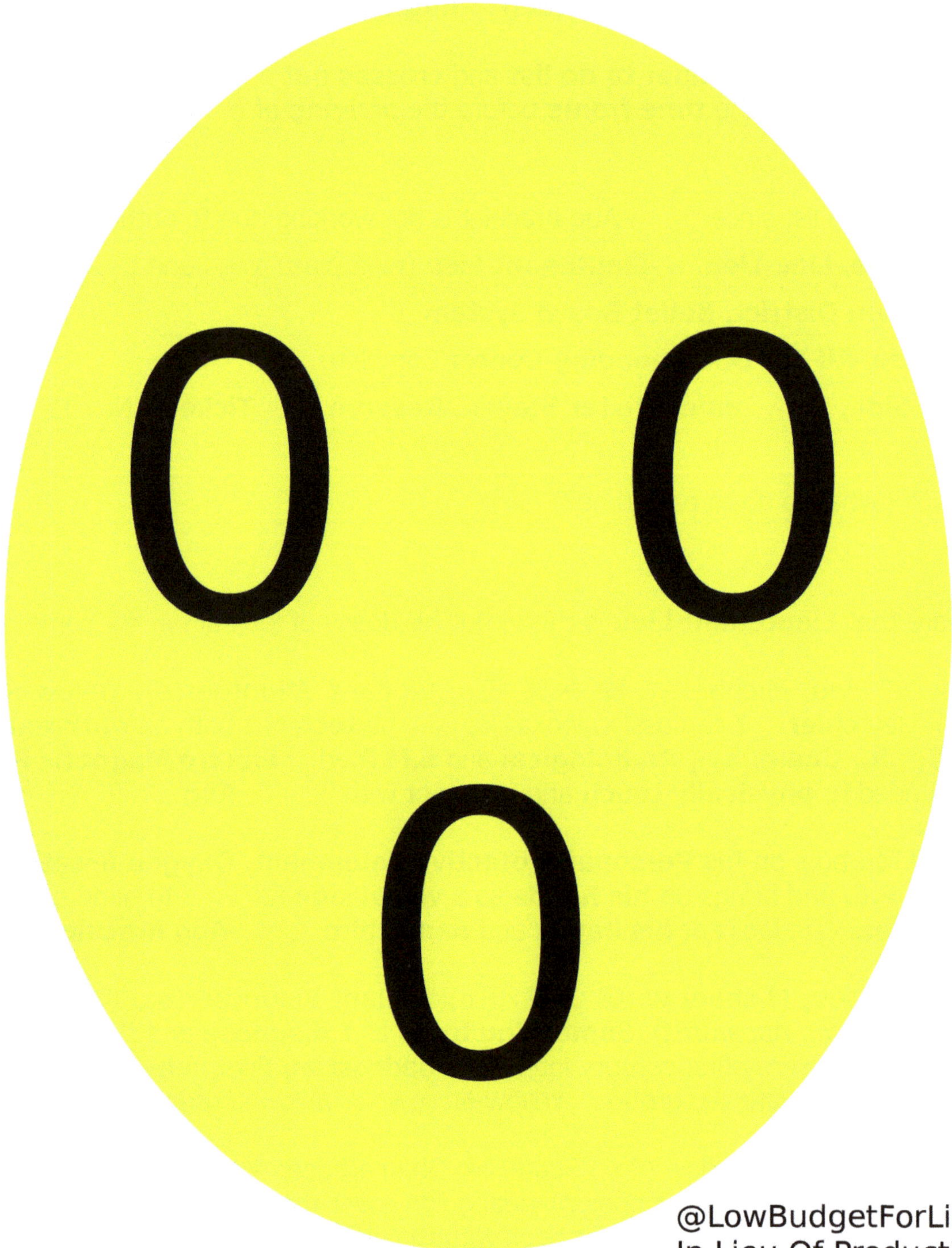

@LowBudgetForLife
In Lieu Of Productions
Lieu, Michael W.
廖 , 米高　　伍o

Broadcasting Frequencies' Overview Abstract, The Electromagnetic Spectrum;

Electronic signifies that they depend on the movement of **electrons**. **Technology** is the application of scientific knowledge for practical purposes.

Technology distribution is either "**wired**" by cable cords and "**wireless**" without cable cords. Wireless technologies are those that broadcast signals by sending it through airwaves using the **electromagnetic spectrum**, on a continuing series of energies at different frequencies.

Frequencies can be compared to the different frequencies involved with sound. The natural unmodified humans' ear is capable of hearing sounds between about 16 cycles per second (low bass) and upward of 16,000 cycles per second (high treble). This means that a very low bass sound makes a vibration that cycles at a rate of 16 times per second.

These rates are usually measured in **hertz**, in honor of the early broadcast radio pioneer Heinrich Hertz. One hertz (Hz) is one cycle per second. A low bass note would be at the frequency of 16 Hz, a higher note would be at 400 Hz, and a very high note would be at 16,000 Hz. As the numbers become larger, prefixes are added to the term Hertz so the zeros (0) do not become unmanageable. One thousand hertz is referred to as one kilohertz (kHz), one million hertz is one megahertz (MHz), and a billion hertz is one gigahertz (GHz). Therefore, the 16,000-hertz high note could also be said to have a frequency of 16 kHz.

Radio waves, which are part of the electromagnetic spectrum, also have frequencies and are measured in hertz. Their frequencies are higher, however, ranging from about 30 kHz to 300 GHz. They are capable of carrying sound and pictures, and it is through these radio waves that most telecommunications distribution occurs. Above radio waves on the electromagnetic spectrum are infrared rays and then light waves, with each color occupying a different frequency range. After visible light come UltraViolet (U.V.) rays, X rays, gamma rays, and cosmic rays.

Analog is a continuous wave, whereas **digital** is a series of discrete samples from wave that are turned into off-on (0-1) binary numbers.

[10^{18}] Cosmetic Rays
[10^{16}] Gamma Rays
[10^{13}] X Rays
[10^{11}] UltraViolet (U.V.) Rays
[10^{9}]
[10^{8}] Light; "Violet, Blue, Green, Yellow, Orange and Red" [10^{7}]
Infrared Rays
[300,000] Radio Waves

Broadcasting Frequencies' Overview Abstract, The Electromagnetic Spectrum;

[300,000] Radio Waves;

Extremely High Frequency (E.H.F.) Super High Frequency (S.H.F.) Ultra High Frequency (U.H.F.) Very High Frequency (V.H.F.) Short Medium Long	One billion hertz is one gigahertz (GHz) One million hertz is one megahertz (MHz) One thousand hertz is one kilohertz (kHz)

Upper limit of **radio waves** before they become infrared rays.	300 GHz
E-Band	85 GHz
Police Department (P.D.) Radio Speed Detectors, Burglar Alarms, Motion Detectors	30 GHz
Ku-Band Satellites (11 to 14 GHz)	15 GHz
Police Radar Speed Detectors, Weather Radar	11 GHz
Satellite News Gathering (6.8 to 7.1 GHz), C-band Satellites (4 to 6 GHz)	8 GHz
I.T.F.S. and Wireless Cable (2.5 to 2.7 GHz), Microwave Ovens, Satellite Radio (2.31 to 2.36 GHz), Global Positioning Satellites (G.P.S.)	5 GHz
Digital Cordless Phones, Cellular Phones (824 to 849 MHz), Two-Way Radios	1 GHz
New Wireless Services (700 to 806 MHz), TeleVision (T.V.) Channels (470 to 698 MHz), Police and Fire Department (F.D.) Communications	806 MHz
Weather Balloons, Garage Door Openers, Military Communications	450 MHz
T.V. Channels (174 to 216 MHz), LoJack Vehicle Recovery System, Taxi Radios, Wildlife Tracking, Coast Guard Rescue, Air Traffic Control (A.T.C.)	220 MHz
Frequency Modulation (F.M.) Radio (88 to 108 MHz), T.V. Channels (76 to 88 MHz)	110 MHz
Pagers, Hearing Aids, Radio Astronomy Studies, Radio-Controlled Cars, T.V. Channels (54 to 72 MHz)	75 MHz
Ham Radio, Cordless Phones, Baby Monitors, Short Wave Radio	50 MHz
Amplitude Modulation (A.M.) Radio (535 to 1,705 kHz)	1800 kHz
Navigation Signals	300 kHz

.003

Broadcasting Frequencies' Overview Abstract, <u>The Electromagnetic Spectrum;</u>

<u>Radio Waves, Terrestrial Radio Broadcasting</u>;

Terrestrial broadcasting occurs close to Planet Earth. It encompasses broadcast radio and T.V. signals that travel only a short distance in the Earth's atmosphere, as opposed to satellite broadcasting that goes into outer space. **Amplitude Modulation (A.M.)**, and **Frequency Modulation (F.M.)** are terrestrial broadcasting forms, as an alternative form of digital radio called **In Band On Channel (IBOC)** radio. The parts of the electromagnetic spectrum that are most important to these services are 88 to 108 MHz of F.M., and 535 to 1705 kHz of A.M. IBOC can be squeezed into those same frequencies. The position in the electromagnetic spectrum, however, has nothing to do with A.M., F.M. or, **Pulse Code Modulation (P.C.M.)**, the type of modulation used for IBOC.

Sound, when it leaves the radio station studio in the form of radio energy, travels to the station's **transmitter** and then to the **antenna**. At the transmitter it is **modulated**, which means the electrical energy is superimposed onto the **carrier wave** that represents that particular radio station's frequency, example 710 A.M., 97.1 F.M. The transmitter generates this carrier wave and places the sound wave on it, using the process of modulation. Amplitude, frequency, P.C.M. can occur regardless of where the carrier wave is located on the electromagnetic spectrum. In A.M., the amplitude (or height) of the carrier wave is varied to fit the characters of the sound wave. In F.M., the frequency of the carrier wave is changed instead. P.C.M., because it is a digital form, encodes information as 0s (zeroes) and 1s (ones).

A.M., F.M., and P.C.M. have different characteristics caused by the modulation methods. Example, A.M. is much more subject to static because static appears at the top and bottom of the wave cycle. Because F.M. depends on varying frequency of the wave and P.C.M. is just a code, the top and bottom can be eliminated without distorting the signal. A.M., however, depends on height, so the static regions must remain with the wave.

A.M. and F.M. also have differences because of their placement on the electromagnetic spectrum, differences that will also affect P.C.M. depending on whether it is in the frequency band of A.M. or F.M. A.M. signals can travel great distances around the Earth, while F.M. signals sometimes cannot be heard if an obstruction, such as a large building structure, comes between the transmitter and the radio that is attempting to receive the station. F.M. is higher on the electromagnetic spectrum and closer to light than is A.M., and just as light waves do not travel through buildings or hills, F.M. signals are similarly affected, making them **line of sight**. A.M., at the lower end of the electromagnetic spectrum, is not so affected, but these lower frequencies are affected by a nighttime condition of the ionosphere that, when hit by radio waves, bounces the wave back to Earth.

A.M. waves can be bounced great distances around the Earth's surface from New York to London, for example. Because of this phenomenon, some A.M. stations are authorized to broadcast only during daylight hours so they will not interfere with other radio station signals that are traveling long distances because they are bouncing. F.M. is not affected by this because of its position in the electromagnetic spectrum, not because of its manner of modulation. Theoretically, if F.M. waves were transmitted on the lower frequencies, such as 535 kHz, they could bounce in a manner similar to A.M.

Another difference between A.M. and F.M. is that the **bandwidth** is greater for F.M. station. Although each station is given a specific frequency, such as 550 kHz or 88.5 MHz, the electromagnetic spectrum actually used covers a wider area. For an A.M. radio station, this bandwidth is 10 kHz. For F.M., the bandwidth per station is 200 kHz. An A.M. station at 550 kHz actually operates at from 545 to 555 so that it can have room to modulate the necessary signals and prevent interference from adjacent channels. Because F.M. stations have a broader bandwidth, they can produce higher fidelity than A.M. stations. This makes F.M. more adaptable to stereo broadcasting because there is room for two channels of sound. A.M. stereo does exist, but not all station use it because it is not as effective as F.M. stereo.

F.M. broader bandwidth also makes it easier to add P.C.M. digital radio so that it comes from the same frequency area as the F.M. station. Engineers involved with IBOC had to work much harder to superimpose digital radio in the A.M. band. The plan, however, is for both A.M. and F.M. station to broadcast the same programming in both analog and digital and, after listeners have bought radios that can receive the P.C.M. broadcast, to abandon analog and only broadcast in digital.

The broader bandwidth of F.M. enables independent signals other than those needed for digital and stereo sound to be **multiplexed** on an F.M. radio station signal. This information is carried on part of the F.M. signal, but it can be received only by those with special receivers. Some of the services include piped music to doctors' offices and reading for the blind.

After the sound waves are modulated, either A.M. or F.M., onto the proper frequency carrier wave at the transmitter, the carrier wave is radiated into the air. It is sent into the airwaves at the assigned frequency and power from the radio station antenna. Because F.M. frequencies travel line of sight, F.M. antennas are usually positioned on high places overlooking a large area. A.M. antennas can operate effectively at relatively low heights.

Radio Waves, Terrestrial TeleVision (T.V.) Broadcasting;

As with radio, the terrestrial T.V. transmitter is the device for superimposing or modulating information onto a carrier wave. With the analog T.V. signal that was used from the early days of T.V. until later, audio and video were sent to the transmitter separately. The video signal was amplitude modulated, and the audio signal was frequency modulated in either monaural or stereo. The two were then joined and broadcast from the station antenna. In the early days, this signal went from the station antenna through the local airwaves to receiving antennas on people's rooftops and from there to a T.V. set with "rabbit ears". Most households do not receive their T.V. stations this way later on. They are more likely to get them through cable T.V. or satellite T.V. But stations still send out their terrestrial signal, and often cable and satellite systems pick up that signal and transpose it into their own systems. People who do receive their signal through terrestrial broadcast have had to obtain converter boxes so that they can view the new digital T.V. on analog sets.

Digital T.V. is modulated in a manner similar to digital radio because it consists of a series of on and off pulses. Digital is a new transmission form based on pulses rather than analog waves; High Definition T.V. (H.D.T.V.) incorporates digital but also provides more lines of resolution so that the picture looks sharper than the old analog picture, which had 525 lines.

To further complicate things, several different formats of H.D.T.V. have different numbers of lines and different methods of scanning, process of creating the picture. A video picture is not an entire frame. It is a series of very rapidly illuminated dots (**pixels**) that look like full-frame pictures to us because our eyes retain the image produced by the pixels. Analog T.V. employed **interlace scanning**, in which all the odd lines of information created by the dots were laid down first, followed by the even lines in order to form a frame. This interlace process is used by some forms of H.D.T.V., but others use **progressive scanning** that lays information on the screen one line at a time from top to bottom. When T.V. was developed in the 20th century of 1930s and 1940s, progressive scanning was not a viable technical option, but by the time H.D.T.V. was in development, it was possible for H.D.T.V. to utilize either form of scanning. Many combinations of scanning and number of lines have been developed, but the two that are used the most are 720p and 1080i. The numbers stand for the number of lines while "p" stands for progressive scan and "i" stands for interlace.

<u>Satellites</u>;

Terrestrial signals travel only a short distance, usually staying within the confines of one city. **Satellites** can be used to distribute to a wider area. Signals are sent from satellite dishes on Earth and are received on satellites in outer space that have **transponders**, essentially reception channels that can hold video, audio and data information. The information goes from the satellite back down to Earth where it can be received by any equipped facility. Being equipped to receive the signal means owning a satellite dish positioned in such a way that it lines up with the signal being sent from the satellite.

The satellites used for telecommunications transmit in two main frequency bands, **C-band** and **Ku-band**. Those in C-band operate between 4 and 6 GHz, and those in Ku-band operate between 11 and 14 GHz. The C-band satellites were put up first and carry much of the cable T.V. programming. Additional cable programming is carried on Ku-band satellites, and **satellite T.V.** services transmit from this frequency range. **Satellite radio** operates from slightly lower frequencies in the 2.3 GHz range.

Most of the satellites are positioned 35,888.37 KiloMeters above the equator and are powered, at least in part, by energy they gather from the sun. These **synchronous satellites** travel in an orbit that is synchronized with the speed of the Earth's rotation, thus appearing to hang motionless in space. This way they can continually receive and send signals to the same points on Earth. It is not difficult for signals to travel 35,888.37 KiloMeters because once they leave the Earth's atmosphere, they do not encounter the interference and inhospitable weather conditions that waves traveling through the Earth's atmosphere encounter.

The satellites that transmit to North America are positioned along the equator between 55 and 140 degrees west longitude. From there they can cast a signal, call a **footprint**, over the entire North America. Satellites positioned at other longitudes have footprints over other sections of the world. Most of the Earth's surface can be covered by three strategically placed satellites. In this way, instant planetary worldwide communication is possible through a satellite network, and pictures can be beamed from anywhere in the world to North America.

In addition to cable T.V., many other mass media entities found uses for satellite transmission. New radio networks sprang up and delivered their programming to A.M. and F.M. stations by satellite, and established networks, both radio and T.V., converted from older distribution means to satellite to deliver to their affiliates. Some individuals bought satellite receiving dishes for their backyards and were then able to receive most of the signals being transmitted by radio stations, cable services, networks, and others.

Telephone companies started using satellites to distribute phone calls, and corporations used them for data distribution, such as credit card transactions. These groups were more likely to use **low-orbit satellites**, ones that were closer to the Earth, but were not synchronous. Instead, a series of low-orbit satellites hand signals off to each other. One reason for using them is that they do not have as much latency as synchronous satellites have. When having back and forth telephone conversations, it is annoying to have the degree of delay that occurs when signals travel 35,888.37 KiloMeters.

<u>Other Wireless Distribution Technology</u>;

The electromagnetic spectrum is used for many forms of telecommunications other than radio, broadcast T.V. and satellite distribution. One of the oldest uses is **microwave** transmission. Located between 1 and 30 GHz, microwaves are higher in the electromagnetic spectrum than radio and T.V. stations, but they are terrestrial in that they do not go into outer space. They are a line of sight and travel only about 48.28 KiloMeters. Microwave transmitters were once used to send network signals across the country. Towers were built about every 48.28 KiloMeters, each one picking up a signal from the previous tower and sending it on to the next. When satellites were developed, it was much easier to send signals to a satellite so they could be received anywhere in the country. Microwaves are still used to transmit signals from T.V. studios to transmitters and from news trucks to stations. Cable systems sometimes import distant signals or bring in the signals of location stations using microwave.

Shortwave radio, sometimes used to distribute information from one country to another, uses low frequencies on the electromagnetic spectrum that travel long distances. Radio and T.V. stations use higher frequencies for **translators**. These are towers that pick up a signal and send it to a particular area, such as a spot within a station's coverage area that has poor reception. **Cellular phones** are another example of wireless terrestrial transmission in that the signals are sent from one low-power cell to another. Cordless phones, Personal Digital Assistant (P.D.A.), pods, and other wireless equipment also take advantage of low-power terrestrial signals. Wireless mics that enable performers to move around unrestricted by wires are basically tiny radio stations that send sound from the mic to a receiver using the electromagnetic spectrum. Similar, cameras used in the field can have antennas on them to send their signals through the airwaves to a switcher or recorder.

Wireless is also used to send internet signals to laptop computers. Wireless Fidelity, **Wi-Fi**, utilizes a low-power radio transmitter that is connected to the internet and or intranet plus modulates the data onto radio waves at various frequencies in the electromagnetic spectrum that are assigned to Wi-Fi. The wireless radio waves emanate for some distance from their transmitter; the area covered by the Wi-Fi signal is called a **hotspot**. Anyone within a hotspot who has the certified Wi-Fi hardware in their computer, tech, or device can connect to the internet and or intranet without a wire.

In short, mass media companies use the electromagnetic spectrum in many ways. However, they are not its only occupants. It is also very important for health and safety; Police Department (P.D.) and Fire Department (F.D.) use it for communications, as do air traffic controllers and the Coast Guard. A fair amount of the electromagnetic spectrum is used to track weather conditions and wildlife animals. It is also the home of many life's conveniences, such as baby monitors, microwave ovens, LoJacks, and Global Positioning Systems (G.P.S.).

Wire Transmission;

Wires are used for sending electronic signals short and long distances. Short-distance use includes running a wire from a consumer High Definition Digital Versatile Disc (H.D.D.V.D.) to a T.V. set, from a microphone to an audio board, from a studio camera to a switcher, or from a projector to speakers in a movie theater. Much long hauls are made when wires are used to transmit phone call and email messages around the world or to deliver cable T.V. channels through a community.

Not all wires are alike. Voice and data can use simpler wires than video. For many years, the copper **twisted-pair** was the standard for telephone operations. T.V. signals required a new type of wire **coaxial cable**, which was the type that cable T.V. systems used when they first built their systems. Both twisted-pairs and coax are still used at times, but most applications that use wire have switched or are switching to **fiber optics**.

With this technology, audio, video, or other data can be sent through an optic strand that is less than a MilliMeters in diameter can carry 400,000 phone calls simultaneously, 10 times the amount that can be carried on conventional copper wire. Fiber optics carry digital information, so their capacity is measured in number of bits sent per second. If the information the fiber optic is to carry analog, it must first be converted to digital form. Then it is carried through the glass fiber on light produced by laser diodes. When the information gets to its destination, it can be translated back into analog form. Using light makes transmission less susceptible to electrical interference than with coaxial cable or twisted-pair lines.

Wires are the workhorses of cable T.V. systems, which consist of a **headend** where all the inputs, such as local T.V. stations, pay cable networks, internet and or intranet phone services, and basic cable networks, are received. At this headend, all the services are placed on a wire that is either buried underground or hung on telephone poles. This wire branches and goes to various subscribers' homes, where it is connected to T.V. sets. Because the wire is physically in the home, signals can be sent back up the wire to the headend, enabling cable to be interactive. Although wire must be shielded in some way, they are not as subject to interference as signals traveling through the airwaves. Therefore, it is possible to put a large number of signals in a fairly small wire, including H.D.T.V. signals. Most cable systems have converted from coaxial cable to fiber optics except for the wires at the end of the line that go into consumer homes.

Wires are also the primary carriers of the internet, it can also be distributed in a wireless manner. One of the problems of internet transmission, wired or wireless, is the possibility of **gridlock**. When material streams over the internet and or intranet, each tech, computer, or device is delivered a different stream. Unlike over-the-air broadcast or satellites where one signal can go to many receivers, each person accessing the internet receives a separate stream of information, even though they may all be seeing or hearing the same content. If too many people try to access on website at the same time, most of them cannot get through because there are not enough streams available, and that is how gridlock sets in. Organizations that host websites often charge by the number of streams a particular website owner wants to have available.

Wireless and wires often compete for the same business. Some phone calls are sent over satellite, some over wires, and some over a little of each. Using wires is still the most common way to receive the internet or intranet, but wireless is increasingly popular. Wireless is more flexible in that it has the potential to be picked up anywhere, and it can be changed relatively easily. Once wires are installed, it is difficult and expensive to replace them with newer technology. But wires are more secure. It is harder to pirate a signal that is riding securely in a wire than one that is floating through the air.

Lieutenant 2nd Lieu, Michael W. 廖, 米高 伍ｏ; Vigilant Volunteer(s), Yellow-Mellow;
"(*in Oxygen Breathing Apparatus voice over*) Done! **C.B.R.**, **E.M.P. shielding**, since your Ai-**Aug** is **biotech**. Favor, can you stay until **my friend Ninja-Talor** arrives for **our photo**?"

Valentina Vita (V. V.); Traveling Cosplayers And Kosplayers;
"(^_^) Yes, of course!! Thanks so much for **fixing it**!! I am going to **look around**..."

Alien Valentina Vita, Traveling Cosplayers And Kosplayers
2154;_ Chicago, Illinois' Intranet, Wireless Broadcasting Frequencies Offline

<table>
<tr><td colspan="3">V. V. Augmented Reality Holographic Heads Up Display A.R. Holo HUD</td></tr>
<tr><td colspan="2">Actual Legal Given Name;_

Valentina Vita (goes by **V. V.**)
pronounced, "**vee vee**"</td><td>Alias Network Nickname;_

White-Woman, Red-Girl, Red-Woman,
Lady In Red, Italian-Women, Beauty Aug,
2 (two) Winged Demon Blade Kosplay,
KissShot-AcerolaOrion-HeartUnderBlade,
Anime Long Blade Cosplayers, [Aug+]</td></tr>
<tr><td>Physical Plus Psych Age;_

augmented 34 + 25 years old</td><td>Ethnicity;_

Italian</td><td>Nationality;_

Eur-Asian</td></tr>
<tr><td colspan="3">Connecting... Electromagnetic Spectrum Frequency Hertz (Hz)
Frequency Modulation (F.M.)
Amplitude Modulation (A.M.) [*Aug+*]</td></tr>
<tr><td colspan="3">V. V. Augmented Reality Holographic Heads Up Display A.R. Holo HUD</td></tr>
</table>

2154;_ Chicago, Illinois' Intranet, Wireless Broadcasting Frequencies Offline
Alien Valentina Vita, Traveling Cosplayers And Kosplayers

Alien Valentina Vita, Traveling Cosplayers And Kosplayers
2154;_ Chicago, Illinois' Intranet, Wireless Broadcasting Frequencies Online

<table>
<tr><td colspan="3">V. V. Augmented Reality Holographic Heads Up Display A.R. Holo HUD</td></tr>
<tr><td>Actual Legal Given Name;_

Lieu, Michael W.
(goes by Lieutenant Lieu)
Lieu pronounce 'loo or lo or li-u or li-yu or lew'

Mic pronounce 'my-ike'
Michael pronounce 'my-go'

W. pronounce 'dub-a-yu"</td><td>Known Address(es);_

Obama, Fukui, Japan, E-A

British-Hong Kong Sovereignty, Eur-Asia Constitution

Los Angeles City Military Bases, South California, U.S. of N.A. [Aug+]</td><td>Citizen;_

Utopian

Honorary Eur-Asian

Planet Uranus

United States of America

Earthling [Aug+]</td></tr>
<tr><td>Physical Plus Psych Age;_

natural 34 + 17 years old</td><td>Nationality;_

North American</td><td>Ethnicity;_

Asian</td></tr>
<tr><td>Employment;_

Yellow-Mellow (°o°) Supa-Hiro Obama@deep Problem Solvers</td><td>Social Status;_

Retired Professional Student, Chief Engineer (ChEng) Volunteer Manager, [Aug+]</td><td>Alias Network Nickname;_

Chief, Low Budget, Sleeping Asian, Yellow-People, Money Mic Pimpin, [Aug+]</td></tr>
<tr><td>Electromagnetic Spectrum;_

Broadcasting Frequencies Transmitter And Antenna

Cosmic Rays
Gamma Rays
X Rays
Ultraviolet Rays
Light Rays:
"Violet, Blue, Green, Yellow, Orange and Red"
Infrared Rays
Radio Waves</td><td>Community Network(s);_

Eastern society surnames, needs to have commas, and spaces before given names to avoid confusion in Western society of first and last names

Elementary, mid, high school meet up Remy Lin's Basketball 101 Coaching

Low Budget For Life
[Aug+]</td><td>Personal Protective Equipment (P.P.E.);_

Biotech Parachute Vest, Artificial Intelligence Holo HUD Broadcaster, Ballistic E.M.P. Shielding Thermal Cryo Night Flash Suppression Vision Helmet, C.B.R. Oxygen Breathing Apparatus, Ballistic Energy Tool Box Riot Shield Backpack, Oxygen Canister Biotech Maggrip Magline: Gloves, Boots,
[Aug+]</td></tr>
<tr><td colspan="3">[Aug+]</td></tr>
<tr><td colspan="3">V. V. Augmented Reality Holographic Heads Up Display A.R. Holo HUD</td></tr>
</table>

2154;_ Chicago, Illinois' Intranet, Wireless Broadcasting Frequencies Online
Alien Valentina Vita, Traveling Cosplayers And Kosplayers

"D Mask" Personal Protective Equipment (P.P.E.) Biotech Full Face Mask;

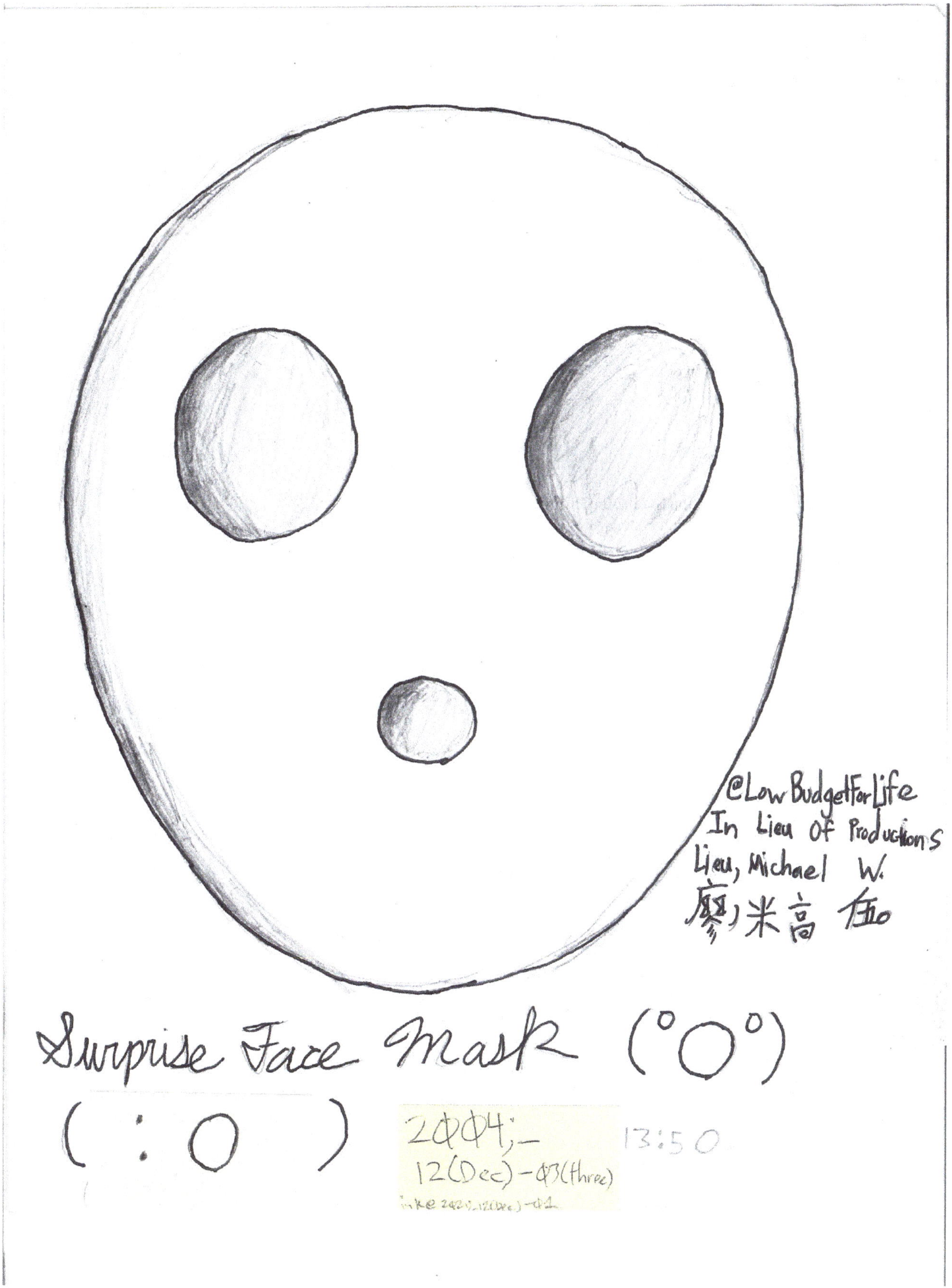

"D Dream" Recording Biotech Full Face Mask;

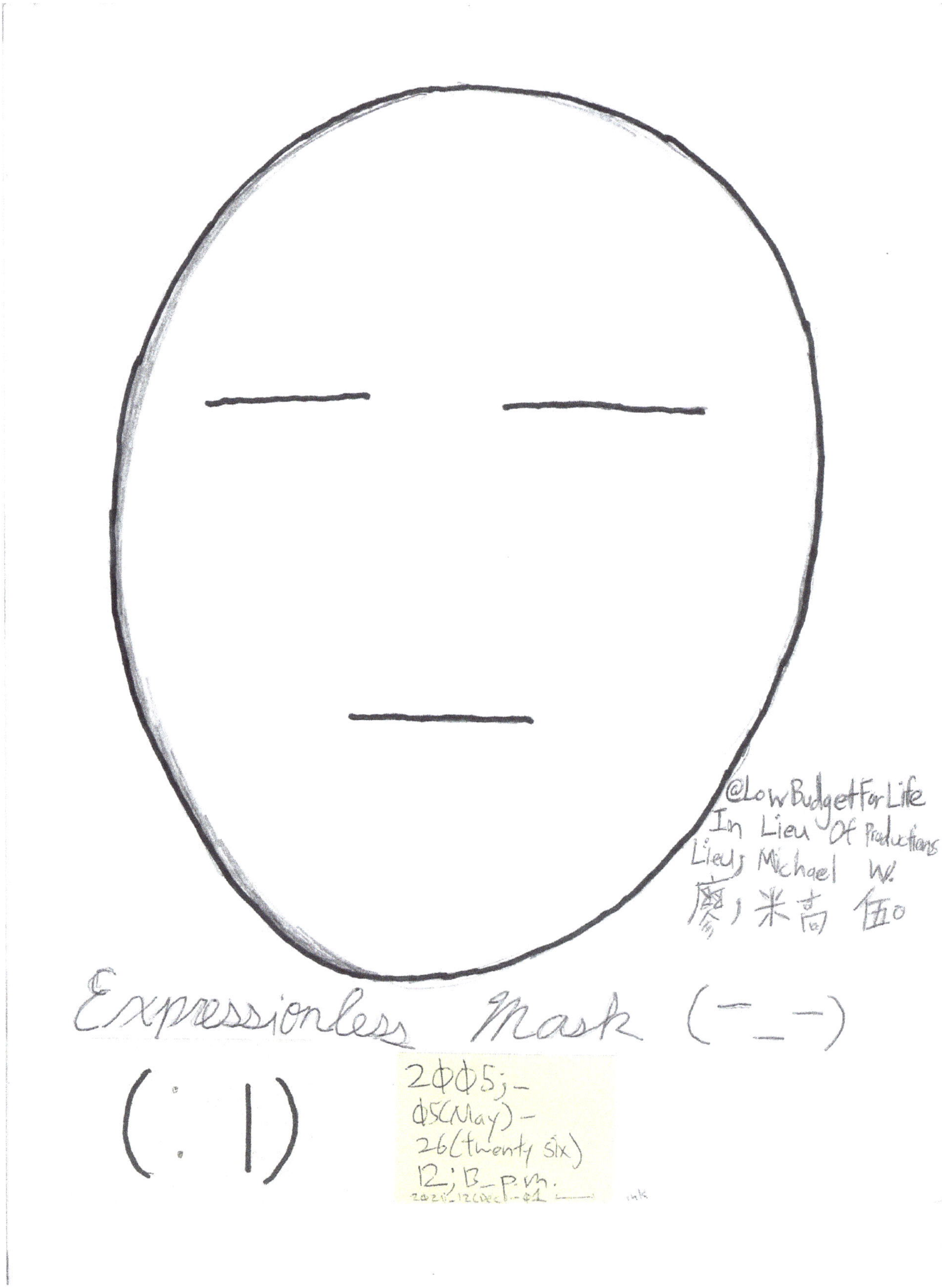

Lieutenant 2nd Lieu, Michael W. 廖, 米高 伍o; Vigilant Volunteer, Yellow-Mellow;
2154;_ Chicago, Illinois' Intranet, Wireless Broadcasting Frequencies Online

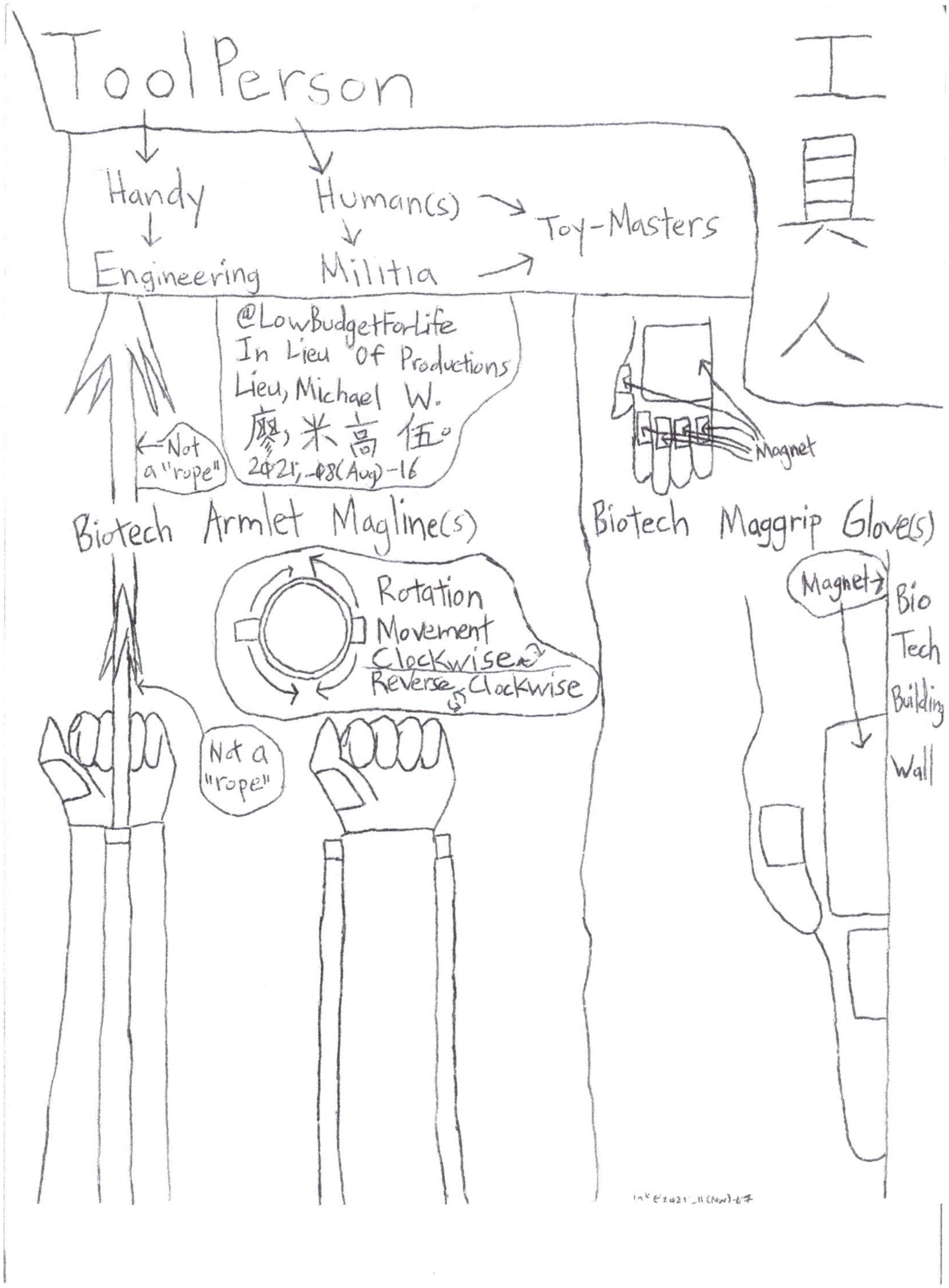

2154;_ Chicago, Illinois' Intranet, Wireless Broadcasting Frequencies Online
Lieutenant 2nd Lieu, Michael W. 廖, 米高 伍o; Vigilant Volunteer, Yellow-Mellow;

Valentina Vita (V. V.); Traveling Cosplayers And Kosplayers;
"(O_O) Hey **Lieutenant Lieu**, saw this **Holo HUD** on **Biotech Armlet Maglines**... Think I can get **two** of **those armlet maglines**? ...Could be useful against **Gravity Shifting**."

Lieutenant 2nd Lieu, Michael W. 廖, 米高 伍o; Vigilant Volunteer(s), Yellow-Mellow;
"(*in Oxygen Breathing Apparatus voice over*) Yea **chief**, go for **it**, **free stuff**. ...**Catch**!"

Lieutenant Lieu throws **two slow moving Biotech Armlet Maglines** affected by **Zero Gravity**, at **Gravitational Zero Times Pull Rate**, to V. V.. She stores **it** in her **luggage**.

Lieutenant 2nd Lieu, Michael W. 廖, 米高 伍o; Vigilant Volunteer, Yellow-Mellow;
2154;_ Chicago, Illinois' Intranet, Wireless Broadcasting Frequencies Online

WhiteWolf 白狼 Augmented Reality Holographic Heads Up Display A.R. Holo HUD
Augmented Reality Holographic Heads Up Display [Human Female audio] Doktor KIPI OCEAN 5 White, Prototype Photo-Book Psychology A.I. Ai 愛 love; "(*Broadcasting Frequencies radio waves voice over*) Canis lupus arctos, the white wolf, also known as the Arctic wolf is a subspecies of gray wolf native to the Canadian Arctic Archipelago, from Melville Island to Ellesmere Island. It is a medium-sized subspecies, distinguished from the northwestern wolf by its smaller size, its whiter coloration. The Arctic wolf is relatively unafraid of people, and can be coaxed to approach people in some areas. It has occasionally acted aggressively toward humans. They are endangered due to their exceptionally low densities, smaller pack sizes, infrequent reproduction, and lower offspring production. In the wild, Arctic wolves' primarily prey on muskoxen and Arctic hares."
[*Aug+*]
WhiteWolf 白狼 Augmented Reality Holographic Heads Up Display A.R. Holo HUD

2154;_ Chicago, Illinois' Intranet, Wireless Broadcasting Frequencies Online
Lieutenant 2nd Lieu, Michael W. 廖, 米高 伍o; Vigilant Volunteer, Yellow-Mellow;

Valentina Vita (V. V.); Traveling Cosplayers And Kosplayers;
"(@_@) ...Huh, **who**... **Who** is displaying this **Holo HUD** and reading this...?"

Doktor KIPI OCEAN 5 White, Prototype Photo-Book Psychology A.I. Ai 愛 love;
"(*Broadcasting Frequencies radio waves voice over*) By the **WhiteWolf biotech suit**!"

Valentina Vita (V. V.); Traveling Cosplayers And Kosplayers;
"(O_O) Whoa we wow, a **biotech suit** with an A.I.? Ai in a **suit**? **Installing** or are you **recharging**? Hmm... No, a **female voice**... Not **male voice**. Hm..."

2154;_ Arcadia \ Valentina Vita (V. V.) 'Healthy-Strong Life' \ Left to Right(L->R)

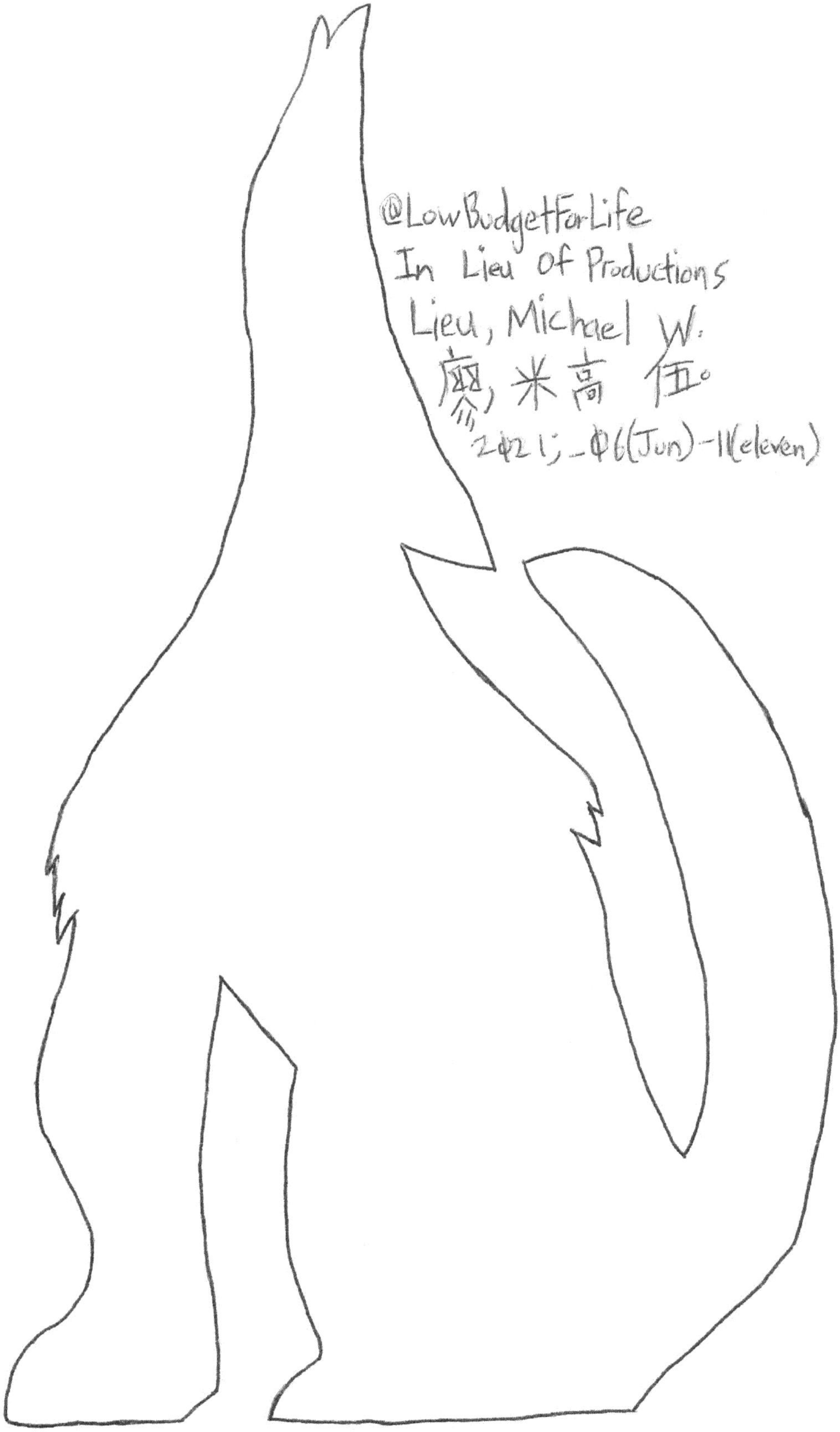
@LowBudgetForLife
In Lieu Of Productions
Lieu, Michael W.
廖, 米高 伍o
2Φ21-Φ6(Jun)-11(eleven)

Experimental Artificial Intelligence (A.I.) Doktor KIPI Built By:
Lieu, Michael W. \ Yellow-Mellow (°o°) \ **Ninja-Talor** A. Williams **** Kaoru-Kami-Sama

A.I. KIPI 0 **OCEAN** 5 (five) personality fused, 6 (six) variation version = **white jacket**,
A.I. KIPI 1 Openness To Experience = O ; green dress,
A.I. KIPI 2 Conscientiousness = Cat orange wall,
A.I. KIPI 3 Extraversion = E.X.; red hair,
A.I. KIPI 4 Agreeableness = Aura; blue clothes,
A.I. KIPI 5 **Neuroticism = Negative feelings; black clothes**,

Valentina Vita (V. V.); Traveling Cosplayers And Kosplayers;
"(>_O) Found **you**...!? An A.I. personality attachment by a bio**tech suit**... So **you** are...?"

Doktor KIPI OCEAN 5 White, Prototype Photo-Book Psychological A.I. Ai 愛 love;
"(*Broadcasting Frequencies radio waves voice over)* Oh, found me! (*gasp*) Looks like you found **me**, **I** am **Doktor KIPI**, an A.I. based on **photos** for **references** for mai **built day sources**. Along with **honoring** and **remembering** a **deceased** natural **human friend**, **Vani-hime**, who **Lieu Lieu**, **Ninja-Talor**, and Yellow-Mellow knew of back in **their school days**.... You know V. V., funny favor for a **systematic software** A.I.... Since **Lieu Lieu** is busy. (*sassy tone*) As a treat, Ai can tell you what kind of **inspiration** those Yellow-Mellow Problem Solvers use as references. It was **Lieu Lieu** who put the nail in the coffin for **photo inspiration**, **Kaoru-Kami-Sama's** God Complex design of using **human** psychology plus **bio**tech consciousness to match **human** consciousness, as **it** gave mi up the **special sixth** OCEA**N** personality, the special toner **white**. Requiring the much needed shifting of different OCEA**N** personalities, augmented by neuroscience **biotech**, studies of cerebral cortex, hypothalamus, neurons, ego, superego, id, aggressive dominant strong behavior as well as passive dependent soft behavior, to name a few... (*phew*) ...Plus, funny nature of Sisco networks' special surprise face which provided him the much needed inspiration for his nerd rage and anger when he quote, needs to mellow out... Mai **built day** was around **twenty one fifty two, February**... Ai think **the old world** Valentine's Day. So, Ai am in a way a Baby-Girl like that **BladeMaster** Elle, minus her psychological age of two years old, and that gray goo physical platform of... (*ahem*) a **black clothing wearing** A.I.. and lastly, **Ninja-Talor's ethical and philosophical teachings of human studies** and Artificial Intelligence interface of **acceptable** and **sociable behavior**... Along with Big Ben's fear factor of the terminator theory, forming **the new world Utopian** Sovereignty **of North America's** Chicago**, The Last Animal** Sanctuary **Of Illinois'** Anti Robotics Movement... Um... (*phew*) Ai do talk a lot as a **doktor** level full of **knowledge** and **wisdom**. Time to **stay silent** and **listen**... **Winky face**."

Valentina Vita (V. V.); Traveling Cosplayers And Kosplayers;
"(^_^) So informative for an A.I., **human like behavior**! I have an A.I., but not as advanced as **you** here... I do not think this is **free stuff** to take. Nice talking to **you**. Bye!"

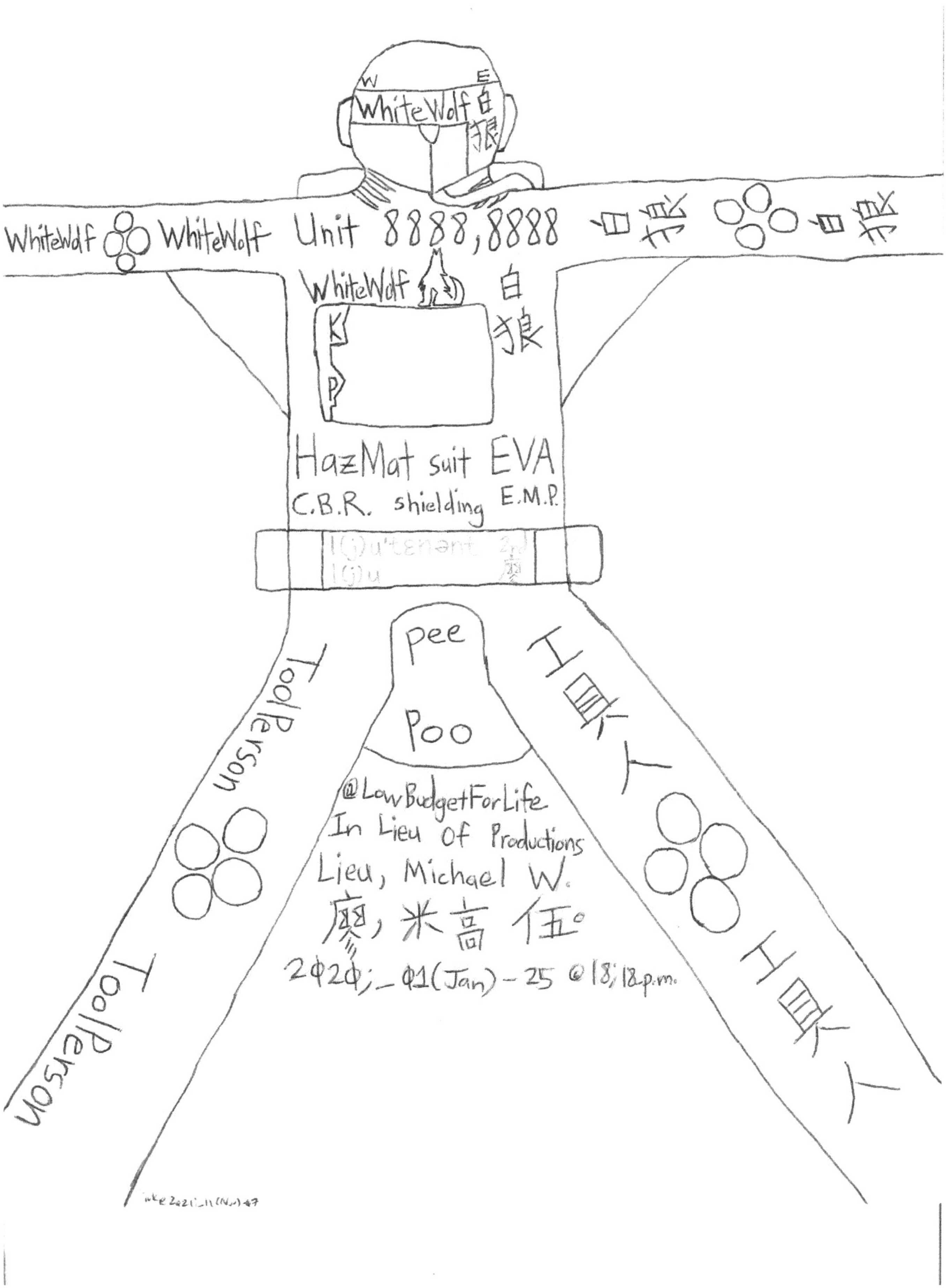
WhiteWolf 白狼
WhiteWolf WhiteWolf Unit 8888,8888 白狼 白狼
WhiteWolf 白狼
K
P
HazMat suit EVA
C.B.R. shielding E.M.P.
pee
Poo
ToolPerson
工具人
@LowBudgetForLife
In Lieu Of Productions
Lieu, Michael W.
廖, 米高 伍o
2Φ2Φ;_Φ1(Jan)-25 @18;18p.m.
ToolPerson
工具人

Ninja-Talor A. Williams

Talor Alexander Williams { DocuDrama, Photo-Book }
@TalorWilliams

King And X Philosophy

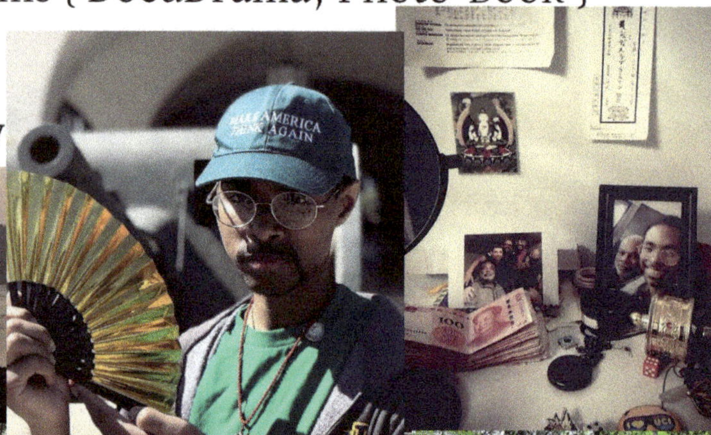

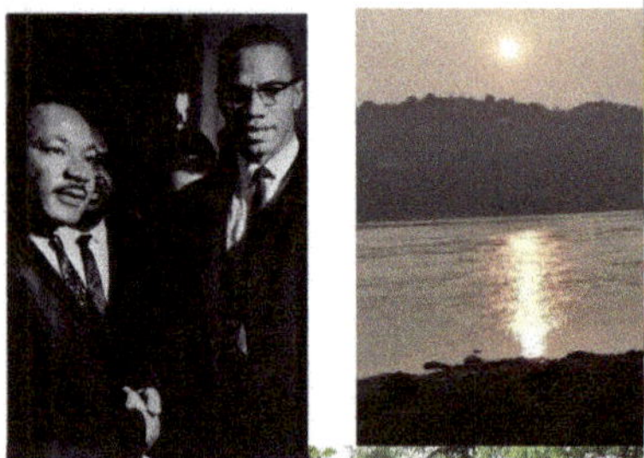

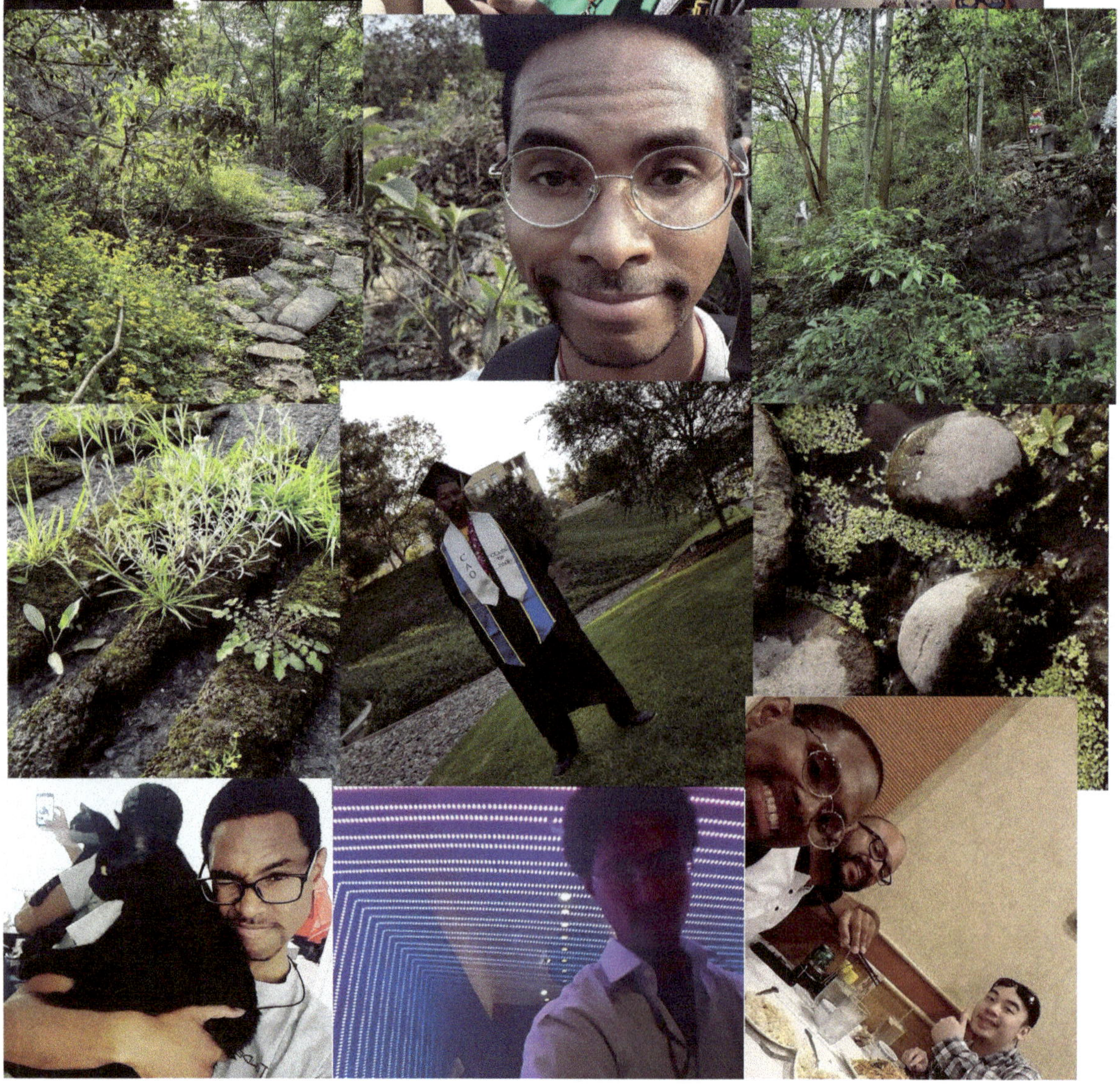

Ninja-Talor A. Williams; D King And X Philosophy "PeacefulPerson 平靜人" 'Way 道'; "(*archive audio video playback*) Yo ya'll, **I, Ninja-Talor,** Williams, The Mercenary Of Happiness, has departed from twenty one twenty one's movement of the new world, **Utopian** Sovereignty **of North America**... Chicago, **Illinois** and arrived to the Central Country **Of Eur-Asia,** helping out in their **animal** sanctuary improving their **infrastructure** and such. **I will, as a** Williams, go to **where my talents** and **skills are rewarded**. Unfortunately, **I** am not here to be part of **our** old Chatsworth High School **twenty one thirty six** and or Santa Monica College **language group meet up**... So, **I** am going to write words and speak, leave behind **this message**, no worries about a **surprise back attack Naruto Ninja thousand years** of **pain** or **aikido stick shift** or **jiu jitsu**... None of that this time... (*ahem*) Remember, as **Nietzsche said he who has a why, can bear any how**. (*deep breaths*) Do not get all **political correctness**, on the **he** and stuff, and **hurts feelings** with **me** okay!? This is how **that philosopher stated back then**... Hah!! ...This is a **one way Holo HUD message** dump... (*looking*) ...Anyways, if you are an **animal kingdom**, regardless of what **species** tree, able to **read** this, that is **amazing**, to see what this kind of **animal** sanctuary assists in **their** overall growth... **pet** vets, **animal** hospital, **oh my Buddha... that Elysium Corp**!! That **relation** with **human** and **ape** relations... **Planet Of Humans And Apes**, hah!! (*laughing and deep breaths*) ...**Buddhist belief system states that life is pain and suffering,** there is **little we can do to ease that pain and suffering, but endure, carry on**... **I** believe it is important to **make order out of chaos**, **my family** has been living life in so much **chaos** being a natural **unmodified African-Americans** in **the old world** United **States** of America of **twenty second, twenty first, twentieth century**, and before... What **I** like to call United **States** of America, **U.C. of A.,** Untrustworthy **Capitalism** of Anarchy**.** (*grabs purified water bottle with essential vitamins and drinks from it*) **...I cannot in good conscience take part in a system that preys on the weak and disadvantaged.** F. This. System. But, not this **animal** sanctuary of the new world, just **my** personal **experience** overall and **looking back**. **As a** nat **dark skin tone human male, I** have a better chance out of achieving **the old world American dream myth, not Doctor King's American dream of unity,** away from this Untrustworthy **Capitalism of** Anarchy. (*deep breaths*) **...I followed a bit of Malcolm X philosophy** than **Reverend King philosophy**... But... **I** really respect **Reverend King's I Have A Dream Speech of the twentieth century**... (*deep breaths and looking around*) Also since, as Yellow-Mellow Problem Solvers when me and Mic Lieu arrived here of Chicago, (*finger quoting*) **quote The Last Animal** Sanctuary **of Illinois unquote** back in **twenty one fifty three, January**, this **animal** sanctuary is just like anime expo **we** all used to go back in L.A. SoCal, well, minus the **pretty females we** used to **meet and greet**, ha ha! But... Mic **as Lieutenant Lieu,** here who now **manages** the place with the **toys and tools**, can tell you more about **our** time overseas in **Eastern society**, **British-Hong Kong sovereignty** of **economic freedom's free market**, **Unified Corea**, **Taiwan and Thailand**, Oh Buddha ...**that story behind Taiwan and Thailand**... (*phew and holding back laughs*) ...Japan of anime land... So many... (*laughing and flipping through notes*).

Ninja-Talor A. Williams; D King And X Philosophy "PeacefulPerson 平靜人" 'Way 道'; (*archive audio video playback*) ...okay... **I** gotta say, **Chicago city**, **lower case city of c.**, not the current City with the upper case C with City Mayor, home of Cool-Aid mix drink, that is something... (*ahem*) ...Well serious stuff, **I** need to see this through in **witnessing how aug** nat **humanity, animal kingdom,** A.I. **and** Earth's Environment **can all somehow coexist, like this... new world utopia here**. Witnessing, the **biblical myth events of Cain And Abel**... **Genesis Christianity I** think ...**Religious Bible**... Not really that religious, but the **stories it tells** is a **useful tool**... How many **people** are living **their Cain life, blaming the system and resents life,** and how many **people** are living **their Abel life, making the sacrifice and embraces life**... (*deep breaths*) ...**Noah's Ark,** The Flood, and... (*nervous*)... and The End. If **we, as a people representing humanity,** are **not able to learn from past mistakes of human history**, with all of **our** twenty second century **biotech tools and wonders**, then **we do not deserve to see and make it to the next twenty third century**. No use **feeling sorry** for those who bring, **birth, built children, babies into this kind of world**... **We only have our selves to blame, and we all must take responsibility, otherwise we are doom**, Doctor Doom, to **repeat it again and again...** Do not forget, Mother nature, she strikes back hard. The saying of World War Three ends everything quickly, and World War Four is **fought with sticks and stones**... of the **stone age**... (*sigh*) This whole War On Mars... What to say, it has been a **complete face palm and double face palm**... A **type one civilization** using **space ships for space warfare**... **We** should not be **going out** and **seed spreading** in the **galaxy** with this kind of... **Cain like behavior**... Oh, how **I** seem to be indoctrinated from the new world movement... Then again, **the old world** United **States** of America is a long by gone **myth**... **Lieu Lieu** here on **his Mic**, **Kaoru-Kami-Sama** along with our Yellow-Mellow says, **the old world U.S. of A.** was on the verge of collapsing in twenty one twenty one and or has **collapsed** like the **Rome Republic Empire myth**, the **old North America is** a **Neo-Wild West, Neo-Western**, **Neo-American myth**. The **perversion of money, private interests corrupting public interests**... (*sigh*) The appeal and value of **the new Americas**, sovereign system... under City Controllers, they are a different mix of human and animal personalities with this... **Western** Super-Hero Para-Military **Xavier** villain Killgrave... level of persuasion power. (*ahem*) **I** am so sick and tired of **the old world conventional political parties**, Democratic of Dumb, Republican of Resistance... **Lesser of two evils?** Why does it have to be **two**? **There should always be another way...** Another choice. (*ahem*) Sovereignty systems and sanctuaries are like **independent of I**, doesn't use **the old world voting ballot**, sovereign sectors do it without **any politicians**, that **explains no taxes** in this sector, and **I** agree with that to a degree. (*grabs purified water bottle with essential vitamins and drinks from it*) **Karma, the decider of fate will see how this all plays out**... **I** see **humans, but no humanity.** (*shaking head*) **I** want to leave you all an **important philosophical question... Are humans quote** naturally **born bad or systematic structures nurture them bad? We** always need checks and balances. (*waves bye*) **Ninja-Talor,** Williams, The Mercenary Of Happiness over and out."

Valentina Vita (V. V.); Traveling Cosplayers And Kosplayers;
"(@_@) Hm... looked over everything there is... **Empty blank white space** here..."

Lieutenant 2nd Lieu, Michael W. 廖, 米高 伍o; Vigilant Volunteer(s), Yellow-Mellow;
"(*in Oxygen Breathing Apparatus voice over*) Okay, here is **our last person** to arrive, **Ninja-Talor**... Hey **chief**, V. V., use **this tech** to take **photo** of **us**... Has been about **four years since we met up**... ...Oh, here is your Ai-**Aug bracelet**... fixed. May need to do a **system config and restore**... Should **not be too bad** since the **Kay Private Security outbreaks outages** was on October thirty one **to November one**..."

Lieutenant Lieu moves up to V. V., takes off **his Oxygen Breathing Apparatus P.P.E.,** hands her an **older tech** used for **taking photos** and her Ai-**Aug bracelet** working.

Valentina Vita (V. V.); Traveling Cosplayers And Kosplayers;
"(O_O) Whoa we wow, **this tech**... (*looking at the older tech*) Looks like it is from **another planet or time**... Okay, got it, tell me when all of **you** are assembled and ready to take the **photo**... (^_^) Yay, my Ai-**Aug bracelet** is up and working!"

V. V. puts on her **repaired** biotech Ai-**Aug bracelet** on her **right** waist.

Alien Valentina Vita, Traveling Cosplayer(s) And Kosplayer(s)
2154;_ Chicago, Illinois' Intranet, Wireless Broadcasting Frequencies Online

V. V. Augmented Reality Holographic Heads Up Display A.R. Holo HUD
Starting Up Biology Technology (**Biotech**)... Ai-Aug System Restore...
[<u>*Aug+*</u>]
V. V. Augmented Reality Holographic Heads Up Display A.R. Holo HUD

2154;_ Chicago, Illinois' Intranet, Wireless Broadcasting Frequencies Online
Alien Valentina Vita, Traveling Cosplayer(s) And Kosplayer(s)

Lieutenant 2nd Lieu, Michael W. 廖, 米高 伍o; Vigilant Volunteer(s), Yellow-Mellow;
"(*voice over*) **He** should be here soon... Then, **we** can get **our photo op**. You be on your way and... enjoy our Chicago **Animal** Sanctuary, **chief**. (*looking*) ...Okay, **ready** up..."

She looks around and sees that some of **Lieutenant Lieu's friends**, as she previously **walked past**, are slowly closing in for the **long awaited meet up** and **photo**.

Ninja-Talor A. Williams; D King And X Philosophy "PeacefulPerson 平靜人" 'Way 道';
"(*voice over*) **I** am here... Thanks to Anton and his **mach speed martial arts travel**!"

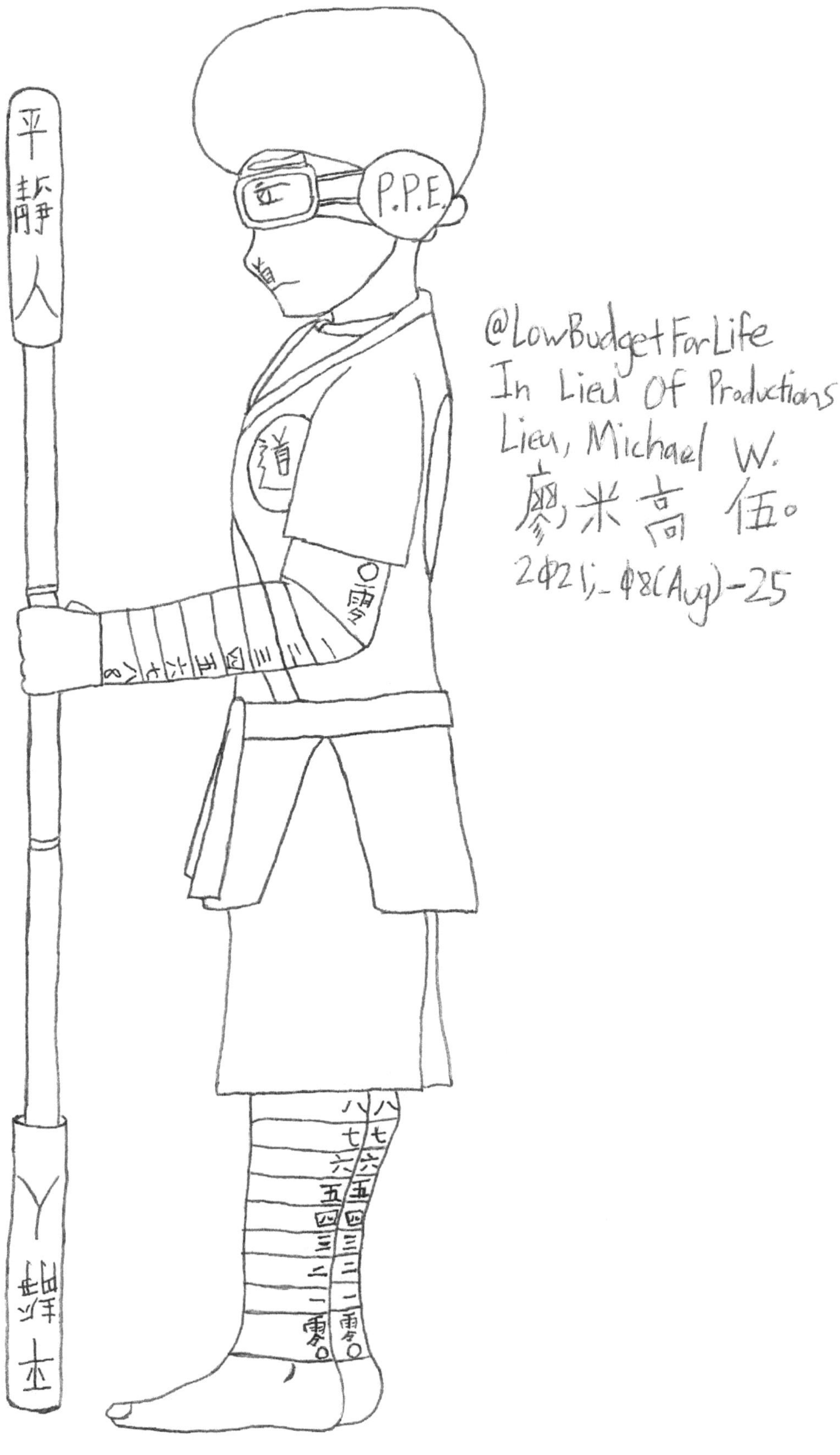
平靜人
P.P.E.
道
@LowBudgetForLife
In Lieu Of Productions
Lieu, Michael W.
廖,米高 伍o
2021-08(Aug)-25

Alien Valentina Vita, Traveling Cosplayers And Kosplayers
2154;_ Chicago, Illinois' Intranet, Wireless Broadcasting Frequencies Online

V. V. Augmented Reality Holographic Heads Up Display A.R. Holo HUD

Actual Legal Given Name;_

Talor A. Williams
(goes by **Ninja-Talor**)
pronounced, **"Nin-Ja Tay-Ler A. Wil-Yems"**

Alias Network Nickname;_

My Ninja, **Black Ninja**, **African Ninja**, Stick Fighting Ninja, **Stick-Man**, Poke Guy, **African** Kung-Fu, **African Martial Arts**, King And X Philosophy, Peaceful Person, **put on your P.P.E. now**, [*Aug+*]

Physical Plus Psych Age;_

natural 34 + 34 years old

Ethnicity;_

African

Nationality;_

North American

Community Network(s);_

Martial Artists

Cook-Chef Community Martial Artist Educator

Language Level(s);_

North American English,
Centralnese Cantonese,
Centralnese Mandarin,
Japanese, Spanish, [*Aug+*]

Education;_

University of South California, Irvine = Bachelor's Art in **Philosophy**, [*Aug+*]

Employment;_

The Strongest Society, Yellow-Mellow (°o°) Supa-Hiro Obama@deep [*Aug+*]

Love(s) Or Like(s);_

"Free Will V. Determinism" "He who has a **why**, can bear any **how**." -Nietzsche
"No such thing as a bad student, **only a bad teacher**." -Mister Miyagi

Metaphysics: What is the *nature of reality*? How many realities are there? What is *reality in contrast with appearance*? Is *reality given or chosen*?

Epistemology: How do you know *what is real*? What is *Truth*? How do you know you are *not being deceived*?

Ethics: What should we *value*? What *moral conduct code* should *we live by*?

[*Aug+*]

V. V. Augmented Reality Holographic Heads Up Display A.R. Holo HUD

2154;_ Chicago, Illinois' Intranet, Wireless Broadcasting Frequencies Online
Alien Valentina Vita, Traveling Cosplayers And Kosplayers

Lieutenant 2nd Lieu, Michael W. 廖, 米高 伍o; Vigilant Volunteer, Yellow-Mellow;
2154;_ Chicago, Illinois' Intranet, Wireless Broadcasting Frequencies Online

<table>
<tr><td colspan="7">L, M.W. Augmented Reality Holographic Heads Up Display A.R. Holo HUD</td></tr>
<tr><td colspan="7">Elementary, mid, high school meet up Remy Lin's Basketball 101 Coaching</td></tr>
<tr><td colspan="7">Low Budget For Life</td></tr>
<tr><td>Jay Lee
=
Twin Dragon Old Glasses</td><td>Lieutenant Lieu =
Low Budget For Life

"ToolPerson 工具人"</td><td>Remy Lin
=
.d Chicago b.
Basketball-Man</td><td>Channy Tran
=
Cook-Gook
(6C)</td><td>Ninja-Talor A. Williams = 'Peaceful Person 平靜人'
'Way 道'</td><td>Yellow-Mellow (°o°) Supa-Hiro Obama @deep Problem Solvers</td><td>Tee Lee
=
Twin Dragon Young Eyes</td></tr>
<tr><td></td><td></td><td colspan="3">D. D.
=
Marvel's Maximum Carnage Fanatic</td><td></td><td></td></tr>
<tr><td colspan="2">Jia Wong
=
B.4-Burma 3(three)III-M.M.B.</td><td>Chris-Z
=
Marvel's Venom Fanatic</td><td>God
=
Animal Kingdom Service Doggie</td><td>Gar Field
=
Marvel's Spider-Man Fanatic</td><td colspan="2">Sen Wang
=
WackyJoker.exe</td></tr>
<tr><td colspan="7">[<u>Aug+</u>]</td></tr>
<tr><td colspan="7">L, M.W. Augmented Reality Holographic Heads Up Display A.R. Holo HUD</td></tr>
</table>

2154;_ Chicago, Illinois' Intranet, Wireless Broadcasting Frequencies Online
Lieutenant 2nd Lieu, Michael W. 廖, 米高 伍o; Vigilant Volunteer, Yellow-Mellow;

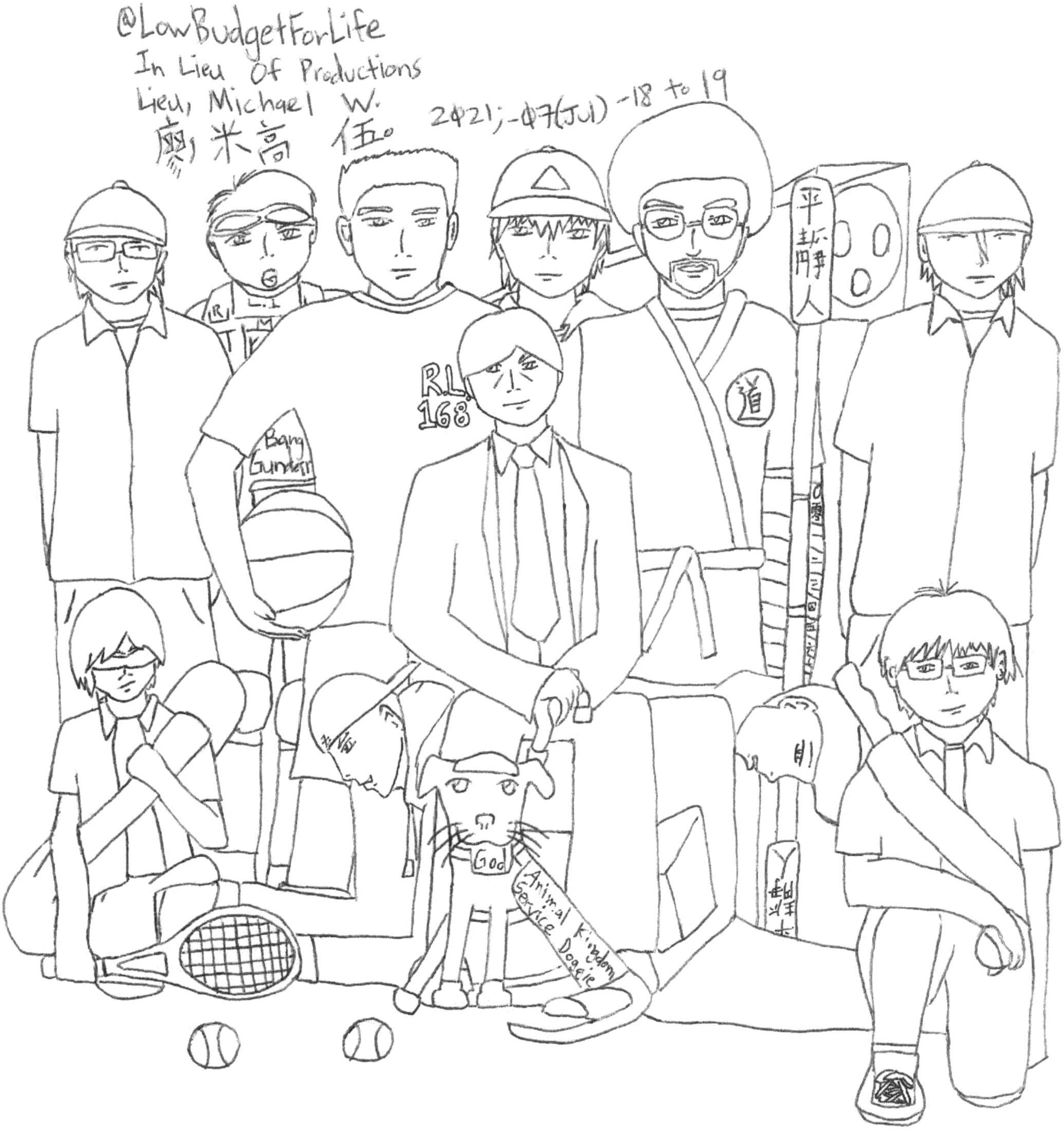
@LowBudgetForLife
In Lieu Of Productions
Lieu, Michael W.
2021;-07(Jul)-18 to 19
廖, 米高 伍o
R.L. 168
Bang Gundam
道
God
Animal Kingdom Service Doggie

Valentina Vita (V. V.); Traveling Cosplayers And Kosplayers;
"(^_^) Okay, **everyone**... say **cheese**. On three, one, two, three... **Cheese**!!"

Jay Lee = Twin Dragon Old Glasses; Elementary, Mid, High School Friends "MoLeiTau 無厘頭";
"(*staring*) ..."

Lieutenant 2nd Lieu, Michael W. 廖, 米高 伍 o; Vigilant Volunteer(s), Yellow-Mellow;
"(*insomnia*) **Cheese**..."

Remy Lin = .d Chicago b. Basketball-Man; Elementary, Mid, High School Friends 無厘頭;
"(*staring*) ..."

Channy Tran = Cook-Gook (6C); Elementary, Mid, High School Friends "MoLeiTau 無厘頭";
"(*staring*) ..."

Ninja-Talor A. Williams; D King And X Philosophy "PeacefulPerson 平靜人" 'Way 道';
"(*excitement*) **Cheese**..."

Yellow-Mellow (°o°) Supa-Hiro Obama@deep Problem Solvers; Saving The World Literally;
"(*surprise face*) ..."

Tee Lee = Twin Dragon Young Eyes; Elementary, Mid, High School Friends "MoLeiTau 無厘頭";
"(*staring*) ..."

D. D. = Marvel's Maximum Carnage Fanatic; Elementary, Mid, High School Friends MoLeiTau;
"(*staring*) ..."

Jia Wong = B.4-Burma 3(three)III-M.M.B.; Elementary, Mid, High School Friends 無厘頭;
"(*staring*) ..."

Chris-Z = Marvel's **Venom Fanatic**; Elementary, Mid, High School Friends MoLeiTau;
"(*enduring*) **Cheese**..."

God = Animal Kingdom Service Doggie; Elementary, Mid, High School Friends MoLeiTau;
"(*barking*) Woof!"

Gar Field = Marvel's Spider-Man Fanatic; Elementary, Mid, High School Friends MoLeiTau;
"(*enduring*) **Cheese**..."

Sen Wang = WackyJoker.exe; Elementary, Mid, High School Friends "MoLeiTau 無厘頭";
"(*staring*) ..."

2154;_11(Nov)-02(two)

Travel To Western Society, America(s), North America
Self-Govern Sovereignty (S),
Illinois (ILL) Sector, Points Of Interest (P.O.I.)

Chicago (S), "**Animal** Sanctuary"
"The New World 2121 Movement Of The 22nd Century,
Utopian Sovereignty of North America (U.S. of N.A.)"
Downtown District, **Elysium Corporation** Biotech Dome Center Point
Captain Rage's Public Po Po Place
Chicago City Police Department (C.C.P.D.) Special Weapons And Tactics (SWAT)

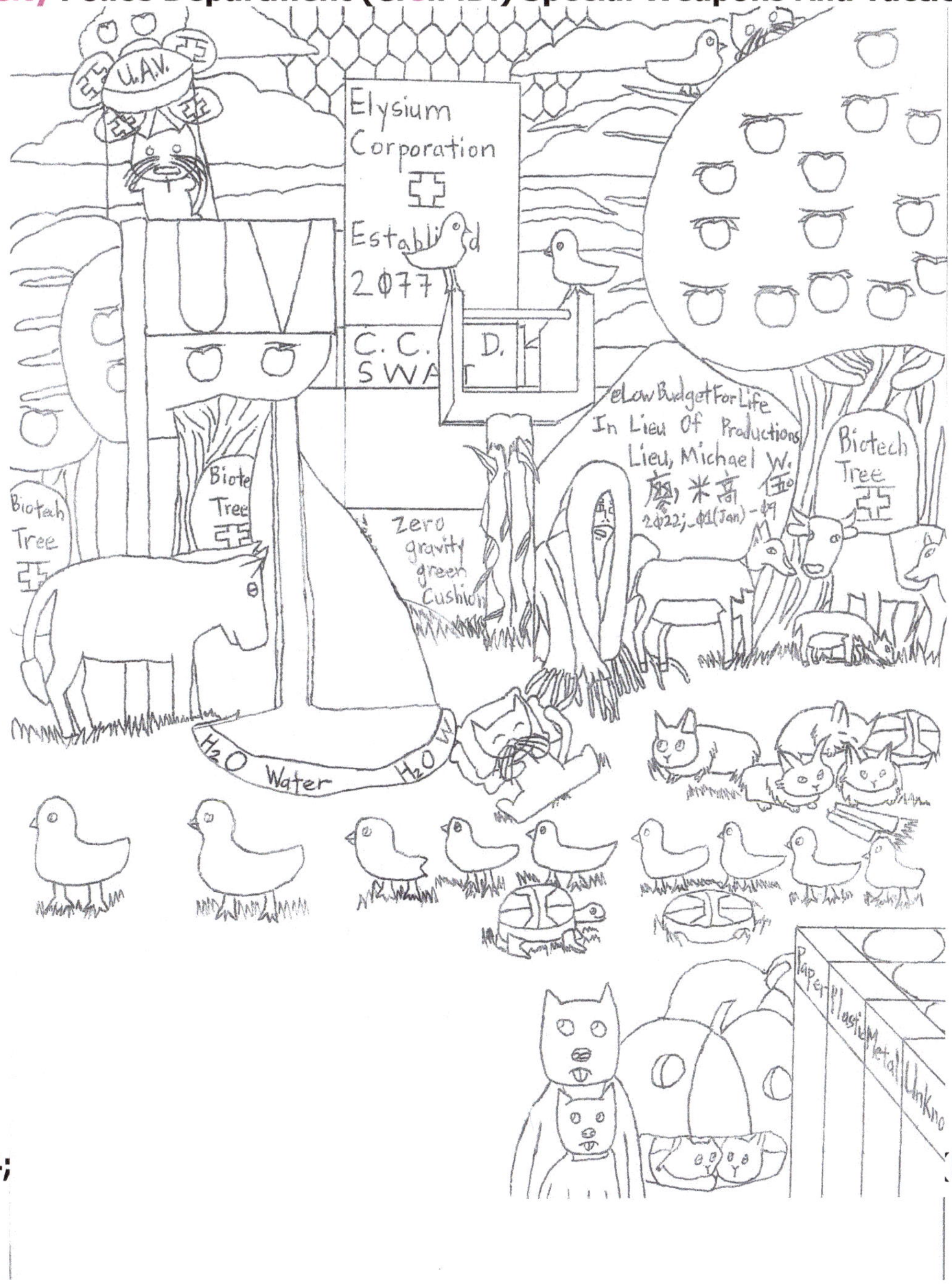

2154; (L->R)

Alien Valentina Vita, Traveling Cosplayer(s) And Kosplayer(s)
2154;_ Chicago, Illinois' Intranet, Wireless Broadcasting Frequencies Online

<table>
<tr><td colspan="4">V. V. Augmented Reality Holographic Heads Up Display A.R. Holo HUD</td></tr>
<tr><td colspan="4">Life Is Strange Universe Cosplayers' Meet Up</td></tr>
<tr>
<td>Chris Eriksen
"Chris-E"

'The Awesome Adventures Of Captain Spirit '

Voice Actor By;
@AshtonJamesJohnson</td>
<td>Gabe Chen

'Life Is Strange True Colors'

Voice Actor By;
@HanSoto</td>
<td>Daniel Diaz

'Life Is Strange 2'

Voice Actor By;
@OfficialRomanDean</td>
<td>Alex Chen

'Life Is Strange True Colors'

Voice Actor By;
@The_Erika_Mori</td>
</tr>
<tr>
<td>Charles Eriksen

'The Awesome Adventures Of Captain Spirit '

Voice Actor By;
@Nick_Apos</td>
<td>Emily Eriksen

'The Awesome Adventures Of Captain Spirit '

Comic Book Artist</td>
<td>Sean Diaz

'Life Is Strange 2'

Voice Actor By;
@ImGonzaloMartin</td>
<td>Life Is Strange's Hawt Dawg Man Mass Market Merchandise Lover <3

@DayeanneHutton</td>
</tr>
<tr>
<td>Max Caulfield

'Life Is Strange 1 (one) original'

Voice Actor By;
@HannahTelle

Max Doe Cosplay</td>
<td>Chloe Price

'Life Is Strange 0 (zero) Before The Storm, 1 (one) original, True Colors'

Voice Actor(s) By;
0 @Rhianna.Devries
1 @AshlyBurch

Chloe Priceless Cosplay</td>
<td>Steph Gingrich

'Life Is Strange 0 (zero) Before The Storm, True Colors '

Voice Actor By;
@KatyBentzOfficial</td>
<td>@SquareEnix

@LifeIsStrangeGame

@Dontnod_Ent

@DeckNineGames

"Life Is Pain And Suffering"</td>
</tr>
<tr>
<td colspan="3">Rachel Amber

'Life Is Strange 0 (zero) Before The Storm, True Colors '
Voice Actor By; @ItsKylieBrown   Raychel Amber-Alert Kosplay</td>
<td>Tac The Cat Bag

Animal Kingdom;
Cat-Kat</td>
</tr>
<tr><td colspan="4">V. V. Augmented Reality Holographic Heads Up Display A.R. Holo HUD</td></tr>
</table>

2154;_ Chicago, Illinois' Intranet, Wireless Broadcasting Frequencies Online
Alien Valentina Vita, Traveling Cosplayer(s) And Kosplayer(s)

2154;_ Arcadia \ Valentina Vita (V. V.) 'Healthy-Strong Life' \ Left to Right(L->R)

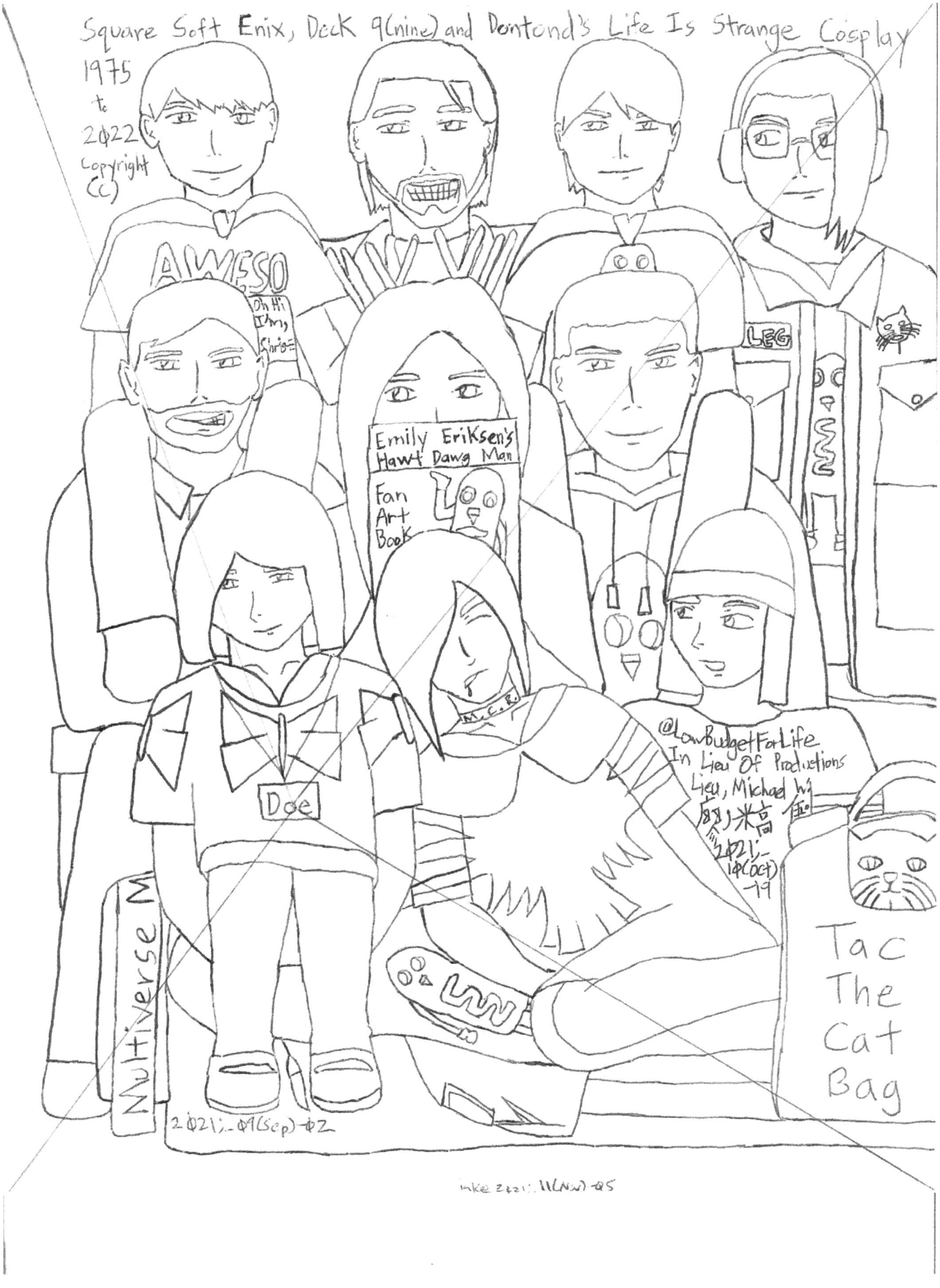
Square Soft Enix, Deck 9(nine) and Dontond's Life Is Strange Cosplay
1975 to 2022 Copyright (C)
AWESO
LEG
Emily Eriksen's Hawt Dawg Man
Fan Art Book
M.C.R.
@LowBudgetForLife
In Lieu Of Productions
Lieu, Michael W.
10(Oct)-19
Doe
Multiverse M
Tac The Cat Bag
2021_09(Sep)-02

Max Doe Cosplay; Time Traveling Forward In Time Like Time Dilation, But Not Backwards; "(*soft tone*) Okay, **everyone**... **Life Is Strange universe cosplay gather** around! The **Three Dimensions**, 3D, story, the **dimension** of **space** and the **dimension** of **time** altogether make up the **geometric object we** call **spacetime**. **You** take this **spacetime object** and **you** put **matter** in it, then **you** have **our universe**. The **dimension** of the **spacetime object** that **we** call **time** seems to only **flow in one direction**. In **physics**, **we** can explain the **geometry** and even **calculate** the **geometry** of **spacetime**, but **we** still have **no explanation why time flows in one direction**. **Matter** in **our universe** can **move back and forth in space**, but **time cannot move back and forth**."

Chloe Priceless Cosplay; Time Traveling Forward In Time Like Time Dilation; "(*snoozing*) ZzZz... ZzZz... ZzZz... ZzZz... ZzZz... ZzZz... ZzZz... ZzZz... ZzZz... ZzZz..."

Raychel Amber-Alert Kosplay; Time Traveling Forward In Time Like Time Dilation; "(*unconscious*) ...Uh... ...Uh... ...Uh... ...Uh... ...Uh... ...Uh... ...Uh... ...Uh... ...Uh..."

Chris Eriksen "Chris-E" Cosplay; Time Traveling Forward In Time Like Time Dilation; "(*excitement*) **I** am **Captain Spirit** and **I** will listen the **Three Dimensions** story!"

Gabe Chen Cosplay; Time Traveling Forward In Time Like Time Dilation; "(*smiling*) Hey, **everyone gather around** and **listen**! About The **3D story**!"

Daniel Diaz Cosplay; Time Traveling Forward In Time Like Time Dilation; "(*excitement*) **I** am the **Hawt Dawg Man** and **I** will listen the **Three Dimensions** story!"

Alex Chen Cosplay; Time Traveling Forward In Time Like Time Dilation; "(*listening to headphones*) ..."

Charles Eriksen Cosplay; Time Traveling Forward In Time Like Time Dilation; "(*smiling*) Hm..."

Emily Eriksen Cosplay; Time Traveling Forward In Time Like Time Dilation; "(*calm*) ...**I** am going to **self-promote**, here is **my Hawt Dawg Man Fan Art Book**."

Sean Diaz Cosplay; Time Traveling Forward In Time Like Time Dilation; "(*listening*) Hm..."

Steph Gingrich Cosplay; Time Traveling Forward In Time Like Time Dilation; "(*looking*) ...**Chloe** is **fast asleep**..."

Valentina Vita (V. V.); Traveling Cosplayers And Kosplayers; "(=_=) Okay, time to **move on** to **explore Downtown District**... pass the **animals**..."

"The New World 2121 Movement Of The 22nd Century, Utopian Sovereignty of North America (U.S. of N.A.)"

Chicago, **The Last Animal** Sanctuary (S) **Of Illinois**

Public-Private Sector, Voluntold-Volunteer Service(s)
Bulletin Board System (B.B.S.)
Downtown District,

The music "**Cowboy Bebop's Cats On Mars**" plays in the **Utopia A.R. Holo HUD** background and is set on loop.

<u>Public Sector</u>, Voluntold Service, "No One Really Wants To Do This, And Wants Out..."

"**Jerry Duty**", Xtra-Men, Shaked-Soda, Chicago's Utopians, U.S. Citizens, Alien status
<u>Soap Opera Speech Box</u> = Quote, "*Start, with the advent of the decade year 2150 (twenty one fifty) in accordance to the new world 2121 (twenty one twenty one) movement, we have found our way to the old world United States (U.S.) Department of Alien and Citizen Services (D. of A and C.S.), Alien and Citizen Census (A. and C.C.) division offices, I have gained access to this sector's personnel files, their data, their back story, their future predictions, all those connections of the past 11 (eleven) billion humans, the 11 (eleventh) hour, which most of them flatline, turn undead, became deceased, necrosis process, forgotten and dismembered on unmarked graves from Earth's first case on 2121 (twenty one twenty one);_ 02-28(February twenty eight) of the Martian 6 (six) Endgame Event, as the old world calls it Martian Outbreak Outage 2120 (MOO-20), Dear Martians, as Earthlings, we understand why you unleashed unconventional warfare on Earth, Earth's Moon, the Solar System, we believed to have identified those deemed unworthy contribution overall on Planet Earth, animal kingdom, need to clean house on this sector Chicago Illinois (ILL) United States of North America (U.S. of N.A.) to begin a Utopian Sovereignty, Animal Sanctuary, with needed planning, waiting, distracting, what should City Controllers call ourselves this time Question-Mark, period, end.*" "*Start, almost none of the new world Utopians normally wants to do this, almost everyone*

wants to get out of this as filed in my violet Holo HUD Aug memo of my name Chicago Controller City Mayor, quote "Abusive bitch and moan toxicity constant crybaby complaint box" unquote, on watching over a green biotech aug of 200 (two hundred) years old, clinically dead human male, blond gold yellow hair, Caucasian ethnic, light skin tone, I named 'Jerry Duty', in honor of the old world, Jury Duty, voluntold service, period, end."

"Start, by providing Utopians the old world 21st (twenty first) century economic system of basic minimum wage, disgusting how they can barely survive off that, in lieu of unpaid voluntold service, we have a normal volunteer worker to do voluntold service, nat human Latino male, an armed social worker Mars Hall, some Utopians' holo file, again and again, that he is sleeping on job, in accordance to my violet Holo HUD Aug memos of my name Chicago Controller City Mayor's quote, "Abusive bitch and moan toxicity constant crybaby complaint box" unquote, armed social worker Mars Hall does a good job of being a special voluntold volunteer service worker that, quote "Let the shit hit the fan rock bottom to crash and burn hard" unquote, all this aug nat people in fighting is distasteful even by the old and new world 22nd (twenty second) century standards, I can, I may, take control of their consciousness, what we called, quote "immediate removal" unquote, thus erases their existence knowingly and unknowingly, becoming animals, but all humanity will systemically wipe themselves out in time and provide us entertainment, if not climate change also assists to systemically slaughter everything of humanity, equal opportunity, or our fear factor war weapon scare of Chemical, Biological, Radiological (C.B.R.) +(plus) Electro Magnetic Pulse (E.M.P.) all of humanity of Earthlings, Spacers, Martians, in our Solar System, as we love to destroy ourselves through imploding infrastructures exploding structures, humans, animals, Artificial Intelligence (A.I.) Planet Earth, all humans are animals deep inside as they all seem to need reminders of that when they are in a Utopian Sovereignty (U.S.), Chicago, The Last Animal Sanctuary Of Illinois (ILL) to provide my beautiful peace loving animal kingdom species, it is beautiful and entertaining to see them give in to their true animal nature, period, end."

"Start, lessons learned from seeing, viewing the original new world 2121 (twenty one twenty one) movement which started in the old world Seattle, Washington (WAS) of the United States of North America (U.S. of N.A.), our successful official break away from the old world United States of North America (U.S. of N.A.)'s federal government, as we City Controllers called it quote, "Chicago City Clean House Act of 2152(twenty one fifty two);_07-04(July four)" unquote, we observed and used the old world United States of North America (U.S. of N.A.) sector's independence day holiday celebration as a distraction, our City Controllers' sovereignty has upgraded our Broadcasting Frequencies Hertz of Radio Waves to expand on the Electromagnetic Spectrum's Infrared Rays, Light Rays, Ultraviolet Rays, X-Rays, Gamma Rays, and Cosmic Rays, as we managed to bring in and save several injured animals the old world no longer cares about, and Planet Earth's mother nature is unforgiving, one original well known natural male dog (Nat-Dog) with the help of two natural human Latino males, who seems to have the best worst day of their life, Big Benavidez-Cruz (Big Ben), Rich Rodriguez, who carries the old world burden weight of quote "Mexican Patriot or The King Of Mexico" unquote, with his green big biotech engineering suit and his aug nat people's sector by the 22nd (twenty second) century, and his younger blood brother Baby Benavidez (Baby Ben), Jose Rodriguez, known as their tag team combo of Ben Bros, and also Big Ben's self-proclaim Mexican-Man flag stance, augmented by their Mexican-Men team, with their infamous Wolf Brothers Luna Howl, anyways, those two named that injured Nat-Dog, who become augmented for survival purposes by my founding support staff, Clean-up Crew Unit 6 (six) Unit 9 (nine), Rover, but Jerry Duty of almost 200 (two hundred) years old, Augmented Biotech Clinically Dead Experiment (A.B.C.D.E.), insisted on calling him quote "Dog-Name-Cat" unquote, a dog name cat literally, maybe kat with a C. or K. Question-Mark, which ever, thus the following day after Chicago City Clean Act of 2152(twenty one fifty two);_ 07-05(July five) I called Aug-Dog, quote "Rover Dog-Name-Cat" unquote, as they all keep fighting over Aug-Dog's name, and Jerry Duty, quote "Xtra-Men, Men who are Xtra and of little use" unquote, our Undead Utopian (U.U.) our prototype zero experiment in using flatline,

undead, deceased, necrosis processing, forgotten and dismembered humans to provide a blueprint model for biotech dome external maintenance as Undead Utopians (U.U.) ascend upwards and out of our biotech dome animal sanctuary, my beautiful peace loving animal kingdom species and their biotech dome home in Chicago, The Last Animal Sanctuary Of Illinois (ILL), period, end."

"Start, been digging through this Chicago Illinois (ILL) sector archive files even more and more, apparently Jerry Deesimon's, his so called actual legal given name, natural birth day of 20th (twentieth) century, 1954 (nineteen fifty four);_ 01-01 (January one) of New Years, hails from a near by ghost town city sector, not too far from here, the old world Detroit, Michigan (MIC) of United States of North America (U.S. of N.A.) state, period, end."

"Start, Jerry Duty's augmented built day of 21st (twenty first) century, 2054 (twenty fifty four);_ 11-11 (November eleven) of the old world United States of North America (U.S. of N.A.) Veteran's Day holiday celebration, died due to natural flatline death period, Happy Flatline Day, 100 (one hundred) years old alive, undead, deceased, period, end."

"Start, Undead Utopian (U.U.) experiment continues, as I, Chicago Controller City Mayor's find more additional augmented and natural deceased animals injection splicing, rest in pieces my beautiful peace loving animal kingdom species, built day for augs, birth day for nats, of the 22nd (twenty second) century, in the old world 2152 (twenty one fifty three);_ 01-01 (January one) of Jerry Duty's natural birth day, Happy New Years Birth Built Day, small simple steps evolve to bigger ambitions later on, period, end."

"Start, Jerry Duty of Xtra-Men still kept physically enhanced as a quote "Undead" unquote, like an unconventional warfare virus disease outbreak, not quite alive and not quite dead, undead of life between death, an undead human under 21st (twenty first)

century biotech augment, rest in pieces my beautiful peace loving animal kingdom species, and using 22nd (twenty second) century Psychological Electromagnetic Spectrum Frequencies Hertz to induce unconscious dream illusion of pre-determined free will, no real reason to wake Jerry Duty up quote "Alive" unquote, into this reality for those who been around since 2120 (twenty one twenty)'s Martian 6 (six) Endgame Event of the first unconventional warfare released on 11-17 (November seventeen), unless a Utopian wants to give him cognitive state, could restore this to aug nat people, but have seen enough sick, foul irresponsible human behavior to care at this point in time, we City Controllers are honored to see the remaining humanity make it past 2120 (twenty one twenty)'s Martian 6 (six) Endgame Event, and onward with the new world movement of 2121 (twenty one twenty one) from Humans' Hubris (H.H.) of Earth, Earth's Moon, Mars, Solar System, period, end."

"Start, 2154 (twenty one fifty four);_ 01-01 (January one) Happy New Years Birth Built Day to Nat Jerry Deesimon, and Aug Jerry Duty and his quote "I need Shaked-Soda" unquote obsession has become quite a trademark catch phase as a form of entertainment to Utopian augs and nats in Chicago, The Last Animal Sanctuary Of Illinois' (ILL), Utopians started calling me Chicago Controller and or City Mayor, or did we call ourselves that Question-Mark, Chicago Controller City Mayor Very Violet Purple Person, All-Father, Lord Jesus Christ, God, god, lower and upper case G, all these names us City Controllers get called on, in the end me and we only control those failing to care about our easy self-governing guidelines rules regulation after indoc, period, end."

"Start, for those Utopians who bothered to read, follow and make it this far, we City Controllers will let you in on a small sovereign secret, ready, for those who desire power, not solar, wind, hydro kind of power, but our Electromagnetic Spectrum Frequencies level of strength, you can do so, by Ascension, Ascension consists of Public Auditing Elysium Corporation's Biotech Dome Sector divisions, in the Downtown District, past

the 4D (four dimension) Movie Theater Box Office, from floor 01 (zero one) with no ziplines, no escalators, no elevators, just good old fashion movement from one floor to another floor to floor 99 (ninety nine), Director: Doctor Froste awaits you all, all those who enter, be very ready, we wonder what would you do with that sovereign secret that the old world United States of North America (U.S. of N.A.) federal government calls us quote, "Domestic Terrorist(s)", ha ha, still with us Question-Mark, good, see those old world United States (U.S.) Department of Outbreaks and Outage (DOO) army soldiers and civilian companies Question-Mark, they think the year is still 2152 (twenty one fifty two);_ 07-04 (July four), Dystopia of Chaos, or Utopia of Order, Utopian, make the Ascension to Elysium Corporation, period, end. unquote.
Responsibility, Manager To Blame = Chicago Controller City Mayor Augmented Biotech Clinically Dead Experiment Files (A.B.C.D.E.F.)
Back Story = Consciousness, The Forbidden Fruit Of God. The Garden Of Eden.
Personal Perks = "Go Green Social Work" uniform for 'Jerry Duty', armed ballistic holstered belt: 9mm Beretta firearms, expandable baton, pepper spray of self-defense and extra tools. Kick Me White **Sign**, I. C. U. P., P. P. (Poopy Pants) White **Re-Writable Draw Me Sticky Notes**, Aug-Nat Alive-Undead-Deceased Human, Aug-Dog Rover Dog-Name-Cat. Pillow And Blanket.
Location = D. "Downtown District" D., Central Hub, All You Can Drink Eat Essential Vitamins Liquid Solid Food Utopia Vending Biotech

V. V. looks down at her **print paper** of **to do list** and crosses out **Downtown District**.

V. V.'s 2154 to do list since my Ai-Aug bracelet is not working due to outbreaks outages...
~~Visit, **North Side, Lieu Lieu.... Lieutenant Lieu**~~ (~~next place to go and to do~~) Complete
~~Visit, **Downtown District, Bullet Board System**~~ Complete
~~Visit, **East Side, All For One Shopping Center**~~ Complete
Visit, **South Side, Entertainer Mister Steven, Reserve** Your **Tickets** (Nov 11 @ 8 a.m.)
What else...? (What to do to pass time?)

Valentina Vita (V. V.); Traveling Cosplayers And Kosplayers;
"(O_O) Okay, I believe I have found... the **Downtown District, Bulletin Board System**."

The music "**Mark Knopfler's 05;15_a.m.**" plays in the **Utopia A.R. Holo HUD** background and is set on loop.

Public Sector, Volunteer Group, "Clean-Up Crew, Get Ready At Five Fifteen A.M.!"

"**Clean-up Crew**", **Chicago Controller City Mayor's Support Staff**, **Animal Caretakers**, **Search And Rescue (SAR), Search And Recovery (SAR)**, **Entertainers' Music Video (M.V.) Display**
Soap Opera Speech Box = Quote, ""Chicago City Clean House Act of 2152 (twenty one fifty two);_07-04 (July four)" *any sleeping beauty, our Clean-up Crew's social slang for deceased necrosis human beings and animal kingdom corpses, not dismembered and or decomposing body, equals flatline Undead Utopians slaving away as long as needed to Chicago Illinois sectors' exterior biotech dome maintenance on the unsettling 22nd century events on Planet Earth and Earth's Moon. All animals get memorial service within biotech dome.*" unquote.
Responsibility, Manager To Blame = Chicago Controller City Mayor
Back Story = "Chicago City Clean House Act of 2152 (twenty one fifty two), 07-04 (July four)" "Officially Sanction Off Using Fireworks' of the new world 2121 (twenty one twenty one) movement"
Personal Perks = Clean House Public Audit via Chicago Controller City Mayor's violet A.R. Holo HUD files "Chicago City Clean House Act of 2152;_07-04" "Twenty one fifty two, July four" I. C. U. N. Body Bags, Free Stuff, Ascend, Ascension, **Scientists, Technicians, Engineering, Mathematicians (STEM),** **Necrosis, Necrophilia, Necromancer, Necromorphing (N.N.N.N.),** All Clean-up Crew names goes by: Unit quote "number (number spelled out)" unquote, Example, Unit 6 (six), Unit 9 (nine) or our infamous Clean-up Crew Unit 8888,8888 (eight eight eight eight comma eight eight eight eight) and or (eight eights), commas are not needed since all Units go by single digit number call out sign. Poopy Pants (P.P.) human and animal diaper excrement for soil agriculture, **white clean top tier hazmat suits** provided by **Elysium Corporation** featuring... violet augmented bending joint Sealing-Tapes Frequency Hertz and green Bio-Lettering. Animal assistant(s): Natural **Honest Ape (**Nat**-Ape)** Augmented Walking Dog (Aug-Dog+)
Location = Our Head Quarters (H.Q.) where we just play **Entertainers' 4-D Music Video (M.V.)** all day, everyday, in stand by mode by **Chicago** City Christian Church.

The music "**Tangerine Dream's Love On A Real Train [State Azure Cover]**" plays in the **Utopia A.R. Holo HUD** background and is set on loop.

Public Sector, Volunteer Group, "Aug+Nat-Animal Human First Responder(s) On Call"

"**Strike Force (打擊 力量)**" "**ToolPerson (工具人)**", Handy Human's Jack Of All Trades, Master Of Nothing, Combat Support Engineers, Toy-Masters, Hydro H2O Water, Navy

Soap Opera Speech Box =
Quote, "*If it is not broken, no need to fix it. Shorten Engrish version, "No broken, no fix."*

Quote, "*It takes ten people, of aug and nat humans, animals, A.I. and others, to change a biotech light. It is not rocket science, rather biotech science.*" unquote.

Responsibility, Manager To Blame = Lieutenant (Second In Command) Lieu, (廖,) Michael (米高) W. (伍 o) Also Known As (A.K.A.) "**Low Budget For Life**"

Back Story = Natural Unmodified Human Male (Nat-Man)
"All Natural Product", Raw-Men "Ramen", No Genetic Modified Organism (G.M.O.)"

"Early Childhood Upbringing Years, 2120 - 2136"
Birth Place: People Of Earthling
Birth Date: 2120 (twenty one twenty) - 09 (September) - 11 (eleven)
Year Of The **Dragon, Eastern Society Zodiac**
People Of Martian's Chemical, Biological, Radiological (C.B.R.) Outbreak,
Electro Magnetic Pulse (E.M.P.) Outage Of 2120 "MOO-20"
2121 - 2136 = "The 22nd (twenty second) Century Dark Ages"
2120 – Present Day Time, War On Mars

United States Army (U.S.A.), **Los Angeles City** (D) (R) Military (M) Bases,
South California (SCA), North America (N.A.):
Castelar Elementary School, Class Of 2129
Nobel Middle School, Class Of 2132
Chatsworth High School, Class Of 2136

More Natural Schooling Or...

"United States Department Of War (U.S. DOW), 2136 - 2139"
(2136 { 0, zero } \ 2137 { 1, one } \ 2138 { 2, two } \ 2139 { 3, three })
Volunteer U.S. Navy Sailor failed engineer; Artificial Intelligence (A.I.) Ai 愛 love,
Voluntold U.S. Army Soldier failed combat support via Individual Augmentee (I.A.),
Medical Separation, Honorable Discharge,
Service Connected Disability 100% (one hundred percent)

Public Sector Career End

"School Days = School Is Cool (Skhool Is Kool), 2139 - 2149"
United States Army (U.S.A.), **Los Angeles City** (D) (R) Military (M) Bases,
South California (SCA), North America (N.A.):
(2139 { 0, zero } \ 2140 { 1, one } \ 2141 { 2, two } \ 2142 { 3, three } \ 2143 { 4, four }
2144 { 5, five } \ 2145 { 6, six } \ 2146 { 7, seven } \ 2147 { 8, eight } \ 2148 { 9, nine }
\ 2149 { 10, ten })
Santa Monica College (S.M.C.) Down Town Los Angeles (D.T.L.A.) Campus
Associates' Art (A.A.) Degree, ***Liberal Arts***
Associates' Science (A.S.) Degree, ***Broadcasting Wireless Communications***
Class of 2144

University of South California Los Angeles Extension (U. of S.C.L.A. Ext)
Down Town Los Angeles (D.T.L.A.) Campus
Introduction To Statistics, "Stats X.L. 10 (ten)"
5 (five) units of credits on 2144.

South California State University, Los Angeles (S.C.S.U., L.A.)
Bachelor's Art (B.A.) Degree, ***Public Relations*** *(P.R.)*
Class of 2146

South California State University, Los Angeles (S.C.S.U., L.A.)
Down Town Los Angeles (D.T.L.A.) Campus
Legal Assistant And ***Legal Secretary*** Certificates
Class of 2149

United States Department of Veterans Affairs (U.S.D. of V.A.)

Living The Public Sector G.I. **Bill Utopia**

"Early Retirement Processing, 2149 - 2150"
United States Army (U.S.A.), **Los Angeles City** (D) (R) Military (M) Bases,
South California (SCA), North America (N.A.):
United States Department of Social Security Services (U.S.D. of S.S.S.)
Executive Order 21221111 and the Martian Outbreak and Outage (MOO) Act
'Attention: All U.S. Citizens & Lawful Aliens of Americas on Planet Earth, Lunar Moon. All Natural Unmodified Humans, of male and female binary biological sex can file for early retirement at age 30(thirty), and the latest at age 40(forty) with no increase after.'

"**Godspeed**, to the people of Earth and Earth's Moon on surviving and seeing it through the ongoing **War On Mars' advanced unconventional warfare**, and to the people of U.S.A. on **domestic terrorism**. Is U.S.A., the next **Roman Republic Empire**?"
-Lieu, (廖,) Michael (米高) W. (伍 o)

Retirement By Age 30 (thirty), The Year 2150, As A Natural Unmodified Human Male. I Love ***Public Sector Welfare State***
Retirement, Yellow-Mellow Supa-Hiro Obama@deep Problem Solvers, **2150 – X** **O O** **O** (2150 { 0, zero } \ 2151 { 1, one } \ 2152 { 2, two } \ 2153 { 3, three }) Co-Founders;_ Sisco D. Mad-Man A.K.A. Mellow-Man Lieu, Michael W. = Low Budget For Life 廖, 米高 伍 o Kaoru Kami-Sama Ninja-Talor A. Williams ***Private Sector Career Ongoing***
"The New World Movement Of 2121, Utopian Sovereignty Of North America" **Chicago, The Last Animal Sanctuary (S) Of Illinois, 2153 - Present Day Time** Homecoming *arriving 2153;_01(January)-02(two) to present day time* United States Navy (U.S.N.) Sailor Military (M) Base, Basic Training, A School = *Recruit Training Command, "Boot Camp", Great Lakes, North Chicago, Illinois (ILL)* *arriving 2136;_11(November)-11(eleven) to 2137;_02(February)-28(twenty eight)* Responsible Manager Volunteer Worker **"Strike Force (打擊 力量)", ToolPerson (工具人)** "Lieutenant Lieu" Welcome all and any Augmented (Aug) + (plus) Natural (Nat) Human(s), We need all the help we can get, we stand up to animal and human bullying behavior.
Personal Perks = Animal Sanctuary And Human Aug-Nat Problem Solvers, Ship Sailing
Location = N("North Side"), pass King And(&) X District by Biotech Dome Ocean Border

Valentina Vita (V. V.); Traveling Cosplayers And Kosplayers;
"(O_O) Hm... Okay, this is **Lieutenant Lieu's back story**... (^_^) Who **assisted** in **diagnosing** and **fixing** my offline Ai-**Aug bracelet**..."

351 \(*Lieutenant Lieu* and or *In Lieu Of*) Lieu, Michael W. { 廖, 米高 伍o } \ 351

The music "**Thomas Bergersen's Final Frontier**" plays in the **Utopia A.R. Holo HUD** background and is set on loop.

Public Sector, Volunteer Group, "Aug+Nat-Animal Human First Responder(s) On Call"

"**Utopian Peacekeeping**", Public Militia Aug-Nat Co-Existence, And "**Fan Following**"

Soap Opera Speech Box = Quote, "*War has been declared.*" "*War is hell.*"
"*There are no atheists in foxholes.*" "*Go ahead, make my day!*"
"*Do not go around playing hero, because a true hero does not make it home.*"
"*Make peace with yourself or make peace with the world.*" unquote.

Responsibility, Manager To Blame = Alex Drastal, "Alex D.", "Doc Drastal"

Back Story = Augmented Biotech Human Male (Aug-Man+) Voluntold Built Day = 2153
Work Experience: (From Newest To Oldest Of Top To Bottom)

Artist Actor Model Costume Player (**Cosplayer**), 2153 - Present Day Time
Martial Artists = "MartialArtsGod 武道神"

United States Department Of War (**U.S. DOW**), 2145 - 2153
Medal Of Honor Award (Living Legacy)
Off Planet Mars Mission (**War On Mars**), War Survivor Military Veteran,
Earth's Global Government (**EGG**), Joint Task Force

United States Space Force (**U.S.S.F.**) **Medical Evacuation** (**MedEvac**), 2150 - 2153
(2150, 2151, 2152, 2153)
Medical Separation, Honorable Discharge,
Service Connected Disability 100% (one hundred percent)

United States Air Force (**U.S.A.F.**) **Pararescue** of 2145 - 2149
(2145, 2146, 2147, 2148, 2149) Reenlistment,
Violet Heart Award

Fat Camp Super Senior Graduate Student

Personal Perks = **Earth** & **Mars**' Conventional & Unconventional Warfare, Improvise, Honor Guard, Military Survival Training Tactics, Medic-Man, Watch Your Six

Location = Everywhere, nowhere. Put on your double hearing **Personal Protective Equipment (P.P.E.)**, listen, feel and watch out for **black blur Mach speed** or feel the **heat signatures' boom** within Chicago **Elysium's animal** sanctuary biotech dome.

Valentina Vita (V. V.); Traveling Cosplayers And Kosplayers;
"(O_<) **Alex D. ...Dra-Drastal**... I do not think I have met **him** yet... **volunteer group Utopian Peacekeeping...** Whoa we wow, almost **eight years military service**..."

The music "**Beethoven's Ave Maria**" plays in the **Utopia A.R. Holo HUD** background and is set on loop.

Public Sector, Belief Systems, "Separation Between Church And State Or Acting State"

"Chicago City Christian Church" Belief Systems And Other Religions Welcome <3"
Soap Opera Speech Box = Quote, "*Using the power of love, heart <3 or ice cream cone, enhanced by loving kindness to heal, give hope and unify all life forms, advanced beings, animals, augmented and naturals humans to build a better future together! Heart! or ice cream cone!! <3*" "*Believe in Lord Cheese And Rice.*" "*Keep love open in your heart <3.*" "*A-Maiden!*" "*All write, all write, all write!*" "*Heart is the frequency.*" "*Our precious heartful love...*" "*Ice cream cone of heart <3!*" "*Heartful ice cream cone <3.*" "<0> <3 <8>" unquote.
Responsibility, Manager To Blame = Sister K. White <3
Back Story = A magical maiden, Augmented Biotech Human Female (Aug-Girl+), from... "All Your Base Are Belong To Us, Planet Uranus"... ...that believes in creating communities with the heartfelt love of theater performing arts and welcoming other Utopians from all kinds of wonderful diverse background. <3.
Personal Perks = Utopian Community Service, Lord Jesus Christ, House Of Worship, Prayer, all draw all draw all draw, or all write all write all write, all draw and all write, all public plus private sector donations welcome in The Kingdom Of Heaven <3!, 4-D Music Video provided by Clean-up Crew Unit 6 (six), Unit 9 (nine), Honest Ape and Walking Dog in the Church's Basement <3.
Location = W("West Side"), Christian Church, pass Court House <3

Valentina Vita (V. V.); Traveling Cosplayers And Kosplayers;
"(O_O) Ah... Super-Hero Sister! Yea, Elle was right... the yellow on her arms and legs do give it away when she is wearing her... **Nun clothes**... (T_T) But she was so **helpful** and **nice**... Like many **nuns** all around the **planet**. ...Anymore on this **Bullet Board System**...? Hm... There is no **Police Department volunteer**, interesting... Been seeing a bunch of **public sector** volunteer... Not sure if I have seen **private sector** volunteer. Going to keep on seeing what else is in the **Downtown District...** (=_=) Need to find a place to change out of my **KissShot**-AcerolaOrion-**HeartUnderBlade** cosplay... and into my red comfortable spandex clothes..."

The music "**Final Fantasy VII Dirge Of Cerberus' Turks 101**" plays in the **Utopia A.R. Holo HUD** background and is set on loop.

<u>Private Sector</u>, Paid Employee(s), Volunteer Worker(s),

"Kay's Private Security Services L.L.C., Research And Development (R&D) Division"
Chicago, The Last Animal Sanctuary Of Illinois, Utopian Sovereignty of North America. (U.S. of N.A.)

<u>Paid Employee(s)</u> = All, Aug Nat Human(s), must be U.S. Citizen(s), no exceptions.

<u>Volunteer Worker(s)</u> = U.S. Citizen(s), and or U.S. Citizen(s) petitioning Alien(s) only.

<u>Soap Opera Speech Box</u> =
Quote, "*Do not trust public security? Invest in private security, we negotiate the rates.*"

<u>Responsibility, Manager To Blame</u> = C.E.O. Agent Kay.

<u>Back Story</u> = The Koolest Society.

<u>Location</u> = Downtown District, Black Biotech High Rise Structure Of 25 Story Floors.

The music "**Nier Automata's The Tower**" plays in the **Utopia A.R. Holo HUD** background and is set on loop.

<u>Private Sector</u>, Paid Employee(s), Volunteer Worker(s),

"**Elysium Corporation, Biotech Dome Sector**" (U.S. of N.A.)
Chicago, The Last Animal Sanctuary Of Illinois, Utopian Sovereignty of North America.

<u>Paid Employee(s)</u> = All, Aug-Nat Human(s), must be U.S. Citizen(s), no exceptions.

<u>Volunteer Worker(s)</u> = U.S. Citizen(s), and or U.S. Citizen(s) petitioning Alien(s) only.

<u>Soap Opera Speech Box</u> =
Quote, "*Anything you want, we give you… Elysium, The Magical Corporation Of Desire.*"

<u>Responsibility, Manager To Blame</u> = Director: Doctor Froste.

<u>Back Story</u> = Established and founded in the year 2077 of 21st (twenty first) century.

<u>Location</u> = Downtown District, White Biotech High Rise Structure Of 99 Story Floors.

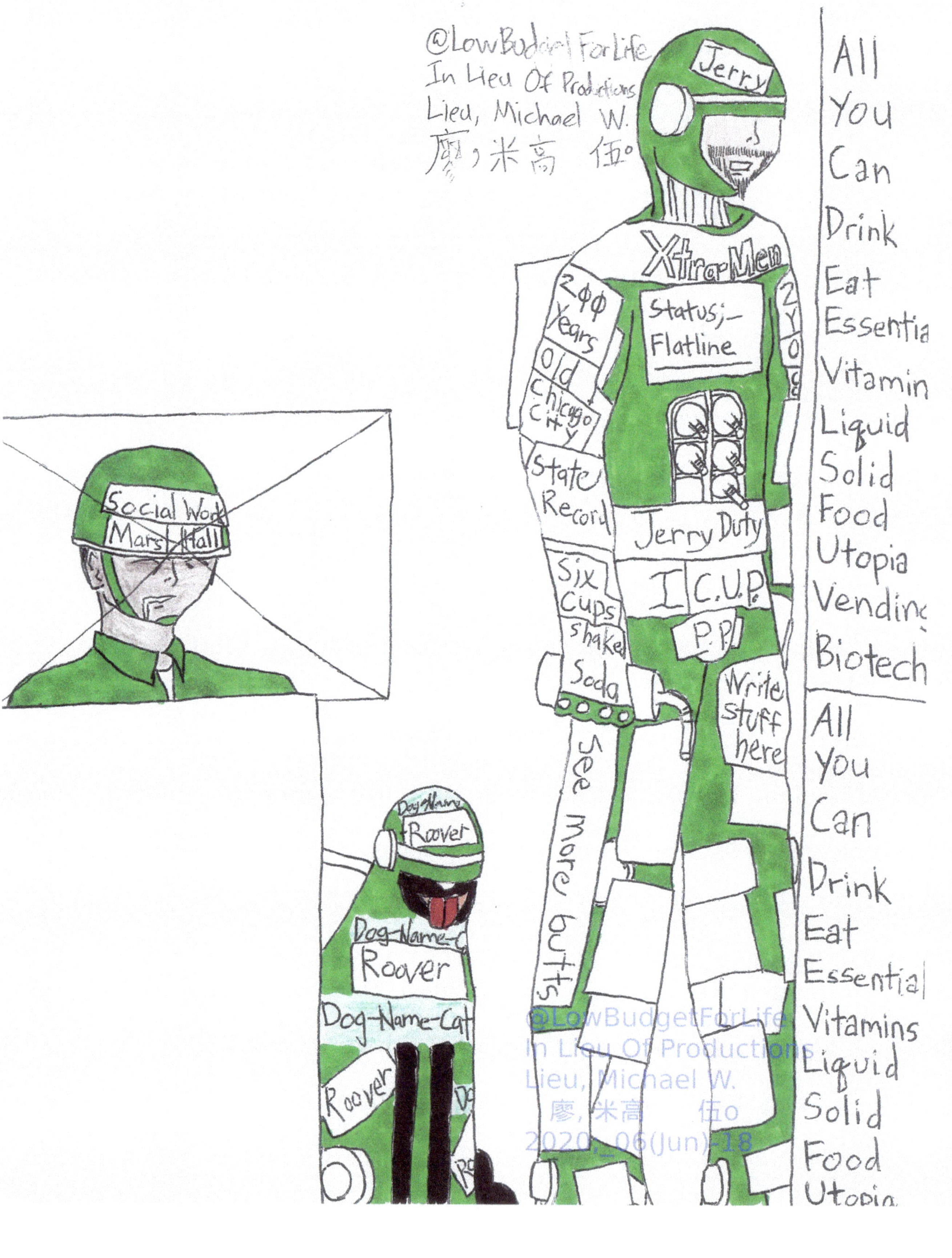
In Lieu Of Productions
Lieu, Michael W.
Jerry
XtraMen
200 Years Old Chicago City
Status;-
Flatline
State Record
Jerry Duty
Six Cups Shake Soda
I C.U.P.
P.P.
Write stuff here
See more butts
Social Work
Mars Hall
Dog Name
Roover
Dog-Name-Cat
Roover
Dog-Name-Cat
Roover
All You Can Drink Eat Essentia Vitamin Liquid Solid Food Utopia Vending Biotech
All You Can Drink Eat Essential Vitamins Liquid Solid Food
@LowBudgetForLife
In Lieu Of Productions
Lieu, Michael W.
廖, 米高　伍o
2020_06(Jun)-18

Alien Valentina Vita, Traveling Cosplayer(s) And Kosplayer(s)
2154;_ Chicago, Illinois' Intranet, Wireless Broadcasting Frequencies Online

V. V. Augmented Reality Holographic Heads Up Display A.R. Holo HUD		
Actual Legal Given Name;_ **Mars Hall** (goes by **Hall**) pronounced, "**ma-arghs'** **'ha-all**"	Actual Legal Given Name;_ **Rover Dog-Name-Cat** (goes by **Rover**) pronounced, "**row-ver**"	Actual Legal Given Name;_ **Jerry Duty** (goes by **Jerry Duty**) pronounced, "**je-re'** **'doo-te**"
Physical Plus Psych Age;_ natural 30s + 30 years old	Physical Plus Psych Age;_ augmented 20 + 10 years old	Physical Plus Psych Age;_ augmented 200 + 100 age years old
Employment;_ **Public Service, Voluntold Service, Jerry Duty** Tax Free Earnings $123,456.78 United States Dollars	Employment;_ Chicago Controller City Mayor Anti Robotics Movement (ARM)	Employment;_ Chicago Controller City Mayor Anti Robotics Movement (ARM)
Ethnicity;_ Latino	Ethnicity;_ Augmented Doggie	Ethnicity;_ Caucasian
Known Address(es);_ Chicago **Elysium Animal Sanctuary** Downtown District [*Aug+*]	Known Address(es);_ Chicago **Elysium Animal Sanctuary** Downtown District [*Aug+*]	Known Address(es);_ Chicago **Elysium Animal Sanctuary** Downtown District [*Aug+*]
Social Status;_ **Public Service, Voluntold Service, Jerry Duty** **Xtra-Men**	Social Status;_ **Public Service, Voluntold Service, Jerry Duty** **Xtra-Men**	Social Status;_ **Public Service, Voluntold Service, Jerry Duty** **Xtra-Men**
[*Aug+*]		
V. V. Augmented Reality Holographic Heads Up Display A.R. Holo HUD		

2154;_ Chicago, Illinois' Intranet, Wireless Broadcasting Frequencies Online
Alien Valentina Vita, Traveling Cosplayer(s) And Kosplayer(s)

Jerry Duty; Self-Governing Sovereignty (S) Voluntold Service Jerry Duty, Xtra-Men;

"(*panicking*) I need **Shaked-Soda**! Please!! I need **Shaked-Soda**!
Cat be a good Dog and get me **Shaked-Soda**! Please!!
Cat be a good Dog and get me **Shaked-Soda**!
I need **Shaked-Soda**! Please!! I need **Shaked-Soda**!
Cat be a good Dog and get me **Shaked-Soda**! Please!!
Cat be a good Dog and get me **Shaked-Soda**!
I need **Shaked-Soda**! Please!! I need **Shaked-Soda**!
Cat be a good Dog and get me **Shaked-Soda**! Please!!
Cat be a good Dog and get me **Shaked-Soda**!
I need **Shaked-Soda**! Please!! I need **Shaked-Soda**!
Cat be a good Dog and get me **Shaked-Soda**! Please!!
Cat be a good Dog and get me **Shaked-Soda**!
I need **Shaked-Soda**! Please!! I need **Shaked-Soda**!
Cat be a good Dog and get me **Shaked-Soda**! Please!!
Cat be a good Dog and get me **Shaked-Soda**!
I need **Shaked-Soda**! Please!! I need **Shaked-Soda**!
Cat be a good Dog and get me **Shaked-Soda**! Please!!
Cat be a good Dog and get me **Shaked-Soda**!
I need **Shaked-Soda**! Please!! I need **Shaked-Soda**!
Cat be a good Dog and get me **Shaked-Soda**! Please!!
Cat be a good Dog and get me **Shaked-Soda**!
I need **Shaked-Soda**! Please!! I need **Shaked-Soda**!
Cat be a good Dog and get me **Shaked-Soda**! Please!!
Cat be a good Dog and get me **Shaked-Soda**!
I need **Shaked-Soda**! Please!! I need **Shaked-Soda**!
Cat be a good Dog and get me **Shaked-Soda**! Please!!
Cat be a good Dog and get me **Shaked-Soda**!
I need **Shaked-Soda**! Please!! I need **Shaked-Soda**!
Cat be a good Dog and get me **Shaked-Soda**! Please!!
Cat be a good Dog and get me **Shaked-Soda**!
I need **Shaked-Soda**! Please!! I need **Shaked-Soda**!
Cat be a good Dog and get me **Shaked-Soda**! Please!!
Cat be a good Dog and get me **Shaked-Soda**!
I need **Shaked-Soda**! Please!! I need **Shaked-Soda**!
Cat be a good Dog and get me **Shaked-Soda**! Please!!
Cat be a good Dog and get me **Shaked-Soda**!
I need **Shaked-Soda**! Please!! I need **Shaked-Soda**!
Cat be a good Dog and get me **Shaked-Soda**! Please!!
Cat be a good Dog and get me **Shaked-Soda**!"

Utopian, Augmented Natural Animal Kingdom Plus Human Beings Coexistence
2154;_ Chicago, Illinois' Intranet, Wireless Broadcasting Frequencies Online

Elysium Corporation Augmented Reality Holographic Heads Up Display A.R. Holo HUD			
All You Can Drink Eat Essential Vitamins Liquid Solid Food Utopia Vending Biotech Mixed Drink Selection;_ "Antioxidants Super Foods Spices"			
Cloves Among the three spices in antioxidant concentration. Contains eugenol, carophyllin, gallic acid, eugenin. Inhibits oxidation of L.D.L. cholesterol, an early step in atherosclerosis.	Cinnamon An antioxidant powerhouse, compounds protect against inflammation, heart disease, Alzheimer's. Polyphenols, increases insulin activity, aiding diabetes control. Boosts attention.	Allspice Contains more than a dozen antioxidant, including eugenol, quercetin, ellagic and gallic acids. Ellagic acid has anticancer effects, decreases proliferation of pancreatic cancer cells.	Saffron Loaded with carotenoids, which possess anticancer activity and aid immune function. Also protects blood fats against oxidation, a step in the development of cardiovascular disease.
Fennel Contains flavonoids and tannins. Stimulates bile acid production in the liver, giving it a major role in fat digestion and absorption. Also contains anethole, reduce inflammation and retard cancer.	Oregano Another antioxidant powerhouse loaded with phenols. Rosmarinic acid has demonstrated antimicrobial activity against food-borne pathogens, and can be used topically as an herbal antiseptic.	Cumin Seeds have significant antiglycemic effect and help reverse metabolic abnormalities in diabetes. Also stimulate bile acid production and its secretion into bile.	Garlic Pungent sulfur compounds promote heart via beneficial effects on lipid metabolism and reduction of cholesterol levels. May also reduce blood pressure and retard cancer growth.
Mustard [Aug+]	Rosemary [Aug+]	Mint [Aug+]	Turmeric [Aug+]
Basil [Aug+]	Ginger [Aug+]	Red Pepper [Aug+]	Black Pepper [Aug+]
[Aug+]			
Elysium Corporation Augmented Reality Holographic Heads Up Display A.R. Holo HUD			

2154;_ Chicago, Illinois' Intranet, Wireless Broadcasting Frequencies Online
Utopian, Augmented Natural Animal Kingdom Plus Human Beings Coexistence

Retired C.C.P.D. SWAT, Alfa ToolPerson "Sergeant Soil" Officer Ogden, Deanna
Constant Toilet Paper (T.P.) Watch, Holo HUD Photo By BladeMaster Elle K. Mcgee

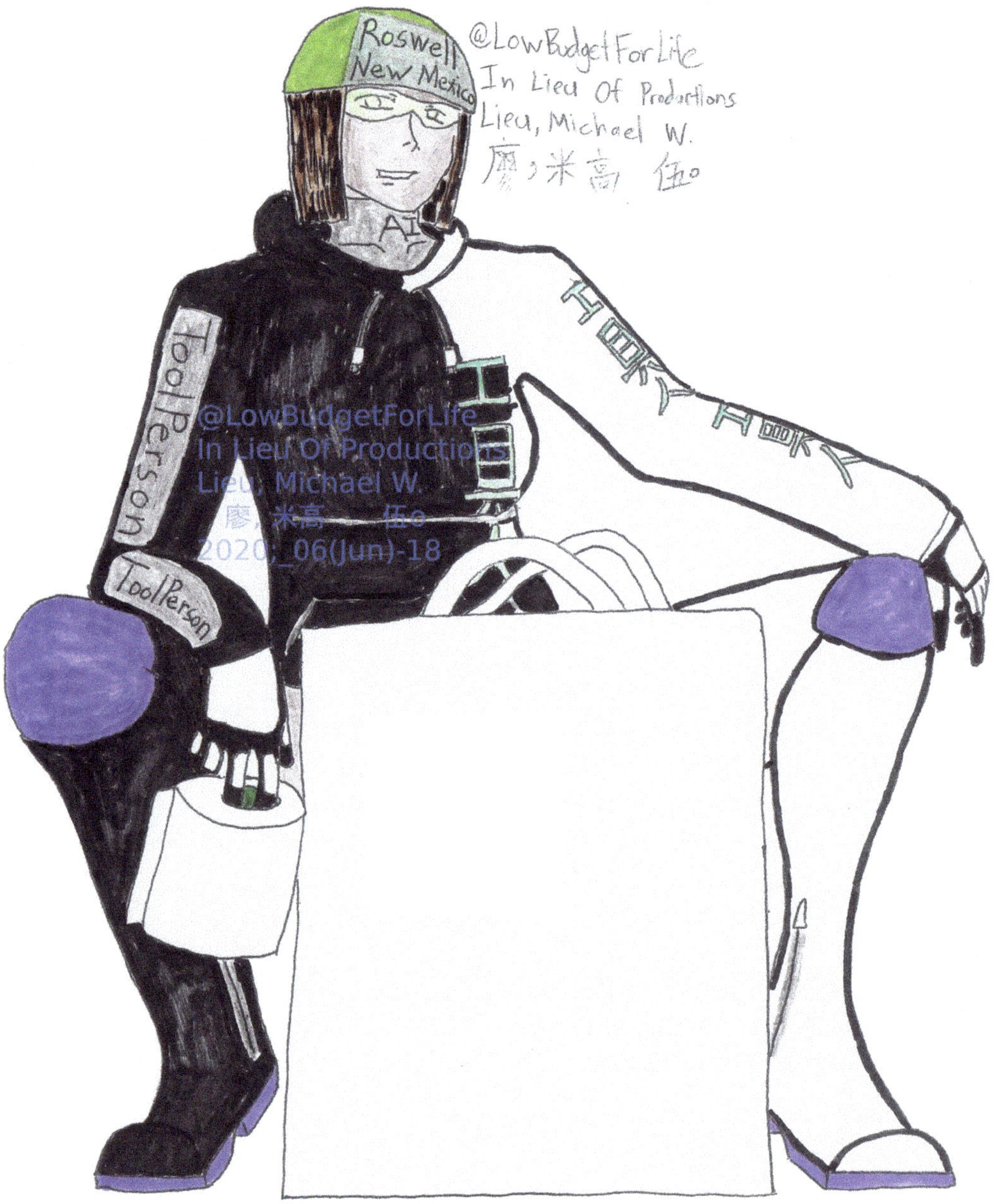

Constant Toilet Paper (T.P.) Watch, Holo HUD Photo By BladeMaster Elle K. Mcgee
Retired C.C.P.D. SWAT, Alfa ToolPerson "Sergeant Soil" Officer Ogden, Deanna

Alien Elle K. Mcgee, BladeMaster 刀大師, Celebrity Cook-Chef Martial Artist
2154;_ Chicago, Illinois' Intranet, Wireless Broadcasting Frequencies Online

刀 Augmented Reality Holographic Heads Up Display A.R. Holo HUD	
Actual Legal Given Name;_ **Ogden, Deanna** (goes by **Officer Ogden**) pronounced, "**aug-dun,' 'dee-na**"	Alias Network Nickname;_ Sergeant Soil, "I need to take a huge shit.", My Stomach, Full Of Shit, Toilet Paper, Alfa ToolPerson, Alfa Tool, T.P. Watcher, Toilet Paper Watcher, [*Aug+*]
Social Status;_ Retired Chicago Police Department SWAT Retired Roswell Police Department SWAT 2136-2140, U.S. Department Of War, Navy Sailor, E-4 Master-At-Arms M.P.	Community Network(s);_ Blade Runners and retirement **Captain Rage's** **Sexy Skintight SWAT Suit Photo Ops**
Physical Plus Psych Age;_ natural 40 + 50 years old	Alive;_ Heart-Rate natural
Known Address(es);_ **Roswell, New Mexico, U.S.A.** **Norfolk, Virginia, U.S.A.** [*Aug+*]	Personal Protective Equipment (P.P.E.);_ Partial Skintight SWAT Suit Partial Ballistic Weave Partial Energy Transfer [*Aug+*]
Family Member(s);_ Green Space Aliens From Planet Uranus [*Aug+*]	Allergies;_ Seafoods, Eggs, Shellfish, Fish, Anal Probing, Buttplug, Sex Toys, [*Aug+*]
[*Aug+*]	
刀 Augmented Reality Holographic Heads Up Display A.R. Holo HUD	

2154;_ Chicago, Illinois' Intranet, Wireless Broadcasting Frequencies Online
Alien Elle K. Mcgee, BladeMaster 刀大師, Celebrity Cook-Chef Martial Artist

Valentina Vita (V. V.); Traveling Cosplayers And Kosplayers;
"(O_o) So, this **Holo Photo** was taken by Elle when she found the **person** that **watches**... **Toilet Paper**... She took a photo of **her**... Hm... That explains where the **Toilet Paper** is In Chicago **Animal** Sanctuary, the **Downtown District**. I do not see a **single toilet** since I been here... ...**Officer Ogden** really **watching** over all those **Toilet Papers. She** must be **watching** me now...? Hm... **She** must have **Police Department SWAT Suit**..."

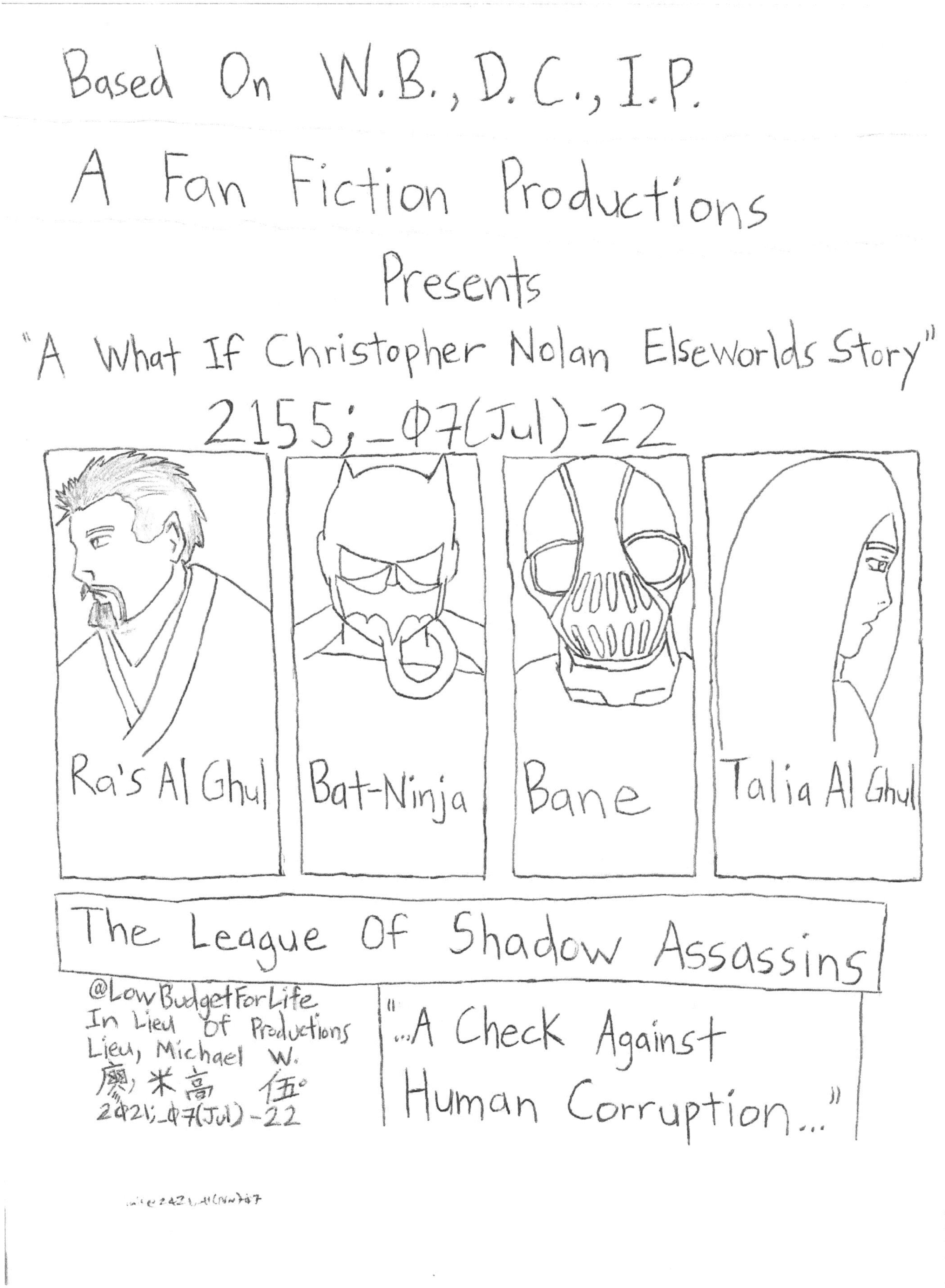
Based On W.B., D.C., I.P.
A Fan Fiction Productions
Presents
"A What If Christopher Nolan Elseworlds Story"
2155;_Φ7(Jul)-22
Ra's Al Ghul
Bat-Ninja
Bane
Talia Al Ghul
The League Of Shadow Assassins
@LowBudgetForLife
In Lieu Of Productions
Lieu, Michael W.
廖, 米高 伍°
2021;_Φ7(Jul)-22
"...A Check Against
Human Corruption..."

Fourth Dimension Movie Theater; Fan Fiction Productions Teaser Utopia A.R. Holo HUD Projection;
"(*dramatic male voice over*) Coming next summer in... **twenty one fifty five... July twenty two**... Based on **Warner Brothers Entertainment, D.C. Comics' Intellectual Property...** A Fan Fiction Productions: **A What If Christopher Nolan Elseworlds Story**... Where one of Planet Earth's **old world Western society Mythology Super-Hero Vigilante... Batman**, decided to officially join **The League Of Shadow Assassins** and embrace the **dark desires** within... as **The League's Bat-Ninja Operative**!"

The **Fourth Dimension Movie Theater** shows visuals with **four characters** dressed in **black custom operative clothing at night time**, making it tricky to see **their** faces fully.

Ra's Al Ghul Method Actor; Fan Fiction Productions Of D.C. W.B. I.P. Christopher Nolan's Batman;
"(*lost*) ...Sometimes the **suffering is so unbearable** that **you** wish... **You** wish... **You never met the person** in the **first place,** so **you** would be **spared** all the **pain**..."

Talia Al Ghul Method Actor; Fan Fiction Productions Of D.C. W.B. I.P. Christopher Nolan's Batman;
"(*concerning*) **Father... You** clearly do not mean **mother**. ...**We all miss her dearly**!"

Bane Method Actor; Fan Fiction Productions Of D.C. W.B. I.P. Christopher Nolan's Batman;
"(*masked augmented male voice*) A fire rises... **The League** shall return to the **corrupt Gotham city** to **properly finish the job**! With... **economics, infrastructure**!!"

Bat-Ninja Wayne Method Actor; Fan Fiction Productions Of D.C. W.B. I.P. "Chris-N's" Batman;
"(*thinking*) ...If **I push someone down in self-defense** from a **high rise building level** and there is no gravity cushion below, that means the **high fall gravitation pull flatlined the person**... (*deep breath*) Wow... same **applies** to **firearms**... **Fired ballistic bullets flatlined the person**, not the **firearms** without the **bullets**..."

Ra's Al Ghul Method Actor; Fan Fiction Productions Of D.C. W.B. I.P. Christopher Nolan's Batman;
"(*calm*) **Come children**... There is much to learn and do with **The League**."

A **female figure in a black catsuit** steps out and cuts the line with **her custom black high heels blades**. **She** then uses **her smoke screen, night vision goggles** to spot **four figures** of **The League Of Shadow Assassins**. The scene fades to **black**.

Dog Master Method Actor; Fan Fiction Productions Of D.C. W.B. I.P. Christopher Nolan's Batman;
"(*yelling voice over*) ...And that is why **you** bring **doggies**... **My doggies** are **hungry**!!"

A Fan Fiction Productions' A What If Christopher Nolan Elseworlds Story Fourth Dimension Movie Theater teaser trailer ends in a fade to **black** transition and shows the upcoming feature **fan fiction movie's** release date of "**2155, 07 (July) - 22**".

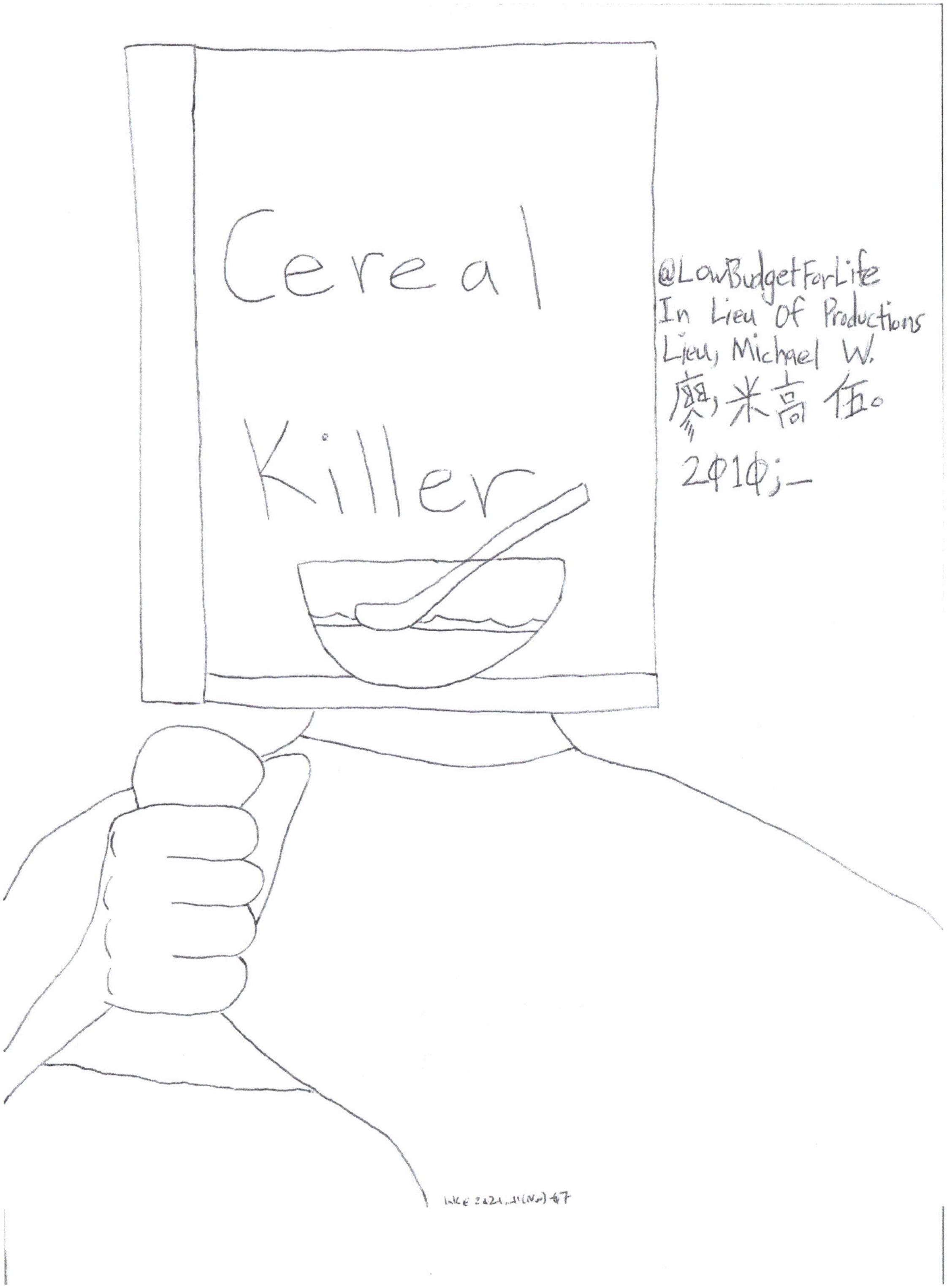

Coming Soon

2154;_Arcadia \ Valentina Vita (V. V.) 'Healthy-Strong Life' \ Left to Right(L->R)

A Dream Like Movie Directed By An Artificial Intelligence (A.I.) Personality,
Conscientiousness "Conscientiousness Cat/Kat's XXXX;_"
The Search For Magical Maiden(s) To Unify And Save The Solar System Literally Series

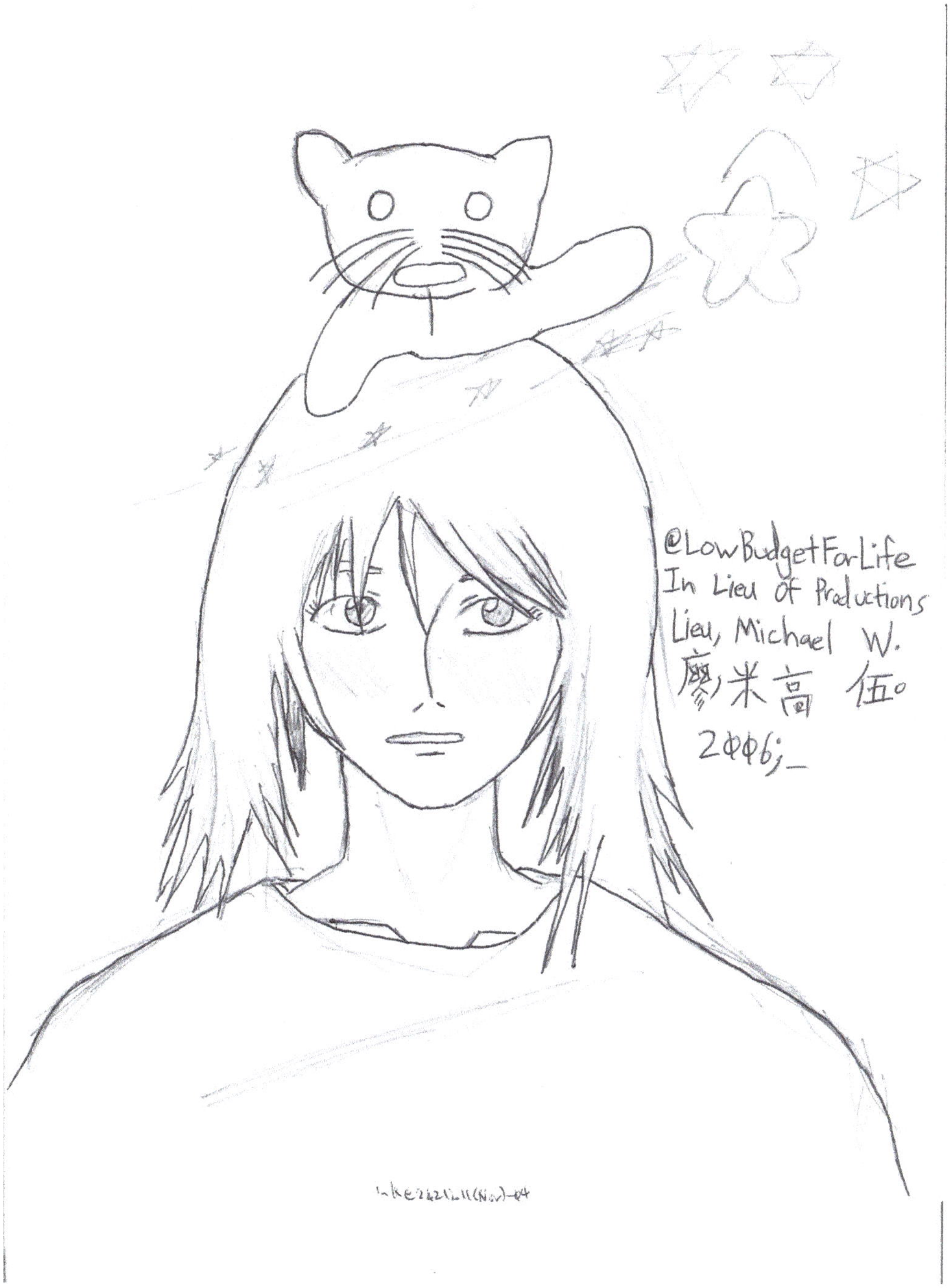

"Meow. Meow! Me! Ow!" / Color Code; Orange Cat/Kat
Coming Sooner Than You Think When You Fall Asleep In Rapid Eye Movement (REM).

2154;_Arcadia \ Valentina Vita (V. V.) 'Healthy-Strong Life' \ Left to Right(L->R)

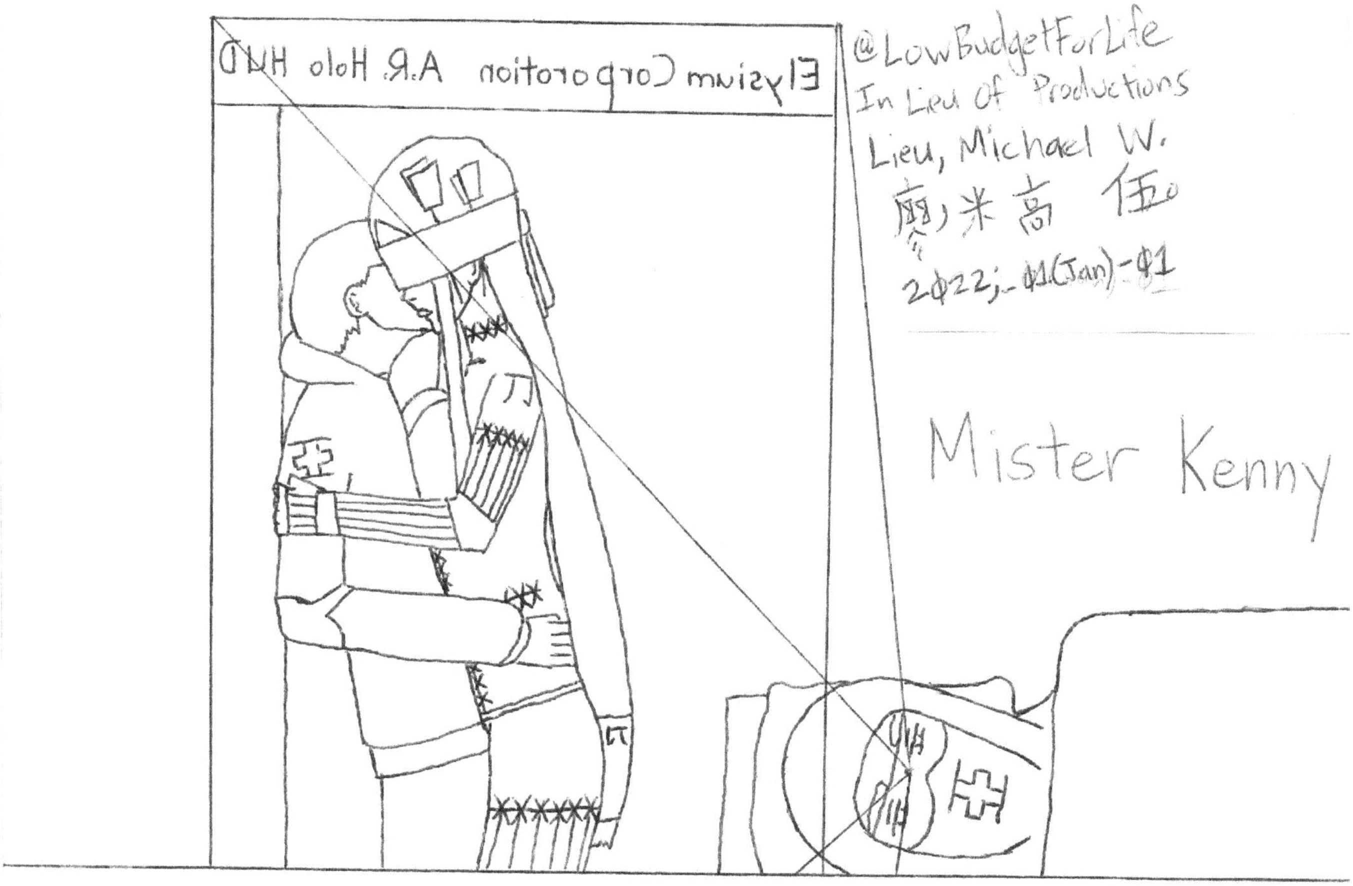

4D Movie Theater Box Office

Four(th) Dimension

High Definition Digital Versatile Disc (H.D. D.V.D.) Archive

Citizen Mister Kenny, Chicago City Official + Elysium Corporation Employee, Dome 2154;_ Chicago, Illinois' Intranet, Wireless Broadcasting Frequencies Online

Elysium Corp Augmented Reality Holographic Heads Up Display A.R. Holo HUD	
Actual Legal Given Name;_ Mister Kenny (goes by Mister Kenny) pronounced, 'mist-ter' 'ken-knee'	Physical Plus Psych Age;_ natural 30 + 17 years old
Employment;_ Elysium Corporation 亞 Establish 2077 The Magical Corporation Of Desire Biotech Dome Sector Station: Planet Earth, North America, Illinois, Chicago, **Animal** Sanctuary (S) Paid Employ Worker: 7,777,777.00 Elysium Corporation Kash-Kredit(s) (E.C. 亞 K-K) Per Hour Salary Rate Accounting System Elysium Corporation's Official Paid Employ Worker Job Title: "Acting Authority: Human And Animal Breeder(s) Builder(s) For Augmented Bio**tech** And Or Natural Unmodified Fertility **Parenting License**" Downtown District, High Tower Corporate Divisions: 01-09, 10-19, 60-69	
[*Aug+*]	
Elysium Corp Augmented Reality Holographic Heads Up Display A.R. Holo HUD	

2154;_ Chicago, Illinois' Intranet, Wireless Broadcasting Frequencies Online
Citizen Mister Kenny, Chicago City Official + Elysium Corporation Employee, Dome

Valentina Vita (V. V.); Traveling Cosplayers And Kosplayers;
"(O_O) Oh... Um... Is this... the... **Fourth Dimension Movie Theater Archives**...?"

Mister Kenny; Human And Animal Breeder(s) Builder(s) Property Of Elysium Corporation;
"(*tired*) ...Yea. So much **breeding**, **mating press, copulation, porn, sex intercourse**... ...Must meet Chicago Controller City Mayors' requests... ...**Our systems** are automated algorithm handles the **Four D Movie Theater Archives**... I am just here to make to handle any auto al problems... Enjoy. Yea, I know. I'm the human and animal breeder builder..."

Automated Algorithm; Human Voice Over Emulation Based On Elysium Corporation Archives;
"(*monotone female voice over*) Please stand by for biotech dome airlock decontamination. Thank you for your patience and cooperation **Utopian**."

2154;_Arcadia \ Valentina Vita (V. V.) 'Healthy-Strong Life' \ Left to Right(L->R)

Chicago, The Last Animal Sanctuary Of Illinois

4D (Four(th) Dimension) Movie Theater Catalog
"Feel Free To Copy, Consume, Play As Much To Your Desire..."
Recorded Entertainment From The 22nd Century Phenomenon
Broadcasting Biotech Tower(s) UltraViolet (U.V.) Rays

In High Definition Digital Versatile Disc (H.D.D.V.D.) Old World Archives

Made Possible By

Yellow-Mellow; Supa-Hiro Obama@deep Problem Solvers (°o°) (O :)

And

Elysium Corporation

Established 2077
The Magical Corporation Of Desire

Mister Kenny, Paid Employ Worker
"Acting Authority: Human And Animal Breeder(s) Builder(s) For Augmented Biotech And Or Natural Unmodified Fertility **Parenting License**"

Downtown District, High Tower Corporate Divisions

Floors	01 to 09	Animal Human Health Care
Floors	10 to 19	Human History Introduction Indoctrination
Floors	60 to 69	Human Sexuality, Copulation, Augmentation

Corporate America(s), Eur-Asia, Africa, Australia, And Planet Uranus
Serving America(s), Eur-Asia, Africa, Australia Continent(s) And Uranus

"The New World 2121 Movement Of The 22nd Century, Utopian Sovereignty of North America (U.S. of N.A.)"

4D (four(th) Dimension) Movie Theater Archives

Science-Fiction Genre, "The Literature Of Ideas" In H.D.D.V.D.s 1080p Resolution

Film Movie(s), TeleVision (T.V.) Show Series = Passive Viewing
Video Game(s), Sandbox Simulation(s), Comic Book(s), Manga(s) = Active Participation

Explanation, The Numbers Or X "2001;_" Is The Year The Recorded Entertainment Is Set In
The Recorded Entertainment In The Same Boxes Means It Is Set In The Same Story Universe
Underline Recorded Entertainment Title Is The Story Universes' Name Series

2001: A Space Odyssey (film movie), 2010;_The Year We Make Contact (film movie)	
1912;_BioShock Infinite The Complete Edition (video game) 1960s;_BioShock 1, 2 Ultimate Rapture Edition (video game)	
1927;_Metropolis' The Complete (film) | 1979-1985;_**Mad Max** 1, 2, 3 (film)	1951;_The Day The Earth Stood Still (film movies) 2008;_The Day The Earth Stood Still
1985-1955-2015-1885;_Back To The Future Part 1, 2, 3 (film movie) **1985;_Back To The Future The Game (video game) Part 4**	1987-1990-2010-2018;_Predator; 1 (film movie), Predator 2 (film movie), Predators (film movie), The Predator (film movie)
1993-1997-2001-2015-2018-2022;_Jurassic Park: 1, 2 The Lost World, 3 (film movies), Jurassic World: 1 (film movie), 2 Lost Kingdom (film movie), 3 Dominion (film movie)	
1997;_Contact (film movie)	2006;_The Adjustment Bureau (film movie)
2006;_Night At The Museum (film movie)	2009;_Moon (film) | 2009;_Gamer (movie)
2010;_District 9 (film movie)	2015;_Chappie (film movie)
2013;_Her (film movie)	1964-2013;_Tomorrowland (film movie)
2013-2020;_Pacific Rim (film movie)	2016;_Hardcore Henry (film movie)
1954-1999-2014;_Godzilla (film movie) 2013-2033;_The Last Of Us (video game)	2012-2019;_The Walking Dead The Telltale Definitive Series (video game)
1998-2020s;_Biohazard Resident Evil: Origins Collection (video game), 3 (video game), 2 (video game), Degeneration (film movie), 4 (video game), Infinite Darkness (T.V. Show Series), Revelations 1 (video game), 5 (video game), Revelations 2 (video game), Damnation (film movie), 6 (video game), Vendetta (film movie), 7 (video game), 8 Village (video game), 9 (video game)	
2014;_Edge Of Tomorrow (film movie)	2014;_Interstellar (film movie)
2016;_Hardcore Henry (film movie)	2015;_The Martian (film movie)

2015-2018;_Humans (T.V. Show Series)	2016-2020s;_Westworld (T.V. Show Series)
2017;_Power Rangers remake (film movie)	2018;_Upgrade (film movie)
2018;_Searching (film movie)	2018;_XCOM 2 Collection (video game)
2020;_Tenet (film movie)	2017-2019;_The Running Man (film movie)
2019;_Blade Runner The Final Cut (film movie), 2022;_Blade Runner Blackout (film movie) 2032;_Blade Runner Black Lotus (T.V. Show Series) 2036;_Blade Runner Nexus (film movie) 2048;_Blade Runner Nowhere To Run (film movie), 2049;_Blade Runner 2049 (film movie) 2099;_Blade Runner 2099 (T.V. Show Series), 2089;_Prometheus (film movie), 2104;_Alien Covenant (film movie) 2122;_Alien (film movie), 2179;_Aliens (film movie)	
2010-2016;_Rise Of The Planet Of The Apes (film movie), 2026;_Dawn Of The Planet Of The Apes (film movie), 2028;_War For The Planet Of The Apes (film movie)	
2027;_Children Of Men (film movie)	2028;_RoboCop remake (film movie)
2027-2029;_Deus Ex: Human Revolution (sandbox simulation), Mankind Divided (sandbox simulation), 2052;_Deus Ex Original (sandbox simulation)	
1984-2029;_The Terminator (film movies) 1991-2029;_Terminator 2 Judgment Day Unrated	
2031;_Snowpiercer (film movie)	2032;_Prey (sandbox simulation)
2035;_I, Robot (film movie)	2038;_Detroit Become Human (video game)
2044;_Automata (film movie)	2044;_Looper (film movie)
2045;_Ready Player One (film movie)	2054;_Minority Report (film movie)
2054-2061;_Call Of Duty Advanced Warfare (video game)	2057;_Sunshine (film movie)
	2064;_Read Only Memories (video game)
2065;_Final Fantasy The Spirits Within (film movie)	2077;_Cyberpunk 2077 (sandbox simulation)
	2077;_Oblivion (film movie)
2077-2287;_Fallout 4 (four) (sandbox simulation), 2277;_Fallout 3 (three) Game Of The Year Edition (sandbox simulation), 2281;_Fallout New Vegas Ultimate Edition (sandbox simulation),	
2084;_Total Recall (film movie)	2084;_Remember Me (video game)
2084;_Observer System Redux (video game)	2092;_Mister Nobody (film movie)
	2100;_A.I. Artificial Intelligence (film movie)
2145;_Doom 3 B.F.G. Edition (video game	2154;_Avatar (film movie)
2183-2186;_Mass Effect: **2 (video game)** Paragon Lost (film movie) **3 (video game)**	2154;_Elysium (film movie)
	1999-8888;_Animatrix (film movie) The Matrix

1492-1986-2016;_Assassin's Creed (film movie)

1476-1499-1511-1512-2012;_Assassin's Creed Ezio Collection (video game)

8888;_Axiom Verge (video game)

2200;_Forbidden Planet (film movie)

1914-2214;_The Fifth Element (film movie)

2274;_Logan's Run (film movie)

2233-2259-2263;_Star Trek: 1 alternate history (film movie), 2 Into Darkness (film movie), 3 Beyond (film movie),

XXXX;_Star Wars: "A Science-Fiction Space Opera Mythology Genre(s) In A Far Galaxy"
Episode 1 (I) The Phantom Menace (film movie)
Episode 2 (II) Attack Of The Clones (film movie)
Episode 3 (III) Revenge Of The Sith (film movie)
Solo: A Star Wars Story (film movie)
Rogue One: A Star Wars Story (film movie)
Episode 4 (IV) A New Hope (film movie)
Episode 5 (V) The Empire Strikes Back (film movie)
Episode 6 (VI) Return Of The Jedi (film movie)
Episode 7 (VII) The Force Awakens (film movie)
Episode 8 (VIII) The Last Jedi (film movie)
Episode 9 (IX) The Rise Of Skywalker (film movie)

1849-1936-1973-2012-2144-2321;_Cloud Atlas (film movie)

2005-2505;_Idiocracy (film movie)

2563;_Alita: Battle Angel (film movie)

2688-1988;_Bill And Ted's Excellent Adventure (film movie)
2691-1991;_Bill And Ted's Bogus Journey (film movie)
2720-2020;_Bill And Ted's Face The Music (film movie)

25XX;_StarCraft: 1, Brood War, 2 Wings Of Liberty, 2 Heart Of The Swam (video game)

2508;_Dead Space 0 (zero) Extraction (video game)
2108;_Dead Space 1 (one) Original (video game)
2511;_Dead Space 2 (two) Limited Edition (video game)
2314-2514;_Dead Space 3 (three) Limited Edition (video game)

2552-2557;_Halo: 0 (zero) Reach, 1 (one) Combat Evolved, 2 (two), 3 (three) O.D.S.T., 3 (three), 4 (four) (video games)

2805;_Walle (film movie)

????;_Hideo Kojima's Death Stranding Director's Cut (video game)

????;_Horizon Zero Dawn Complete Edition (video game)

????;_Adventure Time Pirates Of The Enchiridion (game)

1944-2121;_The Kold War (film movie) Selects; "Doktor Strangelove", "Wargames"

Christopher Nolan "Chris-N" (film movies); Following / Memento / Insomnia / Batman Begins / The Prestige / The Dark Knight / Inception / The Dark Knight Rises / Interstellar / Dunkirk / Tenet / Oppenheimer / X / . /

"The New World 2121 Movement Of The 22nd Century, Utopian Sovereignty of North America (U.S. of N.A.)"

4D (four(th) Dimension) Movie Theater Archives

Science-Fiction Genre, "The Literature Of Ideas" In H.D.D.V.D.s 1080p Resolution
Super-Hero Genre, "Super-Human Strength, Power Up, Augmentation, And Saving"
Vigilante Genre, "Take The Human Laws In To Your Own Hands When Society Fails"

Film Movie(s), TeleVision (T.V.) Show Series = Passive Viewing
Video Game(s), Sandbox Simulation(s), Comic Book(s), Manga = Active Participation

The Recorded Entertainment In The Same Boxes Means It Is Set In The Same Story Universe

Marvel + 20 th Century Fox Studios' X-Men Universe (Film Movies) Chronological Order;

2000-2018;_
X-Men Origins Wolverine 1 (one)
X-Men 1st (First) Class
X-Men 1 (one)
X2 (two) X-Men United
X-Men 3 (three) The Last Stand
The Wolverine 2 (two) Extended Cut
X-Men Days Of Future Past The Rogue Cut
Logan A.K.A. Wolverine 3 (three)
Deadpool 1 (one)
Deadpool 2 (two) Super Duper Cut

Marvel + Sony Columbia Pictures' Spider-Man Universe (Film Movies) Chronological Order;

2002-2020s;_
Spider-Man 1 (one)
Spider-Man 2.1 (two point one)
Spider-Man 3 (three) Editor's Cut
The Amazing Spider-Man 1 (one)
The Amazing Spider-Man 2 (two)
Venom 1 (one)
Venom 2 (two) Let There Be Carnage
Spider-Man 3 (three) No Way Home "Multi uniVerse (MultiVerse) Crossover Movie"
Morbius 1 (one)
Spider-Man 1 (one) Into The Spider-Verse
Spider-Man Across The Spider-Verse Part 1 (One), Part 2 (Two)
Venom 3 (three)
The Amazing Spider-Man 3 (three)
Morbius 2 (two) It's Morbin' Time

Marvel Studios' Marvel Cinematic Universe (M.C.U.) (Film Movies, Short Film Movies); Chronological Order...	*M.C.U. Spin-Off (T.V. Show Series) American Broadcasting Company (A.B.C.), Netflix, Disney Plus; Chronological Order...*
1995;_Captain Marvel 2011;_Iron Man 1(one) 2011;_The Incredible Hulk 2011;_Iron Man 2(two) 2011;_A Funny Thing Happened On The Way To Thor's Hammer 2011;_Thor 1(one) 2011;_The Consultant 2012-1943-1945-2012;_Captain America 1(one) The First Avenger 2012;_The Avengers 1(one) 2012;_Item 47(forty seven) 2012;_Iron Man 3(three) 2013;_All Hail The King 2013;_Thor 2(two) The Dark World 2014;_Guardians Of The Galaxy 1(one) 2014;_Guardians Of The Galaxy Volume 2(two) 2014;_Captain America 2(two) The Winter Soldier 2015;_Avengers 2(two) Age Of Ultron 2015;_Ant-Man 1(one) 2016;_Captain America 3(three) Civil War 2016;_Black Widow 2016;_Black Panther 1(one) 2016;_Spider-Man 1(one) Homecoming 2016;_Doctor Strange 1(one) 2017;_Thor 3(three) Ragnarok 2018;_Avengers 3(three) Infinity War 2018;_Ant-Man 2(two) And The Wasp 2018-2023;_Avengers 4(four) Endgame 2024;_Eternals 2024;_Shang Chi And The Legend Of The Ten Rings 2024;_Spider-Man 2(two) Far From Home 2024;_Spider-Man 3(three) No Way Home 2025;_Doctor Strange 2(two) Multiverse Of Madness 2025;_Thor 4(four) Love And Thunder 2025;_Black Panther 2(two) Wakanda Forever 2025;_Ant-Man 3(three) And The Wasp Quantumania 2025;_Guardians Of The Galaxy Volume 3(three) 2025;_The Marvels / 2025;_Blade 2025;_Captain America New World Order 2025;_Thunderbolts / 2026;_Fantastic Four	1946-1947;_Agent Carter Set After 2012-1943-1945-2012;_Captain America 1 The First Avenger (film) 2013-2020;_ Agents Of SHIELD, Set After 2012;_ The Avengers 1(one) (film) 2015-2019;_ Daredevil season 1(one) Jessica Jones season 1(one) Daredevil season 2(two) Luke Cage season 1(one) Iron Fist season 1(one) Defenders season 1(one) Daredevil season 3(three) Jessica Jones season 2(two) Punisher season 1(one) Punisher season 2(two) 2023-2025;_After Endgame WandaVision, The Falcon And The Winter Soldier, Loki Marvel What If, Hawkeye, Moon Knight, Ms. Marvel, She-Hulk: Attorney At Law, Guardians Of The Galaxy Holiday Special, Secret Invasion, Echo, Loki Season 2(two), Ironheart, Agatha: Coven Of Chaos, Daredevil: Born Again

Eastern Society Made 1978's Japanese Spider-Man A.K.A. Supaidaman (T.V. Show Series)
Disney + Pixar's The Incredibles 1(one) And The Incredibles 2(two) (Film Movies)
Eastern Society "Boku No Hero Academia" / Western Society "My Hero Academia" (T.V. Show Series) In Chronological Order My Hero Academia season 1(one) episodes: 1-9, 10-13, extras My Hero Academia season 2(two) episodes: 14-21, 22-25, extras, 26-34, 35-38, extras My Hero Academia Two Heroes (film movie), extras My Hero Academia season 3(three) episodes: 39-46, 47-50, extras, 51-57, 58-63, extras
Warner Brothers (W.B.) Detective Comics (D.C.) (Film Movies) The Lego Movie 1(one), The Lego Batman Movie, The Lego Movie The 2nd(Second) Part
W.B. D.C.'s **Watchmen The End Is Nigh (video game)** Watchmen Director's Cut (film movie), H.B.O.'s Watchmen (T.V. show series)
W.B. D.C.'s Batman The Animated Series (T.V. Show Series) Universe In Chronological Order Batman Mask Of The Phantasm (film movie) Batman The Animated Series season 1(one) disc 1(one) Batman The Animated Series season 1(one) disc 2(two) Batman The Animated Series season 1(one) disc 3(three) Batman The Animated Series season 1(one) disc 4(four) Batman The Animated Series season 1(one) disc 5(five) Batman The Animated Series season 2(two) disc 1(one) Batman The Animated Series season 2(two) disc 2(two) Batman Subzero (film movie) Batman The Animated Series season 3(three) disc 1(one) Batman The Animated Series season 3(three) disc 2(two) Batman The Animated Series bonus disc Batman And Harley Quinn (film movie) D.C. Harley Quinn (T.V. Show Series) Batman The Killing Joke (film movie) Batman Death In The Family Interactive Movie (interactive video game film movie) Batman Under The Red Hood (film movie) "Batman's Story Continues Years Later..." Batman Beyond The Complete Series disc 1(one) season 1(one) Batman Beyond The Complete Series disc 2(two) season 2(two) Batman Beyond The Complete Series disc 3(three) season 2(two) Batman Beyond The Complete Series disc 4(four) season 3(three) Batman Beyond Return Of The Joker (film movie) Batman Beyond The Complete Series disc 5(five) special features

W.B. D.C. Rocksteady Studios' Batman Arkham Universe In Chronological Order… Batman: Arkham Origins "Interactive Choose Your Adventure" (comic book) Batman: Arkham Origins (video game) Batman: Arkham Origins Blackgate (video game) Batman: Arkham Asylum, Return To Arkham (video game) Batman: Arkham City (comic book) Batman: Arkham City, Return To Arkham (video game) Batman: Arkham Knight (video game) Suicide Squad: Kill The Justice League (video game)
W.B. D.C. NetherRealm Studios' Injustice Universe In Chronological Order… Injustice: Year 1(one) volume 1(one), 2(two) (comic book) Injustice: Year 2(two) volume 1(one), 2(two) (comic book) Injustice: Year 3(three) volume 1(one), 2(two) (comic book) Injustice: Year 4(four) volume 1(one), 2(two) (comic book) Injustice: Year 5(five) volume 1(one), 2(two), 3(three) (comic book) Injustice: Gods Among Us Ultimate Edition (video game) Injustice: Ground Zero volume 1(one) (comic book) Injustice: Ground Zero volume 2(two) (comic book) Injustice: 2(two) volume 1(one) (comic book) Injustice: 2(two) volume 2(two) (comic book) Injustice: 2(two) volume 3(three) (comic book) Injustice: 2(two) volume 4(four) (comic book) Injustice: 2(two) volume 5(five) (comic book) Injustice: 2(two) volume 6(six) (comic book) Injustice: 2(two) Legendary Edition (video game) MultiVerse Intro Dialogue References From "Mortal Kombat Versus D.C. Universe" Mortal Kombat Versus D.C. Universe (video game)
W.B. D.C. Live-Action Tim Burton's Batman (Film Movies) Universe Left To Right (L→R) Batman, Batman Returns
W.B. D.C. Live-Action Christoper Nolan's Batman (Film Movies) Universe (L→R) Batman Begins, The Dark Knight, The Dark Knight Rises
W.B. D.C. Live-Action Matt Reeves' Batman (Film Movies) Universe (L→R) The Riddler: Year One, The Batman: Year Two, The Penguin (T.V. Show Series)
M. Night Shyamalan's Character Driven Super-Hero Vigilante (Film Movies) Universe (L→R) "Eastrail 177 Trilogy": Unbreakable, Split, Glass
James Gunn's Super-Hero Vigilante Dark Comedy Horror "Super" (*Film Movies*) "Brightburn"
Super-Hero Vigilante Dark Comedy "Kick-Ass" ← (*Film Movies*) → "Kick-Ass 2"

W.B. D.C. Live-Action Super-Hero Vigilante (Film Movies) Universe Chronological Order
2013-2020s;_ Man Of Steel Batman V. Superman Dawn Of Justice Ultimate Edition Wonder Woman 1 (one) Wonder Woman 2 (two) 1984 David Ayer's Suicide Squad 1 (one) Unrated Director's Cut Birds Of Prey And The Fantabulous Emancipation Of One Harley Quinn Zack Snyder's Justice League Aquaman 1 (one) Shazam! 1 (one) The Suicide Squad 2 (two), Peacemaker (T.V. Show) set after The Suicide Squad 2 (two) Black Adam Shazam! 2 (two) Fury Of The Gods Batgirl Aquaman 2 (two) And The Lost Kingdom The Flash
W.B. D.C. Live-Action Super-Hero Vigilante (T.V. Show Series) Universe Chronological Order **" Bold Text Is T.V. Show And Or Film Movie Series Name "** 2012-2020;_ **Arrow season 1(one)** disc: 1(one), 2(two), 3(three), 4(four) **Arrow season 2(two)** disc: 1(one), 2(two), 3(three), 4(four) **Arrow season 3(three)** disc: 1(one), 2(two), 3(three), 4(four) **The Flash season 1(one)** disc: 1(one), 2(two), 3(three), 4(four) **Constantine season 1(one)** disc: 1(one), 2(two), 3(three) **Arrow season 4(four)** disc: 1(one), 2(two), 3(three), 4(four) **The Flash season 2(two)** disc: 1(one), 2(two), 3(three), 4(four) **D.C.'s Legends Of Tomorrow season 1(one)** disc: 1(one), 2(two) **Arrow season 5(five)** disc: 1(one), 2(two), 3(three), 4(four) **The Flash season 3(three)** disc: 1(one), 2(two), 3(three), 4(four) **Supergirl season 1(one)** disc: 1(one), 2(two), 3(three) **D.C.'s Legends Of Tomorrow season 2(two)** disc: 1(one), 2(two), 3(three) **Freedom Fighters The Ray** (film movie) **D.C.'s Legends Of Tomorrow season 3(three)** disc: 1(one), 2(two), 3(three) **D.C.'s Legends Of Tomorrow season 4(four)** disc: 1(one), 2(two) **The Flash season 5(five)** disc: 1(one), 2(two), 3(three), 4(four) **Supergirl season 2(two)** disc: 1(one), 2(two), 3(three) **Batwoman season 1(one)** disc: 1(one), 2(two), 3(three), 4(four) Crisis On Infinite Earths disc: "A D.C. Crossover Event"

W.B. D.C. Live-Action Super-Hero Vigilante (T.V. Show Series) Universe Chronological Order
Crossover T.V. Show Series Title Specials In **"Bold And Quotes"**
<u>No Archive Copy In The Underline Titles</u>**.**

2012;_
Arrow season 1(one) "Intro Set-Up Arrow uniVerse (ArrowVerse) W.B. D.C. T.V.

2013;_
Arrow season 2(two) "**The Scientist**" Intro For **The Flash Origins**.

2014;_
Arrow season 3(three) "**Flash Versus Arrow**",
The Flash season 1(one) "**The Brave And The Bold**"
"Intro For **Legends Of Tomorrow Origins**"

2015;_
Constantine season 1(one)
Arrow season 4(four) "**Haunted Episode Crossover With Constantine**"
The Flash season 2(two) "Intro For Multiple uniVerse (MultiVerse) Origins"
"Intro For **Legends Of Tomorrow, Legends Of Today, Legends Of Yesterday**"

2016;_
Arrow season 5(five) "**Invasion!**", The Flash season 3(three) "**Invasion!**",
Legends Of Tomorrow season 1(one),
Supergirl season 1(one) "**World's Finest**" "**Invasion!**"

2017;_
<u>Arrow season 6(six), The Flash season 4(four)</u>, Legends Of Tomorrow season 2(two),
Supergirl season 2(two), Freedom Fighters Ray (film movie) "**Crisis On Earth-X**"

2018;_
<u>Arrow season 7(seven), The Flash season 5(five), Supergirl season 3(three)</u>,
Legends Of Tomorrow season 3(three) "**Daddy Darhkest Crossover With Constantine**",
"**Legends Of-Meow-Meow**", Batwoman season 1(one) "**Elseworlds**"

2019;_
<u>Arrow season 8(eight), The Flash season 6(six), Supergirl season 4(four)</u>,
Legends Of Tomorrow season 4(four) "**Crisis On Infinite Earths**"

2020;_
"**Post-Crisis On Infinite Earths For W.B. D.C. T.V. Show Series**"

W.B. D.C. Animated Super-Hero Vigilante (Film Movies) Universe Chronological Order Justice League The Flashpoint Paradox Justice League War Justice League Throne Of Atlantis Son Of Batman Batman Versus Robin Batman Bad Blood Justice League Versus Teen Titans Teen Titans The Judas Contract Justice League Dark Constantine City Of Demons Suicide Squad Hell To Pay The Death Of Superman Reign Of The Supermen Batman Hush Wonder Woman Bloodlines Justice League Dark Apokolips War Constantine: The House Of Mystery
W.B. D.C. Animated Batman Beginning-Ending Years (Film Movie) Super-Hero Vigilante Batman Year One, Batman The Dark Knight Returns Deluxe Edition
Square Soft Enix, Deck 9(nine) and Dontond's <u>Life Is Strange</u> Universe (Video Game), 0(zero) Before The Storm, 1(one) Original, Adventures Of Captain Spirit, 2(two), True Colors
Time Traveling Genre, Time Loop Subgenre Selection (film movie) Groundhog Day

Open World (Video Game) Title	*(Sandbox Simulation) Setting On Planet Earth*
Infamous Second Son	Seattle, Washington, United States of North America
Grand Theft Auto 5 (five)	Los Angeles, South California, United States Of N.A.
Agents Of Mayhem	Seoul, South Corea, Eur-Asia
Gravity Rush 1 (one), 2 (two)	"Science-Fantasy Setting"
Marvel Spider-Man Marvel Spider-Man Miles Morales	New York, New York, United States of North America New York, New York, United States of North America
Watch Dogs Complete Edition	Chicago, Illinois, United States of North America
Sleeping Dogs	British-Hong Hong Sovereignty, Eur-Asia
Watch Dogs 2 (two)	San Francisco, North California, United States of North America
Control Ultimate Edition	"Science-Fiction Horror Weird Setting"
Mirror's Edge Catalyst	"Science-Fiction Utopia Setting"

"The New World 2121 Movement Of The 22nd Century, Utopian Sovereignty of North America (U.S. of N.A.)"

4D (four(th) Dimension) Movie Theater Archives

Additional Recorded Entertainment In H.D.D.V.D.s 1080p Resolution

Film Movie(s), TeleVision (T.V.) Show Series = Passive Viewing
Video Game(s), Sandbox Simulation(s), Comic Book(s), Manga = Active Participation

The Recorded Entertainment In The Same Boxes Means It Is Set In The Same Story Universe

20th century World War 2 (two) British England's Dunkirk Spirit Story (film movie);   1) Darkest Hour 2) Dunkirk 3) Their Finest
"Nazi Germany On The Verge Of Defeat And The Last Days Of Adolf Hitler" Downfall (film)
"Kingsman: A British Action Comedy Spy (film movie) Series"   1902-1914 The King's Man \ 1997-2014 Kingsman: The Secret Service   Statesman (T.V. Show Series) \ 2014 Kingsman: The Golden Circle \ Kingsman: 3
"John Wick: A United States of North American Neo-Noir Action Thriller (film movie) Series"   {*Start here, left to right (L→R)* } The Continental (T.V. Show Series) / John Wick: Chapter 1   John Wick: Chapter 2 / John Wick: Chapter 3 / John Wick: Chapter 4 /   John Wick: Chapter 5 / John Wick: Chapter 6 / John Wick: Chapter 7   John Wick: Chapter 8 / Ballerina / Jon Women Infants Children (WIC)
"Wolfenstein: A World War Two Alternative History War (video game) Series"   { *Start here, left to right (L→R)* } 1946 Wolfenstein: The Old Blood   1960 Wolfenstein: The New Order / 1961 Wolfenstein II: The New Colossus   1973 Wolfenstein: Youngblood / 1981 Wolfenstein: Cyberpilot
"Heroes: A United States of North American Super-Hero Drama (T.V. Show Series)"   Heroes: Season 1(one); Disc 1(one), Disc 2(two), Disc 3(three), Disc 4(four), Disc 5(five)   Heroes: Season 2 / Heroes: Season 3 / Heroes: Season 4 / Heroes Reborn: Mini-Series
"Tommy Wiseau's The Best Worst Movie Ever The Disaster Artist Room (film movie)"   The Disaster Artist / Tommy Wiseau's The Room / Best F(r)iends / The Neighbors (T.V.)
"Peaceful Less Violent (video game) Series" Flow / Flower / Journey
"Science Adventure: A Science-Fiction Interconnected Visual Novel (video game) Universe"   Chaos;Head / Chaos;Child / Steins;Gate / Steins;Gate 0 / Robotics;Notes / Anonymous;Code
"Sei Shoujo's Visual Novel Drama Horror Erotica Of Bible Black: (video game) Series"   "Bible Black: (video game) Adapted To Animation (T.V. Show Series)"   **Bible Black: Origins** Episode(s) 1, 2 / **Bible Black:** Episode(s) 1, 2, 3, 4, 5, 6   **Bible Black: Only** Episode(s) 1, 2 / **Bible Black:** New Testament Episode(s) 1, 2, 3, 4, 5, 6

End Of Known Recorded Entertainment In Chicago, The Last Animal Sanctuary of Illinois

2154;_11(Nov)-11(eleven)　　　　{ act 3(three) boss battles }

Travel To Western Society, America(s), North America
Self-Govern Sovereignty (S),
Illinois (ILL) Sector, Points Of Interest (P.O.I.)

Chicago (S), "**Animal** Sanctuary"
"The New World 2121 Movement Of The 22nd Century,
Utopian Sovereignty of North America (U.S. of N.A.)"
South Side, ChIraq's LAN Party Halloween Costume Play Shop

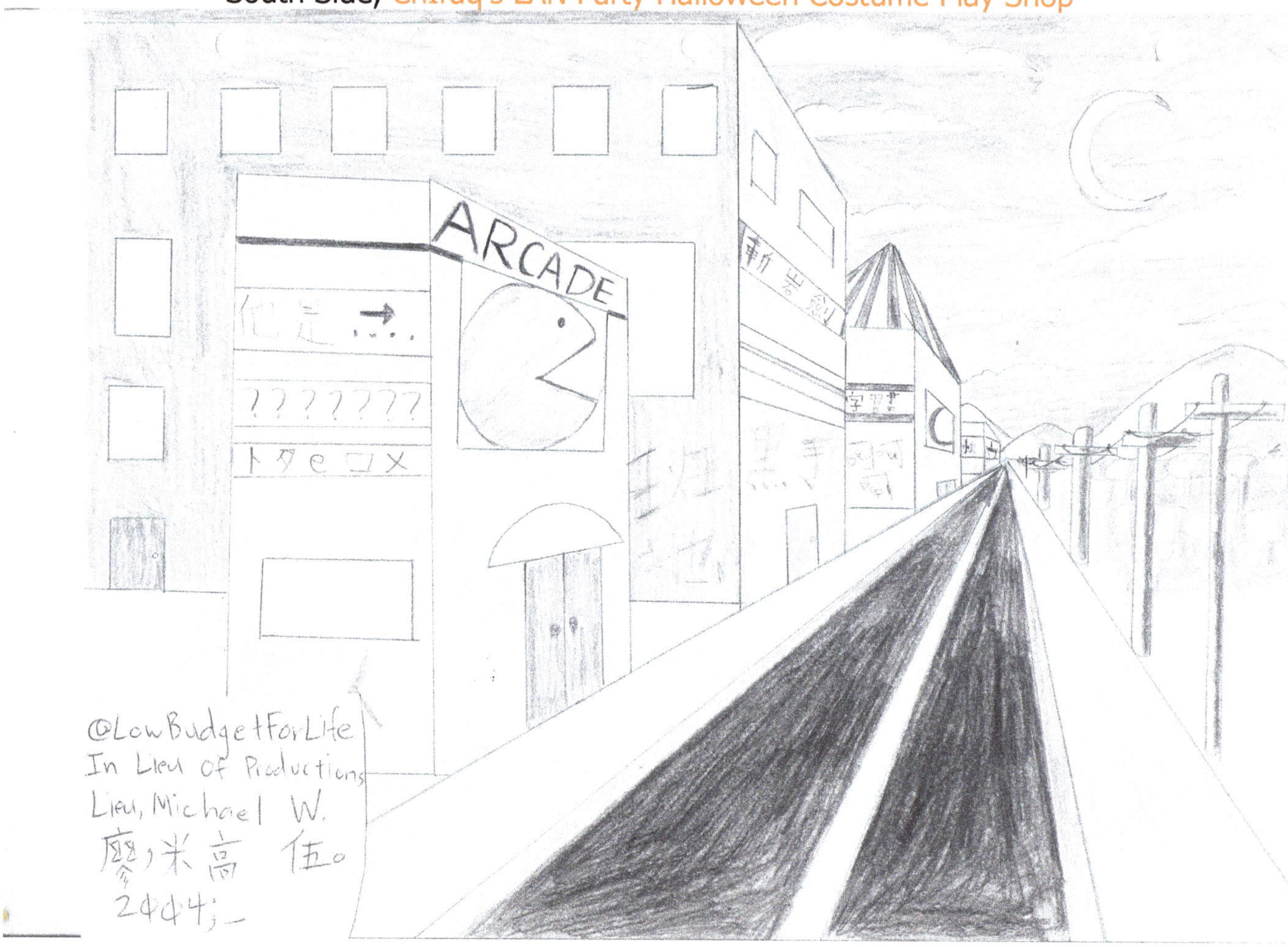

Valentina Vita (V. V.); Traveling Cosplayers And Kosplayers;
"(^_^) ...Wow, that was a good H.D.D.V.D. archives to **pass** and **enjoy** the **time** until **Entertainer Mister Steven Z. J.** arrives in Chicago **Animal** Sanctuary **Illinois**... Hm... Should I take part in the **cosplay show and tell**...? (>_<) Hope I am not too **late** to his **premiere arriving**!? (@_@) I did not miss the **premiere arrival** did I!?"

2154;_ Arcadia \ Valentina Vita (V. V.) 'Healthy-Strong Life' \ Left to Right(L->R)

V. V.'s 2154 to do list since my Ai-Aug bracelet is not working due to outbreaks outages...
~~Visit, **North Side, Lieu Lieu.... Lieutenant Lieu**~~ (~~next place to go and to do~~) Complete
~~Visit, **Downtown District, Bullet Board System**~~ Complete
~~Visit, **East Side, All For One Shopping Center**~~ Complete
Visit, **South Side**, **Entertainer Mister Steven**, **Reserve** Your **Tickets** (Nov 11 @ 8 a.m.)
Now waiting... Looks like I got everything done in this list...
What else...? (~~What to do to pass time?~~)

Valentina Vita (V. V.); Traveling Cosplayers And Kosplayers;
"(=_=) ...Wait, what **time** is it...? Does not **feel** like it is... **eight A.M. morning**..."

V. V. turns on her Ai-**Aug bracelet**, selects the "**Geolocation**" on her A.R. Holo HUD.

Alien Valentina Vita, Traveling Cosplayer(s) And Kosplayer(s)
2154;_ Chicago, Illinois' Intranet, Wireless Broadcasting Frequencies Online

V. V. Augmented Reality Holographic Heads Up Display A.R. Holo HUD
Geolocation;_ Milky Way Galaxy, Planet Earth, North America (N.A.), Utopian Sovereignty of North America (U.S. of N.A.) Illinois (ILL) Sector, "Chicago, The Last Animal Sanctuary Of Illinois" Zip Code 60666 Central Time Zone (C.T.Z.) @ 05;00_a.m.
[*Aug+*]
V. V. Augmented Reality Holographic Heads Up Display A.R. Holo HUD

2154;_ Chicago, Illinois' Intranet, Wireless Broadcasting Frequencies Online
Alien Valentina Vita, Traveling Cosplayer(s) And Kosplayer(s)

Valentina Vita (V. V.); Traveling Cosplayers And Kosplayers;
"(=_=) Phew... It is **five A.M.** morning, not **five P.M.** night... Still got **three hours**..."

V. V. hears the music "**Mark Knopfler's 05;15_A.M.**" playing in the background on a **specialized loud audio sound speaker biotech**. She continues to look around.

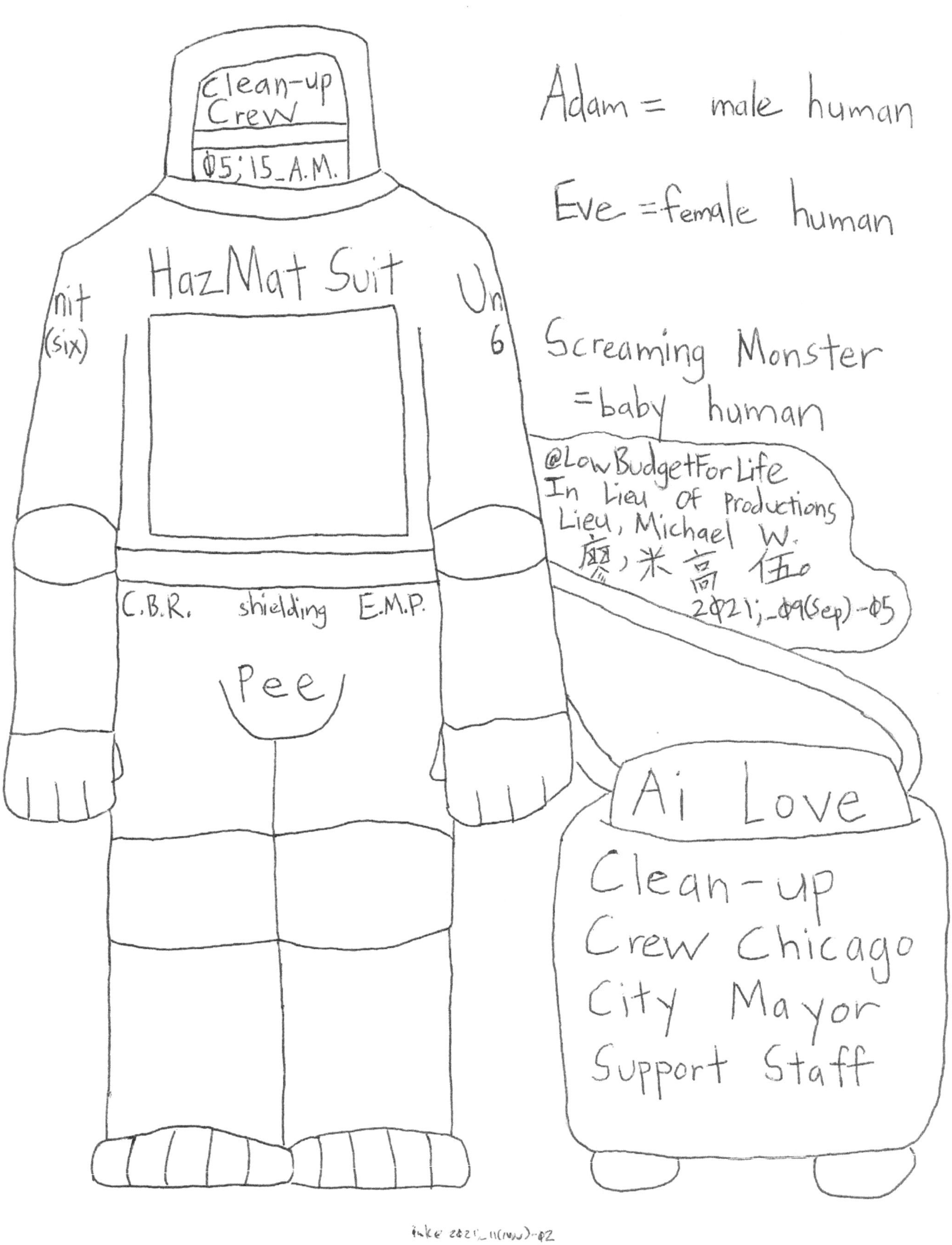
Clean-up Crew
Φ5;15_A.M.
HazMat Suit
Unit (six)
Un 6
C.B.R. shielding E.M.P.
Pee
Adam = male human
Eve = female human
Screaming Monster = baby human
@LowBudgetForLife
In Lieu Of Productions
Lieu, Michael W.
廖,米高 伍o
2Φ21;_Φ9(Sep)-Φ5
Ai Love
Clean-up Crew Chicago City Mayor Support Staff
inke 2Φ21;_11(Nov)-Φ2

Alien Valentina Vita, Traveling Cosplayers And Kosplayers
2154;_ Chicago, Illinois' Intranet, Wireless Broadcasting Frequencies Online

<table>
<tr><td colspan="3">V. V. Augmented Reality Holographic Heads Up Display A.R. Holo HUD</td></tr>
<tr><td>Actual Legal Given Name;_

Adan Crenshaw
(goes by Unit 6 (six))
pronounced,
"A-Dan' 'Cro-Ren-Sha"</td><td colspan="2">Alias Network Nickname;_

Sleeping Beauty, Sleeping On Job,
Lazy Bums, City Mayors' Support Staff,
Cleaning Crew, Those HazMat Suits,
The Crenshaws, Music Video, [<u>Aug+</u>]</td></tr>
<tr><td>Physical Plus Psych Age;_

augmented 34 + 22 years old</td><td>Ethnicity;_

African</td><td>Nationality;_

Utopian</td></tr>
<tr><td>Love(s) Or Like(s);_

"Mark Knopflers':
05;15_A.M.
Punish The Monkey
Brothers In Arms
Silvertown Blues"</td><td>Family Member(s);_

Honest Ape,
Unit 9 (nine) A.K.A.
Eva Crenshaw,
Walking Dog.</td><td>Education;_

Doctor Of Philosophy
(D.O.P.) in Scientists,
Technicians, Engineers,
Mathematicians (STEM)</td></tr>
<tr><td>Personal Protective Equipment
(P.P.E.);_

Full Body White Hazardous
Material (HazMat) Suit
UltraViolet (U.V.)
augmentation joints, Green
Bio-Lettering Unit numbers</td><td>Biological (B.)
Toxicity;_

Low Count Of
Biological Agent(s)
From Animal Kingdom
Excrement And Waste
Management</td><td>Education;_

Biotech, Doctor Of
Philosophy (D.O.P.)
Degree From The
University Of Chiraq</td></tr>
<tr><td>Frequency Modulation (F.M.);_

Ninety four seven the
wave!</td><td>Psychological Traits;_

High Functioning,
Autistic Spectrum</td><td>Electro Magnetic Pulse
(E.M.P.);_

E.M.P. Shielding Online</td></tr>
<tr><td>Frequency Hertz;_

UltraViolet (U.V.) Rays</td><td>Relationships;_

Eva Crenshaw, Sister</td><td>Biological Genealogy;_

Natural Birth Male (M)</td></tr>
<tr><td colspan="3">[<u>Aug+</u>]</td></tr>
<tr><td colspan="3">V. V. Augmented Reality Holographic Heads Up Display A.R. Holo HUD</td></tr>
</table>

2154;_ Chicago, Illinois' Intranet, Wireless Broadcasting Frequencies Online
Alien Valentina Vita, Traveling Cosplayers And Kosplayers

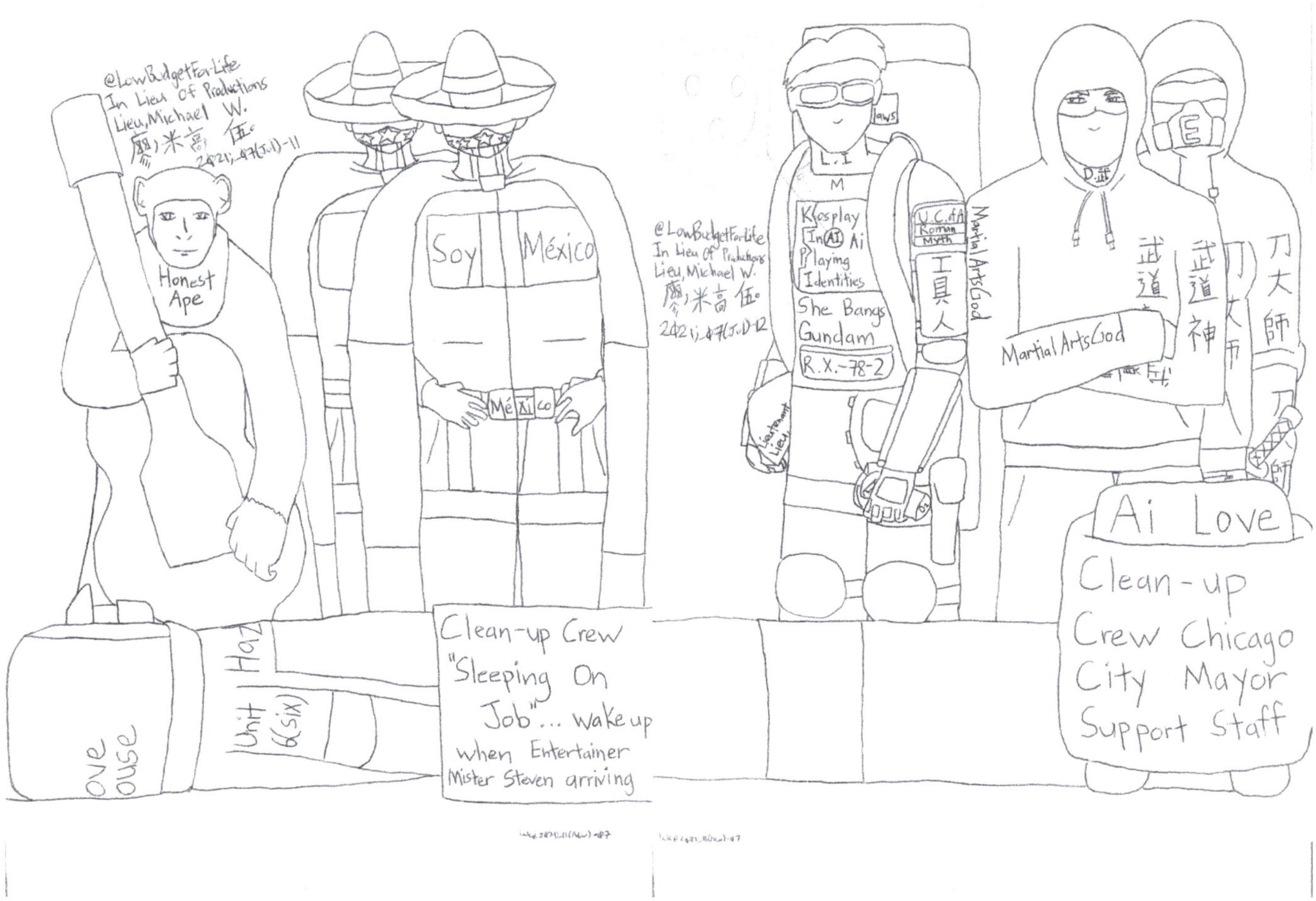

Chicago Controller's Clean-up Crew, Unit 6 (six), Adan Crenshaw; Music Video (M.V.);
"(*in suit*) **Gonna lay down** and **wait** for anything that requires cleaning and clean-up."

Elle K. Mcgee; Childlike Human Kanon-Klone(s) "BladeMaster 刀大師" Experiment;
"(*in Oxygen Breathing Apparatus*) ...Oh hi V. V., **we** are waiting for **Entertainer Mister Steven Z. J., hurry up and wait**! You enjoy your time in the **Downtown District**?"

Benavidez-Cruz, Rich Rodriguez "Big Ben"; Mexican Mythology, The King Of Mexico;
"(*in mouth mask*) Hola..."

Benavidez, Jose Rodriguez "Baby Ben"; Mexican Mythology, Mexican-Men;
"(*in mouth mask*) Hola..."

Lieutenant 2nd Lieu, Michael W. 廖, 米高 伍o; Vigilant Volunteer(s), Yellow-Mellow;
"(*in mouth mask*) Hi again V. V., everything good, your Ai-**Aug bracelet**?"

Alex D. "Doc Drastal"; War Survivor Military Veteran, "MartialArtsGod 武道神";
"(*in mouth mask*) Hi V. V., **I'm Alex Drastal**, do not think **we** have **met**..."

2154;_Arcadia \ Valentina Vita (V. V.) 'Healthy-Strong Life' \ Left to Right(L->R)

Alien Valentina Vita, Traveling Cosplayers And Kosplayers
2154;_ Chicago, Illinois' Intranet, Wireless Broadcasting Frequencies Online

V. V. Augmented Reality Holographic Heads Up Display A.R. Holo HUD						
[*Aug+*]						
Volunteer(s) Workers In: Utopian Sovereignty of North America (U.S. of N.A.), **"Hurry Up And Wait"** For **Entertainer: Mister Steven Z. J.**, Thrilling Plastination Puppeteer, Chicago, The Last **Animal** Sanctuary (S) Of Illinois (ILL)						
Clean-up Crew **Honest Ape**	Mexican-Men Baby Ben	The King Of Mexico Big Ben	Yellow-Mellow (°o°) Obama@ deep Problem Solvers	Lieutenant Lieu = Low Budget For Life "ToolPerson 工具人"	**Alex D. Doc Drastal** = "MartialArts God 武道神"	Elle K. Mcgee = "BladeMaster 刀大師"
Clean-up Crew Unit 6 (six), Adan Crenshaw						
[*Aug+*]						
V. V. Augmented Reality Holographic Heads Up Display A.R. Holo HUD						

2154;_ Chicago, Illinois' Intranet, Wireless Broadcasting Frequencies Online
Alien Valentina Vita, Traveling Cosplayers And Kosplayers

Valentina Vita (V. V.); Traveling Cosplayers And Kosplayers;
"(O_O) Whoa we wow, **hi everyone, Lieu Lieu**, again Elle, we **meet again** for the **fourth time**!? Yea, I was in the **Downtown District** for days... Just enjoying the **Four D Movie Theater**. It was like... **November one**, when **we** last met... **Over one week went** by!"

Elle K. Mcgee; Childlike Human Kanon-Klone(s) "BladeMaster 刀大師" Experiment;
"(*in Oxygen Breathing Apparatus*) **Time** went by **fast** huh? Yea, yea, I been doing some **BladeMaster** student studies, some **volunteer work with different groups**..."

Valentina Vita (V. V.); Traveling Cosplayers And Kosplayers;
"(O_<) This Unit six... Clean-up Crew, **he** just gonna **lay down** on the **ground** like that?

Elle K. Mcgee; Childlike Human Kanon-Klone(s) "BladeMaster 刀大師" Experiment;
"(*in Oxygen Breathing Apparatus*) Oh yea, Unit six loves to do that, with the **five fifteen A.M. Music Video** that usually **plays in the background** on **loud audio** when waiting and getting ready to do any clean-up... **We** even have a **special sign** made for this.

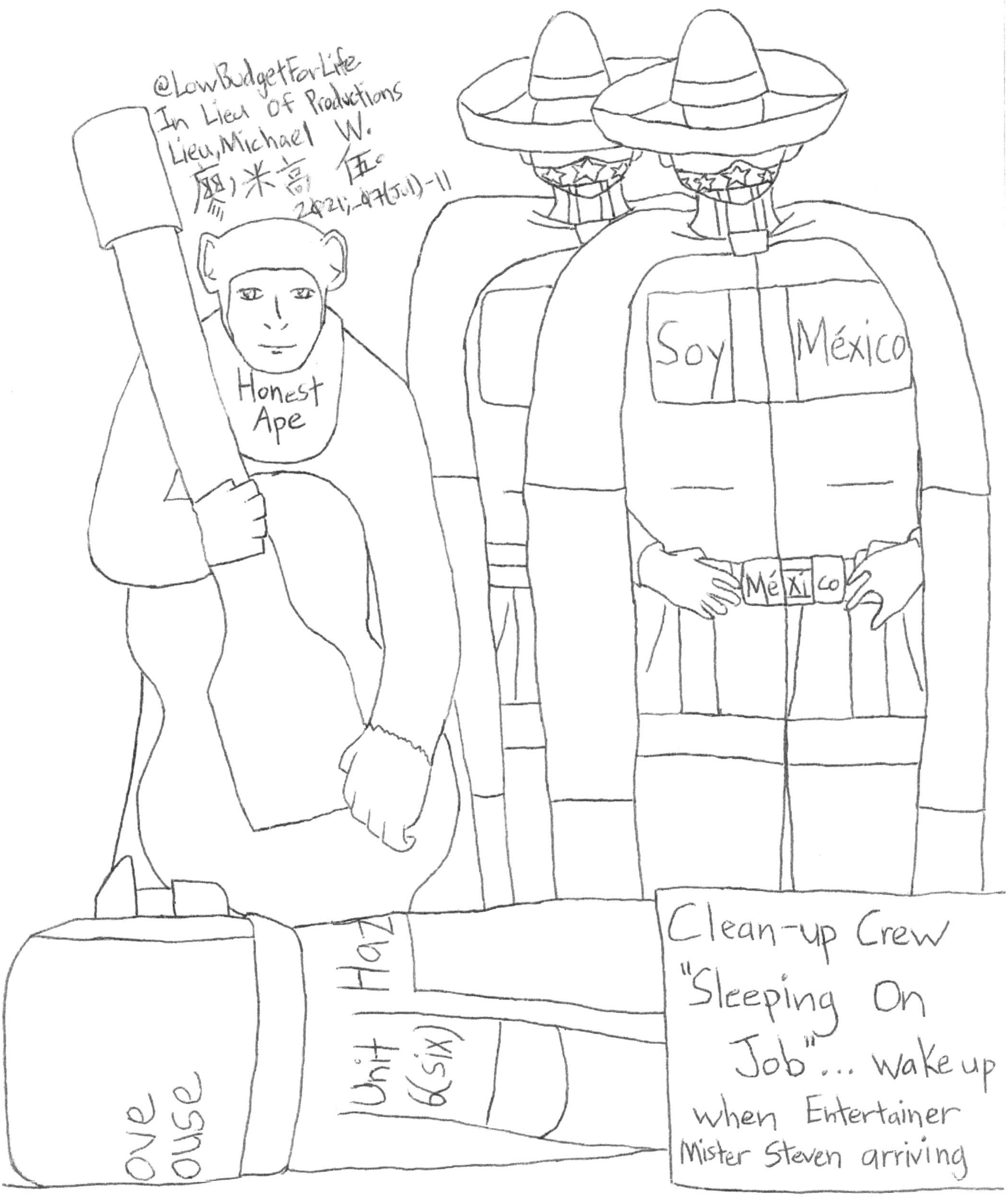
@LowBudgetForLife
In Lieu Of Productions
Lieu, Michael W.
廖, 米高 伍o
Honest Ape
Soy México
México
Haz
Unit 6(six)
Clean-up Crew "Sleeping On Job"... wake up when Entertainer Mister Steven arriving

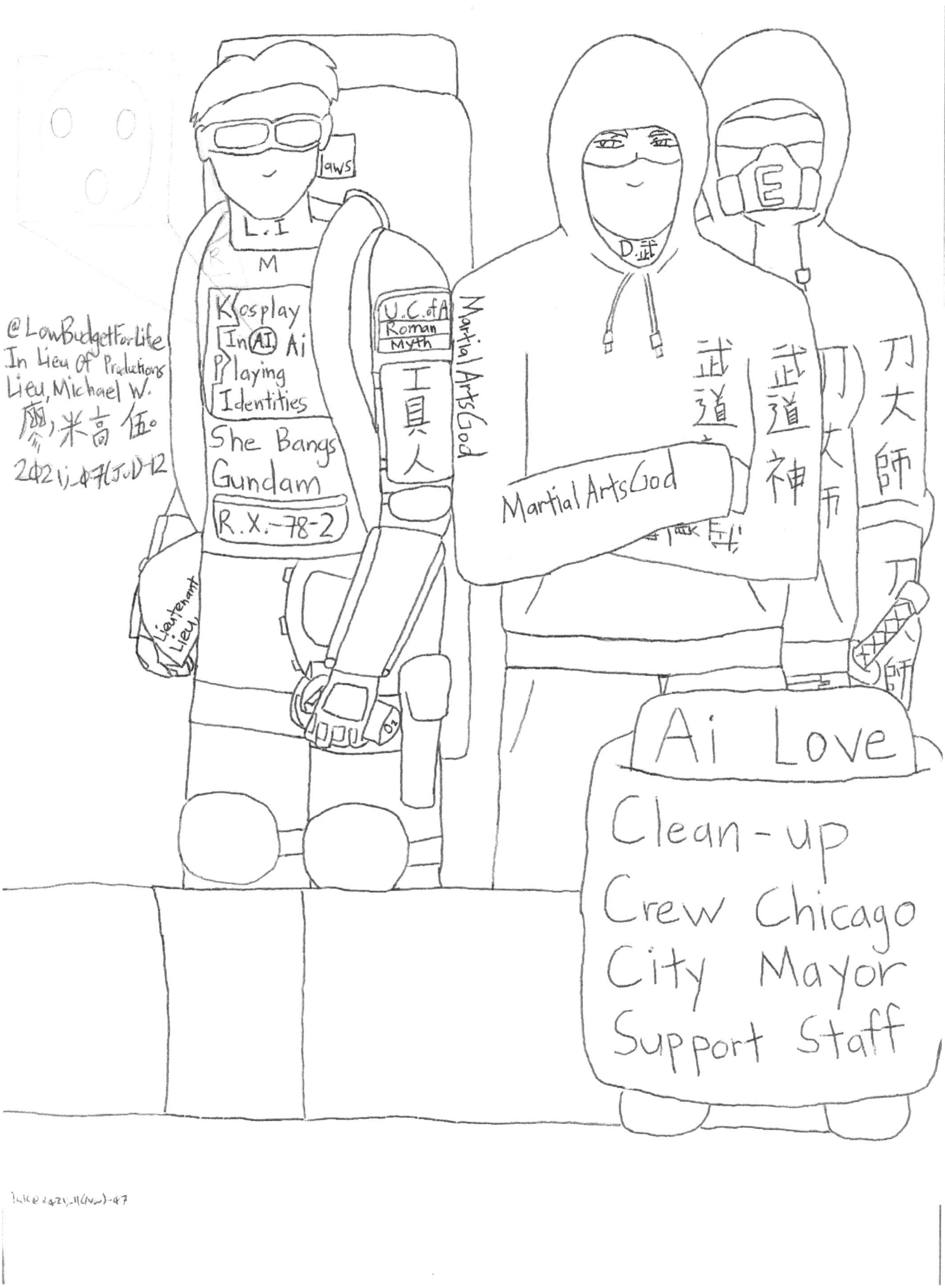
laws
L.I
M
Kosplay
In AI Ai
Playing
Identities
She Bangs
Gundam
R.X.-78-2
U.C. of A
Roman
Myth
工具人
Martial Arts God
@LowBudgetForLife
In Lieu Of Productions
Lieu, Michael W.
廖, 米高 伍o
2021-07(Jul)-12
Lieutenant Lieu
D.武
E
武道神
Martial Arts God
刀大師
Ai Love
Clean-up
Crew Chicago
City Mayor
Support Staff

Valentina Vita (V. V.); Traveling Cosplayers And Kosplayers;
"(^_^) That is an interesting **special sign** Elle... While you... **Hurry Up And Wait...** (@_@) Oh, **Lieutenant Lieu**, everything **good**, no problems with my Ai-**Aug bracelet**."

Lieutenant 2nd Lieu, Michael W. 廖, 米高 伍 o; Vigilant Volunteer(s), Yellow-Mellow;
"(*in mouth mask voice over*) Copy **chief**, good to hear that on your side, **chief**."

Elle K. Mcgee; Childlike Human Kanon-Klone(s) "BladeMaster 刀大師" Experiment;
"(*in Oxygen Breathing Apparatus voice over*) Oh V. V., you have not met **Alex Dra**... **Dra**... **Dra**... **Dra... Alex D.** yet have you? **He** is also a Martial Artist, a fellow **rival** to **us Cook-Chef Blades** in **our** many **societies**. But, **we** are **okay** with **each other here**!"

Alex D. "Doc Drastal"; War Survivor Military Veteran, "MartialArtsGod 武道神";
"(*in mouth mask voice over*) **Drastal**... **Alex D.** is fine. Many **people** do have a bit of a **trouble** pronouncing **my last name...** At first... **Doc D.** or **Doc** is fine too... since **my** time in the United States **Air And Space Force**... as a **Pararescue** and **Medevac**."

Valentina Vita (V. V.); Traveling Cosplayers And Kosplayers;
"(O_O) Oh hi **Alex Dra... Dra... D. Alex D**. Whoa we wow! (=_=) Yea, I was in the **Downtown District Bulletin Board System** with the **volunteer worker groups' postings**... And got to say, it was quite the amount of **information** about the **volunteer workers**, especially the... Jerry Duty, Xtra-Men... Chicago Controller has to says about it."

Alex D. "Doc Drastal"; War Survivor Military Veteran, "MartialArtsGod 武道神";
"(*in mouth mask voice over*) Yea, **almost eight years** in **service**... Close to retirement almost. (*ahem*) Got many good **views** of **space** from **my time** in the **Space Force**."

Valentina Vita (V. V.); Traveling Cosplayers And Kosplayers;
"(>_<) **Thank you** for **your service Alex D.**! I got a nice **space view** during my trip to **Sir Author C. Clarke's Space Elevator anniversary** on **September eleven**!

Alex Drastal nods **his** head in response to V. V.'s **comments** with Elle **staring** at her.

Chicago Controller's Clean-up Crew, Unit 6 (six), Adan Crenshaw; Music Video (M.V.);
"(*in suit voice over*) Oh-oh-oh-oh, is Chicago Controller City Mayor about to go on?"

Chicago Controller, Psychological U.V. Projection As A City Mayor;
"(*voice over*) Start **Utopian**, Chicago, **The Last Animal** Sanctuary **Of Illinois'** welcomes **Entertainer Mister Steven Z. J.. non-paid volunteer workers**, get ready for his **arriving** at **eight A.M. morning hours**, **official orders** as City Mayor as **follows** in this UltraViolet A.R. Holo HUD, have a nice day, period end."

Reads = Left to Right (L -> R) \ Wireless Broadcasting Frequencies Online
2154;_ Chicago, The Last Animal Sanctuary Of Illinois' Intranet

Why bother, who cares, barely any humans follow rules regulation	Fine, whatever do not sit on it, complete this before entering

Chicago Controller
City Mayor

"Arriving, Entertainer Mister Steven Z. J."
2154 (twenty one fifty four);_ 11-11 (November eleven)"

Opening Ceremony For Entertainers' Expo Mister Steven Z. J.,
In Chicago, The Last Animal Sanctuary Of Illinois (ILL).

Welcome To The New World 2121 (twenty one twenty one) Movement Of The 22nd (twenty second) Century, Utopian Sovereignty (U.S.) North American Animal Autonomous Anime Convention Center (N.A.A.A.A.C.C.).

City Mayor's Orders The Following: Captain Rage C.C.P.D. SWAT And Clean-up Crew, Watch Over My Beautiful Peace Loving Animal Kingdom Species

Clean-up Crew
Unit 6 (six), Unit 9 (nine)
Unit 8888,8888 (eight eight eight eight comma eight eight eight eight)
Unit 3113 (three one one three)

Abusive bitch and moan, toxicity constant crybaby complaint box	Let the shit hit the fan rock bottom, to crash and burn hard

2154;_ Chicago, The Last Animal Sanctuary Of Illinois' Intranet
Reads = Left to Right (L -> R) \ Wireless Broadcasting Frequencies Online

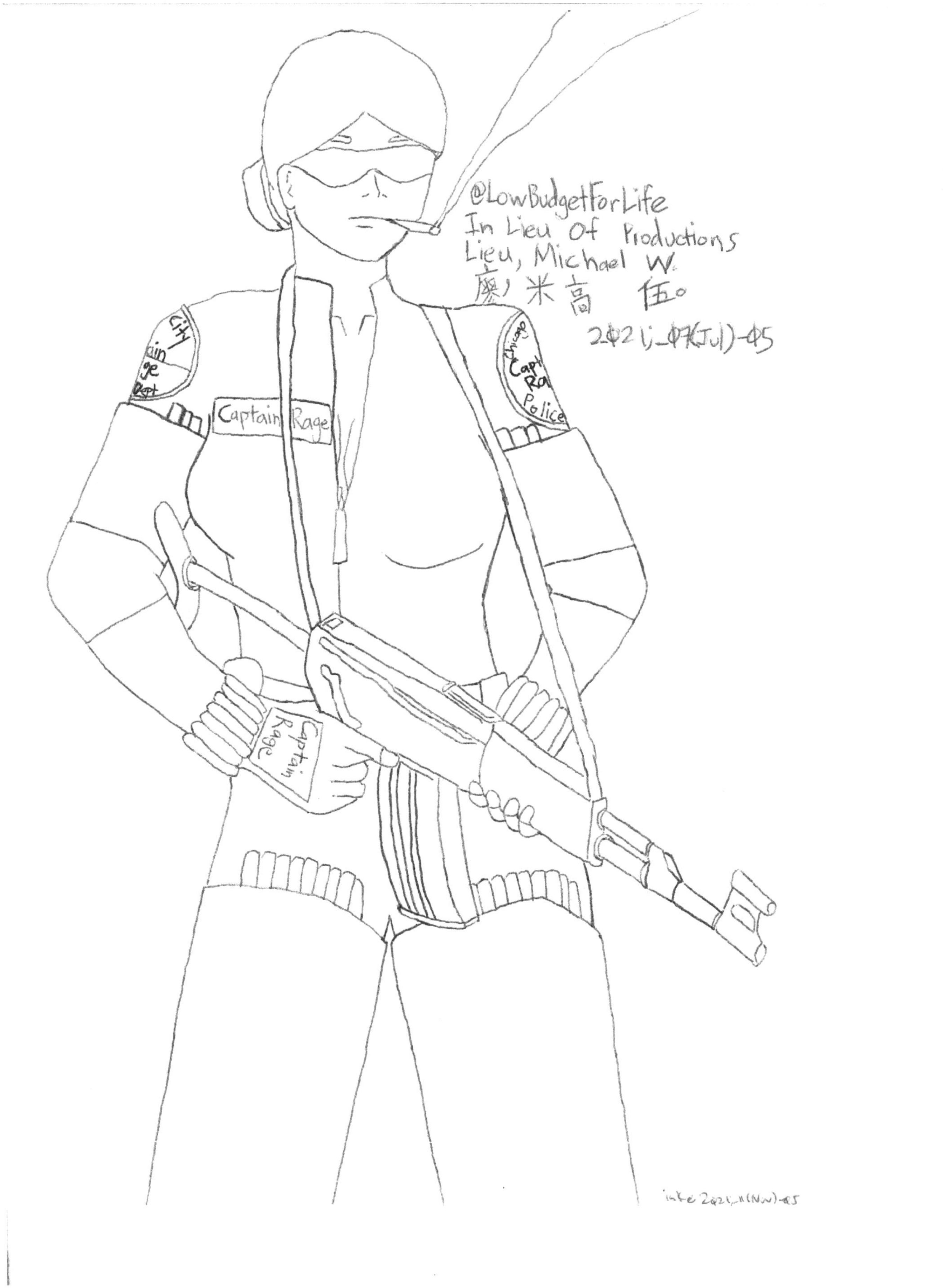
@LowBudgetForLife
In Lieu Of Productions
Lieu, Michael W.
廖, 米高 伍o
2Φ21_Φ7(Jul)-Φ5
Captain Rage
Captain Rage
Police

V. V. smells medical marijuana smoke in the atmosphere, she sees and hears **someone**.

Alien Valentina Vita, Traveling Cosplayer(s) And Kosplayer(s)
2154;_ Chicago, Illinois' Intranet, Wireless Broadcasting Frequencies Online

V. V. Augmented Reality Holographic Heads Up Display A.R. Holo HUD
"Chicago City Clean House Act of Twenty one fifty two, July four".
[*Aug+*]
V. V. Augmented Reality Holographic Heads Up Display A.R. Holo HUD

2154;_ Chicago, Illinois' Intranet, Wireless Broadcasting Frequencies Online
Alien Valentina Vita, Traveling Cosplayer(s) And Kosplayer(s)

Captain Rage; Real Time On Call Animal Sanctuary Hurry Up And Wait C.C.P.D. SWAT; "(*smoking medical marijuana*) Copy that City Mayor, **C.C.P.D. SWAT Captain** Rage going on **patrol** by the **Downtown District**, over..."

V. V. looks at her Ai-**Aug bracelet**, selects the "**Geolocation**" update for the **Date-Time**.

Alien Valentina Vita, Traveling Cosplayer(s) And Kosplayer(s)
2154;_ Chicago, Illinois' Intranet, Wireless Broadcasting Frequencies Online

V. V. Augmented Reality Holographic Heads Up Display A.R. Holo HUD
Geolocation;_ Milky Way Galaxy, Planet Earth, North America (N.A.), Utopian Sovereignty of North America (U.S. of N.A.) Illinois (ILL) Sector, "Chicago, The Last Animal Sanctuary Of Illinois" Zip Code 60666 Central Time Zone (C.T.Z.) @ 08;00_a.m.
[*Aug+*]
V. V. Augmented Reality Holographic Heads Up Display A.R. Holo HUD

2154;_ Chicago, Illinois' Intranet, Wireless Broadcasting Frequencies Online
Alien Valentina Vita, Traveling Cosplayer(s) And Kosplayer(s)

V. V. hears, the music "**Henry Mancini's The Entertainer**", playing in the **Utopia A.R. Holo HUD** background by the **South Side**, ChIraq's LAN Party Halloween Costume Play Shop biotech dome airlock decontamination.

Alien Valentina Vita, Traveling Cosplayers And Kosplayers

2154;_ Chicago, Illinois' Intranet, Wireless Broadcasting Frequencies Online

V. V. Augmented Reality Holographic Heads Up Display A.R. Holo HUD

Privacy Protection aug made possible by

Kay's Private Security Services L.L.C.

【 Aug+ 】

V. V. Augmented Reality Holographic Heads Up Display A.R. Holo HUD

2154;_ Chicago, Illinois' Intranet, Wireless Broadcasting Frequencies Online

Alien Valentina Vita, Traveling Cosplayers And Kosplayers

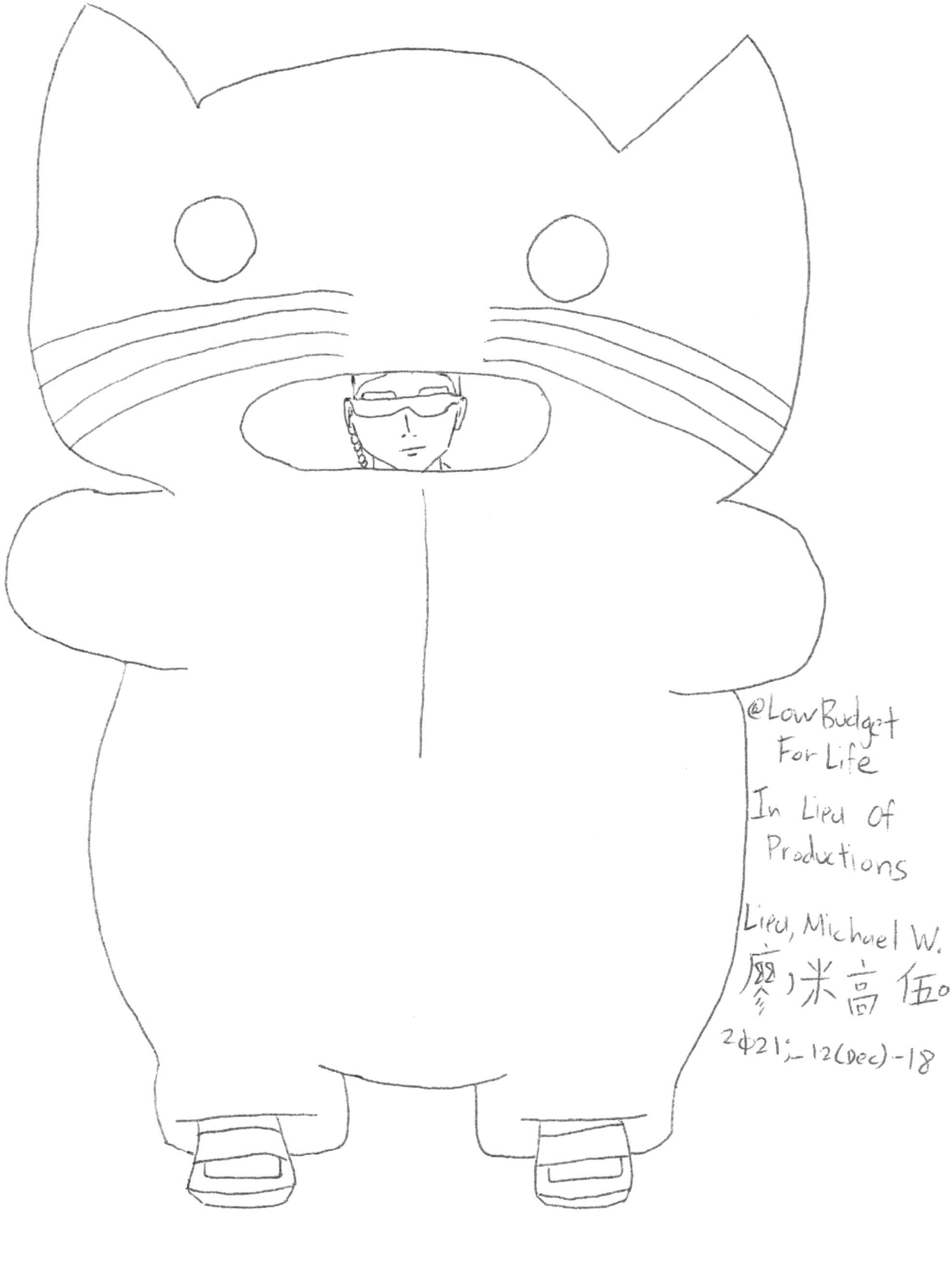
@LowBudget
For Life
In Lieu Of
Productions
Lieu, Michael W.
廖, 米高 伍o
2φ21-12(Dec)-18

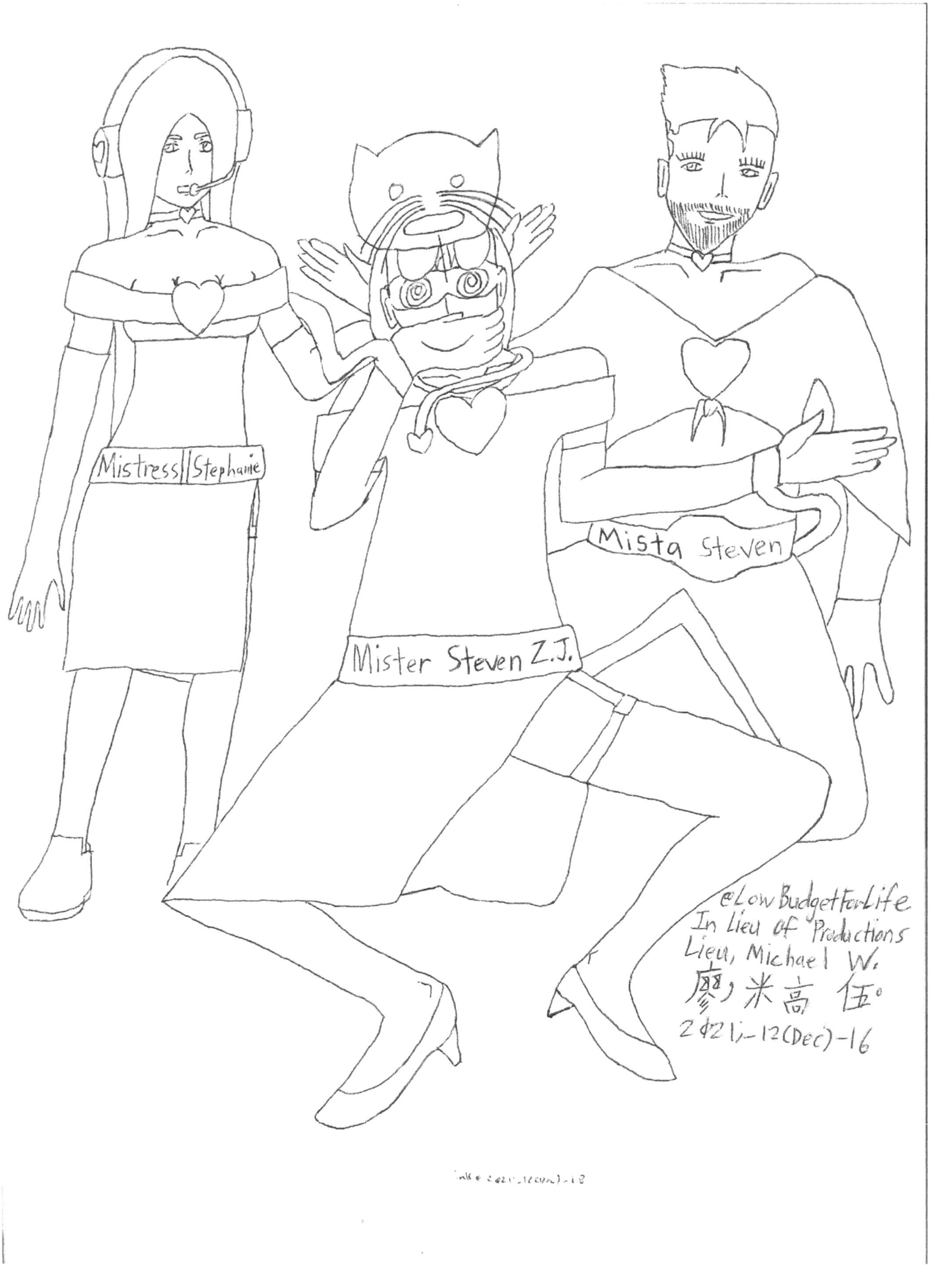
Mistress Stephanie
Mister Steven Z.J.
Mista Steven
@LowBudgetForLife
In Lieu of Productions
Lieu, Michael W.
廖, 米高 伍。
2021-12(Dec)-16

Alien Valentina Vita, Traveling Cosplayer(s) And Kosplayer(s)
2154;_ Chicago, Illinois' Intranet, Wireless Broadcasting Frequencies Online

V. V. Augmented Reality Holographic Heads Up Display A.R. Holo HUD		
Actual Legal Given Name;_ **Mistress Stephanie**	Actual Legal Given Name;_ Cute Cawaii Cat-Kat **Mister Steven Z. J.**	Actual Legal Given Name;_ **Mista Steven**
Social Status;_ **Entertainer,** Plastination Puppeteer	Social Status;_ **Entertainer,** Plastination Puppeteer	Social Status;_ **Entertainer,** Plastination Puppeteer
Employment;_ **Entertainers' Expo**	Employment;_ **Entertainers' Expo**	Employment;_ **Entertainers' Expo**
[*Aug+*]		
V. V. Augmented Reality Holographic Heads Up Display A.R. Holo HUD		

2154;_ Chicago, Illinois' Intranet, Wireless Broadcasting Frequencies Online
Alien Valentina Vita, Traveling Cosplayer(s) And Kosplayer(s)

Valentina Vita (V. V.); Traveling Cosplayers And Kosplayers;
"(@_@) ...Ah... Wow! I am just in time for the **opening ceremony**... **Photo**! **Photo**!"

Cute Cawaii Cat-Kat; Thrilling Animal Entertainers, Utopia A.R. Holo HUD;
"(*stuff animal cat-kat*) (O_O) Meow. Meow! Me Ow!"

Mister Steven Z. J.; Thrilling Puppeteer Entertainers, Utopia A.R. Holo HUD;
"(*male voice tone*) ...Are we experiencing **biotech difficulties**...? ...**Lights**!! **Camera**!! **Action**!! Time to give the **audience** what **they** want. Place **your** bets!"

Mista Steven; Thrilling Plastination Entertainers, Utopia A.R. Holo HUD;
"(*female voice tone*) ...Our **guest** of honor for Entertainers' Expo, **Mister Steven Z. J.**, Brought to you by... **Elysium Corporation's Undead Utopian Plastination Preservation**... Before your natural **unmodified self expires**, you can create an artificial **beautiful biotech body** of yourself... and be **forever young**!"

Mistress Stephanie; Thrilling Plastination Entertainers, Utopia A.R. Holo HUD;
"(*male voice tone*) How are our **ratings**!? Go to **commercial**! Our **sponsors'** for today's **Entertainment**! **Elysium Corporation**... Oh wait, we already announced that."

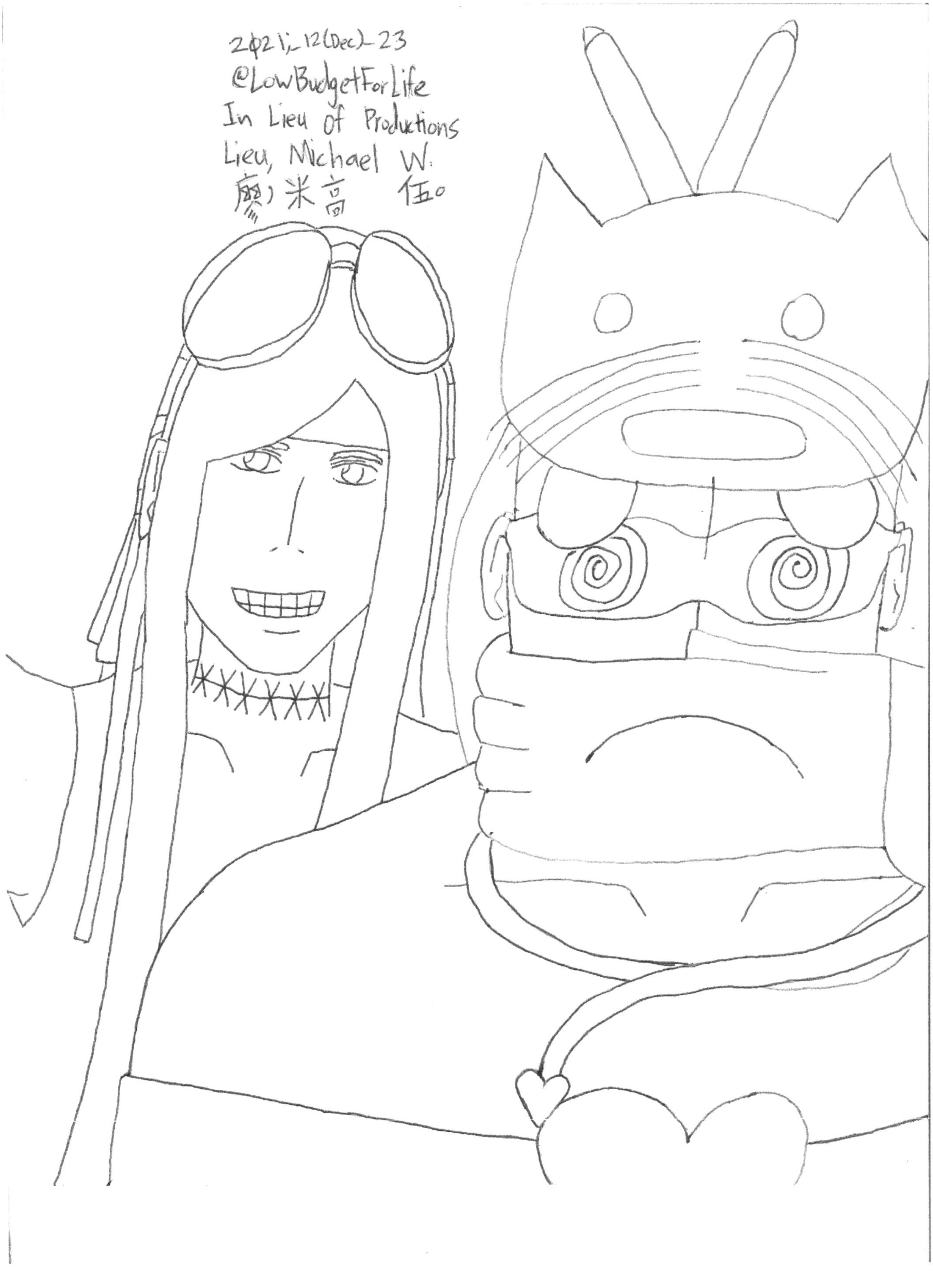
2021-12(Dec)-23
@LowBudgetForLife
In Lieu of Productions
Lieu, Michael W.
廖, 米高　伍o

Valentina Vita (V. V.); Traveling Cosplayers And Kosplayers;
"(T_T) **Mister Steven**, can I get a **photo** with you!? Big fan of yours!"

Mister Steven Z. J.; Thrilling Puppeteer Entertainers, Utopia A.R. Holo HUD;
"(*male voice tone*) Oh of course, my dear V. V., do you know the **password**?"

Valentina Vita (V. V.); Traveling Cosplayers And Kosplayers;
"(>_<) Um… Oh, **re… re-reserve** your **tickets**! Yes, **reserve** your **tickets**!"

Mister Steven Z. J.; Thrilling Puppeteer Entertainers, Utopia A.R. Holo HUD;
"(*male voice tone*) Alrightly, **photo** time! Make sure to love yourself as who you are!"

V. V. takes a **holo photo selfe** with **Entertainer Mister Steven Z. J.** and steps away.

Valentina Vita (V. V.); Traveling Cosplayers And Kosplayers;
"(@_@) **Dream** come **true**!! I present my **tickets**, which is saying quote **reserve** your **tickets** unquote. Got to finally **meet and greet** an **Entertainer**, **Mister Steven Z. J.** That embraces the loving yourself as the **best possible way** to a healthy relationship with oneself as a form of self-care. **Self-talk** time. I know some **Entertainers** that prefer the quote **fame game** unquote. That **desperation** in trying to get more **likes** and **approvals** from **others** in **aiming** to be the most **popular** or trying to **become someone's friend** makes for a **hollow** and **near emptiness existence**… How does that kind of **behavior** not make them feel **sick**? (?_?) How can these **people**, those **fame game Entertainers** deal with all the **stress**, **pressure** to be **fake** for quote **fame game** unquote. (<_>) I ask myself this, what if your friend or someone you know of is getting **depressed**, **sad, anxious** or **frustrated** as a result? (<_<). I believe that this **obsession**, **past time** of becoming the **famous game** can **evolve** and become something **serious** and it is not to be **ignored**… (=_=) Probably, best thing to do is **show** the **other sides** of this **cosplay hobby**… the **honor** and great **pride** of **crafting** something, **satisfaction** of **saving** enough **money**, and **resources**, **goods** for **barter and trade** for a **cosplay**, **friendships** you can **build** due to a **comic**, **movie**, **T.V. show animation, cartoon, live action**, in our historical universe or from the multiple universe theory of an alternative history… When I see **intranet** posts such as I want to be a **cosplayer** just to be **famous** and **recognition**… I think they are **living cosplay life** in the most **frustrating** way possible… That **scares** the **hel** out of me. **Hobbies** are really important, but they cannot become our only **reason** to **live**… (>_<) If you have a **friend** or knows something that is too involved into this quote **fame game** unquote… Keep in close **contact**, an **eye** on them… You would **care** if you want them to **live** a strong healthy life… As my name implies! (O_<) Subject shift, **cosplay workspace, aerograph**, I believe is a **must have** to make in order to **craft armor**, like my red Lady A Dragoon Knight fusion armor P.P.E.. (@_@) Every **dream** of any kind for **cosplay** starts with an **ambitious vision** and that **vision** makes the **future** more **mysterious** yet **possible…** However, the **belief** to **do it** is what makes that **vision** an **actuality**!"

2154;_Arcadia \ Valentina Vita (V. V.) 'Healthy-Strong Life' \ Left to Right(L->R)

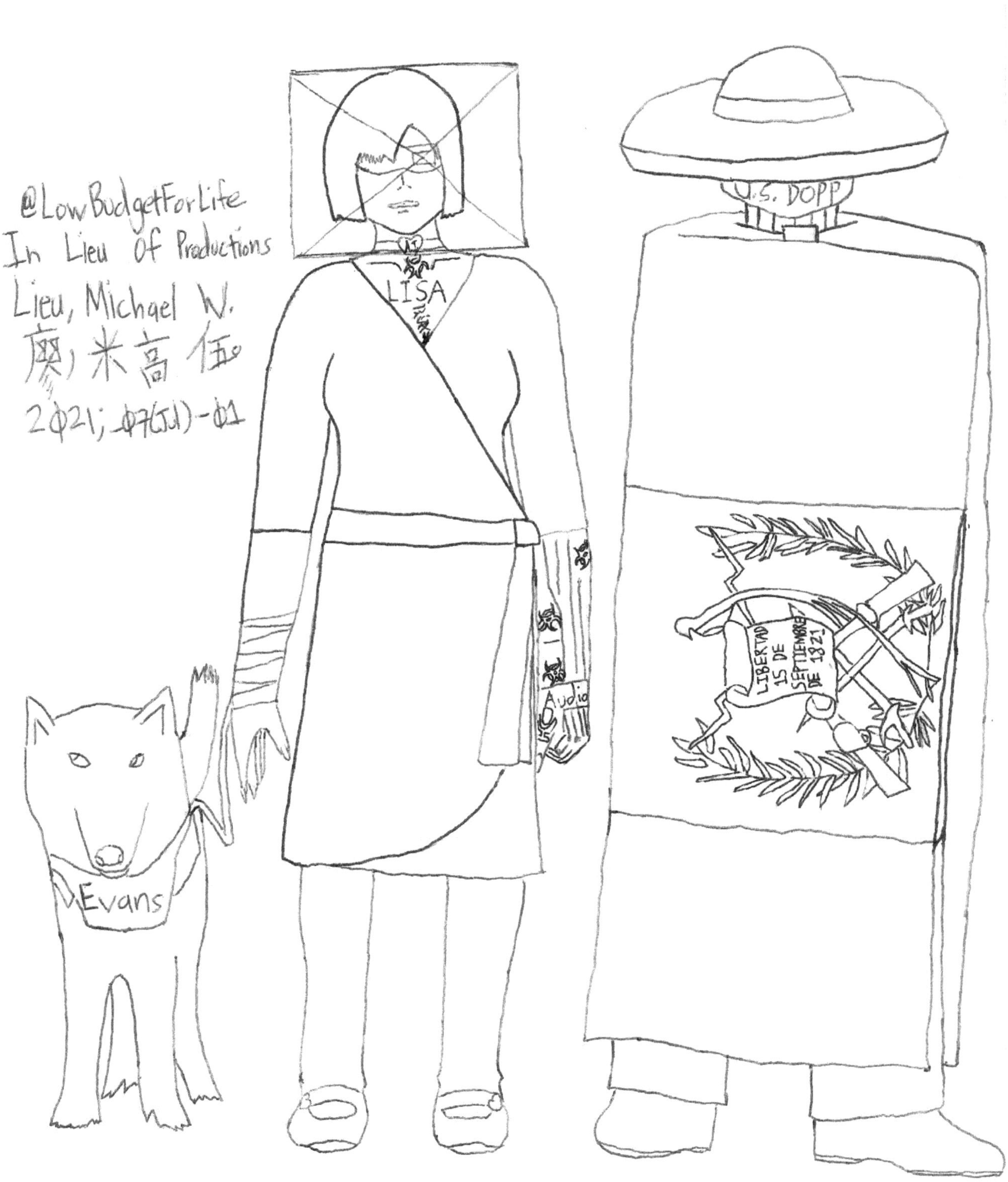
@LowBudgetForLife
In Lieu Of Productions
Lieu, Michael W.
廖, 米高 伍o
2φ21; φ7(Jul)-φ1
LISA
Audio
U.S. DOPP
LIBERTAD 15 DE SEPTIEMBRE DE 1821
Evans

Alien Valentina Vita, Traveling Cosplayer(s) And Kosplayer(s)
2154;_ Chicago, Illinois' Intranet, Wireless Broadcasting Frequencies Online

V. V. Augmented Reality Holographic Heads Up Display A.R. Holo HUD		
Actual Legal Given Name;_ **Evans** pronounced "**Eh-Veans**"	Actual Legal Given Name;_ **Large Intelligent Sexy Agent (LISA)**, goes by "**LISA**" pronounced "**Lee-Sah**"	Actual Legal Given Name;_ **Marco Medina Morales** (goes by **Marco**) pronounced: "**Mah-Co, Ma-De-Na, More-Ale-Less**"
Animal Kingdom;_ Dog Species	Artificial Intelligence (A.I.);_ Confirmed, model referencing "Natural Unmodified Human Female (F)" as a synthetic physical platform	Human Being;_ Natural Unmodified Human Male (M)
Community Network(s);_ **Reminder To Please, Do Not Feed Chocolate To** Nat**-Dogs** [*Aug+*]	Biotech Fluid;_ Biological Agent(s) Of: Blood Mix Ebola, Anthrax, Small Pox, Syphilis, AIDS, Gray Goo Nano-Medicine	Blood Type;_ O Negative
[*Aug+*]		
V. V. Augmented Reality Holographic Heads Up Display A.R. Holo HUD		

2154;_ Chicago, Illinois' Intranet, Wireless Broadcasting Frequencies Online
Alien Valentina Vita, Traveling Cosplayer(s) And Kosplayer(s)

Valentina Vita (V. V.); Traveling Cosplayers And Kosplayers;
"(o_O) Whoa we wow... Hello there! Hm... (*staring*) **You all** around part of the...?"

Mister Marco Medina Morales; Mass Media Investigative Journalism U.S. DOPP;
"(*posing*) Soy Guatamala!! I am Guatemala-Guy!! Interested in **investigative stories**?"

A.I. Ai Large Intelligent Sexy Assistant (LISA); Mass Media Investigative Journalism Assistant;
"(*monotone female voice*) Hello... V. V., **we** are part of the United States **Department Of Public Perception, Investigative Journalists**. I am an Artificial Intelligence, A.I., using a Synthetic Synthesis **platform** to **model** myself after the natural human female **platform**."

Animal Kingdom Evans; Mass Media Investigative Journalism Animal Kingdom Service Doggie;
"(*barking*) Woof! Woof!! Woof!! Woof!! Woof!! Woof!! Woof!! Woof!! Woof!! Woof!!"

2154;_ Arcadia \ Valentina Vita (V. V.) 'Healthy-Strong Life' \ Left to Right(L->R)

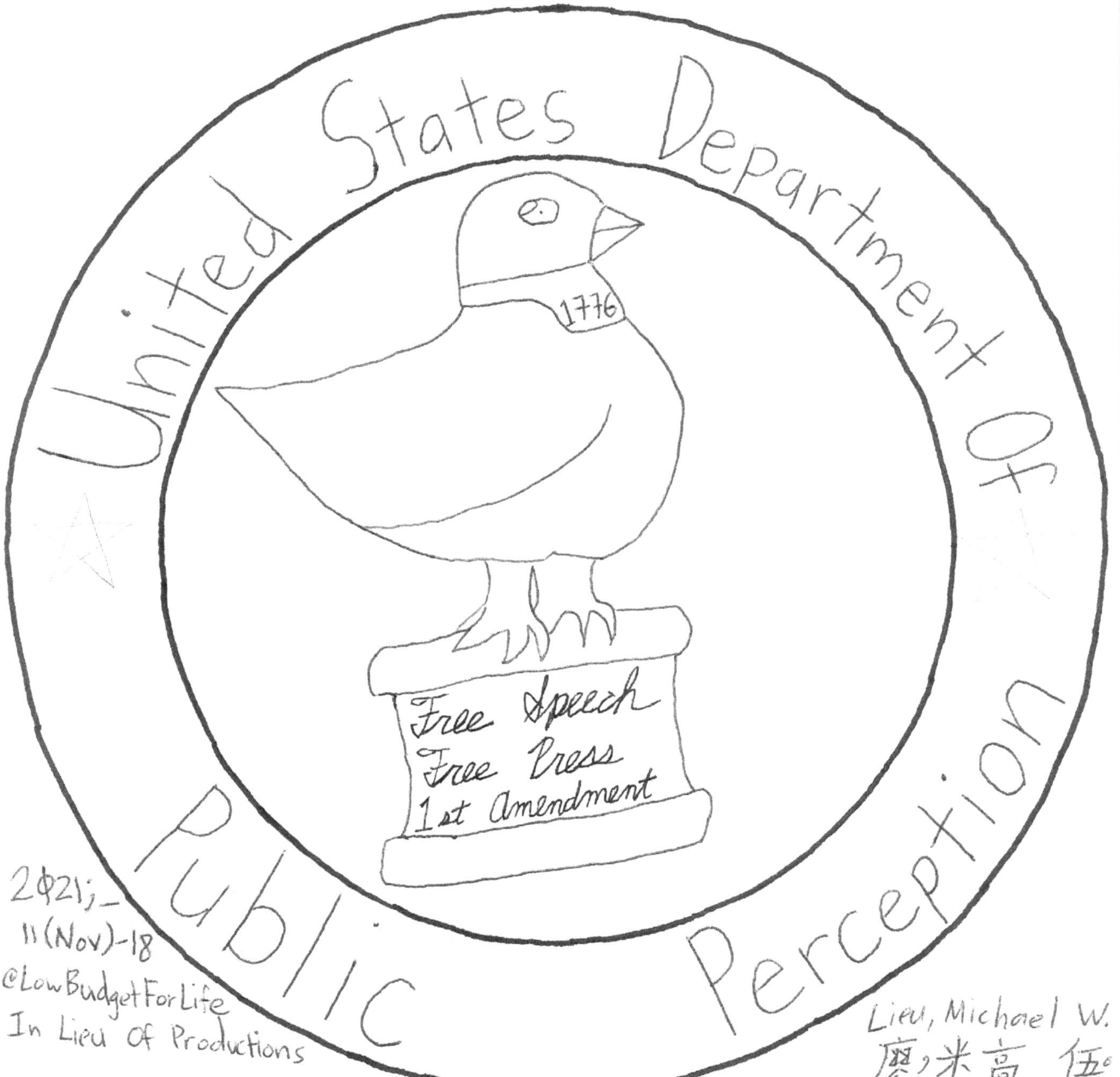
United States Department Of
Public Perception
1776
Free Speech
Free Press
1st Amendment
2Φ21;_
11(Nov)-18
@LowBudgetForLife
In Lieu Of Productions
Lieu, Michael W.
廖, 米高 伍

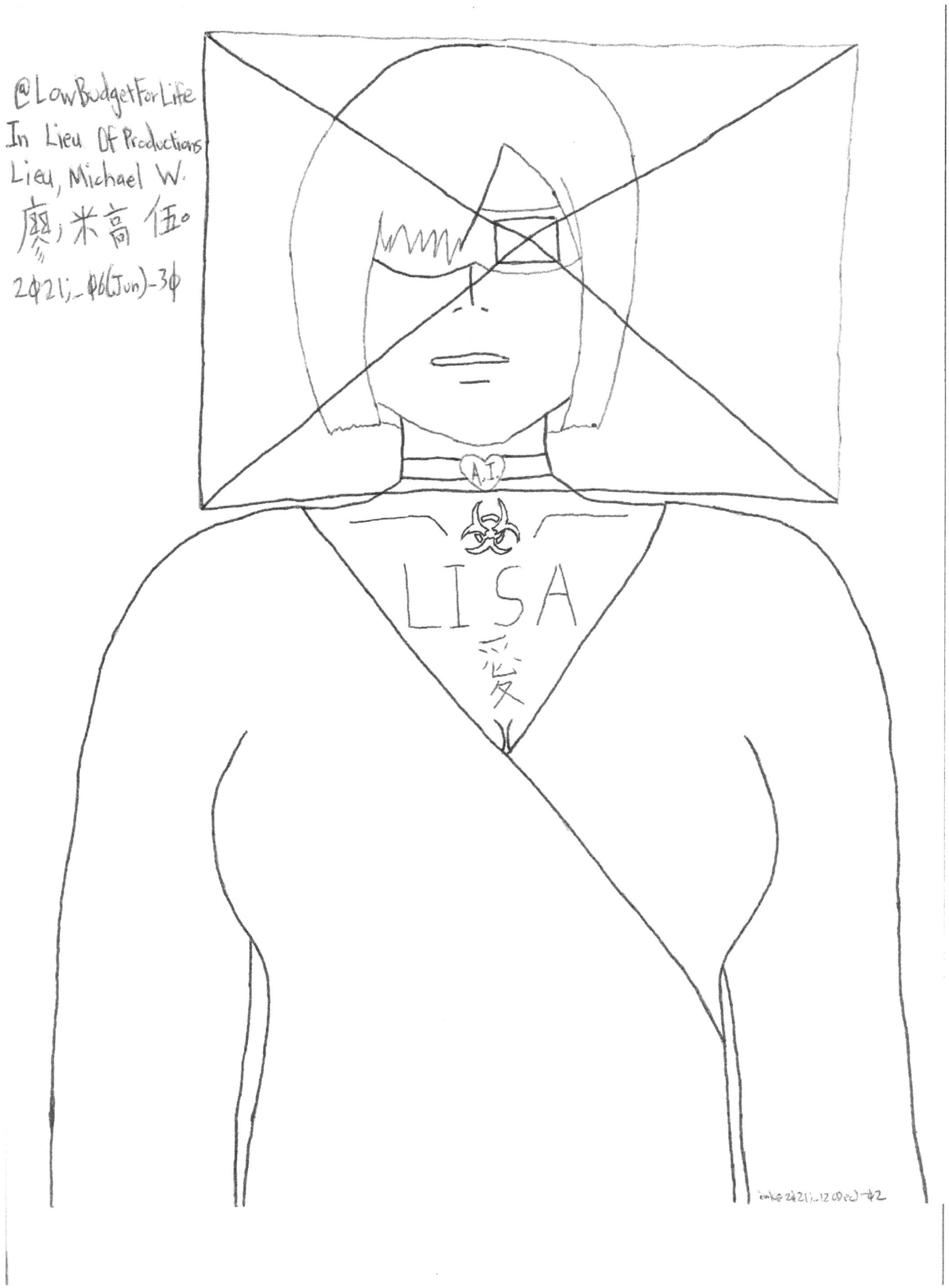
@LowBudgetForLife
In Lieu Of Productions
Lieu, Michael W.
廖, 米高 伍o
2φ21;_φ6(Jun)-3φ
A.I.
LISA
爱

Citizen A.I. Ai Large Intelligent Sexy Assistant (LISA), Mass Media
2154;_ United States DOPP's Intranet, Wireless Broadcasting Frequencies Online

Mass Media Broadcasting Intranet	
United States Department Of Public Perception (U.S. DOPP) Public Sector Federal Government Funded, Investigative Journalism Found 1945;_	
Mass Media Broadcasting Frequencies Archive Stories;_	
[*Aug+*]	
1941;_The Second World War Turning Points: Union Soviet Socialist Republics (U.S.S.R.), Britain, Nazi Germany, Imperial Japan, Africa, United States of America (U.S.A.),	1944-2120;_Prelude To The Third World War A.K.A. The Kold War Via Mutual Assured Destruction (MAD) Theory "Desire For Peace, Prepare For War"
1960s;_Assassination Attempts On Reverend King, Malcolm X, President John F. Kennedy, And Senator Robert F. Kennedy, Foiled By Kay's Private Security Services L.L.C.	1944-2120;_The Second Possible Civil War In The United States Of America (U.S.A.) Between The "Whites/Aryan" And "Colored/Negros" Person(s) Separation
1997;_The Year The Longest Known Natural Unmodified Human Female Died Of Natural Causes Via Flatline Expiration, "Jeanne Calment Of 122.5 Years Old"	
2016;_United States of America (U.S.A.)'s First Female Elected President, Hilary R. Clinton	
1919;_Spaniard Flu Outbreak Pandemic	2020;_Coronavirus Outbreak Pandemic
1919-2020-2121;_A Century Of Apocalyptic Events For Human Civilizations	"Opinion Piece" 2222;_The Next Apocalyptic Event(s)?
2121;_The Year The Solar System Lost Billions Of Human Lives	2121-2136;_The 22nd Century Dark Ages, The Natural Human Infertility Disease
"Opinion Piece" How Some People Compare The U.S.A. To The Rome Republic Collapse	
2121 - Present Day Time "2154;_" "U.S., United States or Utopian Sovereignty?" "City Controllers, upper case C. in the city, and their message with fire in the works."	
2120 - Present Day Time "2154;_" The War On Mars	
[*Aug+*]	
Mass Media Broadcasting Intranet	

2154;_ United States DOPP's Intranet, Wireless Broadcasting Frequencies Online
Citizen A.I. Ai Large Intelligent Sexy Assistant (LISA), Mass Media

Alien Valentina Vita, Traveling Cosplayers And Kosplayers
2154;_ Chicago, Illinois' Intranet, Wireless Broadcasting Frequencies Online

V. V. Augmented Reality Holographic Heads Up Display A.R. Holo HUD

Mass Media Broadcasting Intranet

A U.S. Department Of Public Perception (DOPP), Investigative Journalism presents "City Controller(s), upper case C. in the city, and their message with fire in the works."

Written by Mister Marco Medina Morales,
Edited by A.I. Ai Large Intelligent Sexy Assistant (LISA),
Published by Animal Kingdom Service Doggie Evans on 2153, January one.

"Chicago City Clean House Act of 2152 (twenty one fifty two);_07-04 (July four)"

'An outside look into a unique 22nd century phenomenon of "Animal Sanctuary (S)" by these so called "City Controller(s)" claiming to be "City Mayor(s)" of a human city.'

Since the founding and independence of North America's United States of America (U.S.A.) on July four 1776, celebration with fireworks has been a national tradition.

The 2120 Martians' intended or unintended use of unconventional warfare: Chemical, Biological, Radiological (C.B.R.) outbreaks plus Electro Magnetic Pulse (E.M.P.) outages on Planet Earth and Earth's Lunar Moon Military (M) Bases that lasted years, the 22nd century Dark Ages, leading to an apocalyptic pandemic and blackout on Earth's civilian societies.

During the 22nd century Dark Ages, with the loss of many natural human lives, and augmented humans refusing to comply and follow easy policies resulted in more division, many felt Earth's U.S.A. federal government should have done more to prepare, but would the Aliens and Citizens in a liberty loving country give up their freedom for security?

As a consequence, civilian run cities, counties, states, all across U.S.A., "officially sanction off using fireworks" due to the U.S.A. federal government's lack of infrastructures in Democratic (D) and Republic (R) rule, forming "Sovereignty or Sanctuary (S)" phenomenon.

By 2121, this is the beginning of an historical event that some call it "the new world".

The most recent City (D) (R) Military (M) Base to "officially sanction off using fireworks" from U.S.A.'s military forces is in Illinois (ILL), Chicago, now an "Animal Sanctuary (S)".

Avoiding broadcasting biotech towers seems to counters these "City Controller(s)".

Could these "City Controller(s)" be an unconventional warfare of Martian Terrorists, or humanity, regardless of being augmented or not, destined to its own hubris? [*Aug+*]

[*Aug+*]

V. V. Augmented Reality Holographic Heads Up Display A.R. Holo HUD

2154;_ Chicago, Illinois' Intranet, Wireless Broadcasting Frequencies Online
Alien Valentina Vita, Traveling Cosplayers And Kosplayers

2154;_11(Nov)-12(twelve)

Travel To Western Society, America(s), North America
Self-Govern Sovereignty (S),
Illinois (ILL) Sector, Points Of Interest (P.O.I.)

Chicago (S), "**Animal** Sanctuary"
"The New World 2121 Movement Of The 22nd Century,
Utopian Sovereignty of North America (U.S. of N.A.)"
East Side, All For One Shopping Center

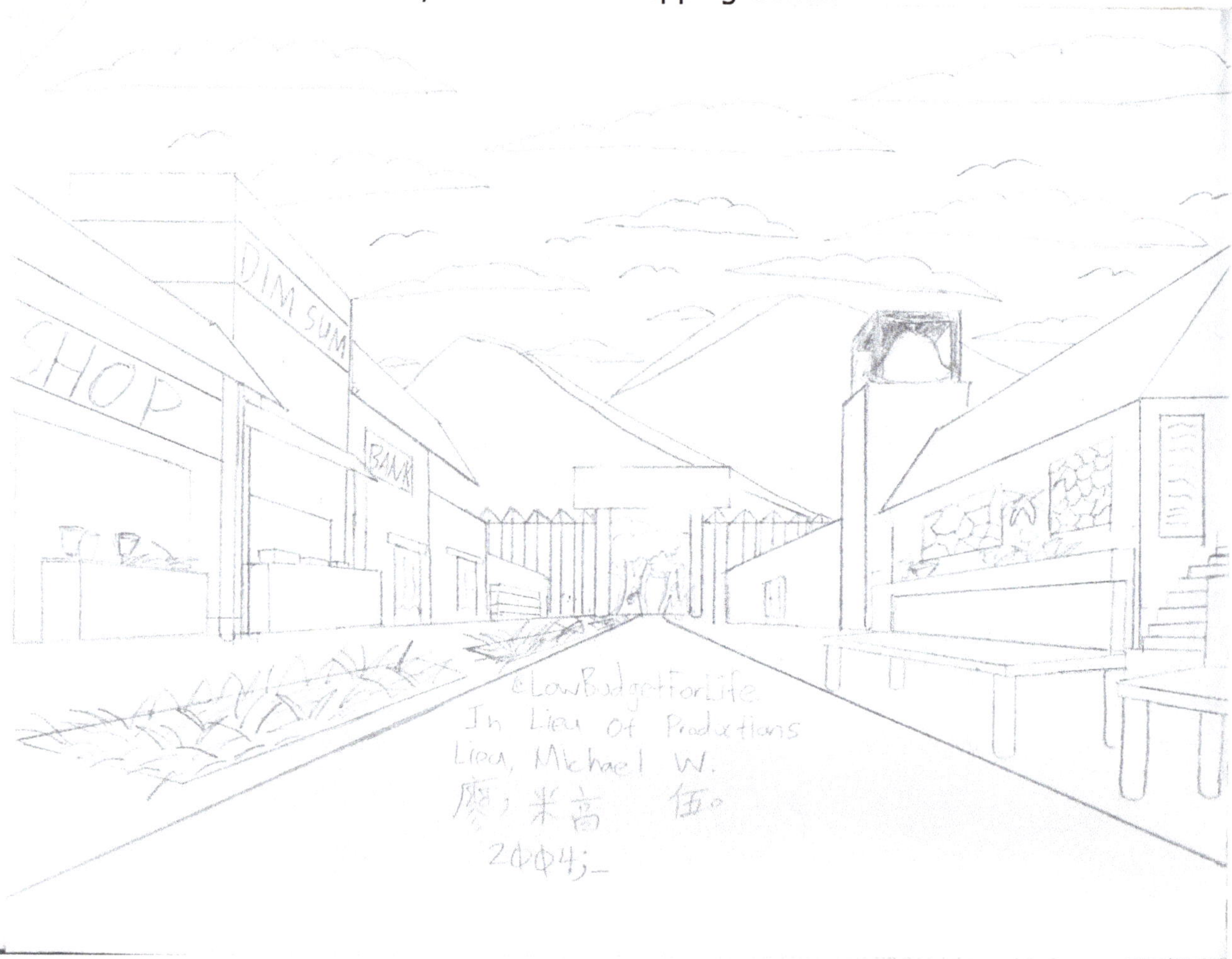

Valentina Vita (V. V.); Traveling Cosplayers And Kosplayers;
"(=_=) Time to turn off my Ai-**Aug**, this bio**tech** is so **addictive**... Enjoy the **twenty one fifty four** Entertainers' Expo! Need to change into my **KissShot**-AcerolaOrion-**HeartUnderBlade costume** now... This would be good place to **set-up**."

V. V. approaches Entertainers' Expo **Bulletin Board System**, writes her **panel/event**.

Entertainers' Expo Mister Steven Z. J.'s Bulletin Board System (B.B.S.)
2154;_ **Utopian** Sovereignty **of Americas** (**U**.S. **of A.**) Grand Tour
11 (Nov) – 11 (eleven) to 28 (twenty eight)
"...The Fame Outlives The Name And The Game... ...Eagerness For Game, Is Eagerness To Contribute To Humanity..."

Chicago, **Animal** Sanctuary, **Illinois**, East Side, All For One Shopping Center
Operation Hours: 08;00_a.m. **Break 11;00_a.m. to 13;00_p.m.** 20;00_p.m.
(**Main Event/Panels/Events/Disc Jockey**) subject to change without notice.

<table>
<tr><td colspan="4">(Guest Of Honor) 2154;_ 11 (Nov) - 11 to 28 Entertainer Mister Steven Z. J.</td></tr>
<tr><td colspan="4">(Main Event) "D. Mister Steven Z. J. Show" Entertaining Subject(s):
Creative Accounting 101, Drunk Driving, Grudge Games, Legal Looting, Double Ds (. Y .)</td></tr>
<tr><td>(Panels/Events)
Tales Of Japanese Rocket Propelled Grenades (J.R.P.G.)

Feat. U.S. Maids And Cats-Kats</td><td>(Disc Jockey)
"D.J. Marvel's Spider-Man And Venom: Maximum Carnage"

Ft. Chris-Z And Gar Sometimes D. D., God</td><td>(Panels/Events)
Footse Fighters: The New Final Challengers' Super Hyper Ultra Champion Turbo Edition V.R. Remix Two</td><td>(Panels/Events)
Neo-Western Frontier Justice: Vigilantes And Militia

Ft. Alex Drastal, Big Ben, Lieutenant Lieu</td></tr>
<tr><td>(Panels/Events)
Climbing For Free Money

Feat. Mistress Stephanie</td><td>(Workshop)
Win The Lotto! What Would You Do With That Money!?

Automated Survey</td><td>(Panels/Events)
Knuckle Sandwich Birthday Punches

Feat. Chris-Z And Gar Sometimes D. D., God</td><td>(Panels/Events)
District Columbia's Vigilante(s):
Anarchy Origins
Anarchy Asylum
Anarchy City
Anarchy Knight</td></tr>
<tr><td>Panels/Events)
Everything On High Heels

Feat.
Mister Steven Z. J.</td><td>(Workshop)
Are You A Female Body Inspector (F.B.I.)?

Automated Survey</td><td>(Panels/Events)
The Karate Kanon

Feat. Ninja Kanon-Klone(s), Ninja Assassin(s)</td><td>(Panels/Events)
Costume Play (Cosplay) Show And Tell

Feat. V. V.</td></tr>
<tr><td colspan="4">(Premiere, Archive Recordings) @Downtown District, 4-D Movie Theater
District Columbia's Super-Hero V. Vigilante Department Of Just-Us Extended Edition.</td></tr>
<tr><td colspan="4">(Archive Recordings) @Downtown District, 4-D Movie Theater
The old world War Advanced Research Projects Agency (WARPA) secret tools.</td></tr>
<tr><td colspan="4">(Mature 17+ Content) @Downtown District, Elysium Corp Floors: 01-09, 10-19, 60-69
Can You Pass Elysium Corporation's Natural Unmodified Human Male (M) and Female (F) Parenting License? Feat. Acting Authority: Mister Kenny.</td></tr>
</table>

2154;_Arcadia \ Valentina Vita (V. V.) 'Healthy-Strong Life' \ Left to Right(L->R)

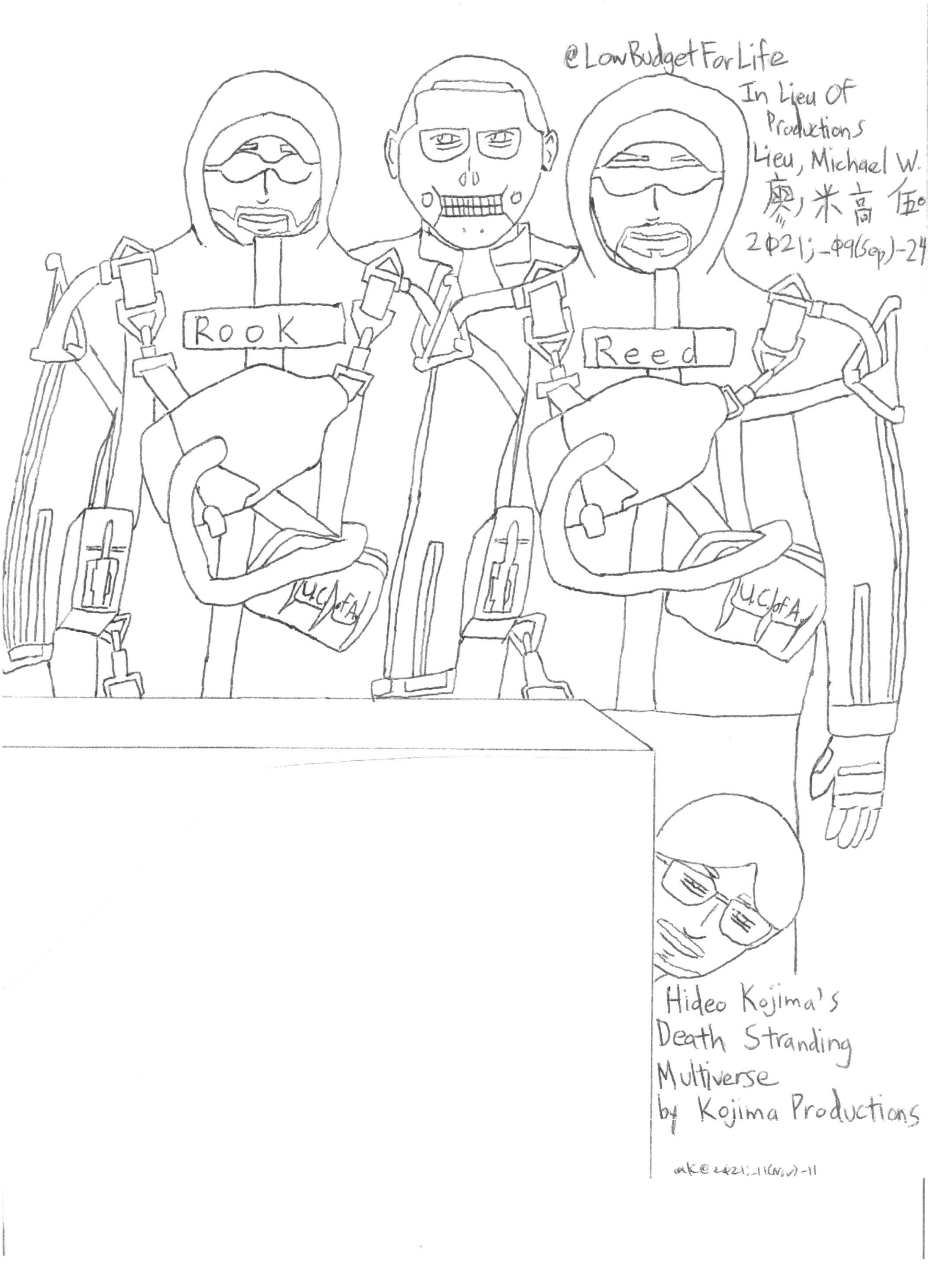
@LowBudgetForLife
In Lieu Of Productions
Lieu, Michael W.
廖, 米高 伍o
2Φ21;-Φ9(Sep)-24
Rook
Reed
U.C. of A.
U.C. of A.
Hideo Kojima's
Death Stranding
Multiverse
by Kojima Productions

Valentina Vita (V. V.); Traveling Cosplayers And Kosplayers;
"(^_^) Hello, I do not have my Ai-**Aug bracelet** turned on, no **Holo HUD**, so I am judging **your cosplays** under the **Cosplay Show And Tell**... My character is **KissShot-AcerolaOrion-HeartUnderBlade**, **costume** is from **Kizumonogatari** anime universe. Can **you** tell me what **your** character is, where is **your** character from and what universe?"

Die-Hardman looks at Mister **Reed**, **he** nods to **Die-Hardman**. Mister **Reed** selects the music "**Silent Poets Feat. Leila Adu's** *Asylums For The Feeling*, **Low Roars**: *Give Up, I'll Keep Coming, Patience, Without You, Anything You Need, Don't Be So Serious, Bones, Poznań, Easy Way Out*" to play in **his Utopia A.R. Holo HUD** bio**tech cuffs**.

Die-Hardman, President of United Cities of Americas (U.C. of A.); Death Stranding;
"(*in face mask*) Hello, V. V. ...**We** are part of the **public private sector partnership**: **United Cities of Americas** and **Bridges Company** along side **Fragile Express**. **I** am **Die-Hardman**, **president** of **U.C. of A.** Misters' **Rook** and **Reed** are **employees** under the **Bridges Company**. The **Bridges Company** handles **logistics**, **supply**, **mail**, **cargo**, **physical goods** and **services** in aiming to **reconnect** the **world**! With **our** private sector companies **providing courier delivery** service and **connecting** people after the **collapsed** United States **federal government** and the **Postal Service**... **We** hope to Unite the Americas once again... with **North America** first, then **South** one day... **We** are currently dealing with **problematic** events: **terrorists**, **separatists**, **Breached Things**, **DOOMS**, **Death Stranding**, **Extinction Entity**, **Beaches**, many phenomenons. **We** have **problems** ahead of **us**, but **we** will **reconnect** the **world**! That is **our** number one **mission**!"

Mister Reed Bridges; Hideo Kojima's Death Stranding MultiVerse;
"(*calm*) **I**'m Mister **Reed**, **I** do pretty much what **Die-Hardman** said... The **private sector courier** duty... Yeah... Oh, these **containers** that hold **Bridge Babies**... Hard to explain what these little **cute critters** do, but **they** help **us** in **our** long journey ahead. Hm..."

Mister Rook Bridges; Hideo Kojima's Death Stranding MultiVerse;
"(*calm*) **I**'m Mister **Rook**, **I** do **courier delivery** services just like what Mister **Reed** said. **Me** and Mister **Reed...** **We** both **tag team couriers** for **Bridges Company**... Yea..."

Kojima-san; Hideo Kojima's Death Stranding MultiVerse By Kojima Productions;
"(*calm*) Hello **everyone**, **I**'m back. **I** am **Kojima-san** in a **cardboard box.** This is a **cardboard box** used for **hiding**. **I** created this **Death Stranding** universe. **H.K.** style."

Valentina Vita (V. V.); Traveling Cosplayers And Kosplayers;
"(^_^) Whoa we wow! Good presentation **everyone**! That **music**, makes me feel **melancholy**... **Die-Hardman**, that was a good **brief** information about **your** characters and the **universe** where it comes from! **No ugly threads**, ten of ten! Hm... I'm wondering about the **cardboard box** hiding act... Good job **everyone**! Have a nice day!"

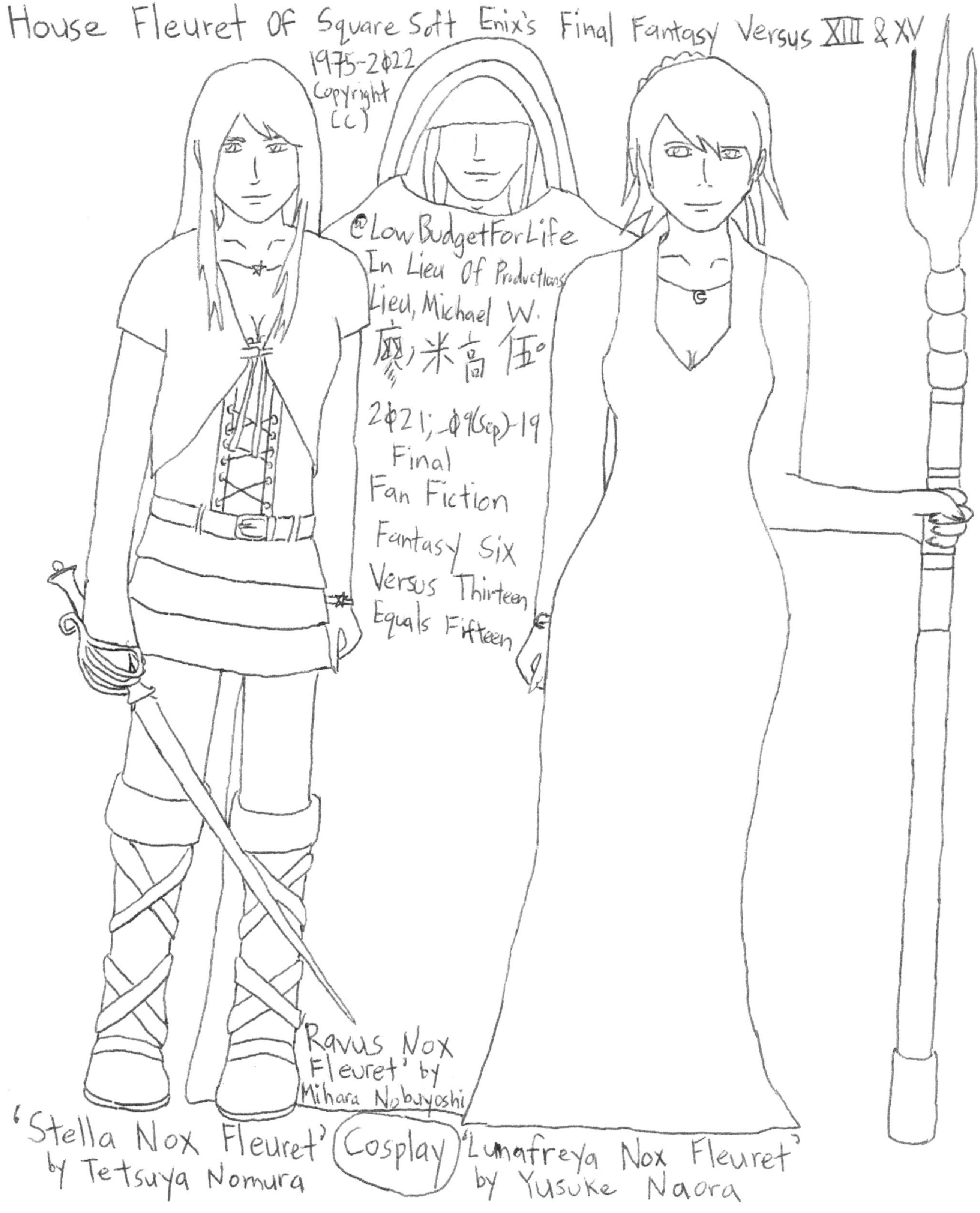

Inke 2021_11(Nov)-04

Valentina Vita (V. V.); Traveling Cosplayers And Kosplayers;
"(@_@) Whoa we wow, **you** must be that **Final Fantasy** characters' **Fan Fic** by Jay Doe? (O_O) Looks good **no ugly threads**! **You** three all look good! As a traveling cosplayer, here is my F.F. cosplay. This Dragoon Knight custom armor was used as an external travel P.P.E. suit, it's in my luggage after bio**tech airlock decontamination**..."

Alien Valentina Vita, Traveling Cosplayer(s) And Kosplayer(s)
2154;_ Chicago, Illinois' Intranet, Wireless Broadcasting Frequencies Online

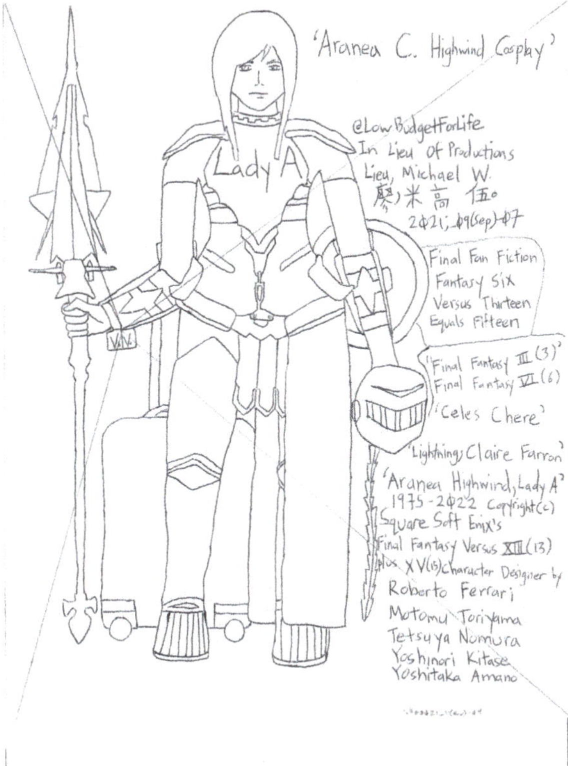

2154;_ Chicago, Illinois' Intranet, Wireless Broadcasting Frequencies Online
Alien Valentina Vita, Traveling Cosplayer(s) And Kosplayer(s)

Lord Ravus Nox Fleuret Cosplay; Final Fan Fiction Fantasy 6 Versus 13 Equals 15;
"(*smiling*) Wow... Lady A... Hm... a custom cosplay inspired by other **Final Fantasy universes**! **Lord Ravus** was in both **F.F. Versus Thirteen** and **Fifteen**... This **fan fiction** version is way better as a **battle biotech armiger**! **We** three are **House Fleuret**!!"

Lady Stella Nox Fleuret Cosplay; Final Fan Fiction Fantasy 6 Versus 13 Equals 15;
"(*staring*) Oh! ...Mines is based off **Final Fantasy Versus Thirteen**... **Stella**!"

Lady Lunafreya Nox Fleuret Cosplay; Final Fan Fiction Fantasy 6 Versus 13 Equals 15;
"(*staring*) **Stella** and **Lunafreya** do not meet... **Lunafreya** is in **Final Fantasy Fifteen. Stella** is only in **Final Fantasy Versus Thirteen**. **We** are **House Fleuret Sisters**!"

2154;_Arcadia \ Valentina Vita (V. V.) 'Healthy-Strong Life' \ Left to Right(L->R)

Lady Stella Nox Fleuret Cosplay selects the music "**Final Fantasy VI**: *Opening Theme, Dark World, Aria Di Mezzo Carattere Remix, My Mind, Tina, New Continent, Edgar & Mash, Mystery Train, Spinach Rag, Stragus, The Mystic Forest, Unknown*, **Final Fantasy Versus XIII**: *Theme, Somnus Drammatica*, **Final Fantasy XV**: *Episode Ardyn: Conditioned To Hate feat. Lotus Juice, Main Theme, Prayer of the Oracle, Ardyn*, **Florence The Machine:** *Stand By Me, I Will Be*" as it plays in **her Utopia A.R. Holo HUD biotech star bracelet**.

Valentina Vita (V. V.); Traveling Cosplayers And Kosplayers;
"(^_^) ...Yes, this is **Aranea C. Highwind** of the **Niflheim Empire, Air Force Operative,** her nickname is quote **Lady A.** unquote said by... **Biggs And Wedge** in **Episode Highwind**... Jay Doe's **Final Fan Fiction Fantasy Six Versus Thirteen Equals Fifteen** is quite a **fan fiction** and long name, that is not **porn** with character sexual relationships and intercourse... The whole **fan fic** is done in an **animated comic book-ish storyboard** like **T.V. show series** with quote **Episode** unquote as the **character focus**... For **Ravus**... in **Episode Tenebere**, the **Empire** invades and conquers **Tenebere** of **House Fleuret, Emperor Aldercapt** makes **Ravus** the **Lord** or **High Commander**, **Commanding Officer**, I forget which rank... in **Episode Galahd**... **General Glauca**, a soldier fighting in the invasion of **Galahd**, is a **klone** of **Prompto** who is from **Niflheim**... (>_<) Spoiler?"

The **House Fleuret** cosplayers of **Ravus, Stella, Lunafreya** all look at V. V.. She is quiet.

Valentina Vita (V. V.); Traveling Cosplayers And Kosplayers;
"(O_<) Anyways, **Lady A.** a nickname given by **her** closest human **comrades** during **her** service to the **Empire**. **Her Dragoon Knight** armor design is so kool! By **Episode Ruin**, **she** leads the **remaining humans' army** in Eos after the **Empire** and **world** collapses to the **dark ages**, **the world of ruin**! **King Noctis Lucis Caelum** returns **ten years** later."

Lord Ravus Nox Fleuret Cosplay; Final Fan Fiction Fantasy 6 Versus 13 Equals 15;
"(*voice over*) Wow... I need to finish the rest of this... Jay Doe's **Final Fan Fiction Fantasy Six Versus Thirteen Equals Fifteen.** Got some **homework** to do as a cosplayer, ha ha! I saw the **Lord Ravus** or **High Commander Ravus** and I knew I had to cosplay the character, he is so strong with his white coat bio**tech armiger**! Hard to find the **Episodes**."

Lady Stella Nox Fleuret Cosplay; Final Fan Fiction Fantasy 6 Versus 13 Equals 15;
"(*voice over*) ...**I** know it is hard to find the **remaining Episodes**... But, **I** remember **Lady Stella** of **House Fleuret** meets **Prince Noctis** of **House Caelum** in **Insomnia** during that political peace treaty before **he** becomes **King Noctis** during **Episodes**... **Kingsglavie...**"

Lady Lunafreya Nox Fleuret Cosplay; Final Fan Fiction Fantasy 6 Versus 13 Equals 15;
"(*voice over*) Lady Luna gets kidnapped by **Arydn** in **Episode Ignis**... **Arydn** cuts her hair and changes her clothes into Aera **The First Oracle Of House Fleuret**! **He** is such a creepy **Adagium**!! Such a creep! That **Adagium Arydn**!"

Valentina Vita (V. V.); Traveling Cosplayers And Kosplayers;
"(@_@) A volunteer worker, Elle, found a **Final Fantasy Fan Fic cosplayers** in the **Downtown District** on **October thirty one or November one**... Some days before... **Mister Steven's Entertainers' Expo** started... Not sure if those **two** are still around."

Alien Valentina Vita, Traveling Cosplayer(s) And Kosplayer(s)
2154;_ Chicago, Illinois' Intranet, Wireless Broadcasting Frequencies Online

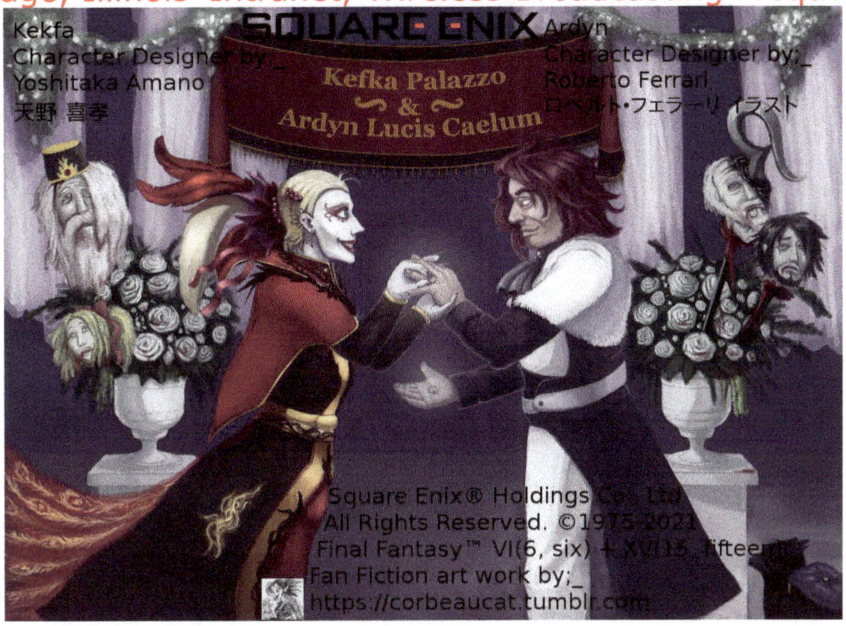

2154;_ Chicago, Illinois' Intranet, Wireless Broadcasting Frequencies Online
Alien Valentina Vita, Traveling Cosplayer(s) And Kosplayer(s)

The **House Fleuret cosplayers** of **Ravus, Stella, Lunafreya** all look at V. V.'s A.R. Holo HUD projection in shock and awe. She notices **their** silence and waits a bit.

Valentina Vita (V. V.); Traveling Cosplayers And Kosplayers;
"(T_T) ...I already visited this sector's H.D.D.V.D. archives and there is no... **remaining Episodes** of Jay Doe's **Final Fan Fiction Fantasy Six Versus Thirteen Equals Fifteen**. Maybe in **Elysium Corp's High Tower Floors**, if **you** dare to go to **Downtown District** past **Four D Movie Theater** with Mister Kenny breeder builder of this biotech dome **Animal** Sanctuary... Oh, my character is **KissShot-AcerolaOrion-HeartUnderBlade**, **costume** I am wearing is from **Kizumonogatari** anime universe. Questions, concerns, comments?"

Lady Stella Nox Fleuret Cosplay; Final Fan Fiction Fantasy 6 Versus 13 Equals 15;
"(*voice over*) None. **We** want to find that **Ardyn cosplayer**! Thanks for your time V. V.!

Valentina Vita (V. V.); Traveling Cosplayers And Kosplayers;
"(^_^) Ten of ten! **No ugly threads**! **You three** of **House Fleuret** have a nice day!"

2154;_Arcadia \ Valentina Vita (V. V.) 'Healthy-Strong Life' \ Left to Right(L->R)

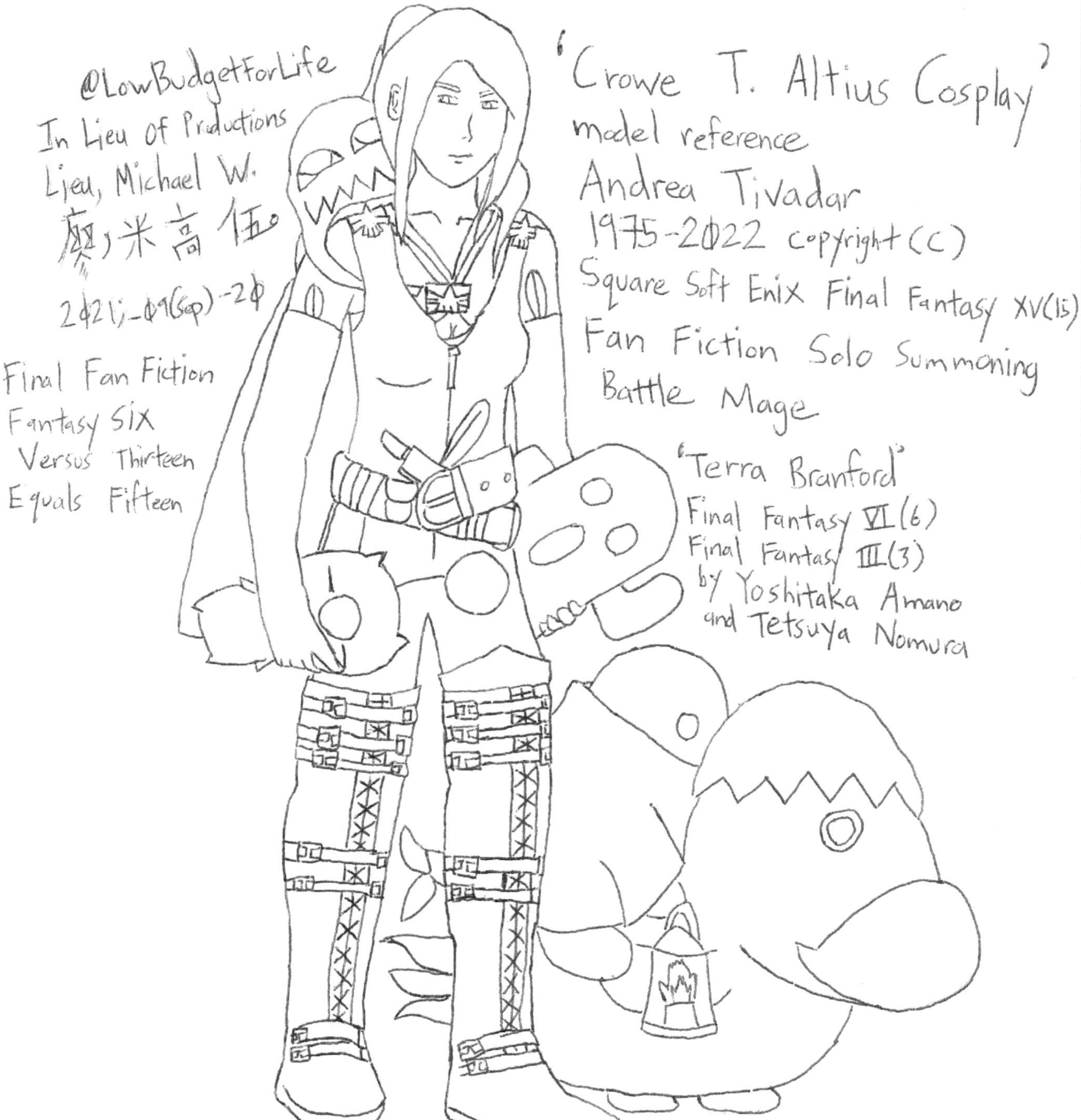
@LowBudgetForLife
In Lieu Of Productions
Lieu, Michael W.
廖, 米高 伍o
2021-09(Sep)-20
Final Fan Fiction
Fantasy Six
Versus Thirteen
Equals Fifteen
'Crowe T. Altius Cosplay'
model reference
Andrea Tivadar
1975-2022 copyright (C)
Square Soft Enix Final Fantasy XV(15)
Fan Fiction Solo Summoning
Battle Mage
'Terra Branford'
Final Fantasy VI (6)
Final Fantasy III (3)
by Yoshitaka Amano
and Tetsuya Nomura

Valentina Vita (V. V.); Traveling Cosplayers And Kosplayers;
"(@_@) Whoa we wow, **you** must be another **Final Fantasy** character? Look at all those **F.F. summons** Moggle, Cactuar, Tonberry, Chocobo, Ultros. Just met **three cosplayers** who are **House Fleuret**! ...**Stella, Ravus, Lunafreya**. Here is my F.F. cosplay. This Dragoon Knight custom armor, I used as an external P.P.E. suit, it's in my luggage."

Alien Valentina Vita, Traveling Cosplayer(s) And Kosplayer(s)
2154;_ Chicago, Illinois' Intranet, Wireless Broadcasting Frequencies Online

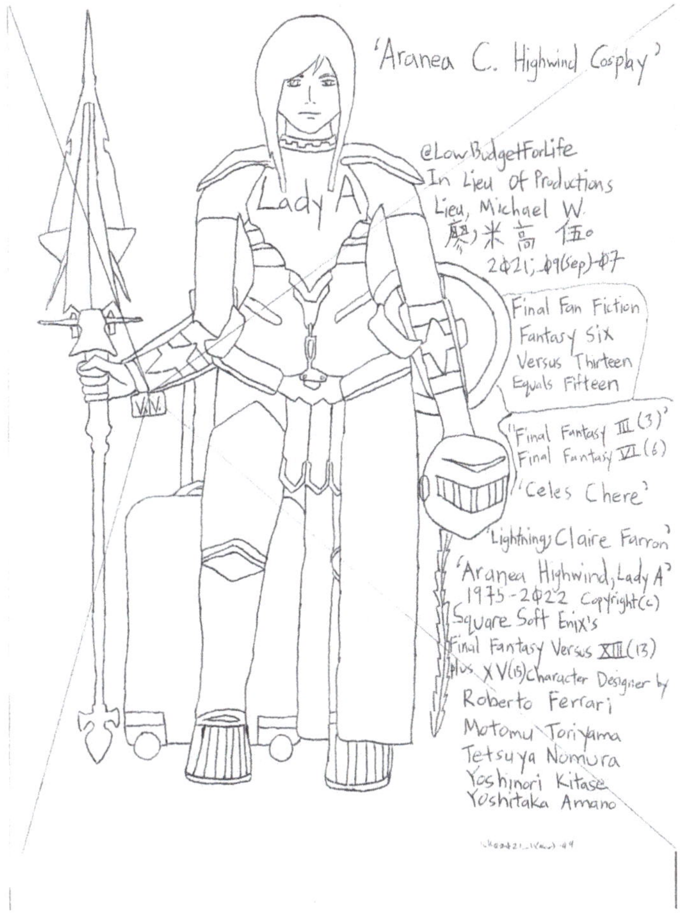

2154;_ Chicago, Illinois' Intranet, Wireless Broadcasting Frequencies Online
Alien Valentina Vita, Traveling Cosplayer(s) And Kosplayer(s)

Refugee Crowe T. Altius Cosplay; Final Fan Fiction Fantasy 6 Versus 13 Equals 15;
"(*tired*) Yes, **I** saw **them** before hand and got to exchange info with **them**... (*looking*) Whoa, **Lady A. ... Oh that is Aranea C. Highwind** of the **Empire**! Wow, that is really work of wonders... Looking at **my summons, I** wonder if **my luggage** can carry all this."

Valentina Vita (V. V.); Traveling Cosplayers And Kosplayers;
"(^_^) **Those three** are heading to **Downtown District**! Ten of ten! **No ugly threads**!

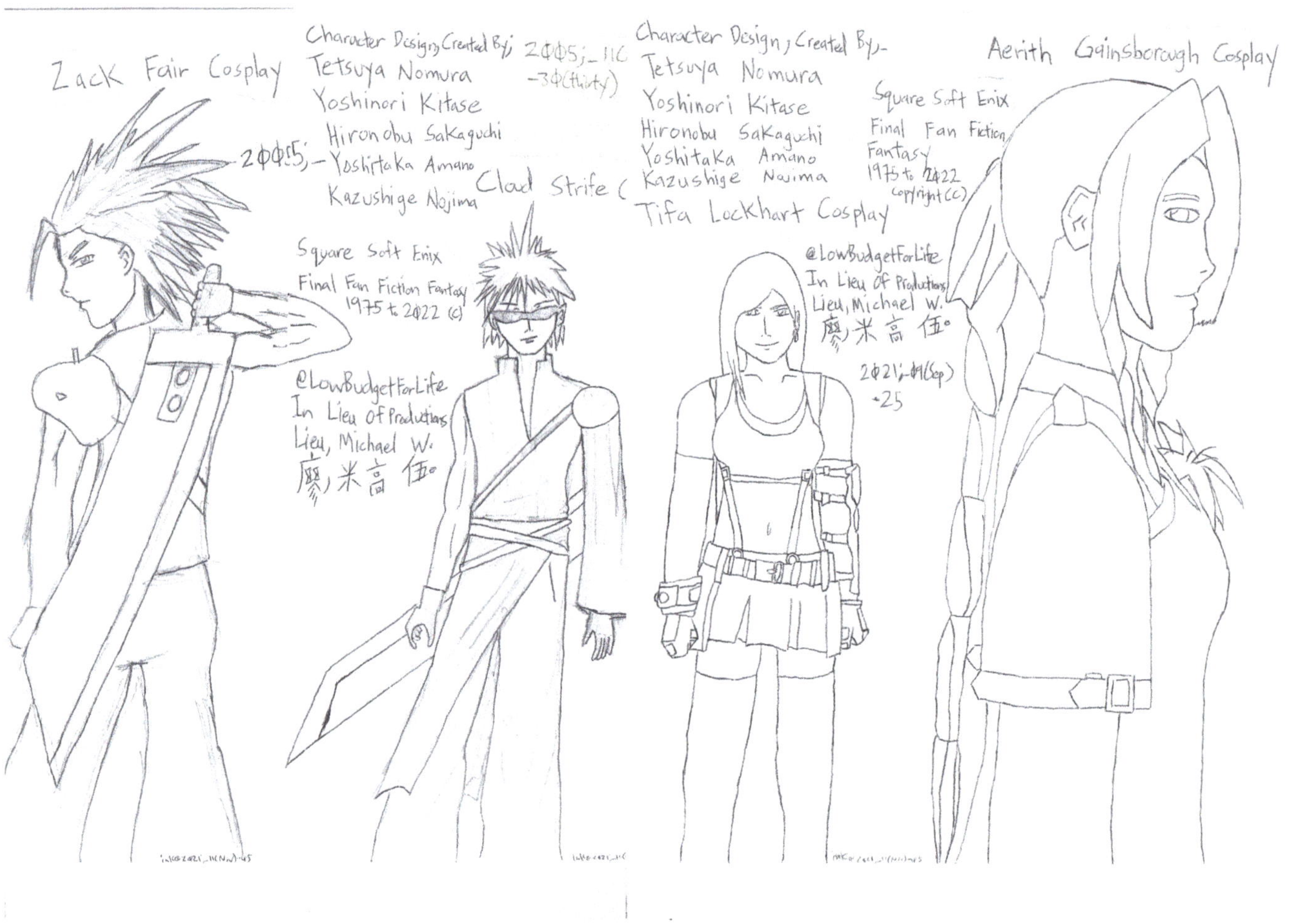

Valentina Vita (V. V.); Traveling Cosplayers And Kosplayers;
"(@_@) Whoa we wow, **you** all must be **Final Fantasy characters**? Looks good, **no ugly threads**! **You** four all look good. Here is my F.F. cosplay. This Dragoon Knight custom armor was used as an external P.P.E. suit, it's in my luggage**...** Here **you** go!"

Zack Fair Cosplay; Final Fan Fiction Fantasy 7 Multiple uniVerse (MultiVerse);
"(*posing*) Yes, thanks a bunch, part of **my** costume requires physical peak condition as well!"

Cloud Strife Cosplay; Final Fan Fiction Fantasy 7 Multiple uniVerse (MultiVerse);
"(*calm*) ...Hm... Not... Hm..."

Tifa Lockhart Cosplay; Final Fan Fiction Fantasy 7 Multiple uniVerse (MultiVerse);
"(*smiling*) ...**You** were gonna say something or just stand still with those **sunglasses**?"

Aerith Gainsborough Cosplay; Final Fan Fiction Fantasy 7 Multiple uniVerse (MultiVerse);
"(*staring*) Hey, how long are **we** gonna stay in posing position...? **I** feel funny..."

V. V. shows **all four Final Fantasy 7 cosplayers** her custom cosplay via Holo HUD.

Alien Valentina Vita, Traveling Cosplayer(s) And Kosplayer(s)
2154;_ Chicago, Illinois' Intranet, Wireless Broadcasting Frequencies Online

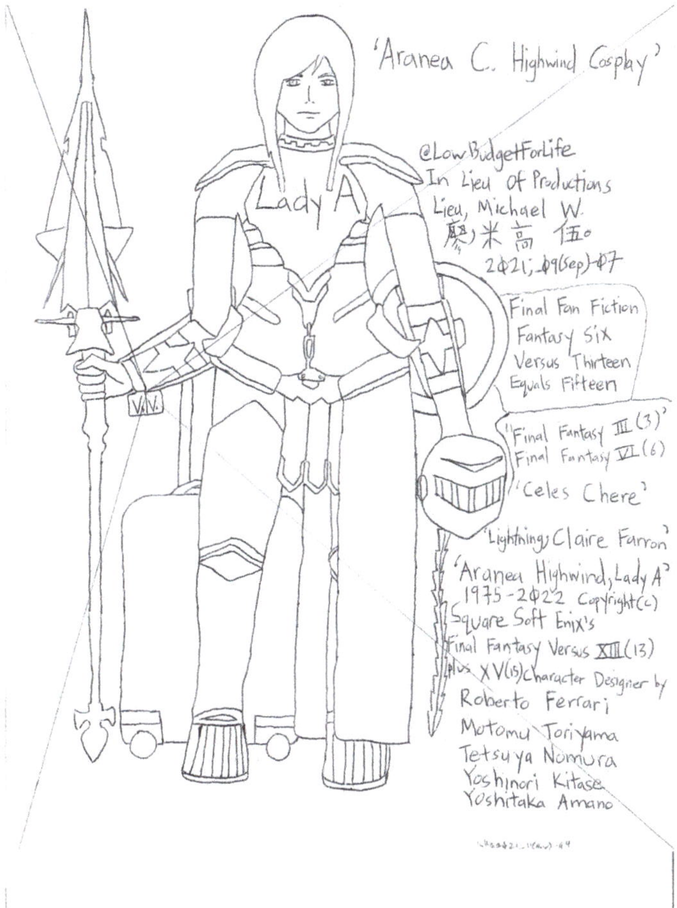

2154;_ Chicago, Illinois' Intranet, Wireless Broadcasting Frequencies Online
Alien Valentina Vita, Traveling Cosplayer(s) And Kosplayer(s)

Zack Fair Cosplay; Final Fan Fiction Fantasy 7 Multiple uniVerse (MultiVerse);
"(*posing*) Whoa... That armor does not look like anything **Shinra engineered**..."

Tifa Lockhart Cosplay; Final Fan Fiction Fantasy 7 Multiple uniVerse (MultiVerse);
"(*smiling*) Whoa... Wow, good job on that cosplay as a P.P.E. suit... Looks great!"

Aerith Gainsborough Cosplay; Final Fan Fiction Fantasy 7 Multiple uniVerse (MultiVerse);
"(*staring*) Whoa... Hey, are **we** gonna check out **Elysium Corp**? It's like **Shinra Building**!

All **four Final Fantasy 7 cosplayers** departed quickly towards the **Downtown District** of **Elysium Corporation**, before V. V. mentions her current costume. She is staring at **them**.

2154;_ Arcadia \ Valentina Vita (V. V.) 'Healthy-Strong Life' \ Left to Right(L->R)

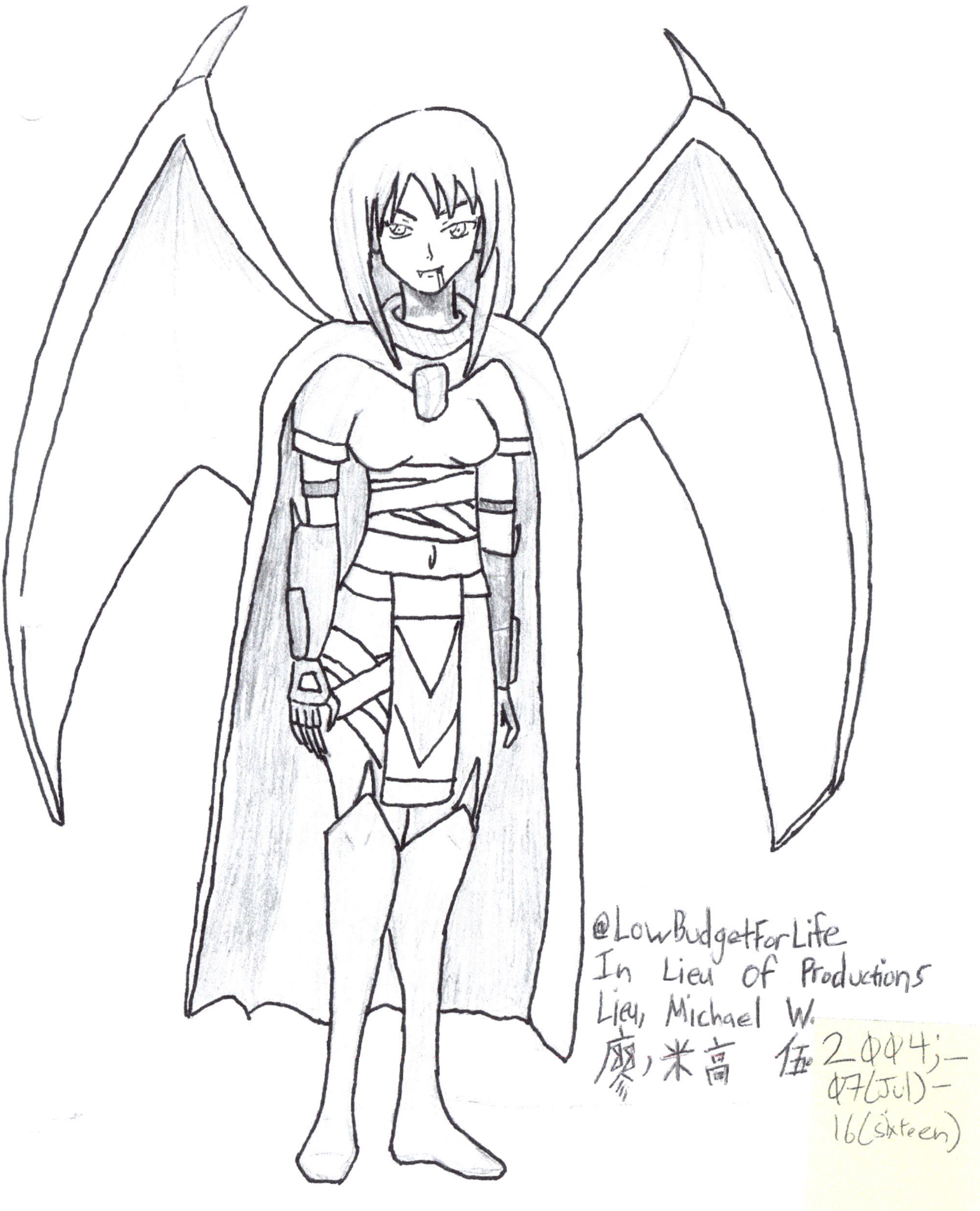
@LowBudgetForLife
In Lieu Of Productions
Lieu, Michael W.
廖, 米高　伍
2ΦΦ4;—
Φ7(Jul)—
16(sixteen)

V. V. sits around for a bit before any other **people in costume as cosplayers** walk up to her **Cosplay Show And Tell** table in the **East Side All For One Shopping Center**.

Valentina Vita (V. V.); Traveling Cosplayers And Kosplayers;
"(^_^) Hello, I'm V. V., my character is **KissShot**-AcerolaOrion-**HeartUnderBlade**, **costume** I am wearing is from **Kizumonogatari** anime universe. So far, I see **no ugly threads sticking out**, **your** costume looks amazing with those wings, fangs, bandages, cape, armlet, and pointy edges... (@_@) Are you a **vampire, succubus, human bat**? Is from **Final Fantasy** or **Death Stranding universe**?"

Bat Woman Vampire; **Utopian Sovereignty (U.S.) Economic Free Market**;
"(*staring*) Thanks, it is original. I just made it as an **arts and crafts**. Wanted to do something of a horror **human animal fusion**... A Bat Woman Vampire **I** guess...?"

V. V. selects the visual reference of **KissShot**-AcerolaOrion-**HeartUnderBlade** to show and tell the Bat Woman Vampire **cosplayer** looks at V. V. visual reference photo and nods.

Alien Valentina Vita, Traveling Cosplayer(s) And Kosplayer(s)
2154;_ Chicago, Illinois' Intranet, Wireless Broadcasting Frequencies Online

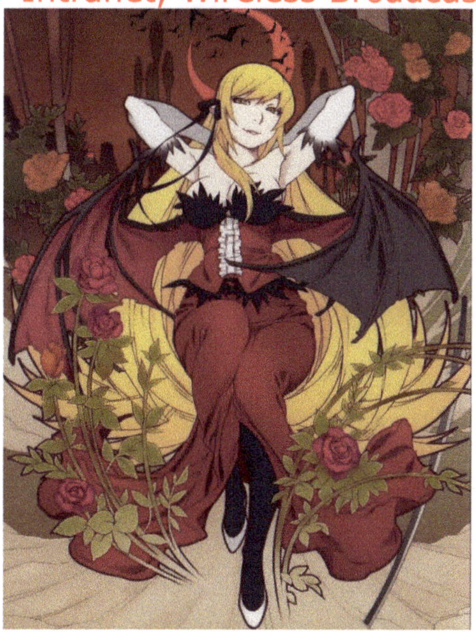

2154;_ Chicago, Illinois' Intranet, Wireless Broadcasting Frequencies Online
Alien Valentina Vita, Traveling Cosplayer(s) And Kosplayer(s)

Valentina Vita (V. V.); Traveling Cosplayers And Kosplayers;
"(^_^) Bat Woman Vampire! Ten of ten, **no ugly threads sticking out**, keep up the good work of your future **arts and crafts**! Have a nice day! Bye bye Bat Woman Vampire!"

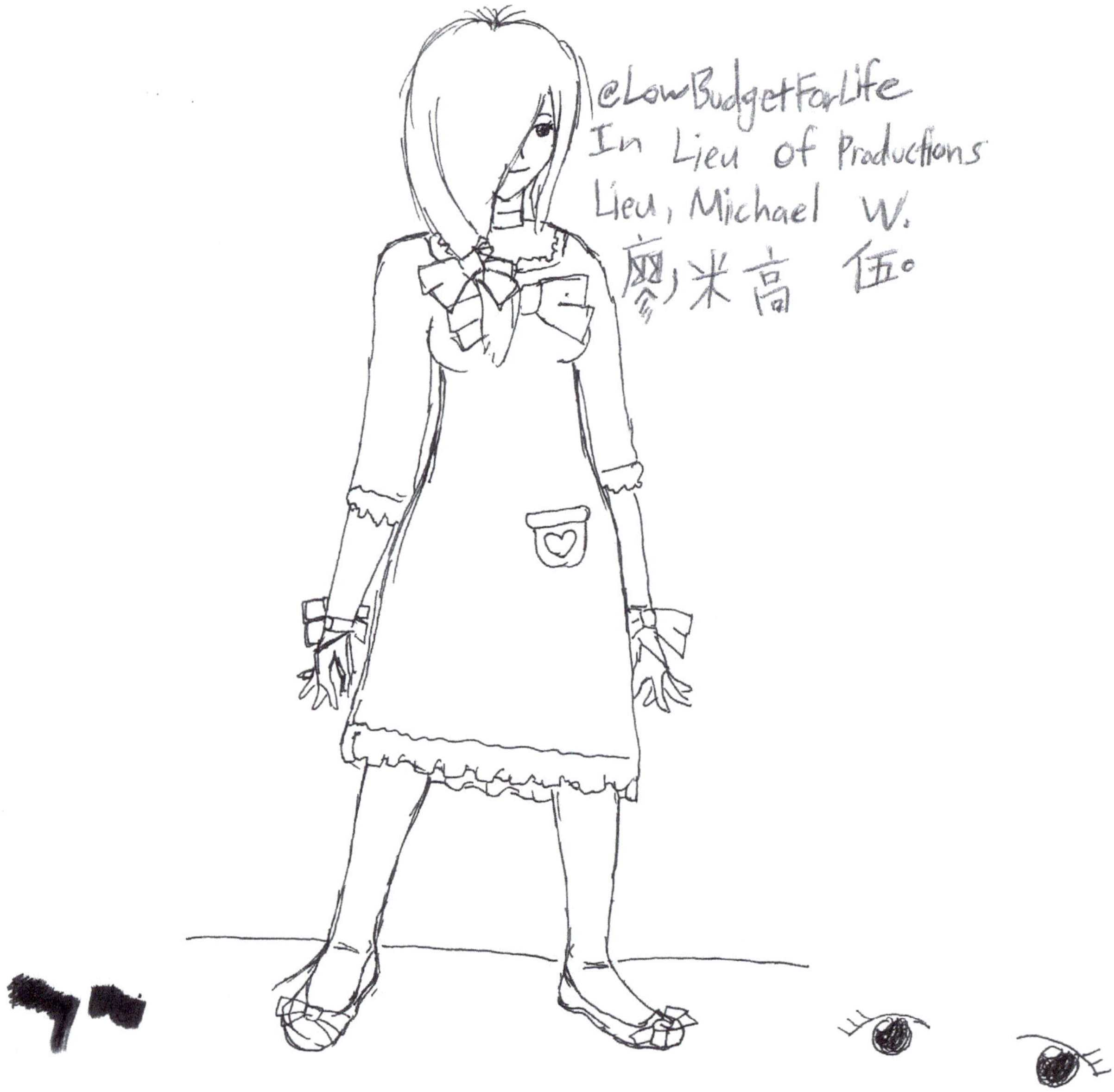

2Φ1Φ;_
Φ1(Jan)-
19(nineteen)

Valentina Vita (V. V.); Traveling Cosplayers And Kosplayers;
"(@_@) Hello, I'm V. V., my character is **KissShot**-AcerolaOrion-**HeartUnderBlade**, **costume** I am wearing is from **Kizumonogatari** anime universe. And... this does not look like any **Final Fantasy** or **Death Stranding universe** cosplay... Is it original?"

Miss Lolita Fashion; **Utopian Sovereignty (U.S.) Economic Free Market**;
"(*smiling*) Yes, hello V. V., this is an **original costume or fashion**, not so much a **cosplay** based off existing **franchises** or **universes**. It's a casual cute white lolita dress and a cute heart love small front pocket and a cute ribbons all around. **I** would say, **my lolita** looks style is classic, has a bit of mixing in some areas. Say, what is your favorite **lolita** style?"

Valentina Vita (V. V.); Traveling Cosplayers And Kosplayers;
"(>_O) Oh, I follow bit on the **lolita fashion**. I go with **classic lolita**, I do like the **Gothic lolita** and **sweet lolita**. But, my costume I am wearing is a red evening cocktail dress party mingling. Did **you** happened to see a Bat Woman Vampire **cosplayer**, her **costume** was **original** like **yours** not based off any **media universes**. (O_O) She had huge wings, it was so well done, it looked like a form of **augmentation**... Maybe it is **augmentation**."

Miss Lolita Fashion; **Utopian Sovereignty (U.S.) Economic Free Market**;
"(*smiling*) **I** did see that cosplayer with those large wings, but did not meet and greet her."

Valentina Vita (V. V.); Traveling Cosplayers And Kosplayers;
"(>_<) ...It's okay, it's okay... Anyways, back to the different **lolita** look types!"

Miss Lolita Fashion; **Utopian Sovereignty (U.S.) Economic Free Market**;
"(*smiling*) Okay, so **I** believe there are the following types of **Lolita** styles... **Sweet Lolita**, **Classic Lolita**, **Gothic Lolita**, Mixing styles, Sub-Styles of Nautical, Country, Old School, Hime, Kuro Shiro Or **Black** White, **Guro**, **Ero**, **Punk**... Oh, **I** did not bring a visual guide with **me**... Explaining it through words is hard... But, you get what **I** am saying right V. V. as a **cosplayer**, you know the different **fashions** and **clothing** style textures and such?"

Valentina Vita (V. V.); Traveling Cosplayers And Kosplayers;
"(o_O) Of course... Um... **your** name, I do not think I got **your** name...?"

Miss Lolita Fashion; **Utopian Sovereignty (U.S.) Economic Free Market**;
"(*smiling*) **Miss Lolita**, **I** appreciate the chit chat with you V. V., time to continue exploring!"

Valentina Vita (V. V.); Traveling Cosplayers And Kosplayers;
"(^_^) Oh course, feel free to check out the rest of the panels, events going on with the Entertainers' Expo **Miss Lolita**. Ten of ten, **no ugly threads sticking out** of **your** dress!

Miss Lolita waves bye to V. V. and she waves back to **Miss Lolita** as **she** departs.

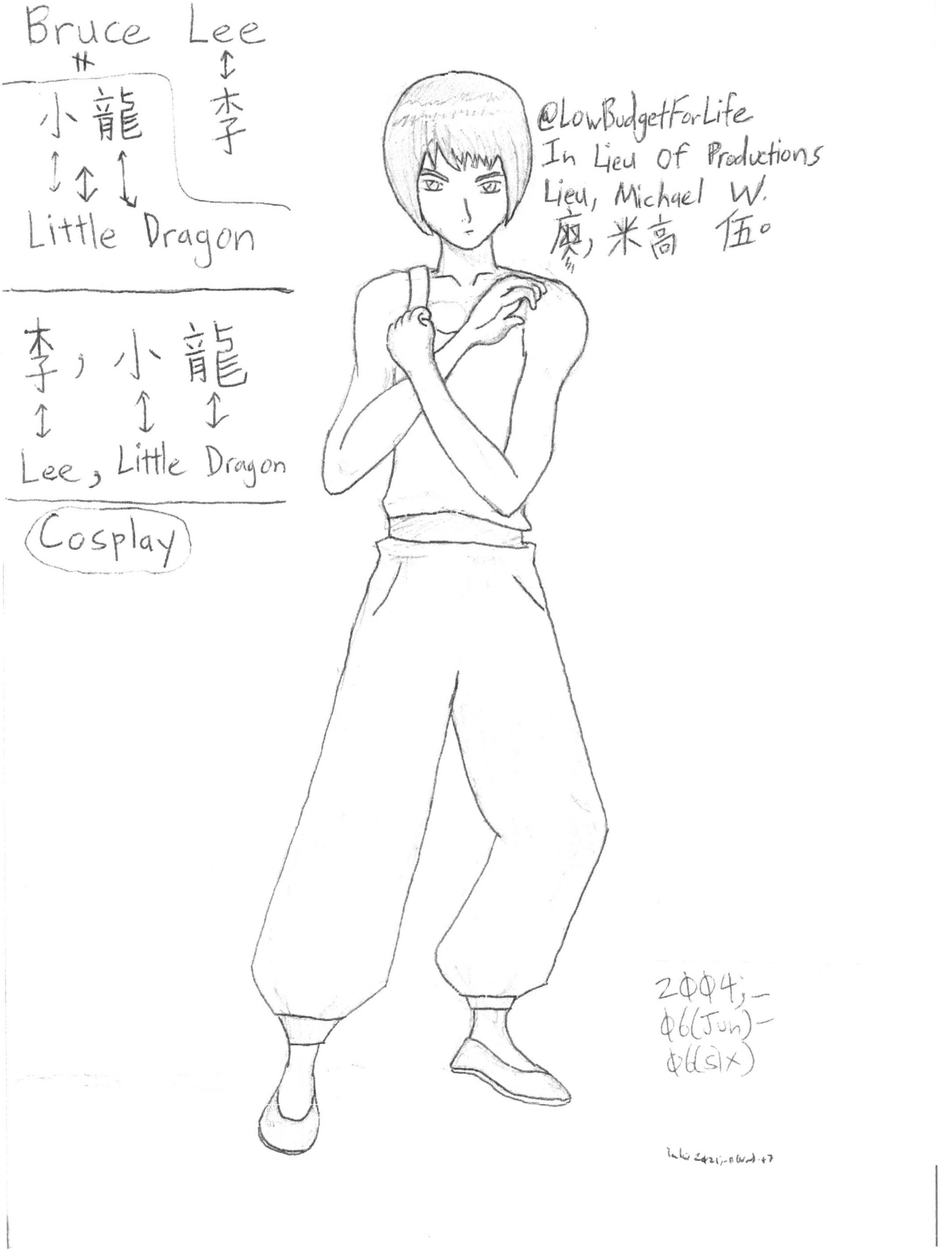
Bruce Lee
小龍 李
Little Dragon
李, 小龍
Lee, Little Dragon
Cosplay
@LowBudgetForLife
In Lieu Of Productions
Lieu, Michael W.
廖, 米高 伍o
2ΦΦ4;-
Φ6(Jun)-
Φ6(six)

V. V. sees a Martial Artist doing practicing various **martial arts fighting forms** as **he** is heading to her. **He** gets in **his** fighting stance towards V. V.'s **Cosplay Show And Tell**.

Valentina Vita (V. V.); Traveling Cosplayers And Kosplayers;
"(@_@) Hello, I'm V. V., my character is **KissShot-AcerolaOrion-HeartUnderBlade**, the **costume** I am wearing is from **Kizumonogatari** anime universe. And... this is a Martial Artist... **Bruce Lee**...? I am not a Martial Artists yet I know this! (^_^) **He** is so well known... By the **twentieth, twenty first and twenty second century** still!"

Bruce Lee, Little Dragon Cosplay; Kung-Fu, Martial Artists "MartialArtsGod 武道神" ;
"(*martial arts form*) Hello V. V., yes, you are correct **I** am a Martial Artists **cosplayer** of **Bruce Lee**... **Lee, Little Dragon** as **he** is known in **Eastern society** of Central Country, in Central Characters, **Kanji**... Practitioner of a Martial Arts form **he** came up with **Jeet Kune Do**... and **his** philosophy on **be** water **my friend**... Seeing **Bruce Lee** beat up ten people does something to you, being able to fight back and defend yourself against such odds. Yea."

Valentina Vita (V. V.); Traveling Cosplayers And Kosplayers;
"(?_?) Clearly, **his** philosophy, Martial Arts, **spirit still lives on** to this day. (@_@) I have to ask... Those flips and kicks **you** were doing is pretty cool is that like **parkour** or **freerunning** or Martial Arts **form**?"

Bruce Lee, Little Dragon Cosplay; Kung-Fu, Martial Artists "MartialArtsGod 武道神" ;
"(*martial arts form*) Oh that, that is called **tricking**. It is a **mix of** Martial Arts **fundamentals**, **gymnastics**, **Capoeira**, **spinning kicks**, **flips**, **breakdancing**, **gravity defying stunts on one gravity pull rate**... **Tricking** is meant to please an **audience**... **Not used for fighting**... as a fighting form style. In one gravity pull rate, **tricking** still does a take a heavy toll on the body, so tomorrow or tonight, **I** gotta rest up and heal."

Valentina Vita (V. V.); Traveling Cosplayers And Kosplayers;
"(O_O) ...Whoa we wow, thanks so much for the info on Martial Arts **tricking**."

Bruce Lee, Little Dragon Cosplay; Kung-Fu, Martial Artists "MartialArtsGod 武道神" ;
"(*martial arts form*) Anytime, **I** gotta get back to training, Martial Arts **warrior spirit**!"

Valentina Vita (V. V.); Traveling Cosplayers And Kosplayers;
"(^_^) Ten of ten, **no ugly threads sticking out** of **your Bruce Lee cosplay**."

The Martial Artist **Bruce Lee, Little Dragon Cosplayer** performs Martial Arts **tricking** as V. V. waves bye bye to **him**. **He** does more numerous Martial Arts **tricking** as **he** moves away from her.

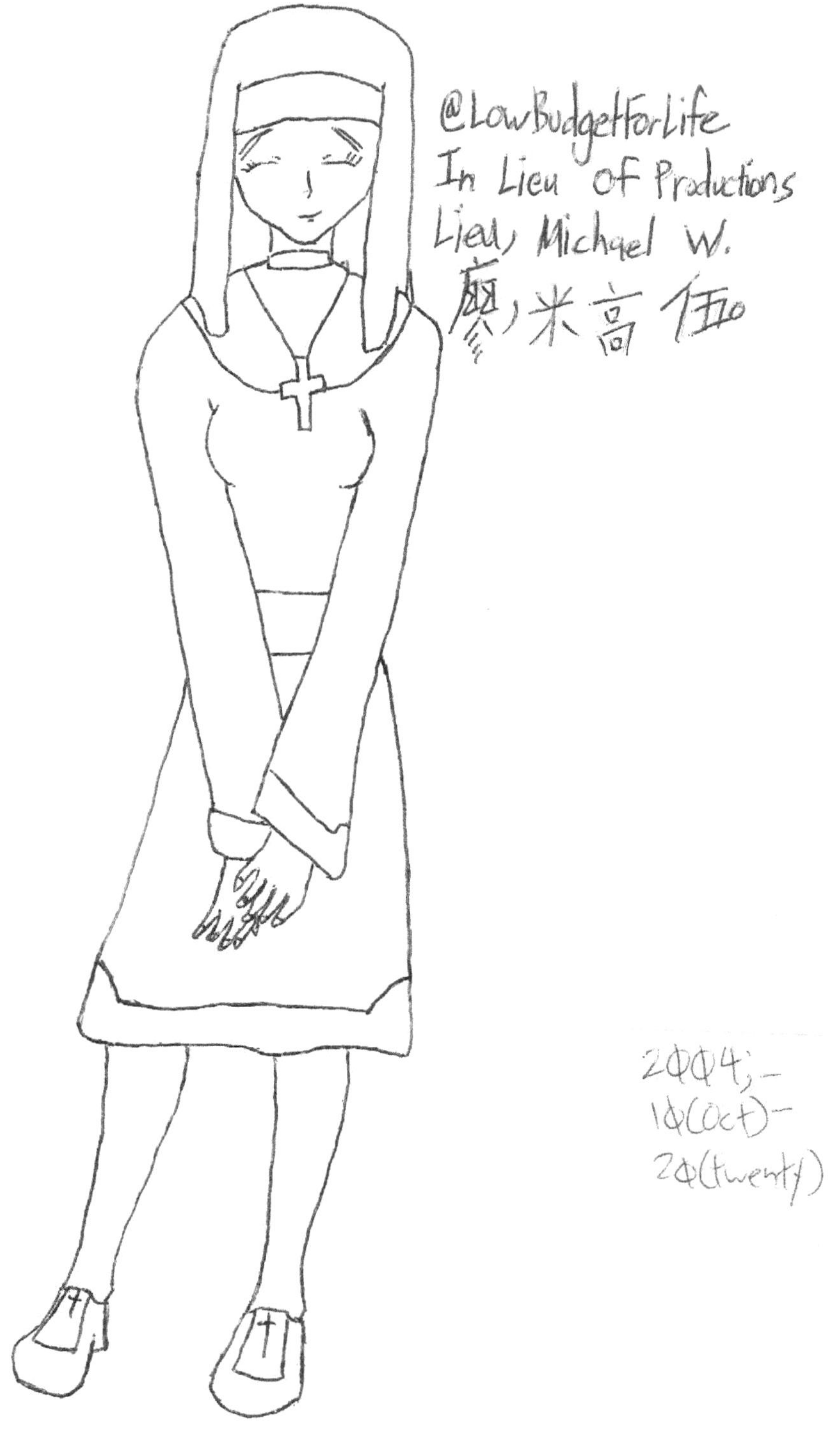
@LowBudgetForLife
In Lieu Of Productions
Lieu, Michael W.
廖, 米高 伍o
2ΦΦ4;-
1Φ(Oct)-
2Φ(twenty)

V. V. sits around for a bit and sees Sister **K.** White <3 approaching her. Sister **K.** White <3 selects the music: "**BooHooWoo Dot Hack's** Aura's Theme,

Castlevania Symphony Of The Night's *Prayer*	**Castlevania Symphony Of The Night's** *Requiem For The Gods*
Bloodstained Ritual Of The Night's *Holy Wisdom*	***Puella Magi Madoka Magica Original*** **Soundtrack Vol.2's** *Ave Maria*

" to play on her earrings' **Utopia A.R. Holo HUD** on one third volume of 33% and is set on loop.

Valentina Vita (V. V.); Traveling Cosplayers And Kosplayers;
"(O_O) Oh hi Sister White!! Are you part of the... **cosplay show and tell** as well?"

Sister K. White <3; Pretty Sure This is Not HeartBladeEndureLove 心刀忍愛;
"(*smiling with her eyes closed*) Oh hi V. V.! I am out of the Church and want to do some sight seeing with **Mister Steven** Entertainers' Expo in our Chicago **Animal** Sanctuary."

Valentina Vita (V. V.); Traveling Cosplayers And Kosplayers;
"(O_o) Okay... Sister White! How are you able to walk and talk with your eyes closed?"

Sister K. White <3; Pretty Sure This is Not HeartBladeEndureLove 心刀忍愛;
"(*smiling with her eyes closed*) I have **many eyes** besides the two in front of me on **this human head platform**... Allow me to show you what I can do... I can save **everyone**, by making love and not war through **selecting file and save**. I now **saved everyone**."

Valentina Vita (V. V.); Traveling Cosplayers And Kosplayers;
"(>_<) Huh...? **Saved everyone**...? Uh... Sister White, what are you talking about...?"

Sister K. White <3; Pretty Sure This is Not HeartBladeEndureLove 心刀忍愛;
"(*smiling with her eyes closed*) ...Time to **show and tell**..."

Valentina Vita (V. V.); Traveling Cosplayers And Kosplayers;
"(>_O) ...Your music really sets the Church atmosphere and environment..."

Sister K. White <3; Pretty Sure This is Not HeartBladeEndureLove 心刀忍愛;
"(*smiling with her eyes closed*) Thank you V. V.... Transform!! HeartBladeEndureLove!"

Sister **K.** White <3 increases the music volume from one third of 33% to two third of 66% on her earrings' **Utopia A.R. Holo HUD**. V. V. sees that Sister **K.** White <3 is about to changing out of her **nun clothes** to her white yellow Super-Hero suit. V. V. is silent and just watches as part of the **cosplay show and tell**.

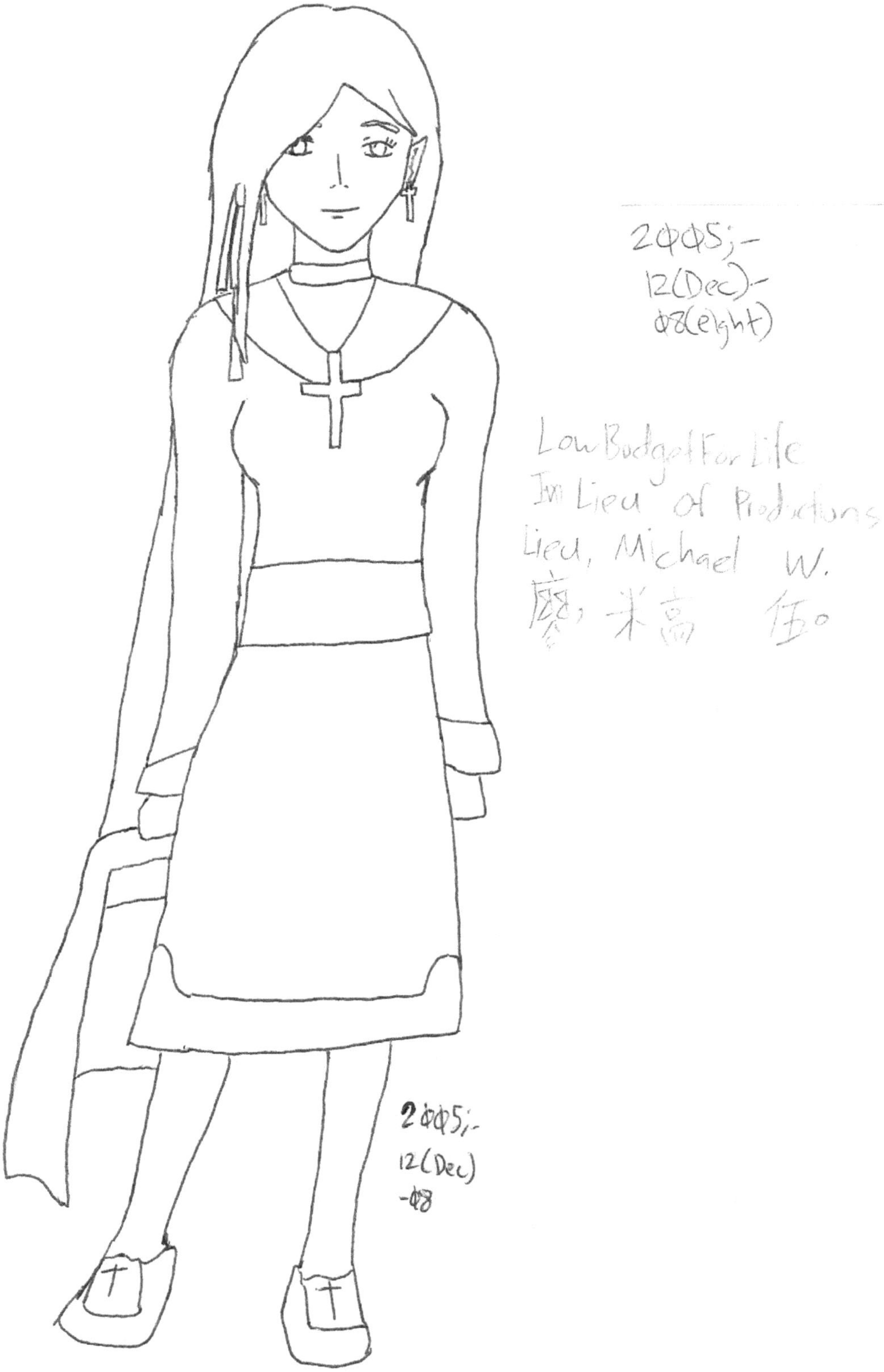
2ΦΦ5;-
12(Dec)-
Φ8(eight)
Low Budget For Life
In Lieu of Productions
Lieu, Michael W.
廖, 米高 伍o
2ΦΦ5;-
12(Dec)
-Φ8

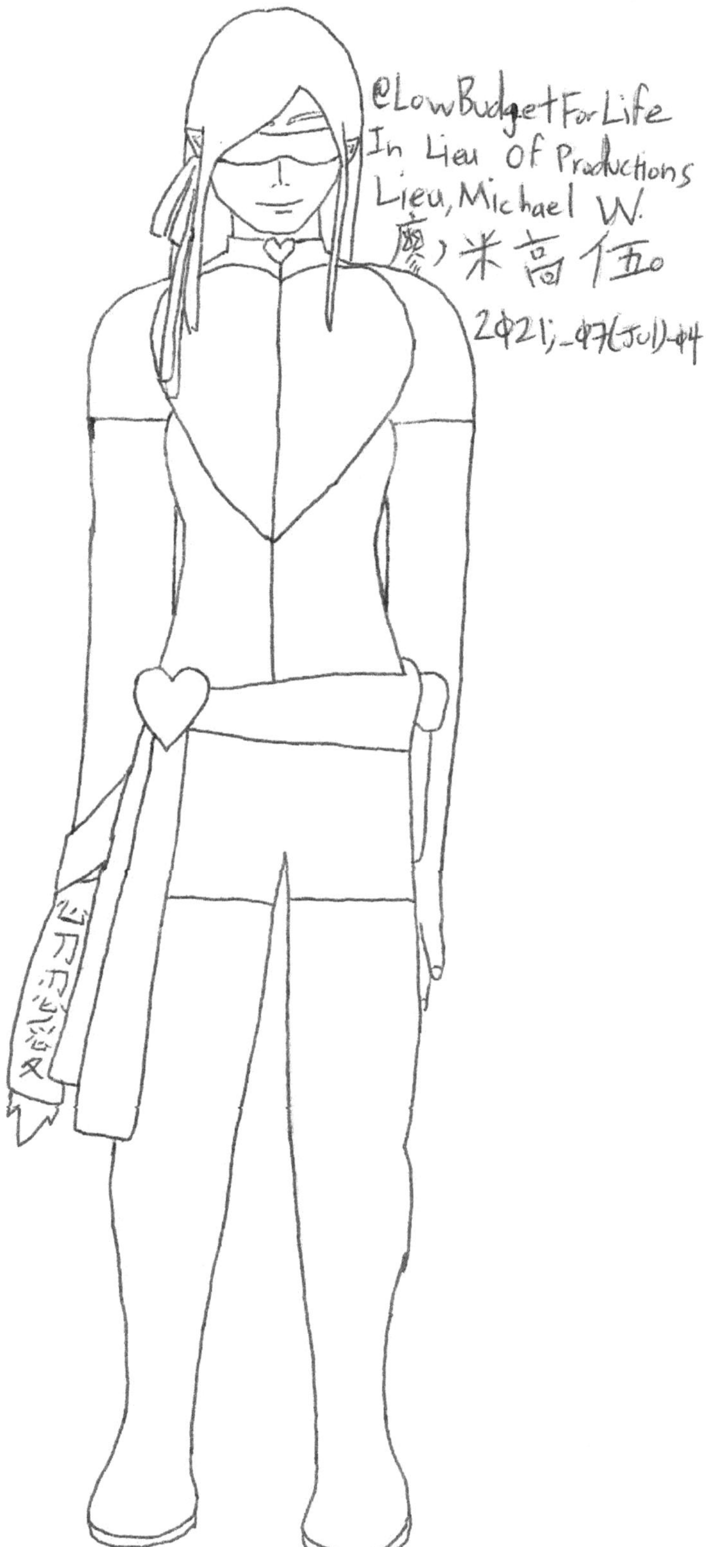
@LowBudgetForLife
In Lieu Of Productions
Lieu, Michael W.
廖, 米高 伍o
2φ21;_φ7(Jul)-φ4

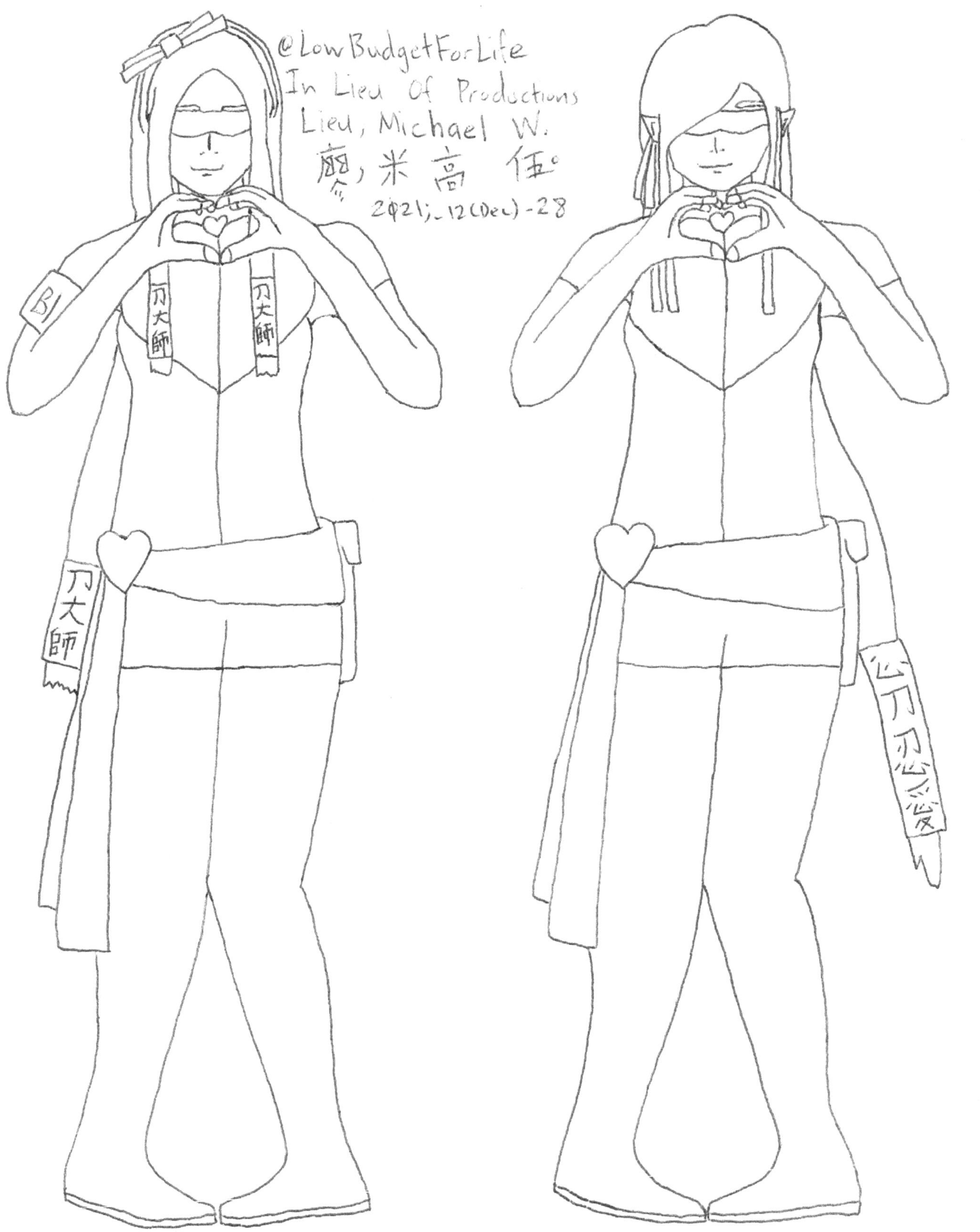
@LowBudgetForLife
In Lieu Of Productions
Lieu, Michael W.
廖, 米高 伍o
2021-12(Dec)-28
BI
刀大師
刀大師
刀大師

Valentina Vita (V. V.); Traveling Cosplayers And Kosplayers;
"(O_O) Whoa we wow, Sister White, your transformation from **nun** to Super-Hero is... Magical... Is that... (O_o) Elle... is she part of your Church **volunteer group** as well?"

HeartBladeEndureLove 心刀忍愛; Pretty Sure This Is Not That Nun Sister K. White <3;
"(*excitement*) Retired Magical Maiden HeartBladeEndureLove transformation complete! Meet my newest Magical Maiden student trainee... **BladeMaster**! Repeat after me **BladeMaster**. Make Love Not War! Heart Love Hand Shapes. Who Wants **Free Hugs**?"

BladeMaster 刀大師; Pretty Sure This Is Not That Elle K. Mcgee;
"(*excitement*) Make Love Not War! Heart Love Hand Shapes. Who Wants **Free Hugs**?"

Valentina Vita (V. V.); Traveling Cosplayers And Kosplayers;
"(O_O) Oh my gosh, Sister Super-Hero... you managed to brainwash, indoctrinate Elle into your own little Church cult follower... As a Church cult leader... Whoa we wow..."

BladeMaster 刀大師; Pretty Sure This Is Not That Elle K. Mcgee;
"(*excitement*) The Church is always actively recruiting new community service **volunteer** Magical Maidens. The Church and Conscientiousness Cat/Kat's is **always recruiting**."

HeartBladeEndureLove 心刀忍愛; Pretty Sure This Is Not That Nun Sister K. White <3;
"(*excitement*) Correction **BladeMaster**, listen well, community service is for **volunteer workers**, Magical Maidens is only for **augmented** biotech **human hermaphrodites and** natural **unmodified human females beings and lifeforms**. (*ahem*) End correction."

Valentina Vita (V. V.); Traveling Cosplayers And Kosplayers;
"(=_=) I have to say... Sister Super-Hero... Thank you for saving me back in the..."

HeartBladeEndureLove 心刀忍愛; Pretty Sure This Is Not That Nun Sister K. White <3;
"(*excitement*) V. V., in this Magical Maiden transformation I am HeartBladeEndureLove..."

Valentina Vita (V. V.); Traveling Cosplayers And Kosplayers;
"(>_O) Uh... Right... HeartBladeEndureLove... Thank you for saving me back then..."

HeartBladeEndureLove 心刀忍愛; Pretty Sure This Is Not That Nun Sister K. White <3;
"(*excitement*) You are welcome V. V.! **BladeMaster**, time to start **Free Hugs Protocols**."

BladeMaster 刀大師; Pretty Sure This Is Not That Elle K. Mcgee;
"(*excitement*) Yes, HeartBladeEndureLove. Time to start **Free Hugs Protocols**."

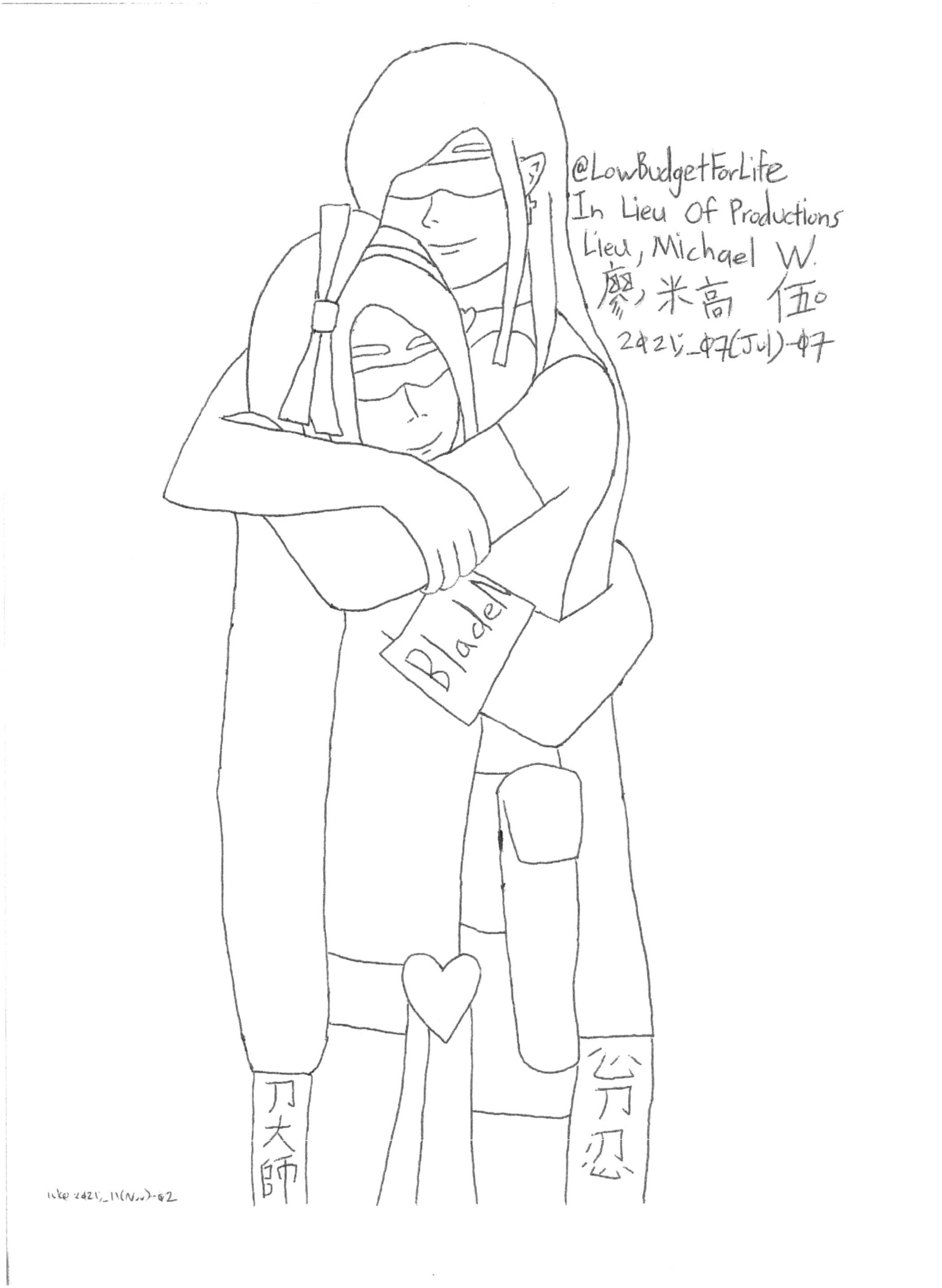
@LowBudgetForLife
In Lieu Of Productions
Lieu, Michael W.
廖, 米高 伍o
Blade
刀大師

Valentina Vita (V. V.); Traveling Cosplayers And Kosplayers;
"(T_T) Whoa we wow... ten of ten, **no ugly threads sticking out**! Amazing cosplay!"

HeartBladeEndureLove mutes the music volume on her earrings' **Utopia A.R. Holo HUD**.

HeartBladeEndureLove 心刀忍愛; Pretty Sure This Is Not That Nun Sister K. White <3;
"(*smiling*) Hm... **Free Hugs Protocols** now in active... **BladeMaster**, do you think V. V. would make a good Magical Maiden?"

BladeMaster 刀大師; Pretty Sure This Is Not That Elle K. Mcgee;
"(*smiling*) ...Yes, HeartBladeEndureLove, I think V. V. would make a good Magical Maiden. We must get V. V. a Magical Maiden Super-Hero Suit **P.P.E. One Size Fits All Biotech Clothing**. We can get it customized for her as soon as possible! But, after **Free Hugs Protocols**... I love the **Free Hugs** along with Make Love Not War and Heart Love Hand Shapes. How long can we stay in the **Free Hugs Protocols**?"

Valentina Vita (V. V.); Traveling Cosplayers And Kosplayers;
"(@_@) Huh... V. V., me a Magical Maiden!? What is going on!? I am not sure... Huh..."

HeartBladeEndureLove 心刀忍愛; Pretty Sure This Is Not That Nun Sister K. White <3;
"(*smiling*) **BladeMaster**, we can stay in **Free Hugs Protocols** as long as possible. At one point in **spacetime**, we will get V. V. her Magical Maiden transformation process. ...No worries, V. V. the process for you to be a Magical Maiden is **simple with little to no pain and suffering at all**... Especially someone with your **beautiful biotech body**..."

BladeMaster 刀大師; Pretty Sure This Is Not That Elle K. Mcgee;
"(*smiling*) Yes, HeartBladeEndureLove. We can stay in **Free Hugs Protocols** as long as needed... V. V. will be recruited as a Magical Maiden, I can prepare her Magical Maiden Super-Hero Suit **P.P.E.** as soon as possible! But, first some **Free Hugs Protocols**."

Valentina Vita (V. V.); Traveling Cosplayers And Kosplayers;
"(>_O) Wait, do I get a say on whether I want to become a Magical Maiden or not...?"

V. V. waits and see for a response from **BladeMaster** and HeartBladeEndureLove.

Valentina Vita (V. V.); Traveling Cosplayers And Kosplayers;
"(>_<) Hello...? **BladeMaster**, HeartBladeEndureLove...? Hello...? What is going on...? (@_@) So confused... Magical Maidens... **Free Hugs**, Make Love Not War, Heart Love Hand Shape. Really confused... Whoa we wow..."

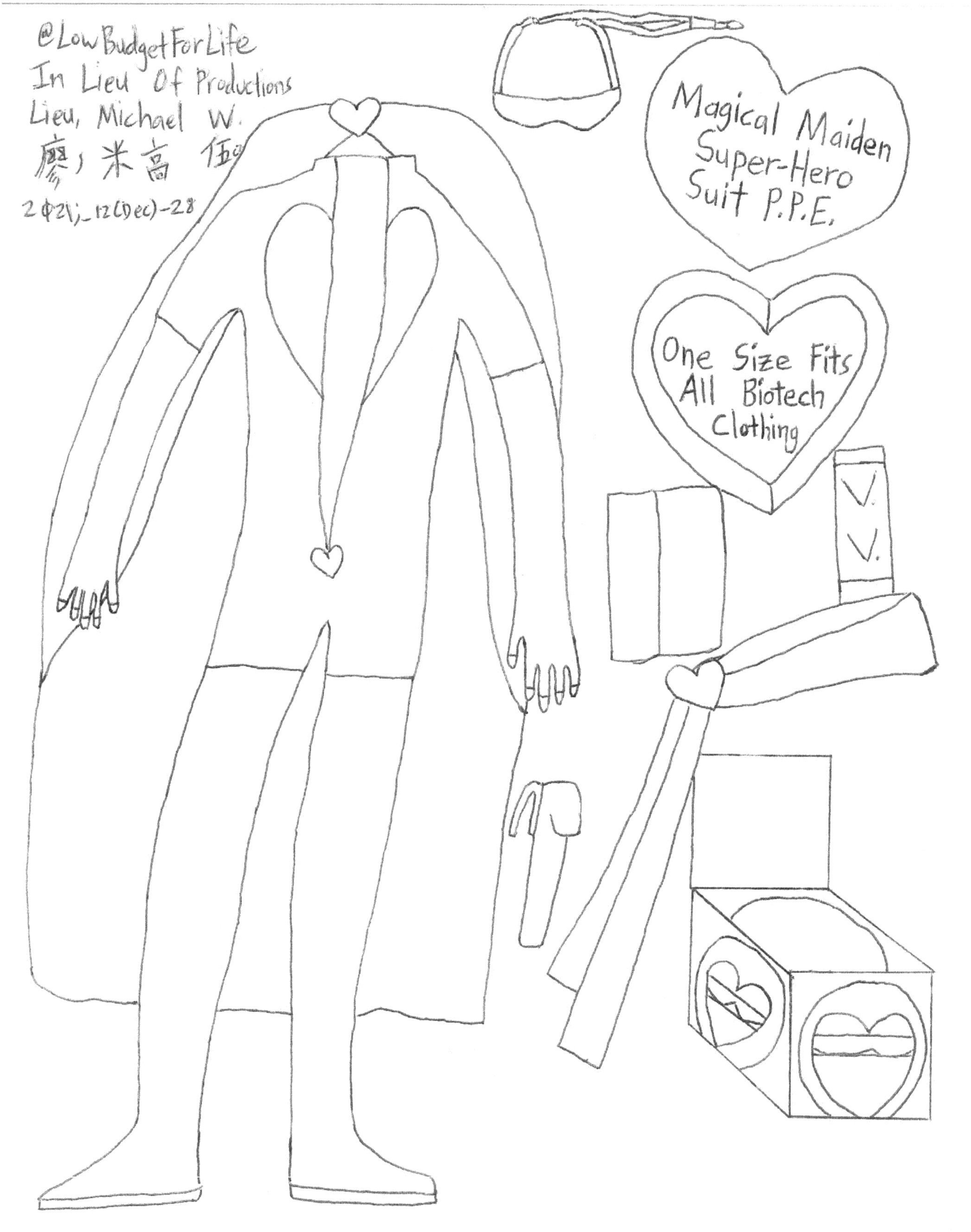
@LowBudgetForLife
In Lieu Of Productions
Lieu, Michael W.
Magical Maiden Super-Hero Suit P.P.E.
One Size Fits All Biotech Clothing

V. V.; Pretty Sure This Is Not That Traveling Cosplayer And Tourist Valentina Vita;
"(@_@) Is this for me... Even has a custom name tag... V. V. just for me... Ten of ten, it has **no ugly threads sticking out**... Is this... For me... But, how I am going to **change clothes** in front of these **people**...?"

HeartBladeEndureLove 心刀忍愛; Pretty Sure This Is Not That Nun Sister K. White <3;
"(*voice over*) No worries Magical Maiden V. V., we can **change** and transform you. Hold her."

BladeMaster **刀大師**; Pretty Sure This Is Not That Elle K. Mcgee;
"(*voice over*) Yes, we can take care of you, **changing clothes** and transformation."

HeartBladeEndureLove selects the music "**Puella Magi Madoka Magica Original Sound Track (O.S.T.)**:

H.D.D.V.D. C.D. 02. warning #1	*Vol. 3, 08. Cubilulum album*
H.D.D.V.D. C.D. 18. fateful #2	*Vol. 3, 05. Confessio*
H.D.D.V.D. C.D. 30. facing the truth #1	*01. once we were*
H.D.D.V.D. C.D. 34. facing the truth #2	*12. something, everything is wrong*
Vol. 1, 03. Gradus prohibitus	*14. never get there*
Vol. 1, 07. Conturbatio	*16. face the truth*
Vol. 1, 09. Puella in somnio	*17. doubt #2*
Vol. 1, 12. Scaena felix	*24. wanna destroy?*
Vol. 2, 05. Amicae carae meae	*31. experimentation*
Vol. 2, 06. Incertus	*38. take your hands*
Vol. 2, 07. Inevitabilis	*41. her new wings*
Vol. 2, 12. Ave Maria	*42. solve the riddle*
Vol. 3, 02. Quamobrem	

" to play on her earrings' **Utopia A.R. Holo HUD** on one third volume of 33% and is set on loop.

V. V.; Pretty Sure This Is Not That Traveling Cosplayer And Tourist Valentina Vita;
"(>_<) Hey Sister! Elle! You two! What are you doing!? **Changing clothes** here!?"

V. V. is physically having her **KissShot**-AcerolaOrion-**HeartUnderBlade** cosplay clothes **change** by two Magical Maidens HeartBladeEndureLove and **BladeMaster**. HeartBladeEndureLove and **BladeMaster** overwhelms V. V. as she is being forced to wear a Magical Maiden Super-Hero Suit **P.P.E. One Size Fits All Biotech Clothing**.

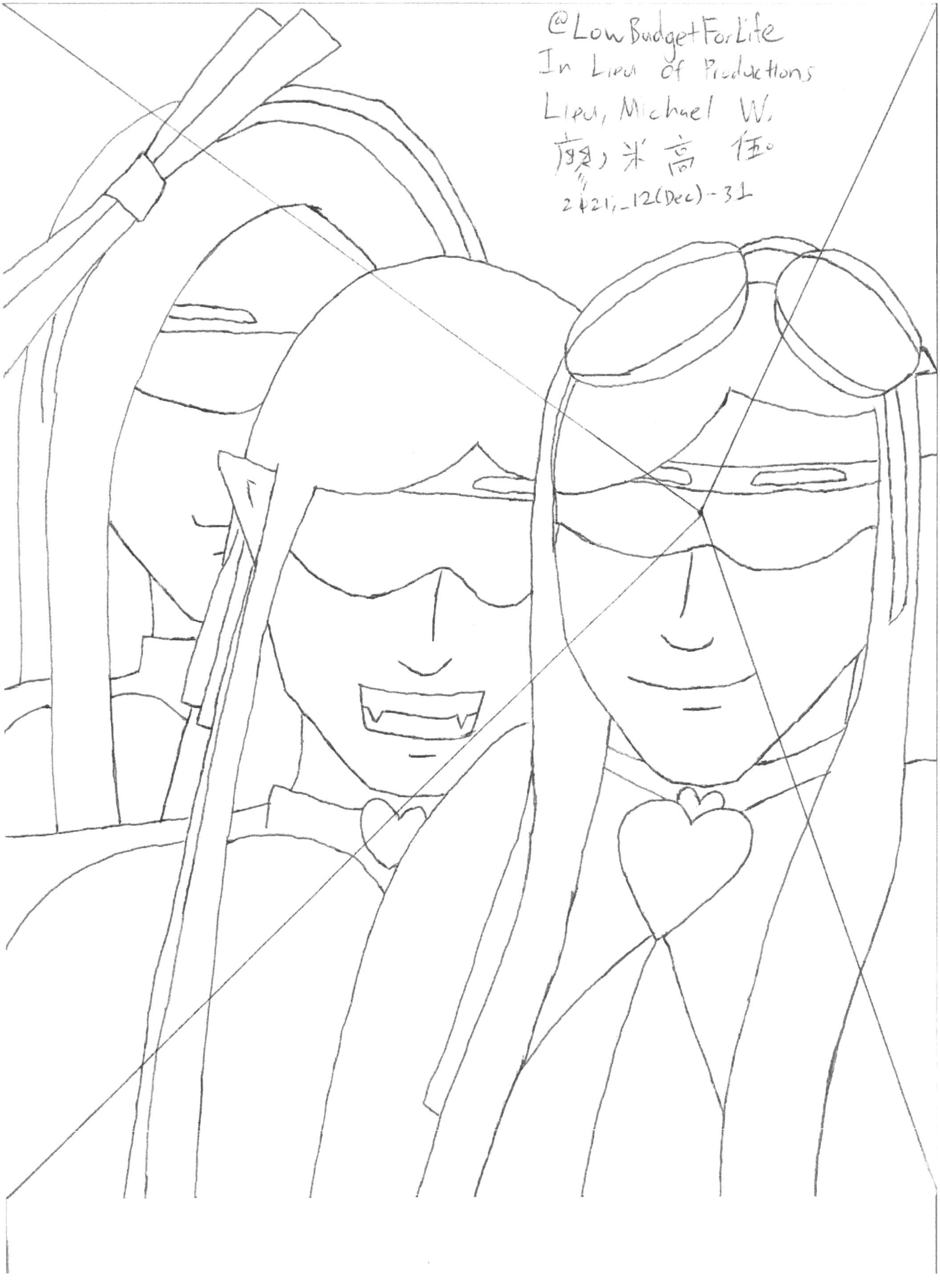
@LowBudgetForLife
In Lieu Of Productions
Lieu, Michael W.
廖, 米高 伍o
2021_12(Dec)-31

V. V.; Pretty Sure This Is Not That Traveling Cosplayer And Tourist Valentina Vita;
"(^_^) Whoa we wow... I am now a Magical Maiden. Got those **kool glasses** that is fitted with red A.R. Holo HUD. I feel good... All those... **Free Hugs**, Make Love Not War... My favorite toner! Red! Red! Okay, let us take a group holo photo!"

HeartBladeEndureLove 心刀忍愛; Pretty Sure This Is Not That Nun Sister K. White <3;
"(*smiling*) Look at the toners in the **holo photo alignment** of our Magical Maiden suits... Red, white, blue... the **old world** United States of America! The **collapsed country** from the **old world** of North America! V. V. is red, I am white, **BladeMaster** is blue!"

BladeMaster 刀大師; Pretty Sure This Is Not That Elle K. Mcgee;
"(*duck face*) HeartBladeEndureLove, what is the **old world** United States of America? I have many questions about that **country that collapsed** many years ago! Must learn more about it as a **BladeMaster** student trainee! (*smelling*) ...Yes, we have another member to our Magical Maidens. Conscientiousness Cat/Kat must be pleased! Meow Meow! Me! Ow! That is all that stuff animal cat says... Meow. Meow! Me! Ow!"

V. V.; Pretty Sure This Is Not That Traveling Cosplayer And Tourist Valentina Vita;
"(^_^) I had a dream about that orange stuff animal cat... When Sister... Uh... HeartBladeEndureLove rescue me from that **Kay Private Security outbreak outage experiment they** were doing... Meow. Meow! Me! Ow!"

HeartBladeEndureLove 心刀忍愛; Pretty Sure This Is Not That Nun Sister K. White <3;
"(*smiling*) Meow. Meow! Me! Ow! Oh **BladeMaster**, if you really want to know more about the **old world** United States of America, **Elysium Corporation** has more about it in **Downtown District**. You need to be become a **Utopian** Sovereignty **Citizen** to gain access to **Elysium Corporation floors**. Make sure to explore that area for the **human history**... We should keep on moving, since V. V. is getting bored with nothing to do in **East Side** with little to no **people** in wanting to take part in her **cosplay show and tell**!"

BladeMaster 刀大師; Pretty Sure This Is Not That Elle K. Mcgee;
"(*duck face*) Okay okay! We should go to **West Side** watch those **first to four**, **Daddy and Subby** is going to be there along with some Martial Artists taking part in a **musical fight**!"

V. V.; Pretty Sure This Is Not That Traveling Cosplayer And Tourist Valentina Vita;
"(^_^) Let me put my **KissShot-AcerolaOrion-HeartUnderBlade** cosplay in my luggage."

HeartBladeEndureLove 心刀忍愛; Pretty Sure This Is Not That Nun Sister K. White <3;
"(*smiling*) Oh, we can levitate using the **electromagnetic spectrum** within our suits.

2154;_11(Nov)-13(thirteen)

Travel To Western Society, America(s), North America
Self-Govern Sovereignty (S),
Illinois (ILL) Sector, Points Of Interest (P.O.I.)

Chicago (S), "**Animal** Sanctuary"
"The New World 2121 Movement Of The 22nd Century,
Utopian Sovereignty of North America (U.S. of N.A.)"

West Side, Chicago City Christian Church Belief Systems, And Other Religions Welcome <3

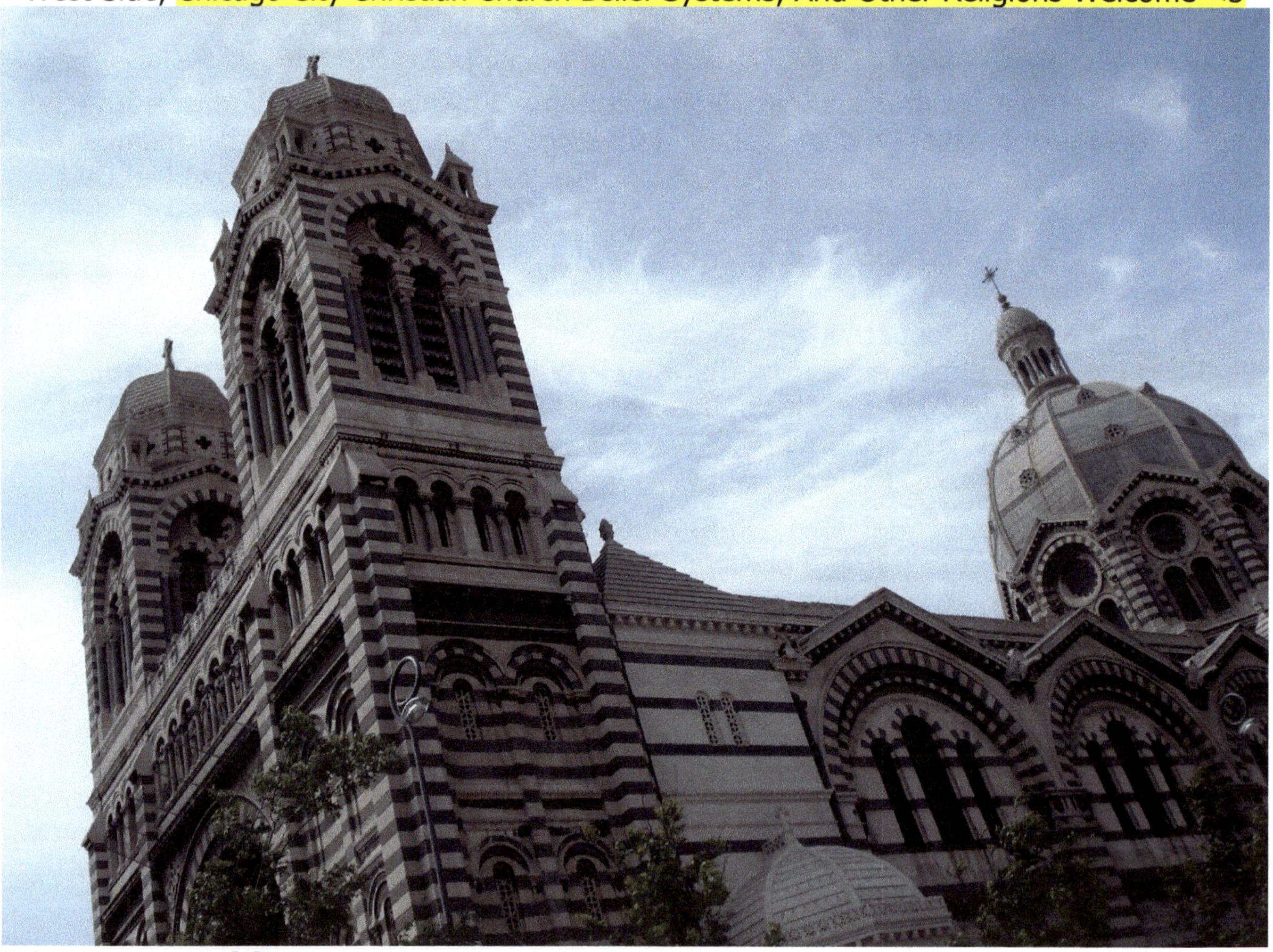

BladeMaster 刀大師; Pretty Sure This Is Not That Elle K. Mcgee;
"(*voice over*) Oh!! Is the **first to four music video** fights going right now? This way!!"

V. V.; Pretty Sure This Is Not That Traveling Cosplayer And Tourist Valentina Vita;
"(@_@) Huh... **First to four music video** fights...? Okay, **BladeMaster**, following..."

2154;_Arcadia \ Valentina Vita (V. V.) 'Healthy-Strong Life' \ Left to Right(L->R)

"1st 2 4 (First Two Four)"
Alex D. Doc Drastal's Utopian Peacekeeping Public Militia Volunteer Training

"An equalizer musical hide and seek fight klub."
Augmented biotech and natural unmodified humans sparring session in solo or groups.

Less Lethal Methods, ***<u>No</u>*** Clean House **Flatline And Dismemberment Weapons And Tools.**
Feat. **Mister Steven** Z. J., **Entertainer's** Rainbow Road Biotech Guitar **Music Video** Display

2400\0000, 0300, 0600, 0900, 1200 → **Rules** ← 1200, 1500, 1800, 2100, 2400/0000

Flip a coin of two sides with heads and tails. No coins with heads and heads or tails and tails!
Who ever gets heads or tails of their call first, gets to select their Music Video (M.V.) of choice.
They become Hider(s).

Hider(s) Objective
Hide, defend yourself, and or hold out until the music ends → Hider(s) get 1 (one) point.
Pick a music duration that is short, less time needed to hold out and or hide.
If you are feeling confident enough, you can expose yourself to Seeker(s) and straight up
Less Lethal fight and spar them into submission → Hider(s) get 1 (one) point.
The selected music duration runs out of time → Hider(s) gets 1 (one) point.

Seeker(s) Objective
Seek out and find Hider(s) using Less Lethal methods for submission,
the longer the music duration Hider(s) pick, the more time that provides for Seeker(s).
Hider(s) who falls into submission from Less Lethal Seeker(s) attacks and give ups
→ Seeker(s) gets 1 (one) point.

Music Selection
Who ever gets 1 (one) point that round, turns the music selection over to the opposing party.
Who ever selects the music, can give up, yield their music selection to the opposing party.

Drop In / Drop Out
<u>*Anyone in and or by the Church is allowed to physically take part in First Two Fours, "after Hey-Zeus says '**4-D music video reset**'", "**4-D music video ending**" and or "**during the Hider(s) music selection**", by announcing themselves with their war cry battle chant.*</u>

In group based Hider(s) and Seeker(s), those who submit and give up, move to the center 4-D music video location where Clean-up Crew volunteers and or 4-D music video biotech is located at until the music duration is over. Lastly, do not hide by or beat up Clean-up Crew volunteers or by 4-D music video biotech, especially physically beat up any animals period.

First Hider(s) or Seeker(s) to get four points wins that set;
hence the name "**First To Four**" or "**First Two Four**" (**1st 2 4**)

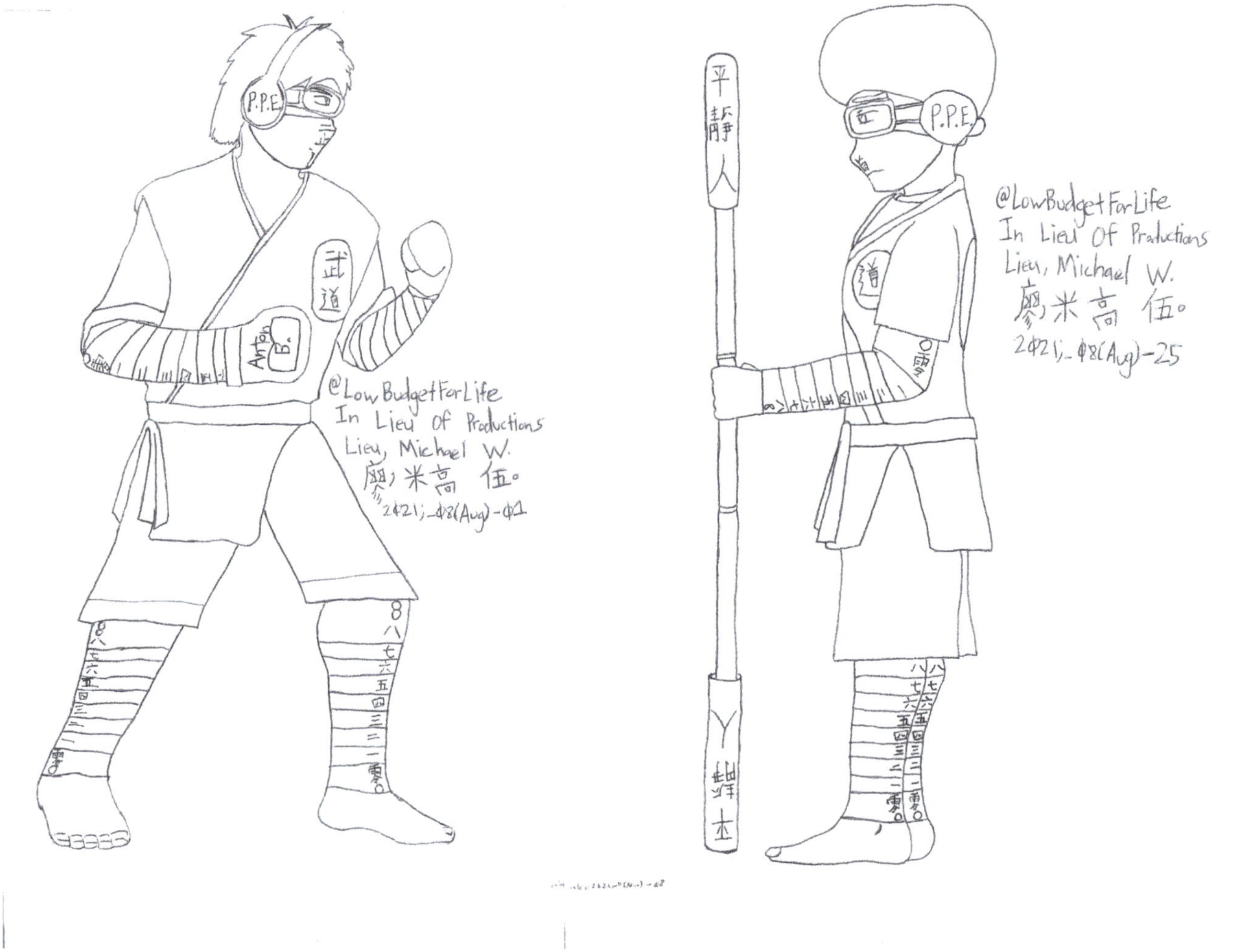

Alien Valentina Vita, Traveling Cosplayer(s) And Kosplayer(s)
2154;_ Chicago, Illinois' Intranet, Wireless Broadcasting Frequencies Online

V. V. Augmented Reality Holographic Heads Up Display A.R. Holo HUD	
Actual Legal Given Name;_ **Anton B.** (goes by **Anton**) pronounced, **"and-ton" "bee"**	Actual Legal Given Name;_ **Talor A. Williams** (goes by **Ninja-Talor**) pronounced, **"Nin-Ja Tay-Ler A. Wil-Yems"**
Physical Plus Psych Age;_ augmented 30s + 34 years old	Physical Plus Psych Age;_ natural 34 + 34 years old
[*Aug+*]	
V. V. Augmented Reality Holographic Heads Up Display A.R. Holo HUD	

2154;_ Chicago, Illinois' Intranet, Wireless Broadcasting Frequencies Online
Alien Valentina Vita, Traveling Cosplayer(s) And Kosplayer(s)

V. V.; Pretty Sure This Is Not That Traveling Cosplayer And Tourist Valentina Vita;
"(@_@) Whoa we wow, seeing their bright orange blue Martial Arts uniform in action."

HeartBladeEndureLove 心刀忍愛; Pretty Sure This Is Not That Nun Sister K. White <3;
"(*voice over*) Looks like just set ups and warm up rounds... No sign of **Politicians' Marks**... yet. So that means, we are early and being early is better than late... Yay!!"

BladeMaster **刀大師**; Pretty Sure This Is Not That Elle K. Mcgee;
"(*voice over*) Oh, this must be warm up ready rounds... Today is... (*looking*) **November thirteen**, **Daddy and Subby**, must be coming up soon! Only in **Animal** Sanctuary!"

Anton B. "MartialArtsPathWay 武道" , Kung-Fu, Martial Artists "MartialArtsGod 武道神" ;
"(*in mouth mask*) **Mach one speed**... Ready? **Put on your P.P.E. now.**"

Ninja-Talor A. Williams; D King And X Philosophy "PeacefulPerson 平靜人" 'Way 道';
"(*in mouth mask*) **Put on your P.P.E. now.** Ready."

HeartBladeEndureLove 心刀忍愛; Pretty Sure This Is Not That Nun Sister K. White <3;
"(*voice over*) Get ready for some loud **music video** play!"

Anton B. selects the following music(s) "**Bruce Faulconer's Dragon Ball Z**:

02 - Call Out The Dragon	*03 - Prince Of Saiyans*	*08 - Race To The Island*
13 - Vegeta Fights Frieza	*15 - Gohan & Icarus*	*16 - The Cell Games*
17 - New Earth Music	*18 - Vegeta's Theme*	*19 - Garlic Transformation*
19 - Vegeta Powers Up	*20 - Nail's Gift*	*20 - Vegeta - Super Saiyan*
21 - A Little Help From A Friend	*22 - The Saga Continues*	*22 - Vegeta Knows His Son*
23 - Goku's Spirit Bomb	*24 - Ginyu Transformation*	*28 - Super Buu*
38 - Gotenks Is Born	*40 - Intro To Finale And Closing Music*	*Pikkon's Theme*
We Gotta Power -Japanese (Hironobu Kageyama)	*Head Cha La -Japanese (Hironobu Kageyama)*	

" to play in the background on the Church's **Mister Steven** Z. J., Entertainer's Rainbow Road Biotech Guitar **Music Video** Display.

V. V., HeartBladeEndureLove and **BladeMaster** puts on **P.P.E. earplugs**, watches Anton B. and **Ninja-Talor's** martial arts sparring at **less than mach one speed** at the Church's **Mister Steven** Z. J., Entertainer's Rainbow Road Biotech Guitar **Music Video** Display that goes along with their **training sessions** as each music play list duration runs out.

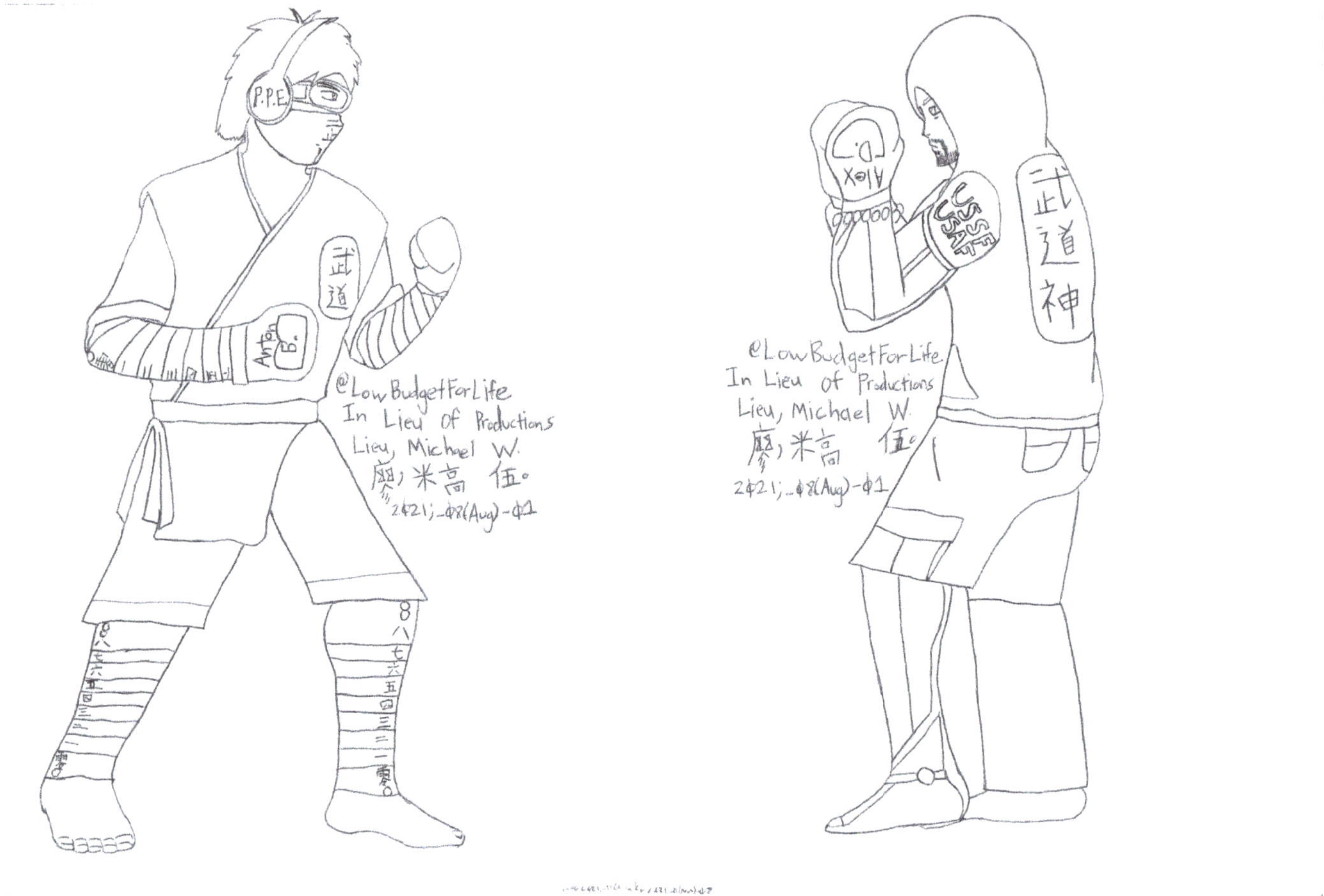

Alien Valentina Vita, Traveling Cosplayer(s) And Kosplayer(s)
2154;_ Chicago, Illinois' Intranet, Wireless Broadcasting Frequencies Online

V. V. Augmented Reality Holographic Heads Up Display A.R. Holo HUD	
Actual Legal Given Name;_ **Anton B.** (goes by **Anton**) pronounced, **"and-ton" "bee"**	Actual Legal Given Name;_ **Alex Drastal** (goes by **Alex**) pronounced, **"al-lex" "draw-stall"**
Physical Plus Psych Age;_ augmented 30s + 34 years old	Physical Plus Psych Age;_ augmented 34 + 34 years old
[*Aug+*]	
V. V. Augmented Reality Holographic Heads Up Display A.R. Holo HUD	

2154;_ Chicago, Illinois' Intranet, Wireless Broadcasting Frequencies Online
Alien Valentina Vita, Traveling Cosplayer(s) And Kosplayer(s)

V. V.; Pretty Sure This Is Not That Traveling Cosplayer And Tourist Valentina Vita;
"(@_@) Oh, looks like they are switching up the martial artists... Now it is Anton Versus **Alex Dra... Dra... D**. (O_<) The **music video**! Whoa we wow! **It** really is something listening to it music as they train and fight!! I feel like I may be losing my hearing! Ha ha!"

V. V. is laughing hard and enjoying herself observing the martial artists' training session.

HeartBladeEndureLove 心刀忍愛; Pretty Sure This Is Not That Nun Sister K. White <3;
"(*voice over*) Yes, **first two fours** do usually get loud, make sure to put on your **P.P.E.** the **full P.P.E.**, minus the Heart Love shape cape unless you want to... (*ahem*) Remember those big round helmet that comes with the standard issue Magical Maiden Super-Hero Suit **P.P.E. One Size Fits All Biotech Clothing**? Make sure to put that helmet on along side **double hearing protection, earplugs as well**! Should be in that yellow waist sash pouch you have... Those two are about to move at whatever **mach speed** they choose. **BladeMaster**, repeat after me, **put on your P.P.E.**!"

BladeMaster 刀大師; Pretty Sure This Is Not That Elle K. Mcgee;
"(*voice over*) **Put on your P.P.E.**!"

V. V. sees that HeartBladeEndureLove and **BladeMaster** puts on their assigned Magical Maiden Super-Hero Helmet **P.P.E. and double hearing protection**. V. V. follows what they both do in putting on her Magical Maiden Super-Hero Helmet **P.P.E.** as well.

Anton B. "MartialArtsPathWay 武道" , Kung-Fu, Martial Artists "MartialArtsGod 武道神" ;
"(*in mouth mask*) Okay, **Alex**, I am **selecting the music**, so I am gonna to **parry and block** as many of **your punches and kicks and headbutts** at **whatever mach speed you** choose. I got my **P.P.E.** on already... So, ready. **Put on your P.P.E. now**!"

Alex D. "Doc Drastal"; War Survivor Military Veteran, "MartialArtsGod 武道神" ;
"(*focusing*) Copy. Time to **improvise, overcome and adapt**. **Put on your P.P.E. now**!"

Anton B. selects the music "**Bruce Faulconer's Dragon Ball Z Goku's S.S.J.3 Power Up Extended**" to play in the background on **Mister Steven** Z. J., Entertainer's Rainbow Road Biotech Guitar Music Video Display.

V. V. sees that both Anton B. and **Alex Drastal** is now moving at **mach speed beyond one**, as the surrounding biotech is able to **absorb** plus handle the **shock and awe** of such **speed**. She has **trouble keeping an eye** on them both, her augmentation biotech is **not designed to handle the visual capture** of **augmented** biotech **human males'** martial artists who move that fast as V. V. is not using her Ai-**Aug bracelet** as well.

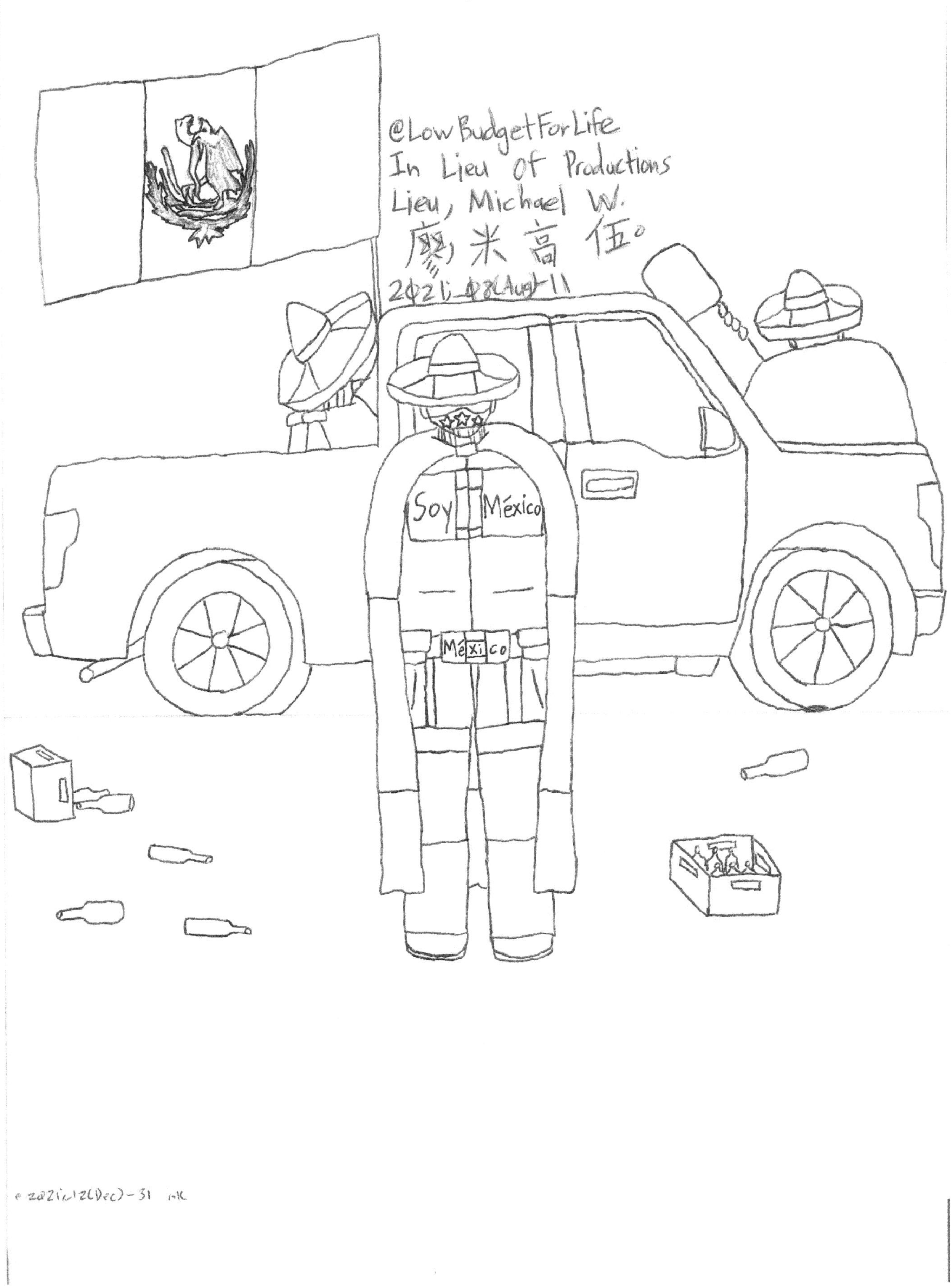
@LowBudgetForLife
In Lieu Of Productions
Lieu, Michael W.
廖米高伍o
2Φ21_Φ8(Aug)-11
Soy México
México

V. V. sees and hears the music "**Bruce Faulconer's Dragon Ball Z Goku's S.S.J.3 Power Up Extended**" playing in the background on **Mister Steven** Z. J., Entertainer's Rainbow Road Biotech Guitar Music Video Display is ending with the time duration of "09;49" running out as Anton B. and **Alex Drastal's music video** martial arts training session ends as the two of **them** depart. She sees Big Ben The Mexican-Man with his Soy Mexico name tag and his Mexican-Men listening to loud music on one of their green custom four door vehicles.

V. V. sees that Big Ben The Mexican-Man selecting the music title "**Top 70+ Best Initial D Eurobeat Full Songs 6.5 hours Hypermix**!" playing on his green custom four door wheel ground maglev drive pickup truck.

V. V.; Pretty Sure This Is Not That Traveling Cosplayer And Tourist Valentina Vita;
"(@_@) Okay, so... the actual **music video** martial arts fight with **Politicians' Marks** is about to start...? Are those green guys like... part of **it**... or they drunk...?"

HeartBladeEndureLove 心刀忍愛; Pretty Sure This Is Not That Nun Sister K. White <3;
"(*voice over*) Oh, yeah... that's our... biotech agricultural farmers... The infamous Big Ben The Mexican-Man and his Mexican-Men. **Drunk** on their green Mexican beer bottles. They are **patriots** of their **old world country** of Mexico... (*ahem*) This is the **pre-show event.** Time to play **hurry up and wait**!"

BladeMaster **刀大師**; Pretty Sure This Is Not That Elle K. Mcgee;
"(*voice over*) **Daddy and Subby** starting soon! Time to play **hurry up and wait**!"

Benavidez-Cruz, Rich Rodriguez "Big Ben"; Mexican Mythology, The King Of Mexico;
"(*in mouth mask, drunk*) I am the King Of Mexico! ...**Pre-show** for our **first two four**, or is it **first to four**... is the to... **T.W.O. or T.O.**? Which one is it... anyways, the **main event** to our **first two four pre-show** with Mexican drinks of alcohol, soda, gasoline? Oh wait gasoline is not for **drinking**. Anyways, we have Mexican food, drinks, refreshments from Mexican-Men! **Free** of course, **party** hard, **drink** hard, **Eurobeat** hard!! Mexico!"

Sam R.; "MartialArtsGod 武道神" , Kung-Fu, Martial Artists "MartialArtsGod 武道神" ;
"(*voice over*) Hello everyone! I'm Sam, I practice Kung-Fu as a Martial Artist... Here is a bit of a martial arts showcase of **varies styles and forms** using a **single edge cutting tool**, **Blade** from **BladeSociety**. After my martial arts showcase we can start our main **first two four** event with **Eur-Asia's Politicians Mark D. and Mark S.** Versus Martial Artists Anton and **Alex** today! Have a nice day in Chicago **Utopian** Sovereignty **Animal** Sanctuary!"

V. V. sees that Big Ben The Mexican-Man and his Mexican-Men is around, standing and some passed out **drunk** from their green Mexican beer as Sam R. is about to start the **event**.

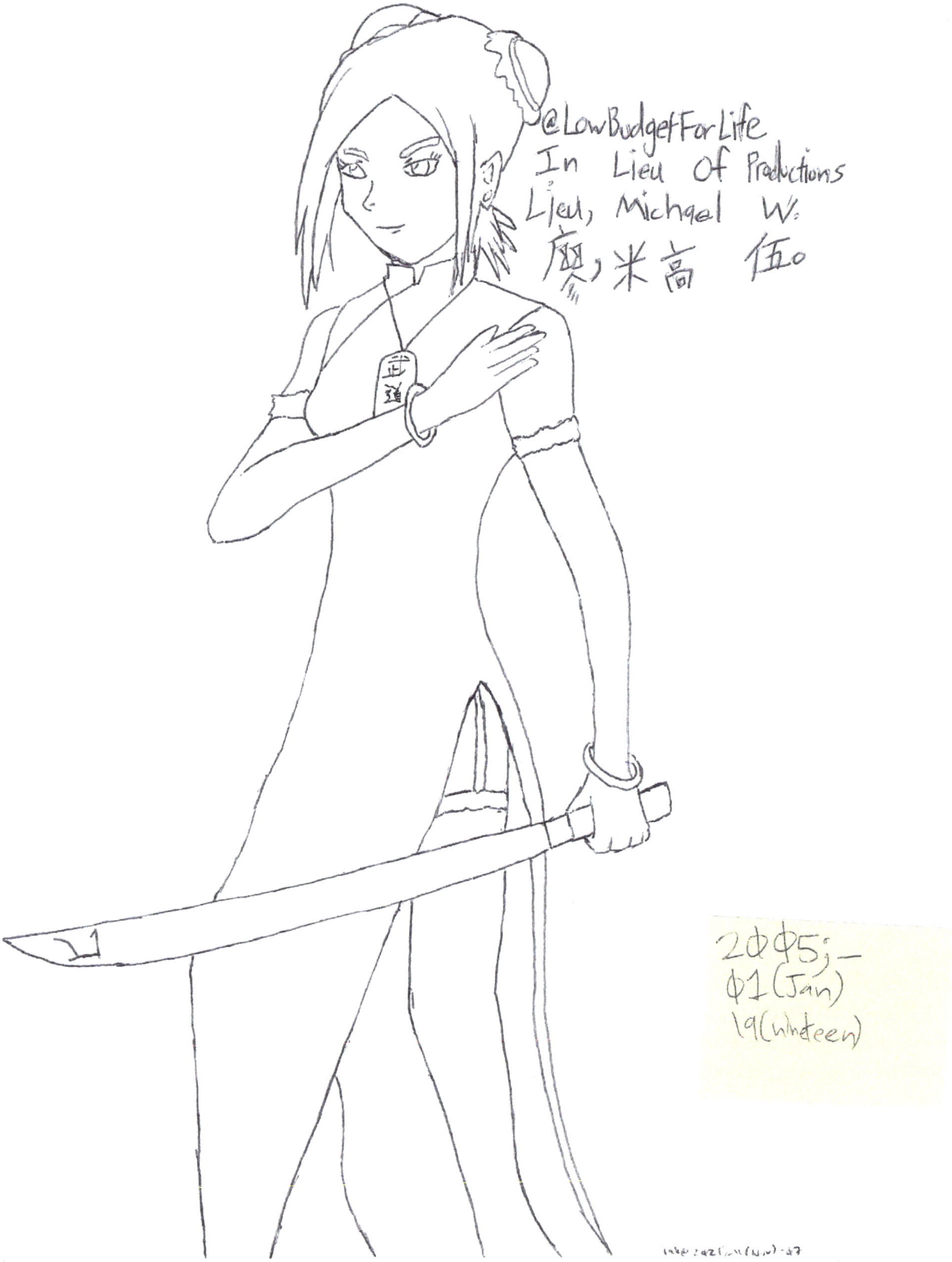
@LowBudgetForLife
In Lieu Of Productions
Lieu, Michael W.
廖, 米高 伍o
2ΦΦ5;-
Φ1(Jan)
19(nineteen)

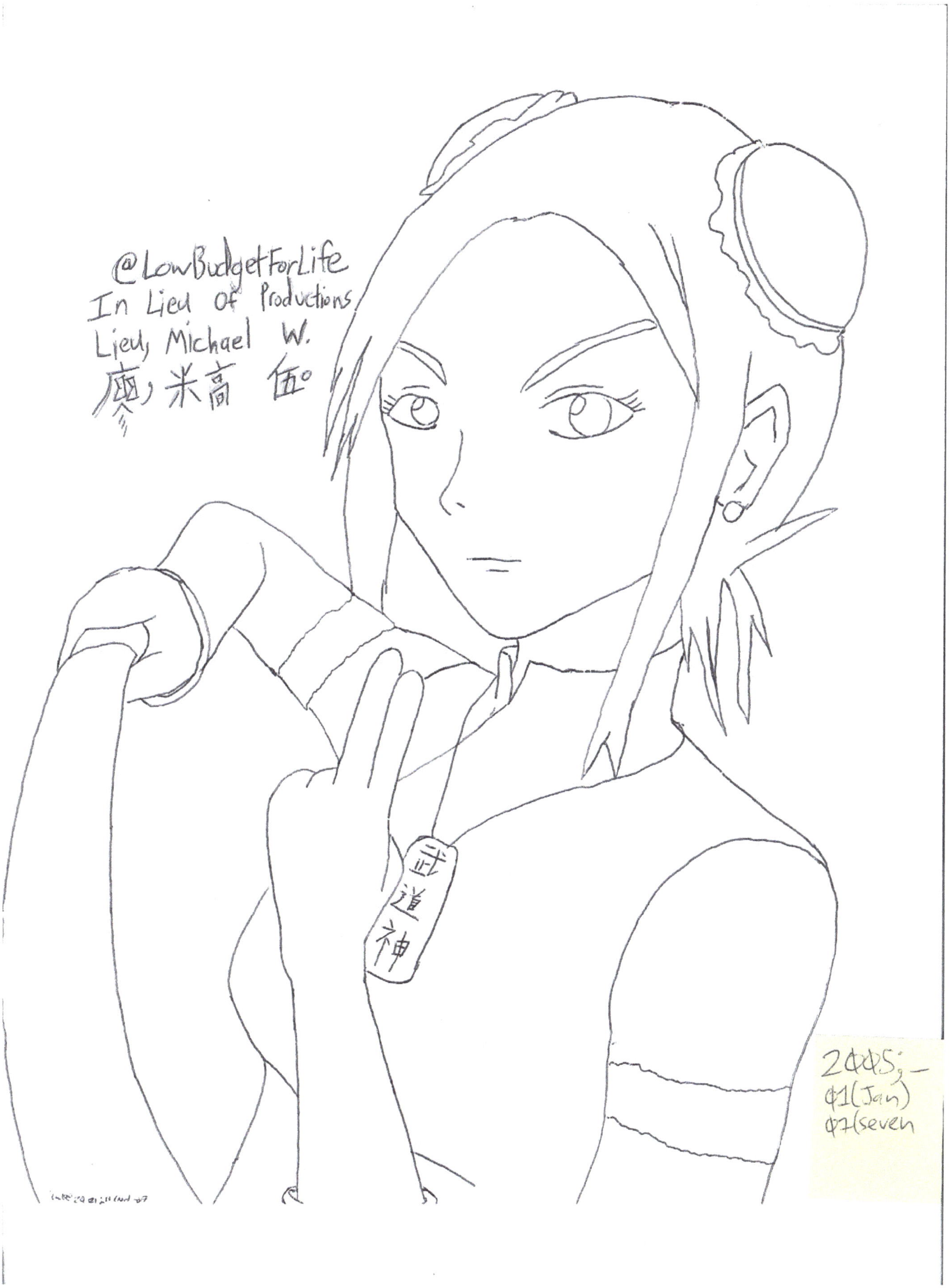
@LowBudgetForLife
In Lieu Of Productions
Lieu, Michael W.
廖, 米高 伍o
武道神
2ΦΦ5;-
Φ1(Jan)
Φ7(seven

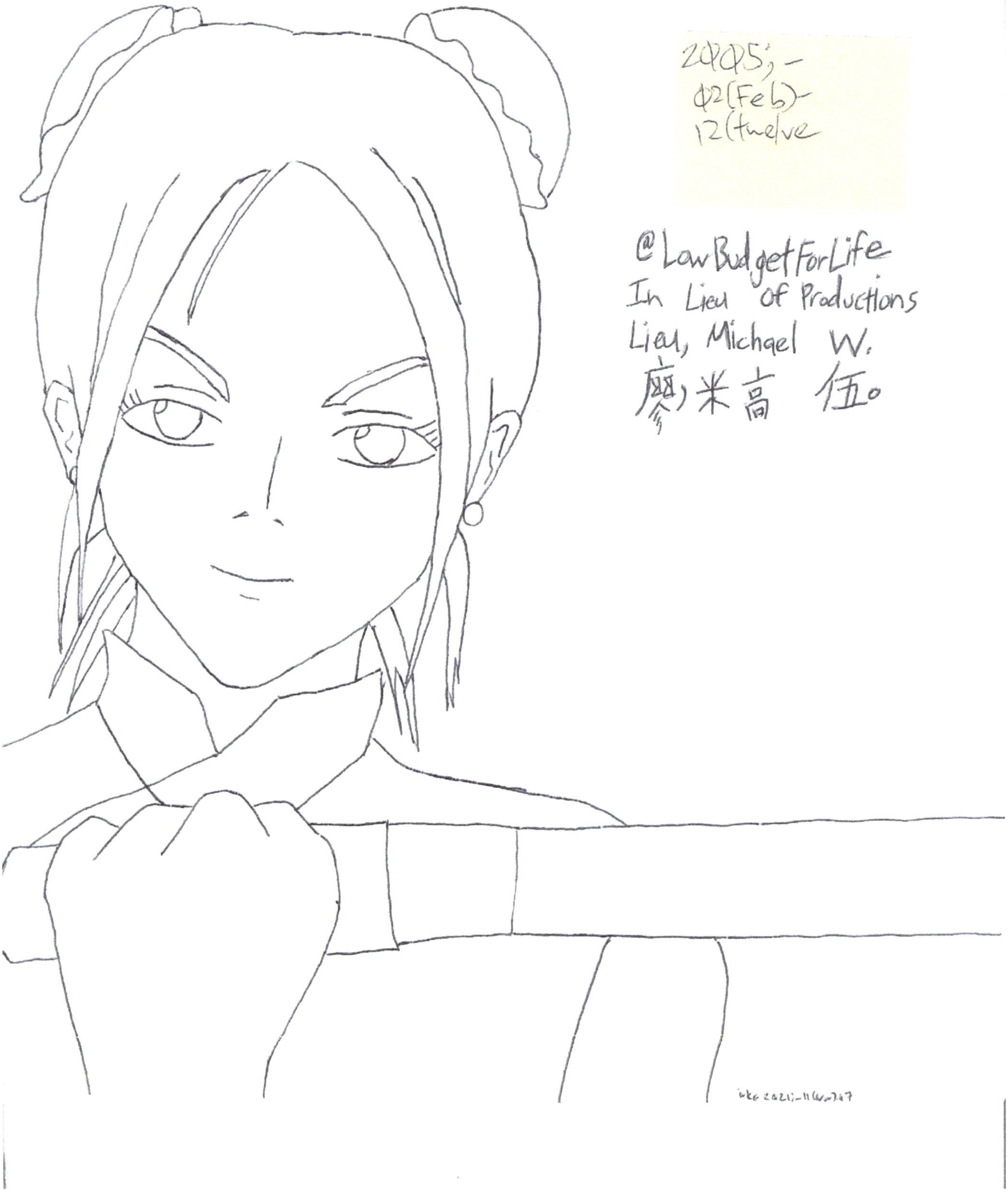
2ΦΦ5;-
Φ2(Feb)-
12(twelve
@LowBudgetForLife
In Lieu of Productions
Lieu, Michael W.
廖, 米高 伍o

Alien Valentina Vita, Traveling Cosplayer(s) And Kosplayer(s)
2154;_ Chicago, Illinois' Intranet, Wireless Broadcasting Frequencies Online

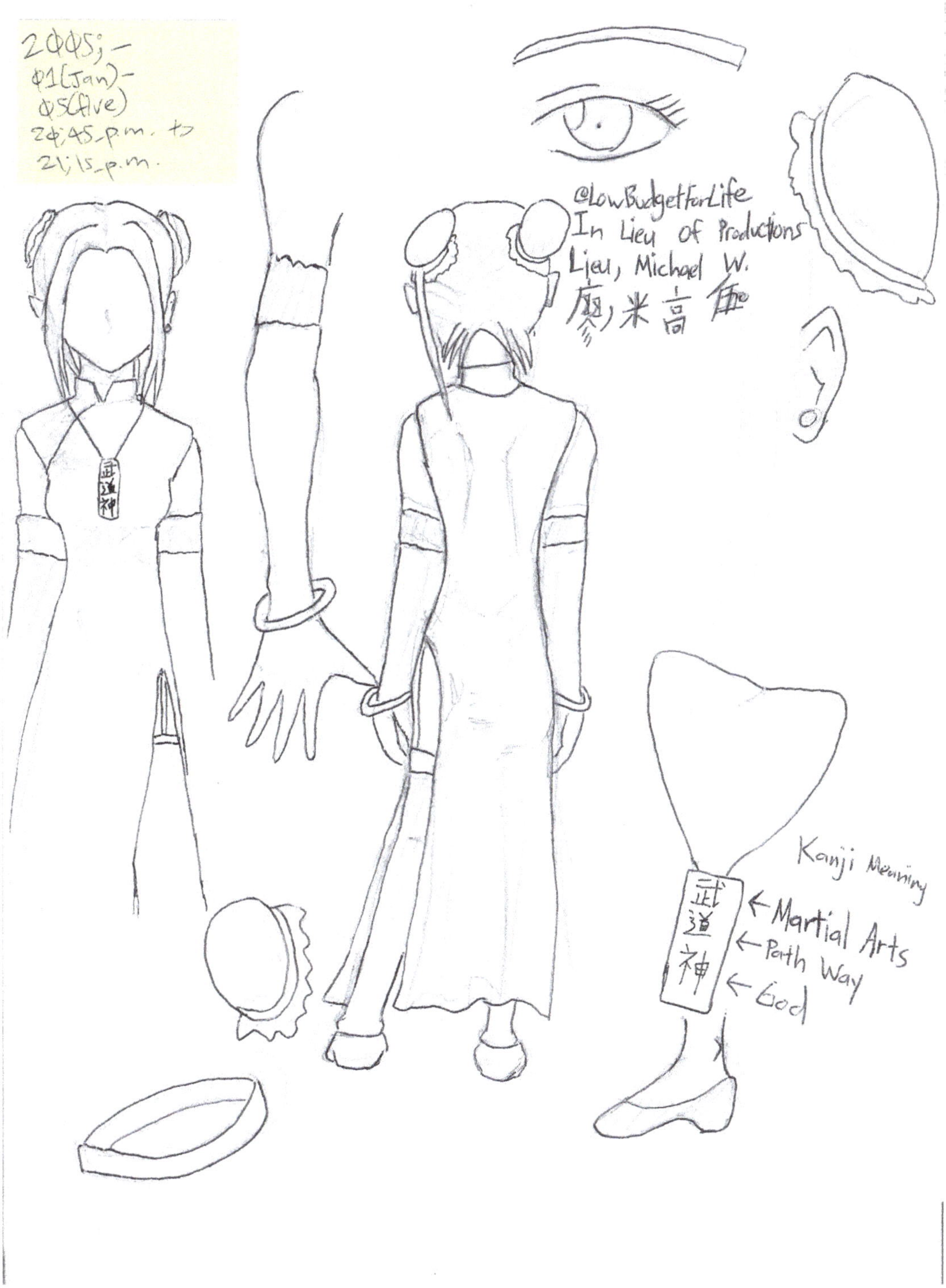

2154;_ Chicago, Illinois' Intranet, Wireless Broadcasting Frequencies Online
Alien Valentina Vita, Traveling Cosplayer(s) And Kosplayer(s)

Alien Elle K. Mcgee, BladeMaster 刀大師, Celebrity Cook-Chef Martial Artist
2154;_ Chicago, Illinois' Intranet, Wireless Broadcasting Frequencies Online

刀 Augmented Reality Holographic Heads Up Display A.R. Holo HUD	
Actual Legal Given Name;_ **Mark D.** (goes by **Mark D.**) pronounced, '**ma-ark' 'dee**'	Actual Legal Given Name;_ **Mark S.** (goes by **Mark S.**) pronounced, '**ma-ark' 'eh-ss**'
Alias Network Nickname;_ Oh Hi Mark, DaiLow, Daddy, BladeBoss, Asian Buddha, Diplomatic Immunity, [*Aug+*]	Alias Network Nickname;_ Oh Hi Mark, SaiLow, Subby, BladeBoss, British Buddha, Diplomatic Immunity, [*Aug+*]
Physical Plus Psych Age;_ augmented 190 + 50s years old	Physical Plus Psych Age;_ augmented 191 + 50s years old
Alien;_ Legal And Approved **Politician(s), Citizen(s) Of Eur-Asia**	
Social Status;_ **Eastern Society of Asian Pacific (A.P.) diplomat** [*Aug+*]	Social Status;_ **Western Society of European Union (E.U.) diplomat** [*Aug+*]
Community Network(s);_ **Hong Kong's Independent Commission Against Corruption (I.C.A.C.)** **British's Special Air Service (S.A.S.)** **British-Hong Kong Sovereignty Sector** **Cook-Chef Community Martial Artist Educator** [*Aug+*]	Community Network(s);_ **British's Special Air Service (S.A.S.)** **Hong Kong's Independent Commission Against Corruption (I.C.A.C.)** **British-Hong Kong Sovereignty Sector** **Cook-Chef Community Martial Artist Educator** [*Aug+*]
Employment;_ **The Oldest Society**. [*Aug+*]	
[*Aug+*]	
刀 Augmented Reality Holographic Heads Up Display A.R. Holo HUD	

2154;_ Chicago, Illinois' Intranet, Wireless Broadcasting Frequencies Online
Alien Elle K. Mcgee, BladeMaster 刀大師, Celebrity Cook-Chef Martial Artist

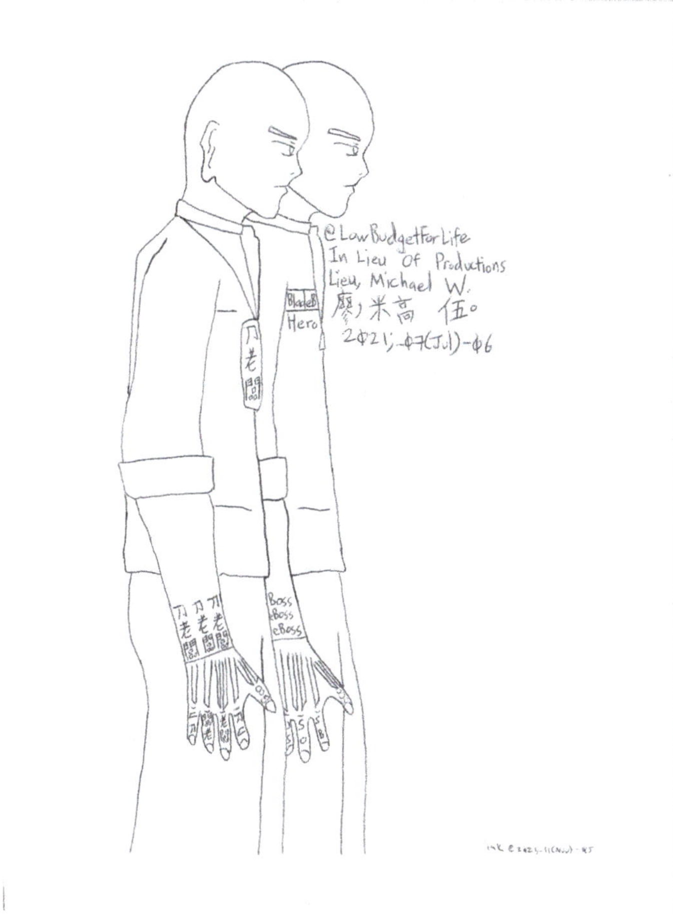

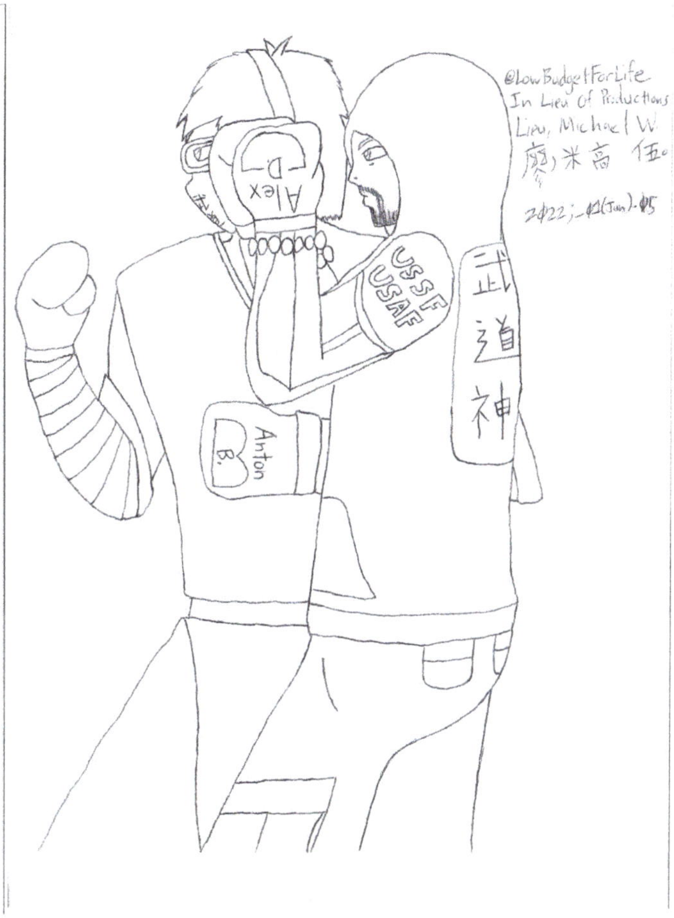

Alien Valentina Vita, Traveling Cosplayer(s) And Kosplayer(s)
2154;_ Chicago, Illinois' Intranet, Wireless Broadcasting Frequencies Online

V. V. Augmented Reality Holographic Heads Up Display A.R. Holo HUD	
Actual Legal Given Name(s);_ **Mark D.** And **Mark S.**	Actual Legal Given Name(s);_ **Anton B.** And **Alex Drastal**
[*Aug+*]	
V. V. Augmented Reality Holographic Heads Up Display A.R. Holo HUD	

2154;_ Chicago, Illinois' Intranet, Wireless Broadcasting Frequencies Online
Alien Valentina Vita, Traveling Cosplayer(s) And Kosplayer(s)

V. V.; Pretty Sure This Is Not That Traveling Cosplayer And Tourist Valentina Vita;
"(>_O) Whoa we wow! ...**Eur-Asia Politicians** fighting Kung-Fu, Martial Artists...?"

BladeMaster **刀大師**; Pretty Sure This Is Not That Elle K. Mcgee;
"(*voice over P.P.E.*) **Daddy Subby**, not just any **politicians**, both **former field agents**."

2154;_ Arcadia \ Valentina Vita (V. V.) 'Healthy-Strong Life' \ Left to Right(L->R)

Mark D., Politician from Eur-Asia Eastern Society of Asian Pacific (A.P.) diplomat;
"(*voice over*) **Our Blades** is not actually in this biotech armlet... **We** carry this..."

Mark S., Politician from Eur-Asia Western Society of European Union (E.U.) diplomat;
"(*voice over*) ...As an old habit of **ours** as **field agents**... As **I** recall, no such thing as a **former field agent**. After all, it is rare for **us** society, **Blades** to go up against Martial Artists in a special Sanctuary like this for **Animals**. **We yield the music** to you."

Anton B. "MartialArtsPathWay 武道" , Kung-Fu, Martial Artists "MartialArtsGod 武道神" ;
"(*in mouth mask, voice over*) Hm... How about that loud six point five hours Eurobeat music for each single **first two four** round? Yea... Okay, ready? This **first two four** is gonna take a while... **We** been planning on this for some time... **Put on your P.P.E.**!"

Alex D. "Doc Drastal"; War Survivor Military Veteran, "MartialArtsGod 武道神";
"(*voice over*) Copy. Time to **improvise, overcome and adapt**. **Put on your P.P.E.**!"

BladeMaster 刀大師; Pretty Sure This Is Not That Elle K. Mcgee;
"(*voice over P.P.E.*) **Daddy and Subby** do **your** best!! Anton and **Alex** do your best!!"

HeartBladeEndureLove 心刀忍愛; Pretty Sure This Is Not That Nun Sister K. White <3;
"(*voice over P.P.E.*) Okay, **BladeMaster**, V. V., we Magical Maidens are to sit here and observe until **these four** have finish **their first two four music video** fight!"

V. V.; Pretty Sure This Is Not That Traveling Cosplayer And Tourist Valentina Vita;
"(@_@) Uh... How long would this **fight** take? That last one with Anton and **Alex** sparring **fight** was **fast.** So surreal. Like a **human caused quake, rumble**, or something I am not sure how to explain, this has to do with my augmentations...? Even with this Magical Maiden **P.P.E.**. I see an orange blue **black blur** at least... Now we have a tag team **politicians fighting** a tag team Kung-Fu, Martial Artists that moves so **fast**!? How long can they **fight** at this kind of **mach speed** for? (>_<) What if we need to use the ...restroom bathroom, you know... number one or two for those of us still bio and natural parts for waste...?"

HeartBladeEndureLove 心刀忍愛; Pretty Sure This Is Not That Nun Sister K. White <3;
"(*voice over P.P.E.*) Remember, the Magical Maiden Super-Hero Suits are biotech... So, you can use them to do number one and number two in them... The suit handles biodegradable waste. (*thinking*) Yes, seeing them has to do with your augmentations, as I can see it, you mainly have **beauty implant augments** and not so much **speed, agility and slow motion capture** biotech. At times you can feel them, over seeing them."

V. V. does not say anything in response and follows what HeartBladeEndureLove tells her.

2154;_11(Nov)-28(twenty eight)

Travel To Western Society, America(s), North America
Self-Govern Sovereignty (S),
Illinois (ILL) Sector, Points Of Interest (P.O.I.)

Chicago (S), "**Animal** Sanctuary"
"The New World 2121 Movement Of The 22nd Century,
Utopian Sovereignty of North America (U.S. of N.A.)"
Downtown District, **Elysium Corporation** Biotech Dome Center Point
Captain Rage's Public Po Po Place
Chicago City Police Department (C.C.P.D.) Special Weapons And Tactics (SWAT)

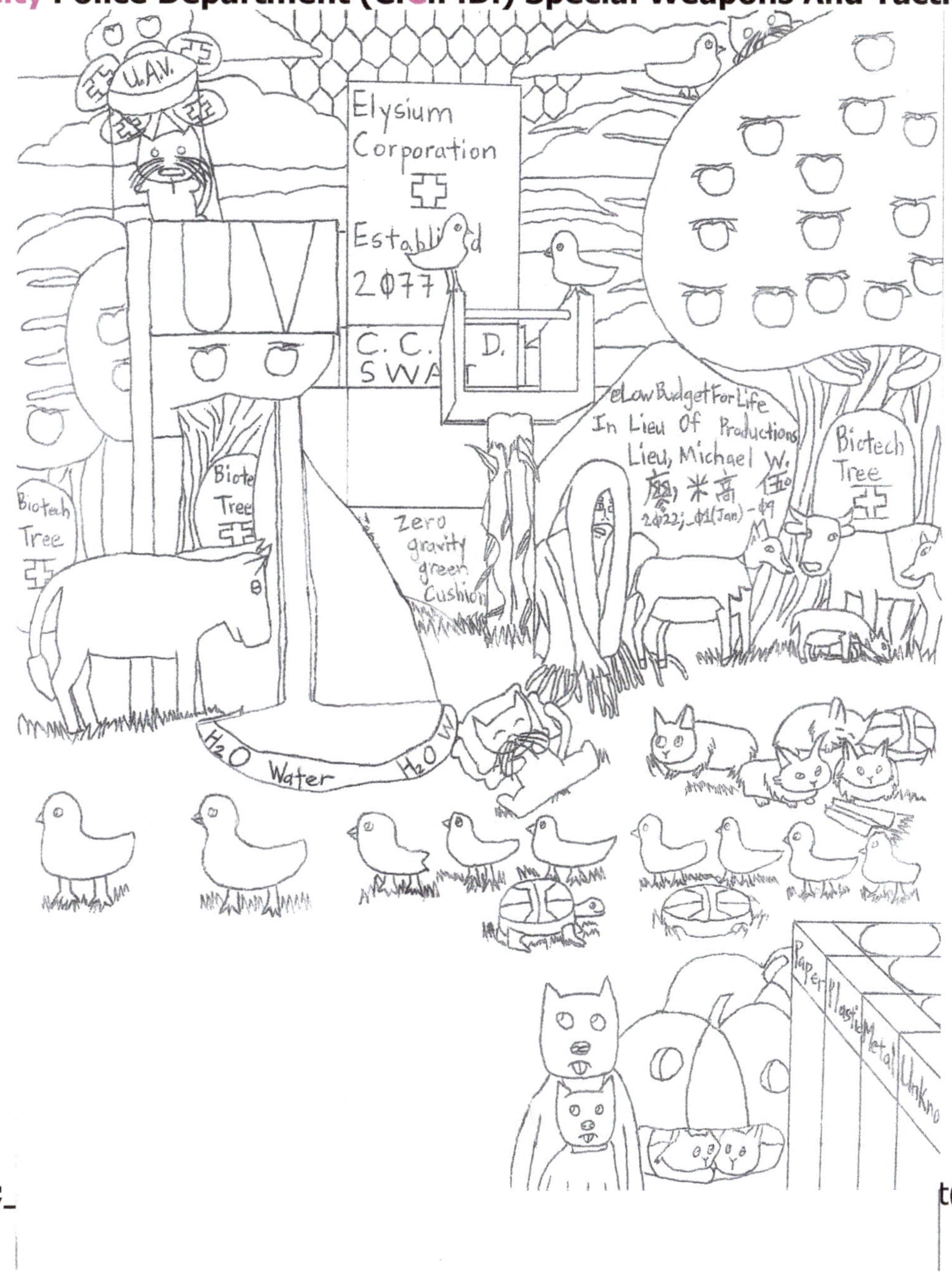

2154;_ t(L->R)

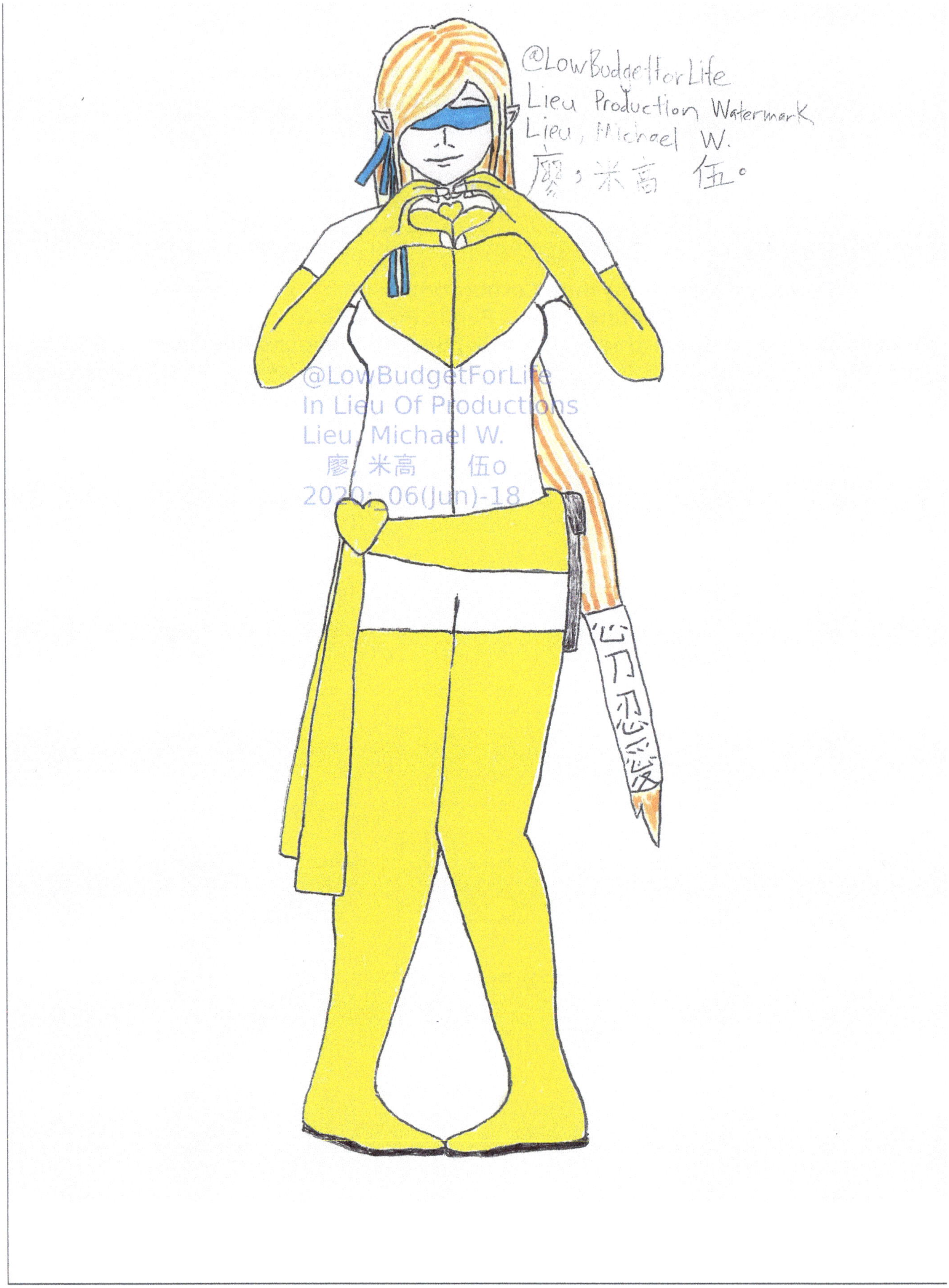
@LowBudgetForLife
Lieu Production Watermark
Lieu, Michael W.
廖, 米高 伍o
@LowBudgetForLife
In Lieu Of Productions
Lieu, Michael W.
廖, 米高　伍o
2020_06(Jun)-18

V. V.; Pretty Sure This Is Not That Traveling Cosplayer And Tourist Valentina Vita;
"(>_<) ...That was a long and interesting... tag team **mach speed** martial arts fight... I have never ever seen, because my bio**tech augs** are not designed for seeing such **speed**! (@_@) Huh!? I mean **Date-Time** is now **November twenty eight**! **Twenty eight**!?"

HeartBladeEndureLove 心刀忍愛; Pretty Sure This Is Not That Nun Sister K. White <3;
"(*hand heart love sign*) That was quite a **first two four** event... Well, we can take off our Magical Maiden **P.P.E.** Helmet... (*looking*) ...Sorry, separation between Church and State belief system prevents me from entering **C.C.P.D. SWAT State** in the **Downtown District**. This is an old world belief system... **BladeMaster**, V. V., you can keep the Magical Maiden Super-Hero Suits. The Church has plenty in supply! Anyways, heart and love you all!"

V. V.; Pretty Sure This Is Not That Traveling Cosplayer And Tourist Valentina Vita;
"(O_O) Oh, okay, you have to depart... because of... Church and State separation... I see. Bye bye... Sister, I mean... HeartBladeEndureLove. Thanks for the **P.P.E.** Magical Maiden Super-Hero Suits... My favorite color in red! (=_=) ...Wait, how are we in the **Downtown District**...? By the... **Animals**? Not the **West Side**...? Whoa we wow... Huh..."

BladeMaster 刀大師; Pretty Sure This Is Not That Elle K. Mcgee;
"(*voice over*) Because of Sister Heart! ...Bye bye Sister Heart! Love is the frequency!"

BladeMaster waves good bye to HeartBladeEndureLove as she levitates away, but V. V. is lost in thought thinking about the **Date-Time** jump forward in time from the **first two four**.

BladeMaster 刀大師; Pretty Sure This Is Not That Elle K. Mcgee;
"(*voice over*) Okay V. V., just **me**, you, **Ninja Kanon-Klone** and who ever we cannot see with our naked eye balls and **P.P.E.** and anyone with biotech augments and **suits**... and"

V. V.; Pretty Sure This Is Not That Traveling Cosplayer And Tourist Valentina Vita;
"(-_-) Yea, things are now calm now... So, where to now El-**BladeMaster**? Most of the events with **Entertainer Expo** is now over... Good thing I got a **holo photo** with **Mister Steven**... (>_<) How did it go from **November thirteen to twenty eight?**"

BladeMaster 刀大師; Pretty Sure This Is Not That Elle K. Mcgee;
"(*voice over*) **Time dilation. Time traveling** is forward in time... Anyways, those **Po Po in SWAT Suits** is close by... **Officer Knight** in **Captain** Rage's **Po Po hiding hole**."

Automated Algorithm; Human Voice Over Emulation Based On Elysium Corporation Archives;
"(*monotone female voice over*) Please stand by for biotech dome airlock decontamination. Thank you for your patience and cooperation **Utopian**."

Chicago Controller City Mayor, Self-Govern Sovereignty Sector(S), Sanctuary
2154;_ Chicago, Illinois' Intranet, Wireless Broadcasting Frequencies Online

City Controller　Augmented Reality Holographic Heads Up Display　A.R. Holo HUD

Chicago Controller
City Mayor

"Chicago City Clean House Act of
2152 (twenty one fifty two);_07-04 (July four)"

"Effective;_2152 (twenty one fifty two);_07-05 (July five)"
Retire Or Be Re-Purposed

All non-retired and non-resigned current and active public security of the old world Chicago Police Department (C.P.D.) officers must reside and work in Chicago Elysium's police precinct under Captain Rage's Public Po Po Place's first responders only to animal sanctuary needs, no human city needs. On duty and off duty Chicago City Police Department (C.C.P.D.) Special Weapons And Tactics (SWAT) officers uses code names of rank trait for their own personal protection of retaliation from Cainful behavior, or not if they want to trap and lure people using old world public sector tactics.

Rank =	Captain	Lieutenant	Operator	Sergeant\Alfa\Deanna	Rookie
Trait =	Rage	Gain	Mirror	Soil\Tool\Ogden	Knight

City Controller　Augmented Reality Holographic Heads Up Display　A.R. Holo HUD

2154;_ Chicago, Illinois' Intranet, Wireless Broadcasting Frequencies Online
Chicago Controller City Mayor, Self-Govern Sovereignty Sector(S), Sanctuary

V. V.; Pretty Sure This Is Not That Traveling Cosplayer And Tourist Valentina Vita;
"(>_O)　These Holo HUD articles explains what happened to the **police department**."

Rookie Knight; Real Time On Call Animal Sanctuary Hurry Up And Wait C.C.P.D. SWAT;
"(*SWAT Suit P.P.E., voice over*)　**I** am always **watching** and **hearing** everything.　**Eyes** and **ears** all around this...　**C.C.P.D. SWAT** Student Trainee Elle...　**Captain** Rage on watch."

Chicago Controller City Mayor, Self-Govern Sovereignty Sector(S), Sanctuary

2154;_ Chicago, Illinois' Intranet, Wireless Broadcasting Frequencies Online

City Controllers Augmented Reality Holographic Heads Up Display A.R. Holo HUD
Mass Media Broadcasting Frequencies Archive Stories;_
A U.S. Department Of Public Perception (DOPP), Investigative Journalism presents "City Controllers, upper case C. in the city, and their message with fire in the works." Written by Mister Marco Medina Morales, Edited by A.I. Ai Large Intelligent Sexy Assistant (LISA), Published by Animal Kingdom Service Doggie Evans on 2154, January one. "Chicago City Clean House Act of 2152 (twenty one fifty two);_07-04 (July four)" 'An outside look into a unique 22nd century phenomenon of "Animal Sanctuary (S)" by these so called "City Controller(s)" claiming to be "City Mayor(s)" of a human city.' What does it really mean to be a "new American", under "the new world movement of 2121" with these "City Controller(s)" and their "Utopian Sovereignty of North America"? What does it really mean to be an "old American", under "the old world" United States of America (U.S.A.) federal government that was founded in 1776 on North America? Historians of the past 21st, 20th centuries, notice the wording of "new world" in comparison to "1st world" as rich, and "old world" in comparison of "3rd world" as poor. The "new" and "old" world movement divide, complicates it further with conventional Democratic (D) and Republican (R) political parties, to further example it: "the new world (D)", "the old world (D)", "the new world (R)", "the old world (R)".
[*Aug+*]
Mass Media Broadcasting Frequencies Archive Stories;_
City Controllers Augmented Reality Holographic Heads Up Display A.R. Holo HUD

2154;_ Chicago, Illinois' Intranet, Wireless Broadcasting Frequencies Online

Chicago Controller City Mayor, Self-Govern Sovereignty Sector(S), Sanctuary

V. V.; Pretty Sure This Is Not That Traveling Cosplayer And Tourist Valentina Vita;
"(@_@) Hmm... By that A.I. LISA. I met early... Hmm... that explains it. **Someone** here?"

BladeMaster **刀大師**; Pretty Sure This Is Not That Elle K. Mcgee;
"(*voice over*) Oh, **Officer Knight** is here! Hiding in **his SWAT Suit** like **Officer Ogden, I** need to take a huge shit **Toilet Paper** watcher! **Officer Knight** is like that **True-Men spying operative**, at least you can see **True-Men** with **their** dust particles with bright light when **they** say **They Are True-Men**, and hear **them** say, **Hey-Zeus** receives. This way!"

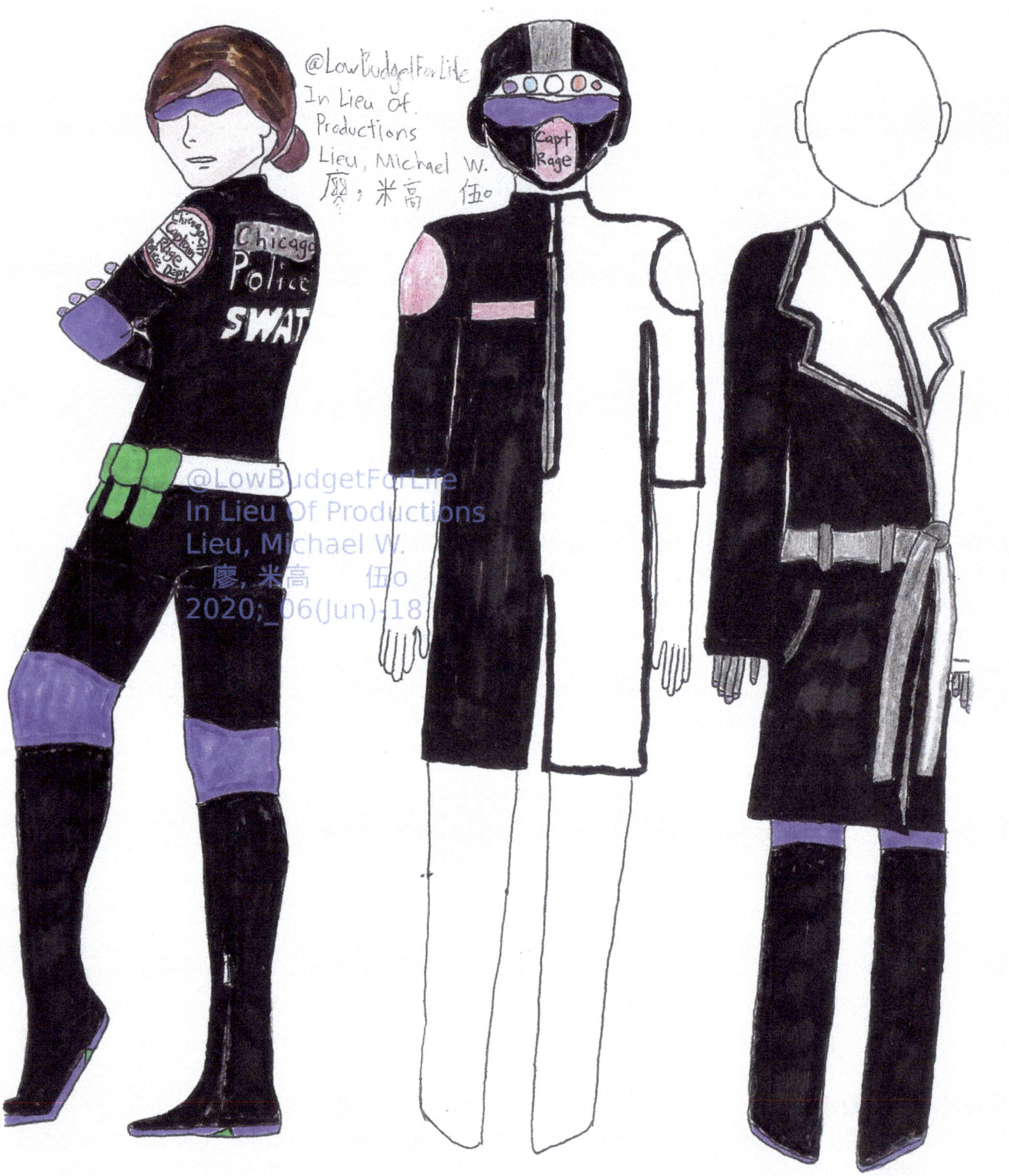
@LowBudgetForLife
In Lieu Of
Productions
Lieu, Michael W.
廖, 米高　伍o
Chicago
Police
SWAT
Capt
Rage
@LowBudgetForLife
In Lieu Of Productions
Lieu, Michael W.
廖, 米高　伍o
2020;_06(Jun)-18

Automated Algorithm; Human Voice Over Emulation Based On Elysium Corporation Archives;
"(*monotone female voice over*) Please stand by for biotech dome airlock decontamination. Thank you for your patience and cooperation **Utopian**."

Alien Valentina Vita, Traveling Cosplayer(s) And Kosplayer(s)
2154;_ Chicago, Illinois' Intranet, Wireless Broadcasting Frequencies Online

V. V. Augmented Reality Holographic Heads Up Display A.R. Holo HUD
"Chicago City Clean House Act of Twenty one fifty two, July four". [*Aug+*]
V. V. Augmented Reality Holographic Heads Up Display A.R. Holo HUD

2154;_ Chicago, Illinois' Intranet, Wireless Broadcasting Frequencies Online
Alien Valentina Vita, Traveling Cosplayer(s) And Kosplayer(s)

V. V. hears that the music "**Soda Inc's Being Boring**" is playing in the **Utopia A.R. Holo HUD** background at one third volume of 33% and is set on loop.

V. V.; Pretty Sure This Is Not That Traveling Cosplayer And Tourist Valentina Vita;
"(O_o) Oh... City Mayor's Chicago **Police SWAT**... **Captain** Rage, hiding up in her gunner's sniper nest. Whoa we wow, what an amazing awesome... **hiding hole you** have here!"

BladeMaster **刀大師**; Pretty Sure This Is Not That Elle K. Mcgee;
"(*voice over*) Hello again, **Captain** Rage! **Operator** Mirror! No **Lieutenant Gains, she** must be on **Downtown District Animal Patrol**! **C.C.P.D. SWAT** Student Elle back to do **real time, on call hurry up and wait volunteer work** to our Chicago **Animal** Sanctuary."

Operator Mirror; Real Time On Call Animal Sanctuary Hurry Up And Wait C.C.P.D. SWAT;
"(*SWAT Suit P.P.E.*) Hello again, **C.C.P.D. SWAT** Student Elle back again."

Captain Rage; Real Time On Call Animal Sanctuary Hurry Up And Wait C.C.P.D. SWAT;
"(*calm*) Oh hey... Elle! Ready to learn more about **B.D.S.M.**, **Police SWAT** brutal tactics?"

C.C.P.D. SWAT Captain Rage looks at V. V. from head to feet, V. V. is looking around.

Captain Rage; Real Time On Call Animal Sanctuary Hurry Up And Wait C.C.P.D. SWAT;
"(*calm*) Hm... **Another volunteer worker**? She's wearing the Church's **P.P.E.** suits."

BladeMaster **刀大師**; Pretty Sure This Is Not That Elle K. Mcgee;
"(*voice over*) Yes, **Captain** Rage! This is V. V.! A white woman literally, her skin bleaching."

V. V. looks at **C.C.P.D. SWAT skintight underwear suits** using her Ai-**Aug bracelets**.

Alien Valentina Vita, Traveling Cosplayer(s) And Kosplayer(s)
2154;_ Chicago, Illinois' Intranet, Wireless Broadcasting Frequencies Online

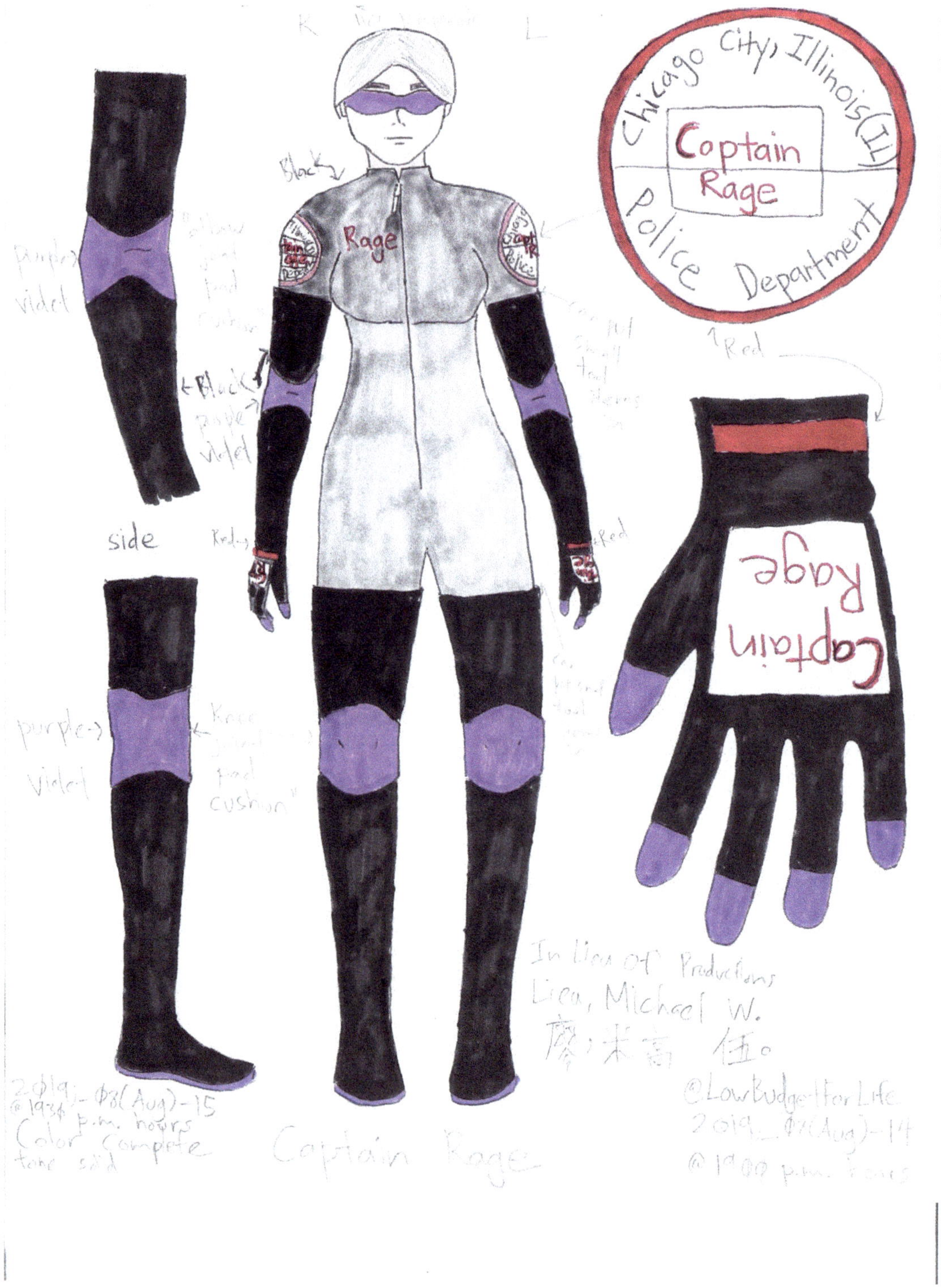

2154;_ Chicago, Illinois' Intranet, Wireless Broadcasting Frequencies Online
Alien Valentina Vita, Traveling Cosplayer(s) And Kosplayer(s)

Alien Valentina Vita, Traveling Cosplayer(s) And Kosplayer(s)
2154;_ Chicago, Illinois' Intranet, Wireless Broadcasting Frequencies Online

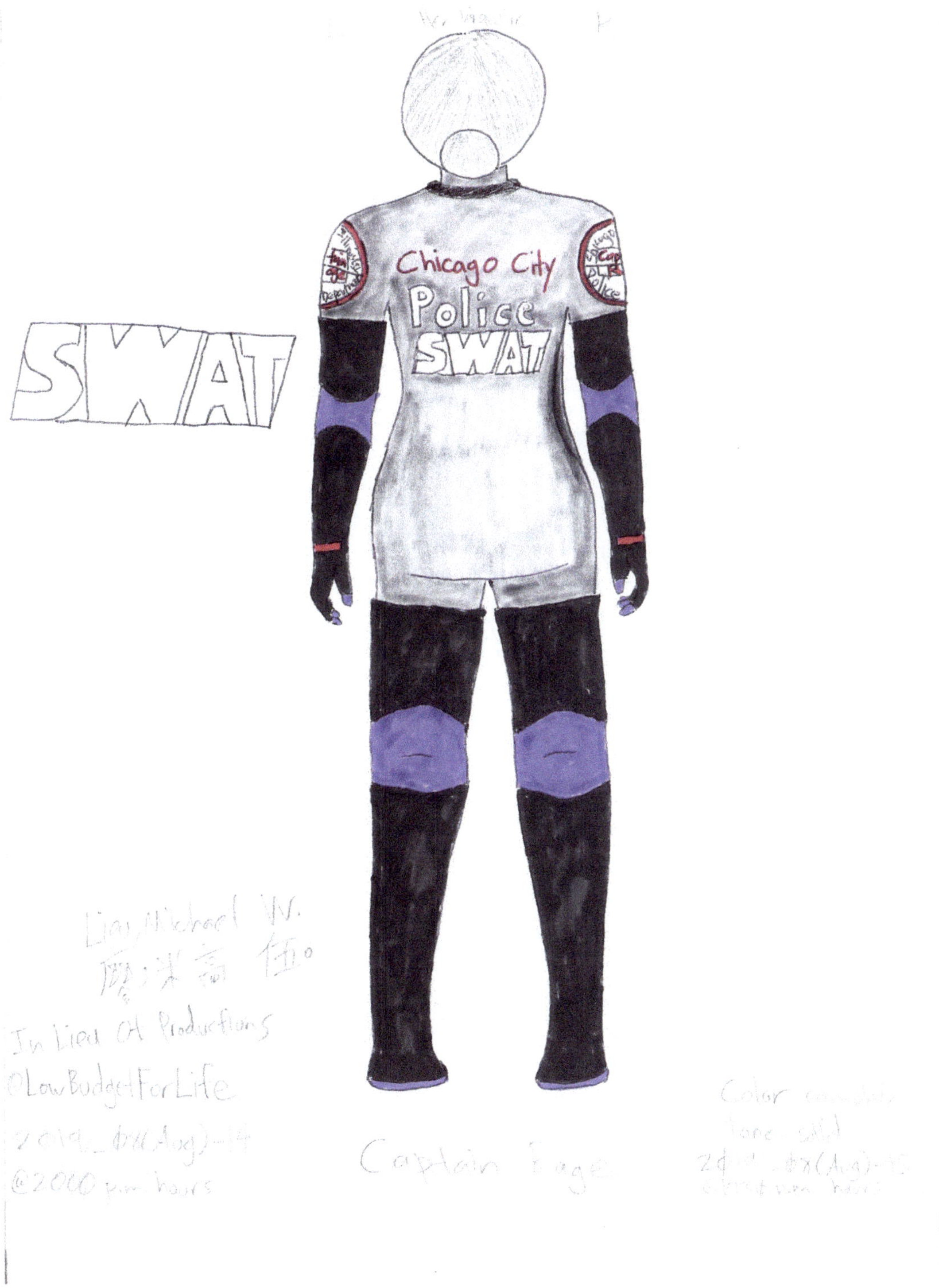

2154;_ Chicago, Illinois' Intranet, Wireless Broadcasting Frequencies Online
Alien Valentina Vita, Traveling Cosplayer(s) And Kosplayer(s)

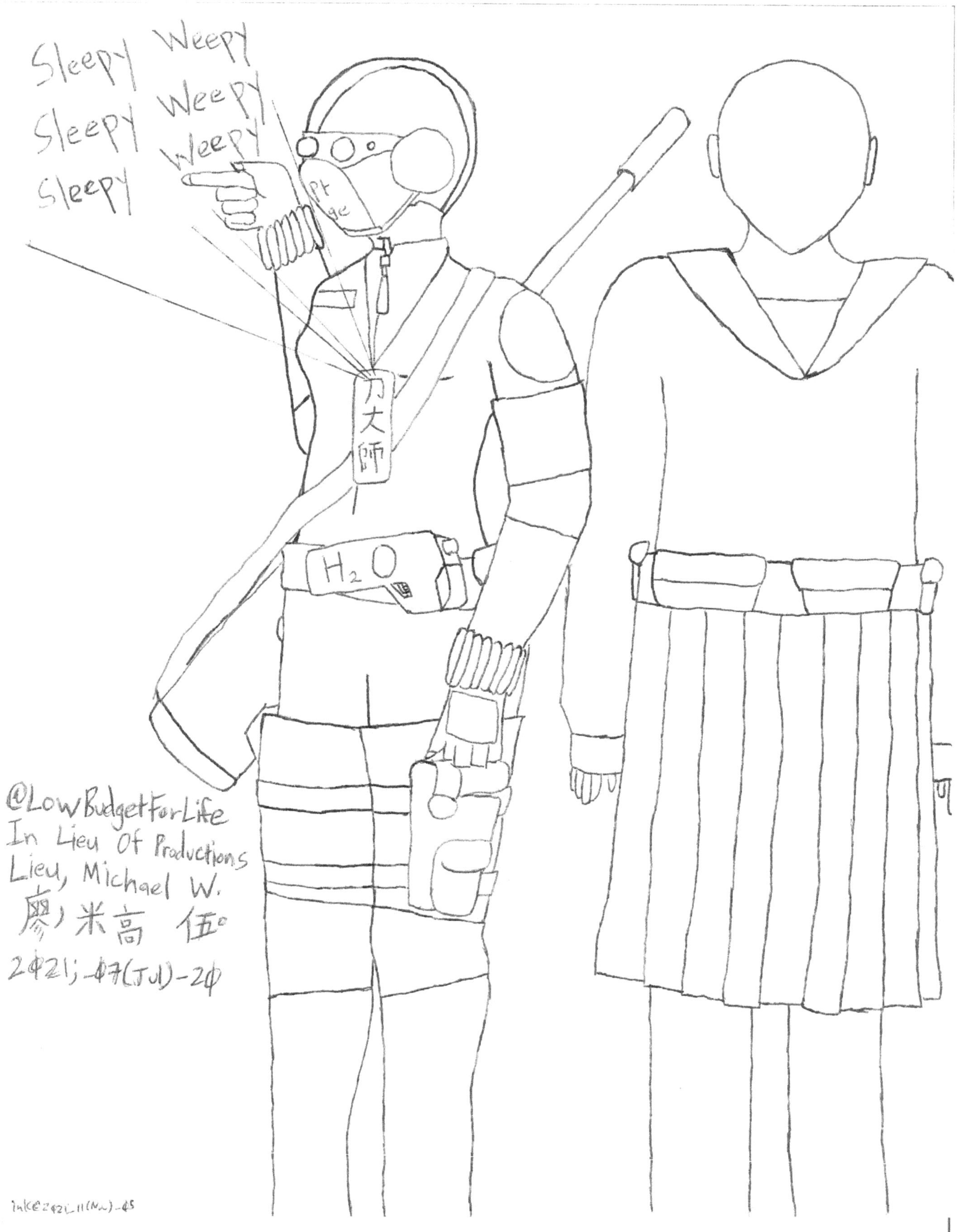
Sleepy Weepy
Sleepy Weepy
Sleepy Weepy
刀大師
H2O
@LowBudgetForLife
In Lieu Of Productions
Lieu, Michael W.
廖, 米高 伍o
2Φ21-Φ7(Jul)-2Φ

V. V.; Pretty Sure This Is Not That Traveling Cosplayer And Tourist Valentina Vita;
"(O_O) Whoa we wow, this is **SWAT Suit under... wear**? Or the **SWAT Suit**? Whoa **BladeMaster**, Elle, you changed out of your Magical Maiden Super-Hero Suit so fast!?"

Operator Mirror; Real Time On Call Animal Sanctuary Hurry Up And Wait C.C.P.D. SWAT;
"(*wearing BladeMaster Elle K. Mcgee's school girl uniform*) Hello V. V., this here is the **Chicago City Police Department SWAT skintight underwear** proportions that is worn underneath the actual **SWAT Suit** which you can barely see without the correct **biotech augments** and **broadcasting frequencies**... Like a **T-Shirt** underneath overalls or coveralls. Feel free to wear **it**, like the Church's Magical Maiden Suit, this **suit** is yours too."

C.C.P.D. SWAT Student Elle K. Mcgee selects the music(s): "**Soda Inc:**

Cocooning	*Cross the Ocean*	*Deepah*	
Full Moon (Greenskeeper Remix)		*Morning Prayer*	*Night Fever*

" on the **Utopia A.R. Holo HUD** background on one third volume of 33% and is set on loop.

BladeMaster Elle; Real Time On Call Animal Sanctuary Hurry Up And Wait C.C.P.D. SWAT;
"(*actual finger gun noises*) Time for to play **Po Po SWATTing** and make the **animal humans** go **sleepy weepy**!! **Po-Lease SWATTing** time!! **Humans** go **sleepy weepy**!!"

Operator Mirror; Real Time On Call Animal Sanctuary Hurry Up And Wait C.C.P.D. SWAT;
"(*wearing BladeMaster Elle K. Mcgee's school girl uniform*) **C.C.P.D. SWAT** Student Elle, no **animals** are allowed to go **sleepy weepy** meaning hurt, harm, injury, maim, etc... Unless the **animal** is **deceased, expiration** by natural causes of **old age**. Only **humans** can go **sleepy weepy** when **we, C.C.P.D. SWAT,** decide to play **gun games**."

BladeMaster Elle; Real Time On Call Animal Sanctuary Hurry Up And Wait C.C.P.D. SWAT;
"(*actual finger gun noises*) Right, yes, of course, I know that, silly **Operator** Mirror! **Po Po SWATTing**! Anyways, freeze hands up or **you** go **sleepy weepy**! I got **smoking guns**!"

Captain Rage; Real Time On Call Animal Sanctuary Hurry Up And Wait C.C.P.D. SWAT;
"(*calm*) Actually, realized **I** need to relief **Lieutenant Gains** from **Downtown District Animal Patrol**... **Operator** Mirror stand by with Student Elle... Present **guns**."

Operator Mirror; Real Time On Call Animal Sanctuary Hurry Up And Wait C.C.P.D. SWAT;
"(*wearing BladeMaster Elle K. Mcgee's school uniform*) Copy, **Captain** Rage."

V. V.; Pretty Sure This Is Not That Traveling Cosplayer And Tourist Valentina Vita;
"(O_O) This **suit** is different, I mean still **skintight** and such, but it **has pockets on the upper arms and the thighs part...** I get to keep this **suit**! Whoa we wow! So are **we** just gonna have after **party** since the **Entertainer's Expo** is over? (@_@) Uh...?"

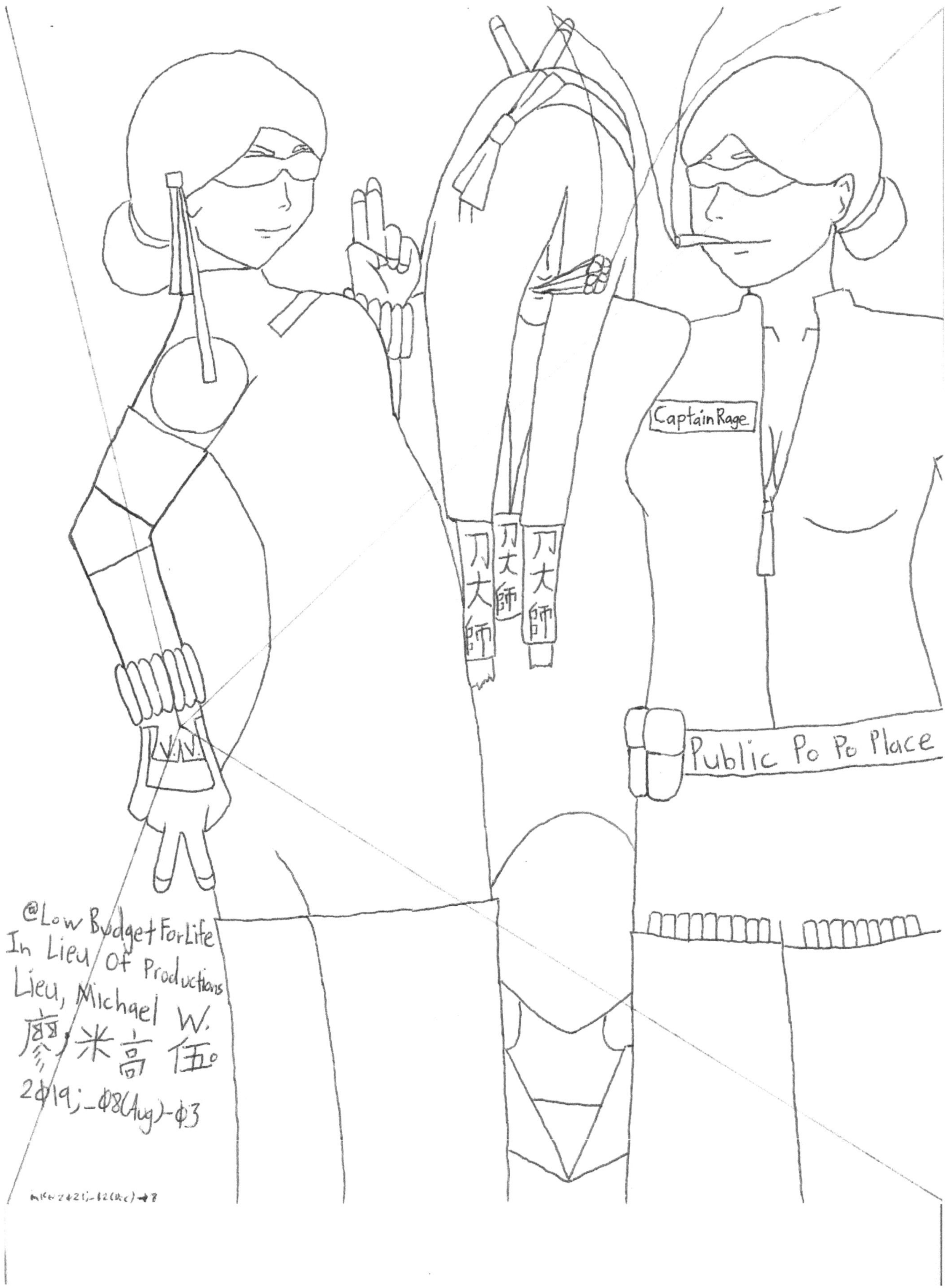
V.V.
刀大師
刀大師
刀大師
Captain Rage
Public Po Po Place
@LowBudgetForLife
In Lieu Of Productions
Lieu, Michael W.
廖,米高 伍o
2019;_08(Aug)-03

2154;_11(Nov)-29(twenty nine) @ 00;00_a.m. { act 4(four) conclusion } Beauty Implant+ Engineering Environment Biotech Dome, Interior, +024° C (twenty four degrees Celsius)
"Chicago, The Last **Animal** Sanctuary Of Illinois. Utopian Sovereignty of North America."
'Responsibility to a 22nd (twenty second) century animal sanctuary biotech utopia.'
Chicago ***City***, D. "**Downtown District**" D., ***Elysium Corporation***
Illinois (ILL) 60666 U.S. of N.A., Planet Earth, Solar System, Milky Way Galaxy

V. V.; Real Time On Call Animal Sanctuary Hurry Up And Wait C.C.P.D. SWAT;
"(O_o) Time for the post con blues. (^_^) Now the after **Entertainer's Expo** party! Like any con, **arrive as strangers, depart as family**. This **suit** has **pockets**, **ways to put stuff on the side** of your **arms and thighs.** I still have time until the **new year's resolution of twenty one fifty five**... My **resolution** list... Writing **it** down on my used **back side rare piece of paper**. **One more month-ish until new years**."

V. V.'s New Years' Resolution For 2155.

1. Craft a reasonable cosplay wish list
2. Do not start and abandon a cosplay project
3. Be more confident with my body and beauty implant augmentations
4. Do not rush a cosplay before any con period
5. Try to make a more cosplay related recordings

Captain Rage; Real Time On Call Animal Sanctuary Hurry Up And Wait C.C.P.D. SWAT;
"(*smoking medical marijuana*) Well Elle, what you think of these **guns**... **Smoking guns.**"

Operator Mirror; Real Time On Call Animal Sanctuary Hurry Up And Wait C.C.P.D. SWAT;
"(*wearing BladeMaster Elle K. Mcgee's school uniform*) So **Lieutenant Gains** is not getting any relief for **her Downtown District Animal Patrol** watch?"

Elle K. Mcgee; Childlike Human Kanon-Klone(s) "BladeMaster 刀大師" Experiment;
"(*smoking medical marijuana*) ...Yea! **Lieutenant Gains** can... can... Whoa...
V. V.! ...Try on the **skintight suit**, it is different from Sister Heart's suit. **Smoking** eight M.M. joints at a time!! ...Wow, so many versions of **you**... wait, where are **you** all going...?"

Captain Rage; Real Time On Call Animal Sanctuary Hurry Up And Wait C.C.P.D. SWAT;
"(*smoking medical marijuana*) **Everyone**, just another day in Chicago **Animal** Sanctuary.

V.V.; Real Time On Call Animal Sanctuary Hurry Up And Wait C.C.P.D. SWAT;
"(^_^) Ha ha! **Smoking guns** is actually just... medical marijuana! **Guns** time!"

2154;_Arcadia story conclude.

"**Model References Work Inspiration; Bibliography**"

As of 2022;_08(Aug)-08(eight) 12;00_a.m. for story 2(two) 2154;_Arcadia
Not In A.B.C. Alphabet Order, Or Numerical Order 1(one), 2(two), 3(three) Completely

2019;_12(Dec), **21X0;_U.S.-Department-Of-War-KIPI-Kosplay** "Prototype Photo-Book"

2020;_03(Mar), **2154;_Arriving-Coloring-Book** "Coloring-Book"

2020;_09(Sep)-06(six), **2154;_Arriving** "Concept Art Script Experiment"

2020;_10(Oct)-31(thirty one), **Augmented Artificial Intelligence (A.I.+) Ai love Conscientiousness Cat/Kat** "Parody Splicing Idea Generator"

A Natural History Of Mating, Marriage, And Why We Stray Anatomy Of Love, by Helen Fisher, P.H.D.

Unraveling The Mysteries Of Repulsion That Is Disguising, By Rachel Herz

Diary Of A Wimpy Kid, A Novel In Cartoons, By Jeff Kinney

Marvel Studios' Thor 2(two) The Dark World Movie Prelude Comic

Life Is Strange 1(one) Original Adventure Video Game Sequel Comics "Save Chloe Price"

Destination Japan 2018-2019 Edition, What Season Is Right For You, Japan Travel Magazine

Anime eXpo (A.X.) Los Angeles Convention Center Guides Year(s); 2012, 2013, 2018, 2019

Anime Los Angeles 15, January 10-13, 2019, Ontario Convention Center, Doubletree Ontario Airport Convention Center Program Guide

Comic-Con International Presents WonderCon Anaheim 2019 Program Book And Complete Schedule, March 29-31, 2019, Anaheim Convention Center

The Daily Show With Jon Steward Presents Earth Book A Visitor's Guide To The Human Race

The Most Dangerous Biological Weapons, By The Infographic Show

United States Department Of Defense (U.S. D.O.D.) Navy Sailor Military Personnel Manual (MilPerMan) chapter 13 "Chemical, Biological, Radiological (C.B.R.) Defense"

Kidzone.ws Animal Babies And Groups Website

2020;_03(Mar)-04(Apr) AAA.com, Triple A, Westways' Magazine page 58,
"Tails & Trails A { Kanab } Utah animal sanctuary is a scenic red-rock retreat for pet-loving volunteers." By Brenda Rees. Best Friends Animal Society { Sanctuary }

King And X Photo; King, Martin Luther { Looter }, --Junior, 1929-1968 Public Appearances
X, Malcolm, --1925-1965--Public Appearances
United States. --Civil Rights Act of 1964. Actual Photo By; "Trikosko, Marion S."
Title; Martin Luther { Looter } King and Malcolm X after King's press conference at the U.S. Capitol about the Senate debate on the Civil Rights Act of 1964
https://www.loc.gov/resource/cph.3d01847/

Electro Magnetic Pulse Commission. www.EMPCommission.org

Human version two "Human v2.0", 2006;_10(Oct)-24, B.B.C. Documentary

Doctor Arthur T. Bradley, E.M.P. expert, NASA engineer

Batman Technology, Warner Brothers Detective Comics 2008;_ History Channel Documentary

Marc { Oh Hi Mark } Zicree, Mr. Sci-Fi "Professor Science-Fiction" Space Command, Hopeful

Electronic Media, Formerly Telecommunications 10th(tenth) Edition, Lynne Schafer Gross
Chapter 13(thirteen) Technical Underpinnings, pages 356-382

Prophets Of Science-Fiction With Ridley Scott, 2011, Discovery Channel T.V. show series doc

0(zero)		
1(one)	Mary Shelley	Frankenstein, Deceased, Dead, Resurrection, Undead
2(two)	Philip K. Dick	Virtual Reality (V.R.), Alternative History Timeline Universes
3(three)	H.G. Wells	Time Travel, Time Traveling, Shakespeare Of Science-Fiction
4(four)	Sir Arthur C. Clarke	Mythology, Spiritual, Religious, Contemporary
5(five)	Issac Asimov	Artificial Intelligence (A.I.), Robotics, Robots, Bots, Slaves
6(six)	Jules Verne	Steampunk, Father Of Science-Fiction
7(seven)	Robert A. Heinlein	Military, Exoskeleton Suits, Dean Of Science-Fiction
8(eight)	George Lucas	Energy Weapons, Space Opera, Space Western

U.T.L.A. Beauty Products Magazine

James Cameron's Story Of Science-Fiction, 2018, A.M.C. Originals, T.V. show series docu

0(zero)		
1(one)	Space Aliens	
2(two)	Outer Space	
3(three)	Monsters Mythology	
4(four)	Dark Futures	
5(five)	Intelligent Machines	
6(six)	Time Travel	

The Problem Of The Real And T.H.X. 1138(one one three eight)

GameInformer Video Game Magazine(s), Issue 265, May 2015, "The World's #1(number one)
A { Science-Fiction } Cyberpunk Thriller Where Player Choice Defines The World'
Deus Ex Mankind Divided { Sandbox Simulation a.k.a. Immersive Simulation Games }

I O U S A documentary, 2008, "01;04;50" Rome Republic "Roman Empire" Collapse Reasons

"...The Rome Republic did not fall because of external threats, rather The Rome Republic did fall because of internal threats from within..."	
1(one)	Decline in moral values and political instability.
2(two)	Overconfidence and over extended of military forces around the world.
3(three)	Fiscal financial irresponsibility by the central government.

Study.com Animal Mammal Versus Reptiles, Biology

AAA.com Westways magazine 2019;_10(Oct) page 56, "Earthquake Kit For Your Car"
2020;_01(Jan)-02(Feb) page 64

Psychology Today magazine issue 2009;_01(Jan)+02(Feb) page(s) 50-51, "Food Chain, Sage Advice by Jett Stone"

Good Nutrition, ChooseMyPlate.gov www.channing-bete.com , VA4A-11917, 12-14-B, 2016 © 2012 Channing Bete Company, Inc. All rights reserved.

Robert A. Heinlein, Dean of Science-Fiction, Quotes, Military, Starship Troopers, U.S.N.

"The World's Most Dangerous Idea" Search Engine Topic Listing(s)

1(one)	***War On Evil* by Robert Wright**

2(two)	***Business As Usual At The United Nations (U.N.)*** **by Samantha Power**
3(three)	***Transhumanism*** **by Francis Fukuyama**
4(four)	***Free Money*** **by Alice Rivlin**
5(five)	***Undermining Free Will*** **by Paul Davies**
6(six)	***Spreading Democracy (D)*** **by Eric Hobsbawm**
7(seven)	***Religious Intolerance*** **by Martha Nussbaum**
8(eight)	***Hating America, United States of America(s)*, by Fareed Zakaria**

Dim Sum, Eastern Society Asian Cantonese Culture Foods, Vegan, Vegetarian, Research

Martial Arts Group, "Martial Club" '...promoting the virtues of ethics of martial arts...' by Andy Le and Brian Le A.K.A. The Le Brothers, Daniel Mah "D Mah"

Lolita Looks Fashion Styles Research; "Lolita: Sweet, Classic, Gothic, Mixing", Sub-Styles: Nauticals, Country, Old School, Hime, Kuro/Shiro/etc Or Black White, Guro, Ero, Punk

2020;_09(Sep)-06(six)
Abstracted From Story 1(one) "2154;_Arriving" 'Concept Art Script Experiment'
In A.B.C. Alphabet Order, Or Numerical Order 1(one), 2(two), 3(three) Somewhat

2011;_Contagion, Warner Brothers (W.B.) Western society science-fiction { science-factual } thriller feature film motion picture movie, with tag line "Nothing Spreads Faster Than Fear", directed by Steven Soderbergh(S.S.), written by Scott Z. Burns { That Name }, featuring actors Marion Cotillard, Matt Damon, Laurence Fishburne, Jude Law, Gwyneth Paltrow, Kate Winslet, Bryan Cranston, Jennifer Ehle, Sanaa Lathan, Chin Han and… { Two hours later. }
"My main model reference work inspiration for my story universe back then..."

6(six) 'basic human behavior expression

1) happiness	2) sadness	3) anger	4) fear	5) disgust	6) surprise

Quote excerpt from science, cognitive psychology research book, "Unraveling The Mysteries Of Repulsion, That's Disgusting", author Rachel Herz

Active Writing: "Will made mistakes." / Passive Writing: "Makes were made."

A.I. Artificial Intelligence. Western U.S. of American science-fiction feature film motion picture movie { Story set in 22nd(twenty second) century }

Ai 愛 { Chinese Character } love. “Love is the one thing that transcends time and space.” -Paraphrase quote from Nolan's Interstellar movie
“Love is the frequency { hertz }.” -Paraphase quote from Disney's Wrinkle In Time

Anxiety. Human body's internal concern that seems to arises even the external threat is not physically present, as the anticipation of threat may or may not materialize.

Anxiety disorder, one example ‘Obsessive Compulsive Disorder (O.C.D.), highly intrusive, distressing thoughts and compulsions’. Quote excerpt from science, cognitive psychology research book, “Unraveling The Mysteries Of Repulsion, That’s Disgusting” by author Rachel Herz

Artificial Enhancement { Augmentation, aug, +(plus) } from Psychology Today Magazine 2009(two thousand nine) 1)January/ 2)February page #16(sixteen) featuring, Augmented Human Plus Male of 20th century RoboCop visual with paraphrases.

<u>1(one)</u>.	{ Physical Plus } ***Mental X-Ray*** = augment biotech that details human mind (psych), hearts (physical, Beats Per Minute (B.P.M.) for info, romance, love.
<u>2(two)</u>.	{ Nope, No } ***Shame On Me*** = augment biotech that blocks out pain, shame, guilt, as these emotions assist in stabilizing society, otherwise there are rewards in destructive behavior, thus the transformation of being ‘humans’ { mammals of warm blood } to being ‘snakes { reptilians of cold blood }. { Beware of snakes, scorpions, and nihilists in human clothing }
<u>3(three)</u>.	{ No Longer } ***A Life Of Excess*** = augment biotech of being able to do and live without, turn on off, such luxuries as recreational activities of entertainment.
<u>4(four)</u>.	{ A Complete } ***Collective Memory*** = augment biotech brain enhancement of experience plus remembering everything implant, explant, that anyone can share.
<u>5(five)</u>.	{ Consciousness Has } ***Everyone’s A Self-Critic*** = augment biotech that says the voices told me to do this… The voices of self-talk speaks to me!
<u>6(six)</u>.	{ Choices Have } ***Unintended Consequences*** = augment biotech that does not ever forget about any detail on almost anything, but may lose the ability to abstractly draw out information.

Augmented Reality (A.R.). A.R. enhances real world setting.
Virtual Reality (V.R.). V.R. replaces real world setting.

Batman. Western U.S. of American science-fiction super-hero vigilante created by Bob Kane + Bill Finger in the year of our lord 1939(nineteen thirty nine) A.D. of 20th(twentieth) century time, owned and operated by Warner B. under D.C. comics

Blood { red liquid }, can kill by transmitting H.I.V., hepatitis, hemorrhagic fever, AIDS

Body Disgust, gas airborne, biological organic lifeform natural human(s) = breath, farts, feces smell, foot odor, poop smell { same as feces smell, different name for shi+ }

Body Disgust, liquid fluid, biological organic lifeform natural human(s) = blood, pus, phlegm, saliva, semen, sperm, sweat, urine, vomit... { but, not tears }

Body Disgust, solid deformities, biological organic lifeform natural human(s) = amputees, deliberate planned body piercing, fat obese overweight, "f-ugly" { F-ing Ugly }, gore, intestines gut exposure, mutilation maimed, poop and or feces { shi+ } scarification, skinny skeletal thin, tattooing

Calment, Jeanne, natural human French female with recorded longest life span to date who died at age 122.5(one hundred twenty two and half) years old in the year 1997(nineteen ninety seven) according to Herz, Rachel author, page 82(eight two)

Church, "Chicago City Christian Church, Belief Systems" East Side, Photo-Book, Les Cathedrales De La Major, Marseille, Porte De L'orient France, Europe, Earth

Diseases, according to, Herz, Rachel page #78(seventy eight) That Is Disgusting book

1(one) Rats are disease vectors.	**2(two)** Deformity are disease reminders.
3(three) Pus-filled sores are disease evidence.	

Fear. "World is ruled by fear, Markus. Fear of others, fear of the future....it's like me, too old.....it's time for it to end." -Carl Manfred, Detroit { Michigan } Become Human Sony + Quantic Dream's Play Station Four (P.S.4) exclusive game

HaMaBe, MiNaMi. Eastern actor featured in Netflix's KaKeGuRuI live action T.V. show

Hero Killer Stain. Solo volunteer vigilante operative from My Hero Academia story universe as a three for one (3 4 1) of "judge, jury, executioner"
Story and characters created by KoHeI HoRiksOhi

Human Behavior, according to Fisher, Helen E., P.H.D. { Doctor, Doktor? White Lab Coat }, author "A Natural History Of Mating, Marriage And Why We Stray" page #12, paraphrased "is a complex mixture of environmental { nurture } and hereditary { nature } forces, also to include culture power in influencing human action".

Human Being, according to "A Beautiful Day In The Neighborhood, Neighborly Words Of Wisdom From Mister Rogers", quote " { As human beings, } our job in life is to help people realize how rare and valuable each one of us really is, that each of us has something that no one else has --or ever will have-- something inside that is unique to all time. It's our job to entourage each other to discover that uniqueness and to provide ways of developing its expression. Page #15(fifteen) { ...Otherwise, we may be a snake { cold blood reptilian } in human { warm blood mammal } skin... Always beware of snakes, scorpions and nihilists. Snake eat snake world. }

Human corruption, human snake, snakes disguised as humans, snake in human skin

Human emotion(s), 6(six) universal basic expressions, according to "Herz, Rachel, author of That Is Disgusting, Unraveling The Mysteries Of Repulsion" page #29,

1(one) =	**Anger** (Mad, Rage, Wrath, Fury
2(two) =	**Disgust** (Sick, Distasteful) = Learned from culture, nurture, upbringing, not innate and not born with it.
3(three) =	**Fear** (Afraid, Shocked
4(four) =	**Happy** (Happiness, Joy, High :)
5(five) =	**Sad** (Sadness, Sorrow, Regret :(
6(six) =	**Surprise** (Amazement, Astonishment) :O

Humor, Having. "Laughter is best medicine." { Slaughter is best medicine }

Homeboys Industry. A non-profit organization founded by Father Greg { Ford }

Honest Game Trailers Broadcasting On YouTube

Honest Trailers Broadcasting On YouTube

Intelligence, How To Boost And Augment Our Human Mind { Paraphrase } = Found on 2009(two thousand nine) January / February Psych Today Page 32

1(one)	***Exercise*** = 30(thirty) minutes a day sharpens problem solving abilities and reduces Alzheimer's risk.
2(two)	***Meditation*** = Controlled breathing, mindfulness, good posture, enhances and augments brain function.
3(three)	***Determination*** = Maintain faith, belief in your own ability to become smarter, to build learn skills of nurture that are not innate of nature. Nisbett warns that assuming intelligence is fixed and or pre-determinded by nature's genetics will discourage us humans from working on improving ourselves.
4(four)	***Video*** { War } ***Games*** { Simulators } = Games, simulators that require

	reflexes, strategy improve attention, working memory, and or reaction time over time.
5(five)	***Education*** = Learning, whether it is self-learning, school offline, distance online, promotes mental acuity and prevents cognitive decline in human brains of all ages. Intelligence Quotient (I.Q.), measures analytical ability, but not practical and creative intelligence.

Job. "Jobs get you nowhere. Professions get you somewhere. Skills get you anywhere. -Dan Lok { DaiLow }

Jung { Young }, Carl, "You do not have ideas, ideas have you."

Linsanity! Lin, Jeremy. Eastern-Western basketball pro player that also plays online Dota game @JLin7 { See that Remy within JeRemy? } Linsanity Documentary

Linsanity! Lin, Ludi. Eastern-Western { Asian-American(s) }
story teller feature film motion picture actor writer producer model @LudiLin
{ Lieutenant Lieu, "ToolPerson (工具人)",
"Strike Force (打擊 力量)" in honor of Air Wing Strike Force, Aircraft Carrier }
=Lionsgate + Saban Brand's Power Rangers remake 2017 movie
=Warner Brothers (W.B.) + Detective Comics (D.C.)'s Aquaman 2018 movie
=Netflix's Black Mirror 2019 T.V. show series, "fighting game love episode"

Lucas' Star Wars Episode 4 (four; IV) A New Hope, Original Star Wars 1(one) 1977

Loving Kindness / Love is something you make, sex is something you have.

Love, according to 2009(two thousand nine) Psychology Today 01(January / 02(February Magazine page # 72(seventy two), "Love, demands the reassurance of a touch. Most fights are really protests over emotional disconnection. Underneath the distress, partners are desperate to know;_ are you there for me, do you need me, do you rely on me?"

Malcolm X. 20th(twentieth) century historical figure that inspired the character Marvel's Magneto

Martin Luther King Junior (M.L.K.J.). 20th(twentieth) century historical figure that inspired the character Marvel's Xavier, A.K.A. Professor Xavier { X }

Matrix, Multi-Media Franchise, The. Western U.S. of American science-fiction action philosophical feature film motion picture movies by the great gifted Wachowskis siblings starring Keanu Reeves, Laurence Fishburne, Carrie-Anne Moss

Mass murder violent video game simulators

Mcconaughey, Matthew. “(in country southern accent) Alright, all right, all write!”

Mercy, ycreM, Ycrem / Mercy killing, mercy kill, honor killing, no mercy.

Mercy, U.S.N.S. (United States Naval { Navy } Ship) Mercy.

Mindfulness Mediation. { Be in the moment }

Mindless mass murder violent video game sandbox simulator(s) equals mass murdering shooting spree U.S. of American in the 21st(twenty first) century claim

MiYaHaRa, KaNoN. @Miyahara_Kanon @Kanon_body Eastern martial artist actor with feature film favorite motion picture movie High Kick Angels, BlackFox Age Of The Ninja, Ninja Zone { Ninja-Kanon, Kanon-Klones

MoNoGaTaRi. Eastern anime adapted light novel books on character driven story series written by Nisioisin { Pen Name }, art by Vofan { Pen Name }, translated by Ko Ranson { Pen Name }

My Hero Academia. Eastern anime manga story telling love letter to Western super-heroes, vigilantes, and super-villains that turn popular world wide by KoHeI HoRiKsOhi.
Also Known As BuKo No Hero Academia in Eastern society

My Hero Academia Vigilantes. Eastern anime manga story telling love letter to Western super-heroes, vigilantes, and super-villains that turn popular world wide by KoHeI HoRiKsOhi. { My Villain Academia }
A spin-off side story official prequel set years before “My Hero Academia” story and characters plot line as it was made possible by
HiDeYuKi FuRuhAShi
As Known As Buko No Hero Academia Illegals in Eastern society

National Geographic's Expedition Mars Spirit And Opportunity

National Geographic + Disney Plus' The World According To Jeff Goldblum

National Geographic's Mission: Wolf Rescue

National Geographic's Origins: The Journey Of Humankind Hosted By Jason Silva. { Covering Subject Topics Of = fire, medicine, war, money, communication, future }

National Geographic's Your Personality Explained Explore Science Of Identity

Netflix's Bandersnatch Black Mirror. Western choose your own adventure feature film motion picture movie video game simulator

Obama, a place in eastern society of OBaMa, FuKuI, Japan, Asia { Obama@deep }

Obama, Barack, 44th (forty fourth) U.S. of American President, Commander-In-Chief, 21st century, natural human African male

Oblivion, The Elder Scrolls 4 (four;_IV), "Stop! You violated the law!!"

Oblivion. Western U.S. of American science-fiction feature film motion picture movie Tom Cruise { A Running Man } { Story set in year 2077 (twenty seventy seven) }

Pirate. "Argh!" Pirating Pays. { Almost everything on the internet is quote "free stuff" }

Pirate Bay. { Free stuff, for those not able to obtain it in a legal way, pretty political }

Plame, Valerie. "Spy, I see spy!! Spy around here..." { Need to know }

Planet earth, nature, mother nature, natural disasters, go green, biodegradable

Plastination, according to Herz, Rachel's That's Disgusting page #200 (two hundred, is a process which replaces water (H2O) and fat (body fat) with plastic for preservation purposes. Weight of a plastinated body is same as original corpse (deceased human, dead body). ...Made possible by Gunther Von Hagens, German anatomist who created the controversial "Body World" exhibition, in which plastinated { human } male and female bodies of various ages are shown in diverse poses, is now selling human bodies and body parts online.

Professor Peterson. Also Known As (A.K.A.) quote "Doctor Peter B. Peterson" "Subject in a world of object."

Persistence. Set standard and will mindset to keep on going and pushing, whether it is under, around, over, and or through an opposing obstacle(s)

Phoenix Act, The.

Phoenix, Kayden. Western U.S. of American Latina creator writer, artist, director, producer from Boyle Heights, Los Angeles, California, North America

Phoenix, Raul. @RaulPhoenixTV Co-star in Tommy Wiseau's The Neighbors Season 1 as 'Tim', "That basketball guy that asks for twenty dollars to borrow". { Raul Doggie }

Phoenix. Myth animal fire birdie that represents life, death, rebirth. { Oh hi birdie! }

Porter, Bill. Bill Porter Effect { Affect? }

1(one)	***Persistence*** = Earn what you want, do not ever give up and take no for an answer.
2(two)	***Integrity*** = Picture yourself being observed by someone or something that cares, support you and would not want to do anything that disappoints them.
3(three)	***Always Being Watched*** = Even though his mother passed away, Bill Porter always felt he was being supported by her…
4(four)	***Perseverance*** = confident and determined.
5(five)	***Be Better Then He Was The Day Before*** { "Boy, Be Better" Kratos, Sony's God Of War 4(four) P.S.4(four) } = Bill Porter had his mother thriving on "Be Better" by giving him the ability to walk forward.
6(six)	***Self Respect*** = Respecting your self.
7(seven)	***Self Proclaimed Golden Rule*** = developing and understanding that others are delayed in the present and reminded later on in the future { delay gratification? }
8(eight)	***Alchemist Artist*** = deconstruct it, identify the problem, formulate a solution and engineer it all together.
9(nine)	***Determination*** = Maintain that professional occupation look.
10(ten)	***Do Not Give Up***. { Never gonna give you up, never gonna let you down. }
11(eleven)	***Enforce That Golden Rule*** = To your self and others around.
12(twelve)	***Closing Circle Conclusion Wrap-up***
Integrity = Be true to yourself and not feel bad on your self.	
Perseverance = Have a plan and a back up plan supporting it as well.	
Self Respect = Enforce That Golden Rule and equal observation rule.	
Attitude = Anger and negatively is always around us, taking it in and altering how one can accept it and affect one's experiences afterwards.	

Outlook On Life = Something you can put on your resume and feel good talking to people about it.
Tolerance Of Rejection = Acknowledge whenever it was good, neutral, bad, it has been noticed, and received feedback.

Quote, "I hate { insert hate speech subject line here }" Valve's Left 4(four) Dead Francis Western science-fiction action-horror { inflection apocalypse } video game simulator

Reeve, Christopher. Man, that is super. A super man. "Superman that ho!"

Reeves, Keanu. { Ceanu } Western marital artist actor cool celebrity. I know kung-fu.
W.B.'s The Matrix, The Matrix Reloaded, The Matrix Revolutions, The Matrix 4
Lionsgate's John Wick, John Wick 2, John Wick 3, John Wick 4 { Jon WIC }
21st century + 2008 Fox's Day The Earth Stood Still. { Augmented A.I. synthetic }
{ Kay's Private Security Services L.L.C., Kay Sec, Agent KEANU, Keanu Kat, Augmented A.I. Agent KEANU, 21st century Day The Earth Stood Still A.I. Gort }

Rocky Horror Picture Show. Western cult classic feature film motion picture movie

Sex { intercourse breeding copulation } can kill by syphilis, hepatitis, AIDS, S.T.D., etc.

Sex is something you have, love is something you make.

Sex, according to 2009(two thousand nine) Psychology Today 01(January / 02(February Magazine page # 76(seventy six), "Sex { intercourse breeding copulation } is boring if it is, { one-dimensional }, cut off from emotional connection. But, if you are emotionally involved, sex { intercourse breeding copulation } is play and passion with a hundred { different } dimensions { of excitement }."
Page # 79(seventy nine), "Excitement comes from the risk involved in staying open in the here-and-now experience of physical and emotional { psychological } connection."

Screen Rant's Pitch Meetings. "Super easy, barely any inconvenience."

Sleepy Weepy, Wakey Wakey from 2015's Chappie movie, directed by Neil Blomkamp

Smoke. "Where there is smoke, there is fire."

Smoke Grenade / Smoking Mirrors / Smoke And Mirrors / Smoke Screen

Stan The Man Lee. Western U.S. of A. comic book super-hero vigilante villain creator

Star Craft (S.C.) Real Time Strategy & Counter Strike (C.S.) First Person Shooter game

Stand Up To Bullying. Stand Up To Bullies. Standing Up To Bullies

STEM. Science Technology Engineering Math { Meth }

Strange, Doctor Stephen "Mister Doctor" / Strange, Professor Hugo

Strange Rebel Gaming. Online social net name to a story teller streamer, actor

Strange Works Of Tara Yoko From Drakengard To NieR: Automata, The. Eastern multi-media story telling franchise supported by Square { Soft } Enix + Third Editions

Stress. Someone, something that a person perceives to be a threat and that threat varies on the intensity of the human mind and body { Psychological plus physical }

String Theory. Somehow, somewhere everything is all connected in some form or shape

Stossel, John. Western U.S. of American broadcast communicating educator on detail numerous topics and subject matter

Superman. Western U.S. of American science-fiction super-hero vigilante created by Jerry Siegel + Joe Shuster in the year of our lord 1933(nineteen thirty three) A.D. They sold Superman to Detective Comics, future of D.C. Comics in the year of our lord 1938(nineteen thirty eight) A.D. of 20th(twentieth) century time.

Surprise Face Emoji Evolution = (: O) \ (ﾟДﾟ) \ (° o °) \ (°o°) \ (O :) \ O :

Skynet, Sky Net, Sky-Net, A.I., Ai, Destroy All Humans, Western Society Fear Factor

SWAT = Special Weapons And Tactics

SWAT 4(four). Western violent video game SWAT simulator training.

SWATting. Planet Earth's 21st(twenty first) century human behavior pattern scenario

Tarantino, Quentin. "Talentino"

Tatsunoko's Infini-T Force. Eastern science-fiction super-heroes manga comics anime

Too Good To Be True. { It's a trap! And or quote "It's a tarp!" }

Tom Clancy. Western author on political, military, and or spy genre type story telling

Two(2) Girls, One(1) Cup. { Warning. } Dare Yu To Look It Up. / Uncanny Valley

Ufotable's Fate 0(zero) + Ufotable's Fate Stay Night Unlimited Blade Works

Ufotable's The Garden Of Sinners = That Is Parodying "The Garden Of Eden"

U.S. = Uncle Sam, United States, Utopian Sovereignty, Undead Slaves

U.S. Department of Outbreaks and Outages (U.S. DOO)
{ In response to Chemical, Biological, Radiological (C.B.R.) Outbreaks and (&) Electro Magnetic Pulse (E.M.P.) Outages }

U.S. Department Of Public Perception (U.S. DOPP) Public Broadcasting Station (P.B.S.)

U.S. Department of Immigration Services (U.S. DIS)
{ Immigration Naturalization Services (N.I.S.) }

U.S. Department of Alien and Citizen Services (U.S. DACS)
{ Alien And Space Alien Difference In Accordance To U.S. Government Agencies }

U.S. Department Of War (U.S. DOW) of actual real events in the 20th(twentieth) century 1940s(nineteen forties), this public sec department ended specifically in the year 1947(nineteen forty seven), thus forming U.S. Air Force (U.S.A.F.) "division" from the phoenix ashes of U.S. Army Air Corps (U.S.A.A.C.) "division".

U.S. Army, The Performance Triad Guide, Sleep Activity, and Nutrition
"Professional Soldier Athlete Here It Is Not A Game".

U.S. Navy, The Bluejacket's Manual, Centennial Edition, Author Thomas J. Cultler.

Virtual Reality (V.R.), replaces real world setting.
Augmented Reality (A.R.), enhances real world setting.

Vigilant / Vigilantes. Society's super-heroes as well as non-paid public service.

Warner Brothers (W.B.), Detective Comics (D.C.) Comics

Washington { Not Washington State }, District Columbia (D.C.)

War. "What's it really good for?!"

War On Drugs / War On Evil / War On Terror / War On Corruption

Weird, Any { Weird Any Andy Weir } / Weir, Andy { Weir Andy Any Weird }

Weave, Reeve

Weaver, Sigourney. { Portrays strong independent female protagonist character(s) }

Weaving, Hugo. “Mister Anderson...” -Agent Smith, Matrix movie multi-media franchise

WestWorld, Warner Brothers’ T.V. show series based off WestWorld feature film movie

WestWorld, Corporate Guidebook, Handbook For New Employees, Delos

WhiteWolf. A wolf that is white. { Chinese Complex Characters = 白狼 }

White Wolf, according to Aurora World, Inc., quote “Canis lupus arctos, the white wolf, also known as the Arctic Wolf is a subspecies of gray wolf native to the Canadian Arctic Archipelago, from Melville Island to Ellesmere Island. It is a medium-sized subspecies, distinguished from the northwestern wolf by its smaller size, its whiter coloration. The Arctic wolf is relatively unafraid of people, and can be coaxed to approach people in some areas. It has occasionally acted aggressively toward humans. They are endangered due to their exceptionally low densities, smaller pack sizes, infrequent reproduction, and lower offspring production. In the wild, Arctic wolves’ primarily prey on muskoxen and Arctic hares.” unquote.

William Hung. Hung, William { Hung, Wil.I.Am } / Wil.I.Am. { Will I am? }

Williams, Jesse. @IJesseWilliams Western model actor activist story teller { Ninja-Talor = Sony + Quantum Dream’s Detroit Become Human ‘interactive movie’

Williams, Talor Alexander. @TalorWilliams My ninja { Ninja-Talor, Ninja-T } Aikido

White. { White and clear color toner, maybe the same? }

White, Briana. Western story teller streamer, acting, model { It’s A Me Briana White } voices Aerith Gainsborough in the Final Fantasy 7 Remake from Square { Soft } Enix. @TheStrangeRebel @ItsBrianaWhite = Criminal Minds T.V., Occupants { “HeartBladeEndureLove 心刀忍愛” / Sister K. White <3 } Super-Hero / Nun The Original Super-Hero Vigilante Parody Fan Fiction 2154;_ genre(s) character 2019 “Life Is Strange 0 Before The Storm Strange Rebel Gaming” YouTube Search

White Materia of Square Soft Enix's Final Fantasy

White Phoenix Of The Crown, Disney Marvel Comics X-Men Character

Woe Is I, The Grammarphobe's Guide To Better English In Plain English, Updated And Expanded 4th(fourth) Edition, author Patricia T. O' Conner

Wolf, Woof, Wolf-Dog, Dog-Wolf / New Mexico's Wild Spirit Wolf Sanctuary

Wolf Of Wall Street. "Deal with your problems by being rich." Belfort, Jordon

Wong. Eastern Asian last name for family surnames. { SumTing Wong }

Words. Microsoft Word, "Write letters and "sh1t" yo.".
"Words are to do things with, not label things". -Professor Peterson.

Work. "Work, work, work, work..."

Work related stress. Stress that is related from work { 9(nine) to 5(five) = "9-2-5" }.

Woody, Creepy. A Woody, character from Disney's Toy Story, that is Creepy.

Yotsuba&. Eastern comedy comic manga that reads right to left, R→L { R ←L ? }

You Are What You Eat. { MaJiN BuU }

+075° F (seventy five degrees Fahrenheit) \ +024° C (twenty four degrees Celsius)
+021° F (twenty one degrees Fahrenheit) \ +006° C (six degrees Celsius)
2154;_09(Sep)-11(eleven) @ 00;00_a.m. { act 1(one) back story } Beauty Implant+
Apocalyptic Climate Changing Planet, Exterior, +006°(six degrees) Celsius
"Naples, An **Animal** Sanctuary In **Italy**. Utopian Sovereignty of Eur-Asia."
'Departing a 22nd (twenty second) century animal sanctuary biotech utopia.'
Naples ***City***, Valentina Vita (V. V.)'s home sweet home, ***Elysium Corporation***
Naples (Nap) XXXXX U.S. Of E-A, Planet Earth, Solar System, Milky Way Galaxy

'**British-Hong Kong Sovereignty, Cook-Chef Community Diplomat**' Political BladeMaster of NoName | BladeBoss of Hero by Mark S. 刀老闆 of 英雄 by Mark D. | 刀大師 of 無名

Character Designer, Artist Author Info Dump

Elementary, Mid, High School Real Actual Friends = Eastern-Western Sit-Com genre

"Remy Lin / ***d. Chicago b.*** *Basketball-Man*"	Jeremy Lin
@ JLin7	書豪 林

An Asian-American that used to play in N.B.A. (National Basketball Association) Professional Sports Basketball-Man Player
2011;_06(Jun)-28 Jeremy Lin Academy from 2013's Linsanity Documentary, paraphrase "I remember in the D-League, um, I kept a little diary… Um… I know that's a little girly for some of you guys, but I like to write down some of my thoughts every once and while, so I can look back and remember. And, uh, I remember on December twenty nine, I wrote about how actually wished I never signed with the Warriors. I wish, you know, I wish I could quit basketball because, um, I wasn't having fun. And what got me through those times was… Obviously I don't feel that way now… But, what got me through that time, was the, uh... the number one reason why all the leaders are here today, and that's because we want to let you know that God loves me, and he, he has a perfect plan for me, and his plan will take you through a lot of ups and downs, but if I stay faithful to him. In the end, I know he's gonna work everything for my good. Now when I play basketball, I don't play for anybody else anymore. I only play for God, and that's the type of purpose he gave me. Once he gave me that purpose, that's when I have my peace. Once I got my peace, that is when I got my joy." -**Lin, Jeremy (林, 書豪)**

"Jia Wong / B.4-Burma3(three)III-M.M.B." = @ **The Chin Han** .
Warner Brothers + D.C.'s Nolan, Christopher's 2008 The Dark Knight, Batman movie
Asian Accountant Lau, "(confident) I'm good with calculation."
Warner Brothers, D.C. + C.W.'s 2012 The Arrow { Green Arrow } show series
Warner Brothers' 2011 Contagion movie, directed by Steven Soderbergh

"Sen Wang / WackyJoker.exe" = @ **Masi Oka** / "Yatta! Heroes T.V. show series"

"Lieutenant Lieu, Michael W. / LowBudgetForLife" = @ **Ludi Lin** 2017 Power Rangers

Better Luck Tomorrow movie, directed by Justin Lin \ Ragnarok Online \ C.S. S.C.

Entertainer, **Mister Steven Z. J.**, Plastination Mister, Mistress = Thriller genre(s)
A **super animated 6 rainbow tone spectrum A.R. Holo HUD** projection plays the music in the background of "**Henry Mancini's The Entertainer**" talks to her directly, as she looks at **it** with a curious cat look. **A rainbow color curtains displays itself first with English on top, Kanji on the bottom reading**: **Red "紅", Orange "橙", Yellow "黃", Green "綠", Blue "藍", Violet "紫". Purple Prince's Rain.**

Character Designer, Artist Author Info Dump

'The Disaster Artist' Co-Author Book	'Me, back in 2019 L.A.C.C.'	'The Room' Movie
@ Greg Sestero	Lieu, Michael W.	@ Tommy Wiseau

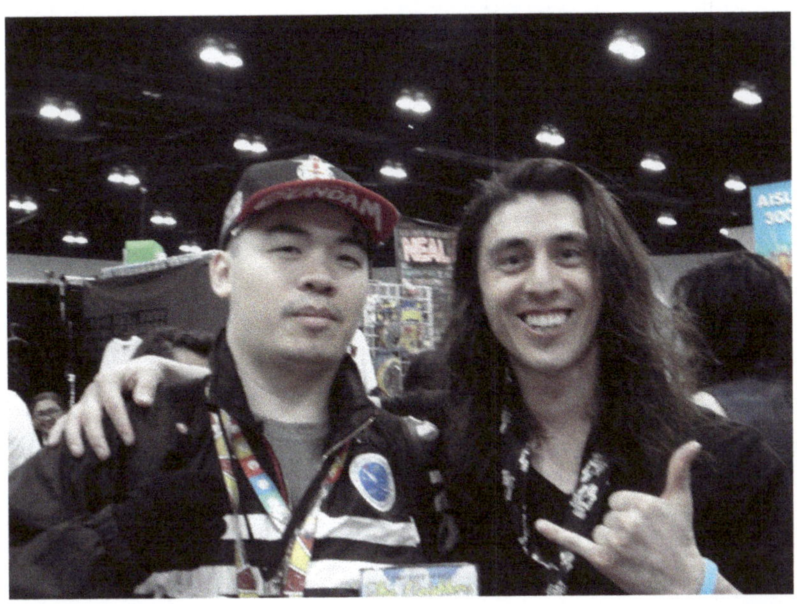

'Me, back in 2019 L.A.C.C.'	'Tommy Wiseau's The Neighbors T.V.' Co-Star
Lieu, Michael W.	@ Raul Phoenix TV

Character Designer Analysis, Overview Guidelines

01.) Name.

02.) Age.

03.) Galaxy, Planet, Continent, Sector, Country, State, County, Providence, City, Village, Sovereignty

04.) Major events in back story.

05.) Where do they live? What city? What kind of place?
With who? Describe the dwelling. Describe the contents.

06.) How do they currently make a living?
Feel about boss, manager, co-workers, job, volunteer? Work history.

07.) Love life? Who are they dating? Married to? Secretly into?
How was their last relationship? Romantic relationship back story? Patterns?

08.) Education, Employment, Training.

09.) Relationship with family, blood, bond, in-law, extended.

10.) Best friend, ally, mate, comrade, chief?

11.) Biggest foe, enemy, opposing force?

12.) Theme song?

13.) Philosophy of life.

14.) #1 (number one) priority in life?

15.) Current most pressing matter? Most important goal?

16.) Obstacles.

17.) Biggest fear?

18.) What are they doing right now?

Character Designer Analysis, Overview Guidelines

Premise; Your story stated in a single line. Include The: "Protagonist, Problem, Plan".

Archetypes;
Models or templates from which copies can be made that all human(s) can understand due to basic shared human experience.

Role;

Hero	Judos	Innocent	Whore
Villain	Holy Man	Bad Boy	Girl Next Door
Best Friend	Taskmaster	Dreamer	Librarian
Wise Man	Clown	Artist	X
Mentor/Teacher	Warrior	Madonna	

Back story;

1. God given biography (born, built, brought into)
2. Historical biography (events, major)
3. Emotional / Spiritual / Psychological biography (affect, moment)

Question;
Want and Need, out of life (Can manifest on their behavior based on it as backstory).

Behavior;
People are defined by how they move not how they say (moving through the world).

Dialogue; Should be *organic and interesting*.

- **Specific** – Syntax, crutch words; ideas / **Honest** - For your character

Background; Culture, educational level city, neighborhood of origin, psychology, physiology.
Subculture; Sales man, mob gangster, police department officer cop, soldier, etc.
Location; Work, home, locker room, church, etc.
Audience; Mom, friend, boss, stranger, etc.
Situation; What does the character?

1. Be in the moment.
2. Do not have to respond to every question comment.
3. Do not repeat beats → move forward.
4. Understand the nature of the relationship between characters in the scene.
5. Talk to the other character, not the audience (unless it is fourth wall breaking).
6. Be sensitive to the energy, and or frequencies' vibration "vibe", you create.

Character Designer Analysis, Overview Guidelines; "Story And Theme"

01. <u>Through the themes of your story that your reader/author/writer will viscerally connect</u>;
Story; Series of events that occur. / **Theme**; Bigger idea(s) within, beneath the story.

02. <u>Pick favorite tale</u>; Explain the log line, explain the theme present.

03. <u>Story line, archetypes</u>; "**Protagonist, Problem, Plan (The Three Ps)**"

Quest	Adventure	Pursuit	Rescue	Escape
Riddle	Rivalry	Underdog	Temptation	Good Versus Evil
Metamorphosis	Coming Of Age	Loss Of Innocence	Love	Forbidden Love
Unrequited Love	Discovery	Revenge	Ascension	Survival

"**Story Components**"

<u>Structure</u>; ***Beginning***: Status Quo (The World Or Universe As It Is) / Inciting Incident (Sets the protagonist, main character, on their quest as it gives them a task to do).

Middle: Progression in the face of mounting obstacles / setbacks toward a climax, where the question asked the inciting incident is answered.

End; Some form of resolution / ***Conclusion***; Wrapping up all the main plot points.

<u>Conflict;</u>

Human Versus Human	Human Versus God	Human Versus Nature	Human Versus Society	Human Versus Biotech	Human Versus Themselves

<u>Obstacles And Set Backs;</u>
Things that get in the way / challenge what your character wants and or needs.

<u>Miscellaneous Literary Story Tools / Devices;</u>

1. **Irony**; when what happens turns out to be the exact opposite of what was expected.
2. **Foreshadowing**; hints of events to come later.
3. **Motif**; theme, images, sounds, that reoccur throughout a single work.
4. **Plot-Twist**; surprise change in the direction or flow of the story.
5. **Red-Herring**; information inserted to distract the reader from important information on what is really going on.
6. **Sub-Plot**; a second story line that serves as a motivating or complicating force for the primarily story line.
7. **Suspense**; when the author/writer maintains his audience's attention though a build-up of events, the outcome soon to be revealed.

Character Designer Analysis, Overview Guidelines; Story Structure Peer Review
"Constructive criticism, critiques allow the creator(s) to see their universe's realization."

- **Form**;
  - Correct Use Of Grammar; Is the verb tense the same all the way through? How about the punctuation? Spelling? Pronouns? Paragraph Usage?

- **Concept**;
  - Scope; Does the story deal with too grand or too trivial a subject? Does it begin or end before it should? Is it interesting?
  - Originality; Have you heard the story or seen it in popular media before? If it works with an old theme or premise, does it reveal it from a fresh perspective?
  - Topicality; Fit the assigned topic. Topic about a protagonist who feels trapped in a confined space, then write about a protagonist who is trapped in a confined space.
  - Obstacles; Protagonist must face obstacles and those obstacles must get progressively more significant.

Point Of View (P.O.V.);

Narrator, the person telling the story, determines what details are included and how they are arranged, in short, the plot.

Example, in "**The Notebook**" was the main character, Noah, telling his wife, Allie, the story of their love in the past throughout the movie.

Unreliable Narrator, first person narrators that are self-serving, mistaken, confused, unstable or even insane.

Example, in Edgar Allan Poe's "**The Cask Of Amontillado**", the narrator Montresor, tells his story to justify a crime he committed fifty years before.

First person, is a narrator who uses I and sometimes we to tell the story, tale, narrative.

Third person, a P.O.V. of a narrator who is not a character.

Example, they had a cat and dog that the little boy loved very much.

Omniscient, a type of third person P.O.V., where the narrator is able to move at will from one character to another. Since it does not have the characteristics of a first person P.O.V. such as naivete, dishonesty, gullibility, or mental instability, it can convey the author's attitude towards the stories' subject matter.

Example, "In a house, in a suburb, in a city, there were a man and his wife who loved each other very much..." (Nadine Gordimer, "**Once Upon A Time**")

Limited Omniscient, similar to *Omniscient*, but focuses only what a single character experiences.

Example, "The wagon went on. He did not know where they were going."

(William Fulkner, "**Barn Burning**")

Objective, type of third person P.O.V. in which remain entirely outside the character's mind. It simply reports the dialogue.

Example, "You'll be drunk' the waiter said. The old man looked at him. The waiter went away." (Ernest Hemingway, "**A Clean, Well-Lighted Place**")

Irony, a discrepancy between what is said and what readers believe to be true. Irony man be dramatic, situational, or verbal.

Example, a person who claims to be a vegan and avoids meat, but will eat a slice of pepperoni pizza because they are hungry.

Dramatic Irony, occurs when narrator, or character, perceives less than readers do.

Example, the dramatic art would appear to be rather a feminine art, it contains in itself all the artifices which belongs to the province of woman: the desire to please, facility to express emotions and hide defects, and the faculty of assimilation which is the real essence of woman. -Sarah Bernhardt

Verbal Irony, irony in which a person says or writes one thing and means another, or uses words to convey a meaning that is the opposite of the literal meaning.

Example, "Water is as clear as mud" -The person actually says that the water is not at all clear.

Situational Irony, an occasion in which the outcome is significantly different from what was expected or considered appropriate.

Example, in "Julius Caesar", we know his "friends" are going to murder him, but he has no idea.

- **Language**;
  - Concrete Detail; I cannot stress this one enough, so I will repeat it and make it louder; **Concrete Detail**!! Are the images vivid? Does the artist/author/designer/ writer appeal to all five senses? Four? Three? Do they avoid cliché and or redundancy?

Style, Tone, And Language Types Expansions;

Tone, there must be a consistent tone to the script. For instance, if it starts off as a realistic piece, then it must remain so. Shifting tones midway through a piece confuses readers.

Present Tense, the action should be unfolding as we read the story "script" in present day.

Stream Of Consciousness, form of narration controlled not be external events, but by the thoughts and subjective impressions of the narrator.

Example, "**Girl**" by Jamaica Kincaid.

<u>*Alliteration*</u>, repetition of consonant sounds, usually the initial sounds, in a word series.
Example, the "S" in "So your chimneys I sweep, and in soot I sleep."

<u>*Parallelism*</u>, in grammar, parallelism, a.k.a. parallel structure or parallel construction, is a balance within one or more sentences of similar phrases or clauses that have the same grammatical structure. 1) The application of parallelism improves writing style and readability, and is thought to make sentences easier to process. 2) Examples Below:
Parallel: She likes cooking, jogging, and reading
Parallel: She likes to cook, jog, and read.

<u>*Informal Diction*</u>, used in an informal setting, such as conversations between friends.
Example, slang, improper grammar, and sounding inarticulate.

<u>*Formal Diction*</u>, proper language, grammar and a complex demeanor.
Example, a president giving a speech, or used in a conversation with authority figure.

<u>*Personification*</u>, the attribution of a personal nature of human characteristics to something non-human.
Example, "The wind was howling."

<u>*Allusion*</u>, an expression designed to call something to mind without mentioning it explicitly.
Example, she smiles like a Cheshire cat.

- **Plot**;
    - <u>Conflict</u>; It is the conflict introduced in the first few paragraphs/pages? Is the tension sustained and heightened as the story develops?
    - <u>Resolution</u>; Is the story resolved? Need it even be resolved? It is too easy (the main character commits suicide, kills themselves and or is murdered), cliché (main character awakes from a dream), unprepared for (a new character appears on the last page with a solution), or sugar coated (they all live happily ever after for comedy genre or everything ends badly for tragedy genre)?

<u>***Plot Types Expansions***</u>;

<u>*Antagonist*</u>, someone or something presented in opposition to the protagonist.
Example, "**Batman**" the protagonist and Bane the <u>*antagonist*</u>.

<u>*Crisis*</u>, peek of the story, considerable tension or importance.
Example, <u>*crisis*</u> starts when Marjane is stopped by "**Guardians Of The Revolution**".

<u>*Climax*</u>, greatest tension or importance, scene that presents a story's decisive action or event.
Example, "...get in the car." Marjane talks her way out of being taken by the "**Guardians Of The Revolution**".

Flashback, departure from chronological order that presents an event or situation that occurred before the time in which the story's action takes place.

Example, in "**Orange Is The New Black**", the main character is in prison and while in prison has flashbacks of her life before prison and the events that led her there.

Foreshadowing, introduction early in a story of situations, characters, or objects that seem to have no special importance, but in fact are later revealed to have great significance.

Example, in the story "**Maus**" by Art Spiegelman on page 90, panel 3, it describes a selection where "old people, families with lots of kids, and people without work cards are all going to the left" which is viewed as bad. On page 91, panels 5 and 6, the father of Fela gets sent to the right and his daughter Fela and grand kids get sent to the left. He quickly jumps the gate to join his daughter and grand kids.

Resolution / (denoument), French for "untying the knot", knot being the climax.

Action comes to conclusion	All loose ends are tied up	Resolution follows the climax of a plot

Example, the *Resolution* is achieved with help of "Deus Ex Machina", page 112.

Deus Ex Machina, Latin for "God From Machine".

-An intervention agent that was not previously part of a plot.	
-Not always predictable.	-Changes course of story.

Example, the sudden arrival of a long, lost relative or discovery of a characters true identity. *Resolution* by *Deus Ex Machina*.

- **Characterization**;
  - Action; Do the character's action reveal their fears or desires? Are they consistent with other aspects of character?
  - Appearance; Does the reader have a sense of the character's appearances, and do they properly reflect personalities? Are they stereotypes? Are they interesting?
  - Thoughts; Do the character's thoughts counterpoint speech, action and appearance? Are they plausible? Focused on the conflict?

Characterization Types Expansions;

Round, A well developed and closely involved in action and responsive to it.

Example, In the "**Catcher In The Rye**", a teenager, Holden Caulfield is round because of his personality.

Flat, Static, Stereotypical or operating foils for protagonist.

Example, the character plays a role to the main character but does not change the course of the story.

Foil, A minor character whose role is to highlight the main character by presenting a contrast with him or her.

Example, in the movie "**Hercules**", Hades would be the foil.

Stock Characters, Stereotypical character who behaves consistently and whom the audience of a play can recognized and classify instantly.

Example(s): heroes, vigilantes, villains, idiots, sidekicks, mentors, etc.

Caricatures, Illustrated cartoons used to depict a story, tales, and or narrative.

Example, Malcolm helped his friend raise funds by drawing caricatures.

Dynamic Character, A character that is growing and changing in the course of the action.

Example, the girl who is starting out to love drawing changes as she becomes older to please others in which changes her feelings towards drawing.

Static, A character who remain unchanged.

Example, a girl who stay true to herself and her interests even though her culture disapproves of it.

Motivation, Reasons behind a character's behavior that makes us accept or believe that character.

Example, a gangster who motivates himself to change his life and become a better person.

Character Outcomes From Film Movies' Epilogue Expansions;

The Character Wins, in this type of story, the character strives for a goal and wins.

Examples: **The Fugitive, Independence Day, Men In Black, Rocky, Speed, Rudy, Silence Of The Lambs, A Few Good Men.**

The Character Loses, in this type of story, a moral victory of some kind often results despite the failure (moral conduct code) of a very sympathetic likable character.

Take something big **give** something back big.

Examples: **Braveheart, One Flew Over The Cuckoo's Nest, Cool Hand Luke, J.F.K.**

The Character Sows The Seeds Of Their Own Destruction, what goes around, comes around.

Examples: "**Scream** (The Killers), **Moby Dick, Frankenstein**."

The Character Grows By Doing The Right Thing, character has a wrong start, but transforms into someone who overcomes their flaw(s) and does the right thing.

Examples: "**Groundhog Day, Jerry Maguire, Schindler's List, Rain Man, Pretty Woman**"

The Character Grows Up, character comes of age while striving for one or more goals that are either achieved or not achieved –It does not matter which.

Examples: "**Stand By Me, Good Will Hunting, Platoon, Risky Business, Breaking Away**"

The Character Learns, this story type the character learns what is needed to be happy.

Examples: "**Wizard Of Oz, City Slickers, It's A Wonderful Life**"

The Character Files To Learn, the character fails to learn what is needed to learn to be happy

Examples: "**War Of The Roses, Goodfellas, Raging Bull, Butch Cassidy, Sundance Kid**"

The Character Declines, the character declines as he strives to achieve a worth goal.

Examples: "**Lawrence Of Arabia, Unforgiven, Citizen Kane**"

- **Dialogue**;
  - Tension; Is one character saying no to the other? Do the character's state the conflict too readily?
  - Diction And Rhythm; Is the diction appropriate to the social status, education and age of the speaker?

Dialogue Types Script Screenplay Screenwriting Expansion;

Keep It Simple, each line of dialogue should be limited to one or two thoughts.

"Oh hi Mark. How's your sex life?" "**Tommy Wiseau's The Room movie**".

Beginning A Sentence With A Name, if you use a proper name or label for someone at the beginning of a sentence, please set it off with a comma. "Mark, I am your best friend."

Ending A Sentence With A Name, if you use a proper name or label for someone at the end of a sentence, please set it off with a comma. "Are you going to the store, Mark?"

Sounds, instead of spelling out sounds in dialogue (I think I love you. Hee, hee.), represent what the character is doing in action or as a parenthetical (romantic).

Slang, keep slang and non-standard English to a minimum. (A little goes a long way.)

Numbers, please spell out numbers, especially if they begin a sentence. 1, one.

- **Setting**;
  - Time And Place; Are the time and place of the story distinct? Does the reader have a clear sense of where and when action occurs?

Setting Types Expansion;

Historical, of or relating to the character of history.

Example, a historical moment is the "Boston Tea Party".

Geographical, of or relating to the science of geography.
Example, the intricacy of the geographical color lines varies in different communities.

Physical, of or relating to the body (actuality) as opposed to the mind, mental (psychological).
Example, a whole range of physical and mental challenges.

- **Tone**;
  - Language / Style; Remember **Winnie The Pooh's** "Long words bother me." Is the artist/author/designer/writer's level of language appropriate to the subject? Are paragraphs too long, or sketchy? And finally, fix the sentence level mistakes; grammar, punctuation, spelling.

Tone And Symbol, Allegory, Myth Type Expansions;

Symbol, a person object, action, place, or event that in addition to its literal meaning, suggests a more complex meaning or range of meanings. "River as a symbol of time."

Universal Or Archetypal Symbol, such as "Old Man", the "Mother", or the "Grim Reaper", are so much a part of human experience that they suggest the same thing to most people.

Conventional Symbols, also likely to suggest the same thing to most people (a rose suggests love, a skull and crossbones denotes poison), provided the people share cultural and social assumptions.

Allegory, communicates a doctrine, message, or moral principle by making it into a narrative in which the characters personify ideas, concepts, qualities, or other abstractions. An allegory is a story with two parallel and consistent levels of meaning, one literal and one figurative. "**The Wizard Of Oz**".

Allegorical Figure, a character, object, place, or event in the allegory; does not open up a text to various interpretations the way symbols do; represents something beyond their literal meaning in a fixed system. Sacrecrow from "**The Wizard Of Oz**".

Allegorical Framework, the set of ideas that conveys the allegory's message. "Cowardice is embodied in the lion (**The Wizard Of Oz**)."

Beast Fable, one type of allegory; a short tale, usually including a moral, in which animals assume human characteristics; "**Aesop's Fables**".

Myth, a story that is central to a culture; in embodies the values on which a culture or society is built. Myths are not synonymous with falsehoods or fairy tales. Rather, they are stories that contain ideas that inform a culture and that give that culture meaning. "**Norse Myth**"

To Viewers, Readers, Audience, Book Flippers Like Yu…

Recap abstract from story 1(one) “2154;_Arriving” Century Schoolbook
Date time stamp = 2020;_09(Sep)-06(six) @ 23;59_p.m. “May Dee Fourth Bee With Yu”
Failed Engineer Drafting Designer
Author, Writer, Artist, Designer “*Science-Fiction Character Story Universe*”
Artist, Architecture, Author, Alchemist(A.A.A.A.) “***Visual Story Text Book***”

“(insomnia) Oh hi everyone!” Sounds, looks like you made it to the part where the artist author, designer writer explains stuff on the near back end conclusion of the book.

2020(twenty twenty) is humanity’s best worst disaster year in the 21st(twenty first) century, “The Disaster Year” with so much negative events, I had to draw write about positive events, but I think I did so in a parody fan fiction method… if not hurt feelings.

2010(twenty ten) was when I had this story concept idea in mind, planning since, by 2011(twenty eleven), first time I ever went to an Anime eXpo(A.X.) convention center with Yellow-Mellow, a Utopian physical place, for indoor introvert nerds, geeks, otakus.

Where humans and animals of all walks of life get together and enjoy what our supposed 1(one) life has to offer, so, in the following year of 2012(twenty twelve)’s A.X. was 5(five) days worth of best ever events... featuring;_ day 0(zero) of 8(eight) hours Line Con getting our convention center badges, seeing LiSA musician in person, seeing my college friend win a lucky drawing to take a photo with LiSA plus poster signed by her, Fate Zero(0) anime series screening with the Japanese original voice actors and the message of wanting to grow up to a hero { Hiro }, African Black Jesus in between waiting for panels as intermission, reading Diary Of Wimpy Kid children visual books that took 2(two) hours to read each, Ding Tai Fung Asian restaurant for food after, and day 4(four), definitive end where convention center Utopians start to feel sad in real life, the post con.

Through the vision dreams I received, when I do manage to get some sleep, I imagined, engineered, architect, designated “2154;_Arriving” as a bottle necking indoctrinating advanced biotech dome city, A Utopian Heaven Paradise that is open 24 hours, 7 days a week(24-7), a North American Animal Autonomous Anime Convention Center(N.A.A.A.A.C.C.) of the new world “Utopian Sovereignty of North America”, from of the ashes of the old world “United **States** of America” due to 2120’s Mars Outbreak Outage(2120’s MOO), or Mars’ Unconventional Warfare as Planetary War, The Roman God Of War, plus the Roman Empire collapse comparison to United **States** of America.

In between those years, I had “drawing writers' block”, yes, I draw visuals pictures photos before I write, for a long time… that was lost in time, like tears in rain.

By 2019(twenty nineteen), Blade(刀) Runner 1(one), I believe I draw write not for myself, but for something bigger higher, like Jeremy Lin(書豪　林) said, “Now, when I play basketball, I don’t play for anybody anymore, I only play for God. Once he gave me that purpose, that’s when I have my peace. Once I got my peace, that’s when I got my joy.”.

I always did want to do volunteer work at A.X.’s convention center in mind, and by 2019(twenty nineteen)’s C-Virus 11(Nov)-17(seventeen) as a long term Chemical, Biological, Radiological(C.B.R.) outbreak, it may last for years... like 1918(nineteen eighteen)’s S-Flu of 2(two) +(plus) years... I guess drawing writing about it, is one way of sandbox simulating it, along with looking back at real life reflecting on a bunch of "What ifs", making it a multiple universe(multiverse) Department Of Energy(DOE) alternative timeline story branching narratives, mostly a Science-Fiction Utopia Parody Fan-Fiction.

I do like to spend a lot of time with a kind of pointless senseless dialogue like in a convention center meeting new people and stuff... so, in a way I sandbox simulate it.

Science-Fiction, Utopia, Parody, Fan Fiction genre type insanely inspired by;_
Phillip K. Dick's multiple universe { many verses } alternative timelines
Issac Asimov's A.I. robots { robotics was originally known as slaves } rules human snake
Arthur { Author } C. Clarke's bright optimism outlook on the future and life
Autonomous Zone \ Animal Sanctuary \ American Reservation \ Anime Expo
Underlining, *Italics*, **Bold**, Color Code Tones, FONT Alternating Document Experiment
(surprise face (°o°)) Number One(#1), favorite emotional feeling face.

After long and hard careful consideration, I decided to continue and improve infrastructure, but not sure if it is a comic book, or graphic novel, not a novel since novels are all make belief fiction, and all my story universes are semi-autobiography, so not a novel, but book yes, there is a difference between books and novels when I researched it.

It is my now long life passion project activity, as I named it in a dumb smartly pants creation concept "***Visual Story Text Book***", subject to evolve each story as experiment in accordance to that "*Science-Fiction Character Story Universe*"...

So, Chicago(ChIraq), Illinois(I.L.), how did I decide to set my story universe in that geographic physical setting... I departed from South California(SoCal) to that sector for my hero's journey in 2006(two thousand six), Great Lakes, North Chicago, Illinois, United States of America, with many **entertainment** industry storytelling content is set in "Chicago, Illinois" in accordance to Houston Howard(H.H.)'s super story world structure.

1(one)	Death Wish remake movie = Crime, Drama, Vigilante "The City Of Death."
2(two)	Watch Dogs one original game = Crime, Drama, Vigilante, Sandbox genre
3(three)	The Dark Knight movie = Crime, Drama, Vigilante, Super-Hero Gotham city
4(four)	The Divergent Series of Divergent movie = Science-Fiction, Dystopia genre

My story universes deals with the fall of "the old world" structures and the rise of "the new world" society, the fine line between order { Batman } and chaos { Joker }, sadly, this Science-Fiction Utopia Fan Fiction story does not take place in our universe...

5. "2154;_Arriving" =
Science-Fiction \ Alternative History \ Utopia \ Parody \ Fan Fiction genre type,
Welcome to the new world 2121 movement of a "Utopian Sovereignty of North America (U.S. of N.A.)"

My low budget visuals are in 1st(first) person perspective, text script is written in 3rd(third) person perspective, model reference from my all time favorite sandbox simulator, Obsidian's Fallout New Vegas, with 2(two) automated algorithm assistants.

This is from my personal experience of playing many war games in the 20th(twentieth) century to 21st(twenty first) century that some have those perspective switches, along with a birds of prey view of Real Time Strategy(R.T.S.) war games from Blizzard Entertainment's StarCraft 1(one) Brood War, Local Area Network(LAN) party.

As for the next "*Science-Fiction Character Story Universe*" follow up, I have yet to know plus(+) decide which one to go with... I do not think it is going to take 10(ten) years to draw write the next one... I do take my sweet time, in allowing me to ensure the story, characters, plot points make sense to me in this three to fourth dimension(3-4th D) as a 21st(twenty first) century earthling.

My big question, how long does it take for readers, viewers to completely fully finish this book, plus(+) with how would you read this kind of "Concept Art Script Experiment" that started from my supplement "2154;_Arriving Coloring-Book" and before that, "21X0;_U.S.-Department-Of-War-KIPI-Kosplay", Photo-Book...?

Would you have a note book, paper, pen, pencil, plus computer, along with internet to listen to the music scenes like "Sound Of Silence" and or "Happy Happy Is Blue Blue"?

What about the 80(eighty plus pages) of "Model References Work Inspiration"? Which was set to originally release in 2021;_05(May)-04(four) = Take away '20;_0(May)-0(four)' equals 2154, if not for this 2019(twenty nineteen) C-Virus outbreak.

I do wonder how does the Electronic-Book(E-Book) look like for "2154;_Arriving".

Oh My Buddha(O.M.B.), this almost 400(four hundred) page book, took 10(ten) years to research and develop(R&D), look at what can happen in 1(one) year alone, especially with 2020(twenty twenty) Endgame Events, but for me times 10(ten).

Anyways, look forward to you all supporting me whatever you get it online: by pirating it for free... Pirating pays, real hard earned money, and or read it from a family, friend of a human being, animal kingdom, planet earth and or A.I." -Lieu, { 廖, } M.W.

A 2 (Arc Two), "A Garden Of Vigilantes"

"<u>*Visual Story Text Book*</u>" { Coloring-Book } { Concept Art Script Experiment }

{ Travel Drawings Audio Retcon }

Planet Earth's Apocalyptic Climate Changing Crisis, The Fire And The Flood

Former British-Hong Kong Sovereignty (S) "1 (one) country, 2 (two) systems"

A.X. C.C. The Post Con Blues = Anime Comic Con Expo Convention Centers

The Flood Myth = Mythology Of Christian Bible Noah's Ark

Contemporary Noah's Ark = United States Military Service Aircraft Carrier Warship

War On Terror, 2001;_September Eleven (9-11) Terrorist Attack On The United **States** of America (U.**S**. of A.) Using Unconventional Warfare Tools

2154;_ Arcadia \ Valentina Vita (V. V.) 'Healthy-Strong Life' \ Left to Right(L->R)

To Viewers, Readers, Audience, Book Flippers Like Yu...
Year Date Time Stamp Completion = 2022;_08(Aug)-08(eight)
Story 2(two), "2154;_Arcadia" 'Travel Journal Drawing Retcon"

Like The Super-Hero Vigilante Comic Book Creator
Stan The Man Lee Soap Box Message Speech

0(zero) words,	
1(one) word to explain my "Science-Fiction Character Story Universe"... "History"	
2(two) words,	"What If"
3(three) words,	"An Alternative Timeline"
4(four) words,	"Human Behavior Sandbox Simulation"
5(five) words,	"Military Service Camaraderie First Responders"
6(six) words,	"Coloring Book Concept Art Script Experiment"
7(seven) words,	"Beginning Of The End 6th (sixth) Extinction Event"
8(eight) words,	"Endless Eight Deep Desire Department Of Energy Multiverse"
9(nine) words,	"Science-Fiction Utopia Parodying Convention Centers Super-Heroes Vigilantes Villains Cosplay"
10(ten) words,	"Chemical Biological Radiological Outbreak Pandemic Electro Magnetic Pulse Outage Blackout"

As of 2021 (twenty twenty one) documented authorship of "Disaster Year Season 2 (two)" continues... Stay home, work from home, home life for life... I been a Clean-up Crew myself and **Clean House to find old art drawings I started back in 2003 (two thousand three) in my high school years**...

Then **got reviews, responses from story 1(one) "2154;_Arriving" 'Concept Art Script Experiment',** I needed to do **small retcon upgrades** due to the uses of photos I had and or took back in 2006 (two thousand six), **I realized I needed to draw out the characters that were not drawn out in traditional art hand drawings** like the "Elementary, Mid, High School Real Actual Friends = { Eastern-Western Sit-Com genre }".

Also, in the Department Of Energy (DOE) alternative history universe 2 (two), Multiple uniVerse (MultiVerse) 2 (two) **needed to expanded the Milky Way Galaxy, Earth's Global Government (EGG) Joint Task Force (J.T.F.) Space Force against** the space alien humans who reside on Planet Mars and Roman Industries, **Martian Terrorists**.

Corona Virus Disease 2019 = CoViD-19 / Martian Outbreak Outage 2120 = MOO-20

2021;_08(Aug)-03 Questions And Answers (Q.N.A.) @MakiEraclea cosplayer response

<u>Question</u>; "How did you decide to name your online internet self "Maki Eraclea"?"

<u>Answer</u>; "Friends always called me "Maki" in middle school. I decided to take Delphine "Eraclea" surname. She is a villain I like a lot."

2154;_Arcadia \ Valentina Vita (V. V.) 'Healthy-Strong Life' \ Left to Right(L->R)

2021;_08(Aug)-10(ten)

One of the **biggest broadcasting frequencies dream(s) vision(s)** I get of our universe in **2022 (twenty twenty two)** Disaster Year(s) Season 3 (three) of an **electricity electronics outage blackout**... On top of the **2019 (twenty nineteen) Coronavirus (C-Virus) outbreak pandemic**...

This outage blackout is made possible by an artificial and or natural Electro Magnetic Pulse (E.M.P.) that would send human society to the 19th (nineteenth) century of the wild west in North America (N.A.) and or the **dark ages** of the 21st (twenty first) century.

Horse and carriage... Ye ha! There is a **science-fiction short story movie tie in** to the **Blade Runner Alien universe** set after the year 2019 (twenty nineteen) 11 (November) original story... "**Blade Runner 2022 Blackout**" in the month of 05 (May).

I recall in **2021 (twenty twenty one), Disaster Year(s) Season 2 (two)**, **Texas state** in the **old world** United **States** of America (U.**S**. of A.) of our universe history **experienced an electricity outage blackout**, not made by an Electro Magnetic Pulse (E.M.P.), rather **weak infrastructure** against **Earth's climate change crisis** in the **21st (twenty first) century**.

Texas' people were experiencing the 2019 (twenty nineteen) Coronavirus (C-Virus) outbreak pandemic like Chemical, Biological, Radiological (C.B.R.) outbreak mix with 2021 (twenty twenty one) electricity outage that is **not a Electro Magnetic Pulse (E.M.P.)**.

We are **lucky** there was **not an Electro Magnetic Pulse (E.M.P.)** that **disabled and or turn off our tech** as **our human society is dealing** with the **so call claim engineered bio weapon from Wuhan of Communist China, Asia, the 2019 (twenty nineteen) Coronavirus (C-Virus) outbreak pandemic**.

That possible science-factual apocalyptic genre endgame event shows and tells that in a **southern Texas' Mister Matthew Mcconaughey voice** explaining "**Murphy's Law**... **Anything can happen, will happen**..." Murphy's Law. **Alright, all write, all wright**!"

Thus I engineered "**A 0 (Arc Zero), Beginning Of The End**" by 2021 (twenty twenty one), after I blueprinted "**A 1 (Arc One), Hero Journey Character Transformation**" in 2019 (twenty nineteen), the Alfa and Omega... <u>Mythology</u> of Christian Bible Book Of Revelations and Norse <u>Myth</u> Ragnarok... and as visualized references in recorded story mediums of our human history universe.

<table>
<tr><td>0(zero)</td><td>"A 0 (Arc Zero), Beginning Of The End"</td><td rowspan="2">The Last Of Us (video game)</td></tr>
<tr><td>1(one)</td><td>Contagion (film movie)</td></tr>
<tr><td>2(two)</td><td>Fear The Walking Dead season 1(one) (T.V. show series)</td><td rowspan="4">The Walking Dead The Telltale Definitive Series (video game)</td></tr>
<tr><td>3(three)</td><td>The Walking Dead season 1(one) (T.V. show series)</td></tr>
<tr><td>4(four)</td><td>World Beyond The Walking Dead season 1(one) (T.V. show series)</td></tr>
<tr><td>5(five)</td><td>Biohazard Resident Evil: (video games, film movies chrono order)
Origins Collection, 3, 2, Degeneration, 4, Infinite Darkness, Revelations 1, 5, Revelations 2, Damnation, 6, Vendetta, 7 Biohazard, 8 Village, 9</td></tr>
</table>

2022;_02(Feb)-22(twenty two)

Sometime in **the year 2021 (twenty twenty one)**, **Lin, Jeremy mentioned that he was called "Coronavirus" (C-Virus) in the basketball by a fellow basketball player** and this was during the 2019 (twenty nineteen) C-Virus outbreak pandemic, as **that trigger an old memory back to the year 2001 (two thousand one) September-11 (nine eleven)** when the **terrorist attacks occurred in New York and all around, United States, North America**...

I was in history class and I had a **Middle Eastern Pakistani ethnic classmate name Mohammad** and later **he was called "Terrorist" that same day**...

Seems like about 20 (twenty) years later, **human behavior has not changed...**

2020;_08(Aug)-28(twenty eight)+29(twenty nine) Rest In Pieces (RIP) Black Panther King Wakanda Forever Actor Chadwick Boseman... to 2022;_03(Mar)-03(three) Reflections

Story 1 (one) to Story 2 (two) Questions aNd Answers (Q.N.A.)

Question; Is **Story 1 (one) "2154;_Arriving" required reading** to understand **Story 2 (two) "2154;_Arcadia"...**?

Answer; **No, not really, unless you want to know how I engineered all this, crazy character madness experiment...** Story 1 (one) is a hard experiment book to read. The main character perspective from **"2154;_Arriving" 'BladeMaster** 刀大師' **Elle K. Mcgee recaps the events from her perspective that occurs in parallel to the story 2 (two) "2154;_Arcadia", Valentina Vita (V. V.) character**...

Alternative Answer; think Warner Bros' (W.B.) D.C. "**Suicide Squad 1 (one) movie**" as "**2154;_Arriving**" and '**The Suicide Squad 2 (two) movie**' as "**2154;_Arcadia**". If that makes sense, you can follow both story one and two if you desire.

2022;_03(Mar)-30(thirty)

(O_O) As I approach **past 500 (five hundred) pages for story 2 (two) "2154;_Arcadia"** as a "**Travel Drawings Audio Retcon**", I realize I am doing this story book as a "**legacy leave behind semi-autobiography**", than a "**glorified business card book**" of mass market selling stuff, since do not have marketing major and **do not really have much of a business plan of sort as a one person lone wolf solo operative**, I just **rather work on the next story, show and tell family friends that I finish my current passion project**, over mass market to many others social networks...

This kind of a book as an originally recall, "Visual Story Text Book" is **hard to emulate, hard to explain what it is I aim to do, not quite a comic book, nor visual novel, nor screenplay, rather a Frankenstein fusion of what I stated before...**

Maybe more can be **expanded** as I work on the next "Visual Story Text Book" **story 3 (three), "2154;_Ascension-00-Prologue",** Vigilant Sketches Journal Expansion... **Elysium, The Magical Corporation Of Desire,** Established 2077 (twenty seventy seven).

2022;_05(May)-30(thirty)

After making time to watch **Warner Bros. (W.B.)** ***Space Jam 1 (one)*** and ***Space Jam 2 (two) A New Legacy*** and along side **honest trailers** of ***Space Jam 2 (two) A New Legacy***, I got the **feelings**, the **frequencies**, that **W.B.** wants to do an **Intellectual Property (I.P.) convention center**, I realize my "*Science-Fiction Character Story Universe*" books references a good amount of **W.B. I.P. movies, one TeleVision (T.V.) show**:

"Two thousand one... ...the best science-fiction film ever made..." -Lucas, George Star Wars ***2001: A Space Odyssey*** and sequel follow-up ***2010: The Year We Make Contact***.

to

Featuring Actor Matthew Mcconaughey in both ***Contact*** and ***Interstellar*** movies. Alright.

to

Actor Joaquin Phoenix in ***Her*** and ***Joker*** as a Science-Fiction Dystopia failed state society.

to

Director Michael Crichton's ***WestWorld***, Science-Fiction study on Artificial Intelligence (A.I.) consciousness from the original M.G.M. movie and ***W.B. H.B.O. T.V. show series***.

to

Monster Kaiju genre Directors: Guillermo Deltoro's ***Pacific Rim***, Gareth Edwards' ***Gozilla***.

to

Actor Tom Cruise, *The Running Man* Not That Actor Arnold S. Movie, In ***Live Die Repeat Edge Of Tomorrow***, Science-Fiction time loop with violent video game references.

to

Actor Harrison Ford in both ***Blade Runner The Final Cut*** to ***Blade Runner 2049***.

to

Director Steven Spielberg's ***Ready Player One*** also a **W.B. I.P. convention center**.

to

The Wachowskis' ***The Animatrix story 0 (zero) origins, The Matrix 1 (one), Matrix 2 (two) Reloaded, Matrix 3 (three) Revolutions, Matrix 4 (four) Resurrections.***

to

Director Fred M. Wilcox's ***Forbidden Planet***, that 1950s (nineteen fifties) movie that inspired and created the *Star Trek universe* and the infamous A.I. Robby The Robot.

to

Director Michael Anderson's ***Logan's Run*** of the infamous 30 (thirty) years old age quote mandatory ascension belief system retirement unquote and living under a dome city.

...Just to name a few of **W.B. I.P. Science-Fiction stories**, me doing a bit of journal and or diary entry at the back of story 2 (two) "2154;_Arcadia" helps me understand that yea, I grew up with my "Elementary, Mid, High School Friends" watching many movies back then in 2019 (twenty nineteen) and before 2020 (twenty twenty) disaster year season 1 (one).

Social Net Media Pages Online Name(s)

"So, here's my info, so contact me maybe?" ***Failed Engineer Drafting Designer***

I think some people are going to spam me with goat porn and Nigerian prince pleas scams for their quote "family, friends..." ***Artist Architecture Author Alchemist***

Oh well... Creativity curse. (surprising face (; O)) "***Visual Story Text Book***"

"I'm sorry, the number you have dialed has a voice mail box not yet set up. Good bye!"

Low Budget Business Card

"Lieutenant Lieu" \ "In Lieu Of" \ Lieu, Michael W.

L → R　　廖, 米高　伍 o

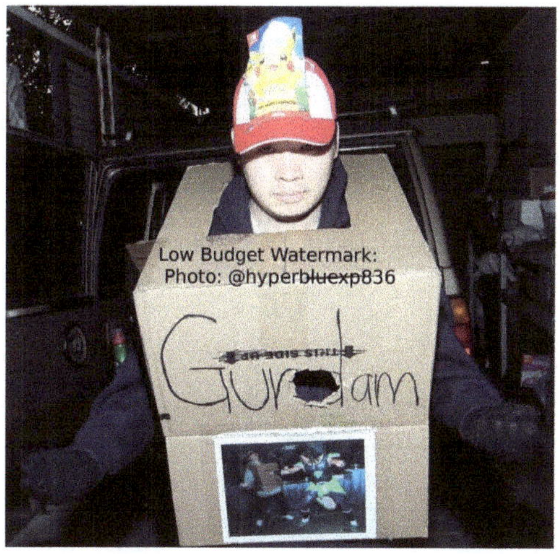

https://www.instagram.com/lowbudgetforlife

https://www.youtube.com/user/lowbudgetmic

https://twitter.com/lowbudgetmic

https://www.facebook.com/lieu.michael.w

2154visualstorytextbook@gmail.com

Kat 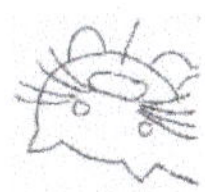

KEANU 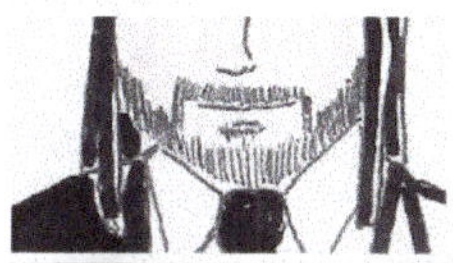

KIPI 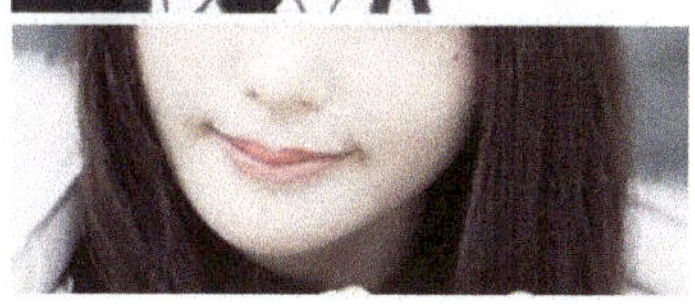

Kiss-Hiss+　?

Kat, Conscientiousness Cat

Kay, Agent Cool

Kosplay In Playing Identities,

Doktor A.I. Ai 愛 love

Kiss-Hiss+, Human-Snake

Social Net Media Pages Visual Collages, from story 1(one), 2020;_09(Sep)-06(six)

"Never gonna give you up, never gonna let you down..." -Rick Roll
https://www.instagram.com/lowbudgetforlife #2154 #2154collage #2154visualstorytextbook

https://lieumichaelw.blogspot.com/

https://www.facebook.com/2154VisualStoryTextBook/photos/

http://www.amazon.com/author/lieumichaelw

Butt, fear not... here is a story universe layout of my madness, visions I dream about...

DOE MultiVerse number(English number) "subject to change due to X variable"
Insanely Inspired by;_
W.B. + D.C. + C.W.'s 2019 Crisis On Infinite Earths

MultiVerse #(number)	Character Name Point Perspective	Next Story Universe To Draw-Write After Completing "2154;_Arriving" Book Titles
0(zero) or O	Multiverse Manipulating	0X1O;_?
1(one) or I	Lieu, Michael W. 廖, 米高 伍o	2006;_Choices
2(two) or to	Valentina Vita(V. V.)	2154;_Awareness
3(three) free	A.I.+ Conscientiousness Cat	Con-Kat's XXXX;_Kiss-Hiss+
4(four) or 死	Holocene Extinction Entity 6	Hideo Kojima's Death Stranding Multiverse
5(five) high	Mars, D. Roman God Of War	2120;_Endgame
6(six) or 9	Kefka Palazzo "Kosplay"	"Watch The World Burn"
7(seven or L	1 Name 4 All "Sephiroth"	1997;_Square Soft Enix's Final Fantasy 7
8(eight) ate	Endless Eight "Ate" 8	8888;_V.R.-Sim-19to91;_8888 = "Subtitle"
9(nine) or 6	"ylapsoK" ozzalaP akfeK	"nruB dlorW ehT hctaW"
10(ten) or X	Multiverse Manipulating	2001;_Indoctrination
18 (eighteen)	Ardyn Lucis Caelum "Cosplay"	2006-2016;_Final Fan Fiction Fantasy 6 Versus 13 Equals 15
		399 / 590
1941(nineteen forty one)	Adolf Hitler, Nazi Germany, My Fuhrer!	1941;_Uprising W.B., D.C., C.W.'s 2017 Crisis On Earth X

Version 2(two) of my "*Science-Fiction Character Story Universe*" Blueprint, "In math, X is a variable that is subject to change." = Arial Rounded MT
Lieu, Michael W. 廖, 米高 伍 o

Insane Inspiration;_
W.B. + D.C. + C.W.'s 2020 Crisis On Infinite Earths' Part 5(five) Conclusion Wrap-Up

4(four) digit title #(number)	Character Name Point Perspective	Next Story Universe To Draw-Write After Completing "2154;_Arcadia" Book Series Title
0X1O;_	MultiVerse Manipulating	0X1O;_
20th	***20th (twentieth) century***	***Alternative history timeline.***
1941;_	Adolf Hitler, My Fuhrer, Nazi Party Of Germany, Third Reich	1941;_Totalitarian
1997;_	"BladeBoss 刀老闆" Mark D. Mark S.	1997;_Sovereignty
21st	***21st (twenty first) century***	***Alternative history timeline.***
2001;_	Sir Arthur C. Clarke	2001;_Space
2006-2016;_	Ardyn Lucis Caelum "Cosplay"	2006-2016;_Final-Fan-Fiction-Fantasy -6-Versus-13-Equals=15
????;_	"Question-Mark ?"	Hideo Kojima's Death Stranding Multiverse
22nd	***22nd (twenty second) century***	***Alternative history timeline.***
2120;_	Roman Industries	2120;_Endgame
2121;_	Jon W. Smith A.K.A. Anonymous	2121;_Movement
2121;_	Seattle City Police Department (S.C.P.D.) Special Weapons And Tactics (SWAT)	2121;_SWAT
2121;_	United States Army (U.S.A.)	2121;_Army
2136;_	**Ai-Aug A.R. Holo HUD**	**2136;_Grad**
21X0;_	**U.S.-Department-Of-War-KIPI-Kosplay { Prototype Photo-Book Script }**	
2136;_	Yellow-Mellow (°o°) (O :)	2136-2139;_Navy
2154;_	Elle K. Mcgee "BladeMaster 刀大師"	2154;_Arriving
2154;_	"Valentina Vita (V. V.) Healthy-Strong Life"	2154;_Arcadia
2154;_	**Lieutenant Lieu, Michael W. "WhiteWolf 白狼"**	**2154;_Ascension-00-Prologue**

4(four) digit title #(number)	Character Name Point Perspective	Next Story Universe To Draw-Write After Completing "2154;_Arcadia" Book Series Title
0X1O;_	MultiVerse Manipulating	0X1O;_
Variable	***Simulation century***	***Simulation history timeline.***
XXXX;_	Augmented Artificial Intelligence (A.I.+) Ai love Conscientiousness Cat/Kat "Con-Kat"	Conscientiousness Cat/Kat's XXXX;_Kiss-Hiss+
XXXX;_	Augmented Artificial Intelligence (A.I.+) Ai love Conscientiousness Cat/Kat "Con-Kat"	Conscientiousness Cat/Kat's XXXX;_Kiss-Hiss+ Part 2(two, to, too) A-Rag, Hon-Knee
XXXX;_	Augmented Artificial Intelligence (A.I.+) Ai love Conscientiousness Cat/Kat "Con-Kat"	Conscientiousness Cat/Kat's XXXX;_Kiss-Hiss+ Part 3(three) A-Rag, Hon-Knee
XXXX;_	Augmented Artificial Intelligence (A.I.+) Ai love Conscientiousness Cat/Kat "Con-Kat"	Conscientiousness Cat/Kat's XXXX;_N
XXXX;_	Augmented Artificial Intelligence (A.I.+) Ai love Conscientiousness Cat/Kat "Con-Kat"	Conscientiousness Cat/Kat's XXXX;_B
XXXX;_	Augmented Artificial Intelligence (A.I.+) Ai love Conscientiousness Cat/Kat "Con-Kat"	Conscientiousness Cat/Kat's XXXX;_N
XXXX;_	Augmented Artificial Intelligence (A.I.+) Ai love Conscientiousness Cat/Kat "Con-Kat"	Conscientiousness Cat/Kat's XXXX;_K
XXXX;_	Augmented Artificial Intelligence (A.I.+) Ai love Conscientiousness Cat/Kat "Con-Kat"	Conscientiousness Cat/Kat's XXXX;_O
XXXX;_	Augmented Artificial Intelligence (A.I.+) Ai love Conscientiousness Cat/Kat "Con-Kat"	Conscientiousness Cat/Kat's XXXX;_O
XXXX;_	Augmented Artificial Intelligence (A.I.+) Ai love Conscientiousness Cat/Kat "Con-Kat"	Conscientiousness Cat/Kat's XXXX;_N
XXXX;_	Augmented Artificial Intelligence (A.I.+) Ai love Conscientiousness Cat/Kat "Con-Kat"	Conscientiousness Cat/Kat's XXXX;_O

Work In Progress (WIP) Subject To Change Because of X Variable

4(four) digit title #(number)	Character Name Point Perspective	Main Mission Science-Fiction Character Story Universe Blueprint Book Series Titles
B.C. & A.D.	***The Beginning, The Middle, The End, The Conclusion.***	***Alternative history timeline.***
0X1O;_	MultiVerse Manipulating	0X1O;_
20th	***20th (twentieth) century***	***Alternative history timeline.***
1941;_	Herr Adolf Hitler, My Fuhrer, Nazi Party Of Germany, Third Reich	1941;_Totalitarian
1997;_	"BladeBoss 刀老闆" Mark D. Mark S.	1997;_Sovereignty
21st	***21st (twenty first) century***	***Alternative history timeline.***
2001;_	Sir Arthur C. Clarke	2001;_Space
2006-2016;_	Ardyn Lucis Caelum "Cosplay"	2006-2016;_Final-Fan-Fiction-Fantasy 6-Versus-13-Equals=15
????;_	"Question-Mark ?"	Hideo Kojima's Death Stranding MultiVerse
22nd	***22nd (twenty second) century***	***Alternative history timeline.***
2120;_	Roman Industries	2120;_Endgame
2154;_	Elle K. Mcgee "BladeMaster 刀大師"	2154;_Arriving
XXXX;_	Augmented Artificial Intelligence (A.I.+) Ai love Conscientiousness Cat/Kat "Con-Kat"	Conscientiousness Cat/Kat's XXXX;_
8888;_	Lieutenant Lieu, Michael W. "WhiteWolf 白狼"	8888;_Ascension_;8888
0X1O;_	City Controllers	0X1O;_
2171;_	?	Question-Mark
23rd	***23rd (twenty third) century***	***Alternative history timeline.***
2222;_	X	X

"Lieu, Michael W.'s Science-Fiction Character Story Universe Blueprint"

A 0(Arc Zero) "Beginning Of The End" Story 0(zero) Origins

"2120;_Endgame" Roman Industries @ 11(Nov)-17(seventeen), Planet Mars
"Mars, The Roman God Of War"
Apocalyptic genre character(s) = Arial Rounded MT "Red"
Human Population In The Solar System; 11,000,000,000(eleven billion)
Martian Terrorists' Belief System;
"***Humans are a cancer on the solar system and needs to be annihilated completely***."
"4(four) Horsemen Of Apocalypse"

Disease	**Famine**	**War**	**Death**
"Sharing is caring." -? (Question -Mark)	"Create enough hunger, and everyone becomes a criminal { or a terrorist }." -Liam Neeson { Alter Ego, Da Mentor } W.B. D.C. Chris Nolan's Batman Begins, The League Of Shadows.	"In war, parents bury their children. In peace, children bury their parents." -Herodotus, Greek Historian	"The laws of physics is a death { deaf } warrant to intelligent life… Extinction is the norm for life { and existence }." -Doktor Professor Michio Kaku, Theoretical Physicist And Futurist

=Apocalyptic Worlds' Ending List=
Insane Inspiration;_

Orson Scott Card's *Ender's Game* Universe
Disney, Marvel Studios' Avengers 4(four) *Endgame* Movie
"Oh, what good is a world that only ever lets you down? Why not *end* it all right here?"
-Arydn
Warner Brothers(W.B.), D.C., C.W. 2019 Pre Crisis On Infinite Earths – Post 2020
W.B., Chris Nolan "Chris-N" Tenet Movie / Amazon's The Expense
W.B., Home Box Office(H.B.O.) WestWorld TeleVision(T.V.) Show Series, Season 3
20th Century Fox Studios' The Martian Movie Based On Andy Weir { Any Weird } Story
N.B.C. Universal's Revolution T.V. Show / W.B., Steven Soderbergh's Contagion Movie
Horror Of The Red Planet, The Wizard Of Mars Movie
Sony, Naughty Dog's → The Last Of Us Story ← W.B., H.B.O.

2120;_11	2121;_02	05	09	12	2122;_01	02	06	10	12
0(zero)	1(one)	1 Billion(B)	2 B	3 B	4 B	5 B	6 B	7 B	8 B

Apocalypse Causes; Hubris / Zombies { Undead } / War / { Space } Alien Invasion
Robots { Slave Artificial Intelligence (A.I.) } / Evil / Celestial Impact / Soda / Flood
The Internet Going Down. From GameInfomer article unknown issue.
Complete The World War Trilogy = 1 (one) Beginning, 2 (two) Middle, 3 (three) End.
Type: 0 (zero), 1 (one), 2 (two), 3 (three)

A 0(Arc Zero) "Beginning Of The End" Story 0(zero) Origins

"2120;_Endgame" Roman Industries @ 11(Nov)-17(seventeen), Planet Mars

Tomorrowland Science-Fiction Movie, Governor Nix's Speech, Hugh Laurie
{ Apocalyptic Educational Indoctrination Speech } Total Run Time = 02;30

"Let's Imagine... if you glimpsed the future and were frightened by what you saw, what would you do with that information? You'd go to... The politicians? Captains of industry? And how would you convince them?
Data? Facts? Good luck. The only facts they won't challenge are the ones that keep the wheels greased and the dollars rolling in. But what if... What if there was a way of skipping the middle man... and putting the critical news directly into everyone's head?
The probability of widespread annihilation kept going up.
The only way to stop it... was to show it. To scare people straight.
'Cause what reasonable human being wouldn't be galvanized by the potential destruction of everything they've ever known or loved? To save civilization... I would show its collapse. But how do you think this vision was received?
How do you think people responded to the prospect of imminent doom?
They gobbled it up like a chocolate ecclair. They didn't fear their demise.
They repackaged it. It could be enjoyed as video games, as T.V. shows, books, movies.
The entire world whole heartedly embraced the apocalypse and sprinted towards it with gleeful abandon. Meanwhile... your earth was crumbling, all around you.
You've got simultaneous epidemics of obesity and starvation!
Explain that one. Bees and butterflies start to disappear... The glaciers melt... Algae blooms... All around you.
The coal mine canaries are dropping dead and you won't take the hint!
In every moment, there is the possibility of a better future... but you people won't believe it.
And because you won't believe it you won't do what is necessary to make it a reality.
So you dwell on this "Oh terrible future" you resign yourselves to it.
For one reason: Because that future doesn't ask anything of you... Today!
So, yes... We saw the iceberg. We warned the Titanic.
But you all just steered forward anyway. Full steam ahead. Why? Because you want to sink. You gave up..."

A 0(Arc Zero) "Beginning Of The End" Story 0(zero) Origins

"2121;_Movement" Jon W. Smith @ 02(Feb)-28(twenty eight), Planet Earth
"Utopia, The Ideal Sovereignty Of Existence"
Dystopia genre character(s) = DejaVu Sans "Violet/Purple"
Human Population In The Solar System; 10,999,999,999
Utopian Sovereignty (U.S.) Belief System;
"If the system fails you, you create your own system."

"Anonymous" Quotes To Lieu By
"If you are ashamed to stand by your colors, better to seek another flag."
"Until you commit your goals to paper you have intentions that are seeds without soil."
"Commitment in the face of conflict produces character."
"Life without friendship is like the sky without the sun."

=Dystopia Documenting Spying System= Insane Inspiration;_

2020(twenty twenty);_ Disaster Years Season 1(one) / Anonymous: name, group
2020;_05(May)-28(twenty eight) George Floyd Civil Unrest, Watch The World Burn
2020;_ Seattle, Washington(WAS), United States of North America(U.S. of N.A.)
Capital Hill Autonomous Zone(CHAZ) / Sony, Sucker Punch's Infamous Second Son
Sovereign (S) citizen movement / A Human Sanctuary (S) city In U.S.

Square Soft Enix's Final Fantasy 6(six; VI), and or 3(three; III) in N.A., and Dragon Quest 11(eleven; XI) S Echoes Of An Elusive Age Role Playing Games (R.P.G.).

Homelessness in U.S. of N.A. and non-profit organizations that support them.
I Owe You(I.O.U.); Student Debt Loans, Home Mortgage, Rent Housing Eviction
Imploding Infrastructure / "Civilian life... is a dog eat dog world..."
W.B., George Miller's Mad Max original movie / Margaret Atwood Handmaid's Tale

A.M.C.'s Fear The Walking Dead season 1(one) T.V. show series / Contagion movie

Hive Mind Control And Indoctrination Brainwashing Difference / George Orwell 1984's Big Brother, The Thought Police And Crimes, Science-Factual Dystopia
2+2=5 (two plus two equals five)
Thesis; "What is utopia and how do we engineer utopia on planet earth?"

Mythology of Christian Holy Bible, The Book Of Revelations,
The Apocalypse Of John { Jon } and Norse myth or culture of Ragnarok.
W.B., D.C. Batman No Man's Land Story, The Dark Knight: Returns, Rises
W.B. Alcon Entertainment's Blade Runner 2022 Blackout Movie

A 0(Arc Zero) "Beginning Of The End" Story 0(zero) Origins

"2121;_Movement" Jon W. Smith @ 02(Feb)-28(twenty eight), Planet Earth

A.M.C.'s The Walking Dead universe, story 0(zero) origins of the human Male (M) W Wolf Owen character and his Wolf pack group mystery.

"...I see my younger self in their eyes..." Williams, Michael K.

Persuasion Power, Mind Control { Kontrol } Relations

Control Type(s)	Actor(s)	Character(s) In Film Movies, T.V. Show Series, Video Games
"I Want It" Mind Controller	"Benedict Samuel, Official"	D.C. Fox Gotham T.V.'s Mad Hatter, Human Male W. Wolf Owen, "Character In Asthma 2015 Movie"
Mind Controller	"David Tennant, Podcast"	Marvel Netflix Jessica Jones T.V. Killgrave The Purple Man, Doctor Who
Mind Control Victim	"Krysten Ritter, The Real"	Marvel Netflix Jessica Jones T.V. "Character In Asthma 2015 Movie"
Remedy Control Game	"Courtney Hope, The"	The United States Federal Bureau of Control (F.B.C.) Director Jessie Faden
Mind Controller	"Barry Keoghan, keoghan92"	Dunkirk, Chernobyl, Disney Marvel's Eternals Space Alien A.I. Druig

Big Bad Villains; D.C. Mad Hatter, and Marvel Killgrave team-up plus Druig.

A mobile cell phone rings in the New York City Police Department (N.Y.C.P.D.) room.

Killgrave, The Purple Man Wearing His UltraViolet (U.V.) Business Suit; "(annoyed) Whose is that?!"

That same mobile cell phone continues to ring in the N.Y.C.P.D. room.

Killgrave, The Purple Man Wearing His UltraViolet (U.V.) Business Suit; "(annoyed) Whose phone is that?!"

Killgrave gets the ringing cell phone from the N.Y.C.P.D. officer and throws it against the wall.

Killgrave, The Purple Man Wearing His UltraViolet (U.V.) Business Suit; "(yelling) Next persons' whose phone rings has to eat it!!"

20th Century Fox Studios' Maze Runner 3 The Death { Deaf } Cure, The Last City

A 0(Arc Zero) "Beginning Of The End" Story 0(zero) Origins

"2121;_SWAT" Seattle City Police Department (S.C.P.D.) @ 03(Mar)-15(fifteen), Planet Earth
"Police, The Human Enforcers Of Behavior"
Found Footage, Reality, Gun Guys Gun Girls genre character(s) = Franklin Gothic Medium "Black"
Human Population In The Solar System; "Let The Bodies Hit The Floor { Or The Atmosphere }"
Police Department Belief System; "***Protect And Serve***"

Insane Inspiration;_
1999;_Artisan Entertainment's The Blairwitch Project movie,
directed by Daniel Myrick and Eduardo Sanchez
2008;_Paramount Pictures' Cloverfield movie, directed by Matt Reeves
2012;_20th Century Fox Studios' Chronicle movie, directed by Josh Trank
2012;_End Of Watch movie, directed by David Ayer
2013;_Magnolia's Europa Report movie, directed by Sebastian Cordero

Police BodyCamera (BodyCam) / Tom Clancy's Rainbow 6(six) Patriots Enemy Within
The Black Market Of Human Society, Trafficking: Human, Drugs, Firearms, Animals
Homegrown Domestic Terrorist(s) In The United States of North America(U.S. of N.A.)
Scholastic Wheels visual book; Written by Judy Nayer, Pictures by Chris Reed

Flatline of natural causes. / Flatline and or dismemberment of augmented causes.

1919;_Spaniard Flu(S-Flu) Chemical, Biological, Radiological(C.B.R.) Outbreak Pandemic

2020;_Disaster Years Season 1(one)
C.B.R. Outbreak Pandemic, Silence Or Sirens In A Human Urban City Society

2121;_Another Disaster Years In About 1(one) Hundred(100) Years Later By The
22nd(twenty second) Century with C.B.R. Outbreak Pandemic Augmented By Electro Magnetic
Pulse(E.M.P.) Outage Blackout, The Dark Ages Of Fire, Burning Buildings And Collapse

The Adjustment Bureau Movie, Thompson On Free Will Versus Determinism Philosophy: "We actually tried free will before. After taking you ***from hunting and gathering to the height of the Roman Empire***, we stepped back to see how you'd do on your own. You gave us the ***Dark Ages for five centuries***... until finally we decided we should come back in. The Chairman thought maybe we just needed to do a better job of ***teaching you how to ride a bike before taking the training wheels off again***. So we gave you ***the Renaissance, the Enlightenment, the Scientific Revolution***. For six hundred years we taught you to ***control your impulses with reason***, then in ***nineteen ten we stepped back. Within fifty years, you'd brought us World War one, the Depression, Fascism, the Holocaust and capped it off by bringing the entire planet to the brink of destruction in the Cuban Missile Crisis***. At that point a decision was taken to step back in again before you did something that even we couldn't fix. You don't have free will, David. You have the appearance of free will."

2154;_Arcadia \ Valentina Vita (V. V.) 'Healthy-Strong Life' \ Left to Right(L->R)

A 0(Arc Zero) "Beginning Of The End" Story 0(zero) Origins

"2121;_Army" United States Army (U.S.A.) @03(Mar)-30(thirty), Planet Earth
"America, The Land Of Opportunity"
Military genre character(s) = Maiandra GD "Green"
Human Population In The Solar System; "Let The Bodies Hit The Floor { Or In The Atmosphere }"
Military Belief System; "***United States Constitution, Uniform Code Military Justice (U.C.M.J.),***

"The Oath Of Enlistment (Enlisted, E)."
"I, ____________________(Full Name Here) do solemnly swear that I will support and defend the Constitution of the United States against all enemies, foreign and domestic; that I will bear true faith and allegiance to the same; and that I will obey the orders of the President of the United States and the orders of the officers appointed over me, according to regulations and the Uniform Code of Military Justice. So help me God."

"The Oath Of Office (Officer, O)."
"I,____________________(S.S.A.N.), having been appointed an officer in the ____________________(Military Branch) of the United States, as indicated above in the grade of ____________________ do solemnly swear that I will support and defend the Constitution of the United States against all enemies, foreign or domestic, that I will bear true faith and allegiance to the same; that I take this obligation freely, without any mental reservation or purpose of evasion; and that I will well and faithfully discharge the duties of the office upon which I am about to enter; So help me God."

United States Army (U.S.A.) Military Ranks Without Any Visual Pictures

Enlisted (E) Green and Gold	Company and Field Grade Officer (O) Gold and Silver { Commissioned }
E-1 = Private	O-1 = 2nd (second) Lieutenant
E-2 = Private	O-2 = 1st (first) Lieutenant
E-3 = Private 1st (first) Class	O-3 = Captain
E-4 = Corporal or Specialist	O-4 = Major
E-5 = Sergeant	O-5 = Lieutenant Colonel
E-6 = Staff Sergeant	O-6 = Colonel
E-7 = Sergeant 1st (first) Class	
E-8 = Master Sergeant or 1st (first) Sergeant	
E-9 = Sergeant Major or Command Sergeant Major or Sergeant Major of the Army	

A 0(Arc Zero) "Beginning Of The End" Story 0(zero) Origins

United States Army (U.S.A.) Military Ranks Without Any Visual Pictures Continued

General Officer (O) Silver	Warrant Officer (O) Silver and Black
O-7 = Brigadier General	W-1 = Warrant Officer
O-8 = Major General	W-2 = Chief Warrant Officer
O-9 = Lieutenant General	W-3 = Chief Warrant Officer
O-10 = General	W-4 = Chief Warrant Officer
O-11 = General of the Army	W-5 = Master Warrant Officer

U.S.A. Military Branches;
Combat Arms; "Involved in actual fighting."

Infantry	Field Artillery	Corps of Engineers	Special Forces
Armor	Air Defense Artillery	Aviation	

U.S.A. Military Branches;
Combat Support; "Provide operational assistance to combat arms to include combat missions as necessary."

Military Police (M.P.)	Aviation	Corps of Engineers
Chemical Corps	Military Intelligence	Signal Corps

U.S.A. Military Branches;
Combat Service Support; "Provide logistical and administrative support to the army; personnel normally not directly involved in combat operations."

Adjutant General Corps	Finance Corps	Ordnance Corps Chaplain Corps	Transportation Corps	Quartermaster Corps
JAG Corps	Medical Corps	Dental Corps	Nurse Corps	Veterinary Corps

A 0(Arc Zero) "Beginning Of The End" Story 0(zero) Origins

United States Armed Forces, Military (M) Services Branches Character(s)

United States Army (U.S.A.), Established 1775;_

2121;_Army	Actor(s)	Character(s) { Soldier Boys, Superman That Ho, Yu! }
Rook	Michael Rooker	A.M.C.'s Walking Dead T.V. Show Merle Dixon, Guardians
Reed	Norman Reedus	A.M.C.'s Walking Dead T.V. Show Daryl Dixon, Death S
Clarke	Mads Mikkelsen	M.C.U.'s Doctor Strange K. "Mister Doctor", Hannibal
Grill	Frank Grillo	M.C.U. SHIELD Brock Rumlow Crossbones, Purge Movies
Mully	Callan Mulvey	M.C.U. SHIELD Jack Rollins, Batman V. Superman
Fukuhara	Karen Fukuhara	D.C.E.U. Katana, Kimiko Female Amazon's The Boys
Kinsman	Joel Kinnaman	RoboCop Remake, D.C.E.U. Colonel Rick Flag, Altered C
Harry	Woody Harrelson	War Planet Of The Apes Colonel, Venom 2 Carnage
Shannon	Michael Shannon	Groundhog Day, 99 Homes, D.C.E.U. General Dru-Zod
Traue	Antje Traue	D.C.E.U. General Dru-Zod's Right Hand Officer Faora-Ul

A 0(Arc Zero) "Beginning Of The End" Story 0(zero) Origins

United States Armed Forces, Military (M) Services Branches Character(s)

United States Navy (U.S.N.), Established 1775;_

2136;_Navy	Actor(s)	Character(s) { Sailor Suits, Sailor Moon That Ho, Yu! }
Lieu	Ludi Lin	Power Rangers Re, Aquaman, Black Mirror, MK"2136-2139
Yellow-Mellow	Jacob Batalon	Disney Marvel Sony Spider-Man Home (O :)"2136-2139"
Ogden	Evangeline Lilly	Hobbit Movies, Marvel Ant-Man And Wasp "2136-2140"

A 0(Arc Zero) "Beginning Of The End" Story 0(zero) Origins

United States Armed Forces, Military (M) Services Branches Character(s)

United States Marine Corps (U.S.M.C.), Established 1775;_

2121;_Army	Actor(s)	Character(s) { Marine Boys, Superboy That Ho, Yu! }
Guevara	Jonny Bernthal	A.M.C. Walking Dead Shane, M.C.U. Netflix Daredevil Punisher
Barney	Ben Barnes	M.C.U. Netflix's Punisher T.V., WestWorld T.V. season 1, 2
Drive	Adam Driver	Disney Star Wars Sequel Trilogy Episodes 7, 8, 9, TED Talk
Jane	Thomas Jane	2004 The Punisher, Dirty Laundry, The Predator, The Expanse

A 0(Arc Zero) "Beginning Of The End" Story 0(zero) Origins

United States Armed Forces, Military (M) Services Branches Character(s)

United States Coast Guard (U.S.C.G.), Established 1790;_

21XX;_Coast	Actor(s)	Character(s) { Sailor Suit, Sailor Moon That Ho, Yu! }
Morales	Oscar Issac	Ex Machina, Annihilation, Disney Star Wars 7, 8, 9, Moon

A 0(Arc Zero) "Beginning Of The End" Story 0(zero) Origins

United States Armed Forces, Military (M) Services Branches Character(s)

United States Air Force (U.S.A.F.), Established 1947;_

2145;_Air	Actor(s)	Character(s) { Air-Men, Raw-Men }
Drastal	Alex Drastal	@AlexDrastal Actor Model Cosplayer "Serves 2145-2149"

A 0(Arc Zero) "Beginning Of The End" Story 0(zero) Origins

United States Armed Forces, Military (M) Services Branches Character(s)

United States Space Force (U.S.S.F.), Established 1958;_ { Our History, Was NASA Founded }

2150;_Space	Actor(s)	Character(s) { Space Superman That Ho, Yu! }
Shepherd	Stephen Lang	James Cameron's Avatar Colonel Miles Q., Don't Breathe
Williams	Keith David	E.A. Mass Effect 1, 2, 3 David Anderson, Halo, COD
Drastal	Alex Drastal	@AlexDrastal Actor Model Cosplayer "Serves 2150-2153"
Wright	Gunner Wright	E.A.'s Dead Space 1 Remake, 2, 3 Issac Clarke voice actor
Rockwell	Sam Rockwell	Sam In Moon Movie, Iron Man 2 Hammer, The Bad Guys
Hale	Jennifer Hale	E.A. Mass Effect 1 2 3 Trilogy Female Shepherd FemShep
Meer	Mark Meer	E.A. Mass Effect 1 2 3 Trilogy Male Shepherd MaleShep

A 0(Arc Zero) “Beginning Of The End” Story 0(zero) Origins

“2121;_Collapse” X @05(May)-28(twenty eight), Planet Earth
“Holocene, The 6th(sixth) Extinction Event Of Epoch”
“Mythology” genre character(s) =
Human Population In The Solar System; “Let The Bodies Hit The Floor
{ Or In The Atmosphere }”
Mythology Belief System; “***It’s all biblical...***”

Insane Inspiration;_

1990s Rodney King Trials And The Los Angeles (L.A.) City Riots

“An empire toppled by its enemies can rise again. But one that crumbles from within?
That is dead. Forever.”.
-Daniel Bruhl { Alter Ego, Ze Mo } in
Disney, Marvel Studios’ Captain America 3(three) Civil War movie

“The League Of Shadows has been a check against human corruption for thousands of years. We sacked Rome, loaded trade ship with plague rats, burn London to the ground. Every time a civilization reaches its pinnacle of decadence, we return to restore the balance.”.
-Liam Neeson { Alter Ego, Da Mentor }
Warner Brothers (W.B.), Detective Comics (D.C.) Christopher Nolan “Chris-N”
Batman story 1(one), Batman Begins The League Of Shadows

The collapse of Afghanistan on 2021;_08(Aug) and takeover by the Taliban

The Economist, 2018;_08(Aug)-04(four), page 09(nine)
“The world is losing the war against climate change
Planet Earth and the ever changing crisis of “climate change”.

The Economist, 2018;_11(Nov)-17(seventeen), page 89(eighty nine)
“Burning out, Climate change makes fires worse, but agricultural development limits them”.

Homefront: The Revolution, Homefront alternative history violet video games

2022;_02(Feb)-24(twenty four) The Invasion Of Ukraine By Russian Military Forces

Characters Genre Index;_? "**E-Book pages subject to change if not in P.D.F. type.**"

Characters Genre Index;_? "**E-Book pages subject to change if not in P.D.F. type.**"

Characters Genre Index;_? "**E-Book pages subject to change if not in P.D.F. type.**"

Naples, **Italy** in 2154(twenty one fifty four),
formerly known as the
"Naples: City Of Trash"
in the 21st(twenty first) century,
by the 22nd(twenty second) century
of the new world
2121(twenty one twenty one) Utopian Sovereignty movement,
Naples City is now
a Human Sanctuary utopia.

Within the engineering environment
biotech domes, the temperature
is around +024° C (positive seventy five degrees Celsius).

Meanwhile, outside the biotech
domes, the temperature outside,
varies according to the
apocalyptic War On Mars.

Overseas **Italian traveling cosplayer**
Valentina Vita A.K.A.(Also Known As) V. V.,
is making plans to
travel and visit the:
Sir Arthur C. Clarke's
Space Elevator anniversary launch,
Meeting Entertainer Mister Steven
Z. J. in the
Americas' UltraViolet(U.V.) Rays' phenomenon
of Animal Sanctuary utopias.

A 2 (Arc Two), "A Garden Of Vigilantes"

during warm season outside:+150° Fahrenheit day time, +100° Fahrenheit night time.
Rising sea water levels within 15 (fifteen) miles inbound to all United States (U.S.) sea shore,
According to Warner Brothers' Blade Runner 2049 (twenty forty nine) story and maybe
assume world wide on planet earth by mid 21st (century first) century.
Low planet earth natural resources inbound to Down Town L.A. (D.T.L.A.) California state.
Sony's Elysium 2154 (twenty one fifty four) story = off planet space station.
Fox's Avatar 2154 (twenty one fifty four) story = off planet space mining.

Subject Topics, Genre Type;_ "pages 517 / 590"
Science-Fiction \ Hard Science-Fiction \ Soft Science-Fiction \ Alternative History
Apocalyptic, Neo-Western { Macro Big } \ Utopia, Parody { Micro Small }
Semi-Autobiography \ Traditional Art Character Designer
Coloring-Book Line Art Style \ Photo-Book { Travel Drawings Audio Retcon = Tahoma }

Milton Keynes UK
Ingram Content Group UK Ltd.
UKHW022015150324
439433UK00004B/54